Relativistic Quantum Mechanics and Field Theory

Relativistic Quantum Mechanics and Field Theory

FRANZ GROSS
College of William and Mary
Williamsburg, Virginia
and
Continuous Electron Beam Accelerator Facility
Newport News, Virginia

WILEY SCIENCE PAPERBACK EDITION PUBLISHED 1999

A Wiley-Interscience Publication
JOHN WILEY & SONS, INC.
New York • Chichester • Weinheim • Brisbane • Singapore • Toronto

This text is printed on acid-free paper. ♾

Wiley Classics Library Edition Published 1999.

Library of Congress Cataloging in Publication Data:

Gross, Franz.
Relativistic quantum mechanics and field theory / Franz Gross.
p. cm.
Includes index.
ISBN 0-471-59113-0
ISBN 0-471-35386-8 Wiley Science Paperback Edition
1. Relativistic quantum theory. 2. Quantum field theory.
3. Symmetry (Physics). 4. Gauge fields (Physics). I. Title.
QC174.24.R4G76 1993
530.1'2—dc20 92-40605

10 9 8 7 6 5 4 3 2

To my parents, Genevieve and Llewellyn

and to the next generation, Glen, Sue, Kathy, Caitlin, and Christina

CONTENTS

PREFACE

Relativistic Quantum Mechanics and Field Theory are among the most challenging and beautiful subjects in Physics. From their study we explain how states decay, can predict the existence of antimatter, learn about the origin of forces, and make the connection between spin and statistics. All of these are great developments which all physicists should know but it is a real challenge to learn them for the first time.

This book grew out of my struggle to understand these topics and to teach them to second year graduate students. It began with notes I prepared for my personal use and later shared with my students. About two years ago I decided to have these notes typed in TEX, little realizing that by so doing I had committed myself to eventually producing this book. My objectives in preparing this text reflect the original reasons I prepared my own notes: to write a book which (i) can be understood by students learning the subject for the first time, (ii) carries the development far enough so that a student is prepared to begin research, and (iii) gives meaning to the study through examples drawn from the fields of atomic, nuclear, and particle physics. In short, the goal was to produce a book which begins at the beginning, goes to the end, and is easy to read along the way.

The first two parts of this book (Part I: Quantum Theory of Radiation, and Part II: Relativistic Equations) assume no previous experience with advanced quantum mechanics. The subjects included here are quantization of the electromagnetic field, relativistic one-body wave equations, and the theoretical explanation for atomic decay, all fundamental subjects which can be regarded as necessary to a well rounded education in physics (even for classical physicists). The presentation is modeled after the first third of a year-long course which I have taught at various times over the past 15 years and these topics are given in the beginning so that those students who must leave the course at the end of the first semester will have some knowledge of these important areas.

To prepare a student for advanced work, the last two parts of this book include an introduction to many of the unique insights which relativistic field theory has contributed to modern physics, including gauge symmetry, functional methods (path integrals), spontaneous symmetry breaking, and an introduction to QCD, chiral symmetry, and the Standard Model. Part III also contains a chapter (Chapter 12) on relativistic bound state wave equations, an important topic frequently overlooked in studies at this level. I have tried to present even these more advanced topics from an elementary point of view and to discuss the subjects in sufficient detail so that the questions asked by beginning students are addressed. The entire book includes a little more material than can comfortably fit into a year long course, so that some selection must be made when used as a text.

To make the book easier to read, most proofs and demonstrations are worked out completely, with no important steps missing. Some topics, such as the quantization of fields, symmetries, and the study of the Lorentz group, are introduced briefly first, and returned to later as the reader gains more experience, and when a greater understanding is needed. This "spiral" structure (as it is sometimes referred to by the educators) is good for beginning students but may be frustrating for more advanced students who might prefer to find all the discussion of one topic in one place. I hope such readers will be satisfied by the table of contents and the index (which I have tried to make fairly complete). Considerable emphasis is placed on applications and some effort is made to show the reader how to carry out practical calculations. Problems can be found at the end of each chapter and four appendices include important material in a convenient place for ready reference.

There are many good texts on this subject and some are listed in the Reference section. Most of these books are either classics, written before the advent of modern gauge theories, or new books which treat gauge theories but omit some of the detail and elementary material found in older books. I believe that most of this elementary material is still very helpful (maybe even necessary) for students, and have tried to cover both modern gauge theories and these elementary topics in a single book. As a result the book is somewhat longer than many, and omits some advanced topics I would very much like to have included. Among these omissions is a discussion of anomalies in field theories.

Many people have helped me in this effort. I am grateful to Michael Frank, Joe Milana, and Michael Musolf for important suggestions and help with individual chapters. I also thank my colleagues Carl Carlson, Nathan Isgur, Anatoly Radyushkin, and Marc Sher. S. Bethke and C. Wohl kindly gave permission to use figures 17.4 and 10.9 (respectively). Many students suffered through earlier drafts, found numerous mistakes, and made many helpful suggestions. Among these are: S. Ananyan, A. Colman, K. Doty, D. Gaetano, C. Hoff, R. Kahler, Z. Li, R. Martin, D. Meekins, C. Nichols, J. Oh, X. Ou, , M. Sasinowski, P. Spickler, Y. Surya, X. Tang, A. B. Wakley, and C. Wang. Roger Gilson did an excellent job transforming my original notes into TEX. And no effort like this would be possible or meaningful without the support of my family. I am especially grateful to my wife, Chris, who assumed many of my responsibilities so I could complete the work on this book in a timely fashion. I could not have done it without her.

FRANZ GROSS

RELATIVISTIC QUANTUM MECHANICS AND FIELD THEORY

PART I

QUANTUM THEORY OF RADIATION

CHAPTER 1

QUANTIZATION OF THE NONRELATIVISTIC STRING

This book will discuss how nonrelativistic quantum mechanics can be extended to describe:

- relativistic systems and
- systems in which particles can be created and annihilated.

The key to both of these extensions is field theory, and we therefore begin with an introduction to this topic. In this chapter we will discuss the quantization of the nonrelativistic, one-dimensional string. This is a many-body system which is also simple and familiar. Quantization of this many-body system leads directly to the (new) concept of a quantum field, and many of the properties of quantum fields can be introduced and illustrated using the nonrelativistic one-dimensional string as an example. The goal of this chapter is to use this simple system to develop an intuition and understanding of the meaning and properties of quantized fields. In subsequent chapters some of these ideas will be developed again in a more general, abstract way, and it is hoped that the intuition gained in this chapter will remove much of the mystery which might otherwise surround those more abstract discussions.

The discussion of relativistic systems begins in the next chapter, where the ideas developed here are immediately extended to the electromagnetic field.

1.1 THE ONE-DIMENSIONAL CLASSICAL STRING

We will approach the treatment of a continuous string by first considering a system of point masses connected together by "springs" and then letting the number of point masses go to infinity, and the distance between them go to zero, in such a way that a continuous system with a uniform density and tension emerges.

Start, then, with a "lumpy" string of overall length L made up of N points, each with mass m, coupled together by springs with a spring constant k. Assume that the oscillators move about their equilibrium positions in a periodic pattern, which is best realized by thinking of the string as closed on itself in a circle, as shown in Fig. 1.1. The oscillators are constrained to vibrate along the circumference of the ring (which has a radius very much greater than the equilibrium separation ℓ so that the system will be treated as a linear system with periodic boundary conditions). The 0th and Nth oscillators are identical, so that if $\bar{\phi}_i$ is the displacement of the ith oscillator from equilibrium, then

$$\left.\begin{aligned} \bar{\phi}_0 &= \bar{\phi}_N \\ \frac{d\bar{\phi}_0}{dt} &= \frac{d\bar{\phi}_N}{dt} \end{aligned}\right\} \quad \text{periodic boundary conditions.}$$

The kinetic energy (KE) and potential energy (PE) are

$$\mathrm{KE} = \frac{1}{2} m \sum_{i=0}^{N-1} \left(\frac{d\bar{\phi}_i}{dt} \right)^2$$

$$\mathrm{PE} = \frac{1}{2} k \sum_{i=0}^{N-1} \left(\bar{\phi}_{i+1} - \bar{\phi}_i \right)^2 .$$

Now, take the continuum limit by letting $\ell \to 0$, $N \to \infty$, such that the length $L = N\ell$, mass per unit length $\mu = m/\ell$, and string tension $T = k\ell$ are fixed. Then the displacement and energy of the string can be defined in terms of a continuous *field* $\bar{\phi}(z,t)$, where

$$\bar{\phi}_i(t) = \bar{\phi}(z_i, t) \to \bar{\phi}(z,t)$$

$$\mathrm{KE} = \frac{1}{2} \frac{m}{\ell} \sum_{i=0}^{N-1} \ell \left(\frac{d\bar{\phi}_i}{dt} \right)^2 \to \frac{1}{2} \mu \int_0^L dz \left(\frac{\partial \bar{\phi}(z,t)}{\partial t} \right)^2$$

$$\mathrm{PE} = \frac{1}{2} k\ell \sum_{i=0}^{N-1} \ell \left(\frac{\bar{\phi}_{i+1} - \bar{\phi}_i}{\ell} \right)^2 \to \frac{1}{2} T \int_0^L dz \left(\frac{\partial \bar{\phi}(z,t)}{\partial z} \right)^2 .$$

The Lagrangian and Hamiltonian are

$$L = \mathrm{KE} - \mathrm{PE} = \int_0^L dz \left\{ \frac{1}{2} \mu \left(\frac{\partial \bar{\phi}}{\partial t} \right)^2 - \frac{1}{2} T \left(\frac{\partial \bar{\phi}}{\partial z} \right)^2 \right\} = \int_0^L dz\, \mathcal{L}(z,t)$$

$$H = \mathrm{KE} + \mathrm{PE} = \int_0^L dz \left\{ \frac{1}{2} \mu \left(\frac{\partial \bar{\phi}}{\partial t} \right)^2 + \frac{1}{2} T \left(\frac{\partial \bar{\phi}}{\partial z} \right)^2 \right\} = \int_0^L dz\, \mathcal{H}(z,t) ,$$

where $\mathcal{L}$ and $\mathcal{H}$ are the Lagrangian and Hamiltonian *densities*. In this example, the field function $\bar{\phi}(z,t)$ is the displacement of an infinitesimal mass from its equilibrium position at z. In three dimensions, $\bar{\phi}$ would be a vector field.

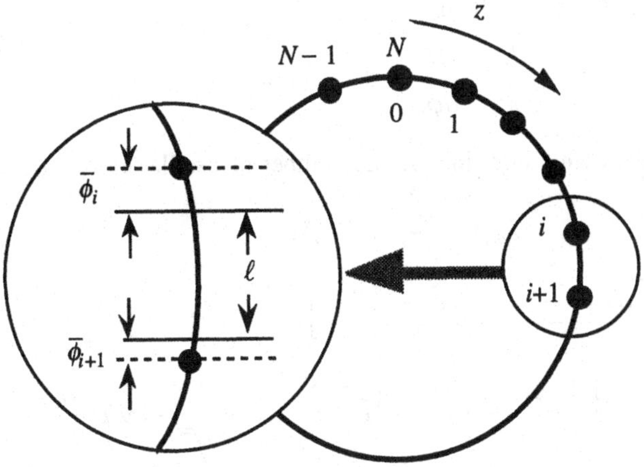

Fig. 1.1 Drawing of the circular string with the location of the oscillators in the interval $[i, i+1]$ enlarged. The equilibrium position of the oscillators are the solid lines separated a distance ℓ.

Anticipating later applications, we redefine $\bar{\phi}$ by absorbing $\sqrt{T}$:

$$\bar{\phi} \to \sqrt{T}\,\bar{\phi} = \phi \quad .$$

Then introducing the wave velocity

$$v^2 = \frac{T}{\mu} \tag{1.1}$$

gives

$$\begin{aligned} \mathcal{L} &= \frac{1}{2}\left[\frac{1}{v^2}\left(\frac{\partial\phi}{\partial t}\right)^2 - \left(\frac{\partial\phi}{\partial z}\right)^2\right] = \mathcal{L}\left(\frac{\partial\phi}{\partial t}, \frac{\partial\phi}{\partial z}\right) \\ \mathcal{H} &= \frac{1}{2}\left[\frac{1}{v^2}\left(\frac{\partial\phi}{\partial t}\right)^2 + \left(\frac{\partial\phi}{\partial z}\right)^2\right] \quad . \end{aligned} \tag{1.2}$$

The equations of motion for the string can be derived from the Lagrangian using the *principle of least action* [for a review, see Goldstein (1977)]. This principle states that the "path" followed by a classical system is the one along which its action $\mathcal{A}$ is an extremum. For the "lumpy" string, made up of discrete coupled oscillators, this condition is

$$\delta\mathcal{A} = \delta\int dt\, L\left(\bar{\phi}_i, \dot{\bar{\phi}}_i, t\right) = 0 \quad ,$$

where $\dot{\bar{\phi}}_i = d\bar{\phi}_i/dt$. Working out the variation gives the Euler–Lagrange equations for the motion of each oscillator:

$$\frac{d}{dt}\frac{\partial L}{\partial \dot{\bar{\phi}}_i} - \frac{\partial L}{\partial \bar{\phi}_i} = 0 \qquad (i = 0 \text{ to } N-1) \ .$$

However, in the continuum limit as the number of oscillators $N \to \infty$,

$$\begin{aligned}
\frac{\partial L}{\partial \bar{\phi}_i} &= -k\left[-\left(\bar{\phi}_{i+1} - \bar{\phi}_i\right) + \left(\bar{\phi}_i - \bar{\phi}_{i-1}\right)\right] \\
&= k\ell\left[\frac{(\bar{\phi}_{i+1} - \bar{\phi}_i)}{\ell} - \frac{(\bar{\phi}_i - \bar{\phi}_{i-1})}{\ell}\right] \\
&\underset{\text{small}\,\ell}{\longrightarrow} \ell\, T\frac{1}{\ell}\left[\frac{\partial\bar{\phi}(z^+,t)}{\partial z} - \frac{\partial\bar{\phi}(z^-,t)}{\partial z}\right] \underset{\text{absorb}\,\sqrt{T}}{\longrightarrow} -\ell\sqrt{T}\,\frac{\partial}{\partial z}\,\frac{\partial \mathcal{L}}{\partial\left(\dfrac{\partial\phi}{\partial z}\right)} \ ,
\end{aligned} \tag{1.3}$$

where $z^+ = z + \frac{1}{2}\ell$ is the midpoint of the interval $z_{i+1} - z_i$, and $z^- = z - \frac{1}{2}\ell$ the midpoint of the interval $z_i - z_{i-1}$. These are appropriate arguments for the two derivatives which arise in the next to the last step of Eq. (1.3). Also:

$$\begin{aligned}
\frac{d}{dt}\frac{\partial L}{\partial \dot{\bar{\phi}}_i} &= m\ddot{\bar{\phi}}_i = \mu\ell\frac{\partial^2\bar{\phi}(z,t)}{\partial t^2} \\
&\underset{\text{absorb}\,\sqrt{T}}{\longrightarrow} \ell\sqrt{T}\,\frac{1}{v^2}\,\frac{\partial^2\phi(z,t)}{\partial t^2} = \ell\sqrt{T}\,\frac{\partial}{\partial t}\,\frac{\partial\mathcal{L}}{\partial\left(\dfrac{\partial\phi}{\partial t}\right)} \ .
\end{aligned}$$

Hence the Euler–Lagrange equations can be expressed directly in terms of $\mathcal{L}$, the Lagrangian density

$$\frac{d}{dt}\left(\frac{\partial L}{\partial \dot{\bar{\phi}}_i}\right) - \frac{\partial L}{\partial \bar{\phi}_i} \underset{\text{small}\,\ell}{\longrightarrow} \ell\sqrt{T}\left\{\frac{\partial}{\partial t}\,\frac{\partial\mathcal{L}}{\partial\left(\dfrac{\partial\phi}{\partial t}\right)} + \frac{\partial}{\partial z}\,\frac{\partial\mathcal{L}}{\partial\left(\dfrac{\partial\phi}{\partial z}\right)}\right\} = 0 \ ,$$

where the $\ell\sqrt{T}$ factor can be discarded.

More generally, $\mathcal{L}$ can also be a function of ϕ as well as $\partial\phi/\partial t$ and $\partial\phi/\partial z$, and for a scalar field in three dimensions, where $r_i = (x, y, z)$ are the three spatial components, we obtain directly

$$\delta\int dt\, d^3r\, \mathcal{L}(r,t) = \int dt\, d^3r\left(\frac{\partial\mathcal{L}}{\partial\dot{\phi}}\delta\dot{\phi} + \frac{\partial\mathcal{L}}{\partial\left(\nabla_i\phi\right)}\delta\left(\nabla_i\phi\right) + \frac{\partial\mathcal{L}}{\partial\phi}\delta\phi\right) = 0 \ ,$$

where summation over repeated indices is implied. Assuming that $\delta\left(\nabla_i\phi\right) = \nabla_i\left(\delta\phi\right)$ and integrating by parts (assuming boundary terms are zero because the boundary conditions are periodic and that the variations $\delta\phi$ at the initial and final times are zero) give

$$0 = \int dt\, d^3r \left\{ -\frac{\partial}{\partial t}\frac{\partial \mathcal{L}}{\partial(\dot{\phi})} - \nabla_i \left(\frac{\partial \mathcal{L}}{\partial(\nabla_i \phi)} \right) + \frac{\partial \mathcal{L}}{\partial \phi} \right\} \delta\phi \, .$$

Using the notation

$$x^\mu = (t, x, y, z) \qquad \mu = 0, 1, 2, 3 \ ,$$

which can be readily generalized to relativistic systems (it will later be the contravariant four-vector), gives the famous Euler–Lagrange equations for a continuous field

$$\boxed{\frac{\partial}{\partial x^\mu}\left(\frac{\partial \mathcal{L}}{\partial\left(\dfrac{\partial \phi}{\partial x^\mu}\right)} \right) - \frac{\partial \mathcal{L}}{\partial \phi} = 0 \ ,} \tag{1.4}$$

where summation over repeated indices is assumed. For the one-dimensional string treated in this chapter, the Euler–Lagrange equations give the familiar wave equation

$$\boxed{\frac{1}{v^2}\frac{\partial^2 \phi}{\partial t^2} - \frac{\partial^2 \phi}{\partial z^2} = 0 \ ,} \tag{1.5}$$

where the wave velocity was defined in Eq. (1.1).

In summary, we have shown in this section how a quantity referred to as a *continuous field* emerges as the natural way to describe a system with infinitely many particles. In this example, the field is $\phi(z, t)$, and it gives the *displacement* of each particle from its equilibrium position at z. Since we absorbed $\sqrt{T}$ into the field, its units (for a one-dimensional system) are $L\sqrt{mL/t^2}$. In the *natural system of units*, where $\hbar = c = 1$, it is dimensionless (for a discussion of the natural system of units, see Sec. 1.3 and Prob. 1.1). In three dimensions, the dimensions of such a field are L^{-1}, which can be deduced directly from the observation that $\int d^3r\ \mathcal{L}$ has units of energy. Regardless of its dimensions, it is useful to remember that a field is always the "displacement" (in a generalized sense) of some dynamical system, and that therefore $\partial\phi/\partial t$ is a generalized velocity.

1.2 NORMAL MODES OF THE STRING

As a preparation to quantizing the string, we find its normal modes. The solutions of the wave equation which satisfy the periodic boundary conditions are

$$\phi \sim e^{\pm i(k_n z - \omega_n t)} \ ,$$

where periodicity requires

$$k_n = \frac{2\pi n}{L} \qquad n = 0,\ \pm 1,\ \pm 2,\ \ldots \tag{1.6}$$

and the wave equation gives

$$\omega_n^2 = v^2 k_n^2 \ .$$

Note that there are both positive and negative frequency solutions. We will adopt the convention that ω_n is *always* positive, and use $-\omega_n$ for negative frequency solutions. The states with positive frequency are written

$$\phi_n(z,t) = \frac{1}{\sqrt{L}} e^{i(k_n z - \omega_n t)} \ . \tag{1.7}$$

The negative frequency states have a time factor $e^{i\omega_n t}$, and since k_n is both positive and negative, it is convenient to denote the negative frequency states by $\phi_n^*(z,t)$. The normalization condition which these states satisfy is

$$\int_0^L dz\, \phi_n^*(z,t)\phi_m(z,t) = \delta_{n,m} \ . \tag{1.8}$$

However, by direct evaluation it is also true that

$$\int_0^L dz\, \phi_n(z,t)\phi_m(z,t) = \delta_{n,-m}\, e^{-2i\omega_n t} \ , \tag{1.9}$$

so the states are not orthogonal in the usual sense. The most general real field can be expanded in normal modes as follows:

$$\begin{aligned} \phi(z,t) &= \sum_{n=-\infty}^{\infty} c_n \left\{ a_n(0)\phi_n(z,t) + a_n^*(0)\phi_n^*(z,t) \right\} \\ &= \sum_{n=-\infty}^{\infty} \frac{c_n}{\sqrt{L}} \left\{ a_n(t)e^{ik_n z} + a_n^*(t)e^{-ik_n z} \right\} \ , \end{aligned} \tag{1.10}$$

where $a_n(0)$ are the coefficients of each normal mode in the expansion (1.10) and the real normalization factor c_n will be chosen later. It will sometimes be convenient to incorporate the time dependence of each normal mode into a generalized $a_n(t)$,

$$a_n(t) = a_n(0)\, e^{-i\omega_n t} \ , \tag{1.11}$$

as was done in the second line of (1.10). The condition that $\phi(z,t)$ is real means that the coefficient of ϕ_n^* must be the complex conjugate of the coefficient of ϕ_n.

Equation (1.11) shows that *each normal mode behaves as an independent simple harmonic oscillator (SHO) satisfying the equation*

$$\ddot{a}_n(t) + \omega_n^2 a_n(t) = 0 \ .$$

To quantize the field, it is only necessary to quantize these oscillators.

Before doing this, however, we evaluate the energy in terms of the dynamical variables $a_n(t)$. Using the "orthogonality" relations (1.8) and (1.9), which can be written

$$\int_0^L dz\, a_n^*(0)\phi_n^*(z,t)a_m(0)\phi_m(z,t) = \delta_{n,m}\,|a_n(0)|^2 = \delta_{n,m}\,|a_n(t)|^2$$

$$\begin{aligned}\int_0^L dz\, a_n(0)\phi_n(z,t)a_m(0)\phi_m(z,t) &= \delta_{-n,m}a_n(0)a_{-n}(0)\,e^{-2i\omega_n t}\\ &= \delta_{-n,m}a_n(t)a_{-n}(t)\ ,\end{aligned}$$

we obtain

$$\begin{aligned}\text{KE} &= \frac{1}{2}\frac{1}{v^2}\int_0^L dz\frac{\partial\phi}{\partial t}\frac{\partial\phi}{\partial t}\\ &= \frac{1}{2}\frac{1}{v^2}\sum_{n=-\infty}^{\infty}\left[2c_n^2 \mid \dot{a}_n(t)\mid^2 + c_n c_{-n}\dot{a}_n(t)\dot{a}_{-n}(t) + c_n c_{-n}\dot{a}_n^*(t)\dot{a}_{-n}^*(t)\right]\\ V &= \frac{1}{2}\int_0^L dz\frac{\partial\phi}{\partial z}\frac{\partial\phi}{\partial z}\\ &= \frac{1}{2}\sum_{n=-\infty}^{\infty}k_n^2\left[2c_n^2 \mid a_n(t)\mid^2 + c_n c_{-n}a_n(t)a_{-n}(t) + c_n c_{-n}a_n^*(t)a_{-n}^*(t)\right]\ .\end{aligned}$$

Using $\dot{a}_n(t) = -i\omega_n a_n(t)$ gives

$$\begin{aligned}H = \text{KE} + V &= \sum_{n=-\infty}^{\infty}\left\{c_n^2\left(\frac{\omega_n^2}{v^2} + k_n^2\right)a_n^*(t)a_n(t)\right.\\ &\quad\left. + \frac{1}{2}c_n c_{-n}\left(-\frac{\omega_n^2}{v^2} + k_n^2\right)\left[a_n(t)a_{-n}(t) + a_n^*(t)a_{-n}^*(t)\right]\right\}\\ &= \sum_{n=-\infty}^{\infty}c_n^2\frac{2\omega_n^2}{v^2}\,a_n^*(t)a_n(t) = \sum_{n=-\infty}^{\infty}c_n^2\frac{2\omega_n^2}{v^2}\,a_n^*(0)a_n(0)\ .\end{aligned}$$

In natural units where $\hbar = 1$, $E = \hbar\omega = \omega$, and the frequency ω has units of energy. It is convenient to choose c_n to make a_n dimensionless. If we choose

$$c_n = \left(\frac{v^2}{2\omega_n}\right)^{1/2}\ ,$$

the Hamiltonian assumes a simple form

$$H = \sum_{n=-\infty}^{\infty}\omega_n\, a_n^*(t)a_n(t)\ .$$

An alternative choice of coordinates will enable us to quantize these oscillators. For this we need generalized positions and momenta, which must be real. Choose

$$q_n(t) = \frac{1}{\sqrt{2\omega_n}} [a_n(t) + a_n^*(t)]$$
$$p_n(t) = \frac{dq_n}{dt} = -\frac{i\omega_n}{\sqrt{2\omega_n}} [a_n(t) - a_n^*(t)] \ .$$

The a's can then be expressed in terms of the real p's and q's

$$a_n = \frac{ip_n + \omega_n q_n}{\sqrt{2\omega_n}} \qquad a_n^* = \frac{-ip_n + \omega_n q_n}{\sqrt{2\omega_n}}$$

and the Hamiltonian becomes

$$H = \sum_{n=-\infty}^{\infty} \frac{1}{2} [p_n^2 + \omega_n^2 q_n^2] \ , \tag{1.12}$$

which is a sum of independent oscillator Hamiltonians. This is confirmed by substituting (1.12) into Hamilton's equations of motion

$$\dot{q}_n = \frac{\partial H}{\partial p_n} = p_n$$
$$\dot{p}_n = -\frac{\partial H}{\partial q_n} = -\omega_n^2 q_n \ ,$$

which gives back the familiar equations of motion for uncoupled oscillators.

1.3 QUANTIZATION OF THE STRING

We now quantize the string by the canonical procedure: the canonical variables are made into operators which are defined by transforming the Poisson bracket relations into commutation relations [for a review of this procedure see, for example, Schiff (1968), Sec. 24]. For the generalized coordinates and momenta this leads to the following commutation relations:

$$\begin{aligned} [q_n, p_m] &= i\hbar\, \delta_{nm} \\ [q_n, q_m] &= [p_n, p_m] = 0 \ . \end{aligned} \tag{1.13}$$

In what follows we will always set $\hbar = c = 1$. This defines the so-called *natural* system of units, which is very convenient. It is important to realize that the correct factors of $\hbar$ and c can always be *uniquely* restored at the end of any calculation, if desired. These units are discussed in Prob. 1.1 at the end of this chapter.

From the commutation relations (1.13) we obtain

$$\boxed{\begin{aligned}\left[a_n, a_m^\dagger\right] &= \delta_{n,m} \\ \left[a_n, a_m\right] &= \left[a_n^\dagger, a_m^\dagger\right] = 0 \,,\end{aligned}} \tag{1.14}$$

where the complex conjugate of a complex number (sometimes called a *c-number*) must be generalized to the Hermitian conjugate of an operator (sometimes called a *q-number*), and the operators a_n are independent of time. The time dependence is in the fields, which are also operators*:

$$\begin{aligned}\phi(z,t) &= v \sum_{n=-\infty}^{\infty} \frac{1}{\sqrt{2\omega_n L}} \left\{a_n\, e^{i(k_n z-\omega_n t)} + a_n^\dagger\, e^{-i(k_n z-\omega_n t)}\right\} \\ &= \phi^{(+)}(z,t) + \phi^{(-)}(z,t) \;,\end{aligned} \tag{1.15}$$

where the positive frequency part, $\phi^{(+)}$, contains the sum over a_n (later to be identified as annihilation operators) and the negative frequency part, $\phi^{(-)}$, is the sum over $a_n^\dagger$ (the creation operators). In this case ϕ is Hermitian because it is associated with a physical observable (the displacement), but in general a field need not be a Hermitian operator. We will study such fields in Part III of this book.

The Hamiltonian is also an operator, and its precise form depends on the order of $a^\dagger$ and a, which was unimportant when these were c-numbers. Perhaps the most "natural" form for H is

$$\begin{aligned}H &= \sum_{n=-\infty}^{\infty} \omega_n \frac{1}{2}\left[a_n^\dagger a_n + a_n a_n^\dagger\right] \\ &= \sum_{n=-\infty}^{\infty} \omega_n \left[a_n^\dagger a_n + \frac{1}{2}\right] \;.\end{aligned}$$

However, the sum over $\frac{1}{2}\omega_n$ gives an infinite contribution to the energy (the zero-point energy), which can be removed simply by redefining the energy. This redefinition will lead to the idea of a *normal ordered product*, which will be defined and discussed in Sec. 1.6 below. For now we will simply adopt the following form for H:

$$H = \sum_{n=-\infty}^{\infty} \omega_n a_n^\dagger a_n \;. \tag{1.16}$$

Note that H is the sum of the dimensionless operators $a_n^\dagger a_n$, each multiplied by the energy ω_n of the normal mode which it describes.

---

*To avoid singularities, we will exclude the state $n = 0$ from this sum. Later, when we take the limit $L \to \infty$ (the continuum limit), the sum will include states of arbitrarily small energy.

1.4 CANONICAL COMMUTATION RELATIONS

The commutation relations between a and $a^\dagger$ also imply relations between the fields ϕ. Suppose we regard ϕ as a canonical coordinate. Then, the canonical momentum is [using $\mathcal{L}$ defined in Eq. (1.2)]

$$\pi(z,t) = \frac{\partial \mathcal{L}}{\partial\left(\dfrac{\partial\phi}{\partial t}\right)} = \frac{1}{v^2}\frac{\partial\phi}{\partial t} \ .$$

Then, generalizing the commutation relations (1.13) to a continuous field, we expect to find relations of the form

$$\boxed{\begin{aligned} &[\phi(z,t),\pi(z',t)] = i\delta(z-z') \\ &[\pi(z,t),\pi(z',t)] = 0 \\ &[\phi(z,t),\phi(z',t)] = 0 \ , \end{aligned}} \tag{1.17}$$

where the $\delta(z-z')$ function is the generalization of the Kronecker δ_{nm} which appears in (1.13). These important commutation relations are known as the *canonical commutation relations*, sometimes referred to as the CCR's.

To prove the relations (1.17), we use the explicit form for π:

$$\pi(z,t) = \frac{1}{v}\sum_{n=-\infty}^{\infty}\frac{1}{\sqrt{2\omega_n L}}\left\{-i\omega_n a_n\, e^{i(k_n z-\omega_n t)} + i\omega_n a_n^\dagger\, e^{-i(k_n z-\omega_n t)}\right\} \ .$$

Then

$$\begin{aligned} [\pi(z,t),\phi(z',t)] &= -\frac{i}{2L}\sum_{n,m}\frac{\sqrt{\omega_n}}{\sqrt{\omega_m}} \\ &\times\Big\{\left[a_n,a_m\right] e^{i(k_n z+k_m z')-i(\omega_n+\omega_m)t} - \left[a_n^\dagger,a_m^\dagger\right] e^{-i(k_n z+k_m z')+i(\omega_n+\omega_m)t} \\ &+\left[a_n,a_m^\dagger\right] e^{i(k_n z-k_m z')-i(\omega_n-\omega_m)t} - \left[a_n^\dagger,a_m\right] e^{-i(k_n z-k_m z')+i(\omega_n-\omega_m)t}\Big\} \\ &= -i\frac{1}{L}\sum_n e^{ik_n(z-z')} \equiv -i\, I(z,z') \ . \end{aligned}$$

However, the functions $\frac{1}{\sqrt{L}}e^{ik_n z}$ are *complete* (i.e., any periodic function can be expanded in terms of them) and orthonormal, and hence

$$\int_0^L dz'\, I(z,z')\, e^{ik_n z'} = e^{ik_n z} \ ,$$

which is the property of the δ-function, and hence

$$I(z,z') = \frac{1}{L}\sum_n e^{ik_n(z-z')} = \delta(z-z') \ . \tag{1.18}$$

This proves the first of the relations (1.17). To prove the others, note that

$$\begin{aligned}[\phi(z,t),\phi(z',t)] =& \frac{v^2}{2L}\sum_{n,m}\frac{1}{\sqrt{\omega_n\omega_m}}\Big[e^{i\left(k_n z-k_m z'\right)-i(\omega_n-\omega_m)t}\left[a_n,a_m^\dagger\right] \\ & \qquad + e^{-i\left(k_n z-k_m z'\right)+i(\omega_n-\omega_m)t}\left[a_n^\dagger,a_m\right]\Big] \\ =& \frac{v^2}{2L}\sum_n \frac{1}{\omega_n} e^{ik_n(z-z')}\left[1-1\right] = 0 \ .\end{aligned}$$

A field theory may be quantized with *either* the CCR's (1.17) *or* the commutation relations (1.14) between the operators a and $a^\dagger$. As we have seen, these two methods are equivalent. Should either be regarded as more fundamental than the other? Many prefer to start from the CCR's because of their close connection with the fundamental relations (1.13), but in this book the relations (1.14) between the a's will be chosen as the starting point for quantizing new field theories. The reason for this choice is that the relations (1.14) are directly related to the oscillators which describe the *independent* dynamical degrees of freedom associated with the field, and therefore always have the same form, while the fields themselves sometimes include degrees of freedom which are *not independent* (such as the vector degrees of freedom of the electromagnetic field) and in these cases the form of the CCR's must be modified so that these dependent degrees of freedom are removed from the commutation relations. This will be apparent in the next chapter where the quantization of the electromagnetic field is discussed.

1.5 THE NUMBER OPERATOR AND PHONON STATES

Next, we find the eigenstates of the Hamiltonian (1.16). The first step is to find the eigenstates of the operator

$$\mathcal{N}_n = a_n^\dagger a_n$$

known as the *number operator*. These are easy to find from the commutation relation for the a's.

Since $\mathcal{N} = a^\dagger a$ is Hermitian (from now on we suppress n), it has a complete set of orthonormal eigenstates. Denote these by $|m\rangle$. Then

$$\begin{aligned}\mathcal{N}|m\rangle &= m|m\rangle \\ \langle m'|m\rangle &= \delta_{m',m} \\ \mathbf{1} &= \sum_m |m\rangle\langle m| \ .\end{aligned}$$

At this point we know only that m is real.

Now consider the state $a^\dagger|m\rangle$. From the commutation relations (1.14) we have

$$\begin{aligned}\mathcal{N} a^\dagger|m\rangle &= \{[\mathcal{N}, a^\dagger] + a^\dagger\mathcal{N}\}\,|m\rangle \\ &= (a^\dagger + a^\dagger m)\,|m\rangle = (m+1)a^\dagger|m\rangle \ .\end{aligned}$$

Hence

$$a^\dagger|m\rangle = C_+|m+1\rangle \ ,$$

where C_+ is a number to be determined. A similar argument gives

$$a|m\rangle = C_-|m-1\rangle \ .$$

The numbers C_+ and C_- can be determined from the norms

$$\langle m|aa^\dagger|m\rangle = |C_+|^2 = \langle m\,|(1+a^\dagger a)|\,m\rangle = m+1 \ .$$

Similarly,

$$|C_-|^2 = \langle m\,|a^\dagger a|\,m\rangle = m \ .$$

The axiomatic development of quantum mechanics requires that all quantum mechanical states lie in a Hilbert space with a positive definite norm. Hence we require that $m > 0$, or if $m = 0$,

$$a|0\rangle = 0 \ .$$

Furthermore, since m can be lowered by integers, all positive m must be integers; otherwise, we could generate negative values for m from positive values by lowering m repeatedly by one unit.

Hence, it is possible to choose phases (signs) so that ($m \geq 0$)

$$\boxed{\begin{aligned}a^\dagger|m\rangle &= \sqrt{m+1}\,|m+1\rangle \\ a|m\rangle &= \sqrt{m}\,|m-1\rangle \\ a^\dagger a|m\rangle &= m|m\rangle \ .\end{aligned}} \tag{1.19}$$

This means that *all* the states can be generated from a "ground state" $|0\rangle$ (sometimes called the "vacuum") by successive operations of $a^\dagger$:

$$|m\rangle = \frac{(a^\dagger)^m}{\sqrt{m!}}|0\rangle \ .$$

For a mechanical system like the string, these states $|m\rangle$ are referred to as *phonon* states, and if $a = a_n$, we will show that m can be interpreted as the number of phonons of energy ω_n, where the quantum of energy carried by the phonon is associated with the *entire system*. This justifies calling $\mathcal{N}$ the number operator and suggests that the operators a and $a^\dagger$ have the following interpretation:

$$\begin{array}{ll} a_n^\dagger & \textit{creates} \text{ a phonon with frequency } \omega_n \\ a_n & \textit{destroys} \text{ a phonon with frequency } \omega_n \quad . \end{array}$$

This description is further supported by the Hamiltonian (1.16) which now has a simple physical interpretation. If $a_n^\dagger a_n$ is an operator which gives the number of phonons of frequency (energy) ω_n, then (1.16) expresses the total energy (H) as a sum of the energy of each phonon (ω_n) times the number of phonons with that energy $\left(a_n^\dagger a_n\right)$. The most general eigenstate of the Hamiltonian can therefore be written*:

$$\boxed{|m_{n_1}, m_{n_2}, m_{n_3}, \ldots\rangle = \frac{\left(a_{n_1}^\dagger\right)^{m_{n_1}}}{\sqrt{m_{n_1}!}} \frac{\left(a_{n_2}^\dagger\right)^{m_{n_2}}}{\sqrt{m_{n_2}!}} \cdots |0\rangle \quad .} \tag{1.20}$$

Since all creation operators commute, these states are completely symmetric and satisfy Bose–Einstein statistics. Such states, with a definite number of phonons of various frequencies, are referred to as *Fock* states.

It is sometimes tempting to try to relate the particles associated with the field (the phonons) to the original mass points from which the string was constructed. However, there is *no* connection between these two kinds of particles. The phonons are associated with frequencies, or normal modes of the string, and hence are related to the motion of the string *as a whole*, its collective motion. They are localized in "frequency," or momentum space, while the particles in the string are localized in position space. Later we will see that there are also particles associated with abstract fields which have no connection with any mechanical system.

1.6 THE QUANTA AS PARTICLES

The quanta associated with a quantum field (the phonons in this example) really are physical particles which carry both momentum and energy. In the previous section we saw how the phonons carry energy. The Hamiltonian tells us that the total energy of a state with a definite number of phonons (a Fock state) is simply the sum of the energy carried by each of the phonons in the state. To complete the description of phonons as particles, we must show that they also carry momentum. This will be done in this section by first finding the momentum operator of the field and then showing that the total momentum of a Fock state is simply the *vector* sum of the momentum of each of the phonons in the state.

*Of course, if more than one state has the same energy (there is a degeneracy), the most general state will be a linear combination of all the states with that energy.

Using Noether's theorem, the momentum (and energy) operators can be determined in an elegant and completely general way for any abstract field theory. This will be discussed later in Chapter 8. In this chapter we exploit the physical properties of the one-dimensional string and determine the momentum operator from the continuity equation.

The energy density carries with it a momentum density which describes how the energy flows. This momentum density is related to the energy density by the continuity equation, which in three dimensions is

$$\frac{1}{v^2}\frac{\partial \mathcal{E}}{\partial t} + \nabla_\imath \mathcal{P}^\imath = 0\,. \tag{1.21}$$

Digression: To recall the origin and physical content of this equation, consider a compressional wave traveling with velocity v in the positive z-direction. This wave has a local mass density $\rho(z-vt)$ different from the average density of the string. Then, the kinetic energy associated with this excess density is

$$\mathcal{E}_{\mathrm{KE}} = \tfrac{1}{2}\rho(z-vt)\,v^2 \quad .$$

By the virial theorem applied to a collection of SHO's, an equal energy also comes from the potential energy, so that

$$\mathcal{E}_{\mathrm{total}} = \rho(z-vt)\,v^2 \quad .$$

But the momentum density associated with the mass flow is

$$\mathcal{P}^z = \rho(z-vt)\,v$$

so that one obtains Eq. (1.21):

$$\frac{1}{v^2}\frac{\partial \mathcal{E}}{\partial t} = -v\rho' = -\frac{\partial}{\partial z}\mathcal{P}^z \quad .$$

∎

Now we will use the continuity equation to find the momentum operator. For the string, $\mathcal{E}$ equals the Hamiltonian density of Eq. (1.2), and hence

$$\begin{aligned}
\frac{1}{v^2}\frac{\partial}{\partial t}\mathcal{E} &= \frac{1}{v^2}\left[\frac{1}{v^2}\frac{\partial \phi}{\partial t}\frac{\partial^2 \phi}{\partial t^2} + \frac{\partial \phi}{\partial z}\frac{\partial^2 \phi}{\partial t \partial z}\right] \\
&= \frac{1}{v^2}\left[\frac{\partial \phi}{\partial t}\frac{\partial^2 \phi}{\partial z^2} + \frac{\partial \phi}{\partial z}\frac{\partial^2 \phi}{\partial t \partial z}\right] \\
&= \frac{1}{v^2}\frac{\partial}{\partial z}\left[\frac{\partial \phi}{\partial t}\frac{\partial \phi}{\partial z}\right] \quad .
\end{aligned}$$

Hence the classical momentum density must be

$$\mathcal{P}^z = -\frac{1}{v^2}\left(\frac{\partial \phi}{\partial t}\frac{\partial \phi}{\partial z}\right) \quad . \tag{1.22}$$

We can turn (1.22) into a quantum mechanical operator by replacing the classical fields by their quantum mechanical operator equivalents. Since the field operators do not in general commute, the order of the terms in any product is important, and it is convenient to choose this order so that (in this example) the expectation value of the momentum of the ground state $|0\rangle$ is zero. To this end we define the *normal ordered product* of two field operators as follows:

$$\begin{aligned}
:\phi_1\phi_2: \equiv & \phi_1\phi_2 - \langle 0|\phi_1\phi_2|0\rangle = \phi_1^{(+)}\phi_2^{(+)} + \phi_1^{(-)}\phi_2^{(+)} + \phi_2^{(-)}\phi_1^{(+)} \\
& + \phi_1^{(-)}\phi_2^{(-)} + \left[\phi_1^{(+)}, \phi_2^{(-)}\right] - \langle 0|\phi_1\phi_2|0\rangle \\
= & \phi_1^{(+)}\phi_2^{(+)} + \phi_1^{(-)}\phi_2^{(+)} + \phi_2^{(-)}\phi_1^{(+)} + \phi_1^{(-)}\phi_2^{(-)} \quad ,
\end{aligned} \tag{1.23}$$

where $\phi_i^{(+)}$ and $\phi_i^{(-)}$ are the positive and negative frequency parts of the field ϕ_i, as defined in Eq. (1.15), and to obtain the last line use the facts that $\phi^{(+)}|0\rangle = 0$, $\langle 0|\phi^{(-)} = 0$, and the commutator $[\phi_1^{(+)}, \phi_2^{(-)}]$ is a *c*-number, so that it is equal to its ground state expectation value. Hence the normal ordered product of operators which satisfy *commutation* relations like (1.17) can be obtained simply by reordering any terms in which creation operators are on the right and the annihilation operators are on the left, so that all the terms have either two annihilation operators, two creation operators, or a creation operator on the left and an annihilation operator on the right.

Using this definition, the total momentum operator of the one-dimensional string is

$$\boxed{P^z = -\int_0^L \frac{dz}{v^2} :\frac{\partial \phi}{\partial t}\frac{\partial \phi}{\partial z}: \quad .} \tag{1.24}$$

The total momentum assumes a simple, clearly interpretable form when expressed in terms of the a's. To obtain it, substitute (1.15) into (1.24), honoring the normal ordered definition (1.23):

$$\begin{aligned}
P^z = & -\sum_{n,m} \frac{k_m\omega_n}{2L\sqrt{\omega_n\omega_m}} \int_0^L dz \\
& \times \left\{ a_n a_m e^{i(k_n+k_m)z - i(\omega_n+\omega_m)t} + a_n^\dagger a_m^\dagger\, e^{-i(k_n+k_m)z + i(\omega_n+\omega_m)t} \right. \\
& \left. - a_n^\dagger a_m\, e^{-i(k_n-k_m)z + i(\omega_n-\omega_m)t} - a_m^\dagger a_n\, e^{i(k_n-k_m)z - i(\omega_n-\omega_m)t} \right\} \\
= & -\frac{1}{2}\sum_n \left\{ k_{-n} a_n a_{-n}\, e^{-2i\omega_n t} + k_{-n} a_n^\dagger a_{-n}^\dagger\, e^{2i\omega_n t} - 2k_n a_n^\dagger a_n \right\} \quad .
\end{aligned}$$

However, the first two terms sum to zero, because they are odd when n is changed to $-n$ (recall $k_n = -k_{-n}$ but $\omega_n = \omega_{-n}$). Hence

$$\boxed{P^z = \sum_n k_n a_n^\dagger a_n \quad ,} \tag{1.25}$$

which expresses the total momentum as a *vector* sum of the momentum of each phonon (k_n) times the number of phonons with that momentum ($a_n^\dagger a_n$). [The full vector character of the momentum operator is only partially illustrated by this one-dimensional example, where P_z has only a z component and all k_n must be in the $\hat{z}$-direction so that k_n can be only positive or negative.] We see that the momentum operator is precisely what we would expect from the interpretation of phonons as *particles* with energy ω_n and momentum k_n.

1.7 THE CLASSICAL LIMIT: FIELD–PARTICLE DUALITY

The Fock states are the quantum mechanical eigenstates of the Hamiltonian. What do these have to do with the classical vibrational states of a string? What is the classical limit? Before giving a full answer to these questions, we make two preliminary observations.

First, note that a state with a *definite number of quanta* corresponds to a case where the *average field is zero,* but otherwise the field is *completely unknown.* To show this, consider a state with n_1 quanta of type 1: $|n_1\rangle$. Then, for quanta of any type m (including $m = 1$)

$$\langle n_1|a_m|n_1\rangle = 0 = \langle n_1|a_m^\dagger|n_1\rangle$$

so that the average field is zero,

$$\langle n_1|\phi(z,t)|n_1\rangle = 0 \ .$$

However, the average of the square of the field is not zero. In fact,

$$\langle n_1|\phi^2(z,t)|n_1\rangle = \langle n_1|v^2 \sum_{i,\,j} \frac{1}{2L\sqrt{\omega_i\omega_j}} \times \left\{a_i a_j^\dagger\, e^{i(k_i-k_j)z-i(\omega_i-\omega_j)t} + a_i^\dagger a_j\, e^{-i(k_i-k_j)z+i(\omega_i-\omega_j)t}\right\} |n_1\rangle \ ,$$

where $\langle n_1|a_i a_j|n_1\rangle = 0 = \langle n_1|a_i^\dagger a_j^\dagger|n_1\rangle$ has been used. Next, note that

$$\langle n_1|a_i a_j^\dagger|n_1\rangle = \delta_{ij} + \langle n_1|a_j^\dagger a_i|n_1\rangle = \delta_{ij} + n_1\delta_{i1}\delta_{j1}$$
$$\langle n_1|a_i^\dagger a_j|n_1\rangle = n_1\delta_{i1}\delta_{j1}$$

and hence

$$\langle n_1|\phi^2(z,t)|n_1\rangle = \frac{v^2}{2L}\left[\frac{2n_1}{\omega_1} + \sum_i \frac{1}{\omega_i}\right] \Rightarrow \infty$$

because $\sum_i |i|^{-1}$ diverges. Hence the uncertainty in ϕ, $\Delta\phi$, is

$$\Delta\phi = \sqrt{\langle\phi^2\rangle - \langle\phi\rangle^2} \Rightarrow \infty$$

and ϕ is completely uncertain, beyond knowing that $\langle\phi\rangle = 0$. [To define the field so that $\Delta\phi \neq \infty$, we may "smear" it, introducing

$$\phi(f) \equiv \int dz\, f(z)\phi(z,t) \ ,$$

where $f(z)$ is strongly peaked in the neighborhood of a point $z = z_0$, and very small elsewhere; see Prob. 1.5 at the end of the chapter.]

For our *second* observation we note that *no state behaves like a classical wave for all z and t.** This would require that the field ϕ and its "velocity" π commute, and the CCR's (1.17) show that this is not the case. Another way to see this is to rewrite ϕ as a sum of traveling waves,

$$\phi(z,t) = v\sum_n \frac{1}{\sqrt{2\omega_n L}}\left\{A_n \cos\left(k_n z - \omega_n t\right) + B_n \sin\left(k_n z - \omega_n t\right)\right\} \ ,$$

where

$$a_n = \tfrac{1}{2}\left(A_n - iB_n\right) \qquad a_n^\dagger = \tfrac{1}{2}\left(A_n + iB_n\right)$$

or, dropping the n

$$A = a + a^\dagger \qquad B = i\left(a - a^\dagger\right) \ .$$

The operators A *and* B *must be simultaneously diagonalized* in order that $\phi(z,t)$ have a definite value for all z and t. But *this is impossible*, because A and B are non-commuting operators:

$$[A,B] = i\left[\left(a + a^\dagger\right), \left(a - a^\dagger\right)\right] = -2i$$

and hence cannot be simultaneously diagonalized (i.e., cannot both have definite values). Furthermore, the above commutator implies an *uncertainty relation*

$$\Delta A\,\Delta B \geq 1 \ . \tag{1.26}$$

These two results give limitations on our ability to define the field and show that it cannot be defined exactly. However, states do exist in which A and B have a very small fractional uncertainty. Such states correspond to a classical field as much

*Thanks to Charles Sommerfield and Alan Chodos for clarification of this point.

as is possible in quantum mechanics. Since an optimization of (1.26) requires that $\Delta A \simeq \Delta B \simeq 1$, small fractional uncertainty in the values of $\langle A\rangle$ and $\langle B\rangle$ is possible only if $\langle A\rangle$ and $\langle B\rangle$ are both very large. However, these quantities are related to the average number of quanta through the relation

$$A^2 + B^2 = 4a^\dagger a + 2 = 4\mathcal{N} + 2$$

and hence such states must have a large average number of quanta $\langle\mathcal{N}\rangle$. If we parameterize $\langle A\rangle$ and $\langle B\rangle$ by

$$\begin{aligned}\langle A\rangle &\cong 2\sqrt{\langle\mathcal{N}\rangle}\cos\delta \\ \langle B\rangle &\cong -2\sqrt{\langle\mathcal{N}\rangle}\sin\delta\end{aligned}$$

and if $\langle\mathcal{N}\rangle \to \infty$, then the fractional uncertainty in $\langle A\rangle$ and $\langle B\rangle$ goes like

$$\left.\begin{aligned}\frac{\Delta A}{\langle A\rangle} &\cong \frac{1}{2\sqrt{\langle\mathcal{N}\rangle}\cos\delta} \to 0 \\ \frac{\Delta B}{\langle B\rangle} &\cong \frac{-1}{2\sqrt{\langle\mathcal{N}\rangle}\sin\delta} \to 0\end{aligned}\right\} \quad \text{as } \langle\mathcal{N}\rangle \to \infty$$

and the fractional uncertainty in A and B is small and the average field $\langle\phi\rangle$ is well-defined in both amplitude and phase (except for exceptional cases where $\sin\delta$ or $\cos\delta = 0$).

An example of a class of states with this property is the *coherent states*, which are the *eigenfunctions of the annihilation operator a*. These states can be written

$$|K\rangle = C\sum_{n=0}^{\infty}\frac{K^n}{\sqrt{n!}}|n\rangle = C\sum_{n=0}^{\infty}\frac{\left(Ka^\dagger\right)^n}{n!}|0\rangle\ , \tag{1.27}$$

where C is a normalization constant, and K is the complex eigenvalue corresponding to the eigenvector $|K\rangle$

$$\begin{aligned}a|K\rangle &= C\sum_{n=0}^{\infty}\frac{K^n}{\sqrt{n!}}a|n\rangle = C\sum_{n=1}^{\infty}\frac{K^n}{\sqrt{n!}}\sqrt{n}|n-1\rangle \\ &= KC\sum_{n=0}^{\infty}\frac{K^n}{\sqrt{n!}}|n\rangle = K|K\rangle\ .\end{aligned}$$

We will parameterize the eigenvalue K by

$$K = \sqrt{N}\,e^{\imath\alpha}$$

and normalize the state

$$\langle K|K\rangle = C^2\sum_{n=0}^{\infty}\frac{\left(|K|^2\right)^n}{n!} = C^2\,e^{|K|^2} = 1\ ,$$

which implies that

$$C = e^{-|K|^2/2} = e^{-N/2} \quad .$$

It is worth noting that the operation of the creation operator on the coherent state is equivalent to differentiating the state with respect to the complex number K,

$$\begin{aligned} a^\dagger|K\rangle &= C\sum_{n=0}^{\infty}\frac{K^n}{\sqrt{n!}}a^\dagger|n\rangle = C\sum_{n=0}^{\infty}\frac{K^n}{\sqrt{n!}}\sqrt{n+1}|n+1\rangle \\ &= C\sum_{n=0}^{\infty} n\,\frac{K^{n-1}}{\sqrt{n!}}|n\rangle = \frac{d}{dK}\,|K\rangle \quad . \end{aligned} \tag{1.28}$$

Using these remarkable results, we can quickly calculate $\langle A\rangle$, $\langle B\rangle$, ΔA, and ΔB. First, using $\langle K|a^\dagger = \langle K|K^*$,

$$\begin{aligned} \langle K|a|K\rangle &= K & \langle K|a^\dagger|K\rangle &= K^* \\ \langle K|a^2|K\rangle &= K^2 & \langle K|a^{\dagger 2}|k\rangle &= K^{*2} \\ \langle K|aa^\dagger|K\rangle &= 1+|K|^2 & \langle K|a^\dagger a|K\rangle &= |K|^2 = N \end{aligned}$$

and hence

$$\begin{aligned} \langle A\rangle &= \langle a+a^\dagger\rangle = 2ReK = 2N^{1/2}\cos\alpha \\ \langle B\rangle &= i\langle a-a^\dagger\rangle = -2ImK = -2N^{1/2}\sin\alpha \\ \langle A^2\rangle &= \left\langle a^2+{a^\dagger}^2+aa^\dagger+a^\dagger a\right\rangle = 2Re(K^2)+1+2|K|^2 \\ &= 2N(\cos 2\alpha+1)+1 = 4N\cos^2\alpha+1 \\ \langle B^2\rangle &= -\left\langle a^2+{a^\dagger}^2-aa^\dagger-a^\dagger a\right\rangle = -2ReK^2+1+2(K)^2 \\ &= 2N(1-\cos 2\alpha)+1 = 4N\sin^2\alpha+1 \quad . \end{aligned}$$

Therefore

$$\begin{aligned} \Delta A &= \left(\langle A^2\rangle-\langle A\rangle^2\right)^{1/2} = \left(4N\cos^2\alpha+1-4N\cos^2\alpha\right)^{1/2} = 1 \\ \Delta B &= \left(\langle B^2\rangle-\langle B\rangle^2\right)^{1/2} = \left(4N\sin^2\alpha+1-4N\sin^2\alpha\right)^{1/2} = 1 \end{aligned}$$

and the fractional uncertainty in A and B does indeed approach 0 if $N\to\infty$. Furthermore,

$$\begin{aligned} \langle\mathcal{N}\rangle &= N \\ \langle\mathcal{N}^2\rangle &= \left\langle\left(a^\dagger a\right)^2\right\rangle = \langle|a^\dagger a\,a^\dagger a|\rangle \\ &= \langle|a^\dagger a+a^\dagger a^\dagger aa|\rangle = N^2+N \end{aligned}$$

so that N is indeed the average number of phonons, but the uncertainty in the number of phonons also approaches zero as $N \to \infty$:

$$\frac{\Delta\mathcal{N}}{\langle\mathcal{N}\rangle} = \frac{\left(\langle\mathcal{N}^2\rangle - \langle\mathcal{N}\rangle^2\right)^{1/2}}{N} = \frac{(N^2 + N - N^2)^{1/2}}{N} = \frac{1}{\sqrt{N}} \to 0 \ .$$

So far our considerations have been limited to a specific frequency. To obtain a well-defined field, we must construct a *coherent state for each frequency*. Hence the general state is of the form

$$|K_1 \ldots K_n \ldots\rangle = C_1 \ldots C_n \ldots \sum_{\substack{n_1 \ldots \\ n_n \ldots = 0}}^{\infty} \frac{(K_1)^{n_1} \ldots (K_n)^{n_n} \ldots}{\sqrt{(n_1!) \ldots (n_n!) \ldots}} |n_1 \ldots n_n \ldots\rangle$$

and there is a field–particle uncertainty relation, or complementarity principle. If $\Delta\mathcal{N} = 0$, then $\Delta\phi = \infty$, while if $\Delta\phi$ is small, $\Delta\mathcal{N}$ must be large.

1.8 TIME TRANSLATION

One of the most fundamental problems in physics is the determination of the time evolution of physical observables. In the language of quantum mechanics, this problem is solved by finding an operator from which it is possible to calculate how matrix elements of quantum mechanical operators evolve in time. We close this chapter with an introductory discussion of how this is done in field theory. We will return to this issue several times in later chapters, but our development will always be very similar to the one presented here.

The time translation operator can be found from the Hamiltonian, which describes how states evolve over an *infinitesimal* period of time. In field theory, this property of the Hamiltonian is described mathematically by the following relations:

$$\boxed{\begin{aligned} [H, \phi(z,t)] &= -i\frac{\partial\phi(z,t)}{\partial t} \\ [H, \pi(z,t)] &= -i\frac{\partial\pi(z,t)}{\partial t} \ . \end{aligned}} \tag{1.29}$$

These fundamental relations are sufficient to establish H as the *generator* of time translations and to permit the construction of the operator for *finite* time translations (for more discussion, see Chapter 8).

To prove the above relations for the one-dimensional string, we ignore the fact that H is normal ordered, since the only difference between a regular product and a normal ordered product is a c-number, which commutes with ϕ and π. Then we use the CCR's to obtain

$$[H, \phi(z,t)] = \left[\int_0^L dz' \frac{1}{2} \left(v^2 \pi^2(z',t) + \left(\frac{\partial \phi(z',t)}{\partial z'} \right)^2 \right), \phi(z,t) \right]$$

$$= \int_0^L dz' v^2 \frac{1}{2} \left[\pi^2(z',t), \phi(z,t) \right] \quad .$$

If $[A, B] = c$, where c is a complex number, then

$$[A^2, B] = 2cA \quad . \tag{1.30}$$

Hence, from the CCR,

$$[H, \phi(z,t)] = -i\, v^2 \int_0^L dz' \pi(z',t) \delta(z'-z) = -i \frac{\partial \phi(z,t)}{\partial t} \quad .$$

For the second relation, use (1.30),

$$[H, \pi(z,t)] = \int_0^L dz' \frac{\partial \phi(z',t)}{\partial z'} \left[\frac{\partial \phi(z',t)}{\partial z'}, \pi(z,t) \right] \quad ,$$

and find the commutator by differentiating the CCR,

$$[\pi(z,t), \phi(z',t)] = -i\delta(z'-z) \Longrightarrow \left[\pi(z,t), \frac{\partial \phi(z',t)}{\partial z'} \right] = -i \frac{d}{dz'.} \delta(z'-z) \quad .$$

Hence, integrating by parts and using the wave equation,

$$[H, \pi(z,t)] = i \int_0^L dz' \frac{d}{dz'} \delta(z'-z) \frac{\partial \phi(z',t)}{\partial z'} = -i \frac{\partial^2 \phi(z,t)}{\partial z^2}$$

$$= -i \frac{1}{v^2} \frac{\partial^2 \phi(z,t)}{\partial t^2} = -i \frac{\partial \pi}{\partial t} \quad .$$

This completes the derivation of the relations (1.29).

The next task is to use these relations to construct the time translation operator, but first we must decide how we are going to describe this operator. In general, in quantum mechanics, there are two choices which can be made. One may choose to have the states change with time and the operators remain fixed (the *Schrödinger picture*) or the states remain fixed and the operators change with time (the *Heisenberg picture*). In the Schrödinger picture, there is a time translation operator which evolves the states from time t_0 to t:

$$\boxed{|t\rangle = U(t, t_0)|t_0\rangle} \qquad \text{Schrödinger picture.} \tag{1.31}$$

In this picture, the operators are fixed at the reference time t_0, and matrix elements at arbitrary time t are written $\langle t|\phi(t_0)|t\rangle$. [For simplicity, in the remainder of this section, we will ignore the dependence of ϕ on z, and write $\phi(z,t) \to \phi(t)$.] The time translation operator is a unitary matrix which operates on the vector space of possible states. It must be unitary because the norm of the state vector, which is the total probability, must be conserved. (If there are several channels, the probability that any particular channel will be occupied may change with time, but the sum of all the probabilities must always add up to unity.)

In the Heisenberg picture, the operators depend on time, and the states are fixed at the reference time t_0. Since all matrix elements must be independent of which picture we use, the relation between the two pictures follows from

$$\underbrace{\langle t|\phi(t_0)|t\rangle}_{\text{Schrödinger}} = \underbrace{\langle t_0|\phi(t)|t_0\rangle}_{\text{Heisenberg}} \; .$$

These are equivalent if the operators in the Heisenberg picture evolve with time according to the following relation:

$$\boxed{\phi(t) = U^{-1}(t,t_0)\phi(t_0)U(t,t_0)} \qquad \text{Heisenberg picture.} \tag{1.32}$$

We will use the Heisenberg picture (which has been employed so far) and the commutation relations Eq. (1.29) to find the form of $U(t,t_0)$. Begin by writing $\partial\phi/\partial t$ in two equivalent ways:

$$\begin{aligned} -i\frac{\partial\phi(t)}{\partial t} &= [H,\phi(t)] = U^{-1}\,[H,\phi(t_0)]\,U \\ &= -i\left\{\frac{d}{dt}U^{-1}(t,t_0)\phi(t_0)U + U^{-1}\phi(t_0)\frac{d}{dt}U(t,t_0)\right\} \; , \end{aligned} \tag{1.33}$$

where, in the first line, we used $U^{-1}HU = H$, which follows from the fact that H is independent of time, and the second line is simply the time derivative of Eq. (1.32). Using $UU^{-1} = 1$, which implies

$$\left(\frac{d}{dt}U\right)U^{-1} + U\frac{d}{dt}U^{-1} = 0 \; , \tag{1.34}$$

Eq. (1.33) can be rearranged as follows:

$$\begin{aligned} [H,\phi(t_0)] &= -iU\left\{\frac{d}{dt}U^{-1}(t,t_0)\phi(t_0)U + U^{-1}\phi(t_0)\frac{d}{dt}U(t,t_0)\right\}U^{-1} \\ &= +i\left\{\left(\frac{d}{dt}U(t,t_0)\right)U^{-1}\phi(t_0) - \phi(t_0)\left(\frac{d}{dt}U(t,t_0)\right)U^{-1}\right\} \\ &= i\left[\left(\frac{d}{dt}U(t,t_0)\right)U^{-1},\ \phi(t_0)\right] \end{aligned}$$

.

or

$$\left[\left(H - i\left(\frac{dU}{dt}\right)U^{-1}\right), \phi(t_0)\right] = 0 \ .$$

Now this must hold for any operator ϕ, and assuming that these operators are a mathematically complete set, so that any operator on the space of Fock states can be expanded in terms of them, the combination $H - i\left(\frac{dU}{dt}\right)U^{-1}$ can commute with all ϕ only if it is a multiple of the identity (this is an application of Schur's Lemma), giving

$$H = i\left(\frac{dU}{dt}\right)U^{-1} + E_0 \ , \tag{1.35}$$

where E_0 is an arbitrary constant. Hence

$$\frac{dU}{dt} = -i(H - E_0)U \ .$$

For H independent of time, this gives

$$U(t, t_0) = \exp\left[-i(H - E_0)(t - t_0)\right] \ , \tag{1.36}$$

where the normalization of the exponential is fixed by the initial condition

$$U(t_0, t_0) = 1 \ . \tag{1.37}$$

This result assumes H is independent of time but can be generalized to cases where H depends on the time, which is normally the case when interactions are included. This will be discussed in Chapter 3.

If we choose E_0 to be the ground state expectation value of H, then $H - E_0$ has a zero ground state expectation value, and that is equivalent to using the normal ordered form for H and taking $E_0 = 0$. With this choice (which we made in the previous sections),

$$\phi(z, t) = e^{iH(t-t_0)}\phi(z, t_0)\, e^{-iH(t-t_0)} \ . \tag{1.38}$$

The form (1.15) for ϕ satisfies this condition (see Prob. 1.4).

In the next chapter we apply these ideas to the quantization of the electromagnetic field.

PROBLEMS

1.1 In this book we are using natural units in which $\hbar = c = 1$. This means that length (L) and time (t) have the same dimensions and that mass (m) has the dimensions of an inverse length.

(a) Using the Fermi (f) as the fundamental unit of length, where $1f = 10^{-15}$ meters, find:

- The mass of the electron.
- The mass of a π meson.
- The radius of the first Bohr orbit of hydrogen.
- The energy of the ground state of hydrogen.

(b) Repeat part (a) using the MeV as the fundamental unit of energy. Find a conversion factor between f and MeV.

(c) An expression in natural units can always be converted *uniquely* into an expression in ordinary units (L, t, m) by inserting $\hbar$ and c in the correct places. Give an argument describing precisely how to do this for any expression and give some examples showing the correctness of your argument.

1.2 The momentum operator of the string is

$$P^z = -\int_0^L \frac{dz}{v^2} : \frac{\partial\phi}{\partial t}\frac{\partial\phi}{\partial z} : \quad .$$

Prove that this is the generator of translation in the z-direction. In particular, prove that

$$[P^z, \phi(z,t)] = i\frac{\partial\phi(z,t)}{\partial z}$$
$$[P^z, \pi(z,t)] = i\frac{\partial\pi(z,t)}{\partial z} \quad .$$

1.3 Consider the Lagrangian density

$$\mathcal{L} = \frac{1}{2}\left[\left(\frac{\partial\phi}{\partial t}\right)^2 - \left(\frac{\partial\phi}{\partial z}\right)^2 - m^2\phi^2\right] \quad ,$$

where $\phi = \phi(z,t)$ is a generalized coordinate.

(a) Find the momentum conjugate to ϕ.

(b) Find the equations of motion for the fields and the solutions. Use periodic boundary conditions.

(c) Suppose the field is expanded in normal modes

$$\phi(z,t) = \sum_n c_n \left\{a_n\phi_n(z,t) + a_n^\dagger\phi_n^*(z,t)\right\} \quad ,$$

where a_n satisfy the commutation relations

$$[a_n, a_{n'}] = \left[a_n^\dagger, a_{n'}^\dagger\right] = 0$$
$$\left[a_n, a_{n'}^\dagger\right] = \delta_{nn'} \quad .$$

Find the coefficients c_n which will insure that the CCR's assume the standard form

$$[\phi(z,t), \pi(z',t)] = i\delta(z-z') \quad .$$

(d) Find the Hamiltonian density, and express the Hamiltonian in terms of the number operators $a_n^\dagger a_n$.

(e) What is the physical significance of this field?

1.4 The Hamiltonian is the generator of time translation. This means that

$$e^{iH(t-t_0)}\phi(z,t_0)\,e^{-iH(t-t_0)} = \phi(z,t) \quad .$$

Prove that this relation holds for the one-dimensional string.

1.5 [Taken from Sakurai (1967).] Consider a three-dimensional scalar field like that introduced in Prob. 3 above:

$$\phi(r,t) = \sum_n \frac{1}{\sqrt{2\omega_n L^3}}\left\{a_n\,e^{-ik_n\cdot x} + a_n^\dagger\,e^{ik_n\cdot x}\right\}$$

where $k_n\cdot x = \omega_n t - \boldsymbol{k}_n\cdot\boldsymbol{r}$ and $\left[a_n, a_{n'}^\dagger\right] = \delta_{nn'}$, $\omega_n = \sqrt{m^2+k_n^2}$.

(a) These fields are singular operators. Show that

$$\begin{aligned}\langle 0|\phi(r,t)|0\rangle &= 0\\ \langle 0|\phi^2(r,t)|0\rangle &= \infty \quad .\end{aligned}$$

(b) To make the fields more regular, we *smear* the fields by averaging them over a small region of space. Suppose we define the average field in the neighborhood of the origin by

$$\bar\phi(0,t) = \left(\frac{1}{2\pi b^2}\right)^{3/2}\int d^3r\,\phi(r,t)\,e^{-r^2/2b^2} \quad .$$

Show that if $b \ll \frac{1}{m}$, then

$$\langle 0|\bar\phi^2(0,t)|0\rangle = (\text{numerical factor})\,\frac{1}{b^2} \quad .$$

Find the precise result if $m = 0$.

CHAPTER 2

QUANTIZATION OF THE ELECTROMAGNETIC FIELD

We now use the techniques developed in Chapter 1 to quantize the electromagnetic (*EM*) field. This system is one of the most important in physics but is also one of the most complicated. The *EM* field appears to be two coupled three-vector fields, but through Maxwell's equations and gauge invariance, it can be reduced to a single four-vector field with only two independent components. The elimination of these redundant components, which are connected with the gauge invariance of the system, poses a new problem unlike any discussed in the previous chapter. The relativistic nature of the *EM* field is also a new feature which needs to be discussed.

This chapter begins with a description of the properties of Lorentz transformations and a discussion of gauge invariance, topics which must be addressed before we can quantize the field. The particles which emerge from this quantization are *photons*, familiar from elementary studies. The vector nature of the *EM* field means that the photons have *spin one* as well as energy and momentum. The appearance of this spin, and its connection to the vector property of the field, will be the last topic covered in the chapter.

The goal of this chapter is to lay the foundation for the treatment of the interaction of the *EM* radiation field with matter, which will be discussed in Chapter 3.

2.1 LORENTZ TRANSFORMATIONS

We begin with a brief discussion of Lorentz four-vectors and transformations. The emphasis here will be on notation; the properties of the Lorentz group will be discussed in more detail in Chapter 5. In the natural system of units, the speed of light, c, is equal to unity, so that the space/time four-vector is denoted

$$\begin{aligned} x^\mu &= (t, \boldsymbol{r}) = (t, x, y, z) = (t, r^i) \\ x_\mu &= (t, -\boldsymbol{r}) = (t, -x, -y, -z) = (t, -r^i) \quad , \end{aligned} \tag{2.1}$$

where x^μ is the *contravariant* and x_μ the *covariant* form. Note that Greek indices on four-vectors (such as μ) vary from 0 to 3, while Roman indices on three-vectors (such as i) vary from 1 to 3.* The invariant length of this four-vector is written

$$x^2 = g_{\mu\nu} x^\mu x^\nu = x_\mu x^\mu = t^2 - \boldsymbol{r}^2 = t^2 - x^2 - y^2 - z^2 \quad , \tag{2.2}$$

where $g_{\mu\nu} = g^{\mu\nu}$ is the *metric tensor* and a sum over repeated indices is always assumed. Note that (2.2) implies that the relation between the contravariant form of x (x^μ) and the covariant form (x_μ) is

$$x_\mu = g_{\mu\nu} x^\nu \quad . \tag{2.3}$$

A *Lorentz transformation* (LT) is any transformation which leaves the *length of four-vectors*, defined in (2.2), *invariant*. In general, a transformation Λ which operates on the space of four-vectors can be written†

$$\begin{aligned} x'^\mu &= \Lambda^\mu{}_\nu x^\nu \\ x^\mu &= \left(\Lambda^{-1}\right)^\mu{}_\nu x'^\nu \quad . \end{aligned} \tag{2.4}$$

In this notation the requirement that the four-vector length remain invariant becomes

$$\begin{aligned} x'^2 &= x'^\mu g_{\mu\nu} x'^\nu = \Lambda^\mu{}_\alpha x^\alpha g_{\mu\nu} \Lambda^\nu{}_\beta x^\beta \\ &= g_{\alpha\beta} x^\alpha x^\beta \quad , \end{aligned} \tag{2.5}$$

which leads to the following condition on Λ:

$$\boxed{g_{\alpha\beta} = \Lambda^\mu{}_\alpha \, g_{\mu\nu} \, \Lambda^\nu{}_\beta \quad .} \tag{2.6}$$

Any transformation which satisfies this relation is an LT. In Sec. 5.8 we will show that all of the transformations which satisfy (2.6) form a group in the mathematical sense.

---

*We will adopt the convention that the *Roman indices on three-vectors will always be written as superscripts*. Be careful to *always include the minus sign* when converting the spatial components of a *covariant* four-vector to a three-vector!

†Free indices on *both sides* of a relativistic equation must always be in the same position (either *up* or *down*), and indices on one side of an equation which are summed (or *contracted*) must always be paired, with one *up* and one *down*. This will insure that both sides of the equation transform in the same way. In *three*-vector equations, the position of the indices is arbitrary, and placement is by convention.

Matrix Notation

It is convenient to introduce a matrix notation for LT's. The following correspondence will be made:

$$G = \{g_{\mu\nu}\} = \left(\begin{array}{c|c} 1 & \\ \hline & -\mathbf{1} \end{array}\right) = \left(\begin{array}{c|ccc} 1 & & & \\ \hline & -1 & & \\ & & -1 & \\ & & & -1 \end{array}\right) \tag{2.7}$$

$$\Lambda = \{\Lambda^\mu{}_\nu\} = \left(\begin{array}{c|c} \Lambda^0{}_0 & \Lambda^0{}_j \\ \hline \Lambda^i{}_0 & \Lambda^i{}_j \end{array}\right)$$

where the matrices have been written in block form, with the upper left element the (0,0) component and the lower right element representing the 3×3 submatrix of spatial components. The Greek indices, μ and ν, always run from 0–3, while the Roman indices, i and j, run over the spatial components 1–3, and the correspondence is $\{\mu\} = (0,\ i)$ and $\{\nu\} = (0,\ j)$. As we have written it, μ labels the *rows* and ν labels the *columns*. Note that therefore $\Lambda^\nu{}_\mu\, g_{\nu\alpha} = \Lambda^{\mathsf{T}} G$.

In this notation, the defining equation (2.6) for the LT's becomes

$$\boxed{G = \Lambda^{\mathsf{T}} G\, \Lambda} \tag{2.8}$$

and, representing the contravariant four-vector by x, and the covariant one by Gx, so that $x' = \Lambda x$, Eqs. (2.2) and (2.4) become

$$\begin{aligned} x' &= \Lambda x \\ Gx' &= G\Lambda G G x = \left(\Lambda^{\mathsf{T}}\right)^{-1} Gx \\ x^2 &= x^{\mathsf{T}} G x \ . \end{aligned} \tag{2.9}$$

Note that the four-gradient

$$\partial_\mu \equiv \frac{\partial}{\partial x^\mu} = \left(\frac{\partial}{\partial t}, \nabla\right) = \left(\frac{\partial}{\partial t}, \frac{\partial}{\partial x}, \frac{\partial}{\partial y}, \frac{\partial}{\partial z}\right)$$

transforms as a covariant four-vector. To prove this easily, use (2.4):

$$\partial'_\mu = \frac{\partial}{\partial x'^\mu} = \frac{\partial x^\alpha}{\partial x'^\mu}\,\frac{\partial}{\partial x^\alpha} = \left(\Lambda^{-1}\right)^\alpha{}_\mu\, \partial_\alpha \ ,$$

which is the same as the transformation law for a general covariant four-vector $\{x_\mu\} = Gx$, as given in Eq. (2.9). If V^μ is a contravariant four-vector, the divergence is

$$\partial_\mu V^\mu = \frac{\partial}{\partial t} V^0 + \nabla \cdot \boldsymbol{V} \ .$$

Note that a *plus sign* appears in this equation, instead of the *minus sign* which might be naively expected.

The LT's are not necessarily orthogonal matrices. The *rotations*, which leave the time component of any four-vector unchanged (and also one direction in space, the rotation axis, unchanged), can be written

$$\Lambda_R = \left(\begin{array}{c|c} 1 & 0 \\ \hline 0 & R \end{array}\right)$$

and are orthogonal. The *boosts*, which leave two directions in space invariant, are *not orthogonal.* A simple example is the boost in the z-direction

$$\Lambda_B = \begin{pmatrix} \cosh\alpha & 0 & 0 & \sinh\alpha \\ 0 & 1 & 0 & 0 \\ 0 & 0 & 1 & 0 \\ \sinh\alpha & 0 & 0 & \cosh\alpha \end{pmatrix} ,$$

which leaves the x- and y-directions invariant. Note that $\Lambda_B^{\mathrm{T}} = \Lambda_B$, *not* Λ_B^{-1}.

2.2 RELATIVISTIC FORM OF MAXWELL'S THEORY

The Maxwell equations (with $c = 1$) are

$$\nabla \cdot \boldsymbol{E} = \rho \qquad \nabla \times \boldsymbol{E} = -\frac{\partial \boldsymbol{B}}{\partial t}$$

$$\nabla \cdot \boldsymbol{B} = 0 \qquad \nabla \times \boldsymbol{B} = \boldsymbol{j} + \frac{\partial \boldsymbol{E}}{\partial t} \ .$$

These are in rationalized Gaussian units where Coulomb's law for a point charge is $V = e^2/4\pi r$ and the fine structure constant is $\alpha = e^2/4\pi$. We replace two of these equations with potentials

$$\boldsymbol{E} = -\nabla\phi - \frac{\partial}{\partial t}\boldsymbol{A} \qquad \boldsymbol{B} = \nabla \times \boldsymbol{A} \ . \tag{2.10}$$

These solve the two homogeneous equations identically, leaving

$$\nabla \cdot \boldsymbol{E} = \rho \qquad \nabla \times \boldsymbol{B} - \frac{\partial \boldsymbol{E}}{\partial t} = \boldsymbol{j} \ . \tag{2.11}$$

To cast these equations into a relativistic form, identify two four-vectors

$$A^\mu = (\phi, \boldsymbol{A}) \qquad j^\mu = (\rho, \boldsymbol{j}) \ . \tag{2.12}$$

Note that the so called "scalar" potential ϕ is now the time component of a four-vector, and it is sometimes convenient to denote it by A^0 instead of ϕ. This potential is still a scalar under rotations, but is no longer a scalar under boosts (and hence is no longer a scalar). Since E and B are coupled, there are six field components which transform into each other. This is just the correct number for an antisymmetric 4×4 tensor, which is denoted $F^{\mu\nu}$. Since three-vectors are always written with their Roman indices as superscripts, we identify

$$\begin{aligned} F^{\mu\nu} &= \{\boldsymbol{E}, \boldsymbol{B}\} \qquad & F^{\mu\nu} &= -F^{\nu\mu} \\ F^{i0} &= E^i \qquad & F^{ij} &= -\epsilon_{ijk}B^k \end{aligned}$$

$$F^{\mu\nu} = \begin{bmatrix} 0 & -E^x & -E^y & -E^z \\ E^x & 0 & -B^z & B^y \\ E^y & B^z & 0 & -B^x \\ E^z & -B^y & B^x & 0 \end{bmatrix} .$$

Here ϵ_{ijk} is the familiar three-dimensional antisymmetric symbol with $\epsilon_{123} = 1$. In this notation, the homogeneous equations (2.10) and the inhomogeneous equations (2.11) become

$$\begin{aligned} F^{\mu\nu} &= \partial^\mu A^\nu - \partial^\nu A^\mu \\ \partial_\mu F^{\mu\nu} &= j^\nu \ . \end{aligned} \tag{2.13}$$

The form of these equations shows immediately that the theory is invariant under Lorentz transformations if A^μ and j^μ are four-vectors and $F^{\mu\nu}$ is a second rank tensor. In this case

$$\begin{aligned} j'^\mu &= \Lambda^\mu{}_\alpha \, j^\alpha \qquad A'^\mu = \Lambda^\mu{}_\alpha \, A^\alpha \\ F'^{\mu\nu} &= \Lambda^\mu{}_\alpha \Lambda^\nu{}_\beta \, F^{\alpha\beta} \ . \end{aligned}$$

Check Eq. (2.13): Noting that $\nabla \to \nabla_i$ and $\boldsymbol{A} \to A^i$ and using the identity $\epsilon_{ijk}\epsilon_{jki'} = 2\delta_{ii'}$, so that $\epsilon_{ijk}\epsilon_{jki'}B^{i'} = -\epsilon_{ijk}F^{jk} = 2B^i$, give

$$\begin{aligned} \boldsymbol{E} \to E^i &= F^{i0} = -\nabla_i A^0 - \partial^0 A^i \to -\nabla\phi - \frac{\partial \boldsymbol{A}}{\partial t} \\ \boldsymbol{B} \to B^i &= -\tfrac{1}{2}\epsilon_{ijk}\left[-\nabla_j A^k + \nabla_k A^j\right] \to \nabla \times \boldsymbol{A} \ . \end{aligned}$$

Next,

$$\begin{aligned}\partial_0 F^{0i} + \partial_j F^{ji} &= -\frac{\partial}{\partial t}E^i + \partial_j \epsilon^{ijk} B^k \\ &\to -\frac{\partial \boldsymbol{E}}{\partial t} + \nabla \times \boldsymbol{B} = \boldsymbol{j} \\ \partial_i F^{i0} &= \nabla \cdot \boldsymbol{E} = \rho \ ,\end{aligned}$$

and Eq. (2.13) is confirmed. ∎

Relativistic Lagrangian

In order to maintain Lorentz invariance, the Lagrangian density for the *EM* theory must be a scalar invariant constructed from the field tensor, the four-vector potential (from now on the four-vector potential will be referred to simply as the "vector" potential), and the currents. This means that all scalar products must be constructed from two quantities, one of which transforms like a contravariant four-vector (or tensor) and one which transforms like a covariant four-vector (or tensor), as in Eq. (2.2). We will show that the Lagrangian density

$$\boxed{\mathcal{L} = -\tfrac{1}{4} F_{\mu\nu} F^{\mu\nu} - j_\mu A^\mu} \tag{2.14}$$

gives the desired equations of motion, and hence is a suitable choice. Discussion of how this Lagrangian density might be uniquely determined from fundamental principles will be deferred to Chapter 13.

To find the equations of motion from (2.14), simplify the expression as follows:

$$\begin{aligned}-\tfrac{1}{4} F_{\mu\nu} F^{\mu\nu} &= -\tfrac{1}{4} \left(\partial_\mu A_\nu - \partial_\nu A_\mu\right)\left(\partial^\mu A^\nu - \partial^\nu A^\mu\right) \\ &= -\tfrac{1}{2} g^{\mu\mu'} g^{\nu\nu'} \left[\partial_\mu A_\nu \partial_{\mu'} A_{\nu'} - \partial_\mu A_\nu \partial_{\nu'} A_{\mu'}\right] \ .\end{aligned}$$

In these expressions, the ∂ in a term like ∂AB operates *only* on A, while in $\partial(AB)$ it operates on *both* A and B. Hence

$$\frac{\partial \mathcal{L}}{\partial(\partial_\alpha A_\beta)} = -\partial^\alpha A^\beta + \partial^\beta A^\alpha = -F^{\alpha\beta}$$

and the Euler–Lagrange equations reduce to

$$\partial_\alpha \left(\frac{\partial \mathcal{L}}{\partial(\partial_\alpha A_\beta)}\right) - \frac{\partial \mathcal{L}}{\partial A_\beta} = -\partial_\alpha F^{\alpha\beta} + j^\beta = 0 \ ,$$

which are the correct equations (2.13). Note that current must be conserved (i.e., its four-divergence is zero), because

$$\partial_\beta j^\beta = \partial_\beta \partial_\alpha F^{\alpha\beta} = 0 \ .$$

Gauge Invariance

The electromagnetic Lagrangian has two special features not encountered before:

(i) The generalized momentum conjugate to the time component of the four-vector potential (which will be denoted by A^0, instead of ϕ, in this subsection) is zero. This follows from the fact that the Lagrangian density (2.14) does not depend on $\partial A^0/\partial t$, and hence

$$\pi^0 = \frac{\partial \mathcal{L}}{\partial \left(\dfrac{\partial A^0}{\partial t} \right)} = 0 \ .$$

Because of this, the Poisson bracket of A^0 with π^0 (or, after we quantize the field, the commutator $[A^0, \pi^0]$) must also be zero. If we attempt to quantize the field component A^0 by turning it into an operator, it would therefore commute with all operators, and by Schur's Lemma would reduce to a c-number. The field component A^0 is special.

(ii) If the current is conserved (and we have seen that consistency requires it), then the Lagrangian is invariant under the gauge transformation

$$A'_\mu = A_\mu - \partial_\mu \Lambda_c \ , \tag{2.15}$$

where Λ_c is a scalar. Note that the Lagrangian *density* is not *locally* gauge invariant [i.e., is not invariant under the transformation (2.15) at every space–time point x], because

$$j_\mu A^{\mu\prime} \to j_\mu A^\mu - \underbrace{j_\mu\, \partial^\mu \Lambda_c}_{\text{not zero}} \ . \tag{2.16}$$

However, the action $\int dt\, L$ (and hence the theory) *is* gauge invariant. To show this, use the fact that the fields are assumed to satisfy periodic boundary conditions, so that when integrating over all space any surface terms which might arise from any integrations by parts can be assumed to vanish or cancel. To justify dropping the surface terms from the *time* integration, assume that $\Lambda_c = 0$ at $t = \pm\infty$. Therefore, integrating the non-zero term in (2.16) by parts gives

$$\int d^4x\, j_\mu\, \partial^\mu \Lambda_c = -\int d^4x\, (\partial^\mu j_\mu)\, \Lambda_c = 0 \ .$$

In order to obtain a definite solution for the EM fields, the arbitrariness associated with the gauge freedom (2.15) must be removed so that the fields can be uniquely specified everywhere. This process is referred to as "gauge fixing" and involves imposing some constraints on the fields which will fix the gauge function Λ_c and remove the gauge freedom. Two popular choices for the constraint, or choice of gauge, are the *Lorentz* and *Coulomb* gauges, defined by the constraints

$$\begin{aligned} &\partial_\mu A^\mu = 0 && \text{Lorentz gauge} \\ &\nabla \cdot \mathbf{A} = \partial_i A^i = 0 && \text{Coulomb gauge} \ . \end{aligned} \tag{2.17}$$

There are advantages and disadvantages which accompany the use of each of these gauges, and the choice of gauge is closely related to how the time component of the four-vector potential, A^0, is to be treated. Since the time derivative of A^0 does not occur naturally in the Lagrangian, and since the gauge transformations give us some freedom to redefine the field in a convenient way, the solution of the electromagnetic problem may be approached in one of two ways:

- The quantity A^0 may be eliminated from the Lagrangian by expressing it in terms of the remaining components of A^μ. This approach is simplified by using the Coulomb gauge.
- A new term may be added to the Lagrangian which contains the time derivative of A^0. In this case, the Lorentz gauge is the preferred constraint.

Each of these approaches will now be discussed briefly.

To see what is involved in eliminating A^0 from the Lagrangian, look at Eq. (2.13) when $\nu = 0$:

$$\partial_\mu \left[\partial^\mu A^0 - \partial^0 A^\mu\right] = -\nabla^2 A^0 - \partial_0 \nabla \cdot \boldsymbol{A} = \rho \ . \tag{2.18}$$

This equation is greatly simplified by imposing the Coulomb gauge, which reduces the equation to Poisson's equation

$$\nabla^2 A^0 = -\rho \ ,$$

and this equation has the unique solution

$$\boxed{A^0(\boldsymbol{r},t) = \frac{1}{4\pi}\int d^3r' \frac{\rho(r',t)}{|r'-r|}} \qquad \text{Coulomb's law} \ . \tag{2.19}$$

This solution is zero if $\rho = 0$. Had we chosen to use the Lorentz gauge, the equation for A^0 which would result from (2.18) is

$$\Box A^0 = \rho \ ,$$

where

$$\Box = \partial_\mu \partial^\mu = \frac{\partial^2}{\partial t^2} - \nabla^2 \tag{2.20}$$

is the familiar wave operator. This equation is manifestly covariant, but the solutions of the wave equation may depend on time and are not zero *even when* $\rho = 0$. For these reasons the Coulomb gauge, which gives Coulomb's law, is used in the study of atomic and other low energy systems, and it will be used in Part I of this book. The disadvantage of this choice is that the Coulomb gauge condition is not manifestly covariant; to maintain this gauge condition in different frames requires that a new gauge function Λ_c be chosen for each frame, so all of the results obtained from this gauge will *look* non-covariant. The final results of any calculation will always turn out to be covariant, but often this is only apparent after the final answer is obtained.

Now consider the second approach to the study of the EM field, in which a term containing the time derivative of A^0 is added to the Lagrangian. This must be done in such a way that the theory is not altered, and a convenient way to do this is to add the following *gauge fixing* term to the Lagrangian density:

$$\mathcal{L}_{\text{gauge}} = -\frac{1}{2\alpha}\left(\partial_\mu A^\mu\right)\left(\partial_\nu A^\nu\right) \quad . \tag{2.21}$$

This extra term can be regarded as a constraint, with the redundant field components related to Lagrange multipliers [see Itzykson and Zuber (1980)]. The parameter α is the *gauge parameter* and may assume any finite value. Two well-known choices are $\alpha = 1$, the *Feynman gauge*, and $\alpha \to 0$, the *Landau gauge.* Note that the overall theory is not affected by the addition of the gauge fixing term because it is zero after the gauge condition $\partial_\mu A^\mu = 0$ is imposed. These gauges are very convenient for the study of high energy scattering processes where it is desirable to maintain manifest Lorentz invariance, and using the method of Gupta [Gu 50] and Bleuler [Bl 50] [see also Bogoliubov and Shirkov (1959)], it is possible to quantize all four components of A^μ as independent degrees of freedom. A modern approach, in which these gauges are used in conjunction with the method of path integrals, will be discussed in Chapter 15.

It is important to realize that the *physics is unaffected by the choice of gauge.* Any gauge may be used, as long as it is used consistently in all parts of the calculation. The intermediate steps may be very different, but the *final result for any physical observable* must be independent of the gauge used to calculate it.

For example, note that a scalar gauge function Λ_c can always be found so that either the Coulomb or Lorentz condition is satisfied. Suppose first that $\nabla \cdot \boldsymbol{A} \neq 0$ and we wish to impose the Coulomb condition. Then change A to A' so that

$$\nabla \cdot \boldsymbol{A}' = \nabla \cdot (\boldsymbol{A} - \nabla \Lambda_c) = 0 \quad .$$

This implies that

$$\nabla^2 \Lambda_c = \nabla \cdot \boldsymbol{A}$$

and we know that this equation (Poisson's equation again) can be solved. Similarly, suppose that $\partial_\mu A^\mu \neq 0$, and we wish to impose the Lorentz gauge. Then change A to A' so that

$$\partial_\mu A'^\mu = \partial_\mu \left(A^\mu - \partial^\mu \Lambda_c\right) = 0 \quad .$$

This implies that

$$\Box \Lambda_c = \partial_\mu A^\mu \quad .$$

This is the inhomogeneous wave equation, which also can be solved. Since the physics does not depend on the scalar Λ_c, the physics also cannot depend on the gauge.

The Lagrangian in the Coulomb Gauge

The next task is to rewrite the Lagrangian density using Coulomb's law to *define* A^0. The resulting Lagrangian will then depend only on the three components of the vector potential A^i (and the charge and current densities, considered sources of the fields and not dependent on them). The three components of A^i will be treated as independent fields, and the Lagrangian will be constructed so that the correct equations of motion for these fields will *emerge naturally.*

To see more clearly what this means, look at the equation for the vector potential. From the field equations (2.13), this equation is

$$\Box A^i + \nabla_i \nabla \cdot \boldsymbol{A} = j^i - \frac{\partial}{\partial t}\nabla_i A^0 = j^i_\perp \ , \tag{2.22}$$

where $j^i_\perp$ is referred to as the *transverse current.* Taking the divergence of both sides of this equation and *assuming that A_0 is given by Poisson's equation* give

$$\begin{aligned}\frac{\partial^2}{\partial t^2}\nabla\cdot\boldsymbol{A} &= \nabla\cdot\boldsymbol{j}_\perp = \nabla\cdot\boldsymbol{j} - \frac{\partial}{\partial t}\nabla^2 A^0 \\ &= \nabla\cdot\boldsymbol{j} + \frac{\partial}{\partial t}\rho = 0 \qquad \text{(by current conservation)} \ .\end{aligned} \tag{2.23}$$

Hence, if we did not know that the Coulomb gauge condition $\nabla\cdot\boldsymbol{A} = 0$ had been used to relate A_0 to ρ, this equation would enable us to recover it in the following sense: if $\nabla\cdot\boldsymbol{A} = 0$ and $\partial\left(\nabla\cdot\boldsymbol{A}\right)/\partial t = 0$ holds at *one* time, it will hold at *all* times. In this sense the Coulomb gauge condition can be regarded as a dynamical consequence of Eq. (2.22) for $\boldsymbol{A}$. Our task is to construct a Lagrangian density which will give this equation.

To find the correct Lagrangian density, we will first separate out the A^0 terms from the Lagrangian density (2.14). All three-vectors will be expressed in a "standard" form, which is taken to be $A^i \to \boldsymbol{A}$ and $\nabla_i \to \nabla$. Hence we use $A_\mu \to (A^0, -A^i)$, $\partial^\mu \to (\partial^0, -\nabla_i)$, and obtain:

$$\begin{aligned}\mathcal{L} =& -\tfrac{1}{4}F_{\mu\nu}F^{\mu\nu} - j_\mu A^\mu \\ =& -\tfrac{1}{2}\partial_\mu A^0\left(\partial^\mu A^0 - \partial^0 A^\mu\right) + \tfrac{1}{2}\partial_\mu A^i\left(\partial^\mu A^i + \nabla_i A^\mu\right) - \rho A^0 + \boldsymbol{j}\cdot\boldsymbol{A} \\ =& -\tfrac{1}{2}\nabla_j A^0\left(-\nabla_j A^0 - \partial^0 A^j\right) + \tfrac{1}{2}\partial_0 A^i\left(\partial^0 A^i + \nabla_i A^0\right) \\ & + \tfrac{1}{2}\nabla_j A^i\left(-\nabla_j A^i + \nabla_i A^j\right) - \rho A^0 + \boldsymbol{j}\cdot\boldsymbol{A} \\ =& \ \ \tfrac{1}{2}\nabla_j A^0 \nabla_j A^0 + \tfrac{1}{2}\partial_0 A^j \partial_0 A^j - \tfrac{1}{2}\boldsymbol{B}^2 + \nabla_j A^0 \partial_0 A^j - \rho A^0 + \boldsymbol{j}\cdot\boldsymbol{A} \ .\end{aligned} \tag{2.24}$$

The third term in the last line was obtained using the identity $\epsilon_{ijk}\epsilon_{i\ell m} = \delta_{j\ell}\delta_{km} - \delta_{jm}\delta_{k\ell}$:

$$\begin{aligned}\boldsymbol{B}^2 &= \epsilon_{ijk}\nabla_j A^k \epsilon_{i\ell m}\nabla_\ell A^m = \nabla_j A^k \nabla_j A^k - \nabla_j A^k \nabla_k A^j \\ &= \nabla_j A^k\left[\nabla_j A^k - \nabla_k A^j\right] \ .\end{aligned}$$

The first term in the last line of (2.24) can be reduced by integrating by parts and dropping the boundary terms, which are guaranteed to vanish because of the periodic boundary conditions imposed on the fields and sources. This procedure has been used several times before and will be used many times again in the following chapters. Stated in general terms, this freedom to integrate by parts means that *two Lagrangian densities which differ by a three-divergence* give the same Lagrangian and hence *are equivalent*. Using $\nabla^2 A^0 = -\rho$ in the last step gives

$$\begin{aligned}\nabla_j A^0 \nabla_j A^0 &= \nabla_j \left(A^0 \nabla_j A^0\right) - A_0 \nabla^2 A_0 \\ &= \begin{matrix}\text{total} \\ \text{divergence}\end{matrix} - A_0 \nabla^2 A_0 \\ &\Longrightarrow A^0 \rho \ .\end{aligned}$$

Similarly, the fourth term is the last line of (2.24) can be replaced by

$$\nabla_j A^0 \partial_0 A^j \Longrightarrow -A^j\, \partial_0 \nabla_j A^0 \ .$$

This replacement is perhaps best justified by noting that both terms are proportional to $\nabla \cdot \boldsymbol{A}$ (plus a total divergence) and hence will give zero after the Coulomb gauge condition is applied.

With these substitutions, the Lagrangian density (2.24) can be written as the sum of two terms:

$$\mathcal{L} = \mathcal{L}_0 + \mathcal{L}_{\text{int}}$$

where

$$\begin{aligned}\mathcal{L}_0 &= \tfrac{1}{2}\boldsymbol{E}_\perp^2 - \tfrac{1}{2}\boldsymbol{B}^2 - \tfrac{1}{2}\rho A_0 \\ \mathcal{L}_{\text{int}} &= \boldsymbol{j} \cdot \boldsymbol{A} - A^j\, \partial_0 \nabla_j A^0 = \boldsymbol{j}_\perp \cdot \boldsymbol{A} \ .\end{aligned}$$

Here $\boldsymbol{E}_\perp$ is defined to be

$$\boldsymbol{E}_\perp = -\frac{\partial}{\partial t}\boldsymbol{A} \ .$$

To see that this Lagrangian density gives the correct equations of motion for the A^i, compute

$$\frac{\partial}{\partial t}\left(\frac{\partial \mathcal{L}_0}{\partial\left(\partial_0 A^i\right)}\right) = \frac{\partial^2 A^i}{\partial t^2}$$

$$\nabla_j \left(\frac{\partial \mathcal{L}_0}{\partial\left(\nabla_j A^i\right)}\right) = -\nabla_j \left(\nabla_j A^i - \nabla_i A^j\right) = -\nabla^2 A^i + \nabla_i \nabla \cdot \boldsymbol{A} \ .$$

Hence the Euler–Lagrange equations implied by $\mathcal{L}$ are

$$\begin{aligned}\frac{\partial}{\partial t}\left(\frac{\partial \mathcal{L}_0}{\partial\left(\partial_0 A^i\right)}\right) + \nabla_j \left(\frac{\partial \mathcal{L}_0}{\partial\left(\nabla_j A^i\right)}\right) - \frac{\partial \mathcal{L}_0}{\partial A^i} &= 0 \\ \Longrightarrow \Box A^i + \nabla_i \nabla \cdot \boldsymbol{A} &= j_\perp^i \ . \qquad (2.25)\end{aligned}$$

These are the desired equations (2.22).

A vector field $\boldsymbol{V}$ which has a zero three-divergence, $\nabla \cdot \boldsymbol{V} = 0$, is said to be *transverse*. Physically, this condition means that the field is perpendicular to its momentum, as will be discussed below. After the equations of motion have been obtained, and the gauge condition applied, $\nabla \cdot \boldsymbol{E}_\perp = 0$, and only transverse radiation fields remain in $\mathcal{L}$. The longitudinal component of $\boldsymbol{E}$, sometimes denoted by $\boldsymbol{E}_\parallel$,

$$\boldsymbol{E}_\parallel = -\nabla A^0 \quad ,$$

is no longer a dynamical variable and is expressed in terms of ρ.

Finally, the Lagrangian derived from $\mathcal{L}$ assumes a nice symmetrical form:

$$L = \int d^3r\, \mathcal{L} = \int d^3r \left\{ \tfrac{1}{2}|\boldsymbol{E}_\perp(r,t)|^2 - \tfrac{1}{2}|\boldsymbol{B}(r,t)|^2 + \boldsymbol{j}_\perp(r,t) \cdot \boldsymbol{A}(r,t) \right\}$$
$$- \frac{1}{8\pi} \int d^3r\, d^3r' \frac{\rho(r,t)\rho(r',t)}{|r'-r|} \quad . \tag{2.26}$$

Note the presence of the instantaneous Coulomb interaction, which makes this approach ideal for application to relativistic atoms. As advertised, L is no longer manifestly covariant.

2.3 INTERACTIONS BETWEEN PARTICLES AND FIELDS

To complete the picture, and to introduce interactions, add two spherically charged particles to the Lagrangian. The location of these particles will be described by generalized coordinates

$$\boldsymbol{q}_a(t) \qquad a = 1 \text{ and } 2 \quad ,$$

and their charge density will be denoted $\rho_a\left(|\boldsymbol{q}_a - \boldsymbol{r}|\right)$, where $|\boldsymbol{q}_a - \boldsymbol{r}|$ is the *length* of the vector which connects the location $\boldsymbol{q}_a$ of the ath particle to the field point $\boldsymbol{r}$. To simplify the notation, the particle coordinates will usually be denoted simply by $\boldsymbol{q}_a$, and the charges by $\rho_a(r)$, although both depend on time. The four-current of each particle is

$$j_a^\mu(r) = \rho_a(r)\left(1\,,\ \dot{\boldsymbol{q}}_a\right) \quad , \tag{2.27}$$

where $\dot{\boldsymbol{q}}_a = d\boldsymbol{q}_a(t)/dt$ and the total charge and current is the sum of the single particle charges and currents:

$$\rho(r) = \rho_1(r) + \rho_2(r)$$
$$j^\mu(r) = j_1^\mu(r) + j_2^\mu(r) \quad .$$

Note that the current of each particle is conserved:

$$\frac{\partial \rho_a(r)}{\partial t} + \nabla \cdot \boldsymbol{j}_a(r) = \rho_a'(r)\left[\frac{(\boldsymbol{q}_a - \boldsymbol{r}) \cdot \dot{\boldsymbol{q}}_a}{|\boldsymbol{q}_a - \boldsymbol{r}|} - \frac{(\boldsymbol{q}_a - \boldsymbol{r}) \cdot \dot{\boldsymbol{q}}_a}{|\boldsymbol{q}_a - \boldsymbol{r}|}\right] = 0 \quad ,$$

where $\rho'_a(r)$ denotes the derivative of $\rho_a(r)$ with respect to its argument, $|\boldsymbol{q}_a - \boldsymbol{r}|$.

The Lagrangian of this field–particle system is composed of three terms:

$$L = L_{\text{particles}} + L_{\text{EM}} + L_{\text{int}}$$

where

$$\begin{aligned}
L_{\text{particles}} &= \frac{1}{2}\sum_{a=1}^{2} m_a |\dot{\boldsymbol{q}}_a|^2 - \frac{1}{2}\int \frac{d^3r\, d^3r'}{4\pi\, |r' - r|}\, \rho(r)\, \rho(r') \\
L_{\text{EM}} &= \frac{1}{2}\int d^3r\, \{\boldsymbol{E}_\perp^2(r,t) - \boldsymbol{B}^2(r,t)\} \\
L_{\text{int}} &= \int d^3r\, [\dot{\boldsymbol{q}}_1\, \rho_1(r) + \dot{\boldsymbol{q}}_2\, \rho_2(r)] \cdot \boldsymbol{A}(r,t) \\
&\quad - \int \frac{d^3r\, d^3r'}{4\pi} A^i(r,t)\, (\nabla_r)_i \frac{\partial}{\partial t}\left\{\frac{\rho(r')}{|r' - r|}\right\} .
\end{aligned} \tag{2.28}$$

Note that the second term in L_{int}, which we will refer to as $L^{(2)}_{\text{int}}$, is zero once the Coulomb gauge is taken into account, and therefore this term will make no contribution to the particle equations or to the total energy.

Equations of Motion

Adding the particle coordinates to the Lagrangian does not change the derivation of the equations for A^i given in Sec. 2.2. We have, as before,

$$\Box A^i + \nabla_i \nabla \cdot \boldsymbol{A} = j^i_\perp \; ,$$

where A^0 is shorthand for the solution of the Poisson equation and the gauge condition $\nabla \cdot \boldsymbol{A} = 0$ is imposed. The only new feature is that the current is now specified in terms of the particle coordinates.

The equations for the motion of the two particles become

$$\begin{aligned}
&\frac{d}{dt}\left(\frac{\partial L}{\partial \dot{q}^j_a}\right) - \frac{\partial L}{\partial q^j_a} = 0 \Longrightarrow \\
&m_a \ddot{q}^j_a + \int d^3r \frac{\partial}{\partial t}\left[\rho_a(r)\, A^j(r,t)\right] + \int d^3r\, (\nabla_r)_j\, \{\dot{\boldsymbol{q}}_a\, \rho_a(r)\} \cdot \boldsymbol{A}(r,t) \\
&-\frac{1}{2}\int \frac{d^3r\, d^3r'}{4\pi |r - r'|}\, \{(\nabla_r)_j\, \rho_a(r)\, \rho(r') + (\nabla_{r'})_j\, \rho_a(r')\, \rho(r)\} = 0 \, ,
\end{aligned} \tag{2.29}$$

where use was made of

$$\frac{\partial}{\partial q^j_a} \rho_a\left(|\boldsymbol{q}_a - \boldsymbol{r}|\right) = -(\nabla_r)_j \rho_a\left(|\boldsymbol{q}_a - \boldsymbol{r}|\right)$$

and the fact that the second interaction term, $L^{(2)}_{\text{int}}$, integrates to zero.

Now simplify the equation (2.29) and extract the Lorentz force law. First, the second and third terms in (2.29) are reduced by integrating by parts and using current conservation:

$$\begin{aligned}
&\int d^3r \left\{ \frac{\partial}{\partial t}(\rho_a(r)A^j(r,t)) + \nabla_j(\dot{q}^i_a\,\rho_a(r))A^i(r,t)\right\} \\
&\quad = \int d^3r \left\{ \frac{\partial \rho_a(r)}{\partial t} A^j(r,t) - \rho_a(r)E^j_\perp(r,t) - \dot{q}^i_a\,\rho_a(r)\nabla_j A^i(r,t)\right\} \\
&\quad = \int d^3r \left\{ -\rho_a(r)E^j_\perp(r,t) + \dot{q}^i_a\,\rho_a(r)\left(\nabla_i A^j - \nabla_j A^i\right)\right\} \ .
\end{aligned}$$

However, the second of these terms is recognized as related to $\boldsymbol{v}\times\boldsymbol{B}$ (where $\boldsymbol{v} = \dot{\boldsymbol{q}}$):

$$\begin{aligned}
\boldsymbol{v}\times\boldsymbol{B} &= \boldsymbol{v}\times(\nabla\times\boldsymbol{A}) \to \epsilon_{jim}v^i\epsilon_{m\ell k}\nabla_\ell A^k \\
&= v^i\left(\nabla_j A^i - \nabla_i A^j\right) \ .
\end{aligned}$$

The fourth term in (2.29) can be simplified by integrating by parts and using

$$\nabla_r \frac{1}{|r-r'|} = -\nabla_{r'}\frac{1}{|r-r'|}$$

to get

$$\begin{aligned}
&\frac{1}{2}\int \frac{d^3r\,d^3r'}{4\pi}\left((\nabla_r)_j \frac{1}{|r-r'|}\right) \\
&\qquad \times \left(\rho_a(r)\left[\rho_1(r') + \rho_2(r')\right] - \rho_a(r')\left[\rho_1(r) + \rho_2(r)\right]\right) \ .
\end{aligned}$$

Note that regardless of the value of a, the only terms in the square brackets which survive are those for particle $b \neq a$, giving

$$\begin{aligned}
&\int \frac{d^3r\,d^3r'}{4\pi}\left((\nabla_r)_j \frac{1}{|r-r'|}\right)\rho_a(r)\rho_b(r') \\
&\qquad = \int d^3r\,\rho_a(r)\,(\nabla_r)_j \underbrace{\int \frac{d^3r'}{4\pi}\frac{\rho_b(r')}{|r-r'|}}_{A^0(r,t)\ \text{from}\ b} \ .
\end{aligned}$$

Combining all terms gives the Lorentz force law:

$$m_a\ddot{\boldsymbol{q}}_a = \int d^3r\,\{\rho_a(r)\,\boldsymbol{E}_a(r,t) + \rho_a(r)\,\dot{\boldsymbol{q}}_a \times \boldsymbol{B}(r,t)\} \ ,$$

where *no summation* over the repeated index a is implied and

$$\boldsymbol{E}_a(r,t) = -\frac{\partial \boldsymbol{A}(r,t)}{\partial t} - \nabla_r \int \frac{d^3r'}{4\pi}\frac{\rho_{b\neq a}(r')}{|r-r'|} \ .$$

Note that the $\boldsymbol{E}_a$ field which enters the force law only includes the longitudinal Coulomb part due to the *other charges*, so there is no self-force.

Hamiltonian

In preparation for calculation of the Hamiltonian, first find the canonical momenta:

$$p_a^i = \frac{\partial L}{\partial \dot{q}_a^i} = m_a \dot{q}_a^i + \int d^3r\, \rho_a(r) A^i(r,t)$$
$$\pi^i(r,t) = \frac{\partial \mathcal{L}}{\partial(\partial_0 A^i)} = \frac{\partial A^i}{\partial t} = -E_\perp^i(r,t)\ . \tag{2.30}$$

Hence the Hamiltonian, dropping the second interaction term $L_{\text{int}}^{(2)}$, is

$$H = \sum_a \boldsymbol{p}_a \cdot \dot{\boldsymbol{q}}_a + \int d^3r\, \boldsymbol{\pi}(r,t) \cdot \frac{\partial \boldsymbol{A}}{\partial t} - L$$
$$= \frac{1}{2}\sum_a m_a \dot{\boldsymbol{q}}_a^2 + \frac{1}{2}\int \frac{d^3r\, d^3r'}{4\pi} \frac{\rho(r)\rho(r')}{|r-r'|} + \frac{1}{2}\int d^3r \left\{\boldsymbol{E}_\perp^2(r,t) + \boldsymbol{B}^2(r,t)\right\}\ .$$

Expressing this in terms of the canonical momenta and the canonical coordinates gives

$$H = \sum_a \frac{1}{2m_a}\left(\boldsymbol{p}_a - \int d^3r\, \rho_a(r) \boldsymbol{A}(r,t)\right)^2 + \frac{1}{2}\int \frac{d^3r\, d^3r'}{4\pi} \frac{\rho(r)\rho(r')}{|r-r'|} + \frac{1}{2}\int d^3r \left\{\boldsymbol{\pi}^2(r,t) + \boldsymbol{B}^2(r,t)\right\}\ . \tag{2.31}$$

The energy can be redefined so that the Coulomb self-energies (the terms proportional to the square of the charge density of a *single* particle) are ignored [choose E_0 from Eq. (1.36) correctly]. The second term is then more familiar:

$$\text{Second term} = \int \frac{d^3r\, d^3r'}{4\pi} \frac{\rho_1(r)\, \rho_2(r')}{|r-r'|}\ .$$

Note that the first term has the familiar $(\boldsymbol{p} - e\boldsymbol{A})^2$ structure.

2.4 PLANE WAVE EXPANSIONS

If there are no charges and currents, the vector potential $\boldsymbol{A}$ in the Coulomb gauge is the solution of the equations

$$\square \boldsymbol{A} = 0$$
$$\nabla \cdot \boldsymbol{A} = 0\ . \tag{2.32}$$

Note that the first of these tells us that *each component* of $\boldsymbol{A}$ satisfies the wave equation, and the second places a restriction on the three components. Therefore the solutions we found in Chapter 1 can be immediately applied to the EM field if they are generalized to:

- three-dimensional space and
- two independent vector degrees of freedom (only two because $\nabla \cdot \boldsymbol{A} = 0$ constrains the third).

In addition, the wave velocity is now that of light, so that $v = c = 1$.

Referring back to Eq. (1.15), the vector field must therefore have the form

$$\boldsymbol{A}(r,t) = \sum_{n,\alpha} \frac{1}{\sqrt{2\omega_n L^3}} \left\{ \epsilon_n^\alpha a_{n,\alpha}\, e^{-ik_n\cdot x} + \epsilon_n^{\alpha\,*} a_{n,\alpha}^\dagger\, e^{ik_n\cdot x} \right\} \quad , \tag{2.33}$$

where the sum is over three integers $n = (n_x, n_y, n_z)$, corresponding to the requirement that the solutions of the wave equation satisfy periodic boundary conditions in *each* of the three space dimensions (referred to as *box* normalization), and the integer $\alpha = 1$ or 2, corresponding to the two independent vector degrees of freedom of the vector potential which are not constrained by the Coulomb gauge. Specifically, the momenta of the plane wave solutions are given by the three-dimensional generalization of Eq. (1.6),

$$k_{n_i} = \frac{2\pi n_i}{L} \qquad n_i = 0, \pm 1, \pm 2, \ldots \qquad i = x, y, z \tag{2.34}$$

and the argument in the exponential of the plane waves is the generalization of (1.7) to three space dimensions

$$k_n \cdot x = \omega_n t - k_{n_x} x - k_{n_y} y - k_{n_z} z = \omega_n t - \boldsymbol{k}_n \cdot \boldsymbol{r} \quad .$$

The plane wave solutions must satisfy the wave equation, which fixes the frequency

$$\boldsymbol{k}_n^2 = k_n^2 = c^2\omega_n^2 = \omega_n^2 \qquad c = 1 \quad . \tag{2.35}$$

As in Chapter 1, the frequency will always be chosen to be positive, so that

$$\omega_n = \frac{2\pi}{L}\sqrt{n_x^2 + n_y^2 + n_z^2} \quad , \tag{2.36}$$

and the negative frequency solutions are the complex conjugates of the positive frequency ones, with the phase $ik \cdot x$.

The vectors ϵ are referred to as *polarization* vectors. They carry the vector direction of $\boldsymbol{A}$ and are dependent on n. The Coulomb gauge condition requires that they must be orthogonal to $\boldsymbol{k}_n$:

$$\nabla \cdot \boldsymbol{A} = 0 \qquad \Longrightarrow \qquad \boldsymbol{k}_n \cdot \epsilon_n^\alpha = 0 \quad ,$$

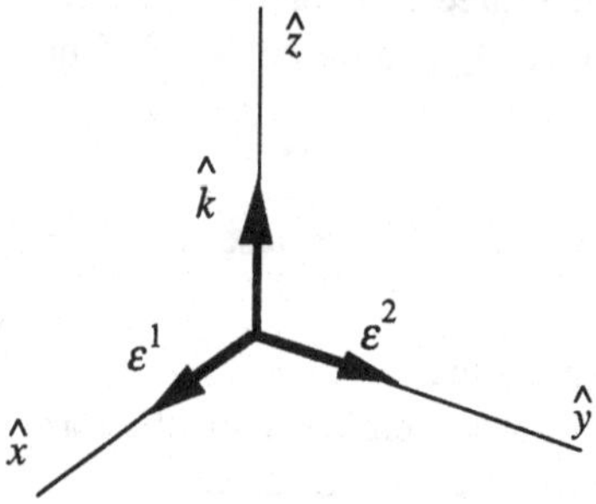

Fig. 2.1 The relative orientation of the two polarization vectors of the photon and its momentum $\hat{k}$.

and hence there can only be two independent vectors for each k_n. To maintain the normalization for the $a_{n,\alpha}$'s introduced in Chapter 1, we require that these vectors be normalized to unity. Since they are in general complex, they will be defined so that

$$\boxed{\epsilon_n^{\alpha'\,*} \cdot \epsilon_n^{\alpha} = \delta_{\alpha,\alpha'}} \quad . \tag{2.37}$$

There are many ways to choose independent ϵ which satisfy (2.37). We will define a *linearly polarized* basis by choosing, for $\boldsymbol{k}_n$ in the $\hat{z}$-direction, $\epsilon^1 = \hat{x}$ and $\epsilon^2 = \hat{y}$, so that

$$\begin{aligned} \epsilon_n^1 \times \epsilon_n^2 &= \hat{k}_n \\ \epsilon_n^2 \times \hat{k}_n &= \epsilon_n^1 \\ \hat{k}_n \times \epsilon_n^1 &= \epsilon_n^2 \quad , \end{aligned} \tag{2.38}$$

where the relative orientation of the two independent polarization vectors is shown in Fig. 2.1. There are only two independent ϵ 's, and they are both perpendicular to $\boldsymbol{k}$. It is this property which leads to the description of the vector potential as *transverse*. There is no simple relation between ϵ_n^1 and $\epsilon_{n'}^1$.

Before we turn to the quantization of the EM field, we will briefly discuss massive vector fields and the differences between massless and massive fields.

2.5 MASSIVE VECTOR FIELDS *

In order to highlight the unique properties of the EM theory, we consider the effect of adding a "mass" term to the Maxwell theory. Massive vector fields play a fundamental role in physics; the $W^{\pm}$ and Z bosons which mediate the electroweak interactions are examples of such fields, and these will be discussed in Sec. 9.10 and in Chapter 15 [see also Appendix D]. For now we are primarily interested in how the massive theory differs from the massless one.

*This section may be omitted on a first reading.

Start by adding a new term to the Lagrangian (2.14):

$$\mathcal{L}_M = -\tfrac{1}{4}F_{\mu\nu}F^{\mu\nu} + \tfrac{1}{2}M^2 A_\mu A^\mu - j_\mu A^\mu \quad , \tag{2.39}$$

where, for now, M is simply regarded as a real parameter. Later we will see that it can be interpreted as the mass of the particles which emerge from the quantization of the field. As before, the four-current j_μ is the source of the field, and it is assumed to be conserved. The equations of motion obtained from this Lagrangian are known as the Proca equations:

$$\partial_\mu F^{\mu\nu} + M^2 A^\nu = j^\nu \quad . \tag{2.40}$$

Taking the four-divergence of both sides and remembering that $F^{\mu\nu}$ is antisymmetric give

$$M^2\, \partial_\nu A^\nu = \partial_\nu j^\nu = 0 \quad . \tag{2.41}$$

*Because the mass is not zero, the Lorentz condition emerges as a necessary constraint.** We no longer have the freedom to choose another constraint (such as the Coulomb gauge condition) because the mass term is *not* gauge invariant. Under a gauge transformation

$$\begin{aligned} M^2 A_\mu A^\mu \to M^2 A'_\mu A'^\mu &= M^2\left(A_\mu A^\mu - \partial_\mu \Lambda_c A^\mu - A_\mu \partial^\mu \Lambda_c + \partial_\mu \Lambda_c\, \partial^\mu \Lambda_c\right) \\ &\neq M^2 A_\mu A^\mu \quad . \end{aligned}$$

Using the Lorentz condition, the equations for the field simplify,

$$\left(\Box + M^2\right) A^\nu = j^\nu \quad . \tag{2.42}$$

If the source is zero, this equation has plane wave solutions

$$A^\nu \sim \epsilon_k^\nu\, e^{-\imath k\cdot x} \quad ,$$

provided the four-vector k satisfies the following equation:

$$k^2 = M^2 \implies \pm k^0 = E_k = \sqrt{M^2 + \boldsymbol{p}^2} \quad . \tag{2.43}$$

This shows that the parameter M is indeed a mass. The Lorentz condition means that the polarization vectors accompanying these plane wave solutions must satisfy

$$k_\mu \epsilon_k^\mu = 0 \quad , \tag{2.44}$$

which is satisfied by *three* independent polarization states (instead of only two as in the massless case). Two of these are the transverse states previously introduced for the EM field, and the third is a *longitudinal* state with a three-vector part

*Note that this constraint must hold for *free* fields, even if the current is not conserved.

in the direction of the particle momentum (for more detail see the discussion in Sec. 9.10).

The most general solution for a *free massive* vector field can therefore be written

$$A^{\mu}(r,t) = \sum_{n}\sum_{\alpha=1}^{3} \frac{1}{\sqrt{2E_n L^3}} \left\{ \epsilon_n^{\alpha\mu} a_{n,\alpha}\, e^{-ik\cdot x} + \epsilon_n^{\alpha\mu\,*} a_{n\alpha}^{\dagger}\, e^{ik\cdot x} \right\} \quad , \tag{2.45}$$

where E_n and ϵ^{μ} satisfy the constraints (2.43) and (2.44), respectively. While this equation appears to be almost identical to the EM field expansion (2.33), it differs in two essential ways. First, the energy and four-momentum are those appropriate to a massive particle and, second, there are *three* independent polarization states instead of only two.

In conclusion, we restate some of the main points of the previous discussion. The Lagrangian for a massive vector theory, Eq. (2.39), still does not depend on $\partial A^0/\partial t$, so that the time component of the field, A^0, must in some sense depend on the sources and other components, as was the case in the massless theory. However, because the massive theory is no longer gauge invariant, the Lorentz condition emerges automatically as the only appropriate constraint on the field, and the Lorentz condition is the constraint which fixes the component A^0 in terms of the other components. Once this condition is taken into account, the free massive field can be expanded in plane waves with three independent polarization degrees of freedom. In the massless case, it is gauge invariance which allows (in fact, requires) us to remove *two* degrees of freedom from the field, which (in Coulomb gauge) amounts to removing the components A^0 *and* A^3 (if the momentum is in the $\hat{z}$-direction), leaving only two independent polarization states.

We now return to a discussion of the quantization of the EM field. Much of the following discussion will be extended to the massive case in Sec. 9.10.

2.6 FIELD QUANTIZATION

We now quantize the theory, described by the Hamiltonian (2.31), for the interaction of the electromagnetic field with nonrelativistic particles, by turning all canonically conjugate variables into operators. The nonrelativistic particles are quantized by the replacements:

$$\begin{aligned} \boldsymbol{r} &\to \boldsymbol{r} \\ \boldsymbol{p} &\to -i\nabla \quad , \end{aligned}$$

where the operator $\boldsymbol{r}$ is simply multiplication by $\boldsymbol{r}$. Similarly, the EM field is quantized by turning $\boldsymbol{A}$ into an operator. The development used in Chapter 1 for the string will be followed again here. This involves two steps. The simple harmonic oscillators which describe the classical field must be found and described, and then they must be quantized.

The plane wave expansion (2.33) for the EM potential $\boldsymbol{A}$ is the solution to the first of these steps; it expresses the field $\boldsymbol{A}$ in terms of *independent* oscillators described by the quantities $a_{n,\alpha}$. The second step, the quantization of the field, is done in *precisely* the same way it was done in Chapter 1; the quantities $a_{n,\alpha}$ are turned into operators by imposing the commutation relations

$$\boxed{\begin{aligned} &[a_{n,\alpha}, a^{\dagger}_{n',\alpha'}] = \delta_{nn'}\delta_{\alpha\alpha'} \\ &[a_{n,\alpha}, a_{n',\alpha'}] = [a^{\dagger}_{n,\alpha}, a^{\dagger}_{n',\alpha'}] = 0 \ . \end{aligned}} \tag{2.46}$$

The only difference between Eq. (2.46) and the corresponding relations for the one-dimensional string is the fact that now there are three space dimensions and two polarizations. This means that Eq. (2.46) must describe many times the number of normal modes, and hence many times the number of independent operators, than were described before. However, this does not really change the result, since operators corresponding to *independent* normal modes still commute, and the commutation relation for a and $a^{\dagger}$ for a *single* normal mode is the same. Thus Eqs. (2.33) and (2.46) give the complete description of the *EM* field and its quantization, and we will now use them to work out several details.

Canonical Commutation Relations

Because of the gauge condition, the forms of the canonical commutation relations for A and π differ from those found for the string. The CCR can be worked out from

$$\begin{aligned} [A^j(r',t)\,,\,\pi^i(r,t)] &= i\sum_{\substack{n,\alpha\\ n',\alpha'}} \frac{\omega_n}{\sqrt{2\omega_n 2\omega_{n'}}\,L^3}\Big\{\epsilon_n^{\alpha i}\epsilon_{n'}^{\alpha' j\,*}[a_{n,\alpha}\,,\,a^{\dagger}_{n',\alpha'}]e^{i(k'\cdot x'-k\cdot x)} \\ &\qquad - \epsilon_n^{\alpha i\,*}\epsilon_{n'}^{\alpha' j}[a^{\dagger}_{n,\alpha}\,,\,a_{n',\alpha'}]\,e^{i(k\cdot x-k'\cdot x')}\Big\} \\ &= i\sum_{n,\alpha}\frac{1}{2L^3}\Big\{\epsilon_n^{\alpha i}\epsilon_n^{\alpha j\,*}\,e^{i\boldsymbol{k}_n\cdot(\boldsymbol{r}-\boldsymbol{r}')} + \epsilon_n^{\alpha i\,*}\epsilon_n^{\alpha j}\,e^{-i\boldsymbol{k}_n\cdot(\boldsymbol{r}-\boldsymbol{r}')}\Big\} \ , \end{aligned}$$

where $x = (t,\boldsymbol{r})$ and $x' = (t,\boldsymbol{r}')$. This can be further reduced using the fact that the polarization vectors, together with $\hat{k}_n$ (the unit vector in the direction of $\boldsymbol{k}_n$), form a complete orthonormal set. Hence, for *each* n,

$$\sum_{\alpha}\epsilon_n^{\alpha i\,*}\epsilon_n^{\alpha j} + \hat{k}_n^i\,\hat{k}_n^j = \delta_{ij}$$

or

$$\sum_{\alpha}\epsilon_n^{\alpha i\,*}\epsilon_n^{\alpha j} = \delta_{ij} - \frac{k_n^i\,k_n^j}{k_n^2} = \delta_{ij}^{\mathsf{T}} \ , \tag{2.47}$$

where δ^{T} is the transverse δ-function. This gives

$$[A^j(r',t),\pi^i(r,t)] = i\sum_n \left(\delta_{ij} - \frac{k_n^i\, k_n^j}{k_n^2}\right)\frac{e^{i\boldsymbol{k}_n\cdot(\boldsymbol{r}-\boldsymbol{r}')}}{L^3}$$

$$= i\left(\delta_{ij} - \frac{\partial_i\partial_j}{\nabla^2}\right)\sum_n \frac{e^{i\boldsymbol{k}_n\cdot(\boldsymbol{r}-\boldsymbol{r}')}}{L^3} \ .$$

Recalling that the sum over the plane wave states gives a delta function for *each* direction in space leads to the following expression for the CCR's for *EM* theory:

$$[A^j(r',t),\pi^i(r,t)] = i\left(\delta_{ij} - \frac{\partial_i\partial_j}{\nabla^2}\right)\delta^3(r-r') \equiv \delta^{3\mathrm{T}}(r-r') \ . \tag{2.48}$$

The extra $\partial_i\partial_j$ term is necessary in order that the CCR's be consistent with the gauge condition:

$$(\nabla_{r'})_j\,[A^j(r',t),\pi^i(r,t)] = i\,[\nabla_i - \nabla_i]\,\delta^3(r-r') = 0$$

$$= [\nabla\cdot A,\pi^i(r,t)] = 0 \ .$$

Note also that

$$\sum_i [A^i(r',t),\pi^i(r,t)] = 2i\delta^3(r-r') \ ,$$

where the factor of 2 appears because there are two independent polarization states.

All of these commutation relations hold at *equal times*, and the commutators are zero if the two points are separated in space. Under Lorentz transformations, the interval $(t-t')^2-(\boldsymbol{r}-\boldsymbol{r}')^2 = (x-x')^2$ is invariant, and thus one consequence of a relativistic generalization of the CCR's is that the *field operators commute when their arguments are separated by a space-like interval* [i.e., one for which the four-vector distance $(x-x')^2 < 0$]. This has a beautiful physical interpretation: it is impossible to exchange information between two points separated by a space-like interval, and hence any physical observables (fields in this case) at two such points must be truly independent of each other. The mathematical expression of this independence is the statement that the operators corresponding to these quantities must commute. This is an important principle, referred to as *local commutativity* or *microscopic causality*, which can be used as a starting point for an axiomatic development of field theory [see Streater and Wightman (1964)].

Perhaps the appearance of the transverse δ-function in the CCR's (2.48) could have been anticipated from the start, but in any case, it follows in a straightforward way from the commutation relations (2.46). These commutation relations between the creation and annihilation operators involve only the *independent* degrees of freedom, and hence are the same for all types of fields. It is for this reason that we have chosen to use them to begin the quantization of any field theory.

Hamiltonian and Momentum Operators

The form of the Hamiltonian and momentum operators can be inferred from the discussion of the string in Chapter 1. Here we will demonstrate that the Hamiltonian does indeed have the expected form. The proof that the momentum operator also has this form is deferred to Prob. 2.1.

For simplicity, assume that the polarization vectors are real. Then, recalling Eq. (2.31), the Hamiltonian for the free EM field is

$$H = \frac{1}{2}\int d^3r \left\{ :\boldsymbol{E}_\perp^2(r,t): + :\boldsymbol{B}^2(r,t): \right\} = \frac{1}{4L^3}\int d^3r \sum_{\substack{n,n' \\ \alpha,\alpha'}} \frac{1}{\sqrt{\omega_n\omega_{n'}}}$$

$$\times \Bigg\{ \left[\omega_n\omega_{n'}\boldsymbol{\epsilon}_n^\alpha\cdot\boldsymbol{\epsilon}_{n'}^{\alpha'} + (\boldsymbol{k}_n\times\boldsymbol{\epsilon}_n^\alpha)\cdot\left(\boldsymbol{k}_{n'}\times\boldsymbol{\epsilon}_{n'}^{\alpha'}\right)\right]$$

$$\times\left(a_{n,\alpha}^\dagger a_{n',\alpha'}\, e^{i(k_n-k_{n'})\cdot x} + a_{n',\alpha'}^\dagger a_{n,\alpha}\, e^{-i(k_n-k_{n'})\cdot x}\right)$$

$$+\left[-\omega_n\omega_{n'}\boldsymbol{\epsilon}_n^\alpha\cdot\boldsymbol{\epsilon}_{n'}^{\alpha'} - (\boldsymbol{k}_n\times\boldsymbol{\epsilon}_n^\alpha)\cdot\left(\boldsymbol{k}_{n'}\times\boldsymbol{\epsilon}_{n'}^{\alpha'}\right)\right]$$

$$\times\left(a_{n,\alpha}a_{n',\alpha'}\, e^{-i(k_{n'}+k_n)\cdot x} + a_{n,\alpha}^\dagger a_{n',\alpha'}^\dagger\, e^{i(k_{n'}+k_n)\cdot x}\right)\Bigg\} \quad ,$$

where the normal ordering prescription has required all cross terms to be written as $a^\dagger a$. Using the fact that

$$\int \frac{d^3r}{L^3} e^{\pm i(k_n-k_{n'})\cdot x} = \delta_{n,n'}$$

$$\int \frac{d^3r}{L^3} e^{\pm i(k_n+k_{n'})\cdot x} = \delta_{n,-n'}\, e^{\pm 2i\omega_n t} \quad ,$$

and $\boldsymbol{k}_n = -\boldsymbol{k}_{-n}$ give

$$H = \frac{1}{4}\sum_{\substack{n \\ \alpha,\alpha'}} \frac{1}{\omega_n}\Bigg\{ \left[\omega_n^2\, \boldsymbol{\epsilon}_n^\alpha\cdot\boldsymbol{\epsilon}_n^{\alpha'} + (\boldsymbol{k}_n\times\boldsymbol{\epsilon}_n^\alpha)\cdot\left(\boldsymbol{k}_n\times\boldsymbol{\epsilon}_n^{\alpha'}\right)\right]\left(a_{n,\alpha}^\dagger a_{n,\alpha'} + a_{n,\alpha'}^\dagger a_{n,\alpha}\right)$$

$$+\left[-\omega_n^2\, \boldsymbol{\epsilon}_n^\alpha\cdot\boldsymbol{\epsilon}_{-n}^{\alpha'} + (\boldsymbol{k}_n\times\boldsymbol{\epsilon}_n^\alpha)\cdot\left(\boldsymbol{k}_n\times\boldsymbol{\epsilon}_{-n}^{\alpha'}\right)\right]$$

$$\times\left(a_{n,\alpha}a_{-n,\alpha'}\, e^{-2i\omega_n t} + a_{n,\alpha}^\dagger a_{-n,\alpha'}^\dagger\, e^{2i\omega_n t}\right)\Bigg\} .$$

Next, use

$$\boldsymbol{\epsilon}_n^\alpha\cdot\boldsymbol{\epsilon}_n^{\alpha'} = \delta_{\alpha\alpha'}$$

$$(\boldsymbol{k}_n\times\boldsymbol{\epsilon}_n^\alpha)\cdot\left(\boldsymbol{k}_n\times\boldsymbol{\epsilon}_{-n}^{\alpha'}\right) = k_n^2\, \boldsymbol{\epsilon}_n^\alpha\cdot\boldsymbol{\epsilon}_{-n}^{\alpha'} - \boldsymbol{k}_n\cdot\boldsymbol{\epsilon}_{-n}^{\alpha'}\ \boldsymbol{k}_n\cdot\boldsymbol{\epsilon}_n^\alpha$$

$$= \omega_n^2\, \boldsymbol{\epsilon}_n^\alpha\cdot\boldsymbol{\epsilon}_{-n}^{\alpha'}$$

$$(\boldsymbol{k}_n\times\boldsymbol{\epsilon}_n^\alpha)\cdot\left(\boldsymbol{k}_n\times\boldsymbol{\epsilon}_n^{\alpha'}\right) = k_n^2\,\delta_{\alpha\alpha'} = \omega_n^2\,\delta_{\alpha\alpha'}$$

to obtain finally

$$\boxed{H = \sum_{n,\alpha} \omega_n a^\dagger_{n,\alpha} a_{n,\alpha} \quad .} \tag{2.49}$$

This is a straightforward generalization of the result for the string. Now there are number operators for photons with momenta in all three spatial directions and with two polarization states.

We leave it as an exercise, Prob. 2.1, to show that the momentum operator is

$$\boldsymbol{P} = \int d^3r : \boldsymbol{E}_\perp(r,t) \times \boldsymbol{B}(r,t) := \sum_{n,\alpha} \boldsymbol{k}_n a^\dagger_{n,\alpha} a_{n,\alpha} \quad . \tag{2.50}$$

This expression shows that the total momentum of the field is a vector sum of the momenta of each photon, as expected from our study of the string.

2.7 SPIN OF THE PHOTON

In the final section of this chapter we show that the particles which emerge from the quantization of the *EM* field (the photons) have *spin one*. Spin can be regarded as an internal degree of freedom of the quanta which is closely connected to the structure of the field from which they emerge. In particular, the quantization of *vector* fields always gives rise to quanta with spin *one*, while the quantization of *scalar* fields gives quanta with spin *zero*.

To obtain these results, it is necessary to discuss the behavior of the field under rotations. Just as the energy of the quanta is displayed by the Hamiltonian (the generator of time translation) and the momenta are displayed by the momentum operator (the generator of space translations), so it is that the spin will emerge from a discussion of the angular momentum operator, the generator of rotations. This section therefore begins with a brief discussion of the rotations of vector fields. A deeper discussion of these topics is postponed until Chapter 8.

Rotations

When transforming vector fields, which are continuous functions of space and time, both the *components* of the vector and the *arguments* of the function must be transformed. For example, a scalar function under rotation transforms in a non-trivial way, as illustrated in Fig. 2.2. In this book, all transformations will be interpreted as *active* transformations, i.e., they transform the state functions, leaving the coordinate system fixed. From examination of the figure, we see that, under an *active* rotation R, the transformed function $\phi_R(r,t)$, where r is a shorthand notation for the three spatial coordinates (x, y, z), is related to the untransformed function $\phi(r,t)$ by

$$\phi_R(r,t) = \phi(R^{-1}r, t) \quad . \tag{2.51}$$

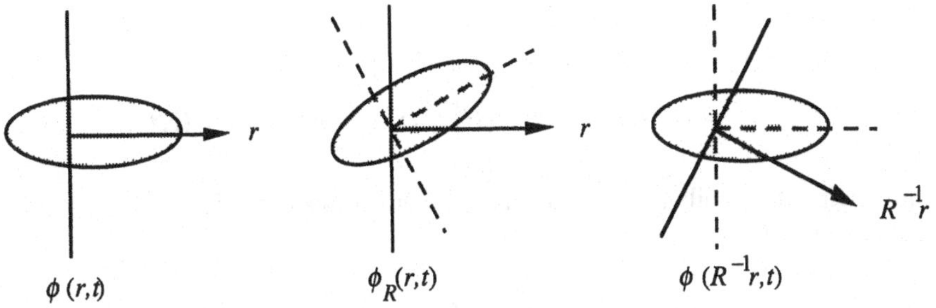

Fig. 2.2 A scalar function ϕ (with contours originally pointing in the $\boldsymbol{r}$-direction) is rotated actively to a new direction. The third figure is identical to the second; only the paper has been turned.

This result can be used to find the transformation law for a vector function. Consider the function $\phi(r,t) = \boldsymbol{a} \cdot \boldsymbol{A}(r,t)$, where $\boldsymbol{a}$ is some reference vector. Under the rotation R this function becomes

$$\boldsymbol{a}_R \cdot \boldsymbol{A}_R(r,t) = \boldsymbol{a} \cdot \boldsymbol{A}(R^{-1}r,t) \quad ,$$

and hence the vector function A satisfies the following transformation law:

$$A^i_R(r,t) = R^{ij} A^j (R^{-1}r,t) \quad . \tag{2.52}$$

Note that we rotate both the *components and the arguments.*

A rotation about the z-axis will be written

$$R_z(\theta) = \begin{pmatrix} \cos\theta & -\sin\theta & 0 \\ \sin\theta & \cos\theta & 0 \\ 0 & 0 & 1 \end{pmatrix} = e^{-iI_z\theta} \quad , \tag{2.53}$$

where I_z is the generator of rotations about the z-axis. The specific form for I_z, and its generalizations to rotations about the axis i, is

$$I_z = \begin{pmatrix} 0 & -i & 0 \\ i & 0 & 0 \\ 0 & 0 & 0 \end{pmatrix} \longrightarrow (I_i)^{jk} = -i\,\epsilon_{ijk} \quad . \tag{2.54}$$

Hence, the change in a vector r under an *infinitesimal* rotation about the i-axis through angle $\delta\theta$ is

$$\delta r^j_i = -\delta\theta \epsilon_{ij\ell} r^\ell = -i\delta\theta\,(I_i)^{j\ell}\, r^\ell \quad .$$

Similarly, the change in a vector field A under the same infinitesimal transformation, using (2.52), is

$$\begin{aligned} A^j_{\delta R_i}(r,t) &= (1 - i\delta\theta I_i)^{j\ell} A^\ell ([1 + i\delta\theta I_i] r, t) \\ &= A^j(r,t) + i\delta\theta_i \left[-I_i^{j\ell} A^\ell(r,t) + I_i^{\ell m} r^m \partial_\ell A^j(r,t) \right] . \end{aligned}$$

Introducing the familiar orbital angular momentum operator, $L^i = -i\,\epsilon_{im\ell} r^m \partial_\ell$, this can be written

$$-i\frac{\delta^i A^j(r,t)}{\delta\theta} = i\,\epsilon_{ij\ell} A^\ell(r,t) - L^i A^j(r,t) \ . \tag{2.55}$$

Now we are ready to apply these ideas to the *EM* field.

Angular Momentum Operator

The angular momentum operator for the *EM* field is

$$\Omega^i = \int d^3r \, [\boldsymbol{r} \times : (\boldsymbol{E}_\perp \times \boldsymbol{B}) :]^i \ . \tag{2.56}$$

This can be reduced to a more tractable form by expanding out the double cross product,

$$\begin{aligned} \Omega^i &= \int d^3r \, \epsilon_{ij\ell} r^j \epsilon_{\ell km} E_\perp^k \epsilon_{mab} \partial_a A^b \\ &= \int d^3r \, \epsilon_{ij\ell} r^j \left(\delta_{\ell a}\delta_{kb} - \delta_{ka}\delta_{\ell b} \right) E_\perp^k \partial_a A^b \\ &= \int d^3r \, \epsilon_{ij\ell} r^j \left\{ E_\perp^b \partial_\ell A^b - \boldsymbol{E}_\perp \cdot \nabla A^\ell \right\} , \end{aligned}$$

and simplifying the last term by integrating by parts and recalling that $E_\perp$ is transverse,

$$\begin{aligned} -\int d^3r \, \epsilon_{ij\ell} r^j \boldsymbol{E}_\perp \cdot \nabla A^\ell &= \int d^3r \, \epsilon_{ij\ell} \left(r^j \nabla \cdot \boldsymbol{E}_\perp \, A^\ell + \delta^{jb} E_\perp^b A^\ell \right) \\ &= \int d^3r \, \epsilon_{ib\ell} E_\perp^b A^\ell \ . \end{aligned}$$

Substituting for $E_\perp$ and using the orbital angular momentum operator $L = -i(r \times \nabla)$ give a more compact form for Ω:

$$\boxed{\Omega^i = -i \int d^3r : \frac{\partial A^b}{\partial t} \left(L^i A^b - i\epsilon_{ib\ell} A^\ell \right) : } \ . \tag{2.57}$$

To interpret each of the two terms in (2.57) it is instructive to compute the commutator of Ω with A. This is easily done using the CCR's:

$$[\Omega^i, A^j(r,t)] = -\int d^3r' \left[\left(\delta_{jb} - \frac{\partial_j \partial_b}{\nabla^2}\right)\delta^3(r-r')\right]\left(L^i A^b - i\epsilon_{ib\ell}A^\ell\right) \quad . \tag{2.58}$$

The $\partial_j \partial_b$ term is zero:

$$\begin{aligned}
&\int d^3r' \frac{\partial_j \partial_b}{\nabla^2}\delta^3(r-r')\left[L^i A^b(r',t) - i\,\epsilon_{ib\ell}A^\ell(r',t)\right] \\
&= -\int d^3x' \frac{\partial_j}{\nabla^2}\delta^3(r-r')\partial^b\left[-i\,\epsilon_{i\ell m}r^\ell \partial_m A^b - i\,\epsilon_{ib\ell}A^\ell\right] \\
&= -\int d^3x' \frac{\partial_j}{\nabla^2}\delta^3(r-r')\left[-i\,\epsilon_{ibm}\partial_m A^b - i\,\epsilon_{i\ell m}r^\ell \partial_m \partial_b A^b + i\,\epsilon_{i\ell b}\partial_b A^\ell\right] = 0
\end{aligned}$$

so that (2.58) simplifies to

$$[\Omega^i, A^j(r,t)] = -L^i A^j(r,t) + i\,\epsilon_{ij\ell}A^\ell(r,t) = -i\frac{\delta^i A^j}{\delta\theta} \quad . \tag{2.59}$$

Note that the right-hand side of this equation describes the infinitesimal rotation of the field around the ith axis [compare with Eq. (2.55)], showing that the angular momentum operator Ω is indeed the generator of rotations for the field theory.

Spin

The spin of a particle, or a field quanta, is an intrinsic property. This suggests separating the total angular momentum operator into two parts:

$$\Omega^i = \Omega_L^i + \Omega_{\text{spin}}^i \quad ,$$

with the spin part associated with transformation of field indices and the orbital part associated with transformation of the field arguments, and

$$\begin{aligned}
\Omega_{\text{spin}}^i &= -\int d^3r \colon \frac{\partial A^j}{\partial t}\epsilon_{ijk}A^k \colon \\
\Omega_L^i &= -i\int d^3r \colon \frac{\partial A^j}{\partial t} L^i A^j \colon \quad .
\end{aligned} \tag{2.60}$$

Note, however, that

$$\begin{aligned}
[\Omega_{\text{spin}}^i, A^j(r,t)] &= i\,\epsilon_{i\ell k}\left[\delta^{j\ell} - \frac{\partial_j \partial_\ell}{\nabla^2}\right]A^k(r,t) \\
[\Omega_L^i, A^j(r,t)] &= -L^i A^j(r,t) + i\,\epsilon_{i\ell k}\frac{\partial_j \partial_\ell}{\nabla^2}A^k(r,t) \quad .
\end{aligned}$$

The extra term in both commutators is required to make each expression consistent with the Coulomb gauge.

From the vector nature of the field we expect it to have spin one, but our task is to see how this comes about naturally in the particle picture. To do this we will express $\Omega^i_{\rm spin}$ in terms of the a's and $a^\dagger$'s, using real polarization vectors for simplicity,

$$\begin{aligned}\Omega^i_{\rm spin} = & -\int d^3r \frac{\epsilon_{ijk}}{2L^3} \sum_{\substack{n,\alpha \\ n',\alpha'}} -\frac{i\omega_n}{\sqrt{\omega_n\omega_{n'}}} \epsilon^{\alpha j}_n \epsilon^{\alpha' k}_{n'} \\ & \times \Big\{ a_{n,\alpha} a_{n',\alpha'} e^{-i(k_n+k_{n'})\cdot x} - a^\dagger_{n,\alpha} a^\dagger_{n',\alpha'} e^{i(k_n+k_{n'})\cdot x} \\ & \qquad - a^\dagger_{n,\alpha} a_{n',\alpha'}\, e^{i(k_n-k_{n'})\cdot x} + a^\dagger_{n',\alpha'} a_{n,\alpha}\, e^{-i(k_n-k_{n'})\cdot x} \Big\} \\ = & i \sum_{\substack{n \\ \alpha,\alpha'}} \frac{\epsilon_{ijk}}{2} \Big\{ \epsilon^{\alpha j}_n \epsilon^{\alpha' k}_{-n} \left[a_{n,\alpha} a_{-n,\alpha'}\, e^{-2i\omega_n t} - a^\dagger_{n,\alpha} a^\dagger_{-n,\alpha'}\, e^{2i\omega_n t} \right] \\ & \qquad + \left[\epsilon^{\alpha j}_n \epsilon^{\alpha' k}_n - \epsilon^{\alpha' j}_n \epsilon^{\alpha k}_n \right] a^\dagger_{n,\alpha'} a_{n,\alpha} \Big\} \ .\end{aligned}$$

The first term in the { } bracket is symmetric in j and k. This can be seen by changing n to $-n$ and $\alpha \leftrightarrow \alpha'$. Hence it is zero when contracted with ϵ_{ijk}. The second term is clearly antisymmetric in j and k and requires $\alpha \neq \alpha'$. Hence

$$\Omega_{\rm spin} = i \sum_{\substack{n \\ \alpha\neq\alpha'}} \left(\epsilon^\alpha_n \times \epsilon^{\alpha'}_n \right) a^\dagger_{n,\alpha'} a_{n,\alpha} \ .$$

Recall that $\epsilon^1 \times \epsilon^2 = \hat{k}$. Hence, carrying out the sum over α, α' gives

$$\Omega_{\rm spin} = i \sum_n \hat{k}_n \left[a^\dagger_{n,2} a_{n,1} - a^\dagger_{n,1} a_{n,2} \right] \ .$$

This is not a convenient form because it is not given in terms of number operators.

We can express $\Omega_{\rm spin}$ in terms of number operators by introducing a new polarization basis referred to as the *circular*, or *helicity*, basis. If $\epsilon^1 \to \hat{x}$, $\epsilon^2 \to \hat{y}$, and $\hat{k} \to \hat{z}$, then the circular polarization basis, in which the states have a definite spin projection along $\hat{z}$, is defined by

$$\begin{aligned}\epsilon^+ &= \text{spin in } +\hat{k}\text{-direction} = -\frac{1}{\sqrt{2}} \left(\epsilon^1 + i\, \epsilon^2 \right) \\ \epsilon^- &= \text{spin in } -\hat{k}\text{-direction} = \frac{1}{\sqrt{2}} \left(\epsilon^1 - i\, \epsilon^2 \right) \ .\end{aligned} \tag{2.61}$$

Note the appearance of the minus sign in the definition of ϵ^+; this is a standard phase convention used in the construction of the spherical harmonics $\mathcal{Y}_{\ell m}$ for

$\ell = 1$ and $m = \pm 1$ from $\hat{x}$ and $\hat{y}$. Then we define $a_\pm$, the annihilation operators corresponding to these circularly polarized states, by the relation (suppress n for now)

$$\epsilon^1 a_1 + \epsilon^2 a_2 = \epsilon^+ a_+ + \epsilon^- a_- \quad .$$

This gives

$$a_1 = -\frac{1}{\sqrt{2}}(a_+ - a_-) \qquad [a_+, a_+^\dagger] = [a_-, a_-^\dagger] = 1$$

$$a_2 = -\frac{i}{\sqrt{2}}(a_+ + a_-) \qquad [a_+, a_-^\dagger] = [a_-, a_+^\dagger] = 0 \quad ;$$

Hence

$$a_2^\dagger a_1 - a_1^\dagger a_2 = -i[a_+^\dagger a_+ - a_-^\dagger a_-]$$

and, restoring n,

$$\boxed{\Omega_{\rm spin} = \sum_n \hat{k}_n [a_{n,+}^\dagger a_{n,+} - a_{n,-}^\dagger a_{n,-}] \quad .} \tag{2.62}$$

The spin operator has now been expressed in terms of number operators for photons with a definite helicity. Note that it is a vector sum of terms which point in the $+\hat{k}$-direction for positive helicity and in the $-\hat{k}$-direction for negative helicity.

In general, the helicity of a particle is the projection of its spin along the direction of its motion, and if a *massive* particle has spin s, its helicity can take on any integer value between s and $-s$ (i.e., $s, s-1, s-2, \ldots, -s$). The direction of motion is simply one special direction in space, and a massive particle of spin s has $2s+1$ states which can always be expanded in terms of states having a definite spin projection along any chosen axis. However, Eq. (2.62) shows that photons do not have this property. It shows that the photon has spin 1, but that out of three possible states (± 1 or 0), only helicity states $+1$ and -1 can occur. The absence of helicity zero is due to the transverse nature of the field (Coulomb gauge) which is due in turn to the absence of a photon rest mass.

The restriction of the photon helicity to its maximum and minimum possible values, ± 1, illustrates a property of any massless particle. In general, if a massless particle has spin s, it may have only two helicity states: $\pm s$. The other possible states are prohibited. This remarkable result is one of the consequences of Wigner's famous analysis of the representations of the *Poincaré group* (which is the group which results from combining the Lorentz transformations with space–time translations). It turns out that the representations of the Poincaré group are characterized by both mass and spin and that the familiar $2s+1$ degeneracy associated with the spin s representations of the $SU(2)$ rotation group occur only when the mass M of the particles described by the representation is non-zero. If

$M = 0$, the spin representations are only two dimensional, explaining why there are only two states with spin projections $\pm s$. For more information see Ryder (1985) or Wigner's original paper [Wi 39].

PROBLEMS

2.1 Prove that

$$\boldsymbol{P} \equiv \int d^3r \ :\boldsymbol{E}_\perp(r,t) \times \boldsymbol{B}(r,t): = \sum_{n,\alpha} \boldsymbol{k}_n a^\dagger_{n,\alpha} a_{n,\alpha} \ .$$

What is the significance of this result?

2.2 **(a)** Compute the following matrix element:

$$\left\langle 1_{n,\alpha}, 1_{n',\alpha'} \right| :\boldsymbol{A}^2(r,t): \left| 0 \right\rangle \ ,$$

where

$$\left| 1_{n,\alpha}, 1_{n',\alpha'} \right\rangle = \frac{1}{\sqrt{2}} a^\dagger_{n,\alpha} a^\dagger_{n',\alpha'} \left| 0 \right\rangle$$

is a two-photon state.

(b) In what physical process might this matrix element play a role?

2.3 **(a)** Compute the following matrix element:

$$\left\langle n_{k_1} \right| \boldsymbol{A}(r,t) \left| m_{k_2} \right\rangle \ ,$$

where

$$\left| m_{k_2} \right\rangle = \frac{\left(a^\dagger_{k_2,\alpha_2} \right)^{m_{k_2}}}{\sqrt{m_{k_2}!}} \ \left| 0 \right\rangle$$

is the state of m_{k_2} photons with momentum $\boldsymbol{k}_2$ and polarization α_2.

(b) Discuss the physical significance of your result.

2.4 The *orbital* part of the angular momentum operator Ω^i_L was defined in Eq. (2.60). Prove that it contains no terms of the form $a^\dagger_{n,\alpha} a^\dagger_{n',\alpha'}$ or $a_{n,\alpha} a_{n',\alpha'}$, and also show that

$$k^i \Omega^i_L \left| 1_k \right\rangle = 0 \ ,$$

where $\left| 1_k \right\rangle$ is the state of one photon with momentum k and polarization α. What is the significance of this result?

CHAPTER 3

INTERACTION OF RADIATION WITH MATTER

In this chapter, the Lagrangian obtained in the last chapter is used to show how atomic decay is explained by field theory. Then the famous *Lamb shift* is calculated and discussed. The Lamb shift is the splitting between atomic levels with the *same* total angular momentum but *different* orbital angular momentum and cannot be explained without the use of field theory. The largest such splitting is between the $2S_{1/2}$ and $2P_{1/2}$ levels and is a noticeable feature of the hydrogen atom spectrum. Finally, we calculate the photodisintegration of the deuteron, one of the first examples of the conversion of energy to mass. To set the stage for these calculations, the chapter begins with a discussion of how to determine the time evolution operator in a case when the Hamiltonian depends on time.

3.1 TIME EVOLUTION AND THE S-MATRIX

Since the interaction Hamiltonian is, in general, time-dependent, we will calculate the interaction between nonrelativistic systems and the quantized EM radiation field using time-dependent perturbation theory.* For definiteness, the nonrelativistic system will be taken to be a heavy atomic nucleus with charge Z at *rest* at the origin and a single electron of mass m with a negative *point* charge located at $\boldsymbol{r}_e$ (other systems will be discussed in Sec. 3.5). The charge distribution for these two particles, in the language of Eq. (2.27), is therefore

$$\begin{aligned}\rho_e(|\boldsymbol{r}_e - \boldsymbol{r}|) &= -e\,\delta^3(r_e - r)\\ \rho_N(|\boldsymbol{r}|) &= Ze\,\delta^3(r) \quad ,\end{aligned}$$

and the only particle coordinates we need to consider are those of the electron. The Hamiltonian given in Eq. (2.31) can therefore be broken up into three parts:

$$H' = H_A + H_{EM} + H_I' \quad , \tag{3.1}$$

*The particles in this chapter are treated nonrelativistically, but the derivation of the time evolution operator is completely general, and the results we obtain here will be applied, in Chapter 9, to relativistic systems.

where

$$\begin{aligned} H_A &= \frac{p_e^2}{2m} - \frac{Z\alpha}{r_e} \\ H_{EM} &= \frac{1}{2}\int d^3r \left\{:\pi^2(r,t): + :B^2(r,t):\right\} \\ H_I' &= \frac{e}{2m}\{p_e \cdot A(r_e,t) + A(r_e,t)\cdot p_e\} + \frac{e^2}{2m}A^2(r_e,t) \ , \end{aligned} \tag{3.2}$$

where $\alpha = e^2/4\pi$ is the fine structure constant. The first two terms, $H_A + H_{EM}$, will be considered the unperturbed Hamiltonian, with H_I' the perturbation. Note that we have included the Coulomb interaction term [the third term in Eq. (2.31)] in H_A because we intend to develop the perturbation theory in terms of atomic wave functions, which include the (nonrelativistic) Coulomb interaction to all orders (exactly). We have omitted the Coulomb self-energies of the atomic nucleus and the electron; for point particles these are infinite constants which may be subtracted by a convenient definition of the energy [as discussed following Eq. (2.31)]. The interaction term H_I' is the expansion of the familiar $(p - eA)^2$ factor and includes a term which is first order in the electron charge e and linear in A and a second order term proportional to A^2.

First, consider the case when the interaction term is zero. Then the EM field coordinates, which are the vector potential operators A, are contained only in H_{EM}, and the electron coordinates, r, are contained only in H_A, which is the usual Schrödinger Hamiltonian. We found the quantum mechanical eigenstates for the free EM field in Chapter 2; the solutions are a Fock space of photon states which are time-independent. The eigenstates of H_A are also known from previous studies of nonrelativistic quantum mechanics; the bound states of hydrogen-like atoms can be described by wave functions

$$\psi_a(r) \ ,$$

where a labels the quantum numbers of the bound state. These states evolve in time by a phase factor only, in the sense that

$$\begin{aligned} \psi_a(r,t) &= e^{-iH_A t}\,\psi_a(r) \\ &= U_A(t)\,\psi_a(r) \ . \end{aligned} \tag{3.3}$$

This expression is similar to Eq. (1.31) with the choice $t_0 = 0$. [Any time t_0 could be chosen, but this choice corresponds to the usual phase convention in which atomic states are real when $t = 0$.]

In Chapters 1 and 2 we used the Heisenberg representation for the fields, while the atomic wave functions are usually given in the Schrödinger representation. It is more convenient to choose a common representation for all fields and operators, and in the remainder of this book we will use the *interaction representation*. In this representation the time dependence of the *free, non-interacting Hamiltonian* is in the operators, and under the influence of the *free* Hamiltonian, the states will not evolve in time. However, under the *full* Hamiltonian, which includes an interaction term, the states *will* evolve in time, and the principal goal of this section is to calculate this evolution. But before we proceed with this calculation, we must give the electron operators the time dependence associated with the free Hamiltonian. This means that, instead of using H' given above, we will use

$$H(t) = U_A^{-1}(t)H'(t)U_A(t) \quad . \tag{3.4}$$

Since H_A commutes with itself and the EM field operators $\boldsymbol{A}$, H_A and H_{EM} are unaffected by this transformation, but H'_I becomes H_I, where

$$\boxed{\begin{aligned} H &= H_A + H_{EM} + H_I(t) = H_0 + H_I(t) \\ H_I(t) &= U_A^{-1}\left\{\frac{e}{2m}\left(\boldsymbol{p}_e \cdot \boldsymbol{A}(\boldsymbol{r}_e, t) + \boldsymbol{A}(\boldsymbol{r}_e, t)\cdot \boldsymbol{p}_e\right) + \frac{e^2}{2m}\boldsymbol{A}^2(\boldsymbol{r}_e, t)\right\}U_A \end{aligned}} \tag{3.5}$$

with $U_A(t)$ defined in Eq. (3.3).

The solutions to the free Hamiltonian $H_0 = H_A + H_{EM}$ are just direct products of atomic wave functions and photon states, which we will write

$$|a, n\rangle = \psi_a(\boldsymbol{r}_e)|n\rangle \quad , \tag{3.6}$$

where a labels the atomic states and $|n\rangle$ the photon Fock states as described (for the string) in Eq. (1.20). The scalar product of the atomic states requires an integration over the coordinate $\boldsymbol{r}_e$,

$$\langle a'|a\rangle = \int d^3r_e\, \psi^*_{a'}(\boldsymbol{r}_e)\psi_a(\boldsymbol{r}_e) \quad .$$

The states $|a, n\rangle$ are a complete set and are stationary under the unperturbed Hamiltonian H_0, which is independent of time. When the interaction H_I is turned on, the states are no longer stationary. The question we ask is: "How do these states evolve in time?" In practice, this may mean "How do excited states $|a, 0\rangle$ decay into other states $|b, n\rangle$ where n photons are emitted?"

To answer this question, we must find the time translation operator for the full Hamiltonian (3.5), which will be written

$$H(t) = H_0 + H_I(t) \quad . \tag{3.7}$$

This Hamiltonian depends on time. We assume that $H_I(t)$ is switched on at time $t = t_0$ so that

$$H(t) = H_0 \quad \text{if} \quad t < t_0 \quad . \tag{3.8}$$

We found the time translation operator for a Hamiltonian which is independent of time in Sec. 1.8, and we therefore know $U_0(t, t_0)$ corresponding to H_0 (it is just the product of U_A and a similar expression for the field). The total time translation operator will be *defined* to be

$$U_{\text{total}}(t, t_0) = U_0(t, t_0) U_I(t, t_0) \quad . \tag{3.9}$$

It is therefore sufficient to find an equation for U_I. Since H_I need not commute with H_0, and since it depends on t, the form of U_I depends on the definition (3.9). Note that our definition differs in important ways from that given in, for example, Fetter and Walecka (1971). From the definition (3.9) the interaction time translation operator is unitary, but note that $U_I(t_1, t_2) U_I(t_2, t_3) \neq U_I(t_1, t_3)$ [see Prob. 3.5].

Now consider any physical observable represented by the operator $\mathcal{O}(t)$. Under the full time translation operator it evolves according to

$$\mathcal{O}(t) = U_{\text{total}}^{-1}(t, t_0) \mathcal{O}(t_0) U_{\text{total}}(t, t_0) \quad . \tag{3.10}$$

Under the free, noninteracting Hamiltonian the same observable evolves according to

$$\mathcal{O}_0(t) = U_0^{-1}(t, t_0) \mathcal{O}(t_0) U_0(t, t_0) \quad . \tag{3.11}$$

Note that the free observable $\mathcal{O}_0(t)$ is not equal to the interacting observable $\mathcal{O}(t)$ because the free time translation operator U_0 is not equal to the full time translation operator U_{total}. This is because when $t > t_0$, the time at which the interaction is turned on, $U_I \neq 1$. However, because of our definition (3.9), there is a simple connection between these two quantities:

$$\begin{aligned} \mathcal{O}(t) &= U_{\text{total}}^{-1} \mathcal{O}_0(t_0) U_{\text{total}} = U_I^{-1} U_0^{-1} \mathcal{O}_0(t_0) U_0 U_I \\ &= U_I^{-1} \mathcal{O}_0(t) U_I \quad . \end{aligned}$$

Hence the connection between the free observable and the interacting observable is

$$U_I(t, t_0) \mathcal{O}(t) U_I^{-1}(t, t_0) = \mathcal{O}_0(t) \quad , \tag{3.12}$$

and the operator U_I converts free observables into interacting observables, and *vice versa*.

We can find the operator U_I from the relations

$$\begin{aligned} [H(t), \mathcal{O}(t)] &= -i \frac{\partial}{\partial t} \mathcal{O}(t) \\ [H_0, \mathcal{O}_0(t)] &= -i \frac{\partial}{\partial t} \mathcal{O}_0(t) \quad , \end{aligned} \tag{3.13}$$

which are the infinitesimal equivalents of Eqs. (3.10) and (3.11). Hence

$$\begin{aligned}[H_0, \mathcal{O}_0(t)] &= -i\frac{\partial}{\partial t}\mathcal{O}_0(t) = -i\frac{\partial}{\partial t}\left[U_I\,\mathcal{O}(t)\,U_I^{-1}\right] \\ &= -i\left[\frac{dU_I}{dt}\underbrace{U_I^{-1}U_I}_{=1}\mathcal{O}(t)\,U_I^{-1} + U_I\,\frac{\partial\mathcal{O}(t)}{\partial t}U_I^{-1} + U_I\mathcal{O}(t)\underbrace{U_I^{-1}U_I}_{=1}\frac{dU_I^{-1}}{dt}\right] \\ &= -i\left[\frac{dU_I}{dt}U_I^{-1}\mathcal{O}_0(t) + iU_I\left[H(t),\mathcal{O}(t)\right]U_I^{-1} - \mathcal{O}_0(t)\,\frac{dU_I}{dt}U_I^{-1}\right] \quad ,\end{aligned} \tag{3.14}$$

where the last term was simplified using Eq. (1.34). However, H is a function of the fields ϕ (a particular subset of the physical observables $\mathcal{O}$),

$$H(t) = H\left[\phi(t)\right] \quad , \tag{3.15}$$

where the square brackets [ ] will be used whenever we wish to express H as a function of field quantities and round brackets ( ) are used to express H as a function of t. Since H can be expanded in powers of ϕ,

$$\begin{aligned} U_I\,H\,U_I^{-1} &= H\left[U_I\,\phi(t)\,U_I^{-1}\right] = H\left[\phi_0(t)\right] \\ &= H_0 + H_I\left[\phi_0(t)\right] \quad . \end{aligned} \tag{3.16}$$

Hence, Eq. (3.14) becomes

$$[H_0, \mathcal{O}_0(t)] = \left[\left(-i\frac{dU_I}{dt}U_I^{-1} + H_0 + H_I\left[\phi_0(t)\right]\right), \mathcal{O}_0(t)\right] \tag{3.17}$$

or

$$\left[\left(\frac{dU_I}{dt}U_I^{-1} + iH_I[\phi_0]\right), \mathcal{O}_0(t)\right] = 0 \quad . \tag{3.18}$$

This is a remarkable equation. Because the H_I in this equation is a function of the *free* fields ϕ_0, the equation allows us to determine the interaction time translation operator *entirely* in terms of the free fields. Since $\mathcal{O}_0$ is any operator in a complete set, and since any operator which commutes with *all* operators in a complete set must be a multiple of the identity (Schur's Lemma), this means that, just as in Eq. (1.35),

$$H_I[\phi_0] = i\frac{dU_I}{dt}U_I^{-1} + E_0(t) \quad , \tag{3.19}$$

where E_0 is a complex number which can depend on time. Hence we obtain an equation for U_I,

$$\frac{dU_I}{dt} = -i\tilde{H}_I U_I \qquad \tilde{H}_I = H_I - E_0 \quad .$$

This can be written as an integral equation

$$U_I(t, t_0) = 1 - i \int_{t_0}^{t} dt' \, \tilde{H}_I(t') U_I(t', t_0) \quad , \tag{3.20}$$

which builds in the initial condition $U_I(t_0, t_0) = 1$. This is a very beautiful result. It gives U_I in terms of $\tilde{H}_I$ only, and $\tilde{H}_I$ is a function of the *free* fields $\phi_0(t)$, which are known.

Now we use perturbation theory to solve Eq. (3.20). If $\tilde{H}_I$ is small, we may solve the equation by iteration:

$$U_I(t, t_0) = 1 - i \int_{t_0}^{t} dt' \, \tilde{H}_I(t') + (-i)^2 \int_{t_0}^{t} dt_1 \int_{t_0}^{t_1} dt_2 \, \tilde{H}_I(t_1) \tilde{H}_I(t_2) + \cdots \quad . \tag{3.21}$$

Note that $\tilde{H}_I(t_1)$ does not necessarily commute with $\tilde{H}_I(t_2)$; the order of terms in the double integral is important, and the later time stands to the left. If we define the *time-ordered product*

$$T\left(\tilde{H}_I(t_1)\tilde{H}_I(t_2)\right) = \tilde{H}_I(t_1)\tilde{H}_I(t_2)\theta(t_1 - t_2) + \tilde{H}_I(t_2)\tilde{H}_I(t_1)\theta(t_2 - t_1) \, , \tag{3.22}$$

where $\theta(x) = 1$ if $x > 0$ and is zero if x is negative, then the double integral may be "symmetrized,"

$$\begin{aligned} \int_{t_0}^{t} dt_1 \int_{t_0}^{t_1} dt_2 \, \tilde{H}_I(t_1)\tilde{H}_I(t_2) &= \int_{t_0}^{t} dt_2 \int_{t_0}^{t_2} dt_1 \, \tilde{H}_I(t_2)\tilde{H}_I(t_1) \\ &= \frac{1}{2} \int_{t_0}^{t} \int_{t_0}^{t} dt_1 \, dt_2 \, T\left(\tilde{H}_I(t_1)\tilde{H}_I(t_2)\right) \end{aligned} \tag{3.23}$$

as shown in Fig. 3.1. Hence,

$$U_I(t, t_0) = 1 - i \int_{t_0}^{t} dt' \, \tilde{H}_I(t') + \frac{(-i)^2}{2!} \int_{t_0}^{t} \int_{t_0}^{t} dt_1 \, dt_2 \, T\left(\tilde{H}_I(t_1)\tilde{H}_I(t_2)\right) + \cdots \quad . \tag{3.24}$$

Since there are $n!$ time orderings for a time-ordered product of n terms, the expansion looks like an exponential and may be formally written

$$U_I(t, t_0) = T \exp\left[-i \int_{t_0}^{t} dt_1 \, \tilde{H}_I\left[\phi_0(t_1)\right]\right] \quad . \tag{3.25}$$

However, because each of the terms in the power series expansion (3.24) of the exponential is time ordered, the terms cannot actually be summed up into a closed form, and (3.25) should be regarded only as a shorthand for the original infinite sum (3.24).

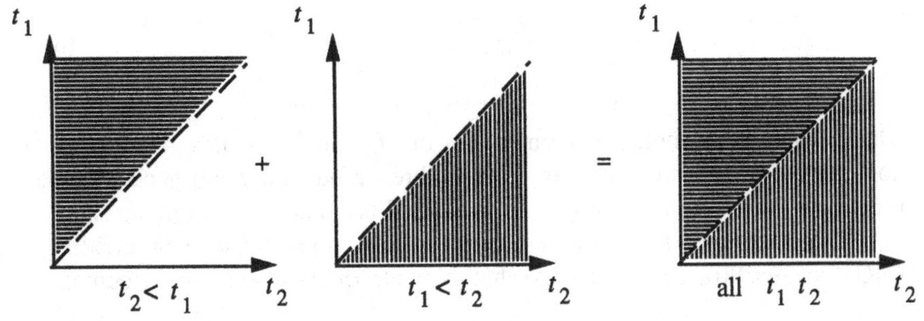

Fig. 3.1 The shaded areas in the two left-hand figures are the regions $t_2 < t_1$ and $t_1 < t_2$, and the integrals over each of these areas are equal, as shown in Eq. (3.22). Adding the two together gives the integral over the total area (shown in the right-hand figure).

The S-Matrix

The time evolution operator gives us the tool necessary to describe scattering and atomic decay. If there are no interactions, the states $|\alpha\rangle$ are eigenstates of the Hamiltonian, and only their phase will change with time. Under the interaction, an initial state $|\alpha\rangle$ will evolve, over time, into a mixture of final states $|\beta\rangle$. In the interaction picture, where the free states do not depend on time, this is expressed in terms of the time translation operator:

$$U_{\text{total}}(T/2,-T/2)\,|\alpha\rangle = N(T)\sum_{\beta} S_{\beta\alpha}(T/2,-T/2)\,U_0(T/2,-T/2)\,|\beta\rangle \quad . \tag{3.26}$$

In words, this equation says that the state which begins as $|\alpha\rangle$ at an initial time $-T/2$ evolves into the state $|\beta\rangle$ at the time $T/2$ with probability $|N(T)\,S_{\beta\alpha}(T/2,-T/2)|^2$. The operator U_0 has been added to the RHS of the equation in order to insure that the "trivial" time-dependent phase factors arising from the time evolution of the unperturbed states will not be included in the expansion coefficients $S(T/2,-T/2)$, and $N(T)$ is a normalization constant to be specified shortly. If the time interval is infinite, described by letting $T\to\infty$, the expansion coefficients $S(T/2,-T/2)\to S$ are referred to as the S-matrix, and calculation of these matrix elements is a central problem in quantum mechanics.

Using Eq. (3.9) and the orthogonality of the states, the S-matrix elements become

$$S_{\beta\alpha} = \frac{1}{N(\infty)}\,\langle\beta|U_I(\infty,-\infty)|\alpha\rangle \quad . \tag{3.27}$$

The S-matrix will be defined by choosing $N(\infty) = \langle 0|U_I(\infty, -\infty)|0\rangle$, so that

$$S_{\beta\alpha} = \frac{\langle\beta|U_I(\infty,-\infty)|\alpha\rangle}{\langle 0|U_I(\infty,-\infty)|0\rangle} , \tag{3.28}$$

where α and β are non-interacting states of H_0 and $|0\rangle$ is the ground state. In nonrelativistic atomic theory, this state is a direct product of the ground state wave function of the atom and the photon vacuum (Fock state with no photons).

There are important reasons why we choose to normalize S by dividing by $\langle 0|U_I(\infty, -\infty)|0\rangle$. First, we show that this number must have unit modulus:

$$\langle 0|U_I|0\rangle = e^{ic} , \tag{3.29}$$

where c is a c-number. To prove this use the facts that U is unitary (which is a consequence of the conservation of probability) and that the ground state is stable (otherwise it would not be the ground state). Stability of the ground state implies that

$$\langle\beta|U_I|0\rangle = 0 \qquad \text{if } \beta \neq 0 , \tag{3.30}$$

where we assume that there is only one vacuum state. Hence, from $U_I^\dagger U_I = 1$, it follows that

$$\begin{aligned} 1 = \langle 0|0\rangle = \langle 0|U_I^\dagger U_I|0\rangle &= \sum_{\beta=0}^{\infty}\langle 0|U_I^\dagger|\beta\rangle\langle\beta|U_I|0\rangle \\ &= \langle 0|U_I^\dagger|0\rangle\langle 0|U_I|0\rangle = |\langle 0|U_I|0\rangle|^2 . \end{aligned} \tag{3.31}$$

This proves the result.

Normalizing the S-matrix elements by this phase factor ensures that they are independent of any *overall* c-number phases. For example, if a c-number is added to $\tilde{H}_I$, then the time translation operator is changed to

$$\begin{aligned} U_I^c(t, t_0) &= T\exp\left[-i\int_{t_0}^{t} dt_1\left[\tilde{H}_I(t_1) + c\right]\right] \\ &= e^{-ic(t-t_0)}U_I(t, t_0) , \end{aligned} \tag{3.32}$$

and this phase becomes infinite as $t - t_0 \to \infty$. But this multiplicative factor cancels in $S_{\beta\alpha}$, since it occurs both in the numerator *and* in the denominator. Thus this cancellation is very useful, since it works even for c's which are *infinite*. An infinite c-number, which might occur order-by-order in U and which would otherwise disturb our concentration, is seen to be irrelevant since it exponentiates and cancels from S, and we may therefore ignore c-number infinities when calculating S. In the context of the time evolution of states, this provides the justification for dropping the electron and nucleus Coulomb self-energies, as discussed in Sec. 2.3. In addition, we now can justify dropping the additional $E_0(t)$ which arose in the derivation of Eq. (3.20) and use H_I instead of $\tilde{H}_I$.

3.2 DECAY RATES AND CROSS SECTIONS

Experimentally, the dynamics of physical systems are studied by preparing an initial state and then following its time evolution in the laboratory. In the broadest terms, it is practical to prepare only two types of initial states: those consisting of a single particle (or a bound state which behaves like a single particle) and those consisting of a beam of two particles (or two bound states) directed toward each other so that a collision is possible. The other logical possibilities, which include the direction of three or more beams at each other so as to produce a three or more body collision, are impractical, except in the most exceptional cases. Hence, our experimental studies are more or less limited to the following types of reactions:

$$\begin{aligned} &\text{one particle} &&\Longrightarrow &&\text{many particles} \\ &\text{two particles} &&\Longrightarrow &&\text{many particles} \;. \end{aligned}$$

In the first instance, if the single particle remains a single particle, then we may measure its mass (or, in some cases, the frequency with which it oscillates into another single particle), while if it decays into two or more particles, we can study the *decay rate* or, when decay into two or more channels occurs, the *branching fraction.* In the second case, we measure the *cross section.* Hence, decay rates and cross sections are very important; they are among the very few physical quantities which can be measured. We now turn to a discussion of how decay rates and cross sections are calculated from the S-matrix.

Decay Rates

The differential decay rate $\Delta W_{\beta\alpha}(T)$ will be defined to be the probability that a state α will decay into state β in the time interval $[T/2, -T/2]$ divided by the total time T (hence a rate). Formally

$$\Delta W_{\beta\alpha}(T) = \frac{|S_{\beta\alpha}(T/2, -T/2)|^2}{T} \;. \tag{3.33}$$

Under most experimental circumstances, when a decaying system is isolated from the apparatus, the measurement is made over a time interval long compared to the internal time scale of the system, and we may therefore take the limit as $T \to \infty$. This limiting rate is denoted $\Delta W_{\beta\alpha}$,

$$\Delta W_{\beta\alpha} = \lim_{T\to\infty} \frac{|S_{\beta\alpha}(T/2, -T/2)|^2}{T} \;. \tag{3.34}$$

Later, we will see that our calculations of the $S_{\beta\alpha}(T/2, -T/2)$ can always be expressed in the form

$$S_{\beta\alpha}(T/2, -T/2) = -i2\pi d_1(T) \left(\frac{1}{\sqrt{L^3}}\right)^{p-1} f_{\beta\alpha} \;, \tag{3.35}$$

where the reduced matrix element $f_{\beta\alpha}$ is independent of T and L^3 is the volume of the box in which the states are normalized, p is the number of particles in the final state (counting the heavy final atomic state as one "particle"), and

$$d_1(T) = \frac{1}{2\pi}\int_{-T/2}^{T/2} dt\, e^{-i(E_\alpha - E_\beta)t} = \frac{T}{\pi}\frac{\sin(\Delta E\, T/2)}{(\Delta E\, T)} , \tag{3.36}$$

where $\Delta E = |E_\alpha - E_\beta|$. Note that

$$\lim_{T\to\infty} d_1(T) = \delta\left(E_\alpha - E_\beta\right) . \tag{3.37}$$

[Proof: As $T \to \infty$, $d_1(T)$ oscillates rapidly around zero unless $\Delta E = 0$, and the integral of $d_1(T)$ over ΔE is unity.] Hence the S-matrix (for decays) can be written

$$S_{\beta\alpha} = -i2\pi\delta\left(E_\alpha - E_\beta\right)\left(\frac{1}{\sqrt{L^3}}\right)^{p-1} f_{\beta\alpha} \qquad \text{Decays} \tag{3.38}$$

which shows that energy is conserved for all decay processes which are allowed to take place over a long time interval.

We now use these results to reduce (3.34) to a convenient form. Returning to (3.35), squaring and dividing by T give

$$dW_{\beta\alpha} = \lim_{T\to\infty}(2\pi)^2\frac{d_1^2(T)}{T}\left(\frac{1}{L^3}\right)^{p-1}\left|f_{\beta\alpha}\right|^2 . \tag{3.39}$$

Now introduce $d_2(T)$:

$$d_2(T) = (2\pi)^2\frac{d_1^2(T)}{T} = 4T\frac{\sin^2(\Delta E\, T/2)}{(\Delta E\, T)^2} \tag{3.40}$$

and observe that

$$\lim_{T\to\infty} d_2(T) = 2\pi\,\delta\left(E_\alpha - E_\beta\right) . \tag{3.41}$$

[Proof: When $T \to \infty$, $d_2(T)$ is zero unless $\Delta E = 0$, and the integral of $d_2(T)$ over ΔE is 2π.] Hence the differential decay rate can be written in the following convenient form:

$$\Delta W_{\beta\alpha} = 2\pi\,\delta\left(E_\alpha - E_\beta\right)\left(\frac{1}{L^3}\right)^{p-1}\left|f_{\beta\alpha}\right|^2 . \tag{3.42}$$

The formula for the total decay rate, sometimes referred to as the *Fermi golden rule*, is found by summing (integrating) $\Delta W_{\beta\alpha}$ over all final states which are detected experimentally. In this formalism where the particles are treated nonrelativistically, the final atomic state is fixed in space, and we sum over all momenta of the light particles produced in the decay (photons in this example), so the total decay rate is

$$W_{\beta\alpha} = \sum_{n_1} \sum_{n_2} \cdots \sum_{n_{p-1}} \left(\frac{1}{L^3}\right)^{p-1} 2\pi\,\delta\left(E_\alpha - E_\beta\right) |f_{\beta\alpha}|^2$$
$$= \int\int\cdots\int \frac{d^3k_1 d^3k_2 \cdots d^3k_{p-1}}{(2\pi)^{(3p-3)}}\, 2\pi\,\delta\left(E_\alpha - E_\beta\right) |f_{\beta\alpha}|^2 \,, \qquad (3.43)$$

where, in the last step, we took the limit $L \to \infty$, referred to as the continuum limit because the spacing between levels $\Delta k = 2\pi/L \to 0$, and for *each* momentum variable,

$$\sum_{n_i} \frac{1}{L}(\quad) = \sum_{n_i} \frac{\Delta k_{n_i}}{2\pi}(\quad) \to \int \frac{dk_i}{2\pi}(\quad) \,. \qquad (3.44)$$

The δ-function insures that energy is conserved in the decay, and the final result is proportional to a density of final states times $|f_{\beta\alpha}|^2$. We will develop these details in applications (below).

If we had worked directly with the S-matrix (3.38), we would encounter the square of a δ-function in the computation of the decay rate. A review of our derivation, which was carried out first for a finite time interval and then followed by taking the limit $T \to \infty$, shows that the final result (3.42) could be obtained directly from the S-matrix by using the substitution

$$\lim_{T\to\infty} \frac{\left[2\pi\,\delta\left(E_\alpha - E_\beta\right)\right]^2}{T} \Rightarrow 2\pi\,\delta\left(E_\alpha - E_\beta\right) \,. \qquad (3.45)$$

This formula, which is a shorthand for the steps we followed, is very convenient and will be used frequently in the subsequent sections.

Cross Section

To treat photon scattering from atoms we will need to calculate the differential cross section. This is defined experimentally as

$$d\sigma = \frac{\#\,\text{particles scattered into solid angle } \Delta\Omega/\text{sec}}{(\#\,\text{particles incident/sec})\,(\#\,\text{scattering centers/area})} \,, \qquad (3.46)$$

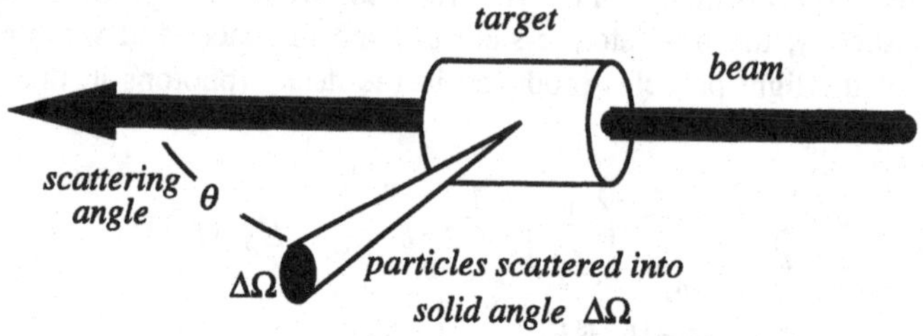

Fig. 3.2 Drawing of an idealized scattering process showing the differential solid angle $\Delta\Omega$ and the scattering angle θ.

where the quantities are defined with the help of Fig. 3.2. Note that the cross section has the units of an area. In most experiments, the target is larger than the beam, as illustrated in Fig. 3.2, so that the number of scattering centers in the path of the beam per unit area is

$$N_c = \frac{\rho\ell}{m_c} = \frac{\text{target density}(\rho)\times\text{ target length}(\ell)}{\text{mass of each scatterer}} \quad . \tag{3.47}$$

If the particles in the beam have charge e_0, then the number of beam particles incident per second can be determined from the beam current

$$n_0 = \frac{j}{e_0} \quad , \tag{3.48}$$

where j is the beam current. For photon beams, the quantity n_0 is determined indirectly from an analysis of how the beam is produced.

Theoretically, we evaluate the cross section assuming one scattering center, and a number of particles incident per second determined by the velocity v and a density derivable from one particle in volume L^3 (consistent with the box normalization introduced in Chapter 2). Hence

$$\begin{aligned} N_c &= \underbrace{\left(\frac{\rho\ell A}{m_c}\right)}_{=1}\frac{1}{A} = \frac{1}{A} \\ n_0 &= \left(\frac{1}{L^3}\right)\frac{Avt}{t} \quad , \end{aligned} \tag{3.49}$$

where A is the area of the beam and Avt is therefore the volume swept out by the beam in time t (see Fig. 3.3). Scattering differs from a decay in that there

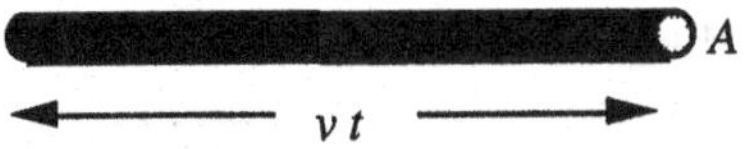

Fig. 3.3 Drawing illustrating the calculation of n_0.

is always an additional incoming plane wave in the initial state, and hence for scattering the reduced matrix element $f_{\beta\alpha}$ is defined

$$S_{\beta\alpha} = -i2\pi\,\delta\left(E_\alpha - E_\beta\right)\left(\frac{1}{\sqrt{L^3}}\right)^p f_{\beta\alpha} \qquad \text{Scattering,} \tag{3.50}$$

where p is again the number of particles in the final state. The extra factor of $L^{-3/2}$ in this equation [compared to Eq. (3.38)] is the normalization factor for the incoming plane wave. The rate at which the scattering takes place is the transition probability divided by the time interval (T) and is equal to

$$\frac{\left|S_{\beta\alpha}\right|^2}{T} = 2\pi\,\delta\left(E_\alpha - E_\beta\right)\left(\frac{1}{L^3}\right)^p \left|f_{\beta\alpha}\right|^2 ,$$

where the convenient substitution (3.45) has been used. Combining all of these factors gives the following result for the differential cross section:

$$\Delta\sigma_{\beta\alpha} = \sum_{\beta\in\Delta\Omega} \frac{\frac{\left|S_{\beta\alpha}\right|^2}{T}}{\left(\frac{1}{L^3}Av\right)\left(\frac{1}{A}\right)} = \frac{1}{v}\sum_{\beta\in\Delta\Omega} 2\pi\,\delta\left(E_\alpha - E_\beta\right)\left(\frac{1}{L^3}\right)^{p-1} \left|f_{\beta\alpha}\right|^2 , \tag{3.51}$$

where the sum is over all final states β which scatter into $\Delta\Omega$ and v is sometimes called the *flux factor*. In the continuum limit ($L \to \infty$) defined above, the cross section is

$$\Delta\sigma_{\beta\alpha} = \frac{1}{v}\int_{\{k_i\}\in\Delta\Omega} \frac{d^3k_1 d^3k_2 \cdots d^3k_{p-1}}{(2\pi)^{(3p-3)}}\, 2\pi\delta\left(E_\alpha - E_\beta\right)\left|f_{\beta\alpha}\right|^2 . \tag{3.52}$$

3.3 ATOMIC DECAY

We are now ready to calculate the electromagnetic decay of an atom! The calculation is so simple, it's almost an anticlimax.

If the initial state α and the final state β are not identical, which is always the case for a decay, $\langle\beta|1|\alpha\rangle = 0$, and to first order in perturbation theory the S-matrix element is

$$S^{(1)}_{\beta\alpha} = -i \int_{-\infty}^{\infty} dt\, \langle\beta|H_I(t)|\alpha\rangle \ , \tag{3.53}$$

where the superscript (1) reminds us that this is the first order expression only and we have assumed that $\langle 0|U(\infty,-\infty)|0\rangle = 1$, which is usually true to first order. The interaction Hamiltonian H_I was given in Eq. (3.5), and the states were defined in Eq. (3.6). For one-photon decay of an initial atomic state a into a final atomic state b and a photon with energy ω_n and polarization λ, the states are

$$\begin{aligned} |\alpha\rangle &= |a,0\rangle = \psi_a(r_e)|0\rangle \\ |\beta\rangle &= |b,1_{n\lambda}\rangle = \psi_b(r_e)|1_{n\lambda}\rangle \ , \end{aligned} \tag{3.54}$$

where $|1_{n\lambda}\rangle = a^\dagger_{n\lambda}|0\rangle$ is the one-photon state with frequency ω_n and polarization λ. Hence

$$S_{ba} = -i \int_{-\infty}^{\infty} dt \langle b, 1_{n\lambda}|U_A^{-1}\left(-\frac{ie}{m}\boldsymbol{A}(r_e,t)\cdot\nabla_e\right) U_A|a,0\rangle \ , \tag{3.55}$$

where $\boldsymbol{p}_e = -i\nabla_e$ and $\nabla_e \cdot \boldsymbol{A}(r_e,t) = 0$ were used to obtain the simplified form (3.55). Since the states are direct products, the matrix element (3.55) reduces immediately to the product of two terms, an atomic matrix element expressed as an integral over r_e and an EM matrix element:

$$S_{ba} = -\int_{-\infty}^{\infty} dt\, e^{i(E_b-E_a)t}\, \frac{e}{m}\int d^3r_e\, \{\psi_b^*(r_e)\nabla_e\psi_a(r_e)\}\cdot\langle 1_{n\lambda}|\boldsymbol{A}(r_e,t)|0\rangle \ , \tag{3.56}$$

where E_a and E_b are the energies of the two atomic states. Now, taking *the matrix element of the field operator between the vacuum and a one-photon state gives a non-zero result! This is the origin of electromagnetic decay.* We get

$$\langle 1_{n\lambda}|\boldsymbol{A}(r_e,t)|0\rangle = \sum_{n'\lambda'} \frac{1}{\sqrt{2\omega_{n'}L^3}}\, \epsilon^{\lambda' *}_{n'}\langle 0|a_{n\lambda}\, a^\dagger_{n'\lambda'}|0\rangle\, e^{ik_{n'}\cdot x_e} \ . \tag{3.57}$$

Only the term with $n' = n$ and $\lambda' = \lambda$ survives, and

$$\langle 1_{n\lambda}|\boldsymbol{A}(r_e,t)|0\rangle = \frac{1}{\sqrt{2\omega_n L^3}}\, \epsilon^{\lambda *}_n\, e^{i(\omega_n t - \boldsymbol{k}_n\cdot\boldsymbol{r}_e)} \ . \tag{3.58}$$

Inserting this expression into (3.56) gives

$$\begin{aligned} S_{ba} &= -\int_{-\infty}^{\infty} dt\, e^{i(E_b+\omega_n-E_a)t}\, \frac{e}{m}\frac{1}{\sqrt{2\omega_n L^3}} \int d^3r_e\, e^{-i\boldsymbol{k}_n\cdot\boldsymbol{r}_e}\psi_b^*(\boldsymbol{r}_e)\, \boldsymbol{\epsilon}_n^{\lambda *}\cdot\nabla\,\psi_a(\boldsymbol{r}_e) \\ &= -i2\pi\,\delta\left(E_b+\omega_n-E_a\right)\frac{1}{\sqrt{L^3}}\, f_{ba} \quad , \end{aligned}$$

where the decay amplitude f_{ba} is

$$f_{ba} = -i\frac{e}{m}\frac{1}{\sqrt{2\omega_n}}\int d^3r_e\, e^{-i\boldsymbol{k}_n\cdot\boldsymbol{r}_e}\, \psi_b^*(\boldsymbol{r}_e)\, \boldsymbol{\epsilon}_n^{\lambda *}\cdot\nabla_e\,\psi_a(\boldsymbol{r}_e) \quad . \tag{3.59}$$

The next step is to reduce f_{ba} and compute the decay rate. In most atomic decays, the energy of the emitted photon, which is equal to $\omega_n = E_a - E_b$, is much less than $1/R$, where R is the size of the atomic system, and hence the maximum range of the integral over r_e. In this case, the dipole approximation

$$e^{-i\boldsymbol{k}_n\cdot\boldsymbol{r}_e} \cong 1 \tag{3.60}$$

is extremely good. Introducing matrix elements of the momentum operator,

$$\boldsymbol{p}_{ba} = -i\int d^3r_e\, \psi_b^*(\boldsymbol{r}_e)\,\nabla_e\,\psi_a(\boldsymbol{r}_e) \quad , \tag{3.61}$$

we can write the dipole approximation to the decay amplitude in the following reduced form:

$$f_{ba} = \frac{e}{m}\frac{1}{\sqrt{2\omega_n}}\,\boldsymbol{\epsilon}_n^{\lambda *}\cdot\boldsymbol{p}_{ba} \tag{3.62}$$

and the differential decay rate becomes

$$\Delta W_{ba} = \frac{2\pi}{L^3}\,\delta\left(E_b+\omega_n-E_a\right)\frac{e^2}{2\omega_n\, m^2}\left|\boldsymbol{\epsilon}_n^{\lambda *}\cdot\boldsymbol{p}_{ba}\right|^2 \quad . \tag{3.63}$$

Summing over all final photon states to get the total $a\to b$ decay rate gives

$$\begin{aligned} W_{ba} =& 2\pi\sum_{n,\lambda}\frac{1}{L^3}\delta\left(E_b+\omega_n-E_a\right)\frac{e^2}{2\omega_n m^2}\left|\boldsymbol{\epsilon}_n^{\lambda *}\cdot\boldsymbol{p}_{ba}\right|^2 \\ \Rightarrow& \frac{e^2}{2m^2}\sum_{\lambda}\int\frac{d^3k}{(2\pi)^2\omega}\delta\left(E_b+\omega-E_a\right)\left|\boldsymbol{\epsilon}_n^{\lambda *}\cdot\boldsymbol{p}_{ba}\right|^2 \quad , \end{aligned} \tag{3.64}$$

where, in the second step, we took the continuum limit (3.44). Eliminating the δ-function by integrating over the magnitude of k and using $k=\omega$ gives

$$W_{ba} = \frac{e^2}{4\pi}\left(\frac{\omega}{2\pi m^2}\right)\sum_{\lambda}\int d\Omega\left|\,\boldsymbol{\epsilon}_n^{\lambda *}\cdot\boldsymbol{p}_{ba}\right|^2 \quad . \tag{3.65}$$

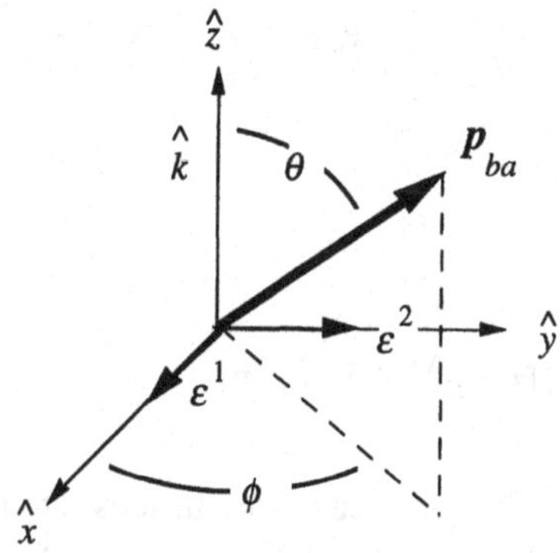

Fig. 3.4 Orientation of the momentum vector with respect to the axes defined by the photon momentum and polarization states.

This integral can be evaluated by integrating over all directions $\hat{k}$ of the outgoing photon. However, since the integrand is rotationally invariant, it is more convenient to fix $\hat{k}$ along the direction of the z-axis and integrate over all directions of the vector $\boldsymbol{p}_{ba}$. This procedure allows us to avoid the problem of defining the directions of the polarization vectors ϵ^1 and ϵ^2, which depend on $\hat{k}$ but can be fixed along $\hat{x}$ and $\hat{y}$ if $\hat{k} = \hat{z}$. The geometry is shown in Fig. 3.4. The integral over the direction of $\boldsymbol{p}_{ba}$ becomes (the polarization vectors are now real)

$$\begin{aligned}\sum_\lambda \int d\Omega\, |\epsilon_n^\lambda \cdot \boldsymbol{p}_{ba}|^2 &= |p_{ba}|^2 \int d\phi \sin\theta\, d\theta \left(\sin^2\theta\cos^2\phi + \sin^2\theta\sin^2\phi\right) \\ &= |p_{ba}|^2 \int d\phi \sin^3\theta\, d\theta = 2\pi\, |p_{ba}|^2 \int_{-1}^{1} dz\,(1-z^2) \\ &= \frac{8\pi}{3}\, |\boldsymbol{p}_{ba}|^2 \ .\end{aligned} \tag{3.66}$$

Then the total rate for the decay of the state a into b is

$$\boxed{W_{ba} = \frac{e^2}{4\pi}\left(\frac{4\omega}{3m^2}\right) |\boldsymbol{p}_{ba}|^2 \ .} \tag{3.67}$$

Finally, the total decay rate for the state a into *any* atomic state b is the sum of the individual decay rates into all states b with $E_b < E_a$:

$$\boxed{W^a_{\text{total}} = \sum_{\substack{b \\ E_b < E_a}} W_{ba} \ .} \tag{3.68}$$

We leave the calculation here, assuming that applications of this result are familiar from previous studies.

Note that quantization of the EM field has given a natural explanation for decay $\left[\ \langle 1_{n\lambda}|\boldsymbol{A}(r_e,t)|0\rangle \neq 0\ \right]$ and the normalization of the decay rate is uniquely predicted by the theory. Also, note how energy conservation ($\omega = E_a - E_b$) arises naturally.

3.4 THE LAMB SHIFT

We search for additional effect due to the quantization of the electromagnetic field. Imagine ourselves back in the late 1940's. The Lamb shift has been discovered.* Everyone believes it is due to field quantization. Can we calculate it? H. A. Bethe did [Be 47], and it is said that he did it on a train, while returning from a conference.

The Lamb shift was measured by W. E. Lamb and W. E. Retherford in 1947 using microwave techniques [LR 47]. It is the splitting between the $2S_{1/2}$ and $2P_{1/2}$ states, which are degenerate to order $(v/c)^2$ (and even exactly to all orders when the Dirac equation is used). The S-state is higher than the P-state by about 1060 MHz. A diagram of the energy levels of hydrogen-like atoms is shown in Fig. 3.5.

To calculate the shift in energy of a bound state, we use second order perturbation theory. The derivation of the energy shift starts from the equation

$$(H_0 + \lambda H_I)\,|\alpha\rangle = \big(E_\alpha^{(0)} + \underbrace{\lambda E_\alpha^{(1)} + \lambda^2 E_\alpha^{(2)} + \ldots}_{\Delta E_\alpha}\big)|\alpha\rangle \quad , \tag{3.69}$$

where λ is a parameter which keeps track of the orders of perturbation theory but is eventually set to $\lambda = 1$. The derivation of the formula for the energy shift in the general case is identical to that from ordinary nonrelativistic, non-degenerate, bound state perturbation theory, so we will not repeat the steps here. We obtain the usual result, valid to second order:

$$\Delta E_\alpha = \langle\alpha|H_I|\alpha\rangle + \sum_{\beta\neq\alpha} \frac{\langle\alpha|H_I|\beta\rangle\langle\beta|H_I|\alpha\rangle}{E_\alpha^{(0)} - E_\beta^{(0)}} \quad . \tag{3.70}$$

The task is to evaluate ΔE_α to second order, i.e., to order e^2.

First, note that $\langle\alpha|H_I|\alpha\rangle = 0$, because the only such term which might be non-zero, the $\boldsymbol{A}^2$ term in H_I, is normal ordered. Hence its vacuum expectation value is zero, and

$$\langle a,0|{:}\boldsymbol{A}^2(r_e,t){:}\,|a,0\rangle = 0 \quad . \tag{3.71}$$

Thus the entire contribution comes from the sum in (3.70).

*For a review of the early experiments see [La 51].

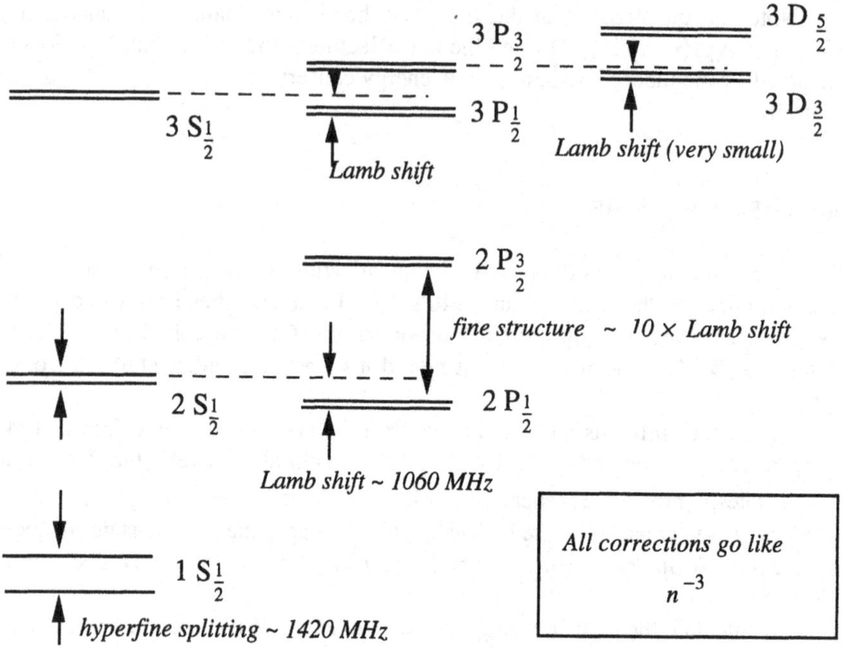

Fig. 3.5 Energy level diagram for a hydrogen-like atom. The splittings are *not* to scale. The Lamb shift is of the same order as the hyperfine splitting and cannot be understood without field theory.

However, if $|\alpha\rangle = |a, 0\rangle$ is a pure atomic state (with no photons present), the only states which can contribute to the sum β are atomic states with one photon present. [In this section we will represent these one-photon states by $|1_k\rangle$, where k is the momentum and the polarization λ will be suppressed for now.] These are the only states which contribute to the sum because only for these states is

$$\langle 1_k|A|0\rangle \neq 0 \ .$$

Furthermore, we have already evaluated these matrix elements. They are just the first order matrix elements of H_I evaluated in Sec. 3.3. In terms of f introduced in Eqs. (3.35) and (3.38),

$$\langle \beta|H_I|\alpha\rangle = \frac{1}{\sqrt{L^3}}\, e^{i(E_\beta - E_\alpha)t}\, f_{ba} \ , \tag{3.72}$$

where $|\beta\rangle = |b, 1_k\rangle$ is a direct product of an atomic state b and a one-photon state with momentum $\boldsymbol{k}$. In this section we denote f by f_{ba} to emphasize that it depends on atomic states a and b and is (nearly) independent of the photon states. Hence ΔE_a reduces to

$$\Delta E_a = \sum_b \sum_{k,\lambda} \frac{1}{L^3} \frac{|f_{ba}|^2}{E_a^{(0)} - E_b^{(0)} - \omega_k} \quad . \tag{3.73}$$

Note that $a = b$ is included in the sum over b. Even when $a = b$, $|\beta\rangle \neq |\alpha\rangle$, because $|\beta\rangle$ has one photon and $|\alpha\rangle$ does not.

The low energy contributions to the sum (3.73) can be estimated using the dipole approximation for f_{ba}, Eq. (3.60). We obtain

$$\begin{aligned}
\Delta E_a &= \sum_b \sum_{k,\lambda} \left(\frac{1}{2\omega_k L^3} \right) \frac{e^2}{m^2} \frac{\left| \epsilon_k^{\lambda *} \cdot \boldsymbol{p}_{ba} \right|^2}{E_a - E_b - \omega_k} \\
&\underset{L\to\infty}{\longrightarrow} \sum_{b,\lambda} \int \frac{d^3k}{2\omega_k (2\pi)^3} \frac{e^2}{m^2} \frac{\left| \epsilon_k^{\lambda *} \cdot \boldsymbol{p}_{ba} \right|^2}{E_a - E_b - \omega_k} \\
&= \frac{1}{2\pi} \sum_b \int_0^\infty \frac{d\omega}{E_a - E_b - \omega} \left[\frac{e^2}{4\pi} \frac{\omega}{2\pi m^2} \sum_\lambda \int d\Omega_k \left| \epsilon_k^{\lambda *} \cdot \boldsymbol{p}_{ba} \right|^2 \right] \quad .
\end{aligned}$$

The term in brackets looks like the transition rate, except that it is not on the "energy shell" defined by $\omega = E_a - E_b$. Introducing a "virtual" transition rate,

$$\overline{W}_{ba}(\omega) \equiv \frac{e^2}{4\pi} \frac{\omega}{2\pi m^2} \sum_\lambda \int d\Omega_k \left| \epsilon_k^{\lambda *} \cdot \boldsymbol{p}_{ba} \right|^2 \quad , \tag{3.74}$$

where $\overline{W}_{ba}(E_a - E_b) = W_{ba}$, permits us to write the energy shift in the following convenient form:

$$\boxed{\Delta E_a = \frac{1}{2\pi} \sum_b \int_0^\infty \frac{d\omega}{E_a - E_b - \omega} \overline{W}_{ba}(\omega) \quad .} \tag{3.75}$$

The integral has a singularity at $\omega = E_a - E_b$, which is defined using the "$i\epsilon$ prescription." With this prescription, the energy denominator $E_a - E_b - \omega$ is replaced by $E_a - E_b - \omega + i\epsilon$, where the limit $\epsilon \to 0$ is understood. The *sign of $i\epsilon$ is determined by causality*. To see this, note that the denominator can be written

$$\frac{1}{E_a - E_b - \omega + i\epsilon} = \mathbb{P} \frac{1}{E_a - E_b - \omega} - i\pi\delta(E_a - E_b - \omega) \quad , \tag{3.76}$$

where $\mathbb{P}$ is the principal value integral. Hence the energy shift is now complex, with

$$\begin{aligned}
\Delta E_a &= \frac{1}{2\pi} \sum_b \int_0^\infty \frac{d\omega}{E_a - E_b - \omega + i\epsilon} \overline{W}_{ba}(\omega) \\
&= \frac{1}{2\pi} \sum_b \mathbb{P} \int_0^\infty \frac{d\omega}{E_a - E_b - \omega} \overline{W}_{ba}(\omega) - i\,\frac{1}{2} \sum_{\substack{b \\ E_b < E_a}} W_{ba} \\
&= Re\Delta E_a + iIm\Delta E_a \quad .
\end{aligned} \tag{3.77}$$

Hence, choosing $+i\epsilon$ gives ΔE_a a negative imaginary part, which equals $\frac{1}{2}$ of the total decay rate of the state a,

$$Im\Delta E_a = -\tfrac{1}{2}W^a_{\text{total}} \quad . \tag{3.78}$$

However, this is just what we should expect from the energy–time evolution factor

$$\begin{aligned}\psi_a(t) &= e^{-iE_at}\psi_a(0) \\ &= e^{-i(E^0_a+Re\Delta E_a)t}\, e^{-\frac{1}{2}W^a_{\text{total}}\,t}\,\psi_a(0) \quad ,\end{aligned} \tag{3.79}$$

which gives

$$|\psi_a(t)|^2 = e^{-W^a_{\text{total}}\,t}\,|\psi_a(0)|^2 \tag{3.80}$$

corresponding to exponential decay of the state a with a half-life equal to the reciprocal of the total decay rate

$$\Gamma^a_{\text{total}} = \frac{1}{W^a_{\text{total}}} \quad . \tag{3.81}$$

The $+i\epsilon$ prescription therefore gives a decay in the probability $|\psi_a(t)|^2$. If we had chosen a $-i\epsilon$ prescription, we would have obtained an exponentially *growing* probability, contrary to causality.

The imaginary part of the energy shift makes the Hamiltonian appear to be non-Hermitian and the norm of ψ_a not conserved. However, when the entire Fock space is considered, it can be shown that the norm of the *total* system is conserved. A decrease in norm of ψ_a is accompanied by an increase in the norm of states with $E_b < E_a$ and with one photon. In detail, the total state is

$$|\beta\rangle = a_0\psi_a|0\rangle + a_1\psi_b|1_\omega\rangle + a_1'\psi_{b'}|1_{\omega'}\rangle + \cdots \tag{3.82}$$

and the total norm

$$|a_0|^2 + |a_1|^2 + |a_1'|^2 + \cdots = 1 \tag{3.83}$$

is conserved.

Now, the real part of ΔE_a gives the shift in energy of the bound state, but it diverges. To see this, insert the expression for the decay rate, Eq. (3.67), into (3.77). Since $|\boldsymbol{p}_{ba}|$ is independent of ω, we obtain

$$Re\Delta E_a = \left(\frac{e^2}{4\pi}\right)\left(\frac{2}{3\pi}\right)\sum_b \frac{|\boldsymbol{p}_{ba}|^2}{m^2}\,\mathbb{P}\int_0^\infty \frac{\omega\,d\omega}{E_a - E_b - \omega} \quad . \tag{3.84}$$

The integral diverges linearly, and we must introduce a high energy cutoff (upper limit) in order to define it. There are physical processes which we have ignored — one is the breakdown of the dipole approximation which is certainly unreliable for $\omega \simeq m$ — which naturally damp out the integral at high energies and help to define such a cutoff. But the sensitivity of the integral (3.84) to the precise choice of the cutoff makes the final result too sensitive to be useful for any reliable estimates. An even greater problem is that the result (3.84) is not physically observable. This leads us to the issue of mass renormalization.

Mass Renormalization

We make the integral more convergent following Bethe's method (first suggested by Kramers) for renormalizing the mass of the bound electron. The idea is to calculate the *observed* energy shift, which must be the *difference* between the shift of a bound electron and the shift of a free electron, each of which is separately not observable. In the process we observe that the *Lamb shift* is interpretable as the *additional shift in the mass of an electron which occurs in the vicinity of a strong electric field.*

Repeating the steps which lead to Eq. (3.75), the energy shift of a *free* electron with momentum p_a is

$$\Delta E_{\text{free}} = \left(\frac{e^2}{4\pi}\right)\frac{2}{3\pi}\frac{1}{m^2}\sum_b \mathbb{P}\int_0^\infty \frac{\omega\, d\omega\, |\boldsymbol{p}_{ba}^{\text{free}}|^2}{\frac{p_a^2}{2m} - \frac{p_b^2}{2m} - \omega} \ , \tag{3.85}$$

where b and a refer to states of a free electron with momenta p_b and p_a. A free electron is described by a plane wave, which in box normalization is

$$\psi_a^{\text{free}}(r_e) = \frac{1}{\sqrt{L^3}}\, e^{i\boldsymbol{p}_a\cdot\boldsymbol{r}_e} \ . \tag{3.86}$$

For such a state, the dipole approximation is not reliable, but the relevant matrix element can be calculated exactly:

$$\begin{aligned}\boldsymbol{p}_{ba}^{\text{free}} &= \int \frac{d^3 r_e}{L^3} e^{-i(p_b+k)\cdot r_e}\left(-i\nabla_{r_e}\right) e^{ip_a\cdot r_e} \\ &= \boldsymbol{p}_a\, \delta^3_{p_a,\, p_b+k} \ .\end{aligned} \tag{3.87}$$

Hence the denominator becomes

$$\frac{p_a^2}{2m} - \frac{p_b^2}{2m} - \omega = \frac{p_a^2}{2m} - \frac{(p_a-k)^2}{2m} - \omega = \frac{\boldsymbol{p}_a\cdot\boldsymbol{k}}{m} - \omega\left(1 + \frac{\omega}{2m}\right) \ .$$

As we did in calculating (3.84), assume $\omega \ll m$, and consider free electron momenta $|\boldsymbol{p}_a|$ which are identical to the average momenta of electrons bound in the atomic state a, which means that $|\boldsymbol{p}_a| < m$. Then the denominator is approximately $-\omega$, and

$$\boxed{\Delta E_{\text{free}} \cong -\left(\frac{e^2}{4\pi}\right)\frac{2}{3\pi}\frac{1}{m^2}\,|\boldsymbol{p}_a|^2\int_0^m d\omega \ ,} \tag{3.88}$$

where the sum over b has been fixed by the $\delta^3_{p_a,\, p_b+k}$.

Note that ΔE_{free} has the form

$$\Delta E_{\text{free}} = -C\, p^2 \ , \tag{3.89}$$

where $p^2 = |\boldsymbol{p}_a|^2$. Hence

$$E_{\text{free}} = \frac{p^2}{2m} + \Delta E_{\text{free}} = \left(\frac{1}{2m} - C\right) p^2$$

and we see that C has the effect of shifting the mass of the electron by

$$\frac{1}{2m_{\text{obs}}} = \frac{1}{2m} - C \qquad \Longrightarrow \qquad m_{\text{obs}} = \frac{m}{1 - 2mC} \ . \tag{3.90}$$

With the cutoff given in Eq. (3.88), $C = 2\alpha/(3\pi m)$ and the size of the correction is small,

$$2mC = \frac{4\alpha}{3\pi} \simeq 0.005 \ . \tag{3.91}$$

Of course, without the cutoff the correction diverges linearly.

Now, the *observed energy shift for the atom* is the *difference between the energy shift of a bound electron and that of a free electron* with the *same* average momentum p_a and is therefore given by

$$\begin{aligned}\Delta E^a_{\text{obs}} &= Re\,(\Delta E_a) - \Delta E_{\text{free}} \\ &= \left(\frac{e^2}{4\pi}\right) \frac{2}{3\pi} \frac{1}{m^2}\, \mathbb{P} \int_0^\infty d\omega \left\{ \sum_b \frac{\omega\, |\boldsymbol{p}_{ba}|^2}{E_a - E_b - \omega} + \langle a|\boldsymbol{p}^2|a\rangle \right\} ,\end{aligned} \tag{3.92}$$

where $|\boldsymbol{p}_a|^2$ has been replaced by $\langle a|\boldsymbol{p}^2|a\rangle$. We can cast the last term in a more convenient form using the completeness of atomic states

$$\langle a|\boldsymbol{p}^2|a\rangle = \sum_b \langle a|\boldsymbol{p}|b\rangle \cdot \langle b|\boldsymbol{p}|a\rangle = \sum_b |\boldsymbol{p}_{ba}|^2 \ . \tag{3.93}$$

Then the two terms can be combined to yield

$$\boxed{\Delta E^a_{\text{obs}} = \left(\frac{e^2}{4\pi}\right) \frac{2}{3\pi} \frac{1}{m^2}\, \mathbb{P} \int_0^\infty d\omega \sum_b \frac{(E_a - E_b)\, |\boldsymbol{p}_{ba}|^2}{E_a - E_b - \omega} \ .} \tag{3.94}$$

This expression now diverges logarithmically, which makes the result far less sensitive to the cutoff (which we still need). If $x > 0$, the principal value is

$$\begin{aligned}\mathbb{P} \int_0^{E_{\max}} \frac{d\omega}{x - \omega} &= \int_0^{x-\epsilon} \frac{d\omega}{x - \omega} + \int_{x+\epsilon}^{E_{\max}} \frac{d\omega}{x - \omega} \\ &= -\log(x - \omega)\,\Big|_0^{x-\epsilon} - \log(\omega - x)\,\Big|_{x+\epsilon}^{E_{\max}} \\ &= -\log\left(\frac{\epsilon}{x}\right) - \log\left(\frac{E_{\max} - x}{\epsilon}\right) \cong -\log\left(\frac{E_{\max}}{x}\right) \ .\end{aligned}$$

If $x < 0$ we have directly

$$\int_0^{E_{\max}} \frac{d\omega}{x - \omega} \cong -\log\left(\frac{E_{\max}}{|x|}\right) .$$

Hence,

$$\Delta E^a_{\text{obs}} = -\left(\frac{e^2}{4\pi}\right)\frac{2}{3\pi}\frac{1}{m^2}\sum_b (E_a - E_b)\left|\boldsymbol{p}_{ba}\right|^2 \log\frac{E_{\max}}{|E_a - E_b|} . \qquad (3.95)$$

The log varies slowly with $E_a - E_b$, so it can be factored out of the sum, replacing $E_a - E_b$ by a mean value

$$\Delta E^a_{\text{obs}} = -\left(\frac{e^2}{4\pi}\right)\frac{2}{3\pi}\frac{1}{m^2}\log\left[\frac{E_{\max}}{\langle|E_a - E_b|\rangle}\right]\sum_b (E_a - E_b)\left|\boldsymbol{p}_{ba}\right|^2 . \qquad (3.96)$$

The rapidly varying sum can be done quickly using a favorite trick from atomic physics:

$$\begin{aligned}
\sum_b (E_a - E_b)\left|\boldsymbol{p}_{ba}\right|^2 &= \sum_b \langle a|H_A p^i - p^i H_A|b\rangle\langle b|p^i|a\rangle \\
&= -\sum_b \langle a|p^i|b\rangle\langle b|H_A p^i - p^i H_A|a\rangle \qquad (3.97)\\
&= \frac{1}{2}\langle a|\left[\left[H_A, p^i\right], p^i\right]|a\rangle .
\end{aligned}$$

But, if $V = -Ze/r_e$, then

$$\left[\left[H_A, p^i\right], p^i\right] = [i\nabla_i V, -i\nabla_i] = -\nabla^2 V = -Z\, e^2 \delta^3(r) . \qquad (3.98)$$

Hence the "final" result is

$$\boxed{\Delta E^a_{\text{obs}} = \Delta E^a_{\text{Lamb}} = Z\,\alpha^2\left(\frac{4}{3m^2}\right)|\psi_a(0)|^2 \ln\left[\frac{E_{\max}}{\langle|E_a - E_b|\rangle}\right] ,} \qquad (3.99)$$

where $\alpha = e^2/4\pi \cong 1/137$ is the fine structure constant.

Discussion

We draw the following conclusions from Eq. (3.99):

- Only for S-states is $\psi_a(0) \neq 0$. Hence the shift is largest for S-states. There is a much smaller shift for other states with $L \neq 0$ arising from small terms which we have not calculated.
- The shift is positive because $E_{\max} > |E_a - E_b|$.

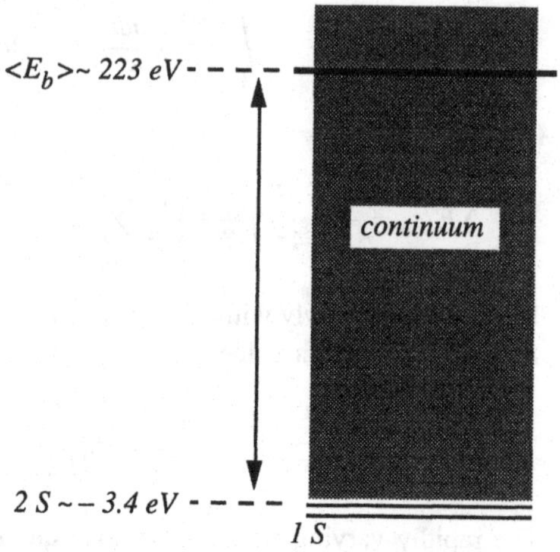

Fig. 3.6 An energy level diagram for hydrogen showing how far into the continuum the mean value (3.100) lies.

To evaluate Eq. (3.99) it is reasonable to take $E_{\max} = m$. However, a good estimate of $\langle |E_a - E_b| \rangle$ is hard to obtain, and the reader is referred to Bethe and Salpeter (1957) for a good discussion. This quantity can be estimated by carrying out the sum over many atomic states, and for the $2S$ state one obtains

$$\langle |E_a - E_b| \rangle_{2s} \cong 16.640\,\text{Ry} = 16.640\frac{\alpha^2 m}{2} = 226.3 \text{ eV} \ . \tag{3.100}$$

This result, illustrated in Fig. 3.6, shows that states *far into the continuum region* are important. The typical excited state contributing to the sum has an excitation energy of $\simeq 200$ eV. With these numbers,

$$\ln \frac{m}{\langle |E_a - E_b| \rangle} \cong 7.72$$

and for hydrogen ($Z = 1$), assuming the $2P_{1/2}$ shift is negligible,

$$\begin{aligned} \Delta E_{2S_{1/2}-2P_{1/2}} &= \frac{4}{3}\frac{\alpha^2}{m^2}(7.72)\left[\frac{1}{4\pi}\frac{1}{(2a_0)^3}\left(2 - \frac{r}{a_0}\right)^2 e^{-r/a_0}\right]_{r=0} \\ &= \frac{4}{3}m\alpha^5\frac{7.72}{8\pi} = m\alpha^5\frac{7.72}{6\pi} \ , \end{aligned} \tag{3.101}$$

where $a_0 = 1/\alpha m$ is the Bohr radius. Hence the transition frequency between the $2S_{1/2}$ and $2P_{1/2}$ states is

$$\begin{aligned} \nu &= \frac{\Delta E}{2\pi} = \frac{mc^2}{\hbar}\left(\frac{\alpha^5}{12\pi^2}7.72\right) \\ &= \frac{0.911 \times 10^{-27}\left(3 \times 10^{10}\right)^2}{1.054 \times 10^{-27}}\left(\alpha^5\frac{7.72}{12\pi^2}\right) \simeq 1051\,\text{MHz} \ . \end{aligned} \tag{3.102}$$

This, of course, is only a rough estimate. There are numerous other terms, and the result can be calculated without a cutoff once the theory is fully renormalized.

The comparison of precise calculations of the Lamb shift (and many other energy shifts) with precise measurements continues to be an active area of research and is a good way to test the validity of the quantum theory of radiation (which becomes *Quantum Electrodynamics* when we also treat the particles as relativistic quantum fields; see Chapter 10). Recent theoretical and experimental results for the Lamb shift in hydrogen are*:

Theory:	1057857(14) kHz	[KS 84]
Experiment:	1057845(9) kHz	[LP 86]
	1057851(2) kHz	[PS 83]

where the numbers in parentheses are an estimate of the errors. So far none of these tests have led to any clear failures; QED is a remarkably successful theory.

3.5 DEUTERON PHOTODISINTEGRATION

As a final application of the quantum theory of radiation, consider the photodisintegration of the deuteron. The deuteron is the only two-body bound system of two nucleons (the neutron and proton) and is therefore the simplest nuclear system. Its binding energy is 2.23 MeV, a very large number when compared with atomic binding energies but quite small on the nuclear scale; it is only about 2% of the mass of a nucleon. Deuteron photodisintegration can occur when a photon (γ) with an energy greater than 2.23 MeV strikes a deuteron (*d*) and breaks it into its constituent nucleons:

$$\gamma + d \rightarrow p + n \ ,$$

where p is the proton with momentum $\boldsymbol{p}_p$ and n the neutron with momentum $\boldsymbol{p}_n$. The observation of this reaction in 1935 was an early confirmation of the conversion of "energy" to "mass" as predicted by relativity [CG 35].

To calculate this reaction, we use the Hamiltonian (3.5), with the proton replacing the electron as the charged particle which interacts with the electromagnetic field (the neutron has a magnetic moment which can also interact with the EM field, but this is a small contribution which can be ignored in a first calculation). Then the interaction Hamiltonian is

$$H_I = U_N^{-1} \frac{e}{m_p} (\boldsymbol{A} \cdot \boldsymbol{p})\, U_N \ , \tag{3.103}$$

where m_p is the proton mass and

$$U_N = e^{-iH_N t} \tag{3.104}$$

*For a recent summary see [BG 87].

with H_N the Hamiltonian of the neutron–proton system. The scattering matrix S and reduced amplitude f, in lowest order perturbation theory, are

$$\begin{aligned} S_{fi} &= -i2\pi\delta\left(E_f - E_i\right)\frac{1}{L^3}f_{fi} \\ &= -i\int_{-\infty}^{\infty} dt\ \langle f\,|H_I(t)|\,i\rangle\ , \end{aligned} \tag{3.105}$$

where Eq. (3.50) has been used to relate the reduced amplitude f to S, $|i\rangle$ is the initial state consisting of a deuteron and a photon, and $|f\rangle$ is the final free neutron and proton.

The nuclear system consists of two particles, the proton at $\boldsymbol{r}_p$ and the neutron at $\boldsymbol{r}_n$. The center of mass (CM) and relative coordinates for these two particles are $\boldsymbol{R} = (\boldsymbol{r}_p + \boldsymbol{r}_n)/2$ and $\boldsymbol{r} = \boldsymbol{r}_p - \boldsymbol{r}_n$. If the initial deuteron is at rest, its wave function is

$$\psi_d(r,R) = \frac{1}{L^{3/2}}\,\phi(r)\ , \tag{3.106}$$

where $\phi(r)$ is the internal wave function of the deuteron and the factor $L^{-3/2}$ is the wave function for the center of mass of the deuteron (obtained from a plane wave with box normalization and zero total momentum). We will discuss the internal wave function shortly. The wave function for the final neutron–proton pair will be approximated by a plane wave

$$\psi_{np}(r,R) = \frac{1}{L^3}\,e^{i(\boldsymbol{p}\cdot\boldsymbol{r}+\boldsymbol{P}\cdot\boldsymbol{R})}\ , \tag{3.107}$$

where $\boldsymbol{P} = \boldsymbol{p}_p + \boldsymbol{p}_n$ is the total momentum and $\boldsymbol{p} = (\boldsymbol{p}_p - \boldsymbol{p}_n)/2$ is the relative momentum of the outgoing pair. In this notation, the incoming and outgoing states are therefore

$$\begin{aligned} |i\rangle &= \frac{1}{L^{3/2}}\,\phi(r)\ \underbrace{a^\dagger_{k,\lambda}|0\rangle}_{\text{one incoming photon}} \\ |f\rangle &= \frac{1}{L^3}\,e^{i(\boldsymbol{p}\cdot\boldsymbol{r}+\boldsymbol{P}\cdot\boldsymbol{R})}\ |0\rangle\ , \end{aligned} \tag{3.108}$$

where $\boldsymbol{k}$ is the momentum of the incoming photon. Hence

$$\begin{aligned} -i\int_{-\infty}^{\infty} dt\ \langle f\,|H_I(t)|\,i\rangle &= -\frac{1}{\sqrt{2\omega}}\,\frac{e}{m}\int_{-\infty}^{\infty} dt\, e^{i(T_n+T_p+\epsilon-\omega)t} \\ &\quad\times\int \frac{d^3r d^3R}{L^6}\,e^{-i(\boldsymbol{p}\cdot\boldsymbol{r}+\boldsymbol{P}\cdot\boldsymbol{R})+i\boldsymbol{k}\cdot(\boldsymbol{R}+\frac{1}{2}\boldsymbol{r})}\left(\epsilon_k^\lambda\cdot\nabla_{r_p}\right)\phi(r) \\ &= -\,i\frac{1}{\sqrt{2\omega}}\,\frac{e}{m}\,2\pi\delta\left(T_n+T_p+\epsilon-\omega\right)\,\epsilon_k^\lambda\cdot(\boldsymbol{p}_p-\boldsymbol{k}) \\ &\quad\times\int \frac{d^3r d^3R}{L^6}\,e^{-i(\boldsymbol{p}\cdot\boldsymbol{r}+\boldsymbol{P}\cdot\boldsymbol{R})+i\boldsymbol{k}\cdot(\boldsymbol{R}+\frac{1}{2}\boldsymbol{r})}\,\phi(r)\ , \end{aligned} \tag{3.109}$$

where T_p and T_n are the kinetic energies of the outgoing proton and neutron, $\epsilon = m_p + m_n - m_d$ is the binding energy of the deuteron, and ω is the energy of the incoming photon. Note that the field is evaluated at the proton point $\boldsymbol{r}_p = \boldsymbol{R} + \frac{1}{2}\boldsymbol{r}$. Integrating over R gives $\boldsymbol{P} = \boldsymbol{k}$, and the reduced amplitude f can be extracted,

$$
\begin{aligned}
f_{fi} &= \frac{e}{\sqrt{2\omega}\, m_p}\, \epsilon_k^\lambda \cdot \left(\boldsymbol{p} - \tfrac{1}{2}\boldsymbol{k}\right) \int d^3r\, e^{-i(\boldsymbol{p}-\frac{1}{2}\boldsymbol{k})\cdot \boldsymbol{r}} \phi(r) \\
&= \frac{e}{\sqrt{2\omega}\, m_p}\, \epsilon_k^\lambda \cdot \boldsymbol{p}\ \tilde{\phi}(p - \tfrac{1}{2}k) \ ,
\end{aligned} \tag{3.110}
$$

where the term $\epsilon_k \cdot \boldsymbol{k}$ is zero because the photon polarization vectors are transverse, and $\tilde{\phi}$ is the momentum space wave function. Now, the energy conservation relation will give

$$
\begin{aligned}
T_p + T_n = \omega - \epsilon &= \frac{\left(\boldsymbol{p} + \frac{1}{2}\boldsymbol{k}\right)^2}{2m_p} + \frac{\left(\boldsymbol{p} - \frac{1}{2}\boldsymbol{k}\right)^2}{2m_n} \\
&\cong \frac{\boldsymbol{p}^2}{m} + \frac{\boldsymbol{k}^2}{4m} = \frac{\boldsymbol{p}^2}{m} + \frac{\omega^2}{4m} \cong \frac{\boldsymbol{p}^2}{m} \\
&\Longrightarrow \boldsymbol{p}^2 = m(\omega - \epsilon) \ ,
\end{aligned} \tag{3.111}
$$

where we neglect the differences in the proton and neutron masses, so that $m_p \cong m_n \cong m$. The neglect of $\boldsymbol{k}^2$ compared to $\boldsymbol{p}^2$ is justified by the last step which gives $\boldsymbol{p}^2 \cong m\omega$, while $k^2 = \omega^2$ is much smaller. Hence $p = |\boldsymbol{p}| \gg |\boldsymbol{k}|$, and we can safely replace $p + \frac{1}{2}k$ by p in the argument of the momentum space wave function of the deuteron (this is just the dipole approximation). The f amplitude therefore reduces to

$$
f_{fi} = \frac{e}{\sqrt{2\omega}\, m}\, \epsilon_k^\lambda \cdot \boldsymbol{p}\ \tilde{\phi}(p) \tag{3.112}
$$

and the cross section is

$$
\frac{d\sigma}{d\Omega} = \frac{1}{(1)} \int_0^\infty \frac{p^2\, dp}{(2\pi)^3}\, 2\pi\delta\left(\frac{p^2}{m} + \epsilon - \omega\right) \left(\frac{e^2}{2\omega m^2}\right) \left|\epsilon_k^\lambda \cdot \boldsymbol{p}\right|^2 |\tilde{\phi}(p)|^2, \tag{3.113}
$$

where the flux factor is unity because $v = c = 1$. Next, assuming the momentum of the photon is in the $\hat{z}$-direction, we average over initial polarization states using Fig. 3.4, which gives

$$
\frac{1}{2} \sum_\lambda \left|\epsilon_k^\lambda \cdot \boldsymbol{p}\right|^2 = \frac{1}{2} p^2 \sin^2\theta \tag{3.114}
$$

and hence the unpolarized cross section reduces to

$$
\frac{d\sigma}{d\Omega} = \left(\frac{e^2}{4\pi}\right) \frac{\sqrt{m\omega}}{8\pi} (\gamma - 1)^{3/2} \sin^2\theta\, |\tilde{\phi}(p)|^2 \ , \tag{3.115}
$$

where we have expressed the answer in terms of the dimensionless quantity $\gamma = \omega/\epsilon$.

To complete the calculation, we must find the wave function of the deuteron. One of the important features of photodisintegration near threshold (the threshold energy is the energy at which the process just becomes physical, which in this case is $\omega = \epsilon$) is that it is insensitive to the details of the deuteron wave function, and hence a reliable prediction is possible without knowing much about the short range structure of the nuclear force.

To see why this is so, we estimate the wave function using the *Hulthén* model for the nuclear potential. This is a crude model which nevertheless is very useful for such estimates. The model assumes that the nuclear force at long range (i.e., for large internucleon separation r) is dominated by the exchange of a single pion (a good approximation) and that spin dependence of the interaction can be neglected (which is not too bad an approximation for interactions which depend only on the charge but overlooks many features of the deuteron, such as the D state).* Under these assumptions the potential is a Yukawa potential with a range of the pion mass (denoted by μ). The *Hulthén* model approximates this potential as follows:

$$V(r) = -\frac{g^2}{4\pi}\frac{e^{-\mu r}}{r} \simeq -\frac{g^2}{4\pi}\,\mu\,\frac{e^{-\mu r}}{1 - e^{-\mu r}} \quad . \tag{3.116}$$

This approximation captures the correct behavior of the potential at both long and short range and permits us to solve the Schrödinger equation for S states exactly. The equation for the relative coordinate is

$$-\frac{1}{m}\frac{d^2}{dr^2}\,\phi(r) - \frac{2}{r}\frac{d}{dr}\,\phi(r) + V(r)\phi(r) = \epsilon\,\phi(r) \quad . \tag{3.117}$$

Substituting a wave function of the form

$$\phi(r) = N\left(\frac{e^{-\alpha r}}{r} - \frac{e^{-\beta r}}{r}\right)$$

into this equation gives a solution, provided

$$\begin{aligned} \alpha \equiv \delta = \sqrt{m\epsilon} &= \frac{1}{2}\left(\frac{g^2}{4\pi}\,m - \mu\right) \\ \beta &= \mu + \delta \quad . \end{aligned} \tag{3.118}$$

The momentum space wave function is the Fourier transform [worked out in Eq. (4.52)], and hence

$$\tilde{\phi}(p) = \int d^3r\, e^{-i\mathbf{p}\cdot\mathbf{r}}\phi(r) = 4\pi N\left(\frac{1}{p^2+\delta^2} - \frac{1}{p^2+(\mu+\delta)^2}\right) \quad . \tag{3.119}$$

*The one-pion exchange force will be derived from field theory in Sec. 9.9.

Note that the second term is very small. For example, if $p = 0$, the two terms are in the ratio of

$$\frac{\delta^2}{(\mu+\delta)^2} \cong \frac{2.2 \times 936}{(139 + \sqrt{2.2 \times 936})^2} \cong 0.061 \cong 6\% \ .$$

Therefore a very good estimate is obtained by using the *asymptotic wave function only* (in which case the answer does not depend on the use of the Hulthén model). The normalization constant for the asymptotic wave function is $N = \sqrt{\delta/2\pi}$, and the square of the asymptotic wave function, evaluated at $p^2 = m(\omega - \epsilon)$, is then

$$\tilde{\phi}^2_{\text{asy}}(p) = \left(\frac{4\pi N}{p^2 + m\epsilon}\right)^2 = \frac{8\pi\sqrt{m\epsilon}}{m^2\omega^2} \ .$$

Substituting this into (3.115) gives finally

$$\frac{d\sigma}{d\Omega} = \left(\frac{e^2}{4\pi}\right) \frac{1}{m\epsilon} \frac{(\gamma-1)^{3/2}}{\gamma^3} \sin^2\theta \ . \tag{3.120}$$

We emphasize that this result only includes the contributions from the proton charge (the *electric dipole* interaction) and that only the contributions from the asymptotic deuteron wave function were retained. It might appear that corrections from the interior part of the deuteron wave function would be uncertain and hard to estimate, but Bethe and Longmire [BL 50] showed that these additional contributions can be expressed in terms of the effective range for the scattering of two nucleons in the $^3S_1$ channel, a quantity which is readily measured. There are also additional contributions from the magnetic interactions of the nucleons which contribute an angular independent background term which dominates at energies within 0.1–0.2 MeV of the threshold but contribute only a few percent to the cross section at higher energies.

Because of the simplicity of this process and its insensitivity to the details of the nuclear force, deuteron photodisintegration has been of considerable interest for many years. Recent precise measurements of the angular distribution at low energies (see, for example, [De 85]) show the large $\sin^2\theta$ dependence expected and are in good agreement with theory.

PROBLEMS

3.1 The ground state wave function of the hydrogen atom is

$$\psi_0 = N_0\, e^{-r/a_0} \ ,$$

where a_0 is the Bohr radius, $a_0 = 1/m\alpha$, and N_0 is a normalization constant. The first excited state is four-fold degenerate. Four linearly independent wave functions which span the space of excited states are

$$\psi_{10} = N_1 \left(2 - \frac{r}{a_0}\right) e^{-r/2a_0}$$
$$\psi_{1x} = N_1 \frac{x}{a_0\sqrt{3}} e^{-r/2a_0}$$
$$\psi_{1y} = N_1 \frac{y}{a_0\sqrt{3}} e^{-r/2a_0}$$
$$\psi_{1z} = N_1 \frac{z}{a_0\sqrt{3}} e^{-r/2a_0} \quad ,$$

where N_1 is another normalization constant, and x, y, and z are the three spatial coordinates of the election and $r = \sqrt{x^2 + y^2 + z^2}$.

(a) Derive a formula for the lifetime of the state ψ_{1z}. Reduce your answer to an integral over the spatial coordinates x, y, z or r, θ, ϕ and constants (N_0, N_1, a_0 and other constants). It is not necessary to fully evaluate the integral, but you should reduce the triple integral to a single integral.

(b) Is the photon which is emitted by the decay polarized? If so, what is its polarization (i.e., in which direction does ϵ^α point)?

(c) What is the lifetime of the states ψ_{1x} and ψ_{1y}? Are the photons emitted by these decays polarized? If so, in which direction?

3.2 A nonrelativistic particle of mass m and charge e is trapped in an infinite one-dimensional square well described by the potential

$$V(z) = 0 \qquad 0 \le z \le \ell$$
$$V(z) = \infty \qquad z < 0 \quad \text{or} \quad \ell < z \ .$$

Calculate the lifetime of the first two excited states. (Suggestion: treat the EM field as one-dimensional.)

3.3 [Taken from Sakurai (1967).] Suppose a photon of energy ω is incident on a hydrogen atom in its ground state. The photon may be absorbed, ionizing the atom. This is a simple model for the photoelectric effect.

(a) Using the formalism developed in this chapter, write the matrix element for the lowest order contribution to this process. (Note that the final state is a scattering state of an electron and a proton.)

(b) If the energy of the incident photon is so large that the final electron–proton scattering state can be approximated by plane waves, show that the differential cross section, defined in Eq. (3.52), is

$$\frac{d\sigma}{d\Omega} = 32 \left(\frac{e^2}{4\pi}\right) \left(\frac{1}{m\omega}\right) \frac{1}{(|\boldsymbol{p}_f|\, a_0)^5} \frac{\sin^2\theta \cos^2\phi}{[1 - v\cos\theta]^4} \quad ,$$

where the spherical coordinate variables θ and ϕ are defined so that the incident photon momentum is along the z-axis, its polarization is along the x-axis, and a_0 is the Bohr radius.

3.4 [Taken from Sakurai (1967).] The phenomenological interaction Hamiltonian responsible for the decay of the spin $\frac{1}{2}$ Σ^0 hyperon ($\Sigma^0 \to \Lambda + \gamma$) located at $\boldsymbol{r} = 0$ can be taken to be

$$\frac{\kappa e}{(m_\Lambda + m_\Sigma)} \tau_{\Lambda\Sigma} \, \boldsymbol{\sigma} \cdot (\nabla \times \boldsymbol{A}) \,|_{\boldsymbol{r}=0}$$

where $\tau_{\Lambda\Sigma}$ is an operator that converts a Σ^0-state into a Λ-state, leaving the spin unchanged, $\boldsymbol{\sigma}$ are the Pauli matrices, which in this case connect the spin $\frac{1}{2}$ spaces of the two hyperons, and κ is a dimensionless constant.

(a) Show that the angular distribution of the decay is isotropic even when the parent Σ^0 is polarized.

(b) Find the mean lifetime (in seconds) for $\kappa = 1$ ($m_\Lambda = 1115$ MeV and $m_\Sigma = 1192$ MeV).

3.5 The interaction time translation operator was defined in Eq. (3.9); rewriting this gives

$$U_I(t, t_0) = U_0^{-1}(t, t_0) U_{\text{total}}(t, t_0) \quad .$$

In the following, assume that $U_{\text{total}}(t, t_0)$ and $U_0(t, t_0)$ are unitary and that they satisfy the multiplicative property $U(t_1, t_2) U(t_2, t_3) = U(t_1, t_3)$.

(a) Prove that U_I is unitary.

(b) Show that U_I does not satisfy the multiplicative property. How would U_I have to be redefined in order to satisfy the multiplicative property?

PART II

RELATIVISTIC EQUATIONS

CHAPTER 4

THE KLEIN–GORDON EQUATION

In the last chapter we brought the subject to the point where the electromagnetic field was quantized, and the production and annihilation of the field quanta (photons in that case) could be treated. The treatment of the EM field was fully relativistic, even if it was not manifestly covariant (because of the Coulomb gauge). However, the treatment of the particles (electrons and protons) remained nonrelativistic, and we had no way to describe the production and annihilation of these particles (which must always occur in particle–antiparticle pairs because both electron and baryon numbers are conserved). With this chapter we begin to develop the tools necessary to describe "classical" particles, such as electrons, covariantly. The development will eventually lead to the construction of field theories for electrons and other classical particles (Chapter 7), with the capability to describe the production of particle–antiparticle pairs. Only then will the distinction between classical particles and fields disappear, with the recognition that the quantum field is the single entity suitable for the description of all matter and energy.

However, before we can introduce these new quantum fields we must first understand how to describe single particles in a covariant fashion. This is the subject of the next three chapters. In this chapter we begin with the simplest relativistic equation, the Klein–Gordon (KG) equation. Then we discuss the Dirac equation (Chapter 5) and applications of the Dirac equation (Chapter 6).

4.1 THE EQUATION

We begin our systematic study of relativistic equations by briefly considering the following equation:

$$i\frac{\partial}{\partial t}\psi(x) = \sqrt{m^2 - \nabla^2}\,\psi(x) \quad , \tag{4.1}$$

where, as in the previous chapters, x represents both the time and space dependence of the wave function, so that $\psi(x) = \psi(\boldsymbol{r}, t)$ is understood. This equation follows the traditional "rules" of quantum mechanics in that it can be obtained from the relativistic energy relation $E = \sqrt{m^2 + \boldsymbol{p}^2}$ by the substitution

$$
\begin{aligned}
E &\to i\frac{\partial}{\partial t} \\
\boldsymbol{p} &\to -i\nabla \ .
\end{aligned}
\tag{4.2}
$$

There is no problem in principle with the operator $E_\nabla = \sqrt{m^2 - \nabla^2}$; this operator is defined by either (i) expanding any function in terms of the eigenfunctions of ∇ (the momentum eigenfunctions), on which the operation E_∇ is easily carried out, or (ii) defining E_∇ by its power series expansion

$$
E_\nabla \cong m\left\{1 - \frac{\nabla^2}{2m^2} - \frac{\nabla^4}{8m^4}\cdots\right\} \ . \tag{4.3}
$$

While this series may not always converge, we may consider its analytic continuation to be the definition of the operation of E_∇ on any function.

The disadvantage of Eq. (4.1) is that it is not *manifestly* covariant. To be manifestly covariant, we must know how to transform the equation not only in time and space, described by the infinitesimal generators H and $\boldsymbol{P}$, but also under the homogeneous Lorentz group, which includes rotations, generated by the angular momentum operators $\boldsymbol{J}$, and the boosts, generated by the operators $\boldsymbol{K}$ (the Lorentz group and its generators will be discussed in Sec. 5.8). While these transformations can sometimes be worked out for equations of the type (4.1), many problems are encountered, and therefore this route was not the one taken in the original developments which led to the quantum field theory of elementary particles [except that the Dirac equation can be regarded as arising from the linearization of the square root in (4.1); see Chapter 5]. Now we know that many particles originally supposed to be "elementary" (such as the proton) are in fact complicated composite structures of valence quarks and a sea of quark–antiquark pairs and hence cannot be described by a single local quantum field. In the search for approximate methods of describing such particles, interest in equations of the type (4.1) has been rekindled and is an active area of current research. Such an approach sometimes is identified as *relativistic Hamiltonian dynamics* and will not be discussed further here.*

The alternative route is to introduce equations which are manifestly covariant, and this method is sometimes referred to as *manifestly covariant dynamics*. It is the route which is traditionally taken to relativistic field theory. A simple, manifestly covariant wave equation is obtained if the substitutions (4.2) are made into the mass energy relation $E^2 = m^2 + \boldsymbol{p}^2$. This gives the *Klein–Gordon* (*KG*) equation for a free particle

$$
\boxed{\left(\Box + m^2\right)\psi(x) = 0 \ ,} \tag{4.4}
$$

*For a review see [Co 89] and [KP 90].

where the wave operator, $\Box$, was previously introduced in Chapter 2, Eq. (2.20), and m is the mass of the particle (to be confirmed below). This equation is manifestly covariant because the wave operator is a scalar, and if the mass m and the wave function ψ are also scalars, the equation and the wave function have the same form in all reference frames. Note that Eq. (2.42) for a free massive vector field (i.e., with $j^\mu = 0$), which we obtained in Sec. 2.5, is just a KG equation for *each* component of the field.

This equation is not first order in the time derivative. This means that it is not sufficient to know the wave function at a particular time in order to determine it at later times; one must also know the time derivative of the wave function at that time. In this sense, Eq. (4.4) appears to depart from one of the basic tenets of quantum mechanics: that knowledge of the wave function at one time is sufficient to determine it at all later times. However, as we shall soon see, this is only an apparent problem and will lead to a reinterpretation of the wave function. Instead, this equation, in common with all manifestly covariant equations, has another problem which is more serious and will be discussed shortly.

To introduce electromagnetic interactions into the KG equation, recall that, in the four-vector notation introduced in Chapter 2, the energy-momentum operator is

$$p^\mu = i\frac{\partial}{\partial x_\mu} = \left(i\frac{\partial}{\partial t}, -i\nabla\right) \equiv i\partial^\mu \quad . \tag{4.5}$$

Using minimal substitution [as encountered in Eq. (2.30)] we are led to the replacement

$$p^\mu \to p^\mu - eA^\mu \qquad \text{minimal substitution,} \tag{4.6}$$

where A^μ is the four-vector potential previously introduced in Chapter 2. This gives

$$\left[-\left(i\frac{\partial}{\partial x^\mu} - eA_\mu\right)\left(i\frac{\partial}{\partial x_\mu} - eA^\mu\right) + m^2\right]\psi(x) = 0$$

or, expanding out the product,

$$\boxed{\left[\Box + m^2 + U(x)\right]\psi(x) = 0 \ ,} \tag{4.7}$$

where the generalized "potential" $U(x)$ consists of a scalar and a vector part

$$\begin{aligned} U(x) &= ie\frac{\partial}{\partial x^\mu}A^\mu + ieA^\mu\frac{\partial}{\partial x^\mu} - e^2A^\mu A_\mu \\ &= \underbrace{i\frac{\partial}{\partial x^\mu}V^\mu + iV^\mu\frac{\partial}{\partial x^\mu}}_{\text{required by hermicity}} + S \quad . \end{aligned} \tag{4.8}$$

In these equations, the $\partial/\partial x^\mu$ operator operates all the way to the right, so that in the first term it operates on *both* A (or V) and ψ. While all the terms in (4.8) are scalars, the first two terms consist of a vector potential which is contracted with the ∂_μ operator to make an overall scalar, while the last term is a scalar by itself. Note that the symmetrized form of the vector term is required in order to maintain the hermiticity of the interaction. In the most general case, the scalar, S, and vector, V^μ, parts of the potential could be independent interactions, but for electromagnetism they are related by

$$S = -e^2 A^\mu A_\mu \qquad V^\mu = eA^\mu \ . \tag{4.9}$$

In some of the following discussion, the vector and scalar parts will be treated as independent interactions, and at other times we will specialize the discussion to electromagnetism, Eq. (4.9).

4.2 CONSERVED NORM

Because of the second time derivative, the conserved norm is no longer $\int d^3r\, \psi^* \psi$. To find the correct norm, consider the following two expressions:

$$\begin{aligned} \psi_b^* \left(\Box + m^2 + U\ \right) \psi_a &= 0 \\ \psi_a \left(\Box + m^2 + U^*\right) \psi_b^* &= 0 \ , \end{aligned}$$

where a and b are the quantum numbers of any two solutions. Subtracting these two expressions gives

$$\psi_b^* \overrightarrow{\Box} \psi_a - \psi_b^* \overleftarrow{\Box} \psi_a + \psi_b^* \left[i \frac{\overrightarrow{\partial}}{\partial x^\mu} V^\mu + i V^\mu \frac{\overrightarrow{\partial}}{\partial x^\mu} + i \frac{\overleftarrow{\partial}}{\partial x^\mu} V^\mu + i V^\mu \frac{\overleftarrow{\partial}}{\partial x^\mu} \right] \psi_a = 0 \ ,$$

where the arrow over the operators tells in which direction they operate. In particular, an arrow pointing to the right ($\rightarrow$) means that the operator operates all the way to the right, including operating on any wave functions which may eventually stand on the right. Similarly, the arrow pointing to the left ($\leftarrow$) operates all the way to the left. [The expressions $\psi_b^* \overrightarrow{\Box} \psi_a$ and $\psi_a \overleftarrow{\Box} \psi_b^*$ are therefore identical.] The above expression can therefore be written as the four-divergence of a four-current j^μ,

$$\frac{\partial}{\partial x^\mu} j^\mu = 0 \ , \tag{4.10}$$

where

$$j^\mu = \psi_b^* i \frac{\overleftrightarrow{\partial}}{\partial x_\mu} \psi_a - 2V^\mu \psi_b^* \psi_a \tag{4.11}$$

and the double arrow is an obvious generalization of the single arrow notation:

$$\psi_b^* \frac{\overleftrightarrow{\partial}}{\partial x^\mu} \psi_a \equiv \psi_b^* \frac{\overrightarrow{\partial}}{\partial x^\mu} \psi_a - \psi_b^* \frac{\overleftarrow{\partial}}{\partial x^\mu} \psi_a \quad . \tag{4.12}$$

If we integrate (4.10) over all space (i.e., over a volume L^3 with periodic boundary conditions), the spatial divergence integrates to zero, and the volume integral of the time component is a constant. In general, for any vector field j^μ which is conserved, so that $\partial_\mu j^\mu = \partial j^0/\partial t + \nabla \cdot \boldsymbol{j} = 0$, and which satisfies periodic boundary conditions, we have

$$\begin{aligned} &\frac{\partial}{\partial t} \int_{L^3} d^3r \; j^0 + \int_{L^3} d^3r \; \nabla \cdot \boldsymbol{j} = 0 \Rightarrow \\ &\frac{\partial}{\partial t} \int_{L^3} d^3r \; j^0 + \underbrace{\int_{\substack{\text{surface} \\ \text{at} \infty}} dS \; \hat{n} \cdot \boldsymbol{j}}_{=0} = 0 \Rightarrow \\ &\int_{L^3} d^3r \; j^0 = \text{constant} \; . \end{aligned} \tag{4.13}$$

For the case given in Eq. (4.10), this becomes

$$\boxed{\int_{L^3} d^3r \left[\psi_b^* \, i \frac{\overleftrightarrow{\partial}}{\partial t} \psi_a - 2V^0 \psi_b^* \psi_a \right] = \text{constant} \quad .} \tag{4.14}$$

If $a = b$, this is the conserved norm for the state a, and it explicitly involves the potential (the time component of the vector part). This result, unusual from the point of view of conventional quantum mechanics, is a general feature of many manifestly covariant relativistic equations (especially two-body equations). Before discussing this norm further, we will obtain the solutions of the free particle equations.

4.3 SOLUTIONS FOR FREE PARTICLES

The solutions of the free particle KG equation (i.e., with $U = 0$) can be obtained by separation of variables. Using box normalization with periodic boundary conditions, the solutions are

$$\phi_n^{(\pm)}(\boldsymbol{r}, t) = N \, e^{i(\boldsymbol{k}_n \cdot \boldsymbol{r} \mp E_n t)} \quad , \tag{4.15}$$

where $E_n = \sqrt{m^2 + \boldsymbol{k}_n^2}$ is always positive and the superscript $(\pm)$ refers to the sign of the energy in the exponential. As in the cases discussed in the previous chapters, solutions with both signs occur; solutions with the superscript $(+)$ will be referred to as "positive" energy solutions (because the factor of $-iE_nt$ in the exponential corresponds to positive energy in nonrelativistic Schrödinger theory), and those with superscript $(-)$ will be referred to as "negative" energy solutions. The periodic boundary conditions, imposed inside of a cube of length L on each side, require

$$\boldsymbol{k}_n = \frac{2\pi}{L}(n_x, n_y, n_z) \qquad n_i = 0, \pm 1, \pm 2, \ldots \ . \tag{4.16}$$

The norms of the positive and negative energy solutions have different sign,

$$i\int_{L^3} d^3r\, \phi_n^{(\pm)*} \frac{\overleftrightarrow{\partial}}{\partial t} \phi_m^{(\pm)} = \pm 2E_n L^3 N^2 \delta_{nm} \ , \tag{4.17}$$

where $\delta_{nm} = \delta_{n_x m_x}\delta_{n_y m_y}\delta_{n_z m_z}$. Using the zeroth component of the conserved current to define invariant scalar products, the different solutions are orthogonal,

$$i\int_{L^3} d^3r\, \phi_n^{(+)*} \frac{\overleftrightarrow{\partial}}{\partial t} \phi_m^{(-)} = 0 \ . \tag{4.18}$$

If we choose $N = \left(2E_nL^3\right)^{-1/2}$, then the two types of solutions, $\phi^{(+)}$ and $\phi^{(-)}$, will each be normalized and orthogonal:

$$\boxed{\phi_n^{(\pm)}(\boldsymbol{r}, t) = \frac{1}{\sqrt{2E_nL^3}}\, e^{i(\boldsymbol{k}_n\cdot\boldsymbol{r} \mp E_n t)}} \ . \tag{4.19}$$

The positive energy solutions $\phi^{(+)}$ have norm $+1$, and the negative energy solutions $\phi^{(-)}$ have norm -1.

Because negative norm solutions exist, $||\psi||^2$ *cannot be a probability density*, and historically this was regarded as a reason for rejecting the KG equation. This point of view is too narrow, but the existence of negative norm solutions is an indication that the quantum mechanics described by such an equation departs from the classical rules of quantum theory. One of these rules is that the states span a vector space with a norm which is positive definite, and this is certainly not the case for the KG equation. Later we will see that if the KG equation is used as the *basis for a field theory* (recall Prob. 1.3), then the states defined by the field theory will all have positive definite norm, and the negative energy states can be reinterpreted as positive energy states of antiparticles. Before developing the field theory, however, it is useful, and maybe even necessary, to study the properties of a quantum mechanics which is based on the use of the KG equation as the equation for single particle states (referred to as the *first quantized* form of the theory). This is the purpose of this chapter, and it is well to realize that even though the first quantized theory can be only partially successful, it is just as important in its own right as the study of the first quantized theory for the Dirac equation (which is taken up in the next chapter). The first quantized theories for both of these equations suffer from the same fundamental disease; they both have negative energy solutions which cannot be treated fully until they are reinterpreted as antiparticles, and this is only fully successful in the second quantized (field theoretic) form.

Keeping these comments in mind, we proceed with our study. If the KG equation is applied to the description of a charged particle, the norm will be interpreted as a charge density, with positive norm states describing + charges and negative norm states describing − charges. The conservation of charge then appears as a consequence of the invariance of the norm. If the particle has no electric charge but has some other quantum number (a generalized charge) which satisfies an additive conservation law, the norm can be interpreted as the density of this generalized charge. In either case, the existence of *two* states, one carrying positive charge and one carrying negative charge, is assumed.

Before we develop these ideas further, it is useful to look at a simple example which illustrates how the KG norm can be consistently interpreted as a charge density and how both particles and antiparticles are described by the equation.

4.4 PAIR CREATION FROM A HIGH COULOMB BARRIER *

Since the norm is conserved, we might suppose that if we start out at some initial time t_0 with a superposition of states with only positive norm, these will evolve at a later time t into a superposition of states with only positive norm, and that therefore in this case we could still interpret the wave function as a probability density. We shall show here that when *interactions* are present, states with negative norm can still appear, and hence they cannot be eliminated from consideration. Our discussion will also enable us to interpret the norm physically.

Consider the reflection of positively charged mesons (π^+ mesons for example) from a high Coulomb barrier. The KG equation (4.7) for this case is

$$\left(\Box + m^2 + 2ieV\frac{\partial}{\partial t} - e^2V^2\right)\psi(x) = 0 \ , \tag{4.20}$$

where V is zero to the left of the barrier and a constant to the right, as shown in Fig. 4.1A. We seek a solution of the form e^{-iEt}, corresponding, in region I, to a free particle with positive energy. We guess the solutions in regions I and II to be of the form

$$\begin{aligned} \psi_{\mathrm{I}}(z,t) &= A\, e^{i(pz-Et)} + B\, e^{-i(pz+Et)} \\ \psi_{\mathrm{II}}(z,t) &= C\, e^{i(Pz-Et)} + D\, e^{-i(Pz+Et)} \ . \end{aligned} \tag{4.21}$$

*Much of the material in this section is discussed in an interesting paper by Winter [Wi 59].

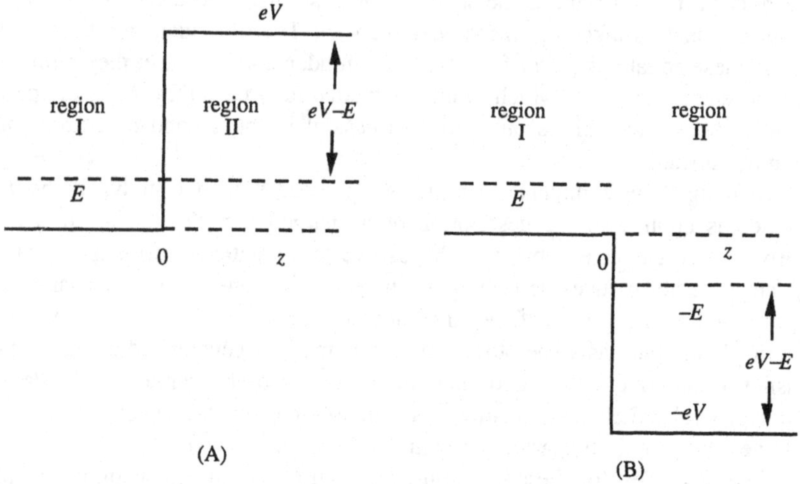

Fig. 4.1 (A) Regions I and II and the high Coulomb barrier as described in the text. (B) The appearance of the "barrier" to a negatively charged particle with total energy $-E$ in region II.

These trial functions will solve the KG equation in the two regions if

$$p = \sqrt{E^2 - m^2} \qquad P = \sqrt{(eV - E)^2 - m^2} \ , \tag{4.22}$$

and the complete solution will later be obtained by requiring that the wave function and its derivative be continuous at the boundary between the two regions. We see that the solution in region I consists of waves moving toward the right (with the coefficient A) and toward the left (with coefficient B). Later we will construct wave packets from these functions and observe that they describe a particle (with positive charge) which approaches the barrier from the left, reflects, and travels back to the right. If $eV < E + m$, the wave number P in region II is complex, and the solutions correspond to damped and growing exponentials, as expected. The particle cannot penetrate into that region, and the correct solution is the one with a purely damped exponential. *However*, if $eV > E + m$ (a high barrier), then P is real and we have oscillating solutions in region II! Before we find all the coefficients, let us investigate the nature of these solutions.

First, we compute the norm of these solutions in region II. If $D = 0$, for example, we have (for box normalization when $-L/2 < z < L/2$)

$$\|\psi_{\rm II}\|^2 = \int_0^{L/2} dz\, |C|^2 (2E - 2eV) < 0 \tag{4.23}$$

if $eV > E + m$ (the high barrier). Hence these oscillating solutions have negative norm and we see that the interaction has forced us to consider such states, even though the solution in region I is the sum of two terms, both with positive norm.

To see how these solutions develop in time, we smear in E to make wave packets. Smearing around $E = E_0$ we have, for ordinary solutions like those in region I,

$$\int_{E_0-\Delta E}^{E_0+\Delta E} dE\, e^{i\left(\sqrt{E^2-m^2}\,z-Et\right)} \cong e^{i(p_0 z-E_0 t)} \int_{-\Delta E}^{+\Delta E} dE\, e^{iE\left(\frac{E_0}{p_0}z-t\right)}$$
$$= e^{i(p_0 z-E_0 t)} f\left(\frac{z-v_0 t}{v_0}\right) , \qquad (4.24)$$

where $v_0 = p_0/E_0$, and the envelope function is

$$f(\eta) \equiv \int_{-\Delta E}^{\Delta E} dE\, e^{iE\eta} = 2\frac{\sin \eta\Delta E}{\eta} . \qquad (4.25)$$

Note that these packets travel in the direction of p with the classical relativistic velocity $v = p/E$.

The packets in region II behave differently, however. We have

$$\int_{E_0-\Delta E}^{E_0+\Delta E} dE\, e^{i\left(\sqrt{(eV-E)^2-m^2}\,z-Et\right)} \cong e^{i(P_0 z-E_0 t)} \int_{-\Delta E}^{\Delta E} dE\, e^{-iE\left(\frac{eV-E_0}{P_0}z+t\right)}$$
$$= e^{i(P_0 z-E_0 t)} f\left(\frac{z+u_0 t}{u_0}\right) , \qquad (4.26)$$

where

$$u_0 = \frac{P_0}{eV-E_0} = \frac{P_0}{\sqrt{m^2+P_0^2}} . \qquad (4.27)$$

Note that this packet propagates to the *left*; its *group* velocity is negative even though its *phase* velocity is positive. It travels with the classical relativistic velocity of a free particle with *kinetic energy* corresponding to $eV - E_0$. Since a π^+ cannot have a positive kinetic energy in this region, the packet must be describing something else. If it were negatively charged, then it would see the potential barrier as a deep hole, and it could have positive kinetic energy. In that case it would have total energy

$$E_{\text{total}} = (eV - E) - eV = -E ,$$

as shown in Fig. 4.1B. A consistent picture of a particle of mass m and charge $-e$ emerges; it is a π^- meson!

To complete the description, we compute the coefficients for a state which is initially a pure π^+ traveling to the right toward the barrier. This means that any π^- meson must be produced by the interaction and will travel toward the right into region II. Hence the boundary condition is that $C = 0$, and it is convenient to choose $D = 1$. Then the continuity of the wave function and its derivative at $z = 0$ require

$$\begin{aligned} A + B &= 1 \\ p(A - B) &= -P , \end{aligned} \qquad (4.28)$$

which gives

$$
\begin{aligned}
A &= \frac{1}{2} - \frac{P}{2p} \\
B &= \frac{1}{2} + \frac{P}{2p}
\end{aligned}
\tag{4.29}
$$

and the solution becomes (renormalizing so $A = 1$)

$$
\begin{aligned}
\psi_{\rm I}(z,t) &= e^{i(pz-Et)} - \frac{P+p}{P-p} e^{-i(pz+Et)} \\
\psi_{\rm II}(z,t) &= -\frac{2p}{P-p} e^{-i(Pz+Et)} \ .
\end{aligned}
\tag{4.30}
$$

To interpret this solution, smear in E and assume that the coefficients are slowly varying functions of E which can be approximated by their value at the central energy E_0. We then get

$$
\begin{aligned}
\bar{\psi}_{\rm I}(z,t) &\cong e^{i(p_0z-E_0t)} f\left(\frac{z-v_0t}{v_0}\right) - \frac{P_0+p_0}{P_0-p_0} e^{-i(p_0z+E_0t)} f\left(\frac{z+v_0t}{v_0}\right) \\
\bar{\psi}_{\rm II}(z,t) &\cong -\frac{2p_0}{P_0-p_0} e^{-i(P_0z+E_0t)} f\left(\frac{z-u_0t}{u_0}\right) \ .
\end{aligned}
\tag{4.31}
$$

Using the fact that the envelope functions are non-zero only when their arguments are small, a moving picture of this state can be constructed as shown in Fig. 4.2.

Note that the norm of the state is a constant of the motion. At $t \to -\infty$, only the incoming packet on the left-hand side exists, and if we take its norm to be one, then the norm of the reflected packet (traveling to the left in region I) is

$$
R = \left(\frac{P_0+p_0}{P_0-p_0}\right)^2
\tag{4.32}
$$

and, recalling Eq. (4.23), the norm of the "transmitted" packet traveling to the right in region II is

$$
T = -\left(\frac{2p_0}{P_0-p_0}\right)^2 \left(\frac{eV-E_0}{E_0}\right) \frac{u_0}{v_0} = -\frac{4p_0P_0}{(P_0-p_0)^2} \ ,
\tag{4.33}
$$

where the ratio $(eV - E_0)/E_0$ comes from the energy factor in (4.23) divided by a similar factor which appears in the norm of the states in region I (which must be divided out because the incoming state is normalized to unity), and the ratio u_0/v_0 is the effect of the fact that the z dependence of the envelope functions is scaled by the velocities in the two regions (see Fig. 4.2). Note that $T < 0$, corresponding to the negative charge of the particle in region I, so that even though $R > 1$,

$$
1 = R + T
\tag{4.34}
$$

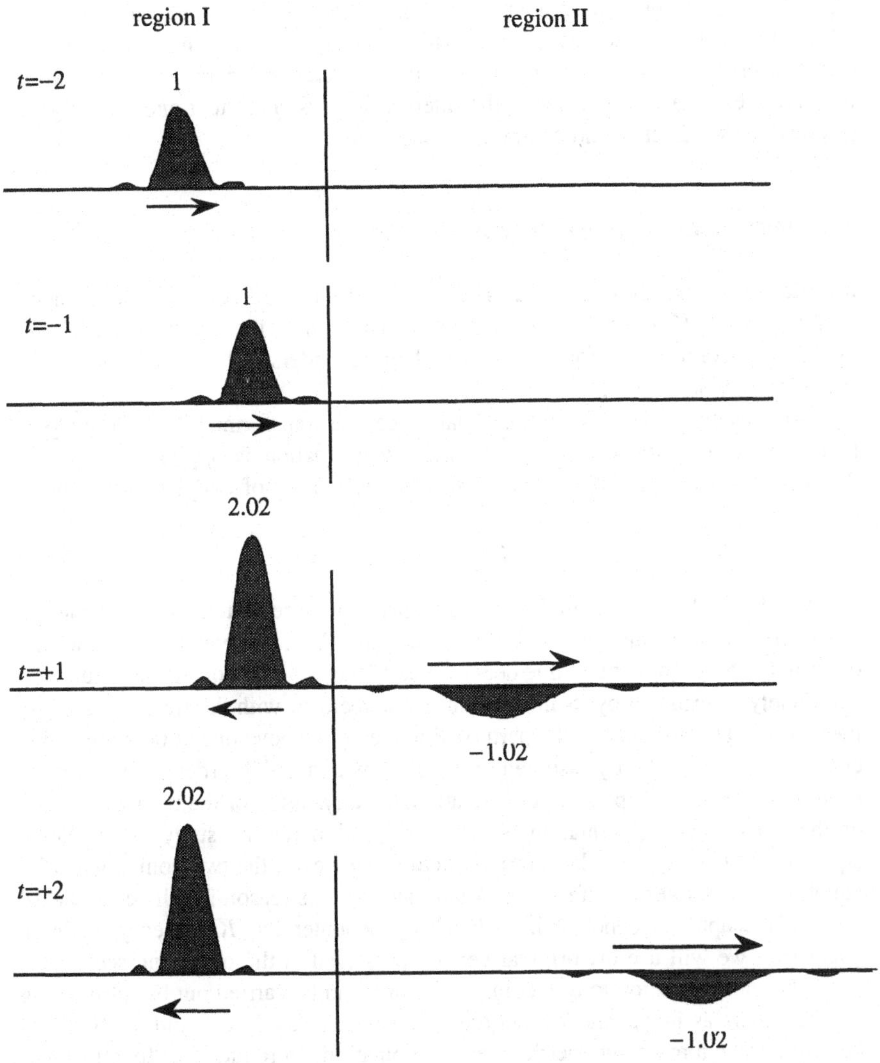

Fig. 4.2 A moving picture of the solution (4.31) for four times around t=0. For this picture we took $E = am$ and $eV - E = bm$, where $a=\frac{13}{12}$ and $b=\frac{13}{5}$. The location and size of the wave packets are to scale for these parameters, and the packets include the energy factor which enters the normalization, as in Eq. (4.23). The normalization of the packets is written above each packet. Note that for $t > 0$, there are 2.02 units of positive charge moving to the left in region I and -1.02 units of negative charge moving to the right in region II, which add up to the original 1 unit of charge incident on the barrier. The incident packet has reflected and produced + and − charged pairs.

as required by the conservation of the norm (the charge).

We interpret this result by saying that the incident pion not only scatters from the barrier but also stimulates the barrier to produce $\pi^+\pi^-$ pairs, and the π^- particles travel into the barrier, which is negative to them, while the π^+ particles produced by the scattering join the scattered π^+ particles. If the norm is interpreted as a charge density, the interpretation is consistent and makes good physical sense. Energy and charge are conserved.

4.5 TWO-COMPONENT FORM *

In order to display these two states explicitly, and to further develop our understanding of the KG equation, we will now discuss how the equation can be cast into a "two-component" form. This will help to understand the equation and to study its nonrelativistic limit.

Any second order differential equation can be transformed into two coupled first order differential equations. If this transformation is applied to the time dependence of the KG equation, we emerge with a set of coupled equations of the form

$$i\frac{\partial \phi}{\partial t} = H\phi \ , \tag{4.35}$$

where ϕ is a vector in a complex two-dimensional space and H is a 2×2 matrix. We gain several advantages from this reduction. First, the equation is now first order in the time, so that the time dependence of the two-component wave function is uniquely determined by its initial value, in agreement with the rules of quantum mechanics. This means that the perturbation theory we developed in the preceding chapters, which implicitly assumed an equation which is first order in time, can be used with the two-component KG equation. Finally, study of the matrix structure of the two-component equation is good preparation for the study of the Dirac equation, which has a similar matrix structure. However, the two-component KG equation is no longer manifestly covariant, and for this reason it will be discarded after this chapter is concluded. When we encounter the KG theory again in Chapter 7, we will use the original version presented in the preceding sections.

The transformation to two-component form can be carried out by introducing ψ and $\partial\psi/\partial t$ as independent functions. However, instead of ψ and $\partial\psi/\partial t$, it is helpful to take a more symmetric linear combination. Introduce two functions ϕ_+ and ϕ_- determined from ψ and $\partial\psi/\partial t$ by

$$\begin{aligned} \phi_+ &= \frac{1}{\sqrt{2m}}\left(i\frac{\partial}{\partial t} - V^0 + m\right)\psi \\ \phi_- &= \frac{1}{\sqrt{2m}}\left(-i\frac{\partial}{\partial t} + V^0 + m\right)\psi \ , \end{aligned} \tag{4.36}$$

*Much of the material in this section is drawn from the interesting review by Feshbach and Villars [FV 58].

where the $\pm$ *subscripts* should not be confused with the $(\pm)$ *superscripts*; they have a completely different meaning. We point out that the choice (4.36) is not unique; it was chosen because it is simple and gives equations with some features suggestive of the Dirac equation. Another choice, interesting because it diagonalizes the Hamiltonian in the absence of interactions, is presented in Prob. 4.4.

The functions (4.36) will be defined to be the upper and lower components of a new wave function, ϕ, and will be organized into a two-component column vector

$$\phi = \begin{pmatrix} \phi_+ \\ \phi_- \end{pmatrix} . \tag{4.37}$$

This vector satisfies the first order differential equation (4.35) with

$$\begin{aligned} H &= \begin{pmatrix} m + V^0 + \dfrac{(\boldsymbol{p}-\boldsymbol{V})^2}{2m} & \dfrac{(\boldsymbol{p}-\boldsymbol{V})^2}{2m} \\[2ex] -\dfrac{(\boldsymbol{p}-\boldsymbol{V})^2}{2m} & -m + V^0 - \dfrac{(\boldsymbol{p}-\boldsymbol{V})^2}{2m} \end{pmatrix} \\ &= \left[m + \frac{(\boldsymbol{p}-\boldsymbol{V})^2}{2m}\right]\tau_3 + V^0 + \frac{(\boldsymbol{p}-\boldsymbol{V})^2}{2m}\, i\tau_2 \ , \end{aligned} \tag{4.38}$$

where τ_i are the Pauli matrices (given explicitly in Appendix A).

Proof: From the definition (4.36) it follows that

$$\begin{aligned} \psi &= \frac{1}{\sqrt{2m}}\,(\phi_+ + \phi_-) \\ i\frac{\partial\psi}{\partial t} &= \sqrt{\frac{m}{2}}\,(\phi_+ - \phi_-) + \frac{V^0}{\sqrt{2m}}\,(\phi_+ + \phi_-) \ . \end{aligned} \tag{4.39}$$

Differentiating (4.36) and using the KG equation and (4.39) give

$$\begin{aligned} i\frac{\partial}{\partial t}\phi_\pm &= \frac{1}{\sqrt{2m}}\left(\mp\frac{\partial^2}{\partial t^2} \mp i\left(\frac{\partial V^0}{\partial t}\right) + (m \mp V^0)\, i\frac{\partial}{\partial t}\right)\psi \\ &= \frac{1}{\sqrt{2m}}\left(\mp\left[-(\boldsymbol{p}-\boldsymbol{V})^2 - i\left(\frac{\partial V^0}{\partial t}\right) - 2iV^0\frac{\partial}{\partial t} + \left(V^0\right)^2 - m^2\right]\psi \right. \\ &\qquad \left. \mp i\frac{\partial V^0}{\partial t}\psi + (m \mp V^0)\left[\sqrt{\frac{m}{2}}(\phi_+ - \phi_-) + V^0\psi\right]\right) \\ &= \left[\frac{1}{2}\left(V^0 \pm m\right) \pm \frac{(\boldsymbol{p}-\boldsymbol{V})^2}{2m}\right](\phi_+ + \phi_-) + \frac{1}{2}(m \mp V^0)(\phi_+ - \phi_-) \ . \end{aligned}$$

Writing this in matrix form gives (4.35) and (4.38). ∎

Substituting (4.39) into (4.14) shows that the two-component form of the conserved norm is

$$\int_{L^3} d^3r\, \phi^\dagger \tau_3 \phi = \int_{L^3} d^3r \left(|\phi_+|^2 - |\phi_-|^2 \right) = \text{constant} \ . \tag{4.40}$$

It is instructive to prove that this is conserved directly from the matrix form of the equation, (4.35). To do this, note first that H is not Hermitian but that

$$\tau_3 H^\dagger \tau_3 = H \ . \tag{4.41}$$

The matrix τ_3 thus plays the role of a "metric tensor," and

$$\begin{aligned} i\frac{d}{dt}\int d^3r\, \phi^\dagger \tau_3 \phi &= i\int d^3r \left\{ \frac{\partial}{\partial t}\phi^\dagger \tau_3 \phi + \phi^\dagger \tau_3 \frac{\partial \phi}{\partial t} \right\} \\ &= -\int d^3r \left\{ \phi^\dagger \left[H^\dagger \tau_3 - \tau_3 H \right] \phi \right\} = 0 \ . \end{aligned} \tag{4.42}$$

We may simplify the notation somewhat by introducing the *KG adjoint*, defined as follows:

$$\overline{\phi} = \phi^\dagger \, \tau_3 \ . \tag{4.43}$$

The conserved norm is then written

$$\int_{L^3} d^3r\, \overline{\phi}\, \phi = \text{constant} \ . \tag{4.44}$$

The solutions of the free KG equation can be found directly from the two-component form of the equation,

$$i\frac{\partial}{\partial t}\phi = H_0 \phi = \begin{pmatrix} m + \dfrac{\boldsymbol{p}^2}{2m} & \dfrac{\boldsymbol{p}^2}{2m} \\ -\dfrac{\boldsymbol{p}^2}{2m} & -\left(m + \dfrac{\boldsymbol{p}^2}{2m} \right) \end{pmatrix} \phi \ . \tag{4.45}$$

If $\phi_{\boldsymbol{k}}^{(\pm)} = \begin{pmatrix} \chi \\ \eta \end{pmatrix} e^{i(\boldsymbol{k}\cdot\boldsymbol{r} \mp Et)}$, then

$$\begin{pmatrix} \pm E - m - \dfrac{\boldsymbol{k}^2}{2m} & -\dfrac{\boldsymbol{k}^2}{2m} \\ \dfrac{\boldsymbol{k}^2}{2m} & \pm E + m + \dfrac{\boldsymbol{k}^2}{2m} \end{pmatrix} \begin{pmatrix} \chi \\ \eta \end{pmatrix} = 0 \ . \tag{4.46}$$

It is easy to see that a solution requires $E^2 = m^2 + \boldsymbol{k}^2$ and that

$$\begin{aligned} \phi_k^{(+)} &= N_+ \begin{pmatrix} 1 \\ \\ \frac{-\boldsymbol{k}^2}{(E+m)^2} \end{pmatrix} e^{i(\boldsymbol{k}\cdot\boldsymbol{r}-Et)} \\ \phi_{-k}^{(-)} &= N_- \begin{pmatrix} \frac{-\boldsymbol{k}^2}{(E+m)^2} \\ \\ 1 \end{pmatrix} e^{-i(\boldsymbol{k}\cdot\boldsymbol{r}-Et)} \quad . \end{aligned} \tag{4.47}$$

Note that $\phi^{(-)}$ has negative norm. If we require

$$\int_{L^3} d^3r\, \overline{\phi}_k^{(+)}\, \phi_k^{(+)} - 1 \qquad \int_{L^3} d^3r\, \overline{\phi}_{-k}^{(-)}\, \phi_{-k}^{(-)} = -1 \ , \tag{4.48}$$

then

$$N_+ = N_- = N = \frac{1}{\sqrt{2EL^3}}\ \frac{E+m}{\sqrt{2m}} \quad . \tag{4.49}$$

We will see later that these solutions bear a striking resemblance to those of the free Dirac equation.

Charge Conjugation

The KG and other manifestly covariant equations (such as the Dirac equation) have a symmetry related to the existence of both positive and negative energy solutions. These solutions can be transformed into each other by an operation referred to as *charge conjugation.* Under this transformation, the interaction term changes sign, but the equation otherwise remains unchanged (in particular, the mass remains the same). This transformation provides a good way to interpret the meaning of the negative energy solutions, $\phi^{(-)}$.

In the two-component theory, the operation of charge conjugation is defined by

$$\phi \Rightarrow \phi^C = \tau_1 \phi^* \quad . \tag{4.50}$$

Now note that ϕ^C satisfies the same equation as ϕ, except with $V^\mu \to -V^\mu$. To see this, remember that $\boldsymbol{p} \to -i\nabla$, so that $\boldsymbol{p}^* = -\boldsymbol{p}$. Then

$$-i\frac{\partial}{\partial t}\phi^* = \left\{ \left[m + \frac{(\boldsymbol{p}+\boldsymbol{V})^2}{2m} \right] \tau_3 + V^0 + \frac{(\boldsymbol{p}+\boldsymbol{V})^2}{2m} i\tau_2 \right\} \phi^* \quad . \tag{4.51}$$

Hence, multiplying both sides of this equation by $-\tau_1$ and using the fact that τ_1 anticommutes with τ_2 and τ_3 give

$$i\frac{\partial}{\partial t}\phi^C = \left\{ \left[m + \frac{(\boldsymbol{p}+\boldsymbol{V})^2}{2m} \right] \tau_3 - V^0 + \frac{(\boldsymbol{p}+\boldsymbol{V})^2}{2m} i\tau_2 \right\} \phi^C \ , \tag{4.52}$$

which completes the proof. Hence, for electromagnetism, ϕ^C describes a particle of opposite charge. Furthermore, note that $\phi^{(-)\,C}$ has positive norm

$$0 < -\int d^3r \left\{ \overline{\phi}^{(-)}\, \phi^{(-)} \right\} = -\int d^3r \left\{ \phi^{(-)\,C\,\mathsf{T}} \tau_1 \tau_3 \tau_1 \phi^{(-)\,C\,*} \right\}$$
$$= \int d^3r \left\{ \overline{\phi}^{(-)\,C}\, \phi^{(-)\,C} \right\} \quad . \tag{4.53}$$

Also, for the free particle states,

$$\phi^{(-)\,C}_{-k} = \tau_1 N \begin{pmatrix} \frac{-k^2}{(E+m)^2} \\ 1 \end{pmatrix} e^{\imath(\boldsymbol{k}\cdot\boldsymbol{r}-Et)} = N \begin{pmatrix} 1 \\ \frac{-k^2}{(E+m)^2} \end{pmatrix} e^{\imath(\boldsymbol{k}\cdot\boldsymbol{r}-Et)}$$
$$= \phi^{(+)}_{\boldsymbol{k}} \quad . \tag{4.54}$$

The charge conjugation operation turns a *negative energy state of momentum* $-\boldsymbol{k}$ *into a positive energy state of the opposite charge and momentum* $\boldsymbol{k}$. This is the origin of the idea that a *negative energy state traveling backward in time is equivalent to an antiparticle state traveling forward in time.* This idea will be developed in considerable detail later in this chapter (see Sec. 4.8).

4.6 NONRELATIVISTIC LIMIT

To gain further insight into the structure of the KG equation, we study its nonrelativistic limit. This is the limit when the mass of the particle is much larger than all momenta or energies and the positive energy solutions have an energy near m. Since the rest mass is not normally included in the nonrelativistic energy, we introduce a difference energy $T = E - m$. Assuming a solution of the form

$$\phi(r,t) = \begin{pmatrix} \chi \\ \eta \end{pmatrix} e^{-\imath Et}$$

and using H given in Eq. (4.38), the coupled equations reduce to

$$\begin{aligned} T\chi &= \left(\frac{(\boldsymbol{p}-\boldsymbol{V})^2}{2m} + V^0 \right) \chi + \frac{(\boldsymbol{p}-\boldsymbol{V})^2}{2m} \eta \\ (2m+T)\eta &= -\frac{(\boldsymbol{p}-\boldsymbol{V})^2}{2m} \chi - \left(\frac{(\boldsymbol{p}-\boldsymbol{V})^2}{2m} - V^0 \right) \eta \quad , \end{aligned} \tag{4.55}$$

where the term $m\tau_3$ in H has been moved to the LHS of the equation, where it cancels a similar term in the first equation but doubles the similar term in the second equation. As $m \to \infty$, the dimensionless quantities $|\boldsymbol{p}|/m$, $|V^0|/m$,

$|\boldsymbol{V}|/m$, and $|T|/m$ are all $\ll 1$, and therefore $\eta \ll \chi$. Expanding the second equation in inverse powers of m and discarding terms of order m^{-3} or higher give

$$\eta \simeq -\frac{(\boldsymbol{p}-\boldsymbol{V})^2}{4m^2}\chi + \mathcal{O}\left(\frac{1}{m^3}\right) \quad . \tag{4.56}$$

Substituting this result into the first equation gives an equation for χ *accurate to order* $1/m^3$,

$$\boxed{T\chi = \left\{\frac{1}{2m}(\boldsymbol{p}-\boldsymbol{V})^2 + V^0 - \frac{1}{8m^3}(\boldsymbol{p}-\boldsymbol{V})^4\right\}\chi \quad .} \tag{4.57}$$

If $\boldsymbol{V} = 0$, the relativistic corrections to the energy up to order m^{-3} are

$$\Delta H_{\rm Rel} = -\frac{\nabla^4}{8m^3} \quad . \tag{4.58}$$

This is the *only* term which can give *fine structure* contributions for a spin zero particle. (See Prob. 4.2 at the end of the chapter.)

Zeeman Effect

The Zeeman effect is the splitting of energy levels which occurs when a bound state (atom) is placed in a weak, magnetic field. In this case the field, $\boldsymbol{B}$, can be assumed to be uniform over the size of the atom, in which case the corresponding vector potential is simply

$$\boldsymbol{A} = -\tfrac{1}{2}\left(\boldsymbol{r}\times\boldsymbol{B}\right) \quad . \tag{4.59}$$

Note that the definition of $\boldsymbol{A}$ is consistent with the Coulomb gauge, and with the identification of $\boldsymbol{B}$ as a constant magnetic field

$$\begin{aligned}(\nabla\times\boldsymbol{A})^i &= -\tfrac{1}{2}\epsilon_{ijk}\nabla_j\epsilon_{k\ell m}r^\ell B^m \\ &= -\tfrac{1}{2}\epsilon_{ijk}\epsilon_{kjm}B^m = B^i \quad .\end{aligned}$$

For a positive charge e, the magnetic interaction term then becomes

$$\begin{aligned}-\frac{1}{2m}\left(\boldsymbol{p}\cdot\boldsymbol{V}+\boldsymbol{V}\cdot\boldsymbol{p}\right) &= \frac{ie}{2m}\left(\nabla\cdot\boldsymbol{A}+\boldsymbol{A}\cdot\nabla\right) = \frac{ie}{m}\boldsymbol{A}\cdot\nabla \\ &= -\frac{ie}{2m}(\boldsymbol{r}\times\boldsymbol{B})\cdot\nabla = \frac{ie}{2m}\boldsymbol{B}\cdot(\boldsymbol{r}\times\nabla) = -\frac{e}{2m}\boldsymbol{B}\cdot\boldsymbol{L} \ ,\end{aligned} \tag{4.60}$$

where L is the familiar angular momentum operator. This interaction gives the "normal" Zeeman effect only (see Prob. 4.3).

Discussion

The KG equation describes the behavior of a spin zero particle and hence would be the correct equation to use for an approximate description of pionic atoms (atomic states with a π^- substituted for an electron). Unfortunately, the pion is very short lived, and these "atomic" states have a very short lifetime. In addition, because the pion is so much more massive than the electron, it is bound in a very small orbit. The orbit is so small that there is a significant probability that the pion will overlap with the nucleus, where it will interact strongly, further broadening the states. These effects make it difficult to study pionic atoms, and direct tests of the applicability of the KG equation to such states is a topic of current research.

In any case, the study of the structure of (perhaps hypothetical) atomic states with spin zero constituents is an interesting intellectual question. Comparing results obtained from the KG equation with those we will obtain later from the Dirac equation will tell us how much of the observed fine structure is due to relativity alone and how much is due to the spin of the electron. Similarly, comparison of the Zeeman effect predicted by each equation helps us separate effects due to the orbital motion of the bound particle (all that we have in the KG theory) from additional effects present in the Dirac theory.

4.7 COULOMB SCATTERING

As an illustration of the usefulness of the two-component theory, we calculate the scattering of a charged spin zero particle from a fixed Coulomb potential

$$V^0 = Z\frac{\alpha}{r} \tag{4.61}$$

which comes from a point charge Ze fixed at the origin.

Because of the fact that the two-component KG theory satisfies a first order differential equation, we may use the formalism for time-dependent perturbation theory developed in Sec. 3.1. The first order S-matrix element is [compare with Eq. (3.53)]

$$S_{fi} = -i\int dt\, \langle f\,|H_I(t)|\,i\rangle \quad , \tag{4.62}$$

where $|i\rangle$ and $|f\rangle$ are initial and final KG free particle states with momenta k_i and k_f. The interaction Hamiltonian in this equation, H_I, must be expressed in the interaction representation, just as we did in Sec. 3.1 in our study of electromagnetism [recall Eqs. (3.4) and (3.5)]. For a pure Coulomb interaction, $\boldsymbol{V} = e\boldsymbol{A} = 0$, and the Schrödinger representation of H_I can be deduced from Eq. (4.38). In the interaction picture, it becomes

$$\boxed{H_I = U_0^{-1}\,\frac{Z\alpha}{r}\begin{pmatrix}1 & \\ & 1\end{pmatrix} U_0} \tag{4.63}$$

with

$$U_0 = e^{-\imath H_0 t} \quad . \tag{4.64}$$

The unperturbed or free particle Hamiltonian, H_0, was introduced in Eq. (4.45). As in Chapter 3, the sole effect of the operators U_0 is to give time-dependent phases when they operate on the initial and final state wave functions:

$$U_0\, \phi^{(+)}_{k_i}(r) = e^{-\imath E_i t}\, \phi^{(+)}_{k_i}(r) \quad .$$

The matrix elements must be put together using the correct scalar product. The matrix element of an operator O, $\langle f|O|i\rangle$, is constructed by inserting the "metric tensor" τ_3 between the final state and $O\,|i\rangle$ or, alternatively, forming the scalar product by multiplying from the left by the adjoint state defined in Eq. (4.43). For the scattering of positive energy states, this gives

$$\begin{aligned}
S_{fi} &= -i\int dt\, d^3r\, \overline{\phi}^{(+)}_{k_f}(r)\, H_I(r,t)\, \phi^{(+)}_{k_i}(r) \\
&= -i\int dt\, d^3r\, \left[\phi^{(+)}_{k_f}(r)\right]^\dagger \underbrace{\tau_3 e^{\imath H_0 t}}_{e^{\imath H_0^\dagger t}\,\tau_3} \frac{Z\alpha}{r}\begin{pmatrix}1 & \\ & 1\end{pmatrix} e^{-\imath H_0 t}\phi^{(+)}_{k_i}(r) \\
&= -i\int dt\, e^{\imath(E_f-E_i)t}\int \frac{d^3r}{L^3}\, e^{\imath(k_i-k_f)\cdot r}\,\frac{Z\alpha}{r}\,\frac{1}{\sqrt{4E_fE_i}} \\
&\qquad \times \underbrace{\frac{(E_f+m)(E_i+m)}{2m}\left(1 \quad \frac{-k_f^2}{(E_f+m)^2}\right)\begin{pmatrix}1 & \\ & -1\end{pmatrix}\begin{pmatrix}1 \\ \frac{-k_i^2}{(E_i+m)^2}\end{pmatrix}}_{=E_f+E_i} \\
&= -i\int dt\, e^{\imath(E_f-E_i)t}\int \frac{d^3r}{L^3} e^{i(k_i-k_f)\cdot r}\,\frac{Z\alpha}{r}\underbrace{\frac{E_f+E_i}{\sqrt{4E_fE_i}}}_{\text{relativistic factors}} \quad .
\end{aligned} \tag{4.65}$$

The time integral gives an energy conserving delta function, and hence $E_f = E_i$ and the new relativistic factors reduce to unity, showing that the S-matrix is identical to the nonrelativistic result. However, the cross section will include relativistic effects which arise from the flux factor.

The cross section becomes [recall Eq. (3.52)]

$$d\sigma_{fi} = \frac{E_i}{|k_i|}\int_{f\in\Delta\Omega} \frac{d^3k_f}{(2\pi)^3}\, 2\pi\,\delta(E_f - E_i)\,|f_{fi}|^2 \quad , \tag{4.66}$$

where f_{fi} is the reduced matrix element, which for this example becomes ($\boldsymbol{q} \equiv \boldsymbol{k}_i - \boldsymbol{k}_f$, and $|\boldsymbol{q}| = q$)

$$\begin{aligned} f_{fi} &= \int d^3r\, e^{i\boldsymbol{q}\cdot\boldsymbol{r}}\, \frac{Z\alpha}{r} e^{-\epsilon r} \\ &= Z\alpha\, 2\pi \int_0^\infty \frac{r^2 dr}{r} \frac{1}{iqr} \left\{ e^{iqr-\epsilon r} - e^{-iqr-\epsilon r} \right\} \\ &= Z\alpha\, 2\pi \frac{1}{iq} \left[-\frac{1}{iq-\epsilon} + \frac{1}{-iq-\epsilon} \right] \\ &= Z\alpha\, 4\pi \frac{1}{q^2+\epsilon^2} = Z\alpha\, 4\pi \frac{1}{q^2} \qquad (\epsilon \to 0) \; . \end{aligned} \tag{4.67}$$

[Note the use of the screening factor $e^{-\epsilon r}$, inserted to insure convergence of the integrals and removed after they have been done by letting $\epsilon \longrightarrow 0$.] Now substitute (4.67) into (4.66), and assume $\Delta\Omega$ is small enough so that all dependence of the integrand on the directions of $\boldsymbol{k}_f$ can be ignored. This gives

$$\begin{aligned} d\sigma_{fi} &= d\Omega \frac{E_i}{k_i} \int \frac{k_f^2 dk_f}{(2\pi)^3} 2\pi\, \delta(E_f - E_i) \frac{(Z\alpha 4\pi)^2}{q^4} \\ \frac{d\sigma}{d\Omega} &= \frac{4(Z\alpha)^2 E^2}{q^4} \; . \end{aligned} \tag{4.68}$$

This is the Coulomb differential scattering cross section for a spin zero particle scattering from a fixed scattering center. Because there is no recoil, the behavior of the cross section is dominated by the familiar q^{-4} factor, where

$$q^2 = (\boldsymbol{k}_f - \boldsymbol{k}_i)^2 = 2k^2 (1 - \cos\theta) = 4k^2 \sin^2 \frac{\theta}{2} \; . \tag{4.69}$$

The scattering is sharply peaked in the forward ($\theta = 0$) direction.

4.8 NEGATIVE ENERGY STATES

The simple example we considered in Sec. 4.4 was sufficient to show that

- negative energy states cannot be ignored and
- they describe the production of particle–antiparticle pairs, which can occur virtually in higher order processes.

The (one-particle) *KG* equation can only do a limited job of describing pair production; a complete description of antiparticles must await the development of field theory (Chapter 7). In this section we lay the background for this study by developing the mathematical description of both positive and negative energy states, to the extent possible without the use of field theory.

To illustrate the techniques, we will calculate the matrix element for Coulomb scattering to *second order* in the electric charge e. From Eq. (3.24) for the time translation operator, the second order S-matrix element is

$$S_{fi}^{(2)} = \frac{(-i)^2}{2} \int_{-\infty}^{\infty} \int_{-\infty}^{\infty} dt_1\, dt_2\, \langle f\,|T\,(H_I(t_2)H_I(t_1))|\, i\rangle$$
$$= - \int_{-\infty}^{\infty} \int_{-\infty}^{\infty} dt_1\, dt_2\, \theta(t_2 - t_1)\, \langle f\,|H_I(t_2)H_I(t_1)|\, i\rangle \quad . \qquad (4.70)$$

The superscript (2) is to remind us that this is the contribution to the infinite sum (3.24) which is second order in the small electric charge e. While this formula was originally obtained for a field theory, it applies equally well, as noted in the previous section, to any quantum mechanical system described by an equation first order in time and which has been separated into an unperturbed Hamiltonian H_0 and an interaction Hamiltonian H_I (written in the interaction picture). We may apply it to the two-component form of the KG theory, which casts the KG equation into a differential equation first order in the time.

The way to evaluate $S^{(2)}$ is to insert a complete set of states between $H_I(t_2)$ and $H_I(t_1)$. Before we can do this, we must discuss the completeness relation for the KG states.

Completeness Relation

We are working in the interaction representation where the free states have been fixed in time (at $t = 0$ for convenience). The completeness relation for the KG states can be written

$$K_{KG}(r, r') = \sum_k \left\{ \phi_k^{(+)}(r)\overline{\phi}_k^{(+)}(r') - \phi_k^{(-)}(r)\overline{\phi}_k^{(-)}(r') \right\} = \mathbf{1}\, \delta^3(r - r') \quad , \qquad (4.71)$$

where $\mathbf{1}$ is a unit 2×2 matrix in the two-component space.

Proof: We can use the orthogonality relations to show that this has the correct properties. For any KG state $\phi(r')$

$$\int d^3r'\, K_{KG}(r, r')\phi(r') = \phi(r) \quad , \qquad (4.72)$$

where the minus sign in the second term compensates for the minus sign which comes from the norm of a negative energy state. However, it is instructive to prove (4.71) directly by construction. Substituting the solutions (4.47) directly into K_{KG}, and remembering that ϕ is a column vector and $\overline{\phi}$ is a row vector give (letting $E_k = E$)

$$K_{KG}(r,r') = \sum_k e^{i\mathbf{k}\cdot(\mathbf{r}-\mathbf{r}')}\,\frac{(E+m)^2}{4mEL^3}$$
$$\times\left\{\begin{bmatrix} 1 \\ \frac{-k^2}{(E+m)^2}\end{bmatrix}\begin{bmatrix}1 & \frac{k^2}{(E+m)^2}\end{bmatrix} - \begin{bmatrix}\frac{-k^2}{(E+m)^2} \\ 1\end{bmatrix}\begin{bmatrix}\frac{-k^2}{(E+m)^2} & -1\end{bmatrix}\right\}$$
$$= \sum_k e^{i\mathbf{k}\cdot(\mathbf{r}-\mathbf{r}')}\,\frac{(E+m)^2}{4mEL^3}$$
$$\times\left\{\begin{bmatrix} 1 & \frac{k^2}{(E+m)^2} \\ \frac{-k^2}{(E+m)^2} & \frac{-k^4}{(E+m)^4}\end{bmatrix} - \begin{bmatrix}\frac{k^4}{(E+m)^4} & \frac{k^2}{(E+m)^2} \\ \frac{-k^2}{(E+m)^2} & -1\end{bmatrix}\right\}$$
$$= \frac{1}{L^3}\sum_k e^{i\mathbf{k}\cdot(\mathbf{r}-\mathbf{r}')}\begin{bmatrix}1 & \\ & 1\end{bmatrix} = \mathbf{1}\,\delta^3(r-r') \ . \qquad \blacksquare$$

Returning to the second order S-matrix and inserting a complete set of states using (4.71) give the following expression for scattering from an initial positive energy state with momentum k_i to a final positive energy state with momentum k_f:

$$S_{fi}^{(2)} = -\int\!\!\int\!\!\int\!\!\int d^3r_1\, d^3r_2\, dt_1\, dt_2\, \theta(t_2-t_1)$$
$$\times\, \overline{\phi}_{k_f}^{(+)}(r_2) H_I(r_2,t_2)\, K_{KG}(r_2,r_1)\, H_I(r_1,t_1)\phi_{k_i}^{(+)}(r_1)\,, \tag{4.73}$$

where the integrals over the spatial coordinates have been written explicitly. Introducing the KG Coulomb matrix elements

$$\langle f\,|H_I(t_1)|\,k\pm\rangle = \int d^3r_1\,\overline{\phi}_{k_f}^{(+)}(r_1)\left(\frac{Z\alpha}{r_1}\right)\phi_k^{(\pm)}(r_1)\, e^{i(E_f\mp E_k)t_1}$$
$$= \frac{1}{L^3} f_{k_f k}^{(\pm)}\, e^{i(E_f\mp E_k)t_1}\ , \tag{4.74}$$

the S-matrix can be reduced to

$$S_{fi}^{(2)} = -\frac{1}{L^6}\sum_k \int\!\!\int dt_1\, dt_2\, \theta(t_2-t_1)$$
$$\times\left\{ f_{k_f k}^{(+)} f_{k_i k}^{(+)*}\, e^{i(E_f-E_k)t_2+i(E_k-E_i)t_1} - f_{k_f k}^{(-)} f_{k_i k}^{(-)*}\, e^{i(E_f+E_k)t_2-i(E_k+E_i)t_1}\right\} \tag{4.75}$$

Next, introduce $T = \frac{1}{2}(t_1 + t_2)$ and $t = (t_2 - t_1)$ to get

$$S_{fi}^{(2)} = -\frac{1}{L^6}\sum_k \int dT\, e^{i(E_f - E_i)T} \int dt\, \theta(t)$$
$$\times \left\{ f_{k_f k}^{(+)} f_{k_i k}^{(+)*}\, e^{i\left[\frac{E_f+E_i}{2} - E_k\right]t} - f_{k_f k}^{(-)} f_{k_i k}^{(-)*}\, e^{i\left[\frac{E_f+E_i}{2} + E_k\right]t} \right\} .$$

The integral over T gives an energy conserving δ-function, and the S-matrix reduces to the standard form

$$S_{fi}^{(2)} = -i2\pi\, \delta\,(E_f - E_i)\, \frac{1}{L^3} f_{fi}^{(2)} \ ,$$

with the reduced amplitude f given by

$$f_{fi}^{(2)} = -i \int dt\, \theta(t) \sum_k \frac{1}{L^3} \left\{ f_{k_f k}^{(+)} f_{k_i k}^{(+)*}\, e^{-i[E_k - E_i]t} - f_{k_f k}^{(-)} f_{k_i k}^{(-)*}\, e^{i[E_k + E_i]t} \right\} . \tag{4.76a}$$

Note that the *negative energy states make a contribution* to this sum, unless $f_{k_i k}^{(-)}$ or $f_{k_f k}^{(-)} = 0$, which is *not* generally the case.*

This confirms our conclusions from Sec. 4.4; the negative energy states cannot be ignored. Even if the initial and final states are restricted to positive energy, the full solution to any problem will usually include virtual contributions from negative energy intermediate states.

The next task is to give a physical interpretation to such contributions. At this point the single particle KG equation does not give a unique answer. First, observe that

$$0 = -i \int dt \sum_k \frac{1}{L^3} f_{k_f k}^{(-)} f_{k_i k}^{(-)*}\, e^{i[E_k + E_i]t}$$

because the integral over t gives $\delta(E_k + E_i)$, which is always zero. Adding this term to Eq. (4.76a) and noting that $\theta(t) + \theta(-t) = 1$ give an alternative equation for the reduced amplitude

$$f_{fi}^{(2)} = -i \int dt \sum_k \frac{1}{L^3} \left\{ f_{k_f k}^{(+)} f_{k_i k}^{(+)*}\, \theta(t)\, e^{-i[E_k - E_i]\,t} \right.$$
$$\left. + f_{k_f k}^{(-)} f_{k_i k}^{(-)*}\, \theta(-t)\, e^{i[E_k + E_i]\,t} \right\} . \tag{4.76b}$$

This equation gives the same mathematical result for the reduced amplitude f but suggests a *very different* physical interpretation. Later, we will see that field

*Note that these $(-)$ matrix elements are zero if energy is conserved but are not zero in second order perturbation theory because the energy of the intermediate state is not the same as the energy of the initial (or final) state.

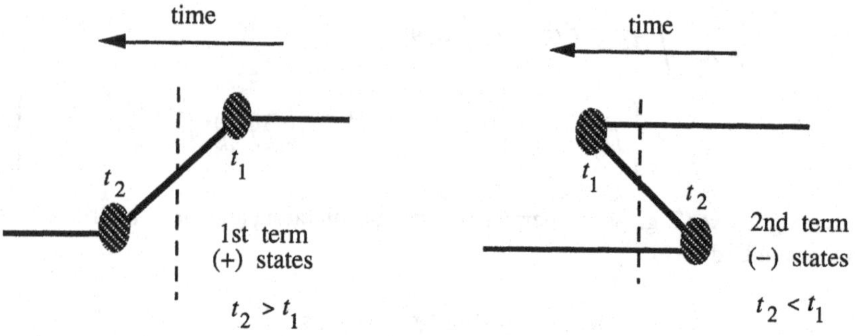

Fig. 4.3 The left diagram illustrates the *forward* propagation of a positive energy intermediate state, while the right is the *backward* propagation of a negative energy intermediate state. The right-hand diagram is *reinterpreted* as the creation of a pair at time t_2, *forward* propagation of the antiparticle to time t_1, followed by annihilation of the antiparticle at time t_1.

theory naturally gives us the interpretation suggested by (4.76b), and this is the only picture which makes sense physically.

In the first of these two descriptions, Eq. (4.76a), both the negative energy and positive energy states propagate forward in time. In the second, Eq. (4.76b), the negative energy states propagate backward in time [because of the $\theta(-t)$ function which implies $t_2 < t_1$]. The meaning of this strange statement is illustrated in Fig. 4.3. The second figure shows that the requirement $t_2 < t_1$ means that the line joining t_1 and t_2 travels backward in time *unless we turn the direction of motion around* and think of a particle–antiparticle pair being created at time t_2 and then annihilated at a later time t_1. Thus the idea that negative energy states propagate backward in time, while at first very strange, actually enables us to reinterpret them as *antiparticle states propagating forward in time.* If the antiparticle states are the charge conjugates of the negative energy states, so that they carry opposite charge, opposite momentum, and have positive energy, then charge is conserved in both descriptions. Reinterpreting the virtual negative energy contributions as virtual antiparticle contributions shows how these contributions describe virtual pair production. This is consistent with the results we obtained in Sec. 4.4.

In order to reduce the amplitude further, we prove an important identity which will be used several times throughout this book:

$$\boxed{\theta(t)\; E^n e^{-iEt} = \frac{1}{2\pi i}\int_{-\infty}^{\infty} d\omega\; \frac{\omega^n\, e^{-i\omega t}}{E-\omega-i\epsilon}\,.} \tag{4.77}$$

In this identity the limit $\epsilon \to 0$ is implied.

Proof: Look at the complex ω plane. The integrand has only one pole at $\omega = E - i\epsilon$ in the lower half plane. If $t > 0$, the contour must be closed in the lower half plane, while if $t < 0$, it must be closed in the upper half plane, in order that, in either case, the exponential has a negative real part and the contribution from the arc at ∞ converges (to zero). Therefore, the integral is e^{-iEt} if $t > 0$ and 0 if $t < 0$. This agrees with the LHS of the identity. ∎

Using this identity (with $n = 0$) for the first term in (4.76b), and using it with $t \to -t$ and $\omega \to -\omega$ in the second term, gives the following reduction of (4.76b):

$$f_{fi}^{(2)} = -\frac{1}{2\pi}\int dt\, d\omega\, e^{-i\omega t}\sum_k \frac{1}{L^3}\left\{\frac{f^{(+)}_{k_f k} f^{(+)*}_{k_i k}}{E_k - E_i - \omega - i\epsilon} + \frac{f^{(-)}_{k_f k} f^{(-)*}_{k_i k}}{E_k + E_i + \omega - i\epsilon}\right\}$$

$$= -\sum_k \frac{1}{L^3}\left\{\frac{f^{(+)}_{k_f k} f^{(+)*}_{k_i k}}{E_k - E_i - i\epsilon} + \frac{f^{(-)}_{k_f k} f^{(-)*}_{k_i k}}{E_k + E_i - i\epsilon}\right\} . \qquad (4.78)$$

Discussion

The main results of this last section are:

- We will define the matrix elements so that *positive energy states propagate forward in time*, associated with $\theta(t_2 - t_1)$, and *negative energy states propagate backward in time*, associated with $\theta(t_1 - t_2)$. This is the *Feynman prescription*. There are two *time-ordered diagrams*, as shown in Fig. 4.3.

- By turning the negative energy line around and reinterpreting it as an antiparticle propagating forward in time, we see how pair production, a multiparticle process, is described by the one-particle KG equation.

For this interpretation to be consistent with the conventional rules of quantum mechanics, all incoming states with energy E must have the usual phase factor e^{-iEt} and outgoing states the complex conjugate phase e^{+iEt}. Using $E_f = E_i$, it is easy to demonstrate that this is indeed true for *both* of the terms in Eq. (4.76b):

$$\begin{aligned} &\text{1st term} \quad e^{-i(E_k - E_i)(t_2 - t_1)} = e^{i(E_f - E_k)t_2}\, e^{+i(E_k - E_i)t_1} \\ &\text{2nd term} \quad e^{i(E_k + E_i)(t_1 - t_2)} = e^{-i(E_i + E_k)t_1}\, e^{i(E_f + E_k)t_2} \ . \end{aligned}$$

Furthermore, the energy denominators given in Eq. (4.78) are consistent with the rules of second order perturbation theory for *positive energy* intermediate states (with one intermediate particle for the first term and three for the second, as required by Fig. 4.3, and with a small *negative* imaginary part assigned to the energy of the intermediate state in cases when the denominator might be zero, as discussed in Sec. 3.4):

$$\text{1st term} \qquad \frac{1}{E_k - E_\imath - i\epsilon}$$

$$\text{2nd term} \qquad \frac{1}{E_i + E_f + E_k - E_i - i\epsilon} = \frac{1}{E_\imath + E_k - i\epsilon} \,.$$

These same features will also arise in our study of the Dirac equation, which is the subject of the next chapter.

PROBLEMS

4.1 Solve the manifestly covariant form of the Klein–Gordon equation for the ground state of the hydrogen atom. Specifically, assume

$$V^0 = -\frac{\alpha}{r} \qquad \mathbf{V} = 0$$

and show that the ground state wave function can be written

$$\psi(r,t) = r^\epsilon \, e^{-\beta r} \, e^{-iEt} \;.$$

Find ϵ, β, and E. Then examine the nonrelativistic limit by projecting out the ϕ_+ and ϕ_- components defined in Eq. (4.36). Interpret your results and compare with the Schrödinger theory.

4.2 Calculate the fine structure splitting of the energy levels for a pion bound in an atom with charge Ze. Draw an energy level diagram showing all the levels up to $n = 3$. (You may use the nonrelativistic form of the Klein–Gordon equation and calculate the splitting in perturbation theory using suitably modified hydrogen atom wave functions.) What are the Bohr radii of these orbits and what is v/c? Estimate the probability that a pion in the S-state will be inside the nucleus.

4.3 Calculate the Zeeman splitting of the levels up to $n = 3$ for a pionic atom.

4.4 Suppose a pion is bound by a scalar potential of the form

$$U(x) = V(r) = -V_0 \, \delta^3(r) \;.$$

Solve the KG equation for the special case when the solution is static (i.e., independent of time). Discuss the significance of your result.

4.5 A pion of mass μ is bound by a scalar one-dimensional square well potential $V(x)$ defined to be:

region I	$R < x$	$V(x) = 0$
region II	$0 < x < R$	$V(x) = -\mu^2 V_0$
region III	$x < 0$	$V(x) = \infty$

(This could be a *very* rough model for a pion inside of a nucleus of radius R.)

(a) Solve the KG equation (4.7) in one space dimension for the positive energy ground state. (Take $U(x) = V(x)$.)

(b) Find the value of R such that the positive energy ground state has energy

$$E_0 = \mu\sqrt{1 - V_0/2} \ .$$

Estimate the size of the pion cloud.

(c) Find the positive and negative energy parts, as defined in Eq. (4.36), of the solution found in part (a). Discuss your result and explain how the negative energy part should be interpreted.

4.6 **New two-component form for the *KG* equation.** One of the features of the two-component form introduced in Eq. (4.36) is that it does not completely decouple positive and negative energy solutions, even if the potentials are zero. In particular, for the free particle solutions

$$\phi^{(+)} \neq \begin{pmatrix} \chi \\ 0 \end{pmatrix} \qquad \phi^{(-)} \neq \begin{pmatrix} 0 \\ \eta \end{pmatrix} .$$

For conceptual purposes, it might be convenient to further diagonalize H so that the non-diagonal terms come from interactions only. This can be done by defining new components:

$$\phi_+ = \frac{E_\nabla^{-1/2}}{\sqrt{2}} \left\{ i\frac{\partial}{\partial t} - V^0 + E_\nabla \right\} \psi$$
$$\phi_- = \frac{E_\nabla^{-1/2}}{\sqrt{2}} \left\{ -i\frac{\partial}{\partial t} + V^0 + E_\nabla \right\} \psi ,$$

where E_∇ is defined by the power series given in Eq. (4.3). Show that:

(a) The conserved norm is identical to (4.40) with τ_3 the "metric tensor."

(b) The equations assume the form (4.35) with the free Hamiltonian being

$$H_0 = \tau_3 \, E_\nabla \ .$$

This completely diagonalizes the $(\pm)$ states for free particles.

(c) The charge conjugation operation and nonrelativistic limits are as before.

(d) The "old" form can be transformed into the "new" form using the following transformation:

$$U = \frac{1}{\sqrt{4mE_\nabla}} \begin{pmatrix} E_\nabla + m & -(m - E_\nabla) \\ -(m - E_\nabla) & E_\nabla + m \end{pmatrix} .$$

In particular, show that this transformation preserves the norm by proving that

$$U^\dagger \tau_3 U = \tau_3 \ .$$

Also, using the explicit forms (4.47), show that

$$U\phi_k^{(+)} = \frac{1}{\sqrt{L^3}} \begin{pmatrix} 1 \\ 0 \end{pmatrix} e^{-ik\cdot x}$$

$$U\phi_{-k}^{(-)} = \frac{1}{\sqrt{L^3}} \begin{pmatrix} 0 \\ 1 \end{pmatrix} e^{+ik\cdot x} \ .$$

Hence U transforms $\phi^{(\pm)}$ into states with only an upper (or lower) component. Finally, show by direct computation that

$$\frac{U}{\sqrt{2m}} \begin{pmatrix} i\frac{\partial}{\partial t} - V^0 + m \\ \\ -i\frac{\partial}{\partial t} + V^0 + m \end{pmatrix} \psi = \frac{1}{\sqrt{2m}} \begin{pmatrix} i\frac{\partial}{\partial t} - V^0 + E_\nabla \\ \\ -i\frac{\partial}{\partial t} + V^0 + E_\nabla \end{pmatrix} \psi \ ,$$

which shows explicitly that U transforms the "old" two-component form into the "new" two-component form.

CHAPTER 5

THE DIRAC EQUATION

In this chapter we continue the discussion of relativistic equations for the first quantization of particles. The Klein–Gordon equation introduced in the last chapter describes spin zero particles. In this chapter we discuss the Dirac equation [Di 28], which describes particles with the two internal degrees of freedom characteristic of a spin $\frac{1}{2}$ particle. Since both electrons and quarks have spin $\frac{1}{2}$, the Dirac equation has many interesting applications, and some of these will be developed in the next chapter.

5.1 THE EQUATION

As discussed in Sec. 4.5, the two-component form of the KG equation could be written

$$i\frac{\partial}{\partial t}\psi = H\,\psi \quad . \tag{5.1}$$

While this equation is first order in the time derivative, the KG Hamiltonian (4.38) is second order in the space derivatives and hence does not treat space and time in an equivalent fashion. Furthermore, because the conserved norm for the KG theory was not positive definite, the two-component KG "Hamiltonian" is not Hermitian. Finally, the covariance of the KG equation is only manifest in its original, one-component form. It is natural to ask: "Is there a relativistic equation which is first order in time, treats space and time in a manifestly symmetric fashion, has a positive definite conserved norm (implying that H is Hermitian), and is manifestly covariant?" The investigation of this question leads directly to the Dirac equation.

To answer this question, we look for an equation which is first order in *both* space and time and which is Hermitian. The equation must have the form

$$i\frac{\partial}{\partial t}\psi = H\,\psi = (\boldsymbol{\alpha}\cdot\boldsymbol{p} + \beta\, m)\,\psi \quad , \tag{5.2}$$

where $\boldsymbol{\alpha}$ and β are Hermitian matrices and $\boldsymbol{p} = -i\nabla$. The relativistic energy momentum relation should emerge naturally, so we require

$$\begin{aligned} E^2\psi \Rightarrow -\frac{\partial^2}{\partial t^2}\psi &= i\frac{\partial}{\partial t}(\boldsymbol{\alpha}\cdot\boldsymbol{p} + \beta m)\psi \\ &= (\boldsymbol{\alpha}\cdot\boldsymbol{p} + \beta m)\, i\frac{\partial\psi}{\partial t} = (\boldsymbol{\alpha}\cdot\boldsymbol{p} + \beta m)^2\psi \\ &= \left(p^2 + m^2\right)\psi \ . \end{aligned} \tag{5.3}$$

Demanding that this relation hold for all ψ gives

$$\begin{aligned} \left(\alpha_i p^i + \beta m\right)^2 &= \beta^2 m^2 + (\alpha_i)^2 \left(p^i\right)^2 + \{\beta, \alpha_i\}\, m\, p^i \\ &\quad + \frac{1}{2}\{\alpha_i, \alpha_j\}_{i\neq j}\, p^i p^j \\ &= \sum_i \left(p^i\right)^2 + m^2 \ , \end{aligned} \tag{5.4}$$

where $\{A, B\} = AB + BA$ is the *anticommutator* of two operators A and B. This equation can hold only if

$$\begin{aligned} \beta^2 = (\alpha_i)^2 &= 1 \\ \{\beta, \alpha_i\} = \{\alpha_i, \alpha_j\} &= 0 \ . \end{aligned} \tag{5.5}$$

To construct such an equation therefore requires a vector space large enough to contain four *anticommuting, Hermitian* matrices.

It is easy to prove that such a space must have a *minimum* of four dimensions and that therefore the matrices α_i and β must be at *least* 4×4. The proof follows in four steps:

Lemma 1: The matrices β and α_i are traceless.
To prove this for the matrices α_i, note that the anticommutation relations imply

$$\beta\alpha_i\beta = -\alpha_i \ .$$

Making use of the fact that the trace of a product of matrices is unchanged by cyclic permutation of the matrices gives

$$\begin{aligned} \mathrm{tr}\,\{\beta\alpha_i\beta\} &= -\,\mathrm{tr}\,\alpha_i \\ &= \mathrm{tr}\,\{\alpha_i\beta\beta\} = \mathrm{tr}\,\alpha_i \ . \end{aligned}$$

Hence $\mathrm{tr}\,\alpha_i = 0$. A similar argument shows that $\mathrm{tr}\,\beta = 0$.

Lemma 2: The eigenvalues of α_i and β must be ± 1.
Since α_i and β are Hermitian, they can be diagonalized, and because $(\alpha_i)^2 = \beta^2 = 1$, their diagonal elements (eigenvalues) can only be ± 1.

Lemma 3: The dimension of the matrices must be even.

In diagonal form the diagonal elements of α_i and β can be only ± 1, and because $\mathrm{tr}\,(\alpha_i) = \mathrm{tr}\,\beta = 0$, all of these matrices must have the same number of $+1$'s as -1's. Hence the dimension can only be even.

Lemma 4: The number of dimensions must be greater than 2.
In general, in n dimensions there are n^2 independent Hermitian matrices; subtracting the identity there are $n^2 - 1$ *Hermitian traceless matrices.* Hence there are only three for $n = 2$ (which can be taken to be the Pauli matrices, σ_i), but for $n = 4$ there are fifteen, more than enough.

We will choose the following representation for the four Dirac matrices:

$$\beta = \begin{pmatrix} 1 & 0 \\ 0 & -1 \end{pmatrix} \qquad \alpha_i = \begin{pmatrix} 0 & \sigma_i \\ \sigma_i & 0 \end{pmatrix} \quad , \tag{5.6}$$

where the matrices are written in 2×2 block form and σ_i are the Pauli matrices. Hence the free particle Dirac equation becomes

$$i\frac{\partial \psi}{\partial t} = (-i\alpha_i \nabla_i + \beta m)\,\psi \ . \tag{5.7}$$

Alternatively, the equation may be written in terms of the γ^μ matrices, defined by

$$\gamma^\mu = (\beta, \beta\alpha_i)$$

$$\gamma^0 = \begin{pmatrix} 1 & 0 \\ 0 & -1 \end{pmatrix} \qquad \gamma^i = \begin{pmatrix} 0 & \sigma_i \\ -\sigma_i & 0 \end{pmatrix} \ . \tag{5.8}$$

Expressed in terms of the γ matrices, the anticommutation relations (5.5) become

$$\{\gamma^\mu, \gamma^\nu\} = 2g^{\mu\nu} \quad , \tag{5.9}$$

and multiplying the Dirac equation by β permits us to write it in the following form (recall $\nabla_i = \partial/\partial x^i$):

$$\left(i\gamma^\mu \frac{\partial}{\partial x^\mu} - m\right)\psi = 0 \qquad \begin{matrix}\text{covariant}\\ \text{Dirac equation.}\end{matrix} \tag{5.10}$$

Electromagnetic interactions may be added to the Dirac equation by using the minimal substitution $p^\mu \to p^\mu - eA^\mu$. This gives

$$\boxed{i\frac{\partial \psi}{\partial t} = H\psi = \left[\alpha_i\left(-i\nabla_i - eA^i\right) + eA^0 + \beta m\right]\psi} \tag{5.11}$$

or, in covariant form,

$$\boxed{\left[\gamma^\mu\left(i\frac{\partial}{\partial x^\mu} - eA_\mu\right) - m\right]\psi = 0 \ .} \tag{5.12}$$

We will use the non-covariant form of the Dirac equation in the next few sections and will return to the covariant form in Sec. 5.9 when we discuss the covariance of the equation.

5.2 CONSERVED NORM

The conserved norm is easily obtained from the equations. Note that, for any two solutions of the Dirac equation,

$$\begin{aligned} \psi_a^\dagger\, i\frac{\partial}{\partial t}\psi_b &= \psi_a^\dagger H\psi_b \\ \left(-i\frac{\partial}{\partial t}\psi_a^\dagger\right)\psi_b &= \left(H\psi_a\right)^\dagger\psi_b \ . \end{aligned} \tag{5.13}$$

Hence, if the electromagnetic interaction (or any other potential) is independent of energy, it will cancel when we subtract the above two equations, and subtracting the first from the second gives

$$i\frac{\partial}{\partial t}\left(\psi_a^\dagger\psi_b\right) = -i\psi_a^\dagger\,\alpha_i\left[\overrightarrow{\nabla}_i + \overleftarrow{\nabla}_i\right]\psi_b \ ,$$

where the arrow over the derivative tells us in which direction in acts, just as in Sec. 4.2. Hence the two terms on the right-hand side become a perfect divergence and

$$i\frac{\partial}{\partial t}\left(\psi_a^\dagger\psi_b\right) + i\nabla_i\left(\psi_a^\dagger\,\alpha_i\psi_b\right) = 0 \ .$$

As in the KG case, we have a four-current which is conserved. The conservation law can be written

$$i\frac{\partial}{\partial x^\mu}\left[\psi_a^\dagger\gamma^0\gamma^\mu\psi_b\right] = 0 \ . \tag{5.14}$$

If we integrate this equation over all space and use the periodic boundary conditions to eliminate the spatial part, just as we did in our discussion of Eq. (4.13), we find that the following quantity is a constant of the motion:

$$\int d^3r\,\psi_a^\dagger\psi_b = \text{constant} \ . \tag{5.15}$$

Note that this expression is positive definite if $a = b$, and hence the states can be normalized as follows:

$$\int d^3r\,\psi_a^\dagger\psi_a = 1 \ . \tag{5.16}$$

This norm is a constant of the motion, and the Dirac equation has *no* states with negative norm. This was first believed to be a great advantage of the Dirac equation, but as we will soon see, the Dirac equation suffers from the same problem as the KG equation; it has negative energy solutions which are difficult to interpret physically.

5.3 SOLUTIONS FOR FREE PARTICLES

As in the KG theory, we will show that the solutions of the free particle Dirac equation have the general form

$$\psi_p^{(\pm)}(r,t) = N_p\, e^{i(\boldsymbol{p}_n\cdot\boldsymbol{r}\mp E_p t)} \begin{pmatrix} \chi \\ \eta \end{pmatrix} \quad , \tag{5.17}$$

where the $(\pm)$ superscript designates the positive $(+)$ or negative $(-)$ energy solutions. We use periodic boundary conditions as before, so that

$$\boldsymbol{p}_n = \frac{2\pi}{L}(n_x, n_y, n_z) \qquad n_i = 0, \pm 1, \pm 2, \ldots \tag{5.18}$$

and $E_p > 0$ always. For simplicity, the subscript n will be frequently ignored, so that $\boldsymbol{p}_n \to \boldsymbol{p}$, and the magnitude of $\boldsymbol{p}$ will be denoted by p.*

Consider the positive energy solutions first. Substituting the ansatz (5.17) into the free particle Dirac equation gives

$$E_p \begin{pmatrix} \chi \\ \eta \end{pmatrix} e^{-ip\cdot x} = \begin{pmatrix} m & -i\boldsymbol{\sigma}\cdot\nabla \\ -i\boldsymbol{\sigma}\cdot\nabla & -m \end{pmatrix} \begin{pmatrix} \chi \\ \eta \end{pmatrix} e^{-ip\cdot x} \quad .$$

Hence

$$\begin{aligned} E_p\chi &= m\,\chi + \boldsymbol{\sigma}\cdot\boldsymbol{p}\,\eta \\ E_p\eta &= \boldsymbol{\sigma}\cdot\boldsymbol{p}\,\chi - m\,\eta \quad . \end{aligned} \tag{5.19}$$

In order for these equations to have a non-zero solution, the determinant of the matrix of coefficients must be zero. Using $(\boldsymbol{\sigma}\cdot\boldsymbol{p})^2 = p^2$ the requirement that the determinant be zero gives the correct energy–momentum relation

$$E_p^2 = p^2 + m^2 \quad .$$

Then, using the second equation to express η in terms of χ gives

$$\eta = \left(\frac{\boldsymbol{\sigma}\cdot\boldsymbol{p}}{E_p + m}\right)\chi \quad . \tag{5.20}$$

*The symbol p will be used to denote either the four-vector or the magnitude of the three-vector. They can be distinguished from each other by the context in which they are used.

This gives the positive energy solution in terms of an arbitrary two-component spinor χ. Choosing $\chi^\dagger\chi = 1$ and normalizing the states to unity determines the normalization constant N_p:

$$\begin{aligned}\int_{L^3} d^3r\,\psi^\dagger\psi = 1 &= N_p^2 L^3 \chi^\dagger \left[1 + \frac{p^2}{(E_p+m)^2}\right]\chi \\ &= N_p^2 L^3 \left[1 + \frac{E_p - m}{E_p + m}\right] = N_p^2 L^3 \frac{2E_p}{E_p+m} \ . \end{aligned} \tag{5.21}$$

The normalization constant is therefore

$$N_p = \sqrt{\frac{E_p + m}{2E_p L^3}} \ . \tag{5.22}$$

It is customary to write the positive energy solution (5.17) in terms of the *positive energy Dirac spinor*, $u(\boldsymbol{p}, s)$, which is defined to be

$$u(\boldsymbol{p}, s) \equiv \sqrt{E_p + m} \begin{pmatrix} 1 \\ \dfrac{\boldsymbol{\sigma}\cdot\boldsymbol{p}}{E_p + m} \end{pmatrix} \chi^{(s)} \ , \tag{5.23}$$

where $\chi^{(s)}$ is a two-component spinor describing the states of a spin $\frac{1}{2}$ particle. If we choose to *quantize the spin in the $\hat{z}$-direction*, the spinors $\chi^{(s)}$ will be eigenvectors of σ_3

$$\chi^{(\frac{1}{2})} = \begin{pmatrix} 1 \\ 0 \end{pmatrix} \qquad \chi^{(-\frac{1}{2})} = \begin{pmatrix} 0 \\ 1 \end{pmatrix} \ . \tag{5.24}$$

Finally, the normalized *positive energy* solutions of the free particle Dirac equation are

$$\psi^{(+)}_{p,s}(x) = \frac{1}{\sqrt{2E_p L^3}}\, u(\boldsymbol{p}, s)\, e^{-ip\cdot x} \ . \tag{5.25}$$

Now find the negative energy solutions. In this case the ansatz (5.17) reduces the coupled Dirac equations to

$$\begin{aligned} -E_p\chi &= m\,\chi + \boldsymbol{\sigma}\cdot\boldsymbol{p}\,\eta \\ -E_p\eta &= \boldsymbol{\sigma}\cdot\boldsymbol{p}\,\chi - m\,\eta \ . \end{aligned} \tag{5.26}$$

As in the positive energy case, the condition $E^2 = p^2 + m^2$ insures that the determinant of the matrix coefficients of (5.26) is zero, so that a non-zero solution exists. Solving for χ in terms of η gives

$$\chi = -\left(\frac{\boldsymbol{\sigma}\cdot\boldsymbol{p}}{E_p + m}\right)\eta \ . \tag{5.27}$$

We choose $\eta^\dagger\eta = 1$, and normalize the state to unity, as before. This gives the same result (5.22) for the normalization constant N_p. For reasons which will be apparent in the next section, we choose phases (signs) such that

$$\boxed{\begin{aligned} \eta^{(-s)} &= -i\sigma_2\chi^{(s)} \\ \eta^{(-\frac{1}{2})} &= -i\sigma_2\chi^{(+\frac{1}{2})} = \begin{pmatrix} 0 \\ +1 \end{pmatrix} \\ \eta^{(+\frac{1}{2})} &= -i\sigma_2\chi^{(-\frac{1}{2})} = \begin{pmatrix} -1 \\ 0 \end{pmatrix} \ , \end{aligned}} \tag{5.28}$$

where σ_2 is the Pauli matrix. Note that this phase convention *differs* from Bjorken and Drell (1964), who choose $\eta^{(\frac{1}{2})} = \binom{1}{0}$. If we introduce the *negative energy Dirac spinor*, $v(\boldsymbol{p}, s)$,

$$\boxed{v(\boldsymbol{p}, s) = \sqrt{E_p + m} \begin{pmatrix} \dfrac{\boldsymbol{\sigma}\cdot\boldsymbol{p}}{E_p + m} \\ 1 \end{pmatrix} \left[-i\sigma_2\chi^{(s)}\right] \ ,} \tag{5.29}$$

the normalized *negative energy* solutions become

$$\psi^{(-)}_{-\boldsymbol{p},-s}(x) = \frac{1}{\sqrt{2E_p L^3}}\, v(\boldsymbol{p}, s)\, e^{ip\cdot x} \ . \tag{5.30}$$

Note that the negative energy solution for momentum $-p$ and spin $-s$ is expressed in terms of the v spinor with momentum p and spin s. This is in accordance with the *hole theory* interpretation to be discussed soon. Note that the (+) and (−) solutions are orthogonal because of the orthogonality of the positive and negative energy spinors:

$$\boxed{v^\dagger(-\boldsymbol{p}, s')u(\boldsymbol{p}, s) = u^\dagger(\boldsymbol{p}, s)v(-\boldsymbol{p}, s') = 0 \ .} \tag{5.31}$$

Comparison with the Two-Component *KG* Solutions

It is instructive to compare these solutions with the two-component KG solutions given in Eq. (4.47). The comparison is presented in Table 5.1.

The principal difference is that the Dirac theory has an extra two-component structure (located in the two-spinors $\chi^{(s)}$), which is identified with the internal degrees of freedom possessed by a spin $\frac{1}{2}$ particle. Otherwise, the structure of the positive and negative energy solutions in the two cases is similar. In both cases the

Table 5.1 Comparison of Dirac and *KG* solutions.

	Dirac $f_p(x)=\sqrt{\frac{E_p+m}{2E_pL^3}}\,e^{-ip\cdot x}$	KG $g_k(x)=\frac{E_k+m}{\sqrt{4mE_kL^3}}\,e^{-ik\cdot x}$
positive energy	$\psi^{(+)}_{p,s}(x)=f_p(x)\begin{bmatrix}1\\ \frac{\sigma\cdot p}{E_p+m}\end{bmatrix}\chi^{(s)}$ norm = 1	$\phi^{(+)}_{k}(x)=g_k(x)\begin{bmatrix}1\\ \frac{-k^2}{(E_k+m)^2}\end{bmatrix}$ norm = 1
negative energy	$\psi^{(-)}_{-p,-s}(x)=f_{-p}(x)\begin{bmatrix}\frac{\sigma\cdot p}{E_p+m}\\ 1\end{bmatrix}\eta^{(-s)}$ norm = 1	$\phi^{(-)}_{-k}(x)=g_{-k}(x)\begin{bmatrix}\frac{-k^2}{(E_k+m)^2}\\ 1\end{bmatrix}$ norm = −1

lower component(s) of the positive energy solutions is (are) smaller than the larger upper component(s), and conversely for the negative energy solutions. In the Dirac theory this suppression of the small components depends on spin (through the appearance of the Pauli spin matrices) and is proportional to the magnitude of $p/(E_p + m)$, while in the KG theory the suppression goes as the *square* of a similar factor and is therefore greater. This leads us to expect (correctly) that the relativistic corrections are spin dependent and larger in the Dirac theory than they were found to be in the KG theory. This will be studied in detail in Sec. 5.7.

5.4 CHARGE CONJUGATION

As in the KG theory there exists a charge conjugation operation which maps the negative energy states into positive energy states.

Consider the following operation on the states:

$$\psi \to \psi^C = C\beta\psi^* \,, \tag{5.32}$$

where

$$C = -i\alpha_2 = \begin{pmatrix} 0 & -i\sigma_2 \\ -i\sigma_2 & 0 \end{pmatrix} . \tag{5.33}$$

Note that $C^2 = -1$ and that, because α_2 is the only imaginary Dirac matrix,

$$C d C^{-1} = -d^* \ , \tag{5.34}$$

where d represents any of the Dirac matrices β or α_i. The operation of C on the covariant γ^μ matrices is

$$C\,\gamma^\mu\, C^{-1} = -\gamma^{\mu\,\mathsf{T}} \ . \tag{5.35}$$

Taking the complex conjugate of the Dirac equation (5.11), multiplying from the left by $C\beta$, and using the relation (5.34) give the equation for ψ^{C}:

$$\begin{aligned}
-i\frac{\partial}{\partial t}\psi^{C} &= C\beta\left[\alpha_i^*\left(i\frac{\partial}{\partial x^i} - eA^i\right) + eA^0 + \beta m\right]\psi^* \\
&= \left[\alpha_i\left(i\frac{\partial}{\partial x^i} - eA^i\right) + eA^0 - \beta m\right]\psi^{C} \ .
\end{aligned}$$

Hence

$$i\frac{\partial}{\partial t}\psi^{C} = \left[\alpha_i\left(-i\frac{\partial}{\partial x^i} + eA^i\right) - eA^0 + \beta m\right]\psi^{C} \tag{5.36}$$

and the charge conjugate amplitude satisfies a Dirac equation with opposite charge from the equation satisfied by ψ. Furthermore, the state $\psi^{(-)\,C}$ has positive energy. To see this, note that

$$\begin{aligned}
C\beta v^*(\boldsymbol{p}, s) &= \sqrt{E_p + m}\begin{pmatrix} 0 & -i\sigma_2 \\ -i\sigma_2 & 0 \end{pmatrix}\begin{pmatrix} \frac{\boldsymbol{\sigma}^*\cdot\boldsymbol{p}}{E_p+m} \\ -1 \end{pmatrix}(-i\sigma_2)\chi^{(s)} \\
&= \sqrt{E_p + m}\begin{pmatrix} 1 \\ -\sigma_2\frac{\boldsymbol{\sigma}^*\cdot\boldsymbol{p}}{E_p+m}\sigma_2 \end{pmatrix}\chi^{(s)} = u(\boldsymbol{p}, s)
\end{aligned} \tag{5.37}$$

because $\sigma_2\boldsymbol{\sigma}^*\sigma_2 = -\boldsymbol{\sigma}$. This simple relation is possible only because of the phase convention introduced in Eq. (5.28) for the two-component spinor η which enters into the definition of v. A similar relation holds for the u spinors; the two relations are

$$\boxed{\begin{aligned} C\beta v^*(\boldsymbol{p}, s) &= u(\boldsymbol{p}, s) \\ C\beta u^*(\boldsymbol{p}, s) &= v(\boldsymbol{p}, s) \ . \end{aligned}} \tag{5.38}$$

Using this result we find

$$C\beta\psi^{(-)\,*}_{-\boldsymbol{p},-s}(x) = \psi^{(+)}_{\boldsymbol{p},s}(x) \ , \tag{5.39}$$

which shows that $\psi^{(-)\,C}$ describes a positive energy particle with *identical mass and spin but opposite charge*. We identify $\psi^{(-)\,C}$ with the physical positive energy state corresponding to $\psi^{(-)}$. It is an antiparticle. To summarize,

$\psi_{p,s}^{(+)}(x)$: is the wave function for a particle with positive energy, momentum p, and spin projection s.

$\psi_{-p,-s}^{(-)}(x)$: is the wave function for a negative energy state with momentum $-p$ and spin projection $-s$, which is *interpreted as an antiparticle state* with positive energy, momentum p, and spin projection s.

In our study of the KG equation in Chapter 4, we also interpreted negative energy states as antiparticles. However, the way in which this interpretation is developed is significantly different for the two equations. First, KG particles (spin zero) obey Bose–Einstein statistics, and there is no limit to the number of negative energy particles which can occupy any negative energy state. Any positive energy KG state is therefore intrinsically unstable; there is no way to prevent it from decaying eventually to a negative energy state. On the other hand, Dirac particles (spin $\frac{1}{2}$) obey Fermi–Dirac statistics (which will be shown in Chapter 7). This means that no more than one particle can occupy any one state (the *Pauli exclusion principle*). If the physical vacuum is assumed to be the state in which *all negative energy states are filled*, a single positive energy state will be stable, since decay to negative energy states will be *Pauli blocked* by the filled *negative energy sea*, and we are able to "explain" why the lowest energy of a single particle is m (and not $-\infty$ as might be expected if the negative energy states were not already occupied). Furthermore, since the energy of the ground state can always be chosen to be zero [by choosing the appropriate constant E_0 in Eq. (3.19)], this picture of the vacuum is physically sensible. In this picture, referred to as *hole theory*, an antiparticle is interpreted as a "hole" in the vacuum, i.e., as the *absence* of one of the particles from the otherwise filled negative energy sea. Being the *absence* of a *negative* energy state, the antiparticle has *positive* energy.

These ideas are illustrated in Fig. 5.1. In Fig. 5.1A the vacuum has no particles, so a single particle with energy $-E < -m$, momentum $-\boldsymbol{p}$, and spin projection $-s$ can exist. In Fig. 5.1B the vacuum is assumed to be filled with negative energy states. The absence of a single negative energy state with quantum numbers $-E < -m$, $-\boldsymbol{p}$, and $-s$ then appears as a hole in this vacuum. Since the vacuum values for these quantum numbers must be zero (by definition), the hole therefore behaves just like a particle with energy $0-(-E)=E$, momentum $\boldsymbol{0}-(-\boldsymbol{p})=\boldsymbol{p}$, and spin projection $0-(-s)=s$, or a *positive* energy antiparticle with energy $E > m$, momentum $\boldsymbol{p}$, and spin projection s. Thus *hole theory* provides a physical picture of how negative energy and antiparticle states are related. Mathematically, this relation is expressed through the charge conjugation transformation.

Hole theory played an important role in the development of relativistic quantum mechanics but is superseded by modern field theory. We no longer think of the vacuum as filled with negative energy particles. In Chapter 7 we will see that a quantum field is equally well suited to the description of either spin zero

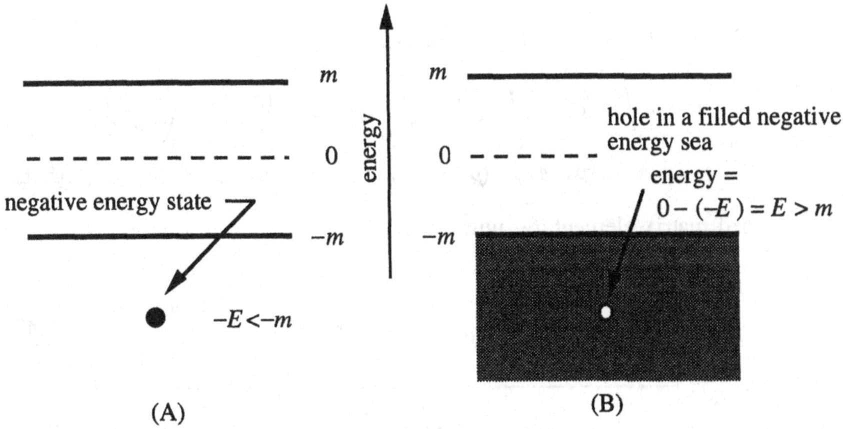

Fig. 5.1 In the model shown in (A), the vacuum is empty and a single negative energy state has energy $-E$. In (B), the vacuum is the state in which the negative energy sea is filled, and an antiparticle is interpreted as a hole in this sea.

particles (which do not satisfy an exclusion principle) or spin $\frac{1}{2}$ particles. In either case both particle and antiparticle degrees of freedom have *positive* energy. However, hole theory still gives us a useful physical picture of the connection between physical antiparticles and the negative energy states which emerge from a one-body relativistic wave equation.

5.5 COULOMB SCATTERING

To illustrate the use of Dirac wave functions and the Dirac formalism, we calculate the lowest order scattering of a Dirac particle by a Coulomb potential. The method is identical to our treatment of Coulomb scattering by a spinless particle, given in Sec. 4.7. The EM potential is assumed to be

$$\mathbf{A} = 0 \qquad eA^0 = \frac{Z\alpha}{r} \ ,$$

and the S-matrix is given by (4.62);

$$S_{fi} = -i \int dt \ \langle f \, | H_I(t) | \, i \rangle$$

except that now $|i\rangle$ and $|f\rangle$ are initial and final Dirac free particle states at $t = 0$ and H_0 and H_I are

$$\begin{aligned} H_0 &= \alpha_i p^i + \beta m \\ H_I &= U_0^{-1} \frac{Z\alpha}{r} \begin{pmatrix} 1 & \\ & 1 \end{pmatrix} U_0 \end{aligned} \tag{5.40}$$

with $U_0 = e^{-iH_0 t}$ as in Eq. (4.64). Hence

$$\begin{aligned} S_{fi} &= -i \int dt\, d^3r\, \psi_{p_f}^{(+)\dagger}(r) H_I \psi_{p_i}^{(+)}(r) \\ &= -i \int dt\, e^{i(E_f - E_i)t} \int \frac{d^3r}{L^3} \frac{1}{\sqrt{4E_f E_i}} e^{i(\boldsymbol{p}_i - \boldsymbol{p}_f)\cdot \boldsymbol{r}} \left(\frac{Z\alpha}{r}\right) \\ &\quad \times u^\dagger(\boldsymbol{p}_f, s_f)\, u(\boldsymbol{p}_i, s_i) \ . \end{aligned} \tag{5.41}$$

The reduced matrix element f_{fi} now becomes

$$\boxed{f_{fi} = \frac{4\pi Z\alpha}{q^2}\left[\frac{1}{2E} u^\dagger(\boldsymbol{p}_f, s_f)\, u(\boldsymbol{p}_i, s_i)\right] \ ,} \tag{5.42}$$

where $q^2 = (\boldsymbol{p}_i - \boldsymbol{p}_f)^2$ as before. Note that the only difference between (5.42) and the KG result (4.68) is the factor in [ ].

The reduced amplitude f_{fi} now depends on the *polarization* of the initial and final particles. We will calculate the *unpolarized* cross section, which will require us to *average* over initial spins (incoming particles are equally likely to have spin up as spin down) and *sum* over final spins.

Part of the calculation of the unpolarized cross section requires the calculation of the following double sum. In evaluating this sum, we use the fact that $E_f = E_i = E$, and hence $p_f^2 = p_i^2 = p^2$, to yield

$$\begin{aligned} &\frac{1}{2}\sum_{s_f, s_i} \left|u^\dagger(\boldsymbol{p}_f, s_f)\, u(\boldsymbol{p}_i, s_i)\right|^2 \\ &= \frac{1}{2}\sum_{s_f, s_i} u^\dagger(\boldsymbol{p}_f, s_f)\, u(\boldsymbol{p}_i, s_i)\, u^\dagger(\boldsymbol{p}_i, s_i)\, u(\boldsymbol{p}_f, s_f) \\ &= \frac{1}{2}\sum_{s_f, s_i} (E+m)^2 \chi^{\dagger(s_f)} \begin{pmatrix} 1 & \frac{\boldsymbol{\sigma}\cdot\boldsymbol{p}_f}{E+m} \end{pmatrix} \begin{pmatrix} 1 \\ \frac{\boldsymbol{\sigma}\cdot\boldsymbol{p}_i}{E+m} \end{pmatrix} \chi^{(s_i)} \chi^{\dagger(s_i)} \\ &\quad \times \begin{pmatrix} 1 & \frac{\boldsymbol{\sigma}\cdot\boldsymbol{p}_i}{E+m} \end{pmatrix} \begin{pmatrix} 1 \\ \frac{\boldsymbol{\sigma}\cdot\boldsymbol{p}_f}{E+m} \end{pmatrix} \chi^{(s_f)} \\ &= \frac{1}{2}(E+m)^2 \sum_{s_f} \chi^{\dagger(s_f)} \left[1 + \frac{\boldsymbol{\sigma}\cdot\boldsymbol{p}_f \boldsymbol{\sigma}\cdot\boldsymbol{p}_i}{(E+m)^2}\right] \left[1 + \frac{\boldsymbol{\sigma}\cdot\boldsymbol{p}_i \boldsymbol{\sigma}\cdot\boldsymbol{p}_f}{(E+m)^2}\right] \chi^{(s_f)} \\ &= \frac{1}{2}(E+m)^2 \ \mathrm{tr}\left\{\left(1 + \frac{\boldsymbol{\sigma}\cdot\boldsymbol{p}_f \boldsymbol{\sigma}\cdot\boldsymbol{p}_i}{(E+m)^2}\right)\left(1 + \frac{\boldsymbol{\sigma}\cdot\boldsymbol{p}_i \boldsymbol{\sigma}\cdot\boldsymbol{p}_f}{(E+m)^2}\right)\right\} \ . \end{aligned}$$

Now use

$$\begin{aligned} \mathrm{tr}\ (\sigma_i \sigma_j) &= 2\delta_{ij} \\ (\boldsymbol{\sigma}\cdot\boldsymbol{p})^2 &= p^2 \end{aligned} \tag{5.43}$$

to get

$$\frac{1}{2}\sum_{s_f,s_i}\left|u^\dagger(\boldsymbol{p}_f,s_f)\,u(\boldsymbol{p}_i,s_i)\right|^2=(E+m)^2\left[1+\frac{2\boldsymbol{p}_f\cdot\boldsymbol{p}_i}{(E+m)^2}+\frac{p^4}{(E+m)^4}\right] .$$

This is further reduced by using $\boldsymbol{p}_f\cdot\boldsymbol{p}_i=p^2\cos\theta$, where θ is the scattering angle

$$(E+m)^2\left[1+\frac{2\boldsymbol{p}_f\cdot\boldsymbol{p}_i}{(E+m)^2}+\frac{p^4}{(E+m)^4}\right]=2\left[2m^2+p^2+p^2\cos\theta\right]$$
$$=4\left[E^2-p^2\sin^2\frac{\theta}{2}\right] . \quad (5.44)$$

Hence, finally

$$\frac{1}{2}\sum_{f_f,s_i}\left|u^\dagger(\boldsymbol{p}_f,s_f)\,u(\boldsymbol{p}_i,s_i)\right|^2=4E^2\left[1-\frac{p^2}{E^2}\sin^2\frac{\theta}{2}\right] . \quad (5.45)$$

The $4E^2$ factor in front is canceled by the $(1/2E)^2$ factor in $|f_{fi}|^2$. The final steps are the same as for the KG case, Sec. 4.7, giving

$$\boxed{\frac{d\sigma}{d\Omega}=\left(\frac{2Z\alpha E}{q^2}\right)^2\left[1-v^2\sin^2\frac{\theta}{2}\right] ,} \quad (5.46)$$

where $v^2=p^2/E^2$. This famous result is the *Mott cross section* for the scattering of a spin $\frac{1}{2}$ particle. Comparing it with the KG result, Eq. (4.68), we see that it differs by the factor

$$\left[1-v^2\sin^2\frac{\theta}{2}\right] . \quad (5.47)$$

For large energy, $v\simeq 1$, and the cross section goes to zero in the backward direction. (See Fig. 5.2.) This difference in the backward direction is due to *magnetic scattering*: the interaction of the magnetic moment of the electron (associated with its spin) with the magnetic field it sees when moving toward the fixed Coulomb field.

5.6 NEGATIVE ENERGY STATES

In this section, the role of negative energy states in the Dirac theory is examined. We will treat second order Coulomb scattering as an example, so the discussion will parallel the development given in Sec. 4.7, where the contribution of negative energy states to second order Coulomb scattering of spinless (KG) particles

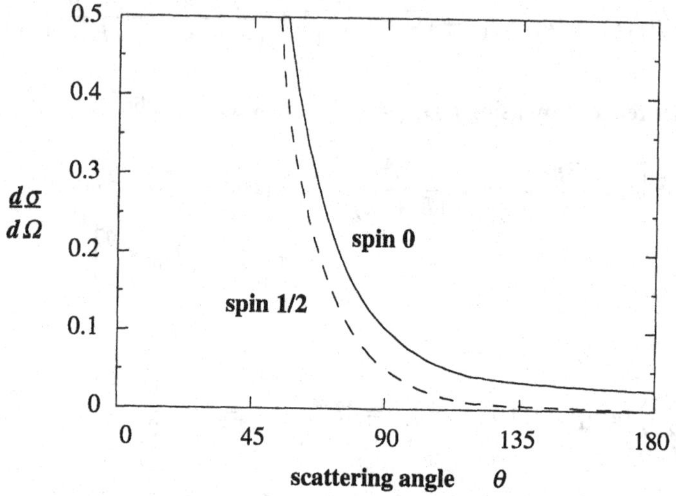

Fig. 5.2 The Coulomb scattering cross section in arbitrary units. The solid line is the cross section for a spin zero particle, and the dashed line is Eq. (5.46) for a spin $\frac{1}{2}$ particle. Note that both cross sections peak strongly in the forward direction but that there is an additional suppression in the backward direction for the spin $\frac{1}{2}$ particle.

was studied. As the results for the Dirac theory are very similar, the discussion here will emphasize the similarities and differences.

Completeness Relation

Recall that the evaluation of the second order matrix element for the S-matrix required the completeness relation. For the two-component KG theory, the needed relation was given in Eq. (4.71). The corresponding relation for the Dirac equation is

$$K_D(r,r') = \sum_{p,s} \left\{ \psi^{(+)}_{p,s}(r)\psi^{(+)\dagger}_{p,s}(r') + \psi^{(-)}_{p,s}(r)\psi^{(-)\dagger}_{p,s}(r') \right\} = \mathbf{1}\,\delta^3(r-r') \ . \tag{5.48}$$

Note that a plus sign stands in front of the negative energy sum; the KG completeness relation, Eq. (4.71), had a minus sign. This difference is due to the different normalization condition satisfied by the positive and negative energy states in the Dirac and KG theories.

Proof: The general proof of (5.48) is identical to the general proof of Eq. (4.71). Even the proof by construction is similar, except that now the matrix is

4×4 instead of 2×2. We have

$$
\begin{aligned}
&K_D(r, r') \\
&= \frac{1}{L^3} \sum_{p,s} \frac{1}{2E_p}\, e^{i\boldsymbol{p}\cdot(\boldsymbol{r}-\boldsymbol{r}')} \left\{ u(\boldsymbol{p}, s) u^\dagger(\boldsymbol{p}, s) + v(-\boldsymbol{p}, -s) v^\dagger(-\boldsymbol{p}, -s) \right\} \\
&= \sum_p e^{i\boldsymbol{p}\cdot(\boldsymbol{r}-\boldsymbol{r}')} \left(\frac{E_p + m}{2E_p L^3} \right) \left\{ \begin{bmatrix} 1 \\ \frac{\boldsymbol{\sigma}\cdot\boldsymbol{p}}{E_p+m} \end{bmatrix} \begin{bmatrix} 1 & \frac{\boldsymbol{\sigma}\cdot\boldsymbol{p}}{E_p+m} \end{bmatrix} + \begin{bmatrix} \frac{-\boldsymbol{\sigma}\cdot\boldsymbol{p}}{E_p+m} \\ 1 \end{bmatrix} \begin{bmatrix} \frac{-\boldsymbol{\sigma}\cdot\boldsymbol{p}}{E_p+m} & 1 \end{bmatrix} \right\} \\
&= \sum_p e^{i\boldsymbol{p}\cdot(\boldsymbol{r}-\boldsymbol{r}')} \left(\frac{E_p + m}{2E_p L^3} \right) \left\{ \begin{pmatrix} 1 & \frac{\boldsymbol{\sigma}\cdot\boldsymbol{p}}{E_p+m} \\ \frac{\boldsymbol{\sigma}\cdot\boldsymbol{p}}{E_p+m} & \frac{p^2}{(E_p+m)^2} \end{pmatrix} + \begin{pmatrix} \frac{p^2}{(E_p+m)^2} & \frac{-\boldsymbol{\sigma}\cdot\boldsymbol{p}}{E_p+m} \\ \frac{-\boldsymbol{\sigma}\cdot\boldsymbol{p}}{E_p+m} & 1 \end{pmatrix} \right\} \\
&= \sum_p \frac{1}{L^3}\, e^{i\boldsymbol{p}\cdot(\boldsymbol{r}-\boldsymbol{r}')} \begin{pmatrix} 1 & 0 \\ 0 & 1 \end{pmatrix} = \mathbf{1}\, \delta^3(r - r') \ . \qquad \blacksquare
\end{aligned}
$$

Now, the Dirac matrix elements of the Coulomb interaction term can be written

$$
\begin{aligned}
\langle f\, | H_I(t_1) |\, n\pm \rangle &= \int d^3r\, \psi^{(+)\dagger}_{p_f, s_f}(r) \left(\frac{Z\alpha}{r} \right) \psi^{(\pm)}_{p_n, s_n}(r)\, e^{i(E_f \mp E_n) t_1} \\
&= \frac{1}{L^3} f^{(\pm)}_{fn}\, e^{i(E_f \mp E_n) t_1} \ ,
\end{aligned}
\tag{5.49}
$$

where n is a shorthand notation for the quantum numbers $\{p_n, s_n\}$ and the reduced matrix elements $f^{(+)}_{fn}$ were given in the preceding section [see Eq. (5.42)]. We have not calculated the reduced matrix element $f^{(-)}_{fn}$, but it could be evaluated from (5.49). All we need to know now is that it is in general not zero. In terms of these reduced matrix elements, the second order S-matrix element for Coulomb scattering can be written

$$
S^{(2)}_{fi} = -i 2\pi \delta\left(E_f - E_i\right) \frac{1}{L^3} f^{(2)}_{fi} \ ,
$$

where

$$
\begin{aligned}
f^{(2)}_{fi} &= -i \int dt \sum_n \frac{1}{L^3} \Big\{ f^{(+)}_{p_f p} f^{(+)*}_{p_i p}\, \theta(t)\, e^{-i(E_p - E_i)\, t} \\
&\qquad - f^{(-)}_{p_f p} f^{(-)*}_{p_i p}\, \theta(-t)\, e^{i(E_p + E_i)\, t} \Big\} \\
&= -\sum_p \frac{1}{L^3} \left\{ \frac{f^{(+)}_{p_f p} f^{(+)*}_{p_i p}}{E_p - E_i - i\epsilon} - \frac{f^{(-)}_{p_f p} f^{(-)*}_{p_i p}}{E_p + E_i - i\epsilon} \right\} \ .
\end{aligned}
\tag{5.50}
$$

The derivation of these results for the Dirac theory is *identical* to that for the KG theory [review the arguments which led from Eq. (4.73) to Eq. (4.78)]. Each of these two expressions for $f^{(2)}$ differs from its KG counterpart [which is (4.76b) for the first and (4.78) for the second] *only in the sign of the negative energy term.* And here, as in the KG theory, the propagation of the negative energy states backward in time is interpreted as the propagation of the corresponding antiparticle states forward in time (recall Fig. 4.3).

The difference in sign of the negative energy contributions to the KG and Dirac expressions for $f^{(2)}$ will appear again in field theory. In that discussion the sign difference will come from the fact that Dirac particles satisfy Fermi-Dirac statistics (i.e., their field operators anticommute) and that when the time ordering of the interactions is changed, as it is for the negative energy states, there is an extra minus sign for fermions.

5.7 NONRELATIVISTIC LIMIT

We now investigate the non-relativistic limit of the Dirac equation. As we did for the Klein–Gordon equation, we will work out the expansion to order $(v/c)^2 \sim (p/m)^2\times$ *leading terms.* In making our estimates, *we assume all potentials* V^0 *and* $\boldsymbol{V}$ *to be of the same order as the kinetic energy term* (justified by the virial theorem). Since all of these leading terms are of order p^2/m, we want all terms up to order p^4/m^3.

Assume a positive energy solution of the form

$$\psi(r,t) = \begin{pmatrix} \chi(r) \\ \eta(r) \end{pmatrix} e^{-iEt} \quad , \tag{5.51}$$

where $E = m + T$. Then, using the Dirac equation, the coupled equations for $\chi(r)$ and $\eta(r)$ become

$$\begin{aligned} T\chi &= V^0\chi + \boldsymbol{\sigma}\cdot(\boldsymbol{p}-\boldsymbol{V})\,\eta \\ (2m+T)\eta &= V^0\eta + \boldsymbol{\sigma}\cdot(\boldsymbol{p}-\boldsymbol{V})\,\chi \quad . \end{aligned} \tag{5.52}$$

In the non-relativistic limit, T, $|p|$, and all components of $|V^\mu| = |eA^\mu|$ are assumed to be very much smaller than m. Hence, the second of the two equations (5.52) shows that the lower components of the Dirac spinor are very much smaller than the upper components, and therefore the equations are solved approximately by eliminating the lower components, as we did for the KG equation. However, if we proceed directly by solving the lower equation for η and substituting the solution into the equation for χ, we obtain

$$T\chi = \left\{ V^0 + \boldsymbol{\sigma}\cdot(\boldsymbol{p}-\boldsymbol{V}) \left(\frac{1}{2m+T-V^0} \right) \boldsymbol{\sigma}\cdot(\boldsymbol{p}-\boldsymbol{V}) \right\} \chi \quad . \tag{5.53}$$

Since T is of the same order as V^0, which is $\simeq p^2/m$, it is necessary to expand the denominator of the second term if we want to collect *all* terms of order p^4/m^3.

This expansion gives

$$T\chi = \left\{ V^0 + \frac{1}{2m}\boldsymbol{\sigma}\cdot(\boldsymbol{p}-\boldsymbol{V})\,\boldsymbol{\sigma}\cdot(\boldsymbol{p}-\boldsymbol{V}) - \frac{1}{4m^2}\boldsymbol{\sigma}\cdot(\boldsymbol{p}-\boldsymbol{V})(T-V^0)\boldsymbol{\sigma}\cdot(\boldsymbol{p}-\boldsymbol{V}) \right\}\chi \ . \tag{5.54}$$

Note the presence of the energy T in the last term on the right-hand side. This means that the effective Hamiltonian defined by Eq. (5.54) is dependent on the energy, and an energy-dependent Hamiltonian leads to many complications which should be avoided, if possible. The explicit dependence on the energy should be eliminated. Since the T dependence occurs only in the highest order term, it might seem that it could be removed by replacing it by an estimate obtained from the solution of the lower order equation, i.e.,

$$T \simeq V^0 + \frac{1}{2m}\boldsymbol{\sigma}\cdot(\boldsymbol{p}-\boldsymbol{V})\,\boldsymbol{\sigma}\cdot(\boldsymbol{p}-\boldsymbol{V}) \ .$$

However, this method will not give a unique answer because T is a number and commutes with $\boldsymbol{\sigma}\cdot(\boldsymbol{p}-\boldsymbol{V})$, while V^0, part of the above estimate for T, does not. It is better to attack the problem from a different direction.

A better method, known as the *Foldy–Wouthuysen (FW) transformation* [FW 50], is to transform the equations to a new form in which the off-diagonal elements of the Hamiltonian are so small that the *leading* order estimate of the lower components (which does not depend on the energy T) is sufficient to get the effective Hamiltonian to the desired order of accuracy. For example, in this problem where we want the Hamiltonian to order p^4/m^3, it would be sufficient to reduce the off-diagonal elements to order p^2/m. If they were that small, the leading contribution from the lower components would be of order p^2/m^2, and their contribution to the equation for χ would therefore be of order p^4/m^3, sufficient for our purposes. In the KG case treated in the last chapter, the off-diagonal elements were initially that small, so we were able to get the desired result immediately. Here, the off-diagonal elements of the Dirac equation are of (larger) order p, so the simplest approach did not work.

To prepare for the application of the FW transformation, return to the matrix equations (5.52), and write them in terms of Dirac matrices

$$T\begin{pmatrix}\chi\\ \eta\end{pmatrix} = \left(-m + V^0 + \boldsymbol{\alpha}\cdot(\boldsymbol{p}-\boldsymbol{V}) + m\beta\right)\begin{pmatrix}\chi\\ \eta\end{pmatrix} = H\begin{pmatrix}\chi\\ \eta\end{pmatrix} \ . \tag{5.55}$$

The off-diagonal terms are those involving the Dirac matrices $\boldsymbol{\alpha}$, and they are large (of order m^0). We want to transform the equation so that they are of order m^{-1}. Then, when the equation is solved, T will not enter into the m^{-3} term.

The equation will be transformed using a general unitary transformation constructed from the Dirac matrices. Since the large off-diagonal terms we wish to reduce depend on $\boldsymbol{\alpha}\cdot\boldsymbol{p}$, it is sufficient to use a transformation of the form

$$U = U^\dagger = A\beta + \frac{\lambda}{m}\boldsymbol{\alpha}\cdot\boldsymbol{p} \qquad A = \sqrt{1 - \frac{\lambda^2 p^2}{m^2}} \;, \tag{5.56}$$

where λ is a parameter which will be chosen later. Using the anticommutation relations satisfied by the Dirac matrices, it is easy to see that

$$U\,U^\dagger = U^\dagger U = 1 \tag{5.57}$$

for any λ. The fact that U is unitary means that the transformed wave function

$$\begin{pmatrix} \chi' \\ \eta' \end{pmatrix} = U \begin{pmatrix} \chi \\ \eta \end{pmatrix} \tag{5.58}$$

has the same norm. Transforming Eq. (5.55) gives

$$T\begin{pmatrix} \chi' \\ \eta' \end{pmatrix} = U\,H\,U^{-1}\begin{pmatrix} \chi' \\ \eta' \end{pmatrix} = H'\begin{pmatrix} \chi' \\ \eta' \end{pmatrix} \;, \tag{5.59}$$

where the individual contributions to the Hamiltonian (5.55) become

$$\begin{aligned}
U(-m)U^{-1} &= -m \\
UV^0U^{-1} &= AV^0A + \beta\frac{\lambda}{m}\left(AV^0\,\boldsymbol{\alpha}\cdot\boldsymbol{p} - \boldsymbol{\alpha}\cdot\boldsymbol{p}\,V^0 A\right) + \left(\frac{\lambda}{m}\right)^2 \boldsymbol{\alpha}\cdot\boldsymbol{p}\,V^0\,\boldsymbol{\alpha}\cdot\boldsymbol{p} \\
U\boldsymbol{\alpha}\cdot(\boldsymbol{p}-\boldsymbol{V})U^{-1} &= -A\,\boldsymbol{\alpha}\cdot(\boldsymbol{p}-\boldsymbol{V})A \\
&\quad + \beta\frac{\lambda}{m}\left[A\,\boldsymbol{\alpha}\cdot(\boldsymbol{p}-\boldsymbol{V})\boldsymbol{\alpha}\cdot\boldsymbol{p} + \boldsymbol{\alpha}\cdot\boldsymbol{p}\,\boldsymbol{\alpha}\cdot(\boldsymbol{p}-\boldsymbol{V})A\right] \\
&\quad + \left(\frac{\lambda}{m}\right)^2 \boldsymbol{\alpha}\cdot\boldsymbol{p}\,\boldsymbol{\alpha}\cdot(\boldsymbol{p}-\boldsymbol{V})\boldsymbol{\alpha}\cdot\boldsymbol{p} \qquad\qquad (5.60) \\
Um\beta U^{-1} &= m\beta A^2 + 2\lambda A\boldsymbol{\alpha}\cdot\boldsymbol{p} - \beta\lambda^2\frac{p^2}{m} \;.
\end{aligned}$$

The off-diagonal terms are those proportional to an odd power of $\boldsymbol{\alpha}$, and they need only be calculated to order m^{-1}. Noting that A can be expanded,

$$A \cong 1 - \frac{\lambda^2 p^2}{2m^2} \;, \tag{5.61}$$

we see that the only off-diagonal terms which survive come from the first term on the RHS of the third of Eqs. (5.60) and the second term on the RHS of the fourth of Eqs. (5.60) and that $A \sim 1$ is sufficient to get all of the $\mathcal{O}\left(m^{-1}\right)$ terms, giving

$$H'_{\text{off-diag}} = -\boldsymbol{\alpha}\cdot(\boldsymbol{p}-\boldsymbol{V}) + 2\lambda\,\boldsymbol{\alpha}\cdot\boldsymbol{p} + \mathcal{O}\left(\frac{1}{m^2}\right) \;. \tag{5.62}$$

Hence, choosing $\lambda = \frac{1}{2}$ gives

$$H'_{\text{off-diag}} \cong \boldsymbol{\alpha} \cdot \boldsymbol{V} \ , \tag{5.63}$$

which is $\mathcal{O}\left(m^{-1}\right)$ by assumption.

With these approximations, the coupled equations (5.52) become

$$\begin{aligned} T\chi' &= H'_{11}\chi' + \boldsymbol{\sigma}\cdot\boldsymbol{V}\,\eta' \\ T\eta' &= \boldsymbol{\sigma}\cdot\boldsymbol{V}\,\chi' - 2m\,\eta' \ , \end{aligned} \tag{5.64}$$

where only the largest (leading) terms have been retained in every element but H'_{11}, which is yet to be reduced. We may now neglect $T\eta'$ in the second equation, giving

$$T\chi' = \left(H'_{11} + \frac{\boldsymbol{\sigma}\cdot\boldsymbol{V}\,\boldsymbol{\sigma}\cdot\boldsymbol{V}}{2m}\right)\chi' = \left(H'_{11} + \frac{V^2}{2m}\right)\chi' \ . \tag{5.65}$$

Note that the $V^2/2m$ term is $\mathcal{O}\left(m^{-3}\right)$.

The remaining task is to reduce H'_{11} using $\lambda = \frac{1}{2}$. Noting that the large terms proportional to m occur in the combination $(-1+\beta)$, which makes no contribution to the H'_{11} matrix element, we have, to $\mathcal{O}\left(m^{-3}\right)$,

$$\begin{aligned} H'_{11} \cong{} & V^0 - \frac{p^2}{8m^2}V^0 - V^0\frac{p^2}{8m^2} + \frac{\boldsymbol{\sigma}\cdot\boldsymbol{p}\,V^0\,\boldsymbol{\sigma}\cdot\boldsymbol{p}}{4m^2} \\ & + \frac{1}{2m}\left(\boldsymbol{\sigma}\cdot(\boldsymbol{p}-\boldsymbol{V})\,\boldsymbol{\sigma}\cdot\boldsymbol{p} + \boldsymbol{\sigma}\cdot\boldsymbol{p}\,\boldsymbol{\sigma}\cdot(\boldsymbol{p}-\boldsymbol{V})\right) - \frac{p^4}{8m^3} - \frac{p^2}{2m} \ , \end{aligned} \tag{5.66}$$

where the first three terms on the RHS are the expansion of $A\,V^0A$, the first two in the second line are the expansion of the contributions from $U\boldsymbol{\alpha}\cdot(\boldsymbol{p}-\boldsymbol{V})U^{-1}$, and the last is the combined contribution from $Um\beta U^{-1}$. To further reduce these terms we will use the identity

$$\sigma_i\sigma_j = \delta_{ij} + i\epsilon_{ijk}\sigma_k \ . \tag{5.67}$$

Using this identity gives

$$\begin{aligned} &\left(\boldsymbol{\sigma}\cdot(\boldsymbol{p}-\boldsymbol{V})\,\boldsymbol{\sigma}\cdot\boldsymbol{p} + \boldsymbol{\sigma}\cdot\boldsymbol{p}\,\boldsymbol{\sigma}\cdot(\boldsymbol{p}-\boldsymbol{V})\right) \\ &\quad = 2p^2 - \boldsymbol{\sigma}\cdot\boldsymbol{p}\,\boldsymbol{\sigma}\cdot\boldsymbol{V} - \boldsymbol{\sigma}\cdot\boldsymbol{V}\,\boldsymbol{\sigma}\cdot\boldsymbol{p} \\ &\quad = 2p^2 - \boldsymbol{p}\cdot\boldsymbol{V} - \boldsymbol{V}\cdot\boldsymbol{p} - i\boldsymbol{\sigma}\cdot(\boldsymbol{p}\times\boldsymbol{V}) - i\boldsymbol{\sigma}\cdot(\boldsymbol{V}\times\boldsymbol{p}) \\ &\quad = (\boldsymbol{p}-\boldsymbol{V})^2 + p^2 - V^2 - \boldsymbol{\sigma}\cdot[\nabla\times\boldsymbol{V}] \\ &\quad = (\boldsymbol{p}-\boldsymbol{V})^2 + p^2 - V^2 - e\,\boldsymbol{\sigma}\cdot\boldsymbol{B} \ , \end{aligned} \tag{5.68}$$

where the use of *square* brackets will mean that $\boldsymbol{p}$ or ∇ operates only *within* the brackets. Note the new term describing a magnetic moment interaction, which was obtained by replacing $\boldsymbol{V}$ by the vector potential $e\boldsymbol{A}$ and using $\boldsymbol{B} = \nabla \times \boldsymbol{A}$. Next, reduce the second through fourth terms in (5.66):

$$\begin{aligned} p^2 V^0 + V^0 p^2 &= \left[p^2\, V^0\right] + 2\left[\boldsymbol{p}\, V^0\right]\cdot \boldsymbol{p} + 2V^0 p^2 \\ \boldsymbol{\sigma}\cdot\boldsymbol{p}\, V^0 \boldsymbol{\sigma}\cdot\boldsymbol{p} &= \boldsymbol{\sigma}\cdot\left[\boldsymbol{p}\, V^0\right]\boldsymbol{\sigma}\cdot\boldsymbol{p} + V^0 p^2 \\ &= \left[\boldsymbol{p}\, V^0\right]\cdot\boldsymbol{p} + i\boldsymbol{\sigma}\cdot\left(\left[\boldsymbol{p}\, V^0\right]\times\boldsymbol{p}\right) + V^0 p^2\ . \end{aligned} \tag{5.69}$$

Thus, in the combination which occurs in H'_{11} the $V^0 p^2$ and $\left[\boldsymbol{p}\, V^0\right]\cdot\boldsymbol{p}$ terms cancel:

$$-\frac{1}{8m^2}\left(p^2 V^0 + V^0 p^2\right) + \frac{\boldsymbol{\sigma}\cdot\boldsymbol{p}\, V^0 \boldsymbol{\sigma}\cdot\boldsymbol{p}}{4m^2} = -\frac{\left[p^2\, V^0\right]}{8m^2} + \frac{i\boldsymbol{\sigma}\cdot\left(\left[\boldsymbol{p}\, V^0\right]\times\boldsymbol{p}\right)}{4m^2}\ . \tag{5.70}$$

Finally, replacing $\boldsymbol{p} = -i\nabla$, $\boldsymbol{V} = e\boldsymbol{A}$, $V^0 = e\phi = e\phi(r)$ gives

$$H'_{11} + \frac{V^2}{2m} = \frac{(\boldsymbol{p} - e\boldsymbol{A})^2}{2m} + e\phi - \frac{e}{2m}\boldsymbol{\sigma}\cdot\boldsymbol{B} - \frac{p^4}{8m^3} + \frac{e\left[\nabla^2\phi\right]}{8m^2} + \frac{e}{4m^2}\boldsymbol{\sigma}\cdot\left(\left[\nabla\phi\right]\times\boldsymbol{p}\right)\ . \tag{5.71}$$

Assuming that the potential is spherically symmetric, so that $\phi = \phi(r)$ where r is the radial coordinate, leads to $\nabla\phi = (\boldsymbol{r}/r)d\phi/dr$, giving finally the effective Hamiltonian

$$\boxed{H_{\text{eff}} = \frac{(\boldsymbol{p} - e\boldsymbol{A})^2}{2m} + e\phi - \frac{p^4}{8m^3} - \frac{e}{2m}\boldsymbol{\sigma}\cdot\boldsymbol{B} + \frac{e\left[\nabla^2\phi\right]}{8m^2} + \frac{e}{4m^2}\frac{1}{r}\frac{d\phi}{dr}\boldsymbol{\sigma}\cdot\boldsymbol{L}\,,} \tag{5.72}$$

where $\boldsymbol{L}$ is the orbital angular momentum operator.

We assume that the reader is familiar with the effective Hamiltonian (5.72) from previous studies, and we will only give a very brief review of these results.* Historically, this effective Hamiltonian was obtained well before Dirac discovered his equation, so that the derivation of the result *from the Dirac equation* can be regarded as a great success and a grand confirmation that the Dirac equation does indeed give the correct description of the interactions of a spin $\frac{1}{2}$ particle with an EM field. Each of the terms in (5.72) was originally derived independently, but using the Dirac equation all of them emerge automatically.

In addition to the $-p^4/8m^3$ term found for the KG equation, there are *three new* corrections connected with the spin of the electron. These give new contributions to the fine structure and Zeeman effect.

Fine Structure (Dirac)

The fine structure now comes from three terms in Eq. (5.72). These are

- $-\dfrac{p^4}{8m^3}$ relativistic mass increase (also from KG theory)

*For an introductory discussion of these topics see, for example, Gottfried (1966) or Sakurai (1985).

- $\frac{e}{8m^2}\nabla^2\phi = -\frac{e}{8m^2}\nabla\cdot\boldsymbol{E} = \frac{Ze^2}{8m^2}\delta^3(r)$ Darwin term

Because of the $\delta^3(r)$, this term is non-zero for S-states only. Physically, it comes from quantum fluctuations in the position of the electron, referred to as *Zitterbewegung* (jittering motion), which make the electron sensitive to the average potential in the vicinity of its average position. The average of the potential is proportional to $\nabla^2\phi \sim \delta^3(r)$, and this accounts for the general structure of the Darwin term.

- $\frac{e}{4m^2}\frac{1}{r}\frac{d\phi}{dr}\,\boldsymbol{\sigma}\cdot\boldsymbol{L} = \frac{e}{2m^2}\left(\frac{1}{r}\frac{d\phi}{dr}\right)\boldsymbol{S}\cdot\boldsymbol{L}$ spin orbit term

where $\boldsymbol{S} = \boldsymbol{\sigma}/2$ is the electron spin operator. This term is due to the interaction of the electron's magnetic moment with the magnetic field it sees due to its motion and automatically includes the *Thomas precession*, which reduces the result naively expected by a factor of 2. It is zero in S-states, because $L = 0$.

The Darwin term contributes only to $L = 0$ states, and the spin orbit term only to states where $L \neq 0$, but when both corrections are taken into account, the spin orbit splitting is given by a single formula which depends only on the principal quantum number and the *total angular momentum* j of the state,

$$T = -m\frac{(Z\alpha)^2}{2n^2} - m\frac{(Z\alpha)^4}{2n^4}\left[\frac{n}{j+\frac{1}{2}} - \frac{3}{4}\right] . \tag{5.73}$$

The first term is the familiar nonrelativistic result, and the second is the fine structure, which splits states with the same n but different j. In the next chapter we will show that the *exact* solutions of the Dirac equation also predict levels which depend on n and j only. This gives a good account of the main features of the hydrogen atom spectrum, but the additional L-dependent Lamb shift can only be explained by field theory, as we discussed in Chapter 3.

Zeeman Effect (Dirac)

The full Zeeman effect comes from two terms. The orbital part is the same as the result obtained from the KG equation and was calculated in Eq. (4.60). The result is

$$-\frac{e}{2m}(\boldsymbol{p}\cdot\boldsymbol{A} + \boldsymbol{A}\cdot\boldsymbol{p}) = -\frac{e}{2m}\boldsymbol{B}\cdot\boldsymbol{L} .$$

Combining this with the spin part, $-e\boldsymbol{B}\cdot\boldsymbol{\sigma}/2m$, gives

$$\boxed{H_{\text{Zeeman}} = -\frac{e}{2m}\boldsymbol{B}\cdot(\boldsymbol{L}+\boldsymbol{\sigma}) = -\frac{e}{2m}\boldsymbol{B}\cdot(\boldsymbol{L}+2\boldsymbol{S}) .} \tag{5.74}$$

Note the *factor of* 2 for the electron's intrinsic gyromagnetic ratio. This factor has no classical explanation but was discovered empirically before the Dirac equation was discovered. Its automatic appearance in the Dirac theory is one of its *major successes* and provides the only "explanation" for this effect that we have.

5.8 THE LORENTZ GROUP

The Dirac space is four-dimensional but is otherwise an abstract space unrelated to physical space–time. To discuss the Lorentz transformation (LT) of a Dirac wave function, the Dirac equation, or a Dirac matrix element requires that we first construct a *representation* of each Lorentz transformation on the Dirac space and then show that the wave functions and matrix elements transform in such a way that the Dirac equation is invariant in form and the matrix elements transform as scalars, four-vectors, or tensors, depending on their structure. In this section the properties of the Lorentz group will be reviewed, and in the next two sections the representation of the Lorentz group on the Dirac space will be worked out and the construction and transformation of Dirac matrix elements will be discussed.

In Sec. 2.1 we discussed how Lorentz transformations change the space–time coordinates. Any transformation which leaves the metric tensor invariant is, by definition, a LT. In the matrix notation, Eq. (2.8), this was written

$$\Lambda^{\mathsf{T}} G \Lambda = G \ .$$

The set of all transformations which satisfy this constraint form a group, which is called the homogeneous Lorentz group. The four group properties are easily demonstrated:

- If Λ_1 and Λ_2 are members of the group, then $\Lambda_1\Lambda_2$ is also, because

$$\Lambda_2^{\mathsf{T}}\Lambda_1^{\mathsf{T}} G \Lambda_1\Lambda_2 = \Lambda_2^{\mathsf{T}} G \Lambda_2 = G \ .$$

- The multiplication law (matrix multiplication in this case) is associative:

$$(\Lambda_1\Lambda_2)\,\Lambda_3 = \Lambda_1\,(\Lambda_2\Lambda_3) \ .$$

- There exists an identity $\Lambda = \mathbf{1}$ which is a Lorentz transformation.
- For each Λ, there exists an inverse Λ^{-1} because

$$\Lambda^{\mathsf{T}} G \Lambda = G \Longrightarrow (\det \Lambda)^2 = 1 \tag{5.75}$$

and hence $\det \Lambda = \pm 1$, and since it is not zero, Λ^{-1} exists. Multiplying Eq. (2.8) by $\left(\Lambda^{\mathsf{T}}\right)^{-1}$ from the left and Λ^{-1} from the right gives

$$G = \left(\Lambda^{-1}\right)^{\mathsf{T}} G \Lambda^{-1}$$

showing that Λ^{-1} is a Lorentz transformation.

The physical Lorentz transformations are real (because they map a real space onto a real space) but complex LT's are very important to the proof of the PCT theorem (see Chapter 8).

The real transformations can be separated into four classes. First, from Eq. (5.75), they may be separated according to whether or not their determinant is $+1$ (proper transformations) or -1 (improper transformations). Next, writing the defining relation (2.8) in block matrix form,

$$\Lambda^{\mathsf{T}} G \Lambda = \left(\begin{array}{c|c} \Lambda_{00} & \Lambda_{j0} \\ \hline \Lambda_{0i} & \Lambda^{\mathsf{T}}_{ji} \end{array}\right) \left(\begin{array}{c|c} 1 & \\ \hline & -1 \end{array}\right) \left(\begin{array}{c|c} \Lambda_{00} & \Lambda_{0i} \\ \hline \Lambda_{j0} & \Lambda_{ji} \end{array}\right) = G\,, \tag{5.76}$$

shows that the 00 component of the LT satisfies the following relation:

$$\left(\Lambda^{\mathsf{T}} G \Lambda\right)_{00} = \Lambda_{00}^2 - \sum_j (\Lambda_{j0})^2 = 1 \;. \tag{5.77}$$

Therefore, for *real* transformations, the values of Λ_{00} must satisfy one of the two conditions

$$\begin{array}{ll} \Lambda_{00} \geq 1 & \text{orthochronous} \\ \Lambda_{00} \leq -1 & \text{non-orthochronous}\;. \end{array} \tag{5.78}$$

Together with the condition on the determinant, there are therefore four classes of real transformations. Since Λ_{00} cannot be changed *continuously* from a value greater than 1 to a value less than 1, and the determinant cannot be continuously changed from $+1$ to -1, these four classes are disconnected, as illustrated in Fig. 5.3. LT's in each class can be continuously deformed into any other LT in that class, and the different classes can therefore be characterized by one of the four basic transformations: 1, T, P, or TP, where T is time inversion and P is space inversion (parity). Explicitly,

$$T = \left(\begin{array}{c|c} -1 & \\ \hline & 1 \end{array}\right) \qquad P = \left(\begin{array}{c|c} 1 & \\ \hline & -1 \end{array}\right) \;. \tag{5.79}$$

The properties of the four classes of Lorentz transformations are summarized in Table 5.2. The restricted group is a subgroup, but none of the others are groups because they have no identity. It can be shown that every $\Lambda \in L_-^{\downarrow}$ can be written as a product $T\Lambda'$, where $\Lambda' \in L_+^{\uparrow}$, and hence T can be viewed as a mapping from $L_+^{\uparrow}$ to $L_-^{\downarrow}$:

$$TL_+^{\uparrow} = L_-^{\downarrow} \;. \tag{5.80}$$

Similarly, P maps $L_+^{\uparrow}$ to $L_-^{\uparrow}$ and TP maps $L_+^{\uparrow}$ to $L_+^{\downarrow}$:

$$\begin{aligned} PL_+^{\uparrow} &= L_-^{\uparrow} \\ TPL_+^{\uparrow} &= L_+^{\downarrow} \;. \end{aligned} \tag{5.81}$$

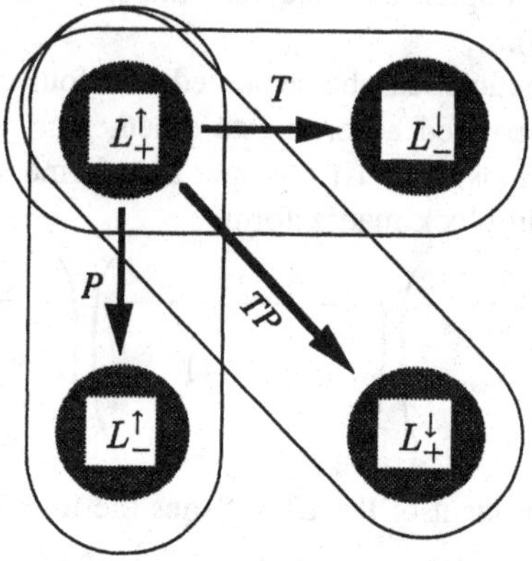

Fig. 5.3 Diagrammatic representation of the four classes of the homogeneous Lorentz group connected by the discrete transformations *T*, *P*, and *TP*.

Figure 5.3 illustrates this continuity by showing the four classes as disconnected regions, with a continuous distribution of transformations within each region (class). The figure and the above equations show that to study the homogeneous Lorentz group, it *is sufficient to study the group of continuous transformations* $L_+^\uparrow$ and the *two discrete transformations* T and P.

The *complex* LT's must also have $\det \Lambda = \pm 1$, but the restriction (5.78) on Λ_{00} no longer holds [because $(\Lambda_{j0})^2$ need no longer be positive]. Therefore

Table 5.2 Properties of the four classes of homogeneous Lorentz transformations.

Label	*Properties*	*Class*	*Continuity with*
$L_+^\uparrow$	$\Lambda_{00} \geq 1$ $\det \Lambda = +1$	orthochronous, proper restricted group	1
$L_+^\downarrow$	$\Lambda_{00} \leq -1$ $\det \Lambda = +1$	non-orthochronous, proper	TP
$L_-^\uparrow$	$\Lambda_{00} \geq 1$ $\det \Lambda = -1$	orthochronous, improper	P
$L_-^\downarrow$	$\Lambda_{00} \leq -1$ $\det \Lambda = -1$	non-orthochronous improper	T

the complex LT's separate into *only two* classes, depending on the sign of the determinant, and the transformations in $L_+^\uparrow$ and $L_+^\downarrow$ can now be connected by a continuous path. As an example of such a "path," consider the transformations

$$\Lambda_\theta = \left(\begin{array}{c|ccc} \cos\theta & & & i\sin\theta \\ \hline & \cos\theta & -\sin\theta & \\ & \sin\theta & \cos\theta & \\ i\sin\theta & & & \cos\theta \end{array}\right) \tag{5.82}$$

which depend on the continuous parameter θ. These transformations satisfy (2.8) for all values of θ and hence map out a continuous path of transformations in the space of complex LT's. By varying θ continuously from 0 to π, we are able to connect the transformations **1** and **−1**. This fact will be of crucial importance to our discussion of the PCT theorem in Sec. 8.7.*

Infinitesimal Transformations in $L_+^\uparrow$

Consider the real LT's in the subgroup $L_+^\uparrow$. Because they can be continuously connected to the identity, they can be written

$$\Lambda = e^{\theta\lambda} \quad , \tag{5.83}$$

where θ is a number and λ is said to be the *generator* of the transformation Λ. [This is assumed without proof. It is a general property of a continuous group.] The structure of the group can be inferred from the structure of the generators λ.

To study this structure, it is sufficient to consider those transformations for which $\theta = \epsilon$ is infinitesimally small. In this case, the transformations can be expanded and only the first order terms retained, so that

$$\Lambda = 1 + \epsilon\lambda \ . \tag{5.84}$$

Since Λ is a LT,

$$\left(1 + \epsilon\lambda^\mathsf{T}\right) G \left(1 + \epsilon\lambda\right) = G \tag{5.85}$$

or, to first order in ϵ,

$$\begin{aligned} \lambda^\mathsf{T} G + G\lambda &= 0 \\ \lambda^\mathsf{T} &= -G\lambda G \ . \end{aligned} \tag{5.86}$$

It is easy to determine the structure of λ from this equation, which looks like

$$\begin{pmatrix} \lambda_{11} & \lambda_{21} & \lambda_{31} & \lambda_{41} \\ \lambda_{12} & \lambda_{22} & \lambda_{32} & \cdots \\ \lambda_{13} & \lambda_{23} & \cdots & \\ \cdots & \cdots & & \end{pmatrix} = \begin{pmatrix} -\lambda_{11} & \lambda_{12} & \lambda_{13} & \lambda_{14} \\ \lambda_{21} & -\lambda_{22} & -\lambda_{23} & \cdots \\ \lambda_{31} & -\lambda_{32} & \cdots & \\ \cdots & \cdots & & \end{pmatrix} \ . \tag{5.87}$$

*For more discussion of these issues see Streater and Wightman (1964).

From this equation we draw the following conclusions:

- All diagonal elements of λ are zero.
- There are three independent λ's which are symmetric. These have space–time components and are the generators of boosts.
- There are three independent λ's which are antisymmetric. These have space–space components and are the generators of rotations.

The generators therefore span a six-dimensional vector space. The six independent generators which will be taken to be basis vectors for this space are denoted $\omega_{\mu\nu}$, where $\mu \neq \nu$ and $\omega_{\mu\nu} = 1$ in the μth column and νth row, is symmetric or antisymmetric depending on the indices, and has zeros for all other elements. Explicitly,

$$\omega_{10} = \omega_{01} = \left(\begin{array}{c|ccc} 0 & 1 & 0 & 0 \\ \hline 1 & & & \\ 0 & & 0 & \\ 0 & & & \end{array}\right) \qquad \omega_{12} = -\omega_{21} = \left(\begin{array}{c|ccc} 0 & & 0 & \\ \hline & 0 & -1 & 0 \\ 0 & 1 & 0 & 0 \\ & 0 & 0 & 0 \end{array}\right)$$

$$\omega_{20} = \omega_{02} = \left(\begin{array}{c|ccc} 0 & 0 & 1 & 0 \\ \hline 0 & & & \\ 1 & & 0 & \\ 0 & & & \end{array}\right) \qquad \omega_{23} = -\omega_{32} = \left(\begin{array}{c|ccc} 0 & & 0 & \\ \hline & 0 & 0 & 0 \\ 0 & 0 & 0 & -1 \\ & 0 & 1 & 0 \end{array}\right)$$

$$\omega_{30} = \omega_{03} = \left(\begin{array}{c|ccc} 0 & 0 & 0 & 1 \\ \hline 0 & & & \\ 0 & & 0 & \\ 1 & & & \end{array}\right) \qquad \omega_{31} = -\omega_{13} = \left(\begin{array}{c|ccc} 0 & & 0 & \\ \hline & 0 & 0 & 1 \\ 0 & 0 & 0 & 0 \\ & -1 & 0 & 0 \end{array}\right) . \tag{5.88}$$

These generators can be written in the following compact form:

$$(\omega_{\mu\nu})^{\alpha}{}_{\beta} = -\tfrac{1}{2}\epsilon_{\mu\nu\lambda\sigma}\epsilon^{\lambda\sigma\alpha}{}_{\beta} \;, \tag{5.89}$$

where $\epsilon_{\mu\nu\alpha\beta}$ is the four-dimensional antisymmetric symbol normalized to

$$\epsilon_{0123} = 1 \;. \tag{5.90}$$

The ω's are the basis for a six-dimensional space of 4×4 traceless matrices, so that any generator is now described by six continuous parameters. The most general infinitesimal LT is then

$$\Lambda = 1 + \xi_i\,\omega_{i0} + \tfrac{1}{2}\theta_i\,\epsilon_{ijk}\omega_{jk} \;. \tag{5.91}$$

The continuous parameters are ξ_i and θ_i. By considering a succession of infinitesimal transformations, we can exponentiate this expression and write the finite transformations as in Eq. (5.83),

$$\Lambda = e^{\left(\xi_i\,\omega_{i0} + \frac{1}{2}\theta_i\,\epsilon_{ijk}\omega_{jk}\right)} \;. \tag{5.92}$$

Equation (5.92) is the explicit characterization of the LT's in $L_+^\uparrow$ which we have been seeking. To better understand this equation, we look at a few examples.

Examples

First, consider a *rotation through angle θ about the z-axis*. In this case $\theta = \theta_3$, and [compare with Eq. (2.92)]

$$R_z(\theta) = e^{\theta\,\omega_{12}} = 1 + \theta\,\omega_{12} + \frac{1}{2}\theta^2\omega_{12}^2 + \frac{1}{3!}\theta^3\omega_{12}^3 + \dots$$

$$= \begin{bmatrix} 1 & 0 & 0 & 0 \\ 0 & 1-\frac{1}{2}\theta^2 & -\theta+\frac{1}{3!}\theta^3 & 0 \\ 0 & \theta-\frac{1}{3!}\theta^3 & 1-\frac{1}{2}\theta^2 & 0 \\ 0 & 0 & 0 & 1 \end{bmatrix} = \begin{bmatrix} 1 & 0 & 0 & 0 \\ 0 & \cos\theta & -\sin\theta & 0 \\ 0 & \sin\theta & \cos\theta & 0 \\ 0 & 0 & 0 & 1 \end{bmatrix} . \tag{5.93}$$

This corresponds to an *active* vector transformation, because a unit vector in the $\hat{x}$-direction is rotated into the first quadrant,

$$\hat{x} = \begin{pmatrix} 0 \\ 1 \\ 0 \\ 0 \end{pmatrix} \quad \Longrightarrow \quad R_z(\theta)\,\hat{x} = \begin{pmatrix} 0 \\ \cos\theta \\ \sin\theta \\ 0 \end{pmatrix} . \tag{5.94}$$

Next, consider a boost in the x-direction. The generator for this boost is ω_{10}, so that

$$L_x(\xi) = e^{\xi\omega_{10}} = 1 + \xi\omega_{10} + \frac{1}{2}\xi^2\omega_{10}^2 + \frac{1}{3!}\xi^3\omega_{10}^3 + \dots$$

$$= \begin{bmatrix} 1+\frac{1}{2}\xi^2 & \xi+\frac{1}{3!}\xi^3 & 0 & 0 \\ \xi+\frac{1}{3!}\xi^3 & 1+\frac{1}{2}\xi^2 & 0 & 0 \\ 0 & 0 & 1 & 0 \\ 0 & 0 & 0 & 1 \end{bmatrix} = \begin{bmatrix} \cosh\xi & \sinh\xi & 0 & 0 \\ \sinh\xi & \cosh\xi & 0 & 0 \\ 0 & 0 & 1 & 0 \\ 0 & 0 & 0 & 1 \end{bmatrix} . \tag{5.95}$$

Hence in coordinate or momentum space

$$\begin{aligned} t' &= t\cosh\xi + x\sinh\xi & \qquad E &= m\cosh\xi \\ x' &= t\sinh\xi + x\cosh\xi & p_x &= m\sinh\xi \\ y' &= y & p_y &= 0 \\ z' &= z & p_z &= 0 \ . \end{aligned} \tag{5.96}$$

These transformations permit us to identify the velocity of the boosted particle with the parameter ξ, which is referred to as the *rapidity*

$$v = v/c = \tanh\xi \ . \tag{5.97}$$

Then, using the familiar relations between the hyperbolic functions leads to the correspondence

$$\cosh\xi = \frac{1}{\sqrt{1-v^2}} \qquad \sinh\xi = \frac{v}{\sqrt{1-v^2}} \tag{5.98}$$

and the familiar form for the *active* boost

$$\begin{aligned} t' &= \frac{1}{\sqrt{1-v^2}}(t+vx) \qquad & E &= \frac{m}{\sqrt{1-v^2}} \\ x' &= \frac{1}{\sqrt{1-v^2}}(x+vt) \qquad & p_x &= \frac{mv}{\sqrt{1-v^2}} \quad . \end{aligned} \tag{5.99}$$

The active boost in the $\hat{x}$-direction propels a particle of mass m from rest into motion along the x-axis with momentum p_x.

5.9 COVARIANCE OF THE DIRAC EQUATION

Now we are ready to study the covariance of the Dirac theory. To establish covariance we must construct a *representation* of the Lorentz group on the four-dimensional Dirac space. In general, a representation of a group is a mapping of each element of the group Λ into a matrix $S(\Lambda)$ which preserves the group multiplication law. This means that if $\Lambda_1\,\Lambda_2 = \Lambda_3$, then $S(\Lambda_1)\,S(\Lambda_2) = S(\Lambda_3)$. Since each group element has an inverse, the matrices which represent the group must also be non-singular, and the identity of the group is represented by the identity matrix.

The representation $S(\Lambda)$ we seek should operate on the four-dimensional Dirac space in such a way that the Dirac equation is invariant in form. For this purpose we use the covariant form, Eq. (5.12), with the γ^μ matrices defined in Eq. (5.8),

$$\left(\gamma^\mu \left[p_\mu - eA_\mu(x)\right] - m\right)\psi(x) = 0 \quad .$$

Then, for any LT Λ which transforms the coordinates and four-vector potential from an unprimed frame to a primed frame,

$$x' = \Lambda x \qquad A'(x') = \Lambda A(x) \quad , \tag{5.100}$$

we seek a representation, $S(\Lambda)$, which transforms the Dirac wave function from the unprimed to the primed coordinate system

$$\boxed{\psi'(x') = S(\Lambda)\psi(x) \quad .} \tag{5.101}$$

Covariance is the requirement that this transformation leave the Dirac equation (5.12) invariant in form, so that in the primed frame,

$$\left(\gamma^\mu \left[p'_\mu - eA'_\mu(x')\right] - m\right)\psi'(x') = 0 \quad .$$

This requirement determines $S(\Lambda)$. To find the equation which defines $S(\Lambda)$, substitute (5.101) into the above equation and multiply by $S^{-1}(\Lambda)$. Recall that $p'^{\mu} = \Lambda^{\mu}{}_{\nu}p^{\nu}$ implies that $p'_{\mu} = \left(\Lambda^{-1}\right)^{\nu}{}_{\mu}\, p_{\nu}$, and obtain

$$S^{-1}(\Lambda)\left\{\gamma^{\mu}\left(\Lambda^{-1}\right)^{\nu}{}_{\mu}\left(p_{\nu} - eA_{\nu}(x)\right) - m\right\} S(\Lambda)\psi(x) = 0 \ .$$

This equation is invariant in form if

$$\left(\Lambda^{-1}\right)^{\nu}{}_{\mu} S^{-1}(\Lambda)\gamma^{\mu} S(\Lambda) = \gamma^{\nu} \ ,$$

which implies

$$\boxed{S^{-1}(\Lambda)\gamma^{\mu} S(\Lambda) = \Lambda^{\mu}{}_{\nu}\gamma^{\nu} \ .} \tag{5.102}$$

This equation will tell us how to construct the $S(\Lambda)$.

Each $\Lambda \in L_{+}^{\uparrow}$ has the form given in Eq. (5.92) and is defined by six numbers $\{\xi_i, \theta_i\}$. The existence of a representation of the Lorentz group on the Dirac space implies that, for *every* choice of the six parameters, there exists a corresponding $S(\Lambda)$ of the form

$$S(\Lambda) = e^{\left(\xi_i B_i + \frac{1}{2}\theta_i \epsilon_{ijk} R_{jk}\right)} \ , \tag{5.103}$$

with the *same* six parameters but with new generators which describe how the transformations act on the Dirac space. To find all of the representations, we need only construct the six generators.

To find the generators, it is sufficient to apply (5.102) to all infinitesimal transformations. If the parameters are infinitesimal, then

$$\begin{aligned} S(\Lambda) &= 1 + \xi_i B_i + \tfrac{1}{2}\theta_i \epsilon_{ijk} R_{jk} \\ \Lambda &= 1 + \xi_i\, \omega_{i0} + \tfrac{1}{2}\theta_i\, \epsilon_{ijk}\omega_{jk} \ , \end{aligned}$$

and Eq. (5.102) becomes

$$\begin{aligned} S^{-1}(\Lambda)\gamma^{\mu} S(\Lambda) &= \left(1 - \xi_i B_i - \tfrac{1}{2}\theta_i \epsilon_{ijk} R_{jk}\right)\gamma^{\mu}\left(1 + \xi_i B_i + \tfrac{1}{2}\theta_i \epsilon_{ijk} R_{jk}\right) \\ &\cong \gamma^{\mu} - \xi_i\left[B_i, \gamma^{\mu}\right] - \tfrac{1}{2}\theta_i \epsilon_{ijk}\left[R_{jk}, \gamma^{\mu}\right] \\ &= \left[1 + \xi_i \omega_{i0} + \tfrac{1}{2}\theta_i \epsilon_{ijk}\omega_{jk}\right]^{\mu}{}_{\nu}\,\gamma^{\nu} \ . \end{aligned} \tag{5.104}$$

Since the parameters ξ_i and θ_i are independent, we must equate their coefficients, giving

$$\begin{aligned} -\left[B_i\, , \gamma^{\mu}\right] &= \left(\omega_{i0}\right)^{\mu}{}_{\nu}\gamma^{\nu} \\ -\left[R_{jk}\, , \gamma^{\mu}\right] &= \left(\omega_{jk}\right)^{\mu}{}_{\nu}\gamma^{\nu} \ . \end{aligned}$$

Substituting the specific forms for the ω's given in Eq. (5.88) gives the following results for the boosts, B_i:

$$\left.\begin{aligned} \left[\gamma^{0}, B_i\right] &= \gamma^{i} \\ \left[\gamma^{j}, B_i\right] &= \delta_{ji}\gamma^{0} \end{aligned}\right\} \Longrightarrow B_i = \tfrac{1}{2}\gamma^{0}\gamma^{i} = \tfrac{1}{2}\alpha_i \ . \tag{5.105}$$

(Remember that only for four-vectors is the placement of the index important; $\gamma^i = -\gamma_i$, but the α's are always written α_i.) To describe the generators of rotations, R_{jk}, we introduce two new Dirac matrices which will be used frequently in the following sections:

$$\sigma_j = \begin{pmatrix} \sigma_j & 0 \\ 0 & \sigma_j \end{pmatrix} \qquad \gamma^5 = \begin{pmatrix} 0 & 1 \\ 1 & 0 \end{pmatrix} \quad . \tag{5.106}$$

In terms of these matrices, the generators of rotations become

$$\left.\begin{aligned} \left[\gamma^0, R_{jk}\right] &= 0 \\ \left[\gamma^\ell, R_{jk}\right] &= -\delta_{\ell j}\gamma^k + \delta_{\ell k}\gamma^j \end{aligned}\right\} \quad \Longrightarrow \quad \begin{aligned} R_{jk} &= \frac{1}{2}\gamma^j\gamma^k = -\frac{i}{2}\epsilon_{jk\ell}\sigma_\ell \\ &= -\frac{i}{2}\epsilon_{jk\ell}\gamma^5\alpha_\ell \quad . \end{aligned} \tag{5.107}$$

Remembering that $j \neq k$, this last equation follows from

$$\begin{aligned} \frac{1}{2}\gamma^j\gamma^k &= \frac{1}{2}\begin{pmatrix} 0 & \sigma_j \\ -\sigma_j & 0 \end{pmatrix}\begin{pmatrix} 0 & \sigma_k \\ -\sigma_k & 0 \end{pmatrix} = -\frac{1}{2}\begin{pmatrix} \sigma_j\sigma_k & 0 \\ 0 & \sigma_j\sigma_k \end{pmatrix} \\ &= -i\frac{1}{2}\epsilon_{jk\ell}\sigma_\ell = -i\frac{1}{2}\epsilon_{jk\ell}\begin{pmatrix} & 1 \\ 1 & \end{pmatrix}\begin{pmatrix} & \sigma_\ell \\ \sigma_\ell & \end{pmatrix} \quad . \end{aligned} \tag{5.108}$$

Digression: The generators can be written in a covariant form. Introduce the matrix

$$\sigma^{\mu\nu} = \frac{i}{2}\left[\gamma^\mu, \gamma^\nu\right] \quad . \tag{5.109}$$

Then

$$\begin{aligned} \sigma^{0i} &= \frac{i}{2}\left[\gamma^0\gamma^i - \gamma^i\gamma^0\right] = i\gamma^0\gamma^i = 2iB_i \\ \sigma^{ij} &= \frac{i}{2}\left[\gamma^i\gamma^j - \gamma^j\gamma^i\right] = i\gamma^i\gamma^j = 2iR_{ij} \qquad i \neq j \ , \end{aligned}$$

and

$$\xi_i B_i + \frac{1}{2}\theta_i\epsilon_{ijk\ell}R_{jk} = -\frac{i}{4}\theta^{\mu\nu}\sigma_{\mu\nu} \ , \tag{5.110}$$

where

$$\begin{aligned} \theta^{\mu\nu} &= -\theta^{\nu\mu} \\ \theta^{i0} = \xi_i \qquad & \theta^{jk} = \theta_i\epsilon_{ijk} \quad . \end{aligned} \tag{5.111}$$

This notation is beautiful but is also cumbersome, and we will not use it. ∎

In the notation we have introduced, the general Lorentz transformation on the Dirac space is

$$\boxed{S(\Lambda) = e^{\frac{1}{2}\boldsymbol{\xi}\cdot\boldsymbol{\alpha} - \frac{i}{2}\gamma^5\boldsymbol{\theta}\cdot\boldsymbol{\alpha}}} \quad . \tag{5.112}$$

We now find the explicit forms of the boosts and rotations.

Boosts in the Dirac Space

The matrices given in Eq. (5.112) for $S(\Lambda)$ can be found in closed form by explicitly summing their power series. For the boosts, the first few terms in this series are

$$S(L_\xi) = 1 + \frac{\boldsymbol{\xi}\cdot\boldsymbol{\alpha}}{2} + \frac{1}{2}\left(\frac{\boldsymbol{\xi}\cdot\boldsymbol{\alpha}}{2}\right)^2 + \frac{1}{3!}\left(\frac{\boldsymbol{\xi}\cdot\boldsymbol{\alpha}}{2}\right)^3 + \cdots \ . \tag{5.113}$$

Since $(\boldsymbol{\xi}\cdot\boldsymbol{\alpha})^2 = \xi^2$, the power series (5.113) is a "repeating" series, with all even powers a multiple of the unit matrix and odd powers a multiple of $\boldsymbol{\alpha}$. Hence

$$\begin{aligned} S(L_\xi) &= \left(1 + \frac{1}{2}\left(\frac{\xi}{2}\right)^2 + \cdots\right) + \hat{\xi}\cdot\boldsymbol{\alpha}\left(\frac{\xi}{2} + \frac{1}{3!}\left(\frac{\xi}{2}\right)^3 + \cdots\right) \\ &= \cosh\frac{\xi}{2} + \hat{\xi}\cdot\boldsymbol{\alpha}\sinh\frac{\xi}{2} \ , \end{aligned} \tag{5.114}$$

where $\xi = |\boldsymbol{\xi}|$ is the magnitude of the vector $\boldsymbol{\xi}$ and $\hat{\xi} = \boldsymbol{\xi}/\xi$ is a unit vector in the direction of $\boldsymbol{\xi}$. This transformation can be expressed in terms of the energy of the particle if we recall from Eq. (5.96) the relation between the rapidity ξ and the energy E imparted by the boost to a particle of mass m,

$$\frac{E}{m} = \cosh\xi \ ,$$

and use relations satisfied by the hyperbolic functions

$$\begin{aligned} \cosh\frac{\xi}{2} &= \sqrt{\frac{\cosh\xi + 1}{2}} = \sqrt{\frac{E+m}{2m}} \\ \sinh\frac{\xi}{2} &= \sqrt{\frac{\cosh\xi - 1}{2}} = \sqrt{\frac{E-m}{2m}} = \sqrt{\frac{E+m}{2m}}\left(\frac{p}{E+m}\right) \ . \end{aligned}$$

Hence, if the boost (5.114) acts on the Dirac wave function of a particle of mass m at rest,

$$\psi_{0,s}(x) = \frac{e^{-imt}}{\sqrt{L^3}}\begin{pmatrix}\chi^{(s)} \\ 0\end{pmatrix} = \frac{e^{-imt}}{\sqrt{2mL^3}}\,u(\mathbf{0}, s) \ ,$$

we obtain

$$\begin{aligned} S(L_\xi)\psi_{0,s}(x) = \psi'_{0,s}(x') &= \frac{e^{-imt}}{\sqrt{L^3}}\sqrt{\frac{E+m}{2m}}\begin{pmatrix} 1 \\ \frac{\boldsymbol{\sigma}\cdot\boldsymbol{p}}{E+m}\end{pmatrix}\chi^{(s)} \\ &= \sqrt{\frac{E}{m}}\,\frac{e^{-i(Et'-\boldsymbol{p}\cdot\boldsymbol{r}')}}{\sqrt{2EL^3}}\,u(\boldsymbol{p}, s) \ . \end{aligned}$$

Recalling the definition of $\psi_{p,s}$, we see that

$$S(L_x)\,\psi_{0,s}(x) = \sqrt{\frac{E}{m}}\,\psi_{p,s}(x') \tag{5.115}$$

and, except for an overall factor $\sqrt{E/m}$, $\psi_{p,s}$ can be obtained from $\psi_{0,s}$ by a Lorentz boost. The non-covariant factor $\sqrt{E/m}$ is present because we have chosen to normalize the Dirac states in a non-covariant manner; this will be discussed further in the next section.

Rotations in Dirac Space

The explicit form for rotations in Dirac space can be found in the same way that the boosts were found. The general rotation about the $\boldsymbol{\theta}$-axis is

$$S(R_\theta) = e^{-\frac{i}{2}\gamma^5\boldsymbol{\theta}\cdot\boldsymbol{\alpha}} = 1 - \frac{i}{2}\gamma^5\boldsymbol{\theta}\cdot\boldsymbol{\alpha} + \frac{1}{2}\left(-\frac{i}{2}\gamma^5\boldsymbol{\theta}\cdot\boldsymbol{\alpha}\right)^2 + \cdots \;. \tag{5.116}$$

Since γ^5 commutes with $\boldsymbol{\alpha}$ and $(\gamma^5)^2 = 1$, it follows that $\left(i\gamma^5\boldsymbol{\theta}\cdot\boldsymbol{\alpha}\right)^2 = -\theta^2$, where $\theta = |\boldsymbol{\theta}|$ is the length of the vector $\boldsymbol{\theta}$, and therefore

$$\begin{aligned} S(R_\theta) &= \left(1 - \frac{1}{2}\left(\frac{1}{2}\theta\right)^2 + \cdots\right) - i\gamma^5\hat{\theta}\cdot\boldsymbol{\alpha}\left(\frac{\theta}{2} - \frac{1}{3!}\left(\frac{\theta}{2}\right)^3 + \cdots\right) \\ &= \cos\frac{\theta}{2} - i\gamma^5\hat{\theta}\cdot\boldsymbol{\alpha}\,\sin\frac{\theta}{2} \;. \end{aligned}$$

As an example, consider a rotation through angle θ about the $\hat{z}$-axis. Recalling that $\gamma^5\boldsymbol{\alpha} = \boldsymbol{\sigma}$, where σ is the diagonal spin matrix (5.106), the action of the rotation about the $\hat{z}$-axis on the Dirac spinor $u(\boldsymbol{p}, s)$ can be written

$$S(R_\theta)\,u(\boldsymbol{p}, s) = \sqrt{E+m}\begin{pmatrix} 1 \\ \frac{s(\theta)\,\boldsymbol{\sigma}\cdot\boldsymbol{p}\,s^\dagger(\theta)}{E+m}\end{pmatrix}\chi^{(s')} \;,$$

where $s(\theta)$ is the representation of the rotation on the 2×2 spinor space, and

$$\chi^{(s')} = s(\theta)\,\chi^{(s)} = \left(\cos\frac{\theta}{2} - i\sigma_z\sin\frac{\theta}{2}\right)\chi^{(s)} = \chi^{(s)}\,d^{(\frac{1}{2})}_{ss'}(\theta) \tag{5.117}$$

is the new two-component spinor which results from the rotation R_θ, familiar from previous studies. The matrix $d^{(\frac{1}{2})}_{ss'}(\theta)$ is the Wigner spin $\frac{1}{2}$ rotation matrix.*

---

*For a discussion of the d functions, see Rose (1957).

The lower components can be further reduced using the properties of the Pauli spin matrices

$$\left(\cos\frac{\theta}{2} - i\sigma_z \sin\frac{\theta}{2}\right) \boldsymbol{\sigma}\cdot\boldsymbol{p} \left(\cos\frac{\theta}{2} + i\sigma_z \sin\frac{\theta}{2}\right)$$
$$= \sigma_z p_z + \sigma_x \left[p_x \cos\theta - p_y \sin\theta\right] + \sigma_y \left[p_y \cos\theta + p_x \sin\theta\right] = \boldsymbol{\sigma}\cdot\boldsymbol{p}' \ .$$

Hence,

$$S(R_z)\psi_{p,s}(x) = \psi_{p',s'}(x') = \psi'_{p,s}(x') \qquad (5.118)$$

and we see that $S(R_z)$ does indeed rotate the state through angle θ about the $\hat{z}$-axis, provided $\chi^{(s)}$ are spin $\frac{1}{2}$ spinors.

Parity

Next, we find the representation of the parity transformation on the Dirac space. Using the defining Eq. (5.102) and the explicit expression for the parity transformation (5.79), we require

$$S^{-1}(P)\gamma^0 S(P) = \gamma^0$$
$$S^{-1}(P)\gamma^i S(P) = -\gamma^i \ .$$

This is satisfied by

$$S(P) = e^{i\phi}\gamma^0 \ , \qquad (5.119)$$

where the phase $e^{i\phi} = \pm 1$ if we require $S(P)^2 = 1$. This phase is related to the intrinsic parity of the particle or state and will be denoted η_P.

Now suppose

$$\psi(r,t) = \begin{pmatrix} F(r,t) \\ G(r,t) \end{pmatrix} \ .$$

Then, under parity

$$\gamma^0 \psi(r,t) = \eta_P \psi'(-r,t) \ , \qquad (5.120)$$

and therefore

$$\psi'(r,t) = \eta_P \begin{pmatrix} F(-r,t) \\ -G(-r,t) \end{pmatrix} \ .$$

If the state has a definite parity, it is unchanged by the transformation, i.e., $\psi' = \psi$. In that case,

$$\begin{aligned} F(r,t) &= \eta_P F(-r,t) \\ G(r,t) &= -\eta_P G(-r,t) \ , \end{aligned} \qquad (5.121)$$

and we find that the upper component of a Dirac wave function has the *same spatial* parity as that of the overall state, while the lower component has the *opposite spatial* parity. This result is due to the action of the γ^0 matrix, which gives an extra phase to the parity transformation on the lower components. We will use this result in the next chapter when we construct the most general solution to the Dirac wave equation.

To complete the discussion of the homogeneous Lorentz group, we need only to find the representation of the time reversal transformation $S(T)$. This will be postponed until Chapter 8, where time reversal will be discussed in some detail.

5.10 BILINEAR COVARIANTS

In this section we discuss the construction matrix elements in Dirac space and the matrix operators from which these matrix elements are constructed. It is convenient to express the most general matrix operator in terms of elementary operators which have definite transformation properties under the homogeneous Lorentz group. These elementary operators are referred to as the *bilinear covariants.*

The study of the covariance of Dirac matrix elements begins with the observation that $S(\Lambda)$ is *not* unitary in every case. This follows from the fact that α_i and $\gamma^5\alpha_i$ are Hermitian operators, but only the generators of rotations have a factor of i in the exponent. Hence $S(R)$ are unitary, while $S(B)$ are not. In fact

$$S(R)^\dagger = S(R)^{-1} \qquad S(B)^\dagger = S(B) \ . \tag{5.122}$$

Hence the density $\psi^\dagger\psi$ is not Lorentz invariant.

To find a Lorentz invariant density, consider a density of the form

$$\rho(x) = \psi^\dagger(x)\theta\psi(x) \ , \tag{5.123}$$

where θ plays the role of a "metric" tensor. Invariance gives

$$\begin{aligned}\rho'(x') = \rho(x) &= \psi'^\dagger(x')\theta\psi'(x') \\ &= \psi^\dagger(x)S^\dagger\theta S\psi(x) \ .\end{aligned} \tag{5.124}$$

Since this must hold for any ψ, we have the requirement

$$S^\dagger\theta S = \theta \ . \tag{5.125}$$

Expanding S in a power series quickly shows that (5.125) is equivalent to the requirement that the "metric" θ commute with the generators of the rotations but *anticommute* with the generators of the boosts,

$$\begin{aligned}\left[\gamma^5\alpha_i, \theta\right] &= 0 \\ \{\alpha_i, \theta\} &= 0 \ .\end{aligned}$$

These conditions are satisfied if $\theta = \gamma^0$, and then, for all Λ

$$\boxed{S^\dagger\gamma^0 S = \gamma^0 \qquad \gamma^0 S^\dagger\gamma^0 = S^{-1} \ .} \tag{5.126}$$

The requirement of covariance has thus led to the introduction of an indefinite metric, similar to the one we encountered in the two-component KG theory (the operator γ^0 plays a role analogous to τ_3). Because this metric must occur in all matrix elements with well-defined covariance properties, it is convenient to introduce the Dirac *adjoint* as follows:

$$\bar{\psi}(x) \equiv \psi^\dagger(x)\gamma^0 \quad . \tag{5.127}$$

This is always a row vector, and a Dirac matrix element will be formed by multiplying from the left by the adjoint spinor $\bar{\psi}(x)$ and from the right by the normal spinor $\psi(x)$. Then

$$\psi'(x') = S(\Lambda)\psi(x)$$
$$\bar{\psi}'(x') = \bar{\psi}(x)S^{-1}(\Lambda)$$

and

$$\rho(x) = \bar{\psi}(x)\psi(x) = \bar{\psi}'(x')\psi'(x') = \rho'(x') \tag{5.128}$$

is a Lorentz invariant scalar density.

All Dirac matrix elements will now be written in the form

$$\bar{\psi}(x)\Gamma\psi(x) \quad ,$$

where Γ is a 4×4 complex matrix. The most general such matrix can always be expanded in terms of 16 independent 4×4 matrices multiplied by complex coefficients. In short, the matrices Γ can be regarded as a *16-dimensional complex vector space* spanned by 16 matrices.

It is convenient to choose the 16 basis matrices, Γ_i, so that they have well-defined transformation properties under LT's. Since the γ^μ's have such properties, we are led to choose the following 16 matrices for this basis:

		# matrices
1	scalar	1
γ^μ	vector	4
$\frac{i}{2}[\gamma^\mu, \gamma^\nu] \equiv \sigma^{\mu\nu}$	antisymmetric tensor	6
$\gamma^5\gamma^\mu$	axial vector	4
$i\gamma^0\gamma^1\gamma^2\gamma^3 \equiv \gamma^5$	pseudoscalar	1
		16

(5.129)

It can be seen by inspection that all of these matrices are linearly independent. Furthermore, their properties under Lorentz transformations are suggested by their labeling. For example

$$\begin{aligned}\rho'^\mu(x') &= \bar{\psi}'(x')\gamma^\mu\psi'(x') = \bar{\psi}(x)S^{-1}(\Lambda)\gamma^\mu S(\Lambda)\psi(x) \\ &= \Lambda^\mu{}_\nu\, \rho^\nu(x) \quad ,\end{aligned} \tag{5.130}$$

the correct transformation law for a vector field. Note that the γ^5 defined above is identical to the one previously introduced in Eq. (5.106) and that an alternative form is

$$\gamma^5 = \frac{i}{24}\epsilon_{\mu\nu\lambda\delta}\gamma^\mu\gamma^\nu\gamma^\lambda\gamma^\delta \ . \tag{5.131}$$

This way of writing γ^5 is useful for proving that γ^5 transforms as a pseudoscalar (see Prob. 5.4). In particular, one can show that

$$S^{-1}(\Lambda)\gamma^5 S(\Lambda) = (\det\Lambda)\,\gamma^5 \tag{5.132}$$

so that

$$\begin{aligned} \rho_5'(x') &= \bar{\psi}'(x')\gamma^5\psi'(x') \\ &= \bar{\psi}(x)S^{-1}(\Lambda)\gamma^5 S(\Lambda)\psi(x) = (\det\Lambda)\,\rho_5(x) \ , \end{aligned} \tag{5.133}$$

which is the correct transformation law for a pseudoscalar if $\Lambda \in L^\uparrow$.

Applications

(1) **Normalization of Dirac wave functions.** Note that the normalization integral, which involves $\psi^\dagger\psi$, can be expressed in terms of the following density:

$$\bar{\psi}(x)\gamma^0\psi(x) \ .$$

This makes it clear that it is the fourth component of a four-current, which is conserved. We already wrote this conservation law in covariant form in Eq. (5.14); in terms of the Dirac adjoint it is

$$\frac{\partial}{\partial x^\mu}\,\bar{\psi}(x)\gamma^\mu\psi(x) = 0 \ . \tag{5.134}$$

The appearance of the factor $\sqrt{E/m}$ in the boost of a Dirac free particle state, Eq. (5.115), can now be understood. The free particle state $\psi_{p,s}(x)$ has been normalized to unity using the normalization condition

$$\int d^3r\,\bar{\psi}_{p,s}(x)\gamma^0\psi_{p,s}(x) = 1 \ . \tag{5.135}$$

Since this condition is the fourth component of a four-vector, the requirement that it be the same in all frames (i.e., behave like a scalar) is inconsistent with its Lorentz nature and must break covariance. We therefore expect a non-covariant factor in the transformation law which carries this state to the rest frame,

$$S^{-1}(B)\psi_{p,s}(x) = N\,\psi_{0,s}(x) \ .$$

The normalization condition (5.135) requires that $N = \sqrt{m/E}$, as already given in Eq. (5.115). To see this, observe that

$$\int d^3r\,\bar{\psi}_{p,s}(x)\gamma^0\psi_{p,s}(x) = \int d^3r\,\bar{\psi}_{p,s}(x)S(L)S^{-1}(L)\gamma^0 S(L)S^{-1}(L)\psi_{p,s}(x)$$
$$= N^2\int d^3r\,\bar{\psi}_{0,s}(x')\Big[\gamma^0\cosh\xi + \gamma\cdot\hat{\xi}\sinh\xi\Big]\psi_{0,s}(x') .$$

But, $\bar{\psi}_{0,s}(x')\gamma^0\psi_{0,s}(x') = 1/L^3$, and because $\psi_{0,s}$ has no lower components and $\gamma\cdot\hat{\xi}$ is off-diagonal, $\bar{\psi}_{0,s}(x')\gamma\cdot\hat{\xi}\psi_{0,s}(x) = 0$. Hence

$$\int d^3r\,\bar{\psi}_{p,s}(x)\gamma^0\psi_{p,s}(x) = N^2\cosh\xi = N^2\frac{E}{m} = 1\ ,$$

which gives the desired result. Thus the *extra non-covariant factor* $1/N = \sqrt{E/m}$ *already incorporated in the definition of* $\psi_{p,s}$ *is just what is needed to insure the state is normalized to 1 in any frame.* Because the normalization condition is non-covariant, the states $\psi_{p,s}$ must include a non-covariant factor.

(2) **Normalization and orthogonality relations for Dirac spinors.** Because of the negative sign in γ^0, the *covariant* normalization and orthogonality relations satisfied by the u and v spinors are:

$$\begin{aligned}\bar{u}(\boldsymbol{p},s)u(\boldsymbol{p},s') &= 2m\,\delta_{ss'}\\ \bar{v}(\boldsymbol{p},s)v(\boldsymbol{p},s') &= -2m\,\delta_{ss'}\\ \bar{u}(\boldsymbol{p},s)v(\boldsymbol{p},s') &= 0\ .\end{aligned} \tag{5.136a}$$

Note that the negative energy v spinors now have *negative norm.* For convenience, the non-covariant versions of (5.136a) are

$$\begin{aligned}u^\dagger(\boldsymbol{p},s)u(\boldsymbol{p},s') &= 2E\delta_{ss'}\\ v^\dagger(\boldsymbol{p},s)v(\boldsymbol{p},s') &= 2E\delta_{ss'}\\ u^\dagger(\boldsymbol{p},s)v(-\boldsymbol{p},s') &= 0\ .\end{aligned} \tag{5.136b}$$

(3) **Energy projection operators.** It is useful to find projection operators which project out the positive and negative energy subspace. The matrices

$$\boxed{\begin{aligned}\Lambda_+(p) &= \frac{1}{2m}\sum_s u(\boldsymbol{p},s)\bar{u}(\boldsymbol{p},s)\\ \Lambda_-(p) &= -\frac{1}{2m}\sum_s v(\boldsymbol{p},s)\bar{v}(\boldsymbol{p},s)\end{aligned}} \tag{5.137}$$

are projection operators with the properties

$$\begin{aligned} \Lambda_+^2 &= \Lambda_+ \qquad & \Lambda_+\Lambda_- &= 0 \\ \Lambda_-^2 &= \Lambda_- \qquad & \Lambda_+ + \Lambda_- &= 1 \ . \end{aligned} \tag{5.138}$$

If any state ψ is expanded in terms of u and v spinors

$$\psi = \sum_s a_s u(\mathbf{p}, s) + \sum_s b_s v(\mathbf{p}, s) \ ,$$

then the operators $\Lambda_\pm$ will project out the separate plus and minus parts

$$\Lambda_+\psi = \sum_s a_s u(\mathbf{p}, s) \qquad \Lambda_-\psi = \sum_s b_s v(\mathbf{p}, s) \ .$$

All of these results follow directly from the orthonormality relations (5.136a).

An alternative form for these projection operators is very useful and is conveniently expressed in terms of the Feynman notation for the scalar product of any four-vector p with the γ matrices,

$$\boxed{\not{p} = p_\mu \gamma^\mu \ .} \tag{5.139}$$

Then, if $p^\mu = (E_p, \mathbf{p})$, the equations satisfied by the u and v spinors, Eqs. (5.19) and (5.26), can be written in the following compact form:

$$\begin{aligned} (\not{p} - m)\, u &= 0 \\ (\not{p} + m)\, v &= 0 \ . \end{aligned} \tag{5.140}$$

Using these equations, it is easy to see that the projection operators can also be written

$$\boxed{\Lambda_\pm = \frac{m \pm \not{p}}{2m} \ .} \tag{5.141}$$

These relations can also be obtained by direct construction from Eqs. (5.137). Using $\not{p}\,\not{p} = p^2 = m^2$, it is a simple matter to prove directly that $\Lambda_\pm^2 = \Lambda_\pm$ and $\Lambda_+\Lambda_- = 0$.

(4) **Spin projection operators.** The spin projection operator for a non-relativistic two-component spinor is

$$\tfrac{1}{2}\,(1 + \boldsymbol{\sigma} \cdot \hat{s})\, \chi^{(s)} = \chi^{(s)} \ , \tag{5.142}$$

where $\hat{s}$ is the unit three-vector in the direction of the spin. For spins in the $\hat{z}$-direction, for example, $\hat{s} = (0, 0, \pm 1)$ for spin up $(+)$ or spin down $(-)$.

Since we use u and v spinors in applications, we define the spin operator so that, for a Dirac particle at rest, $\hat{s} = (0, 0, 1)$ projects out $u(\mathbf{0}, \frac{1}{2})$ and $v(\mathbf{0}, \frac{1}{2})$. Since

$$u\left(\mathbf{0}, \tfrac{1}{2}\right) = \begin{pmatrix} 1 \\ 0 \\ 0 \\ 0 \end{pmatrix} \qquad v\left(\mathbf{0}, \tfrac{1}{2}\right) = \begin{pmatrix} 0 \\ 0 \\ 0 \\ 1 \end{pmatrix} ,$$

this gives

$$\Sigma(\hat{s}) = \frac{1}{2} \begin{pmatrix} 1 + \boldsymbol{\sigma} \cdot \hat{s} & 0 \\ 0 & 1 - \boldsymbol{\sigma} \cdot \hat{s} \end{pmatrix} = \frac{1}{2} \left(1 + \gamma^5 \gamma^\mu \hat{s}_\mu\right) , \tag{5.143}$$

where, *in the rest system of the particle*, $\hat{s}$ is generalized to the four-vector $\hat{s}^\mu = (0, \hat{s})$. Note that this *polarization four-vector* has the same properties as that encountered in Sec. 2.5:

$$\hat{s}^\mu \hat{s}_\mu = -1 \qquad \hat{s} \cdot p_R = \hat{s}_\mu p_R^\mu = 0 , \tag{5.144}$$

where p_R is the four-momentum of the particle at rest. These conditions define the polarization four-vector in any frame, as discussed in Sec. 9.10.

To find the spin projection operator in any frame, use the invariance of the equation

$$\Sigma(\hat{s}) u(\mathbf{0}, \hat{s}) = u(\mathbf{0}, \hat{s}) .$$

If $p = \Lambda p_R$, where $p_R^\mu = (m, \mathbf{0})$, then $s = \Lambda \hat{s}$, and

$$\begin{aligned} S(\Lambda)\Sigma(\hat{s})S^{-1}(\Lambda)S(\Lambda)u(\mathbf{0}, \hat{s}) &= S(\Lambda)u(\mathbf{0}, \hat{s}) \\ &= \tfrac{1}{2}\left[1 + \gamma^5 \left(\Lambda^{-1}\right)^\mu{}_\nu \gamma^\nu \hat{s}_\mu\right] u(p, s) \\ &= \tfrac{1}{2}\left[1 + \gamma^5 \gamma^\nu s_\nu\right] u(p, s) = u(p, s) . \end{aligned}$$

Hence, in general,

$$\boxed{\Sigma(\pm s) = \tfrac{1}{2}\left[1 \pm \gamma^5 \not{s}\right] ,} \tag{5.145}$$

where s^μ is any four-polarization vector satisfying the conditions (5.144). (Check that these are projection operators by direct calculation.)

5.11 CHIRALITY AND MASSLESS FERMIONS

In this last section we discuss some of the special properties possessed by Dirac particles with zero mass. These particles are particularly fascinating and may very well exist in nature. If the masses of the neutrinos (see Appendix D) are not exactly zero, they are certainly very small, and the up (u) and down (d) quarks, which make up the first *generation* (Appendix D), are believed to have a free mass (the mass before interactions are turned on) of only a few MeV, so that for many considerations it is an excellent approximation to regard them as massless.

The spinor for a free massless fermion is ($\hat{p} = \mathbf{p}/|\mathbf{p}|$)

$$u(\mathbf{p}, \lambda) = \sqrt{E} \begin{pmatrix} 1 \\ \sigma \cdot \hat{p} \end{pmatrix} \chi_\lambda = \sqrt{E} \begin{pmatrix} 1 \\ 2\lambda \end{pmatrix} \chi_\lambda \ , \tag{5.146}$$

where χ_λ is the two-component spinor of the fermion quantized in the direction of its motion (the *helicity* spinor), so that $\lambda = \pm\frac{1}{2}$ and

$$\sigma \cdot \hat{p}\, \chi_\lambda = 2\lambda\, \chi_\lambda \ . \tag{5.147}$$

For antiparticles,

$$v(\mathbf{p}, \lambda) = \sqrt{E} \begin{pmatrix} \sigma \cdot \hat{p} \\ 1 \end{pmatrix} (-i\sigma_2 \chi_\lambda^*) = \sqrt{E} \begin{pmatrix} -2\lambda \\ 1 \end{pmatrix} (-i\sigma_2 \chi_\lambda^*) \ , \tag{5.148}$$

where the second step follows immediately if we use $\sigma_i\, \sigma_2 = -\sigma_2\, \sigma_i^*$.

Note that the *helicity states* of the massless spinors have upper and lower components which are equal in magnitude. This means that they are eigenfunctions of the operator γ^5

$$\begin{aligned} \gamma^5\, u(\mathbf{p}, \lambda) &= \ \ 2\lambda\, u(\mathbf{p}, \lambda) \\ \gamma^5\, v(\mathbf{p}, \lambda) &= -2\lambda\, v(\mathbf{p}, \lambda) \ . \end{aligned} \tag{5.149}$$

The eigenvalue of the operator γ^5 is referred to as the *chirality* of the state. Introducing the projection operators

$$\mathcal{P}_\pm = \tfrac{1}{2}\left(1 \pm \gamma^5\right) \tag{5.150}$$

and letting $z(\lambda) = u(\mathbf{p}, \lambda)$ or $v(\mathbf{p}, -\lambda)$, then

$$\begin{aligned} \mathcal{P}_+\, z(\ \tfrac{1}{2}) &= z(\ \tfrac{1}{2}) = z_R \qquad & \mathcal{P}_-\, z(\ \tfrac{1}{2}) &= 0 \\ \mathcal{P}_-\, z(-\tfrac{1}{2}) &= z(-\tfrac{1}{2}) = z_L \qquad & \mathcal{P}_+\, z(-\tfrac{1}{2}) &= 0 \ . \end{aligned} \tag{5.151}$$

Particles with helicity $+\frac{1}{2}$ are referred to as *right-handed*, and those with $-\frac{1}{2}$ are *left-handed* (see Fig. 5.4). In this language, the projection operator $\mathcal{P}_+$ projects out right-handed particles and left-handed antiparticles, denoted collectively by z_R, while the operator $\mathcal{P}_-$ projects out left-handed particles and right-handed antiparticles, denoted by z_L. Note that free right-handed and left-handed states retain their identity under the proper Lorentz transformations only if they are

positive helicity:
clockwise (right-handed) spin

negative helicity:
counter-clockwise (left-handed) spin

Fig. 5.4 Illustration of the relative orientation of spin and momentum for right-handed and left-handed particles.

massless, because only in this case is it impossible to change a particle's helicity by bringing it to rest and reversing its direction of motion.

Right- and left-handed states are not invariant under parity, however. Spin is unchanged by parity (for more discussion see Chapter 8), while momentum changes sign, and hence helicity also changes sign. For massless Dirac particles this result follows from the fact that the parity operator changes the right-handed projection operator into a left-handed one:

$$\gamma^0 \mathcal{P}_{\pm} = \mathcal{P}_{\mp} \gamma^0 \quad . \tag{5.152}$$

For this reason, right- and left-handed states were merely a curiosity until it was discovered in the 1960's that parity in not conserved in the weak interactions. We now know that *only left-handed neutrinos interact weakly*, and in the *Standard Model* of the electroweak interactions *only left-handed neutrinos exist*! We postpone further discussion of these points until Chapters 9 and 15.

This completes our introductory discussion of the Dirac equation. In the next chapter we will use this equation to study some interesting problems.

PROBLEMS

5.1 At $t = 0$ the wave function for an electron (normalized in a volume L^3) is known to be

$$\psi(0, \boldsymbol{r}) = \frac{1}{L^{3/2}} \begin{pmatrix} a \\ b \\ c \\ d \end{pmatrix} e^{i\boldsymbol{p}\cdot\boldsymbol{r}} \quad ,$$

where $\boldsymbol{p} = (0, 0, N)$ (N is an integer) and a, b, c, d are independent of $\boldsymbol{r}$ and t and satisfy

$$|a|^2 + |b|^2 + |c|^2 + |d|^2 = 1 \quad .$$

Find the probabilities that the electron is in the following states:

(a) $E > 0$, spin along z-axis.

(b) $E > 0$, spin along $-z$-axis.

(c) $E < 0$, spin along z-axis.

(d) $E < 0$, spin along $-z$-axis.

5.2 An electron scatters from a repulsive spherical Coulomb potential of the form

$$A^0(r) = \begin{cases} 0 & r > R \\ U = \text{constant} & r < R \ . \end{cases}$$

(a) Calculate the unpolarized cross section in first Born approximation (lowest order in A^0). Use the Dirac formalism.

(b) Compare your relativistic result [from (a) above] with the result you would obtain from the Schrödinger equation in first Born approximation.

5.3 Suppose the Coulomb potential transformed relativistically like a scalar field (rather than like the fourth component of a vector field) so that the interaction of the electron with the Coulomb potential would read

$$e\bar{\psi}(x)\psi(x)A_0(x) \qquad \text{(scalar case)}$$

instead of

$$e\bar{\psi}(x)\gamma^0\psi(x)A_0(x) \qquad \text{(vector case)},$$

where in both cases $e\,A_0(x) = -Ze^2/4\pi|\vec{r}| = -Z\alpha/r$. Calculate the differential cross section in the Born approximation for the scalar case and show that, at high energies, both the angular and energy dependence are completely different from the vector case, even though the two differential cross sections are identical at nonrelativistic energies.

5.4 Prove that $\bar{\psi}\gamma^5\psi$ transforms like a pseudoscalar. If $S(\Lambda)\psi = \psi'$, prove that

$$\bar{\psi}'(x')\gamma^5\psi'(x') = (\det\Lambda)\ \bar{\psi}(x)\gamma^5\psi(x) \ .$$

5.5 Consider the following Dirac matrix element:

$$M^\mu(x) = \bar{\psi}(x)\,\sigma^{\mu\nu}\frac{\overrightarrow{\partial}}{\partial x^\nu}\,\psi(x) \ ,$$

where $\sigma^{\mu\nu}$ was defined in Eq. (5.129).

(a) From the structure of M, guess how it transforms under LT's. Write down the transformation law explicitly, using the notation $x' = \Lambda\,x$.

(b) Using the Lorentz transformation properties of the Dirac wave functions, Eq. (5.101), and the property Eq. (5.102), prove that your transformation law is correct or find the correct one.

5.6 [Taken from Bjorken and Drell (1964).] The Dirac equation describing the interaction of a proton or neutron with an applied electromagnetic field will have an additional magnetic moment interaction representing their observed anomalous magnetic moments:

$$\left(i\gamma^\mu \partial_\mu - e_\imath A_\mu \gamma^\mu - \frac{e\,\kappa_i}{4m_i}\sigma_{\mu\nu}F^{\mu\nu} - m_\imath\right)\psi(x) = 0 \ ,$$

where $F^{\mu\nu}$is the electromagnetic field tensor.

(a) For the proton, $i = p$, $e_p = |e|$; for the neutron $i = n$, $e_n = 0$. Verify that the choice of $\kappa_p = 1.79$ and $\kappa_n = -1.91$ corresponds to the observed magnetic moments and check that the additional interaction does not disturb the Lorentz covariance of the equation. Check also that the Dirac Hamiltonian is Hermitian and that probability is conserved in the presence of the additional interaction.

(b) Make a Foldy–Wouthuysen transformation for the neutron, keeping terms up to order $(v/c)^2$. Give a physical interpretation of the individual terms.

(c) Suppose a negatively charged particle of mass m, charge $-e$, and anomalous moment κ is captured by a nucleus of charge Ze. Suppose that $m \gg m_e$, so that screening by the other electrons can be ignored. Calculate the fine structure splitting of the energy levels, and comment on how the splitting depends on κ.

5.7 New diagonal form for the Dirac equation. Paralleling the discussion following Eq. (5.55), we can introduce a FW transformation which will completely eliminate the lower components from the free positive energy solutions and the upper components from the free negative energy solutions. The advantage of such a representation is that it allows us to regard the mixing of upper and lower components as a dynamical consequence of the interaction; the free Dirac equation is fully diagonalized. A unitary transformation which accomplishes this is

$$U = \sqrt{\frac{E+m}{2E}}\begin{pmatrix} 1 & \frac{\boldsymbol{\sigma}\cdot\boldsymbol{p}}{E+m} \\ \frac{-\boldsymbol{\sigma}\cdot\boldsymbol{p}}{E+m} & 1 \end{pmatrix} .$$

(a) Show that U is unitary by direct computation. Show that

$$U\,u(\boldsymbol{p},s) = \sqrt{2E}\begin{pmatrix} \chi^{(s)} \\ 0 \end{pmatrix}$$

$$U\,v(-\boldsymbol{p},-s) = \sqrt{2E}\begin{pmatrix} 0 \\ -i\sigma_2\chi^{(-s)} \end{pmatrix} .$$

(b) Show that

$$U\,H_0\,U^\dagger = \beta E = \begin{pmatrix} E & & & \\ & E & & \\ & & -E & \\ & & & -E \end{pmatrix} ,$$

which is a diagonal form comparable to the one found for the KG equation in Prob. 4.6.

(c) Show that, in this representation, the Dirac equation with electromagnetic interactions can be written

$$i\frac{\partial}{\partial t}\psi = \beta E\psi + \left(H_1 + \beta H_2 + \boldsymbol{\sigma}\cdot\boldsymbol{H}_3 + \beta\boldsymbol{\sigma}\cdot\boldsymbol{H}_4 \right.$$
$$\left. + \boldsymbol{\alpha}\cdot\boldsymbol{H}_5 + i\beta\boldsymbol{\alpha}\cdot\boldsymbol{H}_6 + \gamma^5 H_7\right)\psi \ .$$

Introducing

$$\boldsymbol{D} = \frac{\boldsymbol{p}}{E+m} = \frac{-i\nabla}{E_\nabla + m}$$
$$\theta = \sqrt{\frac{E+m}{2E}} \ ,$$

show that the H's are

$$H_1 = \theta\left(V^0 + D_j V^0 D_j\right)\theta$$
$$H_2 = -\theta\left(V\cdot D + D\cdot V\right)\theta$$
$$H_3^k = \theta\left(i\epsilon_{ijk} D_i V^0 D_j\right)\theta$$
$$\boldsymbol{H}_4 = -\theta\left(i\left(\mathbf{V}\times\mathbf{D}\right) + i\left(\mathbf{D}\times\mathbf{V}\right)\right)\theta$$
$$\boldsymbol{H}_5 = \theta\left(-\mathbf{V} + (D\cdot V)\mathbf{D} + \mathbf{D}(V\cdot D) - D_j\mathbf{V}D_j\right)\theta$$
$$\boldsymbol{H}_6 = \theta\left(i V^0\mathbf{D} - i\mathbf{D}V^0\right)\theta$$
$$H_7 = \theta\left(i\epsilon_{ijk} D_i V_j D_k\right)\theta \ .$$

CHAPTER 6

APPLICATION OF THE DIRAC EQUATION

This chapter begins with a discussion of the general form of the solutions to the Dirac equation for a potential which is spherically symmetric. Using these results, it is an easy matter to find the solutions for a particle confined by a spherically symmetric square well, a simple model for the treatment of the confinement of quarks in hadrons. The chapter concludes with a discussion of the exact solutions for hydrogen-like atoms.

6.1 SPHERICALLY SYMMETRIC POTENTIALS

In many problems of interest, the potential in the Dirac equation is spherically symmetric, i.e., a function of $r = |\boldsymbol{r}|$ only. For example, if the four-vector potential has the form $V^\mu(r) = (V(r), \mathbf{0})$, which is the case for the Coulomb potential, the equation reduces to

$$i\frac{\partial\psi}{\partial t} = \left[-i\alpha_i\frac{\partial}{\partial x^i} + \beta m + V(r)\right]\psi \ . \tag{6.1}$$

This is an array of four coupled partial differential equations and looks like it would be formidable to solve. However, because of the spherical symmetry, it turns out that these equations can be reduced to only two coupled first order ordinary differential equations, which are comparatively easy to solve. This reduction is a good starting point for the study of many interesting problems, two of which will be treated in the subsequent sections.

The equations are reduced by first finding the symmetries of the system and then using these symmetries to express the solutions in terms of the minimum number of unknown functions which are not determined by symmetry and therefore must be determined from the dynamics. We will see that all solutions can be expressed in terms of *only two scalar functions* and that these can be determined from two coupled first order differential equations.

Symmetries of the Motion

Begin with the introduction of the orbital angular momentum operator

$$\boldsymbol{L} = \boldsymbol{r} \times \boldsymbol{p} = -i\,(\boldsymbol{r} \times \nabla) \tag{6.2}$$

and the spin operator

$$\boldsymbol{S} = \tfrac{1}{2}\gamma^5 \boldsymbol{\alpha} \ . \tag{6.3}$$

Remark: Note that this spin operator is *not* the same as the covariant spin operator $\frac{1}{2}\gamma^5\gamma$ introduced in the last chapter. The difference arises from the fact that, as introduced in Eq. (5.29), the *spin up negative energy state*, for a free particle at rest, is proportional to $v(\mathbf{0}, -\frac{1}{2})$, so that

$$\psi^{(-)}_{0,1/2}(x) = \frac{1}{\sqrt{L^3}} e^{imt} \begin{pmatrix} 0 \\ 0 \\ -1 \\ 0 \end{pmatrix}$$

and using the S_z defined in Eq. (6.3) gives

$$S_z\, \psi^{(-)}_{0,1/2}(x) = +\psi^{(-)}_{0,1/2}(x) \ ,$$

as expected. The operator (6.3) therefore identifies the states $\psi^{(+)}_{p,1/2}(x)$ and $\psi^{(-)}_{p,1/2}(x)$ as "spin up" states, which is the correct definition for use in the first quantized treatment of spin $\frac{1}{2}$ particles. In the last chapter we designed the spin operator so that the "spin up" states were proportional to $u(\boldsymbol{p}, \frac{1}{2})$ and $v(\boldsymbol{p}, \frac{1}{2})$, which is the correct one for use in the second-quantized (field theoretic) treatment. ∎

Neither $\boldsymbol{L}$ nor $\boldsymbol{S}$ are constants of the motion

$$\begin{aligned}
\left[L^i, H\right] &= \left[-i\epsilon_{ijk} r^j \partial_k\, , \, -i\alpha_\ell \partial_\ell\right] = \epsilon_{ijk}\delta_{\ell j}\partial_k \alpha_\ell = \epsilon_{i\ell k}\alpha_\ell \partial_k \\
\left[S^i, H\right] &= \frac{1}{2}\left[\gamma^5\alpha_i, -i\alpha_j\partial_j\right] = -\frac{i}{2}\gamma^5\left[\alpha_i, \alpha_j\right]\partial_j \\
&= -\epsilon_{i\ell k}\alpha_\ell \partial_k \ .
\end{aligned} \tag{6.4}$$

However, note that

$$J^i = L^i + S^i \tag{6.5}$$

are constants of the motion. Furthermore,

$$S^2 = S^i S^i = \tfrac{1}{4}\boldsymbol{\alpha}\cdot\boldsymbol{\alpha} = \tfrac{3}{4} \tag{6.6}$$

is also a constant of the motion but will be suppressed since it always has eigenvalue $S^2 = \frac{3}{4} = (\frac{1}{2})(\frac{3}{2})$ which supports the interpretation that *we are describing* *a spin $\frac{1}{2}$ particle.* We leave it as an exercise to show that $\left[L^2, H\right] \neq 0$. Hence L is not in general a good symmetry, but in this case the states will still have a definite value of ℓ because it turns out that ℓ is fixed uniquely by the parity, which is a good symmetry.

The parity operator is

$$\mathcal{P} = \gamma^0 P \ , \tag{6.7}$$

where γ^0 operates on the Dirac space and P operates on the coordinate space. Note that $\mathcal{P}^2 = 1$ and that

$$\begin{aligned} [\mathcal{P}, H] &= \left[\gamma^0 P\, , -i\alpha_i\partial_i\right] \\ &= -i\left\{\gamma^0 P\alpha_i\partial_i - \alpha_i\partial_i\gamma^0 P\right\} \\ &= -i\left\{-\gamma^0\alpha_i\partial_i - \alpha_i\gamma^0\partial_i\right\} P = 0 \ . \end{aligned} \tag{6.8}$$

Hence, the solutions of the Dirac equation in a spherically symmetric potential are characterized by the following conserved quantities:

$$\boxed{E, \ J^2, \ J_z, \ S^2 = \frac{3}{4}, \ \gamma^0 P = \pm 1 \ .} \tag{6.9}$$

We ignore S^2 from now on.

Structure of the Solutions

Consider a solution of the general form

$$\psi(r) = \begin{pmatrix} F(r) \\ G(r) \end{pmatrix} \ , \tag{6.10}$$

where F and G are two-component spinors which can depend on the quantum numbers which characterize the states. Since parity is a good quantum number, F and G have opposite spatial parity, as shown in Eq. (5.121), and the parity of the overall state is the spatial parity of its upper component. Hence we may define $F^{\pm}$ and $G^{\pm}$ with the following properties:

$$\begin{aligned} F^{\pm}(r) &= \pm F^{\pm}(-r) \\ G^{\pm}(r) &= \mp G^{\pm}(-r) \ . \end{aligned} \tag{6.11}$$

The structure of these functions may be further specified by exploiting the rotational symmetry. The total angular momentum operator has the form

$$\boldsymbol{J} = \begin{pmatrix} \boldsymbol{L} + \frac{1}{2}\boldsymbol{\sigma} & 0 \\ 0 & \boldsymbol{L} + \frac{1}{2}\boldsymbol{\sigma} \end{pmatrix} \ . \tag{6.12}$$

Hence it is clear that both the upper and lower components can be expanded in terms of the generalized spherical harmonics $\mathcal{Y}^{\pm}_{jm}(\hat{r})$, which are constructed by vector addition* from the spatial spherical harmonics $Y_{\ell m}(\hat{r})$ and the spin $\frac{1}{2}$ states

$$\alpha = \begin{pmatrix} 1 \\ 0 \end{pmatrix} \qquad \beta = \begin{pmatrix} 0 \\ 1 \end{pmatrix} \ . \tag{6.13}$$

The $\pm$ superscript on the $\mathcal{Y}$'s denotes the parity. With this notation, the states have the overall structure

$$\psi^{\pm}_{jm}(r) = \begin{pmatrix} f^{\pm}(r)\mathcal{Y}^{\pm}_{jm}(\hat{r}) \\ ig^{\pm}(r)\mathcal{Y}^{\mp}_{jm}(\hat{r}) \end{pmatrix} \ , \tag{6.14}$$

where f and g are now functions of the radial coordinate only and the phase factor i multiplying the lower components is introduced for convenience. Note that (6.14) incorporates the results of (6.11) by explicitly using $\mathcal{Y}$'s with opposite parity to describe the upper and lower components. With the construction (6.14), the states are now eigenstates of angular momentum and parity, with the usual properties

$$\begin{aligned} J^2\psi^{\pm}_{jm}(r) &= j(j+1)\psi^{\pm}_{jm}(r) \\ J_z\psi^{\pm}_{jm}(r) &= m\psi^{\pm}_{jm}(r) \\ \mathcal{P}\psi^{\pm}_{jm}(r) &= \pm\psi^{\pm}_{jm}(-r) \ . \end{aligned} \tag{6.15}$$

The total angular momentum quantum number j is half an odd integer, and the commutation relations between the components of $\boldsymbol{J}$ permit us to introduce raising and lowering operators in the usual way:

$$\begin{aligned} J_{\pm} &= J_x \pm iJ_y \\ J_{\pm}\psi_{jm} &= \sqrt{j(j+1) - m(m\pm 1)}\,\psi_{j,m\pm 1} \ . \end{aligned} \tag{6.16}$$

Hence the $\mathcal{Y}$ states can be explicitly constructed using Clebsch–Gordon (CG) coefficients

$$\begin{aligned} \mathcal{Y}_{jm}(\hat{r}) = {} & \langle \ell\; m{-}\tfrac{1}{2};\, \tfrac{1}{2}\; \tfrac{1}{2} | \, j\, m\rangle\, \alpha\, Y_{\ell,m-\frac{1}{2}}(\hat{r}) \\ & + \langle \ell\; m{+}\tfrac{1}{2};\, \tfrac{1}{2}\; {-\tfrac{1}{2}} | \, j\, m\rangle\, \beta\, Y_{\ell,m+\frac{1}{2}}(\hat{r}) \ , \end{aligned} \tag{6.17}$$

where the CG coefficients come from Table 6.1.

As Eq. (6.17) and Table 6.1 show, there are precisely two $\mathcal{Y}$'s for each j (and m, which we ignore in the following discussion). These have values of the orbital angular momentum ℓ equal to $j + \frac{1}{2}$ or $j - \frac{1}{2}$. The parity of the $\mathcal{Y}$'s depends on whether this value of ℓ is even or odd. Once j and the parity are specified, ℓ is uniquely determined. However, instead of designating these states by parity, which is $\pm$, we introduce a ***new quantum number*** k defined in the following way:

$$k = \pm\left(j + \tfrac{1}{2}\right) \qquad \begin{cases} + & \text{if } \ell = j + \frac{1}{2} \quad \Rightarrow \quad k = \ell \\ - & \text{if } \ell = j - \frac{1}{2} \quad \Rightarrow \quad k = -(\ell + 1) \ . \end{cases} \tag{6.18}$$

---

*A general discussion of angular momentum eigenfunctions and the addition of angular momentum can be found, for example, in Rose (1957).

Table 6.1 Clebsch–Gordon (CG) coefficients.

	$\langle \ell\, m{-}m_2; \frac{1}{2}\, m_2 \vert\, j\, m\rangle$	
	$m_2 = \frac{1}{2}$	$m_2 = -\frac{1}{2}$
$j = \ell + \frac{1}{2}$	$\sqrt{\frac{\ell+\frac{1}{2}+m}{2\ell+1}}$	$\sqrt{\frac{\ell+\frac{1}{2}-m}{2\ell+1}}$
$j = \ell - \frac{1}{2}$	$-\sqrt{\frac{\ell+\frac{1}{2}-m}{2\ell+1}}$	$\sqrt{\frac{\ell+\frac{1}{2}+m}{2\ell+1}}$

The quantum numbers j and k now determine the parity quantum number (and therefore also the correct ℓ corresponding to any particular j) as shown in Table 6.2. It is very convenient to re-express the $\mathcal{Y}^{\pm}$ in terms of this quantum number k. This will simplify all subsequent formulae.

To see how this works, first note that the use of k simplifies the CG Table 6.1. When expressed in terms of k, both coefficients in each column (i.e., both parity states) can be expressed by one algebraic expression:

$$\begin{aligned} \langle \ell\, m{-}\tfrac{1}{2}; \tfrac{1}{2}\; \tfrac{1}{2} \vert\, j\, m\rangle &= -\operatorname{sgn} k \sqrt{\frac{k+\frac{1}{2}-m}{2k+1}} \\ \langle \ell\, m{+}\tfrac{1}{2}; \tfrac{1}{2}\; -\tfrac{1}{2} \vert\, j\, m\rangle &= \sqrt{\frac{k+\frac{1}{2}+m}{2k+1}} \quad , \end{aligned} \tag{6.19}$$

where $\operatorname{sgn} k$ is +1 if $k > 0$ and –1 if $k < 0$. Instead of $\mathcal{Y}^{\pm}$ we will use $\mathcal{Y}^{k}$, where

$$\mathcal{Y}^{k}_{jm}(\hat{r}) = -\operatorname{sgn} k \sqrt{\frac{k+\frac{1}{2}-m}{2k+1}}\, \alpha\, Y_{\ell,m-\frac{1}{2}} + \sqrt{\frac{k+\frac{1}{2}+m}{2k+1}}\, \beta\, Y_{\ell,m+\frac{1}{2}} \quad . \tag{6.20}$$

Note that $\mathcal{Y}^{\pm}$ and $\mathcal{Y}^{k}$ are *identical*, provided the identification of k and $\pm$ is made consistent with Table 6.2, and hence the solution can be written in this new notation:

$$\psi^{k}_{jm}(\mathbf{r}) = \begin{pmatrix} f^{k}_{j}(r)\mathcal{Y}^{k}_{jm}(\hat{r}) \\ i g^{k}_{j}(r)\mathcal{Y}^{-k}_{jm}(\hat{r}) \end{pmatrix} \quad . \tag{6.21}$$

Table 6.2 Relationship between *j*, *k*, *ℓ*, and parity.

j	k	ℓ	parity
$\frac{1}{2}$	+1	1	−
	−1	0	+
$\frac{3}{2}$	+2	2	+
	−2	1	−
$\frac{5}{2}$	+3	3	−
	−3	2	+

Properties of the Angular Functions

Next, we prove some useful properties of the $\mathcal{Y}^k_{jm}(\hat{r})$ functions which will permit us to completely eliminate the angular dependence from the Dirac equation. We prove the following three relations:

$$
\begin{aligned}
-i\boldsymbol{\sigma}\cdot\nabla &= -i\boldsymbol{\sigma}\cdot\hat{r}\,\frac{\partial}{\partial r} + i\boldsymbol{\sigma}\cdot\hat{r}\,\frac{\boldsymbol{\sigma}\cdot\boldsymbol{L}}{r} && \text{(a)}\\
\boldsymbol{\sigma}\cdot\hat{r}\,\mathcal{Y}^k_{jm}(\hat{r}) &= -\mathcal{Y}^{-k}_{jm}(\hat{r}) && \text{(b)} \qquad (6.22)\\
\boldsymbol{\sigma}\cdot\boldsymbol{L}\,\mathcal{Y}^k_{jm}(\hat{r}) &= -(k+1)\,\mathcal{Y}^k_{jm}(\hat{r}) \ . && \text{(c)}
\end{aligned}
$$

Proof: To prove (6.22a), use

$$
\frac{1}{r^2}\left[\boldsymbol{r}\times(\boldsymbol{r}\times\nabla)\right]^i = \frac{1}{r^2}\epsilon_{ijk}r^j\epsilon_{k\ell m}r^\ell\partial_m = \frac{1}{r^2}\left(r^i\boldsymbol{r}\cdot\nabla - r^2\nabla_i\right) \ .
$$

Rearranging this gives

$$
\nabla_i = \frac{r^i}{r}\,\frac{\partial}{\partial r} - i\frac{1}{r^2}(\boldsymbol{r}\times\boldsymbol{L})^i \ .
$$

Hence

$$
-i\boldsymbol{\sigma}\cdot\nabla = -i\boldsymbol{\sigma}\cdot\hat{r}\,\frac{\partial}{\partial r} - \frac{1}{r^2}\,\boldsymbol{\sigma}\cdot(\boldsymbol{r}\times\boldsymbol{L}) \ .
$$

To reduce the last term, use $\sigma_i\sigma_j = \delta_{ij} + i\epsilon_{ijk}\sigma_k$ to obtain

$$
\begin{aligned}
\boldsymbol{\sigma}\cdot(\boldsymbol{r}\times\boldsymbol{L}) &= \epsilon_{ijk}\sigma_i\,r^jL^k\\
&= -i\left(\sigma_j\sigma_k - \delta_{jk}\right)r^jL^k = -i\boldsymbol{\sigma}\cdot\boldsymbol{r}\,\boldsymbol{\sigma}\cdot\boldsymbol{L} \ .
\end{aligned}
$$

This proves (a).

To prove (c), note that

$$J^2 = L^2 + S^2 + \boldsymbol{\sigma}\cdot\boldsymbol{L} \ .$$

Hence

$$\begin{aligned}
&\boldsymbol{\sigma}\cdot\boldsymbol{L}\,\mathcal{Y}^k_{jm}(\hat{r}) \\
&= \left\{\begin{matrix} j(j+1) - \frac{3}{4} - \left(j+\frac{1}{2}\right)\left(j+\frac{3}{2}\right) & \text{if } k>0 \Rightarrow \ell = j+\frac{1}{2} \\ j(j+1) - \frac{3}{4} - \left(j-\frac{1}{2}\right)\left(j+\frac{1}{2}\right) & \text{if } k<0 \Rightarrow \ell = j-\frac{1}{2} \end{matrix}\right\} \mathcal{Y}^k_{jm}(\hat{r}) \\
&= \left\{\begin{matrix} -j-\frac{3}{2} \\ j-\frac{1}{2} \end{matrix}\right\} \mathcal{Y}^k_{jm}(\hat{r}) = \left\{\begin{matrix} -k-1 \\ -k-1 \end{matrix}\right\} \mathcal{Y}^j_{jm}(\hat{r}) \ .
\end{aligned}$$

Because of the definition of k, this result holds for both signs of k, and (c) is proved.

Finally, to prove (b), we will first show that $\boldsymbol{\sigma}\cdot\hat{r}\,\mathcal{Y}^k_{jm}(\hat{r})$ has angular momentum quantum numbers j and m and opposite parity from $\mathcal{Y}^k_{jm}(\hat{r})$. This will establish that

$$\boldsymbol{\sigma}\cdot\hat{r}\,\mathcal{Y}^k_{jm}(\hat{r}) = N_j \mathcal{Y}^{-k}_{jm}(\hat{r}) \ , \tag{6.23}$$

where N_j is a constant of proportionality. Then we will show that $N_j = -1$.

To prove (6.23), it is sufficient to show that J^i *commutes* with $\boldsymbol{\sigma}\cdot\hat{r}$ and that the spatial parity operator *anticommutes* with $\boldsymbol{\sigma}\cdot\hat{r}$:

$$\left[J^i, \boldsymbol{\sigma}\cdot\hat{r}\right] = 0 \qquad \text{and} \qquad \{P, \boldsymbol{\sigma}\cdot\hat{r}\} = 0 \ . \tag{6.24}$$

Then it follows that

$$\begin{aligned}
J^i\,\boldsymbol{\sigma}\cdot\hat{r}\,\mathcal{Y}^k_{jm}(\hat{r}) &= \boldsymbol{\sigma}\cdot\hat{r}\,J^i\,\mathcal{Y}^k_{jm}(\hat{r}) \\
P\,\boldsymbol{\sigma}\cdot\hat{r}\,\mathcal{Y}^k_{jm}(\hat{r}) &= -\boldsymbol{\sigma}\cdot\hat{r}\,P\,\mathcal{Y}^k_{jm}(\hat{r}) \ ,
\end{aligned}$$

and since the $\mathcal{Y}$'s are uniquely specified by j, m, and k and the last expression implies that $\boldsymbol{\sigma}\cdot\hat{r}\mathcal{Y}^k_{jm}(\hat{r})$ has opposite parity from $\mathcal{Y}^k_{jm}(\hat{r})$, it follows that $\boldsymbol{\sigma}\cdot\hat{r}\mathcal{Y}^k_{jm}(\hat{r})$ must be proportional to $\mathcal{Y}^{-k}_{jm}(\hat{r})$.

However, the relations (6.24) are readily proved by direct computation. For the first relation,

$$\begin{aligned}
\left[L^i, \boldsymbol{\sigma}\cdot\hat{r}\right] &= -\frac{i}{r}\epsilon_{ijk}r^j\left[\nabla_k, \sigma_m r^m\right] = -\frac{i}{r}\epsilon_{ijk}r^j\sigma_k \\
\left[S^i, \boldsymbol{\sigma}\cdot\hat{r}\right] &= \frac{1}{2r}\left[\sigma_i, \sigma_j r^j\right] = \frac{i}{r}\epsilon_{ijk}r^j\sigma_k \ .
\end{aligned}$$

Hence $\left[J^i, \boldsymbol{\sigma}\cdot\hat{r}\right] = 0$. The second relation follows from

$$\{P, \boldsymbol{\sigma}\cdot\hat{r}\} = \left[P\,\boldsymbol{\sigma}\cdot\hat{r} + \boldsymbol{\sigma}\cdot\hat{r}\,P\right] = \left[-\boldsymbol{\sigma}\cdot\hat{r} + \boldsymbol{\sigma}\cdot\hat{r}\right]P = 0 \ .$$

This completes the proof of (6.23).

To find N_j, we evaluate both sides of Eq. (6.23) for the special case $\theta = \phi = 0$, or $\hat{r} = \hat{z}$. Using

$$Y_{\ell,m_\ell}(\hat{z}) = \sqrt{\frac{2\ell+1}{4\pi}}\,\delta_{m_\ell,0} \ ,$$

we have, for $m = -\frac{1}{2}$ and $k > 0$ so that $k = \ell$,

$$\mathcal{Y}^k_{j,-1/2}(\hat{z}) = \sqrt{\frac{k}{2k+1}}\sqrt{\frac{2k+1}{4\pi}}\beta = \sqrt{\frac{k}{4\pi}}\beta \ ,$$

and if $k < 0$, so that $\ell = -(k+1)$,

$$\mathcal{Y}^{-k}_{j,-1/2}(\hat{z}) = \sqrt{\frac{k}{2k+1}}\sqrt{\frac{-2k-1}{4\pi}}\beta = \sqrt{\frac{-k}{4\pi}}\beta \ .$$

Hence, for both cases,

$$\boldsymbol{\sigma}\cdot\hat{r}\,\mathcal{Y}^k_{j-1/2}(\hat{z}) = \sigma_3\sqrt{\frac{|k|}{4\pi}}\beta = N_j\sqrt{\frac{|k|}{4\pi}}\beta$$

and $N_j = -1$, which completes the proof of (6.22c). ∎

Using Eqs. (6.22) it is a simple matter to reduce the Dirac equation to two coupled differential equations.

Reduction of the Equations

Assuming a solution of the form $\psi(x) = \psi^k_{jm}(r)\,e^{-iEt}$, the Dirac equation now becomes

$$E\psi^k_{jm}(r) = \begin{bmatrix} m + V(r) & -i\boldsymbol{\sigma}\cdot\hat{r}\,\frac{\partial}{\partial r} + i\boldsymbol{\sigma}\cdot\hat{r}\,\frac{\boldsymbol{\sigma}\cdot\boldsymbol{L}}{r} \\ -i\boldsymbol{\sigma}\cdot\hat{r}\,\frac{\partial}{\partial r} + i\boldsymbol{\sigma}\cdot\hat{r}\,\frac{\boldsymbol{\sigma}\cdot\boldsymbol{L}}{r} & -m + V(r) \end{bmatrix} \psi^k_{jm}(r) \ . \tag{6.25}$$

For the solution with the structure given in Eq. (6.21), the identities (6.22) may now be used to reduce these coupled equations to

$$E\,f^k_j(r) = [\ \ m + V(r)]\,f^k_j(r) - \frac{(1-k)}{r}g^k_j(r) - \frac{dg^k_j(r)}{dr}$$

$$E\,g^k_j(r) = [-m + V(r)]\,g^k_j(r) + \frac{(1+k)}{r}f^k_j(r) + \frac{df^k_j(r)}{dr} \ ,$$

where the first equation is the coefficient of the angular function $\mathcal{Y}^k_{jm}(\hat{r})$ and the second the coefficient of $\mathcal{Y}^{-k}_{jm}(\hat{r})$. The angular variables have thus been completely removed from the equation, leaving only two unknown functions of the radial coordinate. Rearranging terms and dropping the subscripts and superscripts give

$$\boxed{\begin{aligned}\left[E-m-V(r)\right]f(r) &= -\frac{dg(r)}{dr}-\frac{(1-k)}{r}g(r)\\ \left[E+m-V(r)\right]g(r) &= \frac{df(r)}{dr}+\frac{(1+k)}{r}f(r)\,.\end{aligned}} \tag{6.26}$$

This completes the task for this section. In the next section these equations are solved for a constant potential.

6.2 HADRONIC STRUCTURE

As an illustration of the modern use of the Dirac equation, we give a very simple introductory discussion of the structure of hadrons (strongly interacting particles).*

There is strong experimental evidence to support the view that mesons and baryons are composed of elementary spin $\frac{1}{2}$ particles called "quarks." (For a brief summary of the particles of modern physics, see Appendix D.) Mesons are believed to be composed of a "valence" quark (q) and antiquark ($\bar{q}$) pair, surrounded by a "sea" of gluons and other $q\bar{q}$ pairs, and baryons composed of three valence quarks surrounded by a similar sea. Furthermore, quarks are believed to exist only in the combinations of quarks and antiquarks which exist in baryons and mesons. If we attempt to remove a single quark from such a combination, the energy grows with the distance the quark is separated from its neighbors, until it becomes so large that it is energetically favorable to create a $q\bar{q}$ pair and break the "string" connecting the quark to its neighbors. The situation is similar to trying to isolate a single north or south pole of a magnet; if we cut the magnet apart, a new pair of poles is created, defeating our purpose. Because of this property of the forces which bind quarks together, they are said to be *confined*.

The MIT bag model is a very simple model for hadronic structure.† Suppose the hadron occupies a spherical volume of radius R. If a quark is inside this volume, we assume its mass is small, and it may be taken to be zero. If it gets outside, interactions with the neighboring quarks which make up the rest of the hadron are assumed to generate an infinite mass for the quark. Since this implies infinite energy, the quark will not penetrate outside of the hadronic volume, which is designated "Region I" in Fig. 6.1, and is assumed to be spherical.

---

*For a review of modern ideas about hadronic structure, see Bhaduri (1988).

†Two early papers introducing the bag model are [CJ 74]. Additional references can be found in Bhaduri (1988).

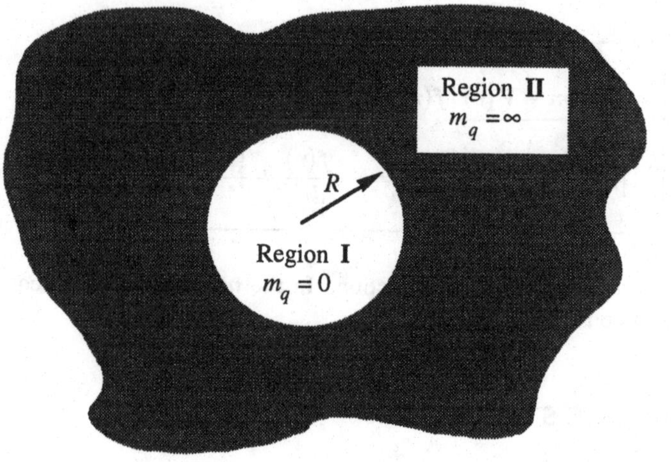

Fig. 6.1 Diagrammatic representation of a hadron in the MIT bag model. Quarks are confined inside a spherical volume (Region I) by the simple *ansatz* that their mass is *infinite* outside the bag (Region II).

To describe this model quantitatively, we must solve the Dirac equation under the assumption that $V = 0$ and that $m = m_q = 0$ inside the volume and $m = m_q \to \infty$ outside. The solutions in the two different regions I and II must be continuous at the surface. (Since the coupled differential equations are first order, and since m is discontinuous at the boundary, we *cannot* require that the derivative of the solution also be continuous at the boundary, as we will see below.)

We therefore begin with a study of the solutions of the spherically symmetric Dirac equation for $V(r) = 0$. We will first obtain solutions for arbitrary quark mass m and later specialize to the limiting cases of interest to the hadronic structure problem. Differentiating the second of Eqs. (6.26) and eliminating $g(r)$ give

$$\frac{d^2 f}{dr^2} + (1+k)\frac{d}{dr}\left(\frac{f}{r}\right) = (E+m)\frac{dg}{dr} = -\left(E^2 - m^2\right) f - \left(\frac{1-k}{r}\right)(E+m)\, g$$
$$= -\left(E^2 - m^2\right) f - \frac{(1-k)}{r}\left[\frac{df}{dr} + \frac{1+k}{r} f\right] .$$

Hence:

$$\boxed{\frac{d^2 f}{dr^2} + \frac{2}{r}\frac{df}{dr} - \frac{k(k+1)}{r^2} f + \left(E^2 - m^2\right) f = 0 .} \qquad (6.27)$$

Recall that the spherical Bessel functions are solutions of the equation*

$$\left[\frac{d^2}{dr^2} + \frac{2}{r}\frac{d}{dr} - \frac{\ell(\ell+1)}{r^2} + k_0^2\right] f_\ell(k_0 r) = 0 \ . \tag{6.28}$$

Hence if $k_0^2 = E^2 - m^2 \geq 0$, the solutions of (6.27) are

$$\begin{array}{ll} j_\ell(k_0 r) & \text{regular as } r \to 0 \\ n_\ell(k_0 r) & \text{singular as } r \to 0 \end{array}$$

while if $k_0^2 = -K_0^2 = E^2 - m^2 < 0$, the solutions are

$$\begin{array}{lll} h_\ell^{(1)}(iK_0 r) & \text{regular as } r \to \infty & \left(h_\ell^{(1)} \to \dfrac{e^{-K_0 r}}{r}\right) \\ h_\ell^{(2)}(iK_0 r) & \text{singular as } r \to \infty & \left(h_\ell^{(2)} \to \dfrac{e^{K_0 r}}{r}\right) \ . \end{array}$$

All of these functions satisfy the recursion relations

$$\begin{aligned} \frac{2\ell+1}{2} f_\ell(x) &= f_{\ell-1}(x) + f_{\ell+1}(x) \\ f_\ell'(x) &= \frac{1}{2\ell+1}\left[\ell f_{\ell-1}(x) - (\ell-1) f_{\ell+1}(x)\right] \ , \end{aligned} \tag{6.29}$$

where f' refers to the derivative of f with respect to its argument x. Finally, for future reference note that

$$\begin{aligned} j_0(x) &= \frac{\sin x}{x} & h_0^{(1)}(ix) &= -\frac{e^{-x}}{x} \\ j_1(x) &= \frac{\sin x}{x^2} - \frac{\cos x}{x} & h_1^{(1)}(ix) &= i\frac{e^{-x}}{x}\left(1 + \frac{1}{x}\right) \ . \end{aligned} \tag{6.30}$$

For the study of the structure of hadrons, we are interested in positive energy solutions (antiquarks will be described by their positive energy charge conjugate states) for which m is both less than and greater than the bound state energy E. There will be two kinds of solutions, depending on the parity (or the sign of k) of the state. If $k > 0$, $\ell = k$, and for solutions in the vicinity of the origin we must choose

$$f(r) = f_k(x) = N j_k(k_0 r) \ , \tag{6.31a}$$

where N is a normalization constant and $x = k_0 r$. The other solution, proportional to n_k, is singular, and hence unacceptable. For a normalizable solution, we must also choose the one which approaches zero as $r \to \infty$, or

$$f(r) = f_k(x) = N h_k^{(1)}(iK_0 r) \ , \tag{6.31b}$$

*A good reference for special functions is Abramowitz and Stegun (1964).

where $x = iK_0 r$. In either case the corresponding g, from Eq. (6.26), is

$$g = \frac{1}{E+m}\left(\frac{df}{dr} + \frac{1+k}{r}f\right)$$
$$= \frac{\kappa}{E+m}\left\{f_k'(x) + \frac{1+k}{x}f_k\right\} \quad ,$$

where in both cases $x = \kappa r$, so that $\kappa = k_0$ for solutions proportional to j and $\kappa = iK_0$ for solutions proportional to $h^{(1)}$. The equation for g may be simplified using a recursion relation obtained by combining the two relations (6.29)

$$f_\ell'(x) = \frac{1}{2\ell+1}\left[\ell f_{\ell-1}(x) - (\ell+1)\left\{\frac{2\ell+1}{x}f_\ell(x) - f_{\ell-1}(x)\right\}\right]$$
$$= -\frac{\ell+1}{x}f_\ell(x) + f_{\ell-1}(x) \quad .$$

Hence, since $k = \ell$,

$$g = \frac{\kappa}{E+m}\, f_{k-1}(x) \quad .$$

The full solution is therefore

$$\boxed{\psi_{jm}^k(r) = \begin{pmatrix} f_k(x) \\ -i\boldsymbol{\sigma}\cdot\hat{\mathbf{r}}\,\frac{\kappa}{E+m}\, f_{k-1}(x)\end{pmatrix}\mathcal{Y}_{jm}^k(\hat{\mathbf{r}}) \quad ,} \tag{6.32}$$

where Eq. (6.22b) has been used to express $\mathcal{Y}^{-k}$ in terms of $\mathcal{Y}^k$, and remember that $k = \ell$, $x = \kappa r$, and f_k is given by Eq. (6.31a) or (6.31b), depending on whether $r < R$ or $r > R$.

The other type of solution occurs when $k < 0$. In that case, $\ell = -(k+1)$, and we must choose

$$f(r) = f_{-(k+1)}(x) = N j_{-(k+1)}(k_0 r) \tag{6.33a}$$

if $r < R$ and

$$f(r) = f_{-(k+1)}(x) = N h_{-(k+1)}^{(1)}(iK_0 r) \tag{6.33b}$$

if $r > R$. In either case, the lower component is

$$g = \frac{1}{E+m}\left(\frac{df}{dr} - \frac{\ell}{r}f\right)$$
$$= \frac{\kappa}{E+m}\left\{f_\ell'(x) - \frac{\ell}{x}\, f_\ell(x)\right\} \quad ,$$

where κ is as before. This suggests manipulating the recursion relations in a different way:

$$\begin{aligned} f'_\ell(x) &= \frac{1}{2\ell+1}\left[\ell\left\{\frac{2\ell+1}{x}f_\ell(x) - f_{\ell+1}(x)\right\} - (\ell+1)f_{\ell+1}(x)\right] \\ &= \frac{\ell}{x}f_\ell(x) - f_{\ell+1}(x) \ . \end{aligned}$$

Hence

$$g = -\frac{\kappa}{E+m}\, f_{-k}(x) \ .$$

The full solution of Type 2, with $k = -(\ell+1) < 0$, is therefore

$$\boxed{\psi^k_{jm}(r) = \begin{pmatrix} f_{-(k+1)}(x) \\ i\boldsymbol{\sigma}\cdot\hat{\boldsymbol{r}}\, \frac{\kappa}{E+m}\, f_{-k}(x) \end{pmatrix} \mathcal{Y}^k_{jm}(\hat{\boldsymbol{r}}) \ ,} \tag{6.34}$$

where now f is given by Eqs. (6.33). Note that the solutions (6.32) and (6.34) involve the same f's (only the ℓ's are different) and identical definitions of κ and x.

We now return to the hadronic structure problem. Only the solution for $k = -1$ will be obtained here. (This is the lowest energy level, or the ground state.) The solution in region I (inside the hadron where $r < R$) with $m = 0$ is

$$\psi_{\rm I}(r) = N_{\rm I} \begin{pmatrix} j_0(Er) \\ i\boldsymbol{\sigma}\cdot\hat{\boldsymbol{r}}\, j_1(Er) \end{pmatrix} \chi^{(s)} \qquad r < R \ , \tag{6.35}$$

where $\chi^{(s)}$ is a two-component spinor. (When $k = -1$, $\mathcal{Y}^k_{jm}$ reduces to a two-component spinor independent of $\hat{\boldsymbol{r}}$.) Outside, in region II, we let $m \gg E$ and neglect E:

$$\psi_{\rm II}(r) = N_{\rm II} \begin{pmatrix} h_0^{(1)}(imr) \\ i\boldsymbol{\sigma}\cdot\hat{\boldsymbol{r}}\, ih_1^{(1)}(imr) \end{pmatrix} \chi^{(s)} \qquad r > R \ . \tag{6.36}$$

Using the forms of h_0 and h_1 given in (6.30), we have

$$\psi_{\rm II}(r) = -\frac{N_{\rm II}}{mr}\, e^{-mr} \begin{pmatrix} 1 \\ i\boldsymbol{\sigma}\cdot\hat{\boldsymbol{r}}\, \left(1 + \frac{1}{mr}\right) \end{pmatrix} \chi^{(s)} \ . \tag{6.37}$$

We want to take the limit $m \to \infty$, but in such a way that the solution is not zero for $r = R$, so that continuity is possible at $r = R$. To accomplish this, choose

$$N_{\rm II} = -mN_0\, e^{mR}$$

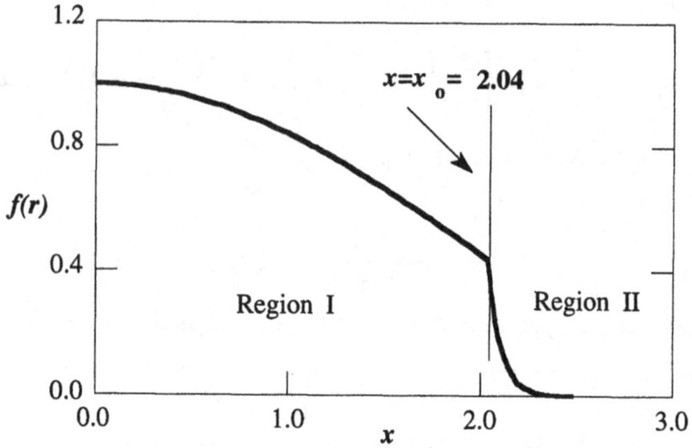

Fig. 6.2 Graph of the upper component of the ground state bag wave function when the quark mass in region II is large but finite. The bag radius is at $x_0 = 2.04$.

where N_0 is finite. Then

$$\psi_{\mathrm{II}}(r) = \frac{N_0}{r}\, e^{-m(r-R)} \begin{pmatrix} 1 \\ i\boldsymbol{\sigma}\cdot\hat{\boldsymbol{r}}\,\left(1+\frac{1}{mr}\right) \end{pmatrix} \chi^{(s)}$$

$$\xrightarrow[m\to\infty]{} \begin{cases} \frac{N_0}{R}\begin{pmatrix} 1 \\ i\boldsymbol{\sigma}\cdot\hat{\boldsymbol{r}} \end{pmatrix}\chi^{(s)} & \text{at } r = R \\ 0 & \text{for all } r > R \quad . \end{cases}$$

The upper component of this solution is shown in Fig. 6.2 (for the case where m in region II is large but still finite). We see that the form of ψ_{II} at $r = R$ *requires that the upper and lower components be equal for* ψ_{I} *at* $r = R$. Specifically, we have

$$\psi_{\mathrm{I}}(R) = \psi_{\mathrm{II}}(R) \Longrightarrow \begin{cases} N_{\mathrm{I}}\, j_0(ER) = \frac{N_0}{R} \\ N_{\mathrm{I}}\, j_1(ER) = \frac{N_0}{R} \quad . \end{cases} \tag{6.38}$$

The eigenvalue condition, for a massless quark confined in a volume of radius R, is therefore

$$\boxed{j_0(ER) = j_1(ER) \quad .} \tag{6.39}$$

This equation is a transcendental equation which can only be solved numerically. Recalling the forms of j_0 and j_1 (see Fig. 6.3) we see that the "cross-over" point must be less than π. It is in fact at

$$x_0 = 2.04 \quad . \tag{6.40}$$

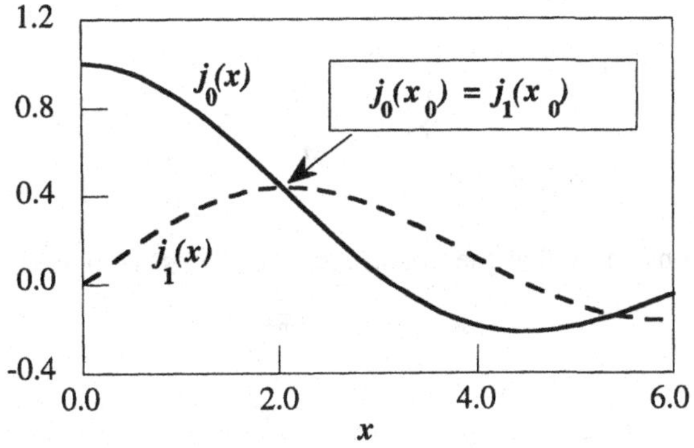

Fig. 6.3 Graph of the first two spherical Bessel functions showing the point at which they are equal.

Hence E is a function of R:

$$\boxed{E = \frac{2.04}{R}} \quad . \tag{6.41}$$

The energy of a massless quark confined in a volume of radius R goes inversely with R. This result can also be obtained (qualitatively) from the uncertainty principle.

We leave it as an exercise (Prob. 6.2) to find some of the other solutions and to prove that the condition (6.41) gives the minimum energy.

6.3 HYDROGEN-LIKE ATOMS

As our final example, we obtain the exact solutions to the Dirac equation for hydrogen-like atoms,* where

$$V = -\frac{Z\alpha}{r} \quad .$$

In this problem, it is convenient to introduce reduced wave functions

$$f = \frac{f_R}{r} \qquad g = \frac{g_R}{r} \tag{6.42}$$

in terms of which the coupled equations (6.26) become

$$\begin{aligned} \frac{df_R}{dr} + \frac{k}{r} f_R - \left(E + m + \frac{Z\alpha}{r}\right) g_R &= 0 \\ \frac{dg_R}{dr} - \frac{k}{r} g_R + \left(E - m + \frac{Z\alpha}{r}\right) f_R &= 0 \quad . \end{aligned} \tag{6.43}$$

*For an elementary discussion see Das (1973).

When $r \to \infty$, the equations reduce to

$$\left.\begin{aligned} \frac{df_R}{dr} - (E+m)g_R = 0 \\ \frac{dg_R}{dr} + (E-m)f_R = 0 \end{aligned}\right\} \Longrightarrow \frac{d^2 f_R}{dr^2} + \left(E^2 - m^2\right) f_R = 0 \ . \tag{6.44}$$

These equations show that the *bound state* solutions, with $E < m$, go like

$$\begin{aligned} f_R &\underset{r\to\infty}{\longrightarrow} e^{-\sqrt{m^2-E^2}\,r} \\ g_R &\underset{r\to\infty}{\longrightarrow} e^{-\sqrt{m^2-E^2}\,r} \ . \end{aligned} \tag{6.45}$$

Hence, it is convenient to scale the equations by introducing

$$\rho \equiv \sqrt{m^2 - E^2}\, r \ . \tag{6.46}$$

Then the coupled equations reduce to

$$\boxed{\begin{aligned} \frac{df_R}{d\rho} + \frac{k}{\rho} f_R - \left(\frac{1}{\epsilon} + \frac{Z\alpha}{\rho}\right) g_R = 0 \\ \frac{dg_R}{d\rho} - \frac{k}{\rho} g_R - \left(\epsilon - \frac{Z\alpha}{\rho}\right) f_R = 0 \ , \end{aligned}} \tag{6.47}$$

where

$$\epsilon = \sqrt{\frac{m-E}{m+E}} \ . \tag{6.48}$$

For weakly bound states (which is the case when Z is near unity), $\epsilon \ll 1$.

The equations (6.47) can be solved by assuming the following power series expansion for the reduced wave functions f_R and g_R:

$$\begin{aligned} f_R &= \rho^\nu \sum_{n=0}^{\infty} A_n \rho^n \, e^{-\rho} \\ g_R &= \rho^\nu \sum_{n=0}^{\infty} B_n \rho^n \, e^{-\rho} \ . \end{aligned} \tag{6.49}$$

Substituting these series into the coupled Eqs. (6.47) and equating the coefficient of the $(n-1)$th power of ρ give:

$$\begin{aligned} (\nu + n + k)\, A_n - A_{n-1} - \frac{1}{\epsilon} B_{n-1} - Z\alpha B_n = 0 \\ (\nu + n - k)\, B_n - B_{n-1} - \epsilon A_{n-1} + Z\alpha A_n = 0 \ . \end{aligned} \tag{6.50}$$

The indicial equation for ν is obtained when $n = 0$:

$$\begin{aligned} (\nu + k)A_0 - Z\alpha B_0 &= 0 \\ Z\alpha A_0 + (\nu - k)B_0 &= 0 \ . \end{aligned} \tag{6.51}$$

To have a non-trivial solution (i.e., A_0 or $B_0 \neq 0$) requires

$$\nu^2 - k^2 + (Z\alpha)^2 = 0 \quad \Rightarrow \quad \nu = \pm\sqrt{k^2 - (Z\alpha)^2} \ . \tag{6.52}$$

Now the *reduced* wave function goes like ρ^ν at the origin and is singular if ν is negative. In order for it to be normalizable, this singularity cannot be stronger than $\rho^{-\frac{1}{2}}$, which implies the condition $\nu > -\frac{1}{2}$. Hence, negative values of ν must be rejected because, even for the smallest value of $|k|$ ($|k| = 1$), negative ν are less than $-\frac{1}{2}$ (unless Z is *very* large, and we will not discuss such extreme cases). Hence

$$\nu = \sqrt{k^2 - (Z\alpha)^2} \ , \tag{6.53}$$

and eliminating A_{n-1} and B_{n-1} from the coupled Eqs. (6.50) gives the following relation for B_n in terms of A_n:

$$B_n = A_n \frac{\epsilon(\nu + n + k) - Z\alpha}{\nu + n - k + Z\alpha\epsilon} \ . \tag{6.54}$$

The recursion relations for A_n can now be found by substituting (6.54) into the coupled Eqs. (6.50),

$$\frac{A_{n+1}}{A_n} = \frac{(\nu + n + 1 - k + Z\alpha\epsilon)\left(2\nu + 2n + Z\alpha\left(\epsilon - \frac{1}{\epsilon}\right)\right)}{(\nu + n - k + Z\alpha\epsilon)\left((\nu + n + 1)^2 - k^2 + (Z\alpha)^2\right)} \ . \tag{6.55}$$

The eigenvalue condition emerges from the recursion relation (6.55). First, note that as $n \to \infty$, the ratio $A_{n+1}/A_n \to 2/(n+1)$. For comparison,

$$e^{2\rho} = \sum_{n=0}^{\infty} \frac{1}{n!}(2\rho)^n = \sum_{n=0}^{\infty} C_n \rho^n \ ,$$

where the ratio of successive terms of this comparison series is

$$\frac{C_{n+1}}{C_n} = \left(\frac{2}{n+1}\right) \ . \tag{6.56}$$

By the ratio test, the series for f_R and g_R will therefore go like

$$f_R \underset{\rho \to \infty}{\longrightarrow} e^{2\rho}\, e^{-\rho} = e^{\rho} \ , \tag{6.57}$$

which is unacceptable as a solution. *Hence the series must terminate.* This produces the eigenvalue condition. Only one of the two terms in the numerator of Eq. (6.55) can be zero and terminate the series. In order for this to happen for some integer N, we require

$$2\nu + 2N - \frac{Z\alpha}{\epsilon}(1-\epsilon^2) = 0 \ . \tag{6.58}$$

Substituting for ϵ

$$\nu + N - \frac{Z\alpha E}{\sqrt{m^2 - E^2}} = 0 \ ,$$

and solving for E gives

$$E_{N,k} = \sqrt{\frac{m^2(\nu+N)^2}{(\nu+N)^2 + (Z\alpha)^2}} = m\sqrt{1 - \frac{(Z\alpha)^2}{(\nu+N)^2 + (Z\alpha)^2}} \ ,$$

where the subscripts remind us that E depends on N and k. Finally, using Eq. (6.53) to eliminate ν and doing some rearranging give

$$E_{N,k} = m\left[1 - \frac{(Z\alpha)^2}{(N+|k|)^2 + 2N\left(\sqrt{k^2 - (Z\alpha)^2} - |k|\right)}\right]^{1/2} \ . \tag{6.59}$$

This is the *exact* solution of the Dirac energy of a hydrogen-like atom.

Before we discuss this result, note that if k is positive, there is *no solution for* $N = 0$. This is because in this case the term in the numerator of the recursion relation (6.55), which would normally terminate the diverging series, is canceled by a zero in the denominator, so the series does not terminate and the would-be "solution" must be discarded. The ratio in question is

$$R = \frac{2\nu + Z\alpha\epsilon - Z\alpha/\epsilon}{\nu - k + Z\alpha\epsilon} \ ,$$

and to see that this can never be zero when k is positive, multiply numerator and denominator by the factor $\nu + k + Z\alpha\epsilon$ (which can have no zeros if k is positive), and substitute $k^2 = \nu^2 + (Z\alpha)^2$ in the denominator, giving

$$\begin{aligned} R &= \frac{(\nu + k + Z\alpha\epsilon)\,(2\nu + Z\alpha\epsilon - Z\alpha/\epsilon)}{(\nu + Z\alpha\epsilon)^2 - \nu^2 - (Z\alpha)^2} \\ &= \frac{(\nu + k + Z\alpha\epsilon)}{Z\alpha\epsilon} \ , \end{aligned}$$

which shows that the expected zero at $2\nu = Z\alpha/\epsilon - Z\alpha\epsilon$ is canceled. In the following discussion we must be careful to exclude the case $N = 0$, $k > 0$ from consideration.

Energy Level Scheme

Since $|k| = j + \frac{1}{2}$ and there is no solution for $N = 0$, $k > 0$, it is convenient to define a new quantum number n as follows:

$$n = N + |k| \geq 1 \qquad -n \leq k < n \ . \tag{6.60}$$

This quantum number is identical to the familiar nonrelativistic principal quantum number. In terms of it, and expressing $|k|$ in terms of j, the energy can be written

$$E_{n,j} = m \left[1 - \frac{(Z\alpha)^2}{n^2 + 2\left(n - (j+\frac{1}{2})\right)\left[\sqrt{(j+\frac{1}{2})^2 - (Z\alpha)^2} - (j+\frac{1}{2})\right]} \right]^{1/2} . \tag{6.61}$$

This is our final expression for the exact energy of a Dirac particle bound by a Coulomb potential. Note that the energies depend on *only two quantum numbers, n and j*. Hence, as we discussed in Chapter 3, *the Lamb shift* (which gives the splitting for different ℓ's associated with the same j) is a physical effect *not described by the Dirac equation.*

It is amusing to expand the exact result to order α^4. This gives

$$E - m \simeq -m\frac{(Z\alpha)^2}{2n^2} - m\frac{(Z\alpha)^4}{2n^4}\left[\frac{n}{j+\frac{1}{2}} - \frac{3}{4}\right] + \mathcal{O}\left((Z\alpha)^6\right) \ . \tag{6.62}$$

As expected, this *agrees exactly with the fine structure results* we obtained previously in Sec. 5.7 using the FW transformation.

The first few Dirac energy levels are tabulated in Table 6.3. The explicit expression for the Dirac wave function for the ground state of a hydrogen-like atom is

$$\psi^{(-1)}_{\frac{1}{2},s}(r) = \begin{pmatrix} f^{(-1)}(r) \\ -i\boldsymbol{\sigma}\cdot\hat{\boldsymbol{r}}\, g^{(-1)}(r) \end{pmatrix} \frac{\chi^{(s)}}{\sqrt{4\pi}} , \tag{6.63}$$

where

$$\begin{aligned} f^{(-1)}(r) &= A_0\, \rho^{\sqrt{1-(Z\alpha)^2}}\, \frac{e^{-\rho}}{\rho} \\ g^{(-1)}(r) &= A_0\, \rho^{\sqrt{1-(Z\alpha)^2}}\, \frac{e^{-\rho}}{\rho} \left[\frac{\epsilon\left(\sqrt{1-(Z\alpha)^2} - 1\right) - Z\alpha}{\sqrt{1-(Z\alpha)^2} + 1 - Z\alpha\epsilon}\right] . \end{aligned} \tag{6.64}$$

Table 6.3 Quantum numbers and exact Dirac energies of the first nine energy levels of hydrogen-like atoms.

n	k	j	ℓ	name	E_{nj}
1	-1	$\frac{1}{2}$	0	$1S_{1/2}$	$m\sqrt{1-(Z\alpha)^2}$
2	-2	$\frac{3}{2}$	1	$2P_{3/2}$	$m\sqrt{1-\frac{1}{4}(Z\alpha)^2}$
	-1	$\frac{1}{2}$	0	$2S_{1/2}$	$\left.\right\} m\sqrt{1-\frac{(Z\alpha)^2}{2+2\sqrt{1-(Z\alpha)^2}}}$
	1	$\frac{1}{2}$	1	$2P_{1/2}$	
3	-3	$\frac{5}{2}$	2	$3D_{5/2}$	$m\sqrt{1-\frac{1}{9}(Z\alpha)^2}$
	-2	$\frac{3}{2}$	1	$3P_{3/2}$	$\left.\right\} m\sqrt{1-\frac{(Z\alpha)^2}{5+2\sqrt{4-(Z\alpha)^2}}}$
	2	$\frac{3}{2}$	2	$3D_{3/2}$	
	-1	$\frac{1}{2}$	0	$3S_{1/2}$	$\left.\right\} m\sqrt{1-\frac{(Z\alpha)^2}{5+4\sqrt{1-(Z\alpha)^2}}}$
	1	$\frac{1}{2}$	1	$3P_{1/2}$	

Note that

$$\epsilon = \sqrt{\frac{m-E}{m+E}} = \frac{m-E}{\sqrt{m^2-E^2}} = \frac{1-\sqrt{1-(Z\alpha)^2}}{Z\alpha} \simeq \frac{1}{2}(Z\alpha) \ ,$$

and to lowest order in $(Z\alpha)^2$, the ground state wave function becomes

$$\begin{aligned} f^{(-1)}(r) &= A_0\, \rho^{-\frac{1}{2}(Z\alpha)^2}\, e^{-\rho} \\ g^{(-1)}(r) &= -\tfrac{1}{2} Z\alpha A_0\, \rho^{-\frac{1}{2}(Z\alpha)^2}\, e^{-\rho} \ . \end{aligned} \tag{6.65}$$

The upper component is very similar to the non-relativistic wave function except for an enhanced (singular) part at small ρ which goes like

$$\frac{1}{\rho^{\frac{1}{2}(Z\alpha)^2}} \ .$$

This singularity is very weak, and the solution is still integrable near the origin. The lower component is very much smaller (by a factor of $\frac{1}{2}Z\alpha$) than the upper component. Hence the relativistic solution differs from the non-relativistic solution only to order $Z\alpha$, or at *very short distances*.

This concludes our study of first quantized relativistic equations. In the next chapter, we begin the study of field theories based on these equations.

PROBLEMS

6.1 Consider quarks confined in a spherical volume, as discussed in Sec. 6.2. Suppose that it requires energy to "make" a volume in which the quarks can move freely, so that the total energy of n non-interacting quarks inside a volume R is

$$E_R = \frac{n\,x_0}{R} + \frac{4\pi}{3}R^3 B \ ,$$

where, for ground state quarks, $x_0 = 2.04$, and B is the energy density of the empty volume.

(a) Minimize the energy with respect to R and show that

$$R_{\min} = \sqrt[4]{\frac{n\,x_0}{4\pi B}}$$

$$E_{\min} = \frac{4}{3}\sqrt[4]{n^3 x_0^3 4\pi B} \ .$$

Suppose the proton (mass 940 MeV) is made of three quarks. What is its radius? What is the mass and radius of a $q\bar{q}$ system? How does this compare with masses of the known mesons ($\pi \cong 140$ MeV, $\rho \cong 770$ MeV, $\omega \simeq 783$ MeV)?

(b) Suppose the confined quark has a rest mass $m_q \neq 0$. Find an equation for its energy if it is confined in a spherical volume of radius R.

6.2 Find the solution for the first excited state of the MIT bag.

6.3 Suppose a massless quark moves under the influence of a SHO potential with both scalar and vector terms,

$$V(r) = \lambda_1 r^2 + \beta\left(V_0 + \lambda_2 r^2\right) \ ,$$

where the term proportional to λ_1 is the fourth component of a vector, the second term (with the Dirac matrix β) is a scalar mass term, and V_0, λ_1, and λ_2 are constants.

(a) Find the correct coupled equations for the upper and lower radial functions $f^k(r)$ and $g^k(r)$ of the Dirac wave function of such a state.

(b) Find the single second order equation for $f^k(r)$.

(c) Choose the constants V_0, λ_1, and λ_2 so that this equation reduces to a Schrödinger equation for a particle moving in a pure simple harmonic potential. Find the ground state energy and the Dirac wave function for the ground state of a massless quark moving in such a potential. Discuss the significance of your result.

6.4 A massless spin $\frac{1}{2}$ particle moves in a one-dimensional *scalar* potential of the form

$$\begin{aligned} V(z) &= 0 & -R \le z \le R \\ &= V_0 & z < -R \quad \text{and} \quad R < z\,, \end{aligned}$$

where the constant V_0 is large and positive.

(a) Write down the correct Dirac equation for the motion of this particle.

(b) Show that the equation found in part (a) is invariant under the parity transformation $z \longrightarrow -z$.

(c) Solve the equation for the ground state energy and wave function of the trapped particle. Take the limit $V_0 \to \infty$, and sketch the solution for this case. Comment on any interesting features which the solution possesses.

6.5 A Dirac particle of mass m and positive charge e scatters from the one-dimensional high barrier shown in Fig. 4.1.

(a) Write down the Dirac equation which correctly describes the scattering if the potential energy eV is the zeroth component of a four-vector (a Coulomb-like interaction).

(b) Write down the Dirac equation which correctly describes the scattering if the potential energy is a scalar (invariant under all Lorentz transformations).

(c) Consider solutions in region II of the form

$$\psi_{\rm II}(z,t) = \begin{pmatrix} \chi \\ \eta \end{pmatrix} \frac{1}{L^{1/2}} e^{i(\pm Kz - Et)} \quad ,$$

where χ and η are two-component spinors, $E > 0$ is fixed, and $eV > E+m$. Solve the Dirac equation in region II for the two cases described above, and discuss the nature of the solutions. Can particles propagate in region II?

(d) Find the full solution for both of the cases described in (a) and (b), and discuss the time evolution of a positive energy state which is localized at large negative z at large negative t and approaches the barrier. Show that the norm is conserved in both cases, and discuss your results.

PART III

ELEMENTS OF QUANTUM FIELD THEORY

CHAPTER 7

SECOND QUANTIZATION

The wave equations discussed in the last three chapters were able to describe the quantum mechanical behavior of *single* particles in a covariant manner. Such a treatment is referred to as *first quantization*. It is suitable for the description of the interactions of massive particles with kinetic energies much less than the particle rest mass, where energy conservation forbids the production of real particle–antiparticle pairs. However, at higher energies where the production of single particles (for cases when particle number is not conserved, such as for neutral pions, π^0's), or particle–antiparticle pairs (in cases where particle number is conserved) is energetically possible, the first quantized form fails completely, and we need to develop a new quantization scheme capable of describing particle production and annihilation fully. Such a quantization scheme is referred to as *second quantization*.

The quantum field theory of the EM field developed in Chapter 2 is just such a theory. In the case of the EM field, we started immediately with the second quantized (field) theory because a useful first quantized theory of photons does not exist. This is because photons have zero rest mass and photon number is not conserved, and therefore photons can always be created, no matter how small the energy. For classical particles the first quantized theory developed in the preceding three chapters was a useful development in itself and an essential first step to a more complete theory.

We are now ready to extend our previous treatment of the EM field to the description of classical particles, such as spin $\frac{1}{2}$ electrons, quarks, or protons or spin zero pions. We proceed by first interpreting the single particle wave functions which emerge from the first quantized theory as "classical" fields and then turning these c-number fields into quantized q-number (operator) fields, just as we did for the photon. The resulting quantum fields have the same general structure as the EM quantum field, showing that classical particles and classical waves (photons) are ultimately described by the same mathematical object, a quantum field.

In this chapter we will discuss the construction of theories which describe free, non-interacting particles. A very important result will emerge. The requirement that the energy of free states be positive leads to the conclusion that the states of spin $\frac{1}{2}$ particles must be antisymmetric and satisfy Fermi–Dirac statistics, while the states of spin 0 particles must be symmetric and satisfy Bose–Einstein statistics. This famous result is referred to as the *connection between spin and statistics* and is one of the great achievements of relativistic quantum field theory. We will conclude this chapter with a brief discussion of how interactions are included in quantum field theories. The study of interacting field theories will resume again in Chapter 9, after a discussion, in Chapter 8, of the role which symmetries play in the development of field theories.

7.1 SCHRÖDINGER THEORY

For comparison, we first discuss the second quantized form of the Schrödinger theory. The first step in this development is to regard the Schrödinger wave function, $\psi(x)$, as a classical field. A Lagrangian density which will yield a Schrödinger equation for this complex field is

$$\mathcal{L} = \frac{1}{2}\psi^*(x) i \frac{\overrightarrow{\partial}}{\partial t}\psi(x) - \frac{1}{2}\psi^*(x) i \frac{\overleftarrow{\partial}}{\partial t}\psi(x) - \frac{1}{2m}\overrightarrow{\nabla}\psi^*(x)\cdot\overrightarrow{\nabla}\psi(x) \; , \tag{7.1}$$

where the arrow over the operator shows in which direction it acts. If ψ and ψ^* are independent (corresponding to two independent real fields), the Euler–Lagrange equations are

$$\frac{\partial}{\partial x^\mu}\frac{\partial\mathcal{L}}{\partial\left(\dfrac{\partial\psi^*}{\partial x^\mu}\right)} - \frac{\partial\mathcal{L}}{\partial\psi^*} = -\frac{i}{2}\frac{\partial\psi(x)}{\partial t} - \frac{i}{2}\frac{\partial\psi(x)}{\partial t} - \frac{1}{2m}\nabla^2\psi(x) = 0 \; ,$$

where one of the two $\partial\psi/\partial t$ terms comes from the derivative of $\mathcal{L}$ with respect to $\partial\psi^*/\partial t$ and the other from the derivative of $\mathcal{L}$ with respect to ψ^*. Combining these terms gives the familiar Schrödinger equation for ψ

$$i\frac{\partial}{\partial t}\psi(x) = -\frac{1}{2m}\nabla^2\psi(x) \; . \tag{7.2}$$

The momentum conjugates to ψ and ψ^* are*

$$\pi^*(x) = \frac{\partial\mathcal{L}}{\partial\left(\dfrac{\partial\psi}{\partial t}\right)} = \frac{i}{2}\psi^*(x) \qquad \pi(x) = \frac{\partial\mathcal{L}}{\partial\left(\dfrac{\partial\psi^*}{\partial t}\right)} = -\frac{i}{2}\psi(x) \; . \tag{7.3}$$

Hence the Hamiltonian density is

$$\begin{aligned}\mathcal{H} = \pi^*(x)\frac{\partial\psi(x)}{\partial t} + \frac{\partial\psi^*(x)}{\partial t}\pi(x) - \mathcal{L} &= \frac{i}{2}\psi^*(x)\frac{\partial\psi(x)}{\partial t} - \frac{i}{2}\psi(x)\frac{\partial\psi^*(x)}{\partial t} - \mathcal{L} \\ &= \frac{1}{2m}\nabla_i\psi^*(x)\,\nabla_i\psi(x) \; ,\end{aligned} \tag{7.4}$$

*In order to agree with the convention we will use later for Dirac fields, the momentum conjugate to ψ will be denoted by π^*, and *not* π.

and the total Hamiltonian, after integrating by parts and assuming the boundary terms vanish (because of the periodic boundary conditions we have always imposed; recall the discussion in Sec. 2.2), becomes

$$H = \int d^3r\, \mathcal{H} = \int d^3r\, \psi^*(x) \left(-\frac{\nabla^2}{2m}\right) \psi(x) \ . \tag{7.5}$$

This is simply the expectation value of the kinetic energy operator, a result familiar from elementary studies.

Note that a popular alternative to the Lagrangian density (7.1) is

$$\mathcal{L} = \psi^*(x)\, i\frac{\overrightarrow{\partial}}{\partial t}\psi(x) - \frac{1}{2m}\overrightarrow{\nabla}\psi^*(x) \cdot \overrightarrow{\nabla}\psi(x) \ .$$

This Lagrangian density will also give the Schrödinger equation for ψ but is not Hermitian and breaks the symmetry which naturally exists between ψ and ψ^*. In particular, it gives $\pi^* = i\psi^*$ and $\pi = 0$ which is inconsistent with other relations. We will always use a Hermitian Lagrangian density.

For the free Schrödinger theory, the eigensolutions of the Schrödinger equation satisfy

$$-\frac{\nabla^2}{2m}\psi_n^{(+)}(x) = E_n^0 \psi_n^{(+)}(x) \ , \tag{7.6}$$

and imposing the same periodic boundary conditions we used in Chapters 2 and 3, they are explicitly

$$\psi_n^{(+)}(x) = \frac{1}{\sqrt{L^3}}\, e^{i\boldsymbol{p}_n\cdot\boldsymbol{r} - iE_n^0 t} = \frac{1}{\sqrt{L^3}}\, e^{-ip\cdot x} \ , \tag{7.7}$$

where $\boldsymbol{p}_n = \frac{2\pi}{L}(n_x, n_y, n_x)$ and $E_n^0 = p_n^2/(2m)$. These states are orthogonal and normalized,

$$\int_{L^3} d^3r\, \psi_n^{(+)*}(x)\, \psi_{n'}^{(+)}(x) = \delta_{nn'} \ . \tag{7.8}$$

We will now quantize this classical field theory. The *general procedure*, which was fully developed in Chapters 1 and 2, is to expand the field in terms of eigensolutions of the field equations and to interpret the expansion coefficients as annihilation and creation operators. For the Schrödinger theory, all $E_n^0 > 0$, so the most general expansion is

$$\psi(x) = \sum_n a_n\, \psi_n^{(+)}(x) \ , \tag{7.9}$$

where $\psi_n^{(+)}(x)$ are positive energy normalized solutions (7.7) of the field equation (7.6) and a_n are annihilation operators with the following interpretation:

a_n destroys a particle of momentum p_n

$a_n^\dagger$ creates a particle of momentum p_n .

These operators satisfy (for now) the commutation relations

$$\left[a_{n'}, a_n^\dagger\right] = \delta_{nn'} \quad . \tag{7.10}$$

[We shall see later that we could also use anticommutation relations with the Schrödinger theory.]

Substituting the field expansion (7.9) into the expression (7.5) for H and using the Schrödinger equation (7.6) and the orthogonality relations (7.8) give immediately

$$H = \sum_n E_n^0 a_n^\dagger a_n \quad . \tag{7.11}$$

This form is familiar from Chapters 1 and 2, and all of the consequences we worked out in those chapters can be carried over to this case. The eigenstates of H are the Fock states, and (7.11) tells us that the total energy of any Fock state is simply the sum of the energies of each of the particles in that state. The equation tells us to first compute the number of particles with momentum p_n (the number operator $a_n^\dagger a_n$), then multiply by the energy of a single particle with momentum p_n, and finally add these contributions together.

The canonical commutation relations for this field theory are

$$\left[\psi(r,t), \pi^\dagger(r',t)\right] = i\,\tfrac{1}{2}\,\delta^3(r-r') \quad , \tag{7.12}$$

which differs by a factor of $\frac{1}{2}$ from what is expected (showing that it is better to use the $[a, a^\dagger]$ commutation relations). To prove this, note that

$$2\left[\psi(r,t), \pi^\dagger(r',t)\right] = i\left[\psi(r,t), \psi^\dagger(r',t)\right]$$

and

$$\begin{aligned}\left[\psi(r,t), \psi^\dagger(r',t)\right] &= \sum_n \psi_n^{(+)}(r,t)\psi_n^{(+)*}(r',t) \\ &= \sum_n \psi_n^{(+)}(r,0)\psi_n^{(+)*}(r',0) = \delta^3(r'-r) \quad .\end{aligned} \tag{7.13}$$

The last relation could be proved from the explicit form of the solutions, but it follows more generally from the *completeness relation*. Because of this factor of 2, the CCR's are usually written, for a theory of this type, in the form

$$\boxed{\left[\psi(r,t), \psi^\dagger(r',t)\right] = \delta^3(r'-r) \quad .} \tag{7.14}$$

Next, observe that the Hamiltonian has the required property of time translation,

$$\boxed{[H, \psi(r,t)] = -i\frac{\partial}{\partial t}\psi(r,t) \quad .} \tag{7.15}$$

The proof is simple and instructive. The commutator is

$$[H, \psi(r,t)] = \left[\sum_n E_n^0 a_n^\dagger a_n \,, \sum_{n'} a_{n'} \psi_{n'}^{(+)}(x)\right] \quad . \tag{7.16}$$

From the commutation relations it follows that

$$\left[a_n^\dagger a_n, a_{n'}\right] = -\delta_{nn'} a_n \quad . \tag{7.17}$$

Hence

$$[H, \psi(r,t)] = -\sum_n E_n^0 a_n \psi_n^{(+)}(x) = -i\frac{\partial}{\partial t}\psi(r,t) \quad . \tag{7.18}$$

Note that the commutation relation (7.17) is a *necessary and sufficient* condition for the result (7.15). *As long as the Hamiltonian has the form (7.11) and the relation (7.17) can be proved, the Hamiltonian will be the generator of time translations.*

7.2 IDENTICAL PARTICLES

As we saw in Chapters 1 and 2 and in the preceding section, the quantization of a classical field leads immediately to creation and annihilation operators and to the introduction of Fock states which describe many particles. The particles associated with the quantization of a *single field are identical*. Quantum mechanically, this means that no *measurement* can be constructed which will distinguish them, and since the results of measurements in quantum theory are expressed as absolute squares of matrix elements, the requirement of indistinguishability takes the mathematical form

$$|\langle f|\mathcal{O}|1_{n_1}1_{n_2}\rangle|^2 = |\langle f|\mathcal{O}|1_{n_2}1_{n_1}\rangle|^2 \quad , \tag{7.19}$$

where $|1_{n_1}1_{n_2}\rangle$ is the Fock state of two identical particles, one with momentum n_1 and the other with momentum n_2 (in general, we will use the notation $|N_{n_1}\rangle$ to denote a state of N particles with momentum n_1), $\mathcal{O}$ is any operator, and $\langle f|$ is any final state. Equation (7.19) is the statement that we can only know that one of the particles has momentum n_1 and the other has n_2, but we cannot know *which* particle has which momentum (this is, in fact, a meaningless question). From Eq. (7.19) we conclude that

$$\langle f|\mathcal{O}|1_{n_1}1_{n_2}\rangle = e^{i\delta}\,\langle f|\mathcal{O}|1_{n_2}1_{n_1}\rangle \quad , \tag{7.20}$$

where the phase factor must be ± 1 if we assume that two interchanges necessarily carry us back to the same state. Since $\mathcal{O}$ and $\langle f|$ are arbitrary, we obtain the result that the Fock states of a quantum field must be either *symmetric* or *antisymmetric*:

$$|1_{n_1}1_{n_2}\;\cdots\rangle = \pm\,|1_{n_2}1_{n_1}\;\cdots\rangle \quad . \tag{7.21}$$

This in turn means that the creation operators (and therefore the annihilation operators as well) must satisfy either *commutation* relations or *anticommutation* relations. We have already discussed quantization with commutation relations and will now discuss anticommutation relations.

Anticommutation Relations

To construct a field theory based on *either commutation or anticommutation* relations, it is sufficient to require the following *commutation* relation:

$$\left[a_n^\dagger a_n, a_{n'}^\dagger\right] = a_n^\dagger \ . \tag{7.22}$$

This one equation is sufficient to both identify the operator $a_n^\dagger a_n$ as the number operator and $a_n^\dagger$ as a creation operator and permit us to proceed with the construction of the Fock states of the theory. [To see this, return to Sec. 1.5 and confirm that the above relation was all we used to construct the states and establish the properties of $a^\dagger$.]

We now assume that the creation operators $a_n^\dagger$ satisfy anticommutation relations (which also implies that the annihilation operators do), and use the required relation (7.22) to find the correct relations *between* $a_n^\dagger$ and a_n. Using the notation $[\ ,\ ]_\mp$ to represent either a commutator or an anticommutator, the implications of (7.22) can be worked out immediately:

$$\begin{aligned}\left[a_n^\dagger a_n, a_{n'}^\dagger\right] &= a_n^\dagger a_n a_{n'}^\dagger - a_{n'}^\dagger a_n^\dagger a_n \\ &= a_n^\dagger \left[a_n, a_{n'}^\dagger\right]_\mp \pm a_n^\dagger a_{n'}^\dagger a_n - a_{n'}^\dagger a_n^\dagger a_n \\ &= a_n^\dagger \left[a_n, a_{n'}^\dagger\right]_\mp \mp a_{n'}^\dagger a_n^\dagger a_n - a_{n'}^\dagger a_n^\dagger a_n \\ &= a_n^\dagger \ ,\end{aligned} \tag{7.23}$$

where the anticommutation of $a_n^\dagger$ and $a_{n'}^\dagger$ is used in going from the second line to the third. Hence, only the lower sign (anticommutation relation) will give the required result and leads to the requirement

$$\left[a_n, a_{n'}^\dagger\right]_+ = \delta_{nn'} \ .$$

We are therefore led to the following set of anticommutation relations:

$$\begin{aligned}\left\{a_n, a_{n'}^\dagger\right\} &= \delta_{nn'} \\ \left\{a_n, a_{n'}\right\} &= \left\{a_n^\dagger, a_{n'}^\dagger\right\} = 0 \ ,\end{aligned} \tag{7.24}$$

where $\{\ ,\ \}$ denotes the anticommutator.

We next discuss a remarkable fact: the Schrödinger theory *can be quantized equally well by imposing commutation* **or** *anticommutation relations* on the operators a and $a^\dagger$. As a demonstration of how this works, we show that the Hamiltonian satisfies (7.15), a condition which *must* hold if the Hamiltonian is to be interpreted as the generator of time translations.

The proof of (7.15) is almost trivial because for either commutation *or* anti-commutation relations we have

$$\begin{aligned}\left[a_n^\dagger a_n, a_{n'}\right] &= a_n^\dagger a_n a_{n'} - a_{n'} a_n^\dagger a_n \\ &= a_n^\dagger \left(\pm a_{n'} a_n\right) - a_{n'} a_n^\dagger a_n \\ &= -\delta_{nn'} a_n + a_{n'} a_n^\dagger a_n - a_{n'} a_n^\dagger a_n \\ &= -\delta_{nn'} a_n \ ,\end{aligned} \tag{7.25}$$

where the upper sign holds for the commutation relations and the lower sign for anticommutation relations. But, as we saw above, this is the necessary and sufficient condition for (7.15), and hence H is the generator of time translations, *regardless* of whether or not the a's satisfy commutation or anticommutation relations. We will not demonstrate it here, but the same holds for the other generators of the Lorentz group.

Implications of Anticommutation Relations

The use of anticommutation relations corresponds to the imposition of Fermi–Dirac statistics and leads to the Pauli exclusion principle. The latter follows from

$$\left\{a_n^\dagger, a_n^\dagger\right\} = 2a_n^\dagger a_n^\dagger = 0 \ . \tag{7.26}$$

Hence the attempt to create a state with two identical particles gives zero. For other states

$$|1_{n_1} 1_{n_2}\rangle \equiv a_{n_1}^\dagger a_{n_2}^\dagger |0\rangle = -a_{n_2}^\dagger a_{n_1}^\dagger |0\rangle = -|1_{n_2} 1_{n_1}\rangle \ . \tag{7.27}$$

Hence these states obey Fermi–Dirac statistics. We will adopt the convention that the order of operators which generate the state from $|0\rangle$ is the same as the order of the indices which label the state:

$$\boxed{|1_{n_1} 1_{n_2} \cdots 1_{n_i} \cdots\rangle = a_{n_1}^\dagger a_{n_2}^\dagger \cdots a_{n_i}^\dagger \cdots |0\rangle \ .} \tag{7.28}$$

Since the states of a particle of momentum n can be either occupied or unoccupied, they can be represented by a two-component vector

$$|0\rangle = \begin{pmatrix} 0 \\ 1 \end{pmatrix} \qquad |1\rangle = \begin{pmatrix} 1 \\ 0 \end{pmatrix} \ , \tag{7.29}$$

where a different vector must be used for *each* momentum state n. In terms of these states, the annihilation and creation operators have the following matrix representation:

$$a^\dagger = \begin{pmatrix} 0 & 1 \\ 0 & 0 \end{pmatrix} \qquad a = \begin{pmatrix} 0 & 0 \\ 1 & 0 \end{pmatrix} \ . \tag{7.30}$$

These matrices have the following properties:

$$a^{\dagger 2} = a^2 = 0 \qquad a^\dagger a = \begin{pmatrix} 1 & 0 \\ 0 & 0 \end{pmatrix} \tag{7.31}$$

as required by their definition.

Finally, we summarize the main result: *The Schrödinger theory is consistent with either commutation relations (Bose–Einstein statistics) or anticommutation relations (Fermi–Dirac statistics) and hence provides no connection between spin and statistics.* The same statement does not hold for relativistic field theories. One of the major triumphs of relativistic quantum field theory is that it does provide such a connection. It can be shown that

$$\begin{aligned} \text{Integer spin} \quad &\Leftrightarrow \quad \text{Bose–Einstein statistics} \\ \text{Half (odd) integer spin} \quad &\Leftrightarrow \quad \text{Fermi–Dirac statistics} \end{aligned}$$

We will show this for spin 0 and spin $\frac{1}{2}$ systems now.

7.3 CHARGED KLEIN–GORDON THEORY

A charged KG field must be complex [otherwise the current defined in Eq. (4.11) would be zero]. The Lagrangian density for a classical complex Klein–Gordon field is

$$\mathcal{L} = \frac{\partial \phi^*}{\partial x^\mu}\frac{\partial \phi}{\partial x_\mu} - m^2 \phi^* \phi \ , \tag{7.32}$$

where ϕ and ϕ^* are regarded as independent fields (corresponding to two independent real fields required to describe the two charge states of the field). The equation of motion which follows from this Lagrangian is the KG equation

$$\frac{\partial}{\partial x^\mu}\left(\frac{\partial \mathcal{L}}{\partial\left(\dfrac{\partial \phi^*}{\partial x^\mu}\right)}\right) - \frac{\partial \mathcal{L}}{\partial \phi^*} \quad \Longrightarrow \quad \Box\phi(x) + m^2\phi(x) = 0 \ . \tag{7.33}$$

The generalized momenta are

$$\pi^* = \frac{\partial \mathcal{L}}{\partial\left(\dfrac{\partial \phi}{\partial t}\right)} = \frac{\partial \phi^*}{\partial t} \qquad \pi = \frac{\partial \phi}{\partial t} \tag{7.34}$$

so that the Hamiltonian density becomes

$$\mathcal{H} = \pi^* \frac{\partial \phi}{\partial t} + \frac{\partial \phi^*}{\partial t}\pi - \mathcal{L} = \pi^* \pi + \nabla_i \phi^* \nabla_i \phi + m^2 \phi^* \phi \ . \tag{7.35}$$

Therefore the total Hamiltonian is

$$H = \int d^3r\, \mathcal{H}(r,t) = \int d^3r\, \left[\pi^*\, \pi + \nabla_i\phi^*\nabla_i\phi + m^2\phi^*\phi\right] \ . \tag{7.36}$$

A convenient formula for the total energy can be obtained from this expression if we substitute for π, integrate by parts (dropping surface terms), and use the KG equation to simplify the final expression:

$$\begin{aligned} H = \int d^3r\, \mathcal{H}(r,t) &= \int d^3r \left[\frac{\partial\phi^*}{\partial t}\,\frac{\partial\phi}{\partial t} + \nabla_i\phi^*\nabla_i\phi + m^2\phi^*\phi\right] \\ &= \int d^3r \left[\frac{\partial\phi^*}{\partial t}\,\frac{\partial\phi}{\partial t} - \phi^*\,\underbrace{\left(\nabla^2\phi - m^2\phi\right)}_{=\frac{\partial^2\phi}{\partial t^2}}\right] \\ &= -\int d^3r\, \phi^*\,\frac{\overleftrightarrow{\partial}}{\partial t}\left(\frac{\partial\phi}{\partial t}\right) \ , \end{aligned} \tag{7.37}$$

where $\overleftrightarrow{\partial}/\partial t = \overrightarrow{\partial}/\partial t - \overleftarrow{\partial}/\partial t$ is familiar from the KG norm, Sec. 4.2.

Now expand the KG field in terms of positive and negative energy solutions of the free KG equation. As before, the field is quantized by imposing commutation (or anticommutation) relations on the expansion coefficients, a step which turns them into particle creation and annihilation operators. As we saw in Chapter 4, the *complete expansion of a relativistic field requires both the positive and negative energy solutions* and therefore has the form

$$\begin{aligned} \phi(x) &= \sum_n \left\{a_n\phi_n^{(+)}(x) + c_n^\dagger\phi_{-n}^{(-)}(x)\right\} \\ \phi^\dagger(x) &= \sum_n \left\{a_n^\dagger\phi_n^{(+)\,*}(x) + c_n\phi_{-n}^{(-)\,*}(x)\right\} \ , \end{aligned} \tag{7.38}$$

where $\phi_n^{(\pm)}(x)$ are the normalized $\pm$ energy states, defined in Eq. (4.19), and we will show that the operators a and c have the following interpretation:

a_n destroys a particle with momentum p_n and positive charge
c_n destroys an antiparticle with momentum p_n and negative charge

$a_n^\dagger$ creates a particle with momentum p_n and positive charge
$c_n^\dagger$ creates an antiparticle with momentum p_n and negative charge .

The interpretation assigned to the a_n's is a straightforward application of our previous study. That the c_n's should destroy and create antiparticles (instead of negative energy particle states) is necessary if the field theory is to describe only positive energy states, which is clearly the goal. That the coefficient of c_n

should be $\phi_{-n}^{(-)}(x)$ instead of $\phi_n^{(-)}(x)$ is required because the charge conjugation transformation shows us that the antiparticle states of momentum p_n are related to negative energy states with momentum $-p_n$. An additional, desirable feature of the expansion (7.38) is that $\phi_{-n}^{(-)}(x)$ has the covariant form of the scalar product $p{\cdot}x$ in its exponent, instead of the clumsy form $E_n t + \boldsymbol{p}_n \cdot \boldsymbol{r}$. The operator $c_n^\dagger$ (instead of c_n) must accompany a_n for two reasons. First, both a_n and $c_n^\dagger$ lower the charge of a state by one unit; a_n does this by *destroying* a particle with + charge, while $c_n^\dagger$ does this by *creating* an antiparticle with − charge. Thus the field operator ϕ always destroys one unit of charge, and by a similar argument, the operator $\phi^\dagger$ creates one unit of charge. Hence the operator $\phi^\dagger\phi$ conserves charge. Had c_n been chosen to accompany a_n, the operator $\phi^\dagger\phi$ would not conserve charge; it would include terms like $a_n^\dagger c_n$, which creates two units of charge, and $c_n^\dagger a_n$, which destroys two units of charge. A second consequence of this assignment is that if the particles were *neutral*, then the particles and antiparticles *might* be identical (but not necessarily); if they were, then $a = c \Rightarrow \phi = \phi^\dagger$. A charged field *requires* that $a \neq c$ and $\phi \neq \phi^\dagger$.

Next we use the orthogonality relations satisfied by free KG wave functions to reduce the Hamiltonian (7.37). These relations, previously given in Sec. 4.3, are

$$\begin{aligned} i\int d^3r\, \phi_n^{(+)*}(x)\overleftrightarrow{\frac{\partial}{\partial t}}\phi_{n'}^{(+)}(x) &= \delta_{nn'} \\ i\int d^3r\, \phi_n^{(-)*}(x)\overleftrightarrow{\frac{\partial}{\partial t}}\phi_{n'}^{(-)}(x) &= -\delta_{nn'} \\ i\int d^3r\, \phi_n^{(-)*}(x)\overleftrightarrow{\frac{\partial}{\partial t}}\phi_{n'}^{(+)}(x) &= 0 \ . \end{aligned} \tag{7.39}$$

Hence the Hamiltonian becomes

$$\begin{aligned} H &= -\int d^3r\, \phi^\dagger \overleftrightarrow{\frac{\partial}{\partial t}}\left(\frac{\partial \phi}{\partial t}\right) \\ &= \int d^3r \sum_n \left\{a_n^\dagger \phi_n^{(+)*}(x) + c_n \phi_{-n}^{(-)*}(x)\right\} \\ &\qquad \times i\overleftrightarrow{\frac{\partial}{\partial t}} \sum_{n'} E_{n'}\left\{a_{n'}\phi_{n'}^{(+)}(x) - c_{n'}^\dagger \phi_{-n'}^{(-)}(x)\right\} \\ &= \sum_n E_n \left\{a_n^\dagger a_n + c_n c_n^\dagger\right\} \ , \end{aligned} \tag{7.40}$$

where we have been careful to preserve the order of the operators a_n, $a_n^\dagger$, c_n, $c_n^\dagger$. Note that the second term is $+c_n c_n^\dagger$, with the + sign coming from the negative norm of the $\phi^{(-)}$ states. This term has the $cc^\dagger$ in the wrong order to be a number operator. If the c's (and hence the a's) satisfy either commutation or anticommutation relations, we may write

$$c_n c_n^\dagger = \pm c_n^\dagger c_n + 1 \ , \tag{7.41}$$

where + is for commutation and − for anticommutation relations. Hence, if the second term is expressed in terms of the number operator $c_n^\dagger c_n$,

$$H = \sum_n E_n \left\{ a_n^\dagger a_n \pm c_n^\dagger c_n \right\} + \langle 0|H|0\rangle \quad , \tag{7.42}$$

where the infinite c-number arising from the sum over the number 1 in (7.41) has been written as a vacuum expectation value $\langle 0|H|0\rangle$. In the general case (i.e., when *either* commutation or anticommutation relations are used), the normal ordered product is defined by the relation

$$:H: \equiv H - \langle 0|H|0\rangle \quad . \tag{7.43}$$

This definition is equivalent to, but more general than, the one used in Chapters 1 and 2. Redefining the Hamiltonian by subtracting its c-number vacuum expectation value, as we discussed following Eq. (3.32) in Sec. 3.1, gives the following:

$$:H: = \sum_n E_n \Big(\underbrace{a_n^\dagger a_n}_{\substack{\text{number of}\\ \text{particles}}} \pm \underbrace{c_n^\dagger c_n}_{\substack{\text{number of}\\ \text{antiparticles}}} \Big) \quad , \tag{7.44}$$

where the plus sign is for commutation relations and the minus sign for anticommutation relations.

We now conclude that the *requirement that the total energy be positive definite can be achieved only if the a's and c's satisfy commutation relations and, hence, Bose–Einstein statistics.* This is because the second sum, which is the energy of all antiparticles, is positive in this case. Since the KG theory describes spin zero particles, we have

spin zero $\Longrightarrow$ Bose–Einstein statistics .

The commutation relations are

$$\left[a_n, a_{n'}^\dagger\right] = \left[c_n, c_{n'}^\dagger\right] = \delta_{n,n'}$$
$$\left[a_n, a_{n'}\right] = \left[a_n^\dagger, a_{n'}^\dagger\right] = \left[a_n, c_{n'}\right] = \left[a_n^\dagger, c_{n'}^\dagger\right] = \left[c_n, c_{n'}\right] = \left[c_n^\dagger, c_{n'}^\dagger\right] = 0 \quad . \tag{7.45}$$

Note finally that our treatment is similar to that of the EM field, which is a real, neutral field with $a = c$. Hence, we also have found that

spin one $\Longrightarrow$ Bose–Einstein statistics .

We now discuss the second quantization of the Dirac theory.

7.4 DIRAC THEORY

It turns out that anticommutation relations are needed in order to keep the energy positive definite in the Dirac theory.

The Lagrangian density for a "classical" Dirac field, which is similar in some ways to the Lagrangian density for the Schrödinger theory, is

$$\mathcal{L} = \bar{\psi}\left[\frac{i}{2}\gamma^\mu \frac{\overleftrightarrow{\partial}}{\partial x^\mu} - m\right]\psi \ , \tag{7.46}$$

where the $\overleftrightarrow{\partial} = \overrightarrow{\partial} - \overleftarrow{\partial}$ as in the KG theory. Note that this is Lorentz invariant. Treating $\bar{\psi}$ and ψ as independent fields, we obtain the Dirac equation:

$$\frac{\partial}{\partial x^\mu}\left(\frac{\partial \mathcal{L}}{\partial\left(\dfrac{\partial \bar{\psi}}{\partial x^\mu}\right)}\right) - \frac{\partial \mathcal{L}}{\partial \bar{\psi}} = \left(-i\gamma^\mu \frac{\partial}{\partial x^\mu} + m\right)\psi = 0 \ . \tag{7.47}$$

The momentum conjugates to the independent fields $\bar{\psi}$ and ψ are

$$\begin{aligned} \bar{\pi} &= \frac{\partial \mathcal{L}}{\partial\left(\dfrac{\partial \psi}{\partial t}\right)} = \frac{i}{2}\bar{\psi}\gamma^0 = \frac{i}{2}\psi^\dagger \\ \pi &= \frac{\partial \mathcal{L}}{\partial\left(\dfrac{\partial \bar{\psi}}{\partial t}\right)} = -\frac{i}{2}\gamma^0\psi \ , \end{aligned} \tag{7.48}$$

where $\bar{\pi}$ is a row vector and π is a column vector. In what follows, we will be careful to always construct scalar products involving $\bar{\pi}$ by multiplying from the left and π by multiplying from the right. Hence, preserving this Dirac matrix structure, the Hamiltonian density is

$$\begin{aligned} \mathcal{H} &= \bar{\pi}\frac{\partial \psi}{\partial t} + \frac{\partial \bar{\psi}}{\partial t}\pi - \mathcal{L} = \frac{i}{2}\bar{\psi}\gamma^0\frac{\overleftrightarrow{\partial}}{\partial t}\psi - \mathcal{L} \\ &= \bar{\psi}\left[-\frac{i}{2}\gamma^i\frac{\overleftrightarrow{\partial}}{\partial x^i} + m\right]\psi = \psi^\dagger\left[-\frac{i}{2}\alpha_i\frac{\overleftrightarrow{\partial}}{\partial x^i} + \beta m\right]\psi \ . \end{aligned} \tag{7.49}$$

Integrating by parts, as we have done frequently, and using the periodic boundary conditions to justify dropping the boundary terms give the following familiar result for the Hamiltonian:

$$H = \int d^3r\,\mathcal{H} = \int d^3r\,\psi^\dagger\left[-i\alpha\cdot\overrightarrow{\nabla} + \beta m\right]\psi \ . \tag{7.50}$$

We now turn this classical theory into a quantum field theory by turning the field ψ into an operator. This is done by expanding ψ in a complete set of positive and negative energy eigenfunctions, introducing annihilation operators $b_{n,s}$ and $d_{n,s}$ as follows:

$$\boxed{\psi(x) = \sum_{n,s} \left\{ b_{n,s}\psi^{(+)}_{n,s}(x) + d^{\dagger}_{n,s}\psi^{(-)}_{-n,-s}(x) \right\} \quad ,} \tag{7.51}$$

where the $\psi^{(\pm)}$ are the positive and negative energy wave functions defined in Sec. 5.3. The structure of (7.51) is similar to the one we introduced for the KG theory and is justified in precisely the same way. The operators are interpreted as follows:

$b_{n,s}$ annihilates a particle of spin projection s, momentum p_n
$d_{n,s}$ annihilates an antiparticle of spin projection s, momentum p_n
$b^{\dagger}_{n,s}$ creates a particle of spin projection s, momentum p_n
$d^{\dagger}_{n,s}$ creates an antiparticle of spin projection s, momentum p_n ,

and, as required by charge conjugation, $\psi^{(-)}_{-n,-s}$ (instead of $\psi^{(-)}_{n,s}$) must go with $d^{\dagger}_{n,s}$.

The Hamiltonian can be re-expressed in terms of annihilation and creation operators following a now standard method. Using the fact that the $\psi^{(\pm)}$ are eigenfunctions of the Dirac operator and orthonormal relations satisfied by the states $\psi^{(\pm)}$, the Hamiltonian reduces to

$$\begin{aligned} H = \int d^3r \sum_{n,s} &\left(b^{\dagger}_{n,s}\psi^{(+)\dagger}_{n,s}(x) + d_{n,s}\psi^{(-)\dagger}_{-n,-s}(x) \right) \\ &\times \sum_{n',s'} E_{n'} \left(b_{n',s'}\psi^{(+)}_{n',s'}(x) - d^{\dagger}_{n',s'}\psi^{(-)}_{-n',-s'}(x) \right) \\ = &\sum_{n,s} E_n \left(b^{\dagger}_{n,s}b_{n,s} - d_{n,s}d^{\dagger}_{n,s} \right) \quad . \end{aligned} \tag{7.52}$$

Note that if d and b were complex numbers instead of operators, the second term would be negative, so H could not be positive definite, and there is *no classical Dirac theory*. Ironically, the positive definite norm gives us trouble with negative energy states since there is no way to change the sign of the energy.

However, if the b's and d's satisfy anticommutation relations, H can be made positive definite. If we require

$$\begin{aligned} \left\{ b_{n,s}, b^{\dagger}_{n',s'} \right\} &= \delta_{nn'}\delta_{ss'} \\ \left\{ d_{n,s}, d^{\dagger}_{n',s'} \right\} &= \delta_{nn'}\delta_{ss'} \end{aligned} \tag{7.53}$$

with all other anticommutators zero, we get

$$H = \sum_{n,s} E_n \left\{ b^\dagger_{n,s} b_{n,s} + d^\dagger_{n,s} d_{n,s} \right\} + \langle 0|H|0\rangle \tag{7.54}$$

or, again,

$$:H: = \sum_{n,s} E_n \left\{ b^\dagger_{n,s} b_{n,s} + d^\dagger_{n,s} d_{n,s} \right\} \quad . \tag{7.55}$$

Note that the *normal ordered product* for *fields which anticommute* still has all creation operators to the left and all annihilation operators to the right but that the *signs which arise when the order of the operators is changed are preserved.* Hence, for example,

$$:d_{n,s} d^\dagger_{n',s'}: = -d^\dagger_{n',s'} d_{n,s} \quad . \tag{7.56}$$

We conclude that the *requirement that the energy be positive definite leads to anticommutation relations for the Dirac theory.* Since this theory describes spin $\frac{1}{2}$ particles, we have

$$\text{spin } \frac{1}{2} \Longrightarrow \text{Fermi–Dirac statistics} \quad .$$

Furthermore, the canonical *anti*-commutation relations for Dirac fields are

$$\left\{ \psi_\alpha(r,t), \psi^\dagger_\beta(r',t) \right\} = \delta^3(r - r')\delta_{\alpha\beta} \quad . \tag{7.57}$$

This is easily shown by writing out the the LHS of the equation:

$$\begin{aligned}
&\left\{ \psi_\alpha(r,t), \psi^\dagger_\beta(r',t) \right\} \\
&= \sum_{\substack{n,s \\ n',s'}} \Bigg\{ \left\{ b_{n,s}, b^\dagger_{n',s'} \right\} \psi^{(+)}_{\alpha,ns}(r,t) \psi^{(+)\dagger}_{\beta,n's'}(r',t) \\
&\qquad + \left\{ d^\dagger_{n,s}, d_{n',s'} \right\} \psi^{(-)}_{\alpha,-n-s}(r,t) \psi^{(-)\dagger}_{\beta,-n'-s'}(r',t) \Bigg\} \\
&= \sum_{n,s} \left\{ \psi^{(+)}_{\alpha,ns}(r,t) \psi^{(+)\dagger}_{\beta,ns}(r',t) + \psi^{(-)}_{\alpha,-n-s}(r,t) \psi^{(-)\dagger}_{\beta,-n-s}(r',t) \right\} \\
&= \quad \delta^3(r - r')\delta_{\alpha\beta} \quad ,
\end{aligned} \tag{7.58}$$

where α and β are the Dirac indices on the spinors in the wave functions, usually suppressed, and the last equation above is recognized as the completeness relation for the Dirac $\pm$ solutions.

This completes our discussion of the construction of free field theories and the connection between spin and statistics. We are now ready to study how interactions are added to these field theories.

7.5 INTERACTIONS: AN INTRODUCTION

The first step in finding the Lagrangian which correctly describes a given physical system is to identify the fundamental degrees of freedom, or particles, which are needed to describe the system. For atomic systems these are electrons and nuclei. The nuclei are complex composites, but at the energy scales probed by atomic physics they may be treated as fundamental. For nuclear systems, the choice of the fundamental constituents is still very much at issue; at energies of a few MeV, most physicists would agree that the neutrons and protons can be chosen to be the constituents, but at higher energies, when the structure of the nucleon begins to become evident, the basic constituents are the quarks and gluons which are the ultimate building blocks from which hadronic matter is formed. Nuclear physicists are currently trying to understand precisely how to incorporate quarks and gluons into the description of nuclei and at what energy scales the structure of nuclei becomes sensitive to the presence of these fundamental constituents. Finally, particle physics now has the very successful *Standard Model*, which includes six flavors of quarks and six leptons organized into three generations as shown in Appendix D, five kinds of gauge bosons, and the Higgs. Particle physicists continue to look for the sixth flavor of quark and for the Higgs and to search for evidence for the existence of possible additional particles, which would signal the breakdown of the Standard Model.

In general, each fundamental constituent is described by a separate quantum field, and the full Lagrangian density is a sum of the free particle Lagrangian densities $\mathcal{L}_i$ for each of the constituents plus an interaction term:

$$\mathcal{L} = \sum_i \mathcal{L}_i + \mathcal{L}_{\rm int} \ . \tag{7.59}$$

The free Lagrangians can describe particles of any type. If the interaction term $\mathcal{L}_{\rm int}$ is zero, the general solutions to the problem are states which are direct products of the free particle states described by each of the free Lagrangians $\mathcal{L}_i$, and the particles are all free particles which do not interact. The interaction term contains fields which enter more than one $\mathcal{L}_i$ and hence couples the fields together and produces the interaction.

How are we to determine the structure of the interaction Lagrangian? Much of the rest of this book will discuss this question. There is no simple answer, and the art of finding the correct interaction is at the heart of modern research in particle physics. In the last 15–20 years, *interactions have been constructed to obey certain symmetries*, and this appears to be the "correct" way to find the interaction. *Gauge invariance* appears to be one of the key symmetries, and the properties of gauge invariant theories will be the subject of much of the latter half of this book. For now we note that $\mathcal{L}_{\text{int}}$ must satisfy the following constraints:

- it must be Lorentz invariant;
- it must be Hermitian;
- it should be local.

The last requirement means that all of the fields in $\mathcal{L}_{\text{int}}$ are evaluated at the *same point in space–time*. Interactions for which this is not the case are said to be *non-local*, and if used, care must be taken to insure that the non-locality is constructed in such a way that Lorentz invariance is not violated. Non-local theories will not be discussed in this book.

In addition to the rules outlined above, the interaction Lagrangian should be simple with as few free parameters as possible. The fewer the number of parameters, the greater the predictive power of the resulting theory. Another criterion, motivated more by simplicity that by any compelling physical requirement, is that $\mathcal{L}_{\text{int}}$ should, if possible, contain no time derivatives. If this is the case, the generalized momenta are not changed by the interaction, and the Hamiltonian can be calculated from

$$\mathcal{H} = \sum_i \mathcal{H}_i + \mathcal{H}_{\text{int}} \ ,$$

where $\mathcal{H}_i$ are the free Hamiltonian densities corresponding to $\mathcal{L}_i$ and $\mathcal{H}_{\text{int}} = -\mathcal{L}_{\text{int}}$. This condition makes the theory simpler, but it is sometimes not possible (for example, the EM interactions of scalar fields and quantum chromodynamics (QCD) involve time derivatives). In practice, interaction Lagrangians are usually *polynomials in the fields*, with a single parameter which defines the strength and is referred to as a coupling constant.

As an example, consider a system with two fundamental constituents described by the Dirac field ψ and a Hermitian (therefore charge zero) scalar field ϕ. A simple interaction between these fields which satisfies the above requirements is

$$\mathcal{L}_{\text{int}} = -\mathcal{H}_{\text{int}} = -g:\overline{\psi}(x)\psi(x)\phi(x): \tag{7.60}$$

where g is a real constant (in order that $\mathcal{L}$ be Hermitian). We will refer to this as a theory with ϕ^3 *structure* because the interaction Lagrangian is a *third order polynomial*; the term ϕ^3 *theory* will be applied exclusively to theories with an interaction involving three *scalar* fields interacting at a point.

It is important to get a physical feeling for the meaning of an interaction like (7.60), and this is fortunately very easily done. First recall that each field operator must create or annihilate only one particle. Hence if n fields act at a point (a theory with ϕ^n structure), the interaction always describes a situation in which $n-\ell$ particles come into the point and ℓ leave (for any $0 \le \ell \le n$). For example, to *first* order in a perturbative treatment of the interaction, a theory with a ϕ^3 structure, such as that described in Eq. (7.60), describes the eight elementary processes shown in Fig. 7.1. The reason for this is that $\overline{\psi}\psi\phi$ contains precisely

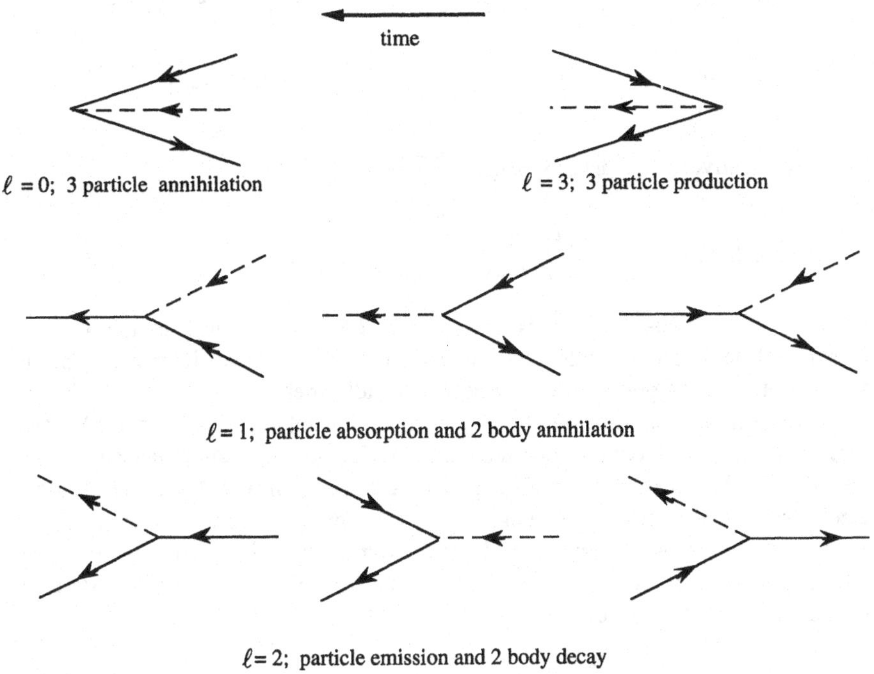

Fig. 7.1 Diagrams showing the possible interactions which result from the single term given in Eq. (7.60). The antifermion lines have arrows pointing in a direction *opposite* to the flow of time (see the discussion in Sec. 10.3).

three annihilation or creation operators and therefore has non-zero matrix elements between the following states:

$$
\begin{array}{llll}
\ell = 0 & \langle 0|\mathcal{H}_{\rm int}|p_1\,\bar{p}_2\,k\rangle & & \\
\ell = 1 & \langle p_2|\mathcal{H}_{\rm int}|p_1\,k\rangle & \langle k|\mathcal{H}_{\rm int}|p_1\,\bar{p}_2\rangle & \langle \bar{p}_2|\mathcal{H}_{\rm int}|\bar{p}_1\,k\rangle \\
\ell = 2 & \langle p_1\,\bar{p}_2|\mathcal{H}_{\rm int}|k\rangle & \langle p_2\,k|\mathcal{H}_{\rm int}|p_1\rangle & \langle \bar{p}_2\,k|\mathcal{H}_{\rm int}|\bar{p}_1\rangle \\
\ell = 3 & \langle p_1\,\bar{p}_2\,k|\mathcal{H}_{\rm int}|0\rangle \quad , & &
\end{array}
$$

where p_1 and p_2 are momenta of fermions, $\bar{p}_1$ and $\bar{p}_2$ are momenta of antifermions (all described by the Dirac field ψ), and k is the momenta of the scalar particle, which is its own antiparticle. All other matrix elements of $\mathcal{H}_{\rm int}$ are zero. In *higher orders*, other processes are possible (as we shall soon see), but they must all be built up out of the eight elementary processes above. Six of these (corresponding to $\ell = 1$ or 2) describe emission and absorption of one particle at a time.

Many other interactions can be constructed. Examples of other types are:

ϕ^4 structure: $\phi_1^\dagger(x)\phi_1(x)\phi_2^\dagger(x)\phi_2(x)$ $\quad [\overline{\psi}(x)\psi(x)]^2$ $\quad \phi^4(x)$

ϕ^{2n} structure: $\left[\phi_1^\dagger(x)\phi_1(x)\right]^n$ $\quad \phi^{2n}(x)$ $\quad \left[\overline{\psi}(x)\psi(x)\right]^{n-\ell}\phi^{2\ell}(x)$

derivative: $\overline{\psi}(x)\gamma^5\gamma^\mu\psi(x)\dfrac{\partial\phi(x)}{\partial x^\mu}$

non-linear: $\dfrac{\lambda\phi^2(x)}{1-\mu\phi^2(x)}$,

where in all of these cases ϕ is a Hermitian field and ϕ_1 is a complex (non-Hermitian) field. The so-called "non-linear" interaction derives its name from the fact that it contains polynomial interactions of all orders.

The simplest and most commonly encountered interactions have a ϕ^3 structure, and these will occupy our attention in Chapter 9. Later, in Chapter 13, we will see that interactions with a ϕ^4 structure occur in QCD and the standard electroweak theory and that the nonlinear interaction arises in the nonlinear sigma model. But before we begin our study of the dynamics of interacting theories, we take a first look at consequences which can be derived solely from the presence of a symmetry of the theory.

PROBLEMS

7.1 Show that the charged KG field discussed in Sec. 7.3 satisfies the CCR

$$[\phi(r,t),\pi^\dagger(r',t)] = i\delta^3(r-r') \tag{7.61}$$

and that all other commutators are zero.

7.2 Neutral *KG* theory. Construct the theory for a neutral KG particle (i.e., where the field ϕ is Hermitian) from the following arguments:

(a) If the charged field $\phi = (\phi_1 + i\phi_2)/\sqrt{2}$, where ϕ_1 and ϕ_2 are commuting Hermitian fields, and if the charged field ϕ satisfies the CCR's worked out in Prob. 1 above, show that

$$[\phi_i(r,t), \frac{\partial\phi_i(r',t)}{\partial t}] = i\delta^3(r-r') \ ,$$

where ϕ_i is either ϕ_1 or ϕ_2.

(b) Show that the Lagrangian density for the charged field ϕ,

$$\mathcal{L} = \frac{\partial\phi^\dagger}{\partial x^\mu}\frac{\partial\phi}{\partial x_\mu} - m^2\phi^\dagger\phi \tag{7.62}$$

[which is just the operator form of Eq. (7.32)], can be written as the sum of two *independent* Lagrangian densities

$$\mathcal{L} = \mathcal{L}_1 + \mathcal{L}_2 \ ,$$

where each density $\mathcal{L}_i$ is multiplied by an overall factor of $\frac{1}{2}$,

$$\mathcal{L}_i = \frac{1}{2}\left(\frac{\partial \phi_i}{\partial x^\mu}\frac{\partial \phi_i}{\partial x_\mu} - m^2\phi_i^2\right) \ , \tag{7.63}$$

compared with the Lagrangian density for its charged counterpart.

(c) Using the density $\mathcal{L}_i$, find the momentum π_i conjugate to ϕ_i, and find the Hamiltonian density $\mathcal{H}_i$ for a neutral theory. Express the Hamiltonian H_1 in terms of the annihilation operators $a_{1n} = (a_n + c_n)/\sqrt{2}$ and the corresponding creation operators $a_{1n}^\dagger = (a_n + c_n)^\dagger/\sqrt{2}$.

(d) Discuss the significance of your results. What is the Lagrangian density for a neutral scalar theory?

7.3 Using the ideas developed in Prob. 2 above, and working from the Lagrangian density for a neutral massive vector field [given in Eq. (2.39) in Sec. 2.5], find the Lagrangian density for a *charged* massive vector field.

7.4 Consider a ϕ^3 theory with a charged scalar field Φ_1 and a neutral scalar field ϕ and an interaction Lagrangian density of the form

$$\mathcal{L}_{\text{int}} = -\lambda : \Phi_1^\dagger(x)\Phi_1(x)\phi(x) : \ .$$

(a) Write out the full Lagrangian density for the theory.

(b) Evaluate the following matrix element:

$$M = \int d^4x \left\langle k\overline{k'}\right| : \Phi_1^\dagger(x)\Phi_1(x)\phi(x) : \left|p\right\rangle \ ,$$

where $|p\rangle$ is the state with one neutral particle with momentum p and $|k\overline{k'}\rangle$ is the state with a charged particle with momentum k and a charged *antiparticle* with momentum k'.

(c) When is the matrix element evaluated in part (b) not equal to zero, and what is its physical significance?

7.5 Construct a theory in which it is possible for two particles to scatter and produce a third ($2 \to 3$) at a single point in space–time.

CHAPTER 8

SYMMETRIES I

This is the first of two chapters devoted to the discussion of symmetries in field theory. Symmetries are important for two reasons. First, if a symmetry is observed in nature, then the Lagrangian must be invariant under the transformations which describe the symmetry, and this imposes a constraint on the form of the interaction Lagrangian. Second, any properties which can be shown to be the consequences of an *exact* symmetry must be *exact* results, regardless of the details of the interactions. It is very difficult to obtain exact results in any other way.

In this chapter we begin with Noether's theorem, which shows that there exists a conserved quantity associated with every *continuous* symmetry and also how to find it. We then study the *discrete* symmetries which the Lagrangian may satisfy. Discrete symmetries are are single, isolated transformations with no infinitesimal form, and the three which are of great importance in field theory are parity (space inversion), charge conjugation (the interchange of particles and antiparticles), and time reversal (or, more correctly, the reversal of the direction of motion). After discussing each of these in turn, we discuss the famous PCT theorem, which states that the product of all three of these transformations must always be a symmetry of the system, even if individual members of this set are not symmetries. The non-trivial consequences of PCT invariance are among the most secure predictions of field theory. In Chapter 13 we continue, with a discussion of *non-Abelian gauge symmetries*, *chiral symmetry*, and *spontaneous symmetry breaking*, all ideas of paramount importance in modern physics.

8.1 NOETHER'S THEOREM

The foundation of all of our discussion of continuous symmetry will be Noether's theorem, which we now state and prove.

> **Theorem:** For every continuous transformation of the field functions and coordinates which leaves the action unchanged, there is a definite combination of the field functions and their derivatives which is conserved (i.e., a constant in time).

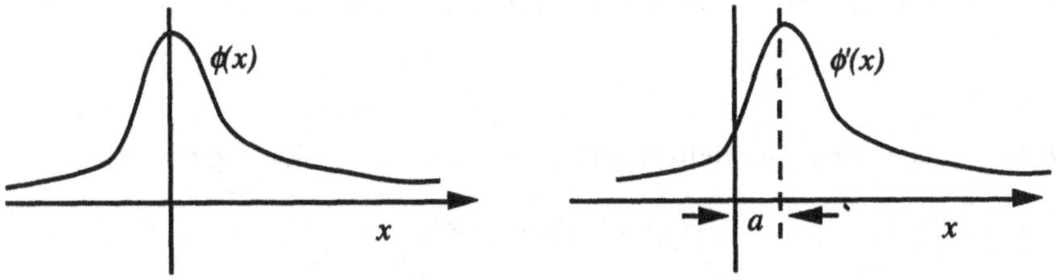

Fig. 8.1 Illustration of the active translation of a scalar function through a distance a in the x-direction.

The infinitesimal transformations of the coordinates and fields will be written

$$\begin{aligned} x'^{\mu} &= x^{\mu} + \lambda^{\mu}{}_{i}(x)\,\epsilon^{i} \\ \psi'_{\alpha}(x') &= \psi_{\alpha}(x) + \Omega_{\alpha i}(x)\,\epsilon^{i} \quad , \end{aligned} \tag{8.1}$$

where λ and Ω are known functions of x and ϵ^i are infinitesimal parameters which describe the transformation. Note that the range of i is not specified. In particular, i need not range from $0 \to 3$ and ϵ^i *may not* be a four-vector.

Examples of Continuous Transformations

Before we proceed with the proof of Noether's theorem, we give some examples of transformations. (Review Sec. 2.7 and 5.8.)

Translations in space and time. The translation of a scalar function $\phi(x)$ through a distance a is illustrated in Fig. 8.1. Here

$$x'^{\mu} = x^{\mu} + g^{\mu}{}_{\nu}\, a^{\nu} \quad , \tag{8.2}$$

where in this case ν [the i of Eq. (8.1)] runs from 0 to 3 and a^ν is a four-vector. If we translate both the function *and* the coordinates, everything is unchanged, so that

$$\phi'(x') = \phi(x) \tag{8.3}$$

or

$$\phi'(x'^{\mu}) = \phi\left(x'^{\mu} - g^{\mu}{}_{\nu}\, a^{\nu}\right) \quad .$$

In this case:

$$\boxed{\begin{aligned} \lambda^{\mu}{}_{\nu} &= g^{\mu}{}_{\nu} \\ \Omega_{\alpha\nu} &= 0 \end{aligned} \qquad \textit{translations.}} \tag{8.4}$$

Homogeneous Lorentz transformations. As an example, consider a vector field A^μ. To see how this transforms, follow steps similar to those taken in Sec. 2.7 in our discussion of rotations. First, form a scalar product

$$\phi(x) = n_\mu A^\mu(x) \ ,$$

where n_μ is a fixed but arbitrary direction in four-dimensional space. Then,

$$\phi'(x') = \phi(x)$$

or, if $x' = \Lambda\, x$, then

$$\begin{aligned} n'_\nu A'^\nu(x') &= n_\mu A^\mu(\Lambda^{-1}x') \\ &= g_{\mu\nu}\, n^\mu A^\nu(\Lambda^{-1}x') \ . \end{aligned}$$

Recalling Eq. (2.6),

$$g_{\mu\nu} = \Lambda^\alpha{}_\mu\, g_{\alpha\beta}\, \Lambda^\beta{}_\nu \ ,$$

this becomes

$$\begin{aligned} g_{\alpha\beta}\, n'^\alpha A'^\beta(x') &= g_{\alpha\beta}\, \Lambda^\alpha{}_\mu\, n^\mu \Lambda^\beta{}_\nu A^\nu(\Lambda^{-1}x') \\ &= g_{\alpha\beta}\, n'^\alpha \Lambda^\beta{}_\nu A^\nu(\Lambda^{-1}x') \ . \end{aligned}$$

Since this holds for any n', we have the transformation law

$$\boxed{A'^\mu(x') = \Lambda^\mu{}_\nu A^\nu(\Lambda^{-1}x') \ .} \tag{8.5}$$

This result holds for *all* Lorentz transformations.

Recalling Eq. (5.89), where the generators of an LT were written

$$(\omega_{\mu\nu})^\alpha{}_\beta = -\tfrac{1}{2}\epsilon_{\mu\nu\lambda\sigma}\, \epsilon^{\lambda\sigma\alpha}{}_\beta \ ,$$

we define the generators of rotations [compare with Eq. (5.91)] by

$$\begin{aligned} (r_i)^\alpha{}_\beta &= \tfrac{1}{2}\epsilon_{ijk}\omega_{jk} = -\tfrac{1}{4}\epsilon_{ijk}\epsilon_{jk\lambda\sigma}\, \epsilon^{\lambda\sigma\alpha}{}_\beta \\ &= -\tfrac{1}{4}\epsilon_{0i}{}^{jk}\, \epsilon_{jk\lambda\sigma}\epsilon^{\lambda\sigma\alpha}{}_\beta \ . \end{aligned} \tag{8.6}$$

The relation $\epsilon_{0123} = -\epsilon^{0123} = 1 = \epsilon_{123}$ was used in the last step to relate the three- and four-dimensional ϵ symbols. Using the identity

$$\epsilon^{\mu\nu\alpha\beta}\, \epsilon_{\mu'\nu'\alpha\beta} = -2(g^\mu{}_{\mu'}\, g^\nu{}_{\nu'} - g^\mu{}_{\nu'}\, g^\nu{}_{\mu'})$$

gives

$$(r_i)^\alpha{}_\beta = -\epsilon_{0i}{}^\alpha{}_\beta \ . \tag{8.7}$$

Hence, for infinitesimal rotations,

$$\Lambda^\mu{}_\nu = g^\mu{}_\nu + \theta_i \, (r_i)^\mu{}_\nu \ ,$$

and

$$\boxed{\begin{array}{lll} \lambda^\mu{}_i = (r_i)^\mu{}_\nu \, x^\nu = -\epsilon_{0i}{}^\mu{}_\nu \, x^\nu & & \\ \Omega_{\mu i} = \epsilon_{0i\mu\nu} A^\nu & \text{vector fields} & \textit{rotations,} \\ \Omega_{\alpha i} = -\dfrac{i}{2} \left(\gamma^5 \alpha^i \psi\right)_\alpha & \text{spinor fields} & \end{array}} \tag{8.8}$$

where the result for spinor fields can be extracted from Eq. (5.116).

Gauge transformations. Gauge transformations leave the coordinates unchanged but transform spinor fields by a phase. Hence $x' = x$, and

$$\psi'(x) = e^{-ig\,\theta}\psi(x) \ . \tag{8.9}$$

Therefore

$$\boxed{\begin{array}{ll} \lambda^\mu{}_\nu = 0 & \textit{gauge} \\ \Omega_\alpha = -ig\psi_\alpha(x) & \textit{transformations.} \end{array}} \tag{8.10}$$

The transformation laws of vector fields under gauge transformations will be given in Chapter 13, where we discuss these important symmetries in detail.

Proof of Noether's Theorem

We return now to the proof of Noether's theorem. We assume that symmetry leaves the Lagrangian density locally invariant (i.e., unchanged in the neighborhood of any point) so that the variation of the action over any finite volume is zero. This means that

$$\begin{aligned} \delta[\mathcal{A}] &= \mathcal{A}' - \mathcal{A} = 0 \\ &= \int_{V'} \mathcal{L}'(x') d^4x' - \int_V \mathcal{L}(x)\, d^4x \\ &\underset{\substack{\text{1st} \\ \text{order}}}{=} \int_{V'} \left(\mathcal{L}'(x') - \mathcal{L}(x')\right) d^4x' + \int \Big[\underbrace{\mathcal{L}(x')d^4x'}_{\substack{\text{volume} \\ V'}} - \underbrace{\mathcal{L}(x)d^4x}_{\substack{\text{volume} \\ V}}\Big] \\ &\underset{\substack{\text{1st} \\ \text{order}}}{=} \int_V \left(\mathcal{L}'(x) - \mathcal{L}(x)\right) d^4x \\ &\quad + \int_V \left(\mathcal{L}(x) + \frac{\partial \mathcal{L}(x)}{\partial x^\mu} \lambda^\mu{}_i \epsilon^i\right) J d^4x - \int_V \mathcal{L}(x) d^4x \ , \end{aligned} \tag{8.11}$$

where J is the Jacobian of the transformation, $\partial\mathcal{L}/\partial x^\mu$ is the total variation of $\mathcal{L}$ with respect to x^μ [with all other $x^\nu (\nu \neq \mu)$ being held constant], and it is

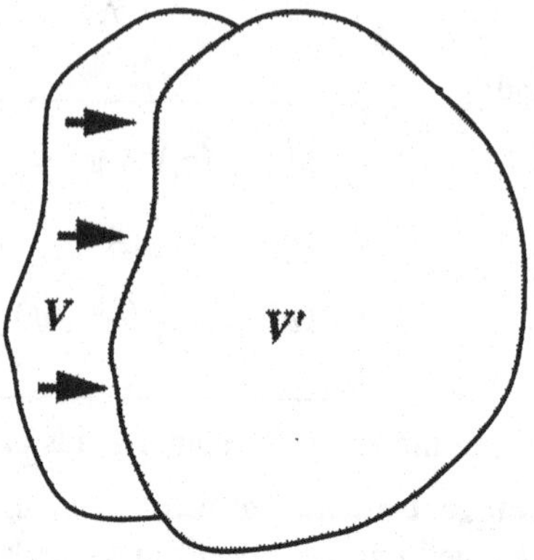

Fig. 8.2 Figure illustrating the transformation of a finite volume of space–time under a symmetry transformation.

assumed that V' is obtained from V by mapping the boundary $f(x) = 0$ to a new boundary $f(x') = 0$ (as suggested in Fig. 8.2). In the last step, all expressions have been expressed in terms of the original volume V (since the difference $\mathcal{L}' - \mathcal{L}$ is already first order, it can be equally well evaluated by integrating over V as V'). Before continuing with the argument, we emphasize that $\delta[\mathcal{A}] = 0$ over *any finite* volume, and it is this feature which will give us a conservation law which will hold for each cell in space–time (i.e., a *local* conservation law). Note that this argument is quite different from the one used to derive the Euler–Lagrange equations; in that case the derivation required integrations by parts, and hence the action was integrated over *all* space so that the boundary terms resulting from those integrations would be zero. In deriving Noether's theorem, boundary terms will play no role because the proof does not require any integrations by parts.

To reduce (8.11), we first calculate the Jacobian of the transformation, which is

$$J = \left|\frac{\partial x'^{\mu}}{\partial x^{\nu}}\right| = \det\left|1 + \frac{\partial \lambda^{\mu}{}_{i}(x)}{\partial x^{\nu}}\,\epsilon^{i}\right| \quad . \tag{8.12}$$

We only want the terms up to first order in ϵ, and hence we need only keep the product of the diagonal terms of the matrix, which when multiplied out to first order in ϵ become

$$J - 1 = \frac{\partial \lambda^{\mu}{}_{i}}{\partial x^{\mu}}\,\epsilon^{i} \qquad \text{(summed over } \mu \text{ and } i) \quad . \tag{8.13}$$

Next, the variation in the *structure* of $\mathcal{L}$ (i.e., variation with x held constant) is

$$\delta[\mathcal{L}] = \mathcal{L}'(x) - \mathcal{L}(x) = \left\{\frac{\partial \mathcal{L}}{\partial \psi_{\alpha}}\,\delta_{i}[\psi_{\alpha}] + \frac{\partial \mathcal{L}}{\partial\left(\dfrac{\partial \psi_{\alpha}}{\partial x^{\mu}}\right)}\,\delta_{i}\left[\frac{\partial \psi_{\alpha}}{\partial x^{\mu}}\right]\right\}\epsilon^{i} \,, \tag{8.14}$$

where the two terms are the variations with respect to the *shape or structure* of ψ_α and $\partial_\mu \psi_\alpha$ (with x held constant). The notation $\delta_i[f]$ refers to the variation of f with respect to the continuous parameter ϵ^i, so that $\delta_i[f]\,\epsilon^i$ is the total variation with respect to all parameters ϵ^i. In this case

$$\delta_i[\psi_\alpha]\,\epsilon^i = \psi'_\alpha(x) - \psi_\alpha(x) = \underbrace{\psi'_\alpha(x) - \psi'_\alpha(x')}_{-\frac{\partial \psi'_\alpha}{\partial x^\mu}\,\delta_i[x^\mu]\,\epsilon^i} + \underbrace{\psi'_\alpha(x') - \psi_\alpha(x)}_{\Omega_{\alpha i}\,\epsilon^i}$$

$$= \left(-\frac{\partial \psi_\alpha}{\partial x^\mu}\lambda^\mu{}_i + \Omega_{\alpha i}\right)\epsilon^i \quad , \tag{8.15}$$

where, in going to the last step, we used $\frac{\partial \psi'}{\partial x} \cong \frac{\partial \psi}{\partial x}$, which is correct to lowest order. In a similar manner,

$$\delta_i\left[\frac{\partial \psi_\alpha}{\partial x^\mu}\right]\,\epsilon^i = \frac{\partial}{\partial x^\mu}\left(\delta_i[\psi_\alpha]\right)\,\epsilon^i \quad . \tag{8.16}$$

Putting all this together gives, to first order in ϵ^i,

$$\begin{aligned}\delta[\mathcal{A}] &= 0 \\ &= \int_V d^4x \left\{ \frac{\partial \mathcal{L}}{\partial x^\mu}\lambda^\mu{}_i + \frac{\partial \mathcal{L}}{\partial \psi_\alpha}\left(\Omega_{\alpha i} - \frac{\partial \psi_\alpha}{\partial x^\nu}\lambda^\nu{}_i\right) \right. \\ &\qquad \left. + \frac{\partial \mathcal{L}}{\partial\left(\frac{\partial \psi_\alpha}{\partial x^\mu}\right)}\frac{\partial}{\partial x^\mu}\left(\Omega_{\alpha i} - \frac{\partial \psi_\alpha}{\partial x^\nu}\lambda^\nu{}_i\right) + \mathcal{L}\frac{\partial \lambda^\mu{}_i}{\partial x^\mu} \right\}\epsilon^i \ . \end{aligned} \tag{8.17}$$

Now use the equations of motion

$$\frac{\partial \mathcal{L}}{\partial \psi_\alpha} = \frac{\partial}{\partial x^\mu}\left(\frac{\partial \mathcal{L}}{\partial\left(\frac{\partial \psi_\alpha}{\partial x^\mu}\right)}\right)$$

to eliminate $\partial \mathcal{L}/\partial \psi_\alpha$. This permits us to extract a perfect differential, and dropping ϵ^i we obtain

$$\int_{\substack{\text{any}\\ V}} d^4x\,\frac{\partial}{\partial x^\mu}\mathcal{O}^\mu{}_i = 0 \ . \tag{8.18}$$

Since this holds for *any* volume, it must be a local relation and we obtain the local form of the conservation law

$$\boxed{\frac{\partial}{\partial x^\mu}\mathcal{O}^\mu{}_i = 0 \ ,} \tag{8.19}$$

where $\mathcal{O}^{\mu}{}_{i}$ is the conserved *current density*

$$\mathcal{O}^{\mu}{}_{i} = \frac{\partial \mathcal{L}}{\partial\left(\dfrac{\partial \psi_\alpha}{\partial x^{\mu}}\right)} \left[\frac{\partial \psi_\alpha}{\partial x^{\nu}} \lambda^{\nu}{}_{i} - \Omega_{\alpha i}\right] - \mathcal{L}\lambda^{\mu}{}_{i} \ . \tag{8.20}$$

Of course this "current" density does not necessarily have anything to do with the usual electric current; the term used here refers to any general four-vector quantity which satisfies a local conservation law.

If the fields fall to zero at spatial infinity, we obtain the constant of the motion by integrating the four-divergence of $\mathcal{O}$ over an infinite slab bounded by any two times t_1 and t_2,

$$\int_{t_1}^{t_2}\int d^3r \frac{\partial}{\partial t}\mathcal{O}^{0}{}_{i} + \int_{t_1}^{t_2}\underbrace{\int d^3r \frac{\partial}{\partial x^j}\,\mathcal{O}^{j}{}_{i}}_{\text{integrates to zero}} = 0 \ . \tag{8.21}$$

Integrating the first term from t_1 to t_2 and noting that t_1 and t_2 could be any two times permit us to conclude that

$$\boxed{\mathcal{Q}_i = \int d^3r\,\mathcal{O}^{0}{}_{i}(\boldsymbol{r},t) = \text{constant}\ .} \tag{8.22}$$

This quantity is the conserved *charge* (i.e., the total time component of the conserved current), and the proof is now complete.

A nice feature of this proof is that it leads to the explicit construction of the conserved quantity, so that we know what is actually conserved as a consequence of the continuous symmetry. We now illustrate the consequences of this important theorem in a number of special cases.

8.2 TRANSLATIONS

Noether's theorem provides the ultimate justification for the definitions of the energy, momentum, and angular momentum operators, which we introduced in Chapters 1 and 2. To obtain these quantum field operators we take the classical c-number quantity (8.21) and substitute the quantum field operators for the classical fields. The order of the terms now matters, and in order to insure that the ground state has zero energy, momentum, and angular momentum, we normal order the terms as discussed in Chapter 1.

As an example, consider the implications of translational invariance for the Dirac theory. From Eq. (8.4), $\lambda^{\mu}{}_{\nu} = g^{\mu}{}_{\nu}$ and $\Omega_{\alpha\nu} = 0$, and the conserved current density, which is called the *stress energy tensor*, is

$$T^{\mu\nu} = \frac{\partial \mathcal{L}}{\partial\left(\dfrac{\partial \psi_\alpha}{\partial x^{\mu}}\right)} \frac{\partial \psi_\alpha}{\partial x_{\nu}} - \mathcal{L}\, g^{\mu\nu} \ . \tag{8.23}$$

The conserved "charges" are therefore

$$P^\mu = \int d^3r\, T^{0\mu} \quad . \tag{8.24}$$

As a consequence, we obtain four conserved quantities associated with invariance under translations in any of the four space–time directions:

(a) Time translations ($\mu = 0$):

$$P^0 \equiv H = \int d^3r\, T^{00} = \int d^3r \left(\frac{\partial \mathcal{L}}{\partial \left(\dfrac{\partial \psi_\alpha}{\partial t} \right)} \frac{\partial \psi_\alpha}{\partial t} - \mathcal{L} \right) \quad , \tag{8.25}$$

which is consistent with the usual definition of the Hamiltonian.

(b) Space translations ($\nu = i = 1, 2, 3$):

$$P^i = \int d^3r\, T^{0i} = -\int d^3r \frac{\partial \mathcal{L}}{\partial \left(\dfrac{\partial \psi_\alpha}{\partial t} \right)} \frac{\partial \psi_\alpha}{\partial x^i} \quad . \tag{8.26}$$

For example, the momentum operator in the Dirac theory is

$$\begin{aligned} P^i &= -\int d^3r \left[\frac{1}{2} \bar{\psi} i \gamma^0 \frac{\partial \psi}{\partial x^i} - \frac{1}{2} \frac{\partial \bar{\psi}}{\partial x^i} i \gamma^0 \psi \right] \\ &= -i \int d^3r\, \psi^\dagger \frac{\partial \psi}{\partial x^i} \quad , \end{aligned} \tag{8.27}$$

where we integrated by parts. Substituting the field expansions

$$\begin{aligned} \psi(r,t) &= \sum_{k,s} \frac{1}{\sqrt{2E_k L^3}} \left\{ u(k,s) b_{k,s} e^{-ik\cdot x} + v(k,s) d^\dagger_{k,s} e^{ik\cdot x} \right\} \\ &= \sum_{k,s} \left\{ b_{k,s} \psi^{(+)}_{k,s}(x) + d^\dagger_{-k,-s} \psi^{(-)}_{k,s}(x) \right\} \end{aligned}$$

and using

$$\frac{\partial}{\partial x^i} \psi^{(\pm)}_{k,s}(x) = ik^i \psi^{(\pm)}_{k,s}(x)$$

and the orthogonality of the Dirac wave functions give

$$P^i = \sum_{k,s} k^i \left\{ b^\dagger_{k,s} b_{k,s} + d_{-k,-s} d^\dagger_{-k,-s} \right\} = \sum_{k,s} k^i \left\{ b^\dagger_{k,s} b_{k,s} - d_{k,s} d^\dagger_{k,s} \right\} \quad . \tag{8.28}$$

Now, using the anticommutation relations to put this into normal order gives

$$P^i = \sum_{k,s} k^i \left[b^\dagger_{k,s} b_{k,s} + d^\dagger_{k,s} d_{k,s} \right] \tag{8.29}$$

as expected.

Note the great power of Noether's theorem. In the above example we were able to find the momentum operator in a case with *no* classical analogue. This provides justification for our heuristic development in Sec. 1.6 and later in Sec. 2.7.

8.3 TRANSFORMATIONS OF STATES AND OPERATORS

In general, a symmetry in quantum mechanics is a group of transformations which preserves matrix elements. Therefore each transformation in a symmetry group can be represented by a unitary matrix which operates on the quantum mechanical states, transforming them under the symmetry

$$U(\theta)|n\rangle = e^{-i\mathcal{Q}\theta}|n\rangle = |n'\rangle \quad , \tag{8.30}$$

where θ is a continuous real parameter and $\mathcal{Q}$ is a Hermitian matrix referred to as the *generator* of the transformations (for simplicity we limit ourselves here to groups with only one continuous parameter and one generator). Each value of θ picks out a different member of the group, which is a *continuous* group because θ can be varied continuously.

The transformation law for the operators can be determined from the transformation law for the states using the following rule:

$$\langle m|\mathcal{O}|n\rangle = \langle m'|\mathcal{O}'|n'\rangle \quad , \tag{8.31}$$

which states that matrix elements are unchanged if *both* the states and the operators are transformed. Hence

$$\langle m'|\mathcal{O}'|n'\rangle = \langle m|U^\dagger(\theta)\,\mathcal{O}'U(\theta)|n\rangle = \langle m|\mathcal{O}|n\rangle \quad , \tag{8.32}$$

and since m and n are arbitrary states, this gives

$$\mathcal{O}' = U(\theta)\,\mathcal{O}\,U^\dagger(\theta) = e^{-i\mathcal{Q}\theta}\mathcal{O}\,e^{i\mathcal{Q}\theta} \quad . \tag{8.33}$$

If the untransformed operator $\mathcal{O}$ does not depend on θ, then differentiating both sides of Eq. (8.33) gives

$$[\mathcal{Q}, \mathcal{O}'(\theta)] = -i\frac{\partial \mathcal{O}'(\theta)}{\partial \theta} \quad . \tag{8.34}$$

Hence, *the infinitesimal change in any operator $\mathcal{O}$ under a symmetry transformation is given by the commutator of the generators of the symmetry group with the operator*. This is a very general relation which we have used several times before.

If the symmetry in question is *also a symmetry of the Lagrangian*, then the transformations (8.30) will commute with the Hamiltonian, and the generator of the transformations will be a *constant of the motion*. In this case, the generator is the conserved "charge" associated with the symmetry, given in Eq. (8.21). In lieu of a proof of this statement, we will show that it is true for the translations.

The finite translations are constructed from the generators of translations, which are the momentum operators, in the following way:

$$U(a) = e^{iP^i a^i} \quad . \tag{8.35}$$

Note that phase in the exponent is $+i$, instead of the $-i$ used for the time translation operator; this is consistent with the covariant scalar product $Ht - P^i r^i$ (and agrees with our definitions in Chapter 1; recall Prob. 1.2). For translations of a scalar operator $\mathcal{O} = \mathcal{O}(x) = \mathcal{O}'(x')$, with $x' = x + a$, and we have

$$\mathcal{O}'(x) = U\mathcal{O}(x)\,U^\dagger = \mathcal{O}(x-a) \quad . \tag{8.36}$$

For infinitesimal a this reduces to

$$\boxed{\left[P^i, \mathcal{O}(x)\right] = i\frac{\partial \mathcal{O}}{\partial x^i}} \quad . \tag{8.37}$$

Equation (8.37) is an example of the general relation (8.34) and is consistent with Eq. (1.38). To see this, note that the time translation operator was $U(\Delta t) = e^{-iH\Delta t}$ (for H independent of time) and that therefore (1.38) could be written

$$\phi(t_0 + \Delta t) = U^\dagger(\Delta t)\phi(t_0)U(\Delta t) \quad . \tag{8.38}$$

Changing $\Delta t \to -\Delta t$ and noting that $U(-\Delta t) = U^\dagger(\Delta t)$ permit us to rewrite (8.38) as

$$U(\Delta t)\phi(t_0)U^\dagger(\Delta t) = \phi(t_0 - \Delta t) \quad , \tag{8.39}$$

which agrees with (8.36).

We will return to the discussion of continuous symmetries in Chapter 13, where we discuss gauge invariance and chiral symmetry. Now we turn to a discussion of the three discrete transformations of great importance to the construction of interactions: space inversion, charge conjugation, and time inversion.

8.4 PARITY

We start with a discussion of space inversion, realized by the parity transformation. Assume that there exists a unitary operator $\mathcal{P}$ which transforms the spatial coordinates from $\boldsymbol{r} \to -\boldsymbol{r}$. Its representation on the Fock space of particle states will be denoted

$$\mathcal{P}|n\rangle = |n'\rangle \ . \tag{8.40}$$

Then the field operators transform according to Eq. (8.32),

$$\langle m'|\phi'|n'\rangle = \langle m|\phi|n\rangle = \langle m'|\mathcal{P}\phi\mathcal{P}^\dagger|n'\rangle \quad ,$$

which implies [recall Eq. (8.33)]

$$\phi' = \mathcal{P}\phi\mathcal{P}^\dagger \ .$$

Since parity is space inversion, the transformation law for the fields is

$$\boxed{\begin{aligned} \mathcal{P}\phi(r,t)\mathcal{P}^\dagger &= \phi' = \eta_\phi^P\,\phi(-r,t) \\ \mathcal{P}\psi(r,t)\mathcal{P}^\dagger &= \psi' = \eta_\psi^P\,S(P)\psi(-r,t) = \eta_\psi\gamma^0\psi(-r,t) \\ \mathcal{P}A^\mu(r,t)\mathcal{P}^\dagger &= A'^\mu = \Lambda(P)^\mu{}_\nu A^\nu(-r,t) \ , \end{aligned}} \tag{8.41}$$

where the phases η_ϕ^P and η_ψ^P are the *intrinsic* parities of the scalar and spinor fields and

$$\Lambda(P)^\mu{}_\nu = \begin{pmatrix} 1 & & & \\ & -1 & & \\ & & -1 & \\ & & & -1 \end{pmatrix} \tag{8.42}$$

is the representation of the parity operator on the four-vector space. Because $\mathcal{P}^2 = 1$, the phases must satisfy the relations $\eta_\phi^{P\,2} = \eta_\psi^{P\,2} = 1$ and hence can only be ± 1. The final determination of their value depends on experiment. Equation (8.41) incorporates the fact that the photon is known to be a vector field (negative parity but positive *intrinsic* parity). The representation of the parity operator as γ^0 (space inversion) for the spinor field was already developed in Sec. 5.9 [recall Eq. (5.119)].

The operation of parity on the annihilation and creation operators can be readily found from the field transformation laws:

$$\begin{aligned} \mathcal{P}\phi(r,t)\mathcal{P}^\dagger &= \eta_\phi^P\,\phi(-r,t) \\ &= \eta_\phi^P \sum_k \frac{1}{\sqrt{2EL^3}}\left(a_{\boldsymbol{k}}e^{-i(Et+\boldsymbol{k}\cdot\boldsymbol{r})} + c_{\boldsymbol{k}}^\dagger e^{i(Et+\boldsymbol{k}\cdot\boldsymbol{r})}\right) \\ &= \eta_\phi^P \sum_k \frac{1}{\sqrt{2EL^3}}\left(a_{-\boldsymbol{k}}e^{-ik\cdot x} + c_{-\boldsymbol{k}}^\dagger e^{ik\cdot x}\right) \ , \end{aligned} \tag{8.43}$$

where k was changed to $-k$ in the sum in the last expression. Hence, equating Fourier coefficients gives

$$\boxed{\begin{aligned} \mathcal{P}a_k\mathcal{P}^\dagger &= \eta_\phi^P a_{-k} \\ \mathcal{P}c_k\mathcal{P}^\dagger &= \eta_\phi^P c_{-k} \quad . \end{aligned}} \tag{8.44}$$

On single particle states, this means that

$$\begin{aligned} \mathcal{P}|k\rangle = \mathcal{P}a_k^\dagger|0\rangle &= \eta_\phi^P a_{-k}^\dagger \mathcal{P}|0\rangle \\ &= \eta_\phi^P |-k\rangle \quad , \end{aligned} \tag{8.45}$$

which shows that parity changes the direction of momentum of a state (as expected) and that particles and antiparticles have the same *intrinsic parity*.

For spinors, a similar argument gives

$$\begin{aligned} \mathcal{P}\psi(r,t)\mathcal{P}^\dagger &= \eta_\psi^P \gamma^0 \psi(-r,t) \\ &= \eta_\psi^P \gamma^0 \sum_k \frac{1}{\sqrt{2EL^3}} \left\{ b_{-k,s} u(-\boldsymbol{k},s)\, e^{-ik\cdot x} + v(-\boldsymbol{k},s) d_{-k,s}^\dagger\, e^{ik\cdot x} \right\} . \end{aligned} \tag{8.46}$$

The Dirac spinors satisfy the relations

$$\begin{aligned} \gamma^0 u(-\boldsymbol{k},s) &= u(\boldsymbol{k},s) \\ \gamma^0 v(-\boldsymbol{k},s) &= -v(\boldsymbol{k},s) \quad , \end{aligned}$$

and therefore

$$\boxed{\begin{aligned} \mathcal{P}b_{k,s}\mathcal{P}^\dagger &= \eta_\psi^P b_{-k,s} \\ \mathcal{P}d_{k,s}\mathcal{P}^\dagger &= -\eta_\psi^P d_{-k,s} \quad . \end{aligned}} \tag{8.47}$$

Hence Fermi particles and antiparticles have opposite parity, and the direction of their spin is unchanged (as we would expect from its cross product nature $\boldsymbol{L} = \boldsymbol{r} \times \boldsymbol{p}$).

For the transverse EM field we have a similar result. Using the *circular polarization* basis,

$$\begin{aligned} \mathcal{P}\boldsymbol{A}(r,t)\mathcal{P}^\dagger &= -\boldsymbol{A}(-r,t) \\ &= -\sum_{k,\alpha} \frac{1}{\sqrt{2\omega L^3}} \left\{ \boldsymbol{\epsilon}_{-k}^{\alpha} a_{-k,\alpha}\, e^{-ik\cdot x} + \boldsymbol{\epsilon}_{-k}^{\alpha\,*} a_{-k,\alpha}^\dagger\, e^{ik\cdot x} \right\} \quad , \end{aligned} \tag{8.48}$$

where the two helicity states are [recall Eq. (2.61)]

$$\begin{aligned} \boldsymbol{\epsilon}_z^+ &= -\frac{1}{\sqrt{2}}\,(\hat{x} + i\hat{y}) \\ \boldsymbol{\epsilon}_z^- &= \frac{1}{\sqrt{2}}\,(\hat{x} - i\hat{y}) \quad . \end{aligned} \tag{8.49}$$

If the transformation from z to $-z$ is achieved by rotating about the y-axis by π, then

$$\begin{aligned}\epsilon^{+}_{-z} &= \frac{1}{\sqrt{2}}(\hat{x}-i\hat{y}) = \epsilon^{-}_{z}\\ \epsilon^{-}_{-z} &= -\frac{1}{\sqrt{2}}(\hat{x}+i\hat{y}) = \epsilon^{+}_{z} \ ,\end{aligned} \tag{8.50}$$

and therefore

$$\boxed{\mathcal{P}a_{k,\pm}\mathcal{P}^{\dagger} = -a_{-k,\mp} \ .} \tag{8.51}$$

Hence the photon has *odd* parity, and its *momentum and helicity* change sign under parity.

Now, if parity is a symmetry of the Lagrangian, then

$$\mathcal{P}\mathcal{L}(x)\mathcal{P}^{\dagger} = \mathcal{L}(-r,t) \ , \tag{8.52}$$

implying that $\mathcal{L}$ is a scalar. This places restrictions on the types of interactions permitted in the Lagrangian. For example, for πNN interactions with no derivatives, we must have

$$\mathcal{L}_p = \bar{\psi}(x)\gamma^5\psi(x)\,\phi(x) \qquad \textit{pseudoscalar} \tag{8.53}$$

if the pion field is pseudoscalar and

$$\mathcal{L}_s = \bar{\psi}(x)\psi(x)\,\phi(x) \qquad \textit{scalar} \tag{8.54}$$

if it is a scalar. To prove this formally, use the transformation laws (8.41),

$$\begin{aligned}\mathcal{P}\mathcal{L}_p(r,t)\mathcal{P}^{\dagger} &= \mathcal{P}\bar{\psi}\mathcal{P}^{\dagger}\gamma^5\mathcal{P}\psi\mathcal{P}^{\dagger}\mathcal{P}\phi\mathcal{P}^{\dagger}\\ &= -\eta_{\psi}^{P\,2}\bar{\psi}(-r,t)\gamma^0\gamma^5\gamma^0\psi(-r,t)\,\phi(-r,t)\\ &= \mathcal{L}_p(-r,t) \ ,\end{aligned} \tag{8.55}$$

where we have used $\eta_{\psi}^{P\,2}=1$ in the last step.

Parity Transformation of Spin and Angular Momentum

Recall that angular momenta (and hence spins) have even parity and are therefore axial vectors (since normal vectors change sign under parity). To show this, recall that the angular momentum vector is a cross product,

$$\boldsymbol{\Omega} = \boldsymbol{r}\times\boldsymbol{p} \ .$$

In terms of its components,

$$\Omega^i = \epsilon_{ijk}\,r^j p^k \ ,$$

and under any transformation $r'^j = \Lambda^j{}_\ell r^\ell$,

$$\Omega'^i = \epsilon_{ijk} r'^j p'^k = \epsilon_{ijk} \Lambda^j{}_\ell \Lambda^k{}_m r^\ell p^m \quad . \tag{8.56}$$

But

$$\epsilon_{ijk} \Lambda^i{}_n \Lambda^j{}_\ell \Lambda^k{}_m = (\det \Lambda) \epsilon_{n\ell m} \quad ,$$

and hence, multiplying this expression by $\Lambda^{i'}{}_n$ and summing over n (for orthogonal transformations Λ where $\Lambda^{i'}{}_n \Lambda^i{}_n = \delta_{i'i}$) give

$$\epsilon_{i'jk} \Lambda^j{}_\ell \Lambda^k{}_m = (\det \Lambda) \Lambda^{i'}{}_n \epsilon_{n\ell m} \quad .$$

Hence the transformation (8.56) can be written

$$\begin{aligned} \Omega'^i &= (\det \Lambda) \Lambda^i{}_n \epsilon_{n\ell m} r^\ell p^m \\ &= (\det \Lambda) \Lambda^i{}_n \Omega^n \quad . \end{aligned} \tag{8.57}$$

The extra factor of $\det \Lambda$ shows that Ω does not change sign under the parity transformation and hence is an axial vector.

8.5 CHARGE CONJUGATION

The charge conjugation transformation, as we saw in Chapters 4 and 5, is associated with a symmetry of a relativistic wave equation which allows us to transform the negative energy solutions into positive energy solutions which satisfy the same wave equation but with the *sign of the charge reversed*. Such a transformation can also be defined when a particle has no charge but has some other quantum number which changes sign under the transformation (for example, the K_0, $\bar{K}_0$ system). Hence charge conjugation might be more appropriately referred to as "field conjugation."

To extend this idea to field theory, we postulate the existence of a unitary operator $\mathcal{C}$ which transforms the fields according to the relations

$$\boxed{\begin{aligned} \mathcal{C}\phi(x)\mathcal{C}^\dagger &= \eta_\phi^C \phi^\dagger(x) \\ \mathcal{C}\psi(x)\mathcal{C}^\dagger &= \eta_\psi^C C \bar{\psi}^\mathsf{T}(x) \\ \mathcal{C}\boldsymbol{A}(x)\mathcal{C}^\dagger &= -\boldsymbol{A}(x) \quad , \end{aligned}} \tag{8.58}$$

where again the η's are phases which can be ± 1 and C (to be distinguished from $\mathcal{C}$) is the Dirac conjugation matrix introduced in Eq. (5.32) with the following properties:

$$C = -C^\mathsf{T} = -C^{-1} \qquad C\gamma^\mu = -\gamma^{\mu\mathsf{T}} C \quad . \tag{8.59}$$

The form of the transformation laws (8.58) is obtained from the corresponding transformation of free KG particles discussed in Sec. 4.5, the transformation of Dirac particles discussed in Section 5.4, and the observation that the charge changes sign under $\mathcal{C}$ and hence the EM field must also.

The proof that the free Dirac Lagrangian is invariant under $\mathcal{C}$ begins in the same way as it did for parity,

$$\begin{aligned}
\mathcal{C}\mathcal{L}(x)\mathcal{C}^\dagger &= \frac{1}{2}\mathcal{C}\bar{\psi}\mathcal{C}^\dagger \left(i\gamma^\mu \frac{\overrightarrow{\partial}}{\partial x^\mu} - m \right) \mathcal{C}\psi\mathcal{C}^\dagger + \text{h.c.} \\
&= \frac{1}{2}\psi^\mathsf{T} \underbrace{\gamma^0 C^\dagger \gamma^0}_{-C^{-1}} \left(i\gamma^\mu \frac{\overrightarrow{\partial}}{\partial x^\mu} - m \right) C\bar{\psi}^\mathsf{T} + \text{h.c.} \\
&= \frac{1}{2}\psi^\mathsf{T} \left(i\gamma^{\mu\mathsf{T}} \frac{\overrightarrow{\partial}}{\partial x^\mu} + m \right) \bar{\psi}^\mathsf{T} + \text{h.c.}
\end{aligned} \tag{8.60}$$

Since this is a scalar, we may take the transpose in Dirac space. However, if we interchange $\bar{\psi}$ and ψ, which we need to do when taking the transpose, we are exchanging $\bar{\psi}$ and ψ not only in Dirac space (which we are entitled to do), but *also in the Fock space*, and since $\bar{\psi}$ and ψ anticommute, this will introduce *an extra minus sign*. (The c-number anticommutator which emerges from interchange will be ignored because we normal order the final answer anyway.) Hence the transpose of (8.60) becomes

$$\begin{aligned}
\mathcal{C}\mathcal{L}(x)\mathcal{C}^\dagger &= -\frac{1}{2}\bar{\psi} \left(i\gamma^\mu \frac{\overleftarrow{\partial}}{\partial x^\mu} + m \right) \psi + \text{h.c.} \\
&= \frac{1}{2}\bar{\psi} \left(-i\gamma^\mu \frac{\overleftarrow{\partial}}{\partial x^\mu} - m \right) \psi + \text{h.c.}
\end{aligned} \tag{8.61}$$

Note that this term is just the Hermitian conjugate (h.c.) of what we started with, and it is easy to verify that the h.c. term transforms into the original term. Under $\mathcal{C}$ the two terms interchange places. Hence

$$\mathcal{C}\mathcal{L}(x)\mathcal{C}^\dagger = \mathcal{L}(x) \; , \tag{8.62}$$

proving that $\mathcal{L}$ is invariant under $\mathcal{C}$.

Now we find the effect of $\mathcal{C}$ on the annihilation and creation operators and hence on the states. For scalar fields,

$$\begin{aligned}
\mathcal{C}\phi(x)\mathcal{C}^\dagger &= \eta_\phi^C \phi^\dagger(x) \\
&= \eta_\phi^C \sum_k \frac{1}{\sqrt{2E_k L^3}} \left\{ a_k^\dagger e^{ik\cdot x} + c_k e^{-ik\cdot x} \right\} \\
&= \mathcal{C} \sum_k \frac{1}{\sqrt{2E_k L^3}} \left\{ a_k e^{-ik\cdot x} + c_k^\dagger e^{ik\cdot x} \right\} \mathcal{C}^\dagger \; .
\end{aligned} \tag{8.63}$$

Since the exponentials are independent, and since $\mathcal{C}$ is unitary, we have

$$\boxed{\begin{aligned} \mathcal{C}a_k\mathcal{C}^\dagger &= \eta_\phi^C\, c_k \\ \mathcal{C}c_k\mathcal{C}^\dagger &= \eta_\phi^C\, a_k \quad . \end{aligned}} \tag{8.64}$$

Thus $\mathcal{C}$ interchanges particles and antiparticles, with the same phase. For single particle states, assuming $\mathcal{C}|0\rangle = |0\rangle$, we have

$$\begin{aligned} \mathcal{C}|k\rangle &= \mathcal{C}a_k^\dagger|0\rangle = \mathcal{C}a_k^\dagger\mathcal{C}^\dagger\mathcal{C}|0\rangle \\ &= \eta_\phi^C\, c_k^\dagger|0\rangle = \eta_\phi^C|\bar{k}\rangle \ , \end{aligned} \tag{8.65}$$

where $|\bar{k}\rangle$ is the state of a single antiparticle with momentum k. The same relation holds for antiparticles,

$$\mathcal{C}|\bar{k}\rangle = \eta_\phi^C|k\rangle \quad . \tag{8.66}$$

For the Dirac field, the derivation requires use of Eq. (5.38), which can be written

$$\boxed{\begin{aligned} C\bar{u}^\mathsf{T}(p,s) &= v(p,s) \\ C\bar{v}^\mathsf{T}(p,s) &= u(p,s)\ . \end{aligned}} \tag{8.67}$$

Using these relations,

$$\begin{aligned} \mathcal{C}\psi(x)\mathcal{C}^\dagger &= \eta_\psi^C\, C\sum_{p,s}\frac{1}{\sqrt{2E_pL^3}}\left\{b_{p,s}^\dagger\bar{u}^\mathsf{T}(p,s)\,e^{ip\cdot x} + d_{p,s}\bar{v}^\mathsf{T}(p,s)\,e^{-ip\cdot x}\right\} \\ &= \eta_\psi^C\sum_{p,s}\frac{1}{\sqrt{2E_pL^3}}\left\{b_{p,s}^\dagger v(p,s)\,e^{ip\cdot x} + d_{p,s}u(p,s)\,e^{-ip\cdot x}\right\} \quad . \end{aligned} \tag{8.68}$$

Hence

$$\boxed{\begin{aligned} \mathcal{C}b_{p,s}\mathcal{C}^\dagger &= \eta_\psi^C d_{p,s} \\ \mathcal{C}d_{p,s}\mathcal{C}^\dagger &= \eta_\psi^C b_{p,s} \end{aligned}} \tag{8.69}$$

and the effect of $\mathcal{C}$ on fermions is the same as it is on bosons.

Finally, for photons

$$\mathcal{C}\boldsymbol{A}(x)\mathcal{C}^\dagger = -\boldsymbol{A} \ , \tag{8.70}$$

which gives

$$\mathcal{C}\sum_{k,\alpha}\frac{1}{\sqrt{2\omega L^3}}\left\{a_{k,\alpha}\vec{\epsilon}_k^{\,\alpha}\,e^{-ik\cdot x} + a_{k,\alpha}^\dagger\vec{\epsilon}_k^{\,\alpha\,*}\,e^{ik\cdot x}\right\}\mathcal{C}^\dagger = -\boldsymbol{A} \tag{8.71}$$

and this requires

$$\mathcal{C}a_{k,\alpha}\mathcal{C}^\dagger = -a_{k,\alpha} \quad . \tag{8.72}$$

Photons are *odd* under charge conjugation.

Positronium Decay

To illustrate the usefulness of charge conjugation invariance, we consider positronium decay. Positronium is a bound state of the e^+e^- system, which can decay through the annihilation of the e^+e^- pair into photons. This decay is most likely to happen when the e^+ and e^- are very close to each other, and this in turn is only probable in S-states, where the wave function at the origin is not zero. There are two types of S-states: the spins of the particles may be aligned, so that the total spin is one (a spin *triplet* state), or they may be antialigned in a spin zero (a spin *singlet* state). The standard spectroscopic notation for these states is ${}^{2S+1}L_J$, where J, L, and S are the total angular momentum, orbital angular momentum (in the S, P, D, $\cdots$ notation), and total spin, respectively. Now, experimentally, it is observed that the ${}^3S_1$-states of positronium decay *only* into 3γ's (2γ decay is not observed and 1γ decay is forbidden by energy momentum conservation) while the ${}^1S_0$-states decay *only* into 2γ's. Since 3γ decay is *much* less probable that 2γ decay (because the decay rate is smaller by an extra factor of α and is further suppressed by the small size of the three-body phase space), the ${}^3S_1$-state is metastable.

To understand these results, we will represent the positronium states by the vector

$$|B\rangle = \sum_{s,s'} \int d^3p\, f(\boldsymbol{p}; s, s') b^\dagger_{\boldsymbol{p},s} d^\dagger_{-\boldsymbol{p},s'} |0\rangle \ , \tag{8.73}$$

where $\boldsymbol{p}$ is the relative momentum of the e^+e^- pair (the total momentum being zero by assumption) and s and s' are the possible spin states of the electron and positron, respectively. The wave function of the state is related to f, which for S states depends only on the magnitude of $\boldsymbol{p}$, so that $f(\boldsymbol{p}; s, s') = f(-\boldsymbol{p}; s, s') = f(p; s, s')$. Angular momentum conservation insures that f separates into a triplet part, symmetric in s and s', and a single part, antisymmetric in s and s'. We can therefore distinguish two different functions f, one symmetric under $s \leftrightarrow s'$ and the other antisymmetric,

$$f_\epsilon(p; s, s') = \epsilon f_\epsilon(p; s', s) \ , \tag{8.74}$$

where $\epsilon = +1$ for ${}^3S_1$-states and -1 for ${}^1S_0$-states. Hence,

$$|B, \epsilon\rangle = \sum_{s,s'} \int d^3p\, f_\epsilon(p; s, s') b^\dagger_{\boldsymbol{p},s} d^\dagger_{-\boldsymbol{p},s'} |0\rangle \ . \tag{8.75}$$

Now note that

$$\begin{aligned}
\mathcal{C}|B, \epsilon\rangle &= \sum_{s.s'} \int d^3p\, f_\epsilon(p; s, s') d^\dagger_{\boldsymbol{p},s} b^\dagger_{-\boldsymbol{p},s'} |0\rangle \\
&\underset{\substack{p\to -p \\ s\leftrightarrow s'}}{=} \sum_{s,s'} \int d^3p\, f_\epsilon(p; s', s) d^\dagger_{-\boldsymbol{p},s'} b^\dagger_{\boldsymbol{p},s} |0\rangle \\
&= -\epsilon \sum_{s,s'} \int d^3p\, f_\epsilon(p, s, s') b^\dagger_{\boldsymbol{p},s} d^\dagger_{-\boldsymbol{p},s'} |0\rangle \\
&= -\epsilon |B, \epsilon\rangle \ ,
\end{aligned} \tag{8.76}$$

where the minus sign in the next to the last step arises when the two creation operators are interchanged. Since the EM interactions conserve $\mathcal{C}$, and remembering that the photon is odd under $\mathcal{C}$, we obtain the following correspondence:

$$\begin{array}{llll} {}^3S_1 & \epsilon = 1 & \text{odd under } \mathcal{C} & 3 \text{ (or } 5, 7, \ldots) \ \gamma\text{'s only} \\ {}^1S_1 & \epsilon = -1 & \text{even under } \mathcal{C} & 2 \text{ (or } 4, 6, \ldots) \ \gamma\text{'s only.} \end{array} \tag{8.77}$$

The decays of positronium are explained by $\mathcal{C}$ invariance. Note that the anticommutation of the Dirac creation operators played an essential role in the argument.

8.6 TIME REVERSAL

We now turn to time reversal, the last of the discrete symmetries. This discussion leads us to a new consideration: in order that the time reversal operator leave the Hamiltonian (energies) positive, *and* be an invariance of the theory, it must be an *antiunitary* operator.

To understand the precise difference between a unitary and antiunitary operator, return to basic definitions. A linear operator $\mathcal{O}$ has the property

$$\mathcal{O}\left(a|x\rangle + b|y\rangle\right) = a\mathcal{O}|x\rangle + b\mathcal{O}|y\rangle \qquad \textit{linear} \ , \tag{8.78}$$

while an antilinear operator $\mathcal{O}_A$ has the property

$$\mathcal{O}_A\left(a|x\rangle + b|y\rangle\right) = a^*\,\mathcal{O}_A|x\rangle + b^*\,\mathcal{O}_A|y\rangle \qquad \textit{antilinear} \ . \tag{8.79}$$

Complex conjugation is an example of an antilinear operator. Finally, a norm preserving operator $\mathcal{N}$ has the following property:

$$\langle \mathcal{N}x|\mathcal{N}x\rangle = \langle x|x\rangle \qquad \textit{norm preserving} \ . \tag{8.80}$$

Now, we use these definitions to obtain some derived relations. By definition, a unitary operator $\mathcal{U}$ is both *linear* and *norm preserving*. From this it follows that

$$\boxed{\langle \mathcal{U}x|\mathcal{U}y\rangle = \langle x|y\rangle} \qquad \textit{unitary} \ . \tag{8.81}$$

Proof: Consider the vector $|z\rangle = |x\rangle + i|y\rangle$. Then

$$\begin{aligned} \langle \mathcal{U}z|\mathcal{U}z\rangle = \langle z|z\rangle &= \langle x|x\rangle + \langle y|y\rangle + i\,\langle x|y\rangle - i\,\langle y|x\rangle \\ &= \langle x|x\rangle + \langle y|y\rangle + i\,\langle \mathcal{U}x|\mathcal{U}y\rangle - i\,\langle \mathcal{U}y|\mathcal{U}x\rangle \ . \end{aligned}$$

Next, consider $|w\rangle = |x\rangle + |y\rangle$. Then

$$\begin{aligned} \langle \mathcal{U}w|\mathcal{U}w\rangle = \langle w|w\rangle &= \langle x|x\rangle + \langle y|y\rangle + \langle x|y\rangle + \langle y|x\rangle \\ &= \langle x|x\rangle + \langle y|y\rangle + \langle \mathcal{U}x|\mathcal{U}y\rangle + \langle \mathcal{U}y|\mathcal{U}x\rangle \ . \end{aligned}$$

Hence

$$\langle \mathcal{U}x|\mathcal{U}y\rangle - \langle \mathcal{U}y|\mathcal{U}x\rangle = \langle x|y\rangle - \langle y|x\rangle$$
$$\langle \mathcal{U}x|\mathcal{U}y\rangle + \langle \mathcal{U}y|\mathcal{U}x\rangle = \langle x|y\rangle + \langle y|x\rangle \ ,$$

and from this it follows that

$$\langle \mathcal{U}x|\mathcal{U}y\rangle = \langle x|y\rangle \ .$$

∎

An antiunitary operator $\mathcal{U}_A$ is defined to be both *antilinear* and norm preserving, and it is now quite easy to see that such an operator has a somewhat different property:

$$\boxed{\langle \mathcal{U}_A x|\mathcal{U}_A y\rangle = \langle x|y\rangle^* = \langle y|x\rangle} \qquad \textit{antiunitary} \ . \tag{8.82}$$

Proof: The proof is almost identical to the above, except

$$\langle \mathcal{U}_A z|\mathcal{U}_A z\rangle = \langle x|x\rangle + \langle y|y\rangle - i\,\langle \mathcal{U}_A x|\mathcal{U}_A y\rangle + i\,\langle \mathcal{U}_A y|\mathcal{U}_A x\rangle \ ,$$

where the change in sign of the imaginary terms is due to the *antilinear* nature of the operator $\mathcal{U}_A$. Hence

$$\langle \mathcal{U}_A x|\mathcal{U}_A y\rangle - \langle \mathcal{U}_A y|\mathcal{U}_A x\rangle = -\,\langle x|y\rangle + \langle y|x\rangle$$
$$\langle \mathcal{U}_A x|\mathcal{U}_A y\rangle + \langle \mathcal{U}_A y|\mathcal{U}_A x\rangle = \ \ \langle x|y\rangle + \langle y|x\rangle \ .$$

which gives the result (8.82),

$$\langle \mathcal{U}_A x|\mathcal{U}_A y\rangle = \langle y|x\rangle = \langle x|y\rangle^* \ .$$

∎

Transformation of Operators under Antiunitary Transformations

We now investigate how a typical operator $\mathcal{O}$ transforms under an antiunitary transformation $\mathcal{U}_A$. The operator is defined by its effect on a typical vector $|x\rangle$, which is transformed by $\mathcal{O}$ into another state $|z\rangle$:

$$\mathcal{O}|x\rangle = |z\rangle \ .$$

The antiunitary operator U_A transforms these two vectors as follows:

$$\mathcal{U}_A|x\rangle = |x'\rangle$$
$$\mathcal{U}_A|z\rangle = |z'\rangle \ .$$

The transformed $\mathcal{O}$, denoted by $\mathcal{O}'$, is defined by the requirement

$$|z'\rangle = \mathcal{O}'|x'\rangle \ .$$

Hence

$$\begin{aligned}\mathcal{U}_A|z\rangle &= \mathcal{O}'\mathcal{U}_A|x\rangle \\ &= \mathcal{U}_A\mathcal{O}|x\rangle \ .\end{aligned}$$

This equality must hold for any $|x\rangle$, which gives the relation

$$\mathcal{O}' = \mathcal{U}_A \mathcal{O} \mathcal{U}_A^{-1} \ . \tag{8.83}$$

Furthermore, under $\mathcal{U}_A$ the matrix element $\langle y|z\rangle$ transforms as

$$\langle y'|z'\rangle = \langle y|z\rangle^* \ ,$$

from which we obtain the general result for the transformation of operators under antiunitary transformations,

$$\langle y|\mathcal{O}|x\rangle = \langle y'|\mathcal{O}'|x'\rangle^* = \langle x'|\mathcal{O}'^{\dagger}|y'\rangle \ . \tag{8.84}$$

Now we will discuss why the physics requires that the operation of time inversion, $\mathcal{T}$, must be an antiunitary operator. If $\mathcal{O}$ is a scalar operator which depends on time, $\mathcal{O}'(t) = \mathcal{O}(-t)$. Since the Hamiltonian is the time component of a four-vector, we would expect the transformation law to be $H'(t) = -H(-t)$, but this would have the undesirable effect of changing the sign of all energies. We thus *require* that the correct transformation for H be $H'(t) = H(-t)$. If this is the correct transformation law for the Hamiltonian, the effect of time inversion on the interaction time translation operator can be found by examining its effect on the nth term in its expansion, Eq. (3.24), which is

$$U_I\bigg|_{n\text{th term}} = \frac{(-i)^n}{n!}\int_{-\infty}^{\infty}\cdots\int_{-\infty}^{\infty} dt_1\cdots dt_n\, T\left(H_I(t_1)\cdots H_I(t_n)\right) \ . \tag{8.85}$$

Hence the operation of time inversion gives

$$\begin{aligned}\mathcal{T}\,U_I\bigg|_{n\text{th term}}\mathcal{T}^{-1} &= \frac{(+i)^n}{n!}\mathcal{T}\int_{-\infty}^{\infty}\cdots\int_{-\infty}^{\infty} dt_1\cdots dt_n\, T\left(H_I(t_1)\cdots H_I(t_n)\right)\mathcal{T}^{-1} \\ &= \frac{(+i)^n}{n!}\int_{-\infty}^{\infty}\cdots\int_{-\infty}^{\infty} dt_1\cdots dt_n\, T_A\left(H_I(-t_1)\cdots H_I(-t_n)\right) \\ &\underset{t_i\to -t_i}{=} \frac{(+i)^n}{n!}\int_{-\infty}^{\infty}\cdots\int_{-\infty}^{\infty} dt_1\cdots dt_n\, T_A\left(H_I(t_1)\cdots H_I(t_n)\right) \\ &= \frac{(+i)^n}{n!}\int_{-\infty}^{\infty}\cdots\int_{-\infty}^{\infty} dt_1\cdots dt_n\left[T\left(H_I(t_n)\cdots H_I(t_1)\right)\right]^{\dagger} ,\end{aligned} \tag{8.86}$$

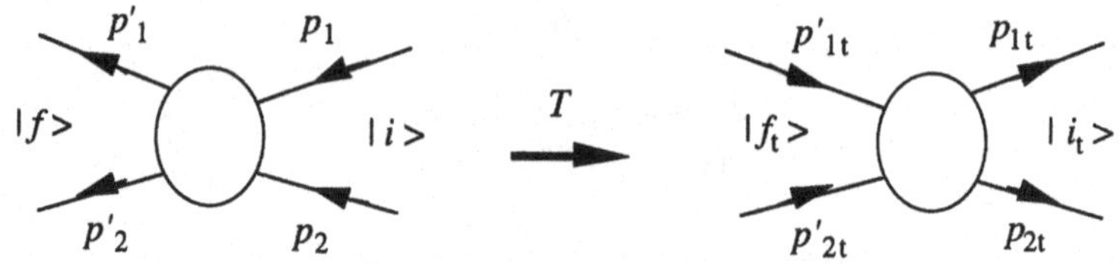

Fig. 8.3 Under time inversion, which is interpreted as the reversal of the direction of motion, the scattering process on the left is transformed into the process on the right.

where in the first line we used the assumed antiunitarity of $\mathcal{T}$ when we changed the factor of $(-i)^n$ to $(+i)^n$, in the second line noted that when the action of $\mathcal{T}$ on H changed the sign of the time, terms which were time ordered became "anti-time-ordered" (denoted by T_A), in the third line changed $t \to -t$ in each of the time integrals, and finally in the last line replaced the anti-time-ordering with the Hermitian conjugate. Hence we conclude from (8.86) that

$$\mathcal{T}\, U_I\, \mathcal{T}^{-1} = U_I' = U_I^\dagger = U_I^{-1} \quad . \tag{8.87}$$

This is a very elegant result. Taking matrix elements and using Eq. (8.84) give

$$\langle f|U_I|i\rangle = \left\langle i_t|{U_I'}^\dagger|f_t\right\rangle = \langle i_t|U_I|f_t\rangle \quad , \tag{8.88}$$

where $|f\rangle$ and $|i\rangle$ are the final and initial state, respectively, and $|i_t\rangle$ and $|f_t\rangle$ are time-reversed initial and final states. As illustrated in Fig. 8.3, this gives a beautiful interpretation of the $\mathcal{T}$ operation as the *reversal of direction of motion.* This interpretation would not emerge without the mapping $|i\rangle \to |f_t\rangle$ and $|f\rangle \to |i_t\rangle$ and without the change in the sign of i which insures that it is the same (correct) time translation operator U_I which is involved in both cases. And both of these properties are a consequence of the assumption that $\mathcal{T}$ is antiunitary.

The necessity for time reversal to be antiunitary can also be demonstrated with two simple arguments:

- If the energy E is to be kept positive under time inversion, then the exponent in the factor e^{-iEt} must not be allowed to change sign. If $t \to -t$, the only way to insure this is to change the sign of i, which requires that $\mathcal{T}$ be antiunitary.

- If we wish to preserve the meaning of H as a time translation operator, we must preserve the relation

$$-i\frac{\partial\phi}{\partial t} = [H, \phi] \quad , \tag{8.89}$$

and this requires that $\mathcal{T}$ be antiunitary.

To see why the preservation of Eq. (8.89) requires that $\mathcal{T}$ be antiunitary, consider a scalar field, which satisfies the following transformation law:

$$\mathcal{T}\phi(x)\mathcal{T}^{-1} = \phi'(x) = \phi(r, -t) \ . \tag{8.90}$$

If $\mathcal{T}$ were unitary, (8.89) would transform as follows:

$$\begin{aligned} \mathcal{T}\left(-i\frac{\partial\phi}{\partial t}\right)\mathcal{T}^{-1} &= -i\frac{\partial\phi'}{\partial t} = i\frac{\partial\phi'}{\partial t'} \\ &= \mathcal{T}[H,\phi]\mathcal{T}^{-1} = [H',\phi'] \ , \end{aligned} \tag{8.91}$$

and hence its time-inverted form would become

$$i\frac{\partial\phi'}{\partial t'} = [H',\phi'] \ , \tag{8.92}$$

which can be restored to the original form only by letting $H' = -H$. In this case we would obtain the unphysical result that $\mathcal{T}$ changes the sign of all energies. However, if $\mathcal{T}$ is antiunitary, $i \to -i$, and

$$\mathcal{T}\left(-i\frac{\partial\phi}{\partial t}\right)\mathcal{T}^{-1} = i\frac{\partial\phi'}{\partial t} = -i\frac{\partial\phi'}{\partial t'} \ . \tag{8.93}$$

Now we preserve the result if $H' = H$, which is desired. We conclude that *the physics requires that $\mathcal{T}$ be an antiunitary operator.*

We now apply this to the various fields we have been studying previously. Their transformation laws under time inversion assume the following form:

$$\boxed{\begin{aligned} \mathcal{T}\phi(r,t)\mathcal{T}^{-1} &= \eta_\phi^T\,\phi(r,-t) \\ \mathcal{T}\psi(r,t)\mathcal{T}^{-1} &= \eta_\psi^T\,T\psi(r,-t) \\ \mathcal{T}A^\mu(r,t)\mathcal{T}^{-1} &= -\Lambda(T)^\mu{}_\nu A^\nu(r,-t) \ , \end{aligned}} \tag{8.94}$$

where the η's are intrinsic phases which can only be ± 1. The vector potential must change sign to compensate for the change in sign of the current under reversal of the direction of motion. The Dirac matrix T which generates time inversion for the Dirac equation is yet to be determined. It will turn out to be

$$T = C\gamma^5 \ , \tag{8.95}$$

where C is the charge conjugation matrix. We turn now to a discussion of the implications of these transformations.

First, look at scalar fields. The transformation law gives

$$\begin{aligned} \mathcal{T}\phi(r,t)\mathcal{T}^{-1} &= \eta_\phi^T \sum_k \sqrt{\frac{1}{2E_k L^3}} \left(e^{ik\cdot x} a_{-k} + e^{-ik\cdot x} c_{-k}^\dagger\right) \\ &= \sum_k \sqrt{\frac{1}{2E_k L^3}} \left(e^{ik\cdot x} \mathcal{T} a_k \mathcal{T}^{-1} + e^{-ik\cdot x} \mathcal{T} c_k^\dagger \mathcal{T}^{-1}\right) , \end{aligned} \tag{8.96}$$

where in the first line the combined effect of changing $t \to -t$ and $k \to -k$ (in the sum) leads to a change in sign of the exponents and in the second line, because of the antilinear nature of $\mathcal{T}$, the sign of i in the exponents was changed as the operator $\mathcal{T}$ passed by them. Hence

$$\begin{aligned} \mathcal{T} a_k \mathcal{T}^{-1} &= \eta_\phi^T a_{-k} \\ \mathcal{T} c_k \mathcal{T}^{-1} &= \eta_\phi^T c_{-k} \quad . \end{aligned} \tag{8.97}$$

Note that these are consistent with the notion of reversal in direction of motion. We have, for arbitrary $|y\rangle$,

$$\begin{aligned} \langle y|k\rangle &= \left\langle y|a_k^\dagger|0\right\rangle = \left\langle y_t|a_{-k}^\dagger|0\right\rangle^* \\ &= \langle 0|a_{-k}|y_t\rangle = \langle -k|y_t\rangle \quad . \end{aligned} \tag{8.98}$$

The momentum of the particle changes sign, and its position in the matrix element changes from an initial to a final state.

Next, consider Dirac fields. First, determine T by requiring that the free Dirac Lagrangian be invariant under $\mathcal{T}$, or that

$$\mathcal{T}\mathcal{L}(x)\mathcal{T}^{-1} = \mathcal{L}(r,-t) \quad . \tag{8.99}$$

Transforming the Lagrangian gives

$$\begin{aligned} \mathcal{T}\bar{\psi}\left(\frac{i}{2}\gamma^\mu \frac{\overleftrightarrow{\partial}}{\partial x^\mu} - m\right)\psi\mathcal{T}^{-1} &= \mathcal{T}\bar{\psi}\mathcal{T}^{-1}\left(-\frac{i}{2}\gamma^{\mu *} \frac{\overleftrightarrow{\partial}}{\partial x^\mu} - m\right)\mathcal{T}\psi\mathcal{T}^{-1} \\ &= \bar{\psi}(r,-t)\gamma^0 T^\dagger \gamma^0 \left(\frac{i}{2}\gamma^{\mu\mathsf{T}} \frac{\overleftrightarrow{\partial}}{\partial x'^\mu} - m\right) T\psi(r,-t) , \end{aligned} \tag{8.100}$$

where $x'^\mu = (-t, \boldsymbol{r})$ and the hermiticity properties of the γ matrices, namely $\gamma^{0\dagger} = \gamma^0$ and $\gamma^{i\dagger} = -\gamma^i$, were used to make the replacement,

$$\gamma^{\mu *}\frac{\partial}{\partial x^\mu} = -\gamma^{\mu\mathsf{T}}\frac{\partial}{\partial x'^\mu} \quad .$$

The requirement (8.99) then leads to the following conditions:

$$\begin{aligned} \gamma^{\mu\mathsf{T}} T &= T\gamma^\mu \\ \gamma^0 T^\dagger \gamma^0 T &= 1 \quad . \end{aligned} \tag{8.101}$$

The first is satisfied by $T = \lambda C\gamma^5$, where λ is a constant. The second determines the square of λ,

$$|\lambda|^2\gamma^5 C^\dagger C\gamma^5 = 1 \quad \Rightarrow \quad |\lambda|^2 = 1 \ . \tag{8.102}$$

Since the overall phase can only be ± 1, as previously discussed, we are free to choose

$$T = C\gamma^5 \ .$$

It is left as a problem (Prob. 8.7) to show that

$$\boxed{\begin{aligned} Tu(\mathbf{p}, s) &= (-1)^{\frac{1}{2}-s} u^*(-\mathbf{p}, -s) \\ Tv(\mathbf{p}, s) &= (-1)^{\frac{1}{2}-s} v^*(-\mathbf{p}, -s) \ . \end{aligned}} \tag{8.103}$$

Results such as these are expected, because under $\mathcal{T}$ both the direction of momentum and the direction of spin should change. The origin of the phase will be discussed shortly.

Now we find the implications of time reversal for Dirac fields. If $\eta_\psi^T = 1$, the transformation law (8.94) becomes

$$\begin{aligned} \mathcal{T}\psi(r,t)\mathcal{T}^{-1} &= T\sum_{\mathbf{p},s} \frac{1}{\sqrt{2EL^3}} \left\{ e^{ip\cdot x} u(-\mathbf{p}, s)\, b_{-\mathbf{p},s} + e^{-ip\cdot x} v(-\mathbf{p}, s)\, d^\dagger_{-\mathbf{p},s} \right\} \\ &= \sum_{\mathbf{p},s} \frac{1}{\sqrt{2EL^3}} \left\{ e^{ip\cdot x} u^*(\mathbf{p}, s)\, \mathcal{T} b_{\mathbf{p},s} \mathcal{T}^{-1} \right. \\ &\qquad\qquad \left. + e^{-ip\cdot x} v^*(\mathbf{p}, s)\, \mathcal{T} d^\dagger_{\mathbf{p},s} \mathcal{T}^{-1} \right\} , \end{aligned} \tag{8.104}$$

where the complex conjugation of the spinors in the second line arises because of the antilinear nature of $\mathcal{T}$. Using the relations (8.103) and changing the sign of s in the first sum give

$$\boxed{\begin{aligned} \mathcal{T} b_{\mathbf{p},s} \mathcal{T}^{-1} &= (-1)^{\frac{1}{2}+s}\, b_{-\mathbf{p},-s} \\ \mathcal{T} d^\dagger_{\mathbf{p},s} \mathcal{T}^{-1} &= (-1)^{\frac{1}{2}+s}\, d^\dagger_{-\mathbf{p},-s} \ . \end{aligned}} \tag{8.105}$$

The spin-dependent phase has a very simple physical origin. For spin $\frac{1}{2}$ particles it says that

$$\begin{aligned} \mathcal{T}|p, +\tfrac{1}{2}\rangle &= -|-p, -\tfrac{1}{2}\rangle \\ \mathcal{T}|p, -\tfrac{1}{2}\rangle &= \ \ |-p, +\tfrac{1}{2}\rangle \ . \end{aligned} \tag{8.106}$$

We will now show that these two equations are consistent with, and required by, the normal phase convention used to define angular momentum states. To see this, work with a particle at rest, $\boldsymbol{p} = 0$, and recall that the states are related to each other by the raising and lowering operators,

$$\begin{aligned} J_+|0,-\tfrac{1}{2}\rangle &= (J_x + iJ_y)\,|0,-\tfrac{1}{2}\rangle = |0,\ \tfrac{1}{2}\rangle \\ J_-|0,\ \tfrac{1}{2}\rangle &= (J_x - iJ_y)\,|0,\ \tfrac{1}{2}\rangle = |0,-\tfrac{1}{2}\rangle \ . \end{aligned} \tag{8.107}$$

If we choose the phase so that

$$\mathcal{T}|0,-\tfrac{1}{2}\rangle = |0,\tfrac{1}{2}\rangle \ ,$$

and use the fact that $\mathcal{T}J_i = -J_i\mathcal{T}$, we obtain

$$\begin{aligned} \mathcal{T}|0,\tfrac{1}{2}\rangle &= \mathcal{T}\,(J_x + iJ_y)\,|0,-\tfrac{1}{2}\rangle = (-J_x + iJ_y)\,\mathcal{T}|0,-\tfrac{1}{2}\rangle \\ &= -J_-|0,\tfrac{1}{2}\rangle = -|0,-\tfrac{1}{2}\rangle \ . \end{aligned} \tag{8.108}$$

Hence, the negative phase for the spin "up" state is *required* by all of the definitions and choices previously imposed!

We conclude this section by discussing time inversion for the photon. Using the helicity representation, Eq. (8.50),

$$\epsilon^{\pm}_{\hat{z}} = \frac{1}{\sqrt{2}}(\mp\hat{x} - i\hat{y}) \ ,$$

the photon polarization vectors satisfy the following relations:

$$\epsilon^{\pm\,*}_{-\hat{z}} = \epsilon^{\mp\,*}_{\hat{z}} = -\epsilon^{\pm}_{\hat{z}} \ . \tag{8.109}$$

Therefore, under time inversion

$$\begin{aligned} \mathcal{T}A(x)\mathcal{T}^{-1} &= -\sum_{k,\alpha}\frac{1}{\sqrt{2\omega L^3}}\left\{\epsilon^{\alpha}_{-k}a_{-k,\alpha}\,e^{ik\cdot x} + \epsilon^{\alpha\,*}_{-k}a^{\dagger}_{-k,\alpha}\,e^{-ik\cdot x}\right\} \\ &= \sum_{k,\alpha}\frac{1}{\sqrt{2\omega L^3}}\left\{\epsilon^{\alpha\,*}_{k}\mathcal{T}a_{k,\alpha}e^{ik\cdot x}\mathcal{T}^{-1} + \epsilon^{\alpha}_{k}\mathcal{T}a^{\dagger}_{k,\alpha}\mathcal{T}^{-1}\,e^{-ik\cdot x}\right\} \end{aligned} \tag{8.110}$$

and the transformation law for the annihilation operators is

$$\mathcal{T}a_{k,\alpha}\mathcal{T}^{-1} = a_{-k,\alpha} \ . \tag{8.111}$$

Note that the phase is positive, and the helicity does *not* change sign, because both the momentum and the spin do.

8.7 THE *PCT* THEOREM

We conclude this chapter with a discussion of one of the more interesting theorems in the subject of symmetry.

> **Theorem:** If the Lagrangian density is a Hermitian, normal-ordered, Lorentz invariant operator constructed from fields quantized with the usual connection between spin and statistics, then the product of the $\mathcal{PCT}$ transformations is always a symmetry of the theory.

Interactions can readily be constructed which violate $\mathcal{P}$, $\mathcal{C}$, or $\mathcal{T}$ individually (often it is only necessary to change the phase of a coupling constant to achieve this), but the *PCT* theorem says that it is possible to *choose the phases* η [which enter into the transformation laws Eqs. (8.41), (8.58), and (8.94)] so that the product of all of these symmetries, $\mathcal{O} = \mathcal{PCT}$, *is always a symmetry of the theory*, regardless of how the interactions are constructed (provided only that they conform to the restrictions stated in the theorem) and regardless of the phases of the coupling constants. In this section, a proof of this theorem will be sketched and implications discussed.

If $\mathcal{O} = \mathcal{PCT}$, then from the previous sections,

$$\boxed{\begin{aligned} \mathcal{O}\phi(x)\mathcal{O}^{-1} &= \eta_\phi\, \phi^\dagger(-x) \\ \mathcal{O}\psi(x)\mathcal{O}^{-1} &= \eta_\psi\, \gamma^5\gamma^0\bar{\psi}^{\mathsf{T}}(-x) \\ \mathcal{O}A^\mu(x)\mathcal{O}^{-1} &= -A^\mu(-x)\ , \end{aligned}} \tag{8.112}$$

where, as in the previous cases, the phase η_ψ is free to be chosen, subject to the condition $\eta_\psi^2 = 1$. However, as we will discuss below, the *PCT* symmetry will emerge *only if the phase* η_ϕ *accompaning the transformation of spin zero fields is chosen to be* +1. Since the Lagrangian density is a scalar, invariance under this transformation implies

$$\mathcal{O}\mathcal{L}(x)\mathcal{O}^{-1} = \mathcal{L}^\dagger(-x) = \mathcal{L}(-x) \tag{8.113}$$

because $\mathcal{L}$ is also Hermitian.

Now consider the most general Lagrangian density imaginable. Lorentz invariance requires that all terms involving the Dirac ψ fields be constructed from the bilinear covariants discussed in Sec. 5.10, which must in turn be contracted with other tensors constructed from other fields which have the same Lorentz invariance properties. We have denoted these bilinear covariant matrices by Γ, and without loss of generality we may assume that the Γ's are Hermitian (any non-Hermitian terms we might want to consider can be constructed from Hermitian Γ's multiplied by complex coefficients). The ψ fields will therefore enter the

Lagrangian through combinations like $\bar{\psi}(x)\Gamma\psi(x)$. Consider the transformation of this quantity under $\mathcal{O}$. Because $\mathcal{O}$ is antiunitary,

$$\begin{aligned}\mathcal{O}\bar{\psi}(x)\Gamma\psi(x)\mathcal{O}^{-1} &= \mathcal{O}\bar{\psi}(x)\mathcal{O}^{-1}\mathcal{O}\Gamma\psi(x)\mathcal{O}^{-1} \\ &= \eta_\psi^2\psi^{\mathsf{T}}(-x)\gamma^5\gamma^0\Gamma^*\gamma^5\gamma^0\bar{\psi}^{\mathsf{T}}(-x) \;. \end{aligned} \tag{8.114}$$

Now, we can remove the transpose in the Dirac space by re-ordering the Dirac fields (as we did in Sec. 8.5), remembering that this will give a minus sign. Hence, since $\eta_\phi^2 = 1$,

$$\mathcal{O}\bar{\psi}(x)\Gamma\psi(x)\mathcal{O}^{-1} = -\bar{\psi}(-x)\gamma^0\gamma^5\Gamma^\dagger\gamma^0\gamma^5\psi(-x) \;. \tag{8.115}$$

[The normal ordering which is implied means that we can drop all anticommutators which would otherwise emerge when the $bb^\dagger$ terms and $dd^\dagger$ terms were exchanged under interchange of $\bar{\psi}$ and ψ.] Now, the implied hermiticity of the Γ's gives the requirement

$$\left[\bar{\psi}(x)\Gamma(x)\psi(x)\right]^\dagger = \bar{\psi}(x)\gamma^0\Gamma^\dagger\gamma^0\psi(x)$$

so that

$$\gamma^0\Gamma^\dagger\gamma^0 = \Gamma \;. \tag{8.116}$$

Using $\gamma^0\gamma^{\mu\dagger}\gamma^0 = \gamma^\mu$, the specific (Hermitian) forms of the 16 Γ's are

$$\Gamma = \left\{1,\ \gamma^\mu,\ \sigma^{\mu\nu},\ \gamma^5\gamma^\mu,\ i\gamma^5\right\} \;. \tag{8.117}$$

Hence, recalling that γ^5 anticommutes with all γ^μ, under $\mathcal{O}$ the Γ's transform to

$$\Gamma' = \gamma^5\Gamma\gamma^5 \;. \tag{8.118}$$

Under PCT the Γ's therefore fall into two classes:

$$\begin{aligned}\Gamma'_S &= \ \ \Gamma_S \quad \text{for} \quad \Gamma_S = \left\{\mathbf{1}, \sigma^{\mu\nu}, i\gamma^5\right\} \\ \Gamma'_A &= -\Gamma_A \quad \text{for} \quad \Gamma_A = \left\{\gamma^\mu, \gamma^5\gamma^\mu\right\} \;. \end{aligned} \tag{8.119}$$

Note that Γ's with one Lorentz vector index change sign; those with an even number (0 or 2) do not. However, Γ's with one vector index must necessarily be multiplied by

$$A^\mu(x) \quad \text{or} \quad \frac{\partial}{\partial x_\mu} \quad \text{or} \quad \text{another} \quad \Gamma_A \;.$$

All of these also change sign under $\mathcal{O}$, so that their product with Γ_A does not. If a Γ has an even number of vector indices, it must be contracted with other terms which have the same (even) number of indices, and there will again be no sign change. This result can be immediately extended to include interaction terms with complex coupling constants; if the Lagrangian includes the term $\lambda\Gamma \times \Gamma$, where λ is complex, the hermiticity of the Lagrangian density guarantees that the Hermitian conjugate of this term is also present, and the combined term $\lambda\Gamma \times \Gamma + \lambda^*\Gamma \times \Gamma$ does not change sign under PCT. Hence, while individual factors involving the fermion fields may change sign (or phase) under $\mathcal{O}$, this change of sign (or phase) is always balanced by other terms with a compensating change of sign (or phase) and therefore *all terms involving fermions are invariant under* $\mathcal{O}$.

It now remains to examine interactions involving ϕ. The Lagrangian density could contain (for example) an interaction term of the form

$$\mathcal{L} = -\lambda\phi^3(x) - \lambda^*\phi^{\dagger\,3}(x) \quad ,$$

which is both Lorentz invariant and Hermitian. Under PCT, $\phi(x)$ is transformed to $\eta_\phi\phi^\dagger(-x)$ and $\phi^\dagger(x)$ to $\eta_\phi\phi(-x)$, and remembering that $\mathcal{O}$ is antilinear, these terms transform into each other, *with an overall phase change of* η_ϕ^3. If we chose the phase $\eta_\phi = -1$, these terms would be odd under PCT, and the theorem would not hold. However, choosing $\eta_\phi = +1$ guarantees that all such terms are even. This choice also insures that any terms involving products of Dirac and scalar fields are even. We conclude that $\mathcal{L}$ is invariant under $\mathcal{O}$ [i.e., it transforms like Eq. (8.113)].

As we have seen, the freedom to choose $\eta_\phi = +1$ is central to the proof of the PCT theorem, and it must be demonstrated that this choice is always permitted by the physics. A full discussion of this point requires techniques which we have not developed in this book. Briefly, it can be shown that a field theory can be fully defined by the vacuum expectation values of all products of its field operators, and that these matrix elements, initially defined for real space–time points, can be *analytically continued into complex space-time*. The concept of analytic continuation is familiar from elementary studies of complex functions. It is a remarkable fact that a smooth function which is initially known only over an interval of the real axis can be continued into the entire complex plane and that this continuation is unique. In a similar fashion, the matrix elements of a Hermitian scalar field (a simple example), initially defined for real space–time points, can be defined uniquely in complex space–time. Now, for points in real space–time, the matrix elements of a scalar field are invariant under the real Lorentz group, and their unique extension into complex space–time is necessarily invariant under the group of *complex Lorentz transformations*. However, as we discussed briefly in Sec. 5.8, for the group of complex Lorentz transformations, TP is continuously connected to the identity, so that any function must transform under TP with the same phase as it does under the restricted group $L_+^\uparrow$ (recall Fig. 5.3 and Table 5.2). A scalar field is invariant under $L_+^\uparrow$, and because TP is connected to the identity through the process of analytic continuation, the phase η_ϕ for a Hermitian scalar field must therefore equal +1. Hence there can be no physical impediment to this choice, which is, in this case, *required* by the interpretation of $\mathcal{TP}$ as space–time inversion. For further discussion, the interested reader is referred to Streater and Wightman (1964) and to the literature (see, for example, [Lu 57]) .

Implications of *PCT* Invariance

We begin our examination of the implications of PCT invariance by considering the effect of $\mathcal{PCT}$ on meson annihilation and creation operators. Since $\eta_\phi = 1$, we have

$$\begin{aligned} \mathcal{O}\phi(x)\mathcal{O}^{-1} &= \sum_k \frac{1}{\sqrt{2E_k L^3}} \left\{a_k^\dagger e^{-ik\cdot x} + c_k e^{ik\cdot x}\right\} \\ &= \sum_k \frac{1}{\sqrt{2E_k L^3}} \left\{\mathcal{O}a_k\mathcal{O}^{-1} e^{ik\cdot x} + \mathcal{O}c_k^\dagger \mathcal{O}^{-1} e^{-ik\cdot x}\right\} \quad .(8.120) \end{aligned}$$

Hence

$$\mathcal{O}a_k\mathcal{O}^{-1} = c_k \tag{8.121}$$

and

$$\mathcal{O}|k\rangle = |\bar{k}\rangle \ , \tag{8.122}$$

where, as before, $|\bar{k}\rangle$ is the state of an antiparticle with momentum k. The operator $\mathcal{O}$ turns particles into antiparticles. We can therefore use PCT invariance to prove the following theorem:

> **Theorem:** The masses of particles and antiparticles are equal provided they are stable or cannot decay into any other single particle state.

Proof: For stable particles, the proof is simple because they are eigenstates of the total Hamiltonian, and therefore

$$\begin{aligned} H|k\rangle &= E_k|k\rangle \\ H|\bar{k}\rangle &= \bar{E}_k|\bar{k}\rangle \ , \end{aligned} \tag{8.123}$$

where $E_k = \sqrt{m^2 + k^2}$ and $\bar{E}_k = \sqrt{\bar{m}^2 + k^2}$ are the energies of the particle and antiparticle states of momentum k. But because of the invariance under PCT, $\mathcal{O}$ commutes with H, and

$$\begin{aligned} \mathcal{O}H|k\rangle &= E_k\mathcal{O}|k\rangle = E_k|\bar{k}\rangle \\ &= H\mathcal{O}|k\rangle = H|\bar{k}\rangle = \bar{E}_k|\bar{k}\rangle \ . \end{aligned} \tag{8.124}$$

Hence $E_k = \bar{E}_k$, which implies $m = \bar{m}$. An example is the proton; $m_p = \bar{m}_p$.

For unstable particles the energy is, by definition, the expectation value of the Hamiltonian,

$$\begin{aligned} E_k &= \langle k|H|k\rangle \\ \bar{E}_k &= \langle\bar{k}|H|\bar{k}\rangle \ . \end{aligned} \tag{8.125}$$

Therefore, under PCT

$$\begin{aligned} E_k &= \langle k|H|k\rangle = \langle\mathcal{O}k|H|\mathcal{O}k\rangle \\ &= \langle\bar{k}|H|\bar{k}\rangle = \bar{E}_k \ . \end{aligned} \tag{8.126}$$

If these particles do not couple to another single particle state, the Hamiltonian on the subspace of single particle states is diagonal,

$$H = \begin{pmatrix} E_k & 0 \\ 0 & \bar{E}_k \end{pmatrix} \tag{8.127}$$

and the equality of the diagonal elements implies equality of the masses. An example is the π^+ and π^- mesons. ∎

Any violation of these predictions would imply a breakdown of PCT invariance and would have profound implications for field theory and for physics.

The K^0, $\bar{K}^0$ System

We conclude this discussion by noting that there are particles and antiparticles which can couple to each other. This can happen only when they are neutral, have a sufficiently long lifetime to be observed as particles with a well-defined mass, and have some non-zero quantum number which is violated by the weak interactions but conserved by the strong interactions which produces them. Hence, under the strong interactions they are distinct particles with a well-defined mass. Their coupling through the weak interactions will mix the two states, however, leading to a mass splitting in apparent (but not real) violation of the equality of masses of particles and antiparticles. Such a system is the K^0, $\bar{K}^0$ system.

The K^0 and $\bar{K}^0$ states can be represented by a two-component vector

$$|K_0\rangle = \begin{pmatrix} 1 \\ 0 \end{pmatrix} \qquad |\bar{K}_0\rangle = \begin{pmatrix} 0 \\ 1 \end{pmatrix} \ . \tag{8.128}$$

In this subspace, the Hamiltonian has the form

$$H = \begin{pmatrix} A & B \\ B^* & A \end{pmatrix} , \tag{8.129}$$

where PCT invariance insures us that the diagonal elements are equal,

$$A = \langle K_0|H|K_0\rangle = \langle \bar{K}_0|H|\bar{K}_0\rangle \ ,$$

but says nothing about the off-diagonal elements,

$$B = \langle K_0|H|\bar{K}_0\rangle = \langle \bar{K}_0|H|K_0\rangle^* \ .$$

Because B is non-zero (even though it is small), the eigenstates of the matrix H are no longer $\binom{1}{0}$ and $\binom{0}{1}$. The eigenvalues (for $\mathbf{k} = 0$) become

$$m_\pm = A \pm |B| \ , \tag{8.130}$$

corresponding to eigenstates

$$
\begin{aligned}
m_-: \qquad & K_S = \frac{1}{\sqrt{B+B^*}} \left\{ \sqrt{B}\,|K_0\rangle - \sqrt{B^*}\,|\bar{K}_0\rangle \right) \\
m_+: \qquad & K_L = \frac{1}{\sqrt{B+B^*}} \left(\sqrt{B}\,|K_0\rangle + \sqrt{B^*}\,|\bar{K}_0\rangle \right) \ .
\end{aligned}
\tag{8.131}
$$

Note that the masses are no longer equal. If CP is conserved, it turns out that $B = B^*$, and K_S decays into an even number of π's (mainly 2π), while K_L decays into an odd number (only 3π's since single pion decay is forbidden by energy–momentum conservation). It was the observation that K_L sometime decays into 2π's, which led to the discovery of CP violation.

PROBLEMS

8.1 Find the momentum operator P^i for the Klein–Gordon field (using Noether's theorem) and prove that

$$
\frac{\partial \phi}{\partial x^i} = -i\left[P^i, \phi\right] \ .
$$

8.2 **(a)** Show explicitly that

$$
i\colon \bar{\psi}(x)\gamma^5\psi(x)\colon \phi(x) + i\colon \bar{\psi}(x)\gamma^5\psi(x)\colon \phi^\dagger(x)
$$

is Hermitian. (Here ψ is a Dirac field, ϕ a *scalar* field.) Also show (assuming all phases are unity) that

(i) this interaction does *not* conserve parity, but

(ii) the interaction does conserve $\mathcal{PCT}$.

(b) Assuming all phases are unity, what other transformation ($\mathcal{T}$ or $\mathcal{C}$) is not conserved by this interaction? Compute the effect of $\mathcal{T}$ and $\mathcal{C}$ on the interaction.

8.3 Construct a field theory with spin zero particles *only*, in which

(i) 3 particles interact at a point

(ii) 4 particles interact at a point

Assume that parity is conserved. What is the parity of the particles in each case? [To "construct a field theory," it is sufficient to write down $\mathcal{L}$ and $\mathcal{H}$.]

8.4 Show that the following Lagrangian density is invariant under charge conjugation:

$$\mathcal{L} = \bar{\psi}(x)\left(\frac{i}{2}\overleftrightarrow{\not\partial} - m\right)\psi(x) - ig\bar{\psi}(x)\gamma^5\psi(x)\phi(x)$$
$$+ \frac{1}{2}\left(\partial_\mu\phi(x)\partial^\mu\phi(x) - m_\phi^2\phi^2(x)\right) ,$$

where ϕ is a Hermitian scalar field with a positive phase under charge conjugation and ψ is a Dirac field.

8.5 Consider a field theory with the following interaction:

$$\mathcal{L}_{\text{int}} = -g\left[\bar{\psi}(x)\gamma^\mu\psi(x)\right]\left[\bar{\psi}(x)\gamma_\mu\gamma^5\psi(x)\right] ,$$

where g is a real constant.

(a) Prove that $\mathcal{L}_{\text{int}}$ is Hermitian.

(b) Prove that $\mathcal{L}_{\text{int}}$ does not conserve parity.

(c) What are the *simplest* physical interactions described by $\mathcal{L}_{\text{int}}$?

8.6 Suppose the electron had a static *electric* dipole moment analogous to the magnetic moment. Write a Hamiltonian density that represents the interaction of the electric dipole moment with the electromagnetic field and prove that it is not invariant under parity.

8.7 Show that

$$Tu(-p, -s) = (-1)^{\frac{1}{2}+s}u^*(p, s)$$
$$Tv(-p, -s) = (-1)^{\frac{1}{2}+s}v^*(p, s) ,$$

where $T = C\gamma^5$.

CHAPTER 9

INTERACTING FIELD THEORIES

With this chapter we begin a systematic study of interacting field theories in which all particles are described by relativistic quantum fields. This means that all particles are handled in a similar way and that the annihilation or creation of particles or particle pairs can be treated in a way consistent with the description of their scattering. The particles are isolated from their surroundings and interact only with each other, so that momentum (as well as energy) is conserved, and the recoil of the target in a collision process can be properly described. One of the great successes of this treatment is that it leads naturally to a description of particle forces.

9.1 ϕ^3 THEORY: AN EXAMPLE

We often want a simple interacting field theory to use as an example when we begin the discussion of a new subject. In this book we will use ϕ^3 theory for this purpose. While this theory has very few applications, it has the virtue of being one of the more simple theories which can be constructed and is also rich enough to illustrate the general techniques used to treat any interacting theory.

In its most complete form, our illustrative ϕ^3 theory will include three kinds of scalar particles: two charged scalar particles with masses m_i, $i = 1, 2$, and a neutral scalar particle with mass μ. As discussed in Sec. 7.5, each particle will be described by a separate quantum field, which is Φ_i for particles 1 and 2 and ϕ for the neutral particle. The total Lagrangian is the sum of four terms,

$$\mathcal{L} = \mathcal{L}_1 + \mathcal{L}_2 + \mathcal{L}_0 + \mathcal{L}_{\text{int}} \quad . \tag{9.1}$$

The free Lagrangians for the charged fields were discussed in Sec. 7.3 and are

$$\mathcal{L}_i =: \frac{\partial \Phi_i^\dagger}{\partial x^\mu} \frac{\partial \Phi_i}{\partial x_\mu} - m_i^2 \Phi_i^\dagger \Phi_i : \tag{9.2}$$

where i is *not summed over*, and for the neutral scalar field, where $\phi^\dagger = \phi$, the Lagrangian is

$$\mathcal{L}_0 = \frac{1}{2}\colon\left\{\frac{\partial\phi}{\partial x^\mu}\frac{\partial\phi}{\partial x_\mu} - \mu^2\phi^2\right\}\colon \quad , \tag{9.3}$$

as discussed in Prob. 7.2. All three of these Lagrangians describe spin 0 Klein–Gordon particles, and hence their corresponding fields satisfy *commutation relations*. The interaction term will have a ϕ^3 structure and consist of three possible terms,

$$\mathcal{L}_{\rm int} = -\lambda_1 \colon\Phi_1^\dagger(x)\Phi_1(x)\phi(x)\colon -\lambda_2 \colon\Phi_2^\dagger(x)\Phi_2(x)\phi(x)\colon -\frac{\lambda}{3!}\colon\phi^3(x)\colon \quad , \tag{9.4}$$

where the coupling constants of the theory are λ_1, λ_2, and λ, all of which are real because (9.4) must be Hermitian. In some references, the term "ϕ^3 theory" is reserved exclusively for a purely neutral self-interaction term like the last term given in (9.4), but we will refer to any of the interaction terms in (9.4) as a "ϕ^3 theory." The reason for dividing the neutral ϕ^3 term by 3! will be discussed later. In some applications we will take some of these constants to be zero, giving simpler theories. According to the discussion in Sec. 7.5, this interaction Lagrangian describes elementary processes in which three particles interact at a point.

The fields $\Phi_i(x)$ and $\phi(x)$ have the following structure:

$$\begin{aligned}\Phi_i(x) &= \sum_p \left\{a_{ip}\Phi_{ip}^{(+)}(x) + c_{ip}^\dagger\Phi_{i\,-p}^{(-)}(x)\right\}\\ \phi(x) &= \sum_k \left\{a_k\phi_k^{(+)}(x) + a_k^\dagger\phi_{-k}^{(-)}(x)\right\} \quad ,\end{aligned} \tag{9.5}$$

where the subscripts on the annihilation and creation operators, a_{ip}, label both the particle type, i, and the momentum, p, and the corresponding operators for the neutral particle will be distinguished by having only one subscript, the momentum k. The subscripts on the wave functions Φ label both the particle type and the momentum.

The interaction Hamiltonian corresponding to the ϕ^3 theory proposed in Eq. (9.2) is

$$\mathcal{H}_{\rm int} = -\mathcal{L}_{\rm int} = \lambda_1\colon\Phi_1^\dagger(x)\Phi_1(x)\phi(x)\colon + \lambda_2\colon\Phi_2^\dagger(x)\Phi_2(x)\phi(x)\colon + \frac{\lambda}{3!}\colon\phi^3(x)\colon \quad . \tag{9.6}$$

The perturbation theory for the time translation operator, worked out in Sec. 3.1, can also be applied to a perturbative treatment of interacting second quantized theories, and we will carry it over without further discussion. Expressed in terms of Hamiltonian *densities*, the time translation operator to second order is

$$\boxed{U_I = 1 - i\int d^4x\,\mathcal{H}_{\rm int}(x) + \frac{(-i)^2}{2}\int d^4x_1\,d^4x_2\,T\left(\mathcal{H}_{\rm int}(x_1)\mathcal{H}_{\rm int}(x_2)\right) + \dots ,} \tag{9.7}$$

where the T operation is the time-ordered product of the Hamiltonian densities, with the later times on the left.

In the next few sections, we will discuss particle decay (which happens in first order in ϕ^3 theory) and particle scattering (which happens in second order). Our discussion here parallels the discussion in Sec. 3.2 but is more general because *all particles are treated relativistically, with the possibility of particle production and annihilation, and there is no longer any fixed center of force.* The latter allows us to conserve momentum as well as energy.

9.2 RELATIVISTIC DECAYS

The Hamiltonian (9.6) will permit the neutral particle to decay into a particle and antiparticle of type-1 if $\mu > 2m_1$. In this section we will calculate this decay rate and the corresponding lifetime of the neutral particle. While the details of this calculation are given for this example, nearly all of our results apply to any decay process and thus are very general. We will extract the general features of the calculation as we go along.

The S-matrix for decay is

$$S = \langle p\bar{p}' \left| U_I \right| k\rangle \quad , \tag{9.8}$$

where k is the momentum of the neutral, heavy particle and p and p' are the momenta of particles and antiparticles of type-1. The bar over a momentum variable will be used to denote an antiparticle. To lowest order in the interaction, this matrix element is

$$S = -i\left\langle 0 \left| a_{1p} c_{1p'} \int d^4x\, \lambda_1 \colon \Phi_1^\dagger(x)\Phi_1(x)\phi(x)\colon a_k^\dagger \right| 0 \right\rangle \quad , \tag{9.9}$$

where, if $\lambda_2 \neq 0$, the term with $\Phi_2^\dagger\Phi_2$ must be zero because the interaction term is normal ordered and there are no annihilation or creation operators for particles of type-2 in either the initial or the final state to prevent these operators from acting directly on the vacuum state and giving zero. Now, recalling the expansions for Φ_1 and ϕ given in Eq. (9.5), and the commutation relations satisfied by these operators, we see that the only non-zero term can come from the a terms in ϕ, the $a_1^\dagger$ terms in $\Phi_1^\dagger$, and the $c_1^\dagger$ terms in Φ_1. Furthermore, using the result

$$\left\langle 0 \left| a_{1p} c_{1p'} a_{1q}^\dagger c_{1q'}^\dagger a_{k'} a_k^\dagger \right| 0 \right\rangle = \delta_{k,k'}\delta_{p,q}\delta_{p',q'} \quad , \tag{9.10}$$

the momentum sums in these three fields must collapse to a single term, giving simply

$$\begin{aligned} S &= -i\lambda_1 \int d^4x\, \Phi_{1p}^{(+)\,*}(x)\Phi_{1\,-\bar{p}'}^{(-)}(x)\phi_k^{(+)}(x) \\ &= -i\lambda_1 \int \frac{d^4x\, e^{i(p+p'-k)\cdot x}}{\sqrt{8E_1(p)E_1(p')\omega(k)\, L^9}} \end{aligned}$$

where $E_i(p) = \sqrt{m_i^2 + p^2}$ for $i = 1$ or 2 and $\omega(k) = \sqrt{\mu^2 + k^2}$ for the neutral particle. Integrating over all x and going to continuum limit ($L \to 2\pi$) show that this result can be cast into the following form:

$$\boxed{S = -i\frac{(2\pi)^4\delta^4(p + p' - k)}{\sqrt{(2\pi)^9 8E_1(p)E_1(p')\omega(k)}}\,\mathcal{M}\ ,} \tag{9.11}$$

where the relativistic $\mathcal{M}$-matrix has been introduced. In the *general case*, when the total number of particles entering *and* leaving the interaction is n, the relativistic $\mathcal{M}$-matrix is *defined* by the relation

$$\boxed{S = -i(2\pi)^4\delta^4(p_f - p_\imath)\prod_{\imath=1}^{\imath=n} f_\imath\,\mathcal{M}\ ,} \tag{9.12}$$

where, if $E_\imath$ is the energy of the $\imath$th particle (either E_1 or ω in this example), then

$$\boxed{f_\imath = \frac{1}{\sqrt{(2\pi)^3 2E_\imath}}}$$

for each *particle* entering *or* leaving the interaction. In this simple ϕ^3 example, the $\mathcal{M}$-matrix is simply the coupling constant

$$\mathcal{M} = \lambda_1\ . \tag{9.13}$$

Decay Rate

To obtain the decay rate the calculations in Sec. 3.1 must be generalized. Since there is now no potential, or fixed center of force, decay (or scattering) takes place throughout *all* space. In this case, the decay rate per unit volume is calculated, and the result is summed over all space and all momenta. Hence, for a box of volume L^3, the differential decay rate is

$$dW_{\beta\alpha} = \frac{|S_{\beta\alpha}|^2}{TL^3} \times L^3\ , \tag{9.14}$$

where the first term is the rate *per unit volume* and L^3 is the volume. The rate per unit volume is calculated in much the same way as the time average was treated in Eq. (3.34). If the volume integral over the wave functions is denoted by Δ,

$$\Delta = \int_{L^3} d^3r\, e^{-i(p_f - p_\imath)\cdot r} = L^3\delta_{p_f,p_\imath} \underset{\text{continuum}}{\longrightarrow} (2\pi)^3\delta^3(p_f - p_\imath)\ , \tag{9.15}$$

then the volume average of the square of the integral, which is what is needed for the calculation of the decay rate, is

$$\frac{\Delta^2}{L^3} = \frac{L^6 \delta_{p_f, p_i}}{L^3} = L^3 \delta_{p_f, p_i} \underset{\text{continuum}}{\longrightarrow} (2\pi)^3 \delta^3(p_f - p_i) \ . \tag{9.16}$$

Hence, in the continuum limit, we have a formal relation completely analogous to the time average relation Eq. (3.45),

$$\frac{\left[(2\pi)^3 \delta^3(p_f - p_i)\right]^2}{L^3} = (2\pi)^3 \delta^3(p_f - p_i) \ . \tag{9.17}$$

Using this, and Eq. (3.45) for the time average, gives, in the continuum limit,

$$\begin{aligned} dW_{\text{decay}} &= \frac{d^3p\, d^3p'\, (2\pi)^4 \delta^4(p + p' - k)}{(2\pi)^9\, 8E_1(p)E_1(p')\omega(k)} |\mathcal{M}|^2 \times (2\pi)^3 \\ &= \left[\frac{d^3p}{(2\pi)^3 2E_1(p)}\right]\left[\frac{d^3p'}{(2\pi)^3 2E_1(p')}\right]\left(\frac{1}{2\omega(k)}\right)(2\pi)^4\delta^4(p + p' - k)\, |\mathcal{M}|^2 . \end{aligned} \tag{9.18}$$

The generalization of this formula to an n-body decay satisfies the following rules:

- Define $\mathcal{M}$ as in Eq. (9.12).
- Construct the decay rate dW as follows:

> - a factor of $(2\pi)^4 \delta^4(p_f - p_i)$,
> - a factor of
>
> $$\frac{d^3p}{(2\pi)^3\, 2E_p}$$
>
> for each *particle in the final state*,
> - a factor of $1/2E$ for the initial particle which is decaying,
> - the absolute square of the $\mathcal{M}$-matrix.

These rules are also recorded in Appendix B for future reference.

The total decay rate for the ϕ^3 example under consideration (in which $\mathcal{M}$ is independent of angles) is

$$\begin{aligned} W &= \int \frac{d^3p\, d^3p'}{(2\pi)^6} \frac{(2\pi)^4 \delta^4(p + p' - k)\, |\mathcal{M}|^2}{8E_1(p)E_1(p')\omega(k)} \\ &= \frac{4\pi}{(2\pi)^2} \int \frac{pE_1(p) dE_1(p)}{8E_1^2(p)\mu} \delta(2E_1(p) - \mu)\, |\mathcal{M}|^2 \\ &= \frac{p^*}{8\pi\, \mu^2} |\mathcal{M}|^2 = \left(\frac{\lambda_1^2}{4\pi}\right) \frac{p^*}{2\mu^2} \ , \end{aligned} \tag{9.19}$$

where the calculation was carried out in the rest system of the neutral particle, so $\omega(k) = \mu$ and $E_1(p) = E_1(p')$, and the relative momenta of the two decay products is obtained from the equation

$$E_1^2(p^*) = m_1^2 + p^{*\,2} = \tfrac{1}{4}\mu^2 \quad ,$$

so the final result can be written

$$W = \frac{1}{\tau} = \left(\frac{\lambda_1^2}{4\pi}\right) \frac{\sqrt{\mu^2 - 4m_1^2}}{4\mu^2} \quad , \tag{9.20}$$

where τ is the lifetime.

Remarks

Decays exhibit the following features:

- The decay rate is proportional to λ_1^2, and hence, if λ_1 is small so that the decay is well described by the lowest order perturbative result, the *magnitude* of the coupling constant λ_1 can be determined from the decay.
- The decay will not take place unless $\mu > 2m_1$.
- The decay rate is proportional to the *phase space*, which for this simple S wave two-body decay is

$$\rho(\mu; m_1, m_1) = \frac{p^*}{4\pi\mu} = \frac{1}{8\pi}\sqrt{1 - \frac{4m_1^2}{\mu^2}} \quad . \tag{9.21}$$

If μ is very close to $2m_1$, the rate is low (the lifetime long), *even if* λ_1 *is large.*

The *phase space* of an n-particle decay is defined to be

$$\rho(\mu; m_1, \cdots, m_n) = \int \cdots \int \frac{d^3p_1 \cdots d^3p_n}{(2\pi)^{3n} 2E_1 \cdots 2E_n} \,(2\pi)^4\, \delta(p_f - k) \quad , \tag{9.22}$$

where k is the four-momentum of the decaying particle and $k^2 = \mu^2$. The phase space ρ is an integral operator, but if it acts on a *constant* $\mathcal{M}$-matrix, all the integrals can be carried out and it can be reduced to a known function of the masses. Evaluation the phase space integral is a useful way to estimate many body decay rates and cross sections. For two-body decays it is easy to work out, and the task of obtaining the general result is left to the reader (Prob. 9.4).

9.3 RELATIVISTIC SCATTERING

To describe scattering, we need to go to second order in the ϕ^3 interaction Hamiltonian (a ϕ^4 term would be needed for scattering to occur in lowest order). As our first example, consider the case of a particle of type-1 scattering from a particle of type-2. The process is represented diagrammatically in Fig. 9.1, where the

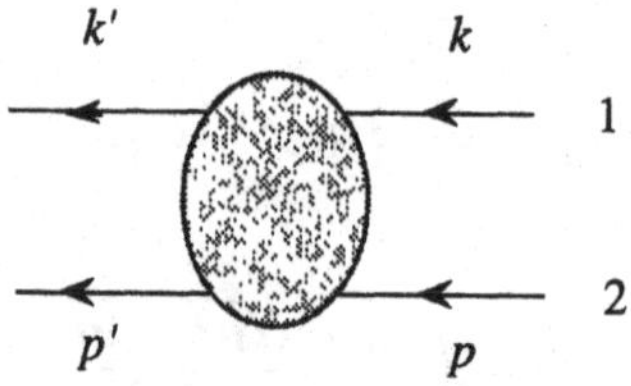

Fig. 9.1 Diagram of the scattering of particles of type 1 and 2.

momenta of particle 1 are k and k' and of particle 2 are p and p'. We will assume $k \neq k'$, $p \neq p'$, which means the scattering is *not* in the forward direction.

The S-matrix is (the factor of $\frac{1}{2}$ is compensated by two identical terms)

$$S = (-i)^2 \lambda_1 \lambda_2 \left\langle k' p' \left| \int d^4x_1\, d^4x_2 \right.\right.$$
$$\left.\left. \times\, T\left\{ : \Phi_1^\dagger(x_1)\Phi_1(x_1)\phi(x_1) : : \Phi_2^\dagger(x_2)\Phi_2(x_2)\phi(x_2) : \right\} \right| k\, p \right\rangle \ . \tag{9.23}$$

This is the only term to second order which can contribute to the matrix element. The $a_1^\dagger$ in $\Phi_1^\dagger$, a_1 in Φ_1, $a_2^\dagger$ in $\Phi_2^\dagger$, and a_2 in Φ_2 are the terms which survive. Using

$$\left\langle k' p' \left| a_{1n}^\dagger a_{1m} a_{2i}^\dagger a_{2j} \mathcal{O} \right| k\, p \right\rangle = \left\langle 0 \left| a_{1k'} a_{2p'} a_{1n}^\dagger a_{1m} a_{2i}^\dagger a_{2j} a_{1k}^\dagger a_{2p}^\dagger \mathcal{O} \right| 0 \right\rangle$$
$$= \delta_{k'n}\delta_{p'i}\delta_{mk}\delta_{jp} \langle 0|\mathcal{O}|0\rangle \ , \tag{9.24}$$

which holds for any operator $\mathcal{O}$ which commutes with a_1, $a_1^\dagger$, a_2, and $a_2^\dagger$, gives

$$S = -\lambda_1\lambda_2 \int d^4x_1\, d^4x_2\, \Phi_{1k'}^{(+)\,*}(x_1)\Phi_{2p'}^{(+)\,*}(x_2)\Phi_{1k}^{(+)}(x_1)\Phi_{2p}^{(+)}(x_2)$$
$$\times \langle 0|T(\phi(x_1)\phi(x_2))|0\rangle \ , \tag{9.25}$$

where $\Phi_{1k'}^{(+)}(x)$ is the KG wave function for a free particle of type-1 with momentum k', etc. Now, to carry out the integrals, introduce

$$R = \tfrac{1}{2}(x_1 + x_2) \qquad r = (x_1 - x_2) \tag{9.26}$$

and use the fact, which we will show later, that

$$\langle 0\,|T\,(\phi(R + \tfrac{1}{2}r)\,\phi(R - \tfrac{1}{2}r))|\,0\rangle = \langle 0|T\,(\phi(r)\phi(0))\,|0\rangle \tag{9.27}$$

is a function of r only, and go to continuum normalization to obtain

$$S = -\lambda_1\lambda_2 \int \frac{d^4R\, d^4r\; e^{i[(k'-k)\cdot(R+\frac{1}{2}r)+(p'-p)\cdot(R-\frac{1}{2}r)]}}{(2\pi)^6\sqrt{16E_1(k)E_1(k')E_2(p)E_2(p')}} \langle 0|T(\phi(r)\phi(0))|0\rangle \; . \tag{9.28}$$

Defining the reduced matrix $\mathcal{M}$ as in Eq. (9.12) above,

$$\boxed{S = -i\frac{(2\pi)^4\delta^4(k'+p'-k-p)}{(2\pi)^6\sqrt{16E_1(k)E_1(k')E_2(p)E_2(p')}}\,\mathcal{M}\;,} \tag{9.29}$$

we obtain

$$\mathcal{M} = -i\lambda_1\lambda_2 \int d^4r\, e^{i(k'-k)\cdot r}\; \underbrace{\langle 0|T(\phi(r)\phi(0))|0\rangle}_{\text{propagator}} \; . \tag{9.30}$$

In this example, the $\mathcal{M}$-matrix is proportional to the Fourier transform of the *vacuum expectation value of the time-ordered product of two field operators.* This vacuum expectation value is referred to as the *propagator* and is a very important concept in field theory.

Instead of calculating the propagator of the *neutral* field (9.30), we will calculate the propagator for a *charged* field with the same mass. Since a charged field has antiparticles which are distinct from its particles, as discussed in Sec. 7.3, this calculation will enable us to track the role of antiparticles more easily. At the end, we will see that the two propagators have the same mathematical form. In coordinate space the charged propagator is

$$i\Delta(x_1 - x_2) = \langle 0|T(\phi(x_1)\phi^\dagger(x_2))|0\rangle \;, \tag{9.31}$$

where the charged field expansion is

$$\phi(x) = \sum_k \left\{ a_k\phi_k^{(+)}(x) + c_k^\dagger\phi_{-k}^{(-)}(x)\right\} \; . \tag{9.32}$$

The charged field differs from the neutral field only in that $c_k \neq a_k$; the wave functions $\phi^{(\pm)}$ are identical in the two cases.

The propagator is easily calculated. Only the $a\,a^\dagger$ and $c\,c^\dagger$ terms can contribute, and these contribute to different time orderings. Using the (by now) familiar result (which also holds for the c's)

$$\left\langle 0|a_n a_{n'}^\dagger|0\right\rangle = \delta_{nn'} \tag{9.33}$$

gives

$$
\begin{aligned}
&\langle 0|T\left(\phi(x_1)\phi^\dagger(x_2)\right)|0\rangle \\
&= \left\langle 0 \left| \left(\sum_n a_n\phi_n^{(+)}(x_1)\right)\left(\sum_{n'} a_{n'}^\dagger \phi_{n'}^{(+)\,*}(x_2)\right)\right| 0\right\rangle \theta(t_1-t_2) \\
&\quad + \left\langle 0 \left| \left(\sum_n c_{n'}\phi_{-n'}^{(-)\,*}(x_2)\right)\left(\sum_n c_n^\dagger \phi_{-n}^{(-)}(x_1)\right)\right| 0\right\rangle \theta(t_2-t_1) \\
&= \sum_n \phi_n^{(+)}(x_1)\phi_n^{(+)\,*}(x_2)\theta(t_1-t_2) + \sum_n \phi_{-n}^{(-)\,*}(x_2)\phi_{-n}^{(-)}(x_1)\theta(t_2-t_1) \ .
\end{aligned}
\tag{9.34}
$$

The particle part of this expansion is the first term, which propagates positive energy solutions to times $t_1 > t_2$, while the antiparticles contribute to the second term, which has the form of *negative energy solutions propagating to times* $t_1 < t_2$. Thus the field theory propagator automatically gives us the interpretation we developed in Sec. 4.8 (recall Fig. 4.3).

Having shown that the antiparticles are responsible for the $t_1 < t_2$ part of the propagator, we return to our discussion and observe that the neutral propagator gives a result identical to (9.34) because $\phi_{-n}^{(-)} = \phi_n^{(+)}$. Substituting for $\phi_n^{(+)}$, letting $\omega_k = \sqrt{\mu^2 + k^2}$, and taking the continuum limit give

$$
\begin{aligned}
&\langle 0|T\left(\phi(x_1)\phi(x_2)\right)|0\rangle \\
&= \int \frac{d^3k}{(2\pi)^3}\,\frac{e^{i\boldsymbol{k}\cdot(\boldsymbol{r}_1-\boldsymbol{r}_2)}}{2\omega_k}\left\{e^{-i\omega_k(t_1-t_2)}\theta(t_1-t_2) + e^{i\omega_k(t_1-t_2)}\theta(t_2-t_1)\right\} ,
\end{aligned}
\tag{9.35}
$$

where the integration was changed from $\boldsymbol{k} \to -\boldsymbol{k}$ in the second term. Recalling the identity (4.77),

$$
\theta(t)\, e^{-i\omega_k t} = \frac{1}{2\pi i}\int_{-\infty}^{\infty} \frac{dk_0\, e^{-ik_0 t}}{\omega_k - k_0 - i\epsilon} ,
$$

we can write the propagator in the following form:

$$
\begin{aligned}
&\langle 0|T\left(\phi(x_1)\phi(x_2)\right)|0\rangle \\
&= \int \frac{d^3k}{(2\pi)^3}\,\frac{e^{i\boldsymbol{k}\cdot(\boldsymbol{r}_1-\boldsymbol{r}_2)}}{2\omega_k}\,\frac{1}{2\pi i}\left\{\int_{-\infty}^{\infty}\frac{dk_0\, e^{-ik_0(t_1-t_2)}}{(\omega_k - k_0 - i\epsilon)} + \int_{-\infty}^{\infty}\frac{dk_0\, e^{-ik_0(t_1-t_2)}}{\omega_k + k_0 - i\epsilon}\right\} \\
&= \int \frac{d^3k}{(2\pi)^3}\,\frac{e^{i\boldsymbol{k}\cdot(\boldsymbol{r}_1-\boldsymbol{r}_2)}}{2\omega_k}\,\frac{1}{2\pi i}\int \frac{dk_0\, e^{-ik_0(t_1-t_2)}}{\underbrace{(\omega_k^2 - k_0^2 - 2i\epsilon\,\omega_k - \epsilon^2)}_{\to\, \mu^2 - k_0^2 + \boldsymbol{k}^2 - i\epsilon'}}\, 2\omega_k ,
\end{aligned}
\tag{9.36}
$$

where the identity (4.77) with $t \to -t$ and $k_0 \to -k_0$ was used for the second term. Since the role of the $i\epsilon$ term in the denominators is *only to tell how to go around the pole in the complex plane*, the ϵ^2 term in the last denominator can be ignored, and $i\epsilon' = 2i\epsilon\,\omega_k$ is completely equivalent to $i\epsilon$ (because $\omega_k > 0$), and in *all subsequent calculations we will assume this equivalence by taking* $\epsilon' = \epsilon$ *without comment.* Putting this all together gives

$$\langle 0|T\left(\phi(x_1)\phi(x_2)\right)|0\rangle = \int \frac{d^4k}{(2\pi)^4}\, e^{-ik\cdot(x_1-x_2)} \left[\frac{-i}{\mu^2 - k^2 - i\epsilon}\right] \ . \tag{9.37}$$

This is a beautifully simple, covariant formula for the configuration space propagator of a spin zero particle. It shows that the propagator depends only on $x_1 - x_2$, and hence justifies Eq. (9.27). It displays the propagator as the four dimensional Fourier transform of a simple function of the *square of the virtual four-momentum* of the particle. The integration is over all four components of k^2, and $k^2 \neq \mu^2$. We say that the four-momentum of the propagating particle is "off-mass-shell," or simply "off-shell." Inserting this result into Eq. (9.30) gives the following second order result for the $\mathcal{M}$-matrix which describes $1+2 \to 1+2$ scattering in this simple theory:

$$\mathcal{M} = \frac{-\lambda_1\lambda_2}{\mu^2 - (k-k')^2 - i\epsilon} \ . \tag{9.38}$$

This calculation of the $\mathcal{M}$-matrix was more lengthy that the calculation of the decay amplitude in Sec. 9.2 but was still reasonably easy. The calculation of more complicated processes can be very tedious if done in this way (as we shall see!), and in the late 1940's Feynman discovered a very beautiful diagrammatic way to organize perturbative calculations so the results can be obtained much more easily. In the next section we begin our discussion of the famous Feynman rules.

9.4 INTRODUCTION TO THE FEYNMAN RULES *

One of our principal objectives as we continue to study interactions will be to extract the Feynman rules for the calculation of the $\mathcal{M}$-matrix. For the rest of this chapter and during the next two, we will calculate results directly from field theory the "hard" way, but as we do so, we will pause at the end and introduce new Feynman rules illustrated by the particular calculation just completed. In this way we will demonstrate explicitly how most of the Feynman rules arise in lowest order calculations and also develop an understanding of how results can be obtained directly from the field operators using perturbation theory. A summary of the Feynman rules we will introduce is included in Appendix B, and the reader is encouraged to refer to this frequently.

*Two of the original papers, [Fe 49], are reprinted in a nice introductory volume by Feynman (1961).

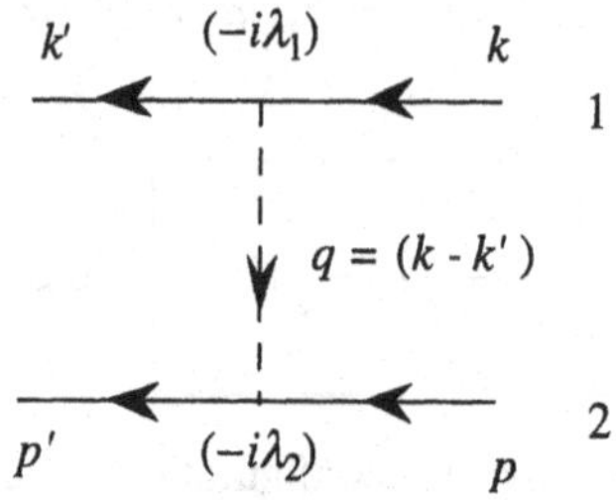

Fig. 9.2 Feynman diagram for the scattering of two scalar particles.

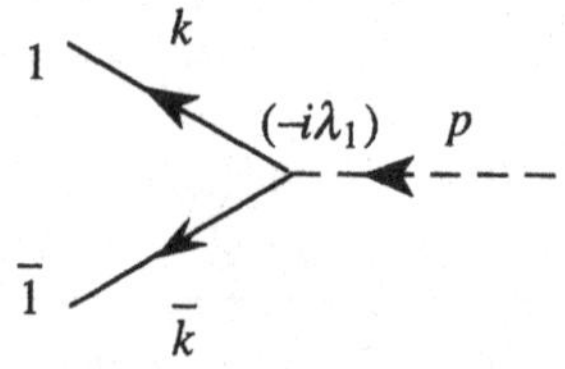

Fig. 9.3 Feynman diagram for the decay of a heavy scalar particle.

The approach we are taking in the next few chapters will *not* demonstrate that the Feynman rules we extract from our lowest order calculations will also work to *all orders in perturbation theory.* This proof is rather difficult and clumsy to carry out with the operator formalism we are using in this part of the book and will be defered until Chapter 14, where we introduce the path integral formalism.

Returning to the ϕ^3 scattering calculation just completed in the last section, we write the final result for the $\mathcal{M}$-matrix in the following way:

$$\mathcal{M} = \underbrace{i}_{Rule\ 0} \underbrace{(-i\lambda_1)(-i\lambda_2)}_{Rule\ 1} \underbrace{\left[\frac{-i}{\mu^2 - (k-k')^2 - i\epsilon}\right]}_{Rule\ 2} . \tag{9.39}$$

This illustrates the first three Feynman rules for the construction of the $\mathcal{M}$-matrix in ϕ^3 theory:

- First, draw all *Feynman diagrams* which describe a given scattering process to a given order. These are topologically distinct drawings which describe possible mechanisms for particle production, annihilation, and propagation which can lead to the final scattering process. For this example there is only one such diagram (shown in Fig. 9.2) which describes the exchange of the neutral particle between particles 1 and 2.

- Next, find the mathematical factor associated with each diagram, and add the factors for all diagrams together to get the total result. The rules which enter in this first simple example are

Rule 0: a factor of i.

Rule 1: a factor of $-i\lambda_j$ for each vertex where the neutral particle is emitted or absorbed by particle j.

Rule 2: a factor of

$$\frac{-i}{\mu^2 - k^2 - i\epsilon}$$

for each internal line of four-momentum k corresponding to the virtual propagation of a scalar particle.

- Four-momentum is to be conserved at each vertex, like current flowing through a circuit.

Note that the decay calculated in the last section also followed these rules (refer to Fig. 9.3).

The Feynman rules fall into two general categories. There are rules which are very general and apply to all theories and rules which are specific to a particular theory. Rules 0 and 2 above are examples of the first type, and Rule 1 is of the second. Each theory has its own characteristic interactions, and for each type of interaction there is a Rule 1 with a unique structure. Many of these are summarized in Appendix B. Rule 0 may seem frivolous; if the factor of i is omitted from *every* amplitude, it will be of no consequence for any prediction since all observables depend on the absolute square of an amplitude. However, omitting this factor is inconvenient from a theoretical point of view, since many amplitudes known to be real (or imaginary) will not have the expected property unless it is included. Finally, the propagator is a key building block in the construction of Feynman amplitudes, and the propagator given above (Rule 2) is the correct propagator to use in any Feynman amplitude which includes an internal *scalar* particle, and in this sense it is a general result. We will discuss some of the properties of the propagator now.

The Propagator

We will first rederive the explicit form for the propagator given in Eq. (9.37) using a different technique. This new derivation will give us experience with the treatment of matrix elements of field operators and insight into the physics which goes into its definition.

We begin by showing that the propagator $i\Delta(x) = \langle 0|T\left(\phi(x)\phi(0)\right)|0\rangle$ [recall Eq. (9.31)] satisfies the following inhomogeneous KG equation:

$$\left(\Box + \mu^2\right)\Delta(x) = -\delta^4(x) \quad . \tag{9.40}$$

To prove this, note that the field operators satisfy the KG equation (they are constructed from its plane wave solutions), so the only non-zero contributions must come from the time derivatives of the θ-functions associated with the time-ordered product. Recalling that

$$\frac{d}{dt}\theta(t) = \delta(t) \quad , \tag{9.41}$$

we have

$$\begin{aligned}
\frac{\partial^2}{\partial t^2}\Delta(t) &= \frac{\partial^2}{\partial t^2}\Big[\theta(t)\ \langle 0|\phi(x)\phi(0)|0\rangle + \theta(-t)\ \langle 0|\phi(0)\phi(x)|0\rangle\Big] \\
&= \frac{\partial}{\partial t}\left(\underbrace{\delta(t)\ \langle 0|\left[\phi(x),\phi(0)\right]|0\rangle}_{=0} + \theta(t)\left\langle 0\left|\frac{\partial\phi(x)}{\partial t}\phi(0)\right|0\right\rangle\right. \\
&\qquad\qquad \left. + \theta(-t)\left\langle 0\left|\phi(0)\frac{\partial\phi(x)}{\partial t}\right|0\right\rangle\right) \\
&= \delta(t)\left\langle 0\left|\left[\frac{\partial\phi(x)}{\partial t},\phi(0)\right]\right|0\right\rangle + \left(\frac{\partial^2}{\partial t^2}\phi(x) \quad \text{terms}\right) \\
&= \delta(t)\ \langle 0|\left[\pi(r,0),\phi(0)\right]|0\rangle = -i\delta^4(x) \ ,
\end{aligned} \tag{9.42}$$

where the canonical commutation relations were used to discard the commutator in the second line and to obtain the additional $\delta^3(r)$ term in the last line and the $\partial^2\phi/\partial t^2$ terms are discarded because they are canceled by the remaining $\nabla^2 - \mu^2$ from the KG equation. Hence we have proved Eq. (9.40).

If the propagator and the δ-function are expanded in plane waves, Eq. (9.40) can be used to obtain the following solution for the propagator:

$$i\Delta(x) = \int \frac{d^4k}{(2\pi)^4}\, e^{-ik\cdot x}\left[\frac{-i}{\mu^2 - k^2}\right] + \text{homogeneous solution} \ , \tag{9.43}$$

where the homogeneous solution is fixed by the *boundary conditions* we wish to impose. Before we discuss this, note that the denominator of the inhomogeneous term in (9.43) still has the same two zeros at $k_0 = \pm\omega_k$ that we encountered before. The integral is defined by regarding these as poles in the complex k_0 plane, and there are four possibilities: both poles could be in the upper half plane, both in the lower half plane, or we could have one in each. We *choose* to put the positive energy pole in the lower half plane and the negative energy one in the upper half plane, as illustrated in Fig. 9.4, and to set the homogeneous terms to zero. This choice is referred to as the *Feynman prescription* and is equivalent to imposing the requirement that positive energy states propagate forward in time and negative energy states propagate backward in time and is, as we have seen, the correct prescription for a theory in which negative energy states are to be the antiparticles. We leave it to the reader to show (Prob. 9.5) that any other choice of propagator differs from the Feynman choice only by a homogeneous term, and hence the Feynman prescription is equivalent to imposing boundary conditions on the theory. With the Feynman prescription, our solution (9.43) becomes identical to the original solution (9.37) and our rederivation of the result is complete.

We conclude this discussion by showing, directly from Eq. (9.37), that the Feynman prescription propagates positive energy states forward in time and negative energy states backward in time. [In other words, we show that the steps which

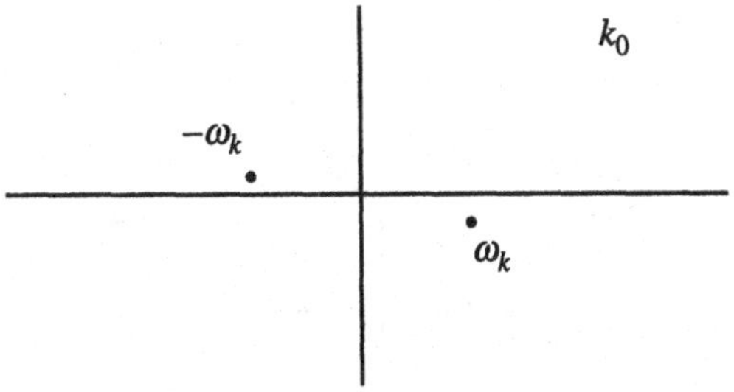

Fig. 9.4 The complex k_0 plane showing the location of the two poles of the propagator.

lead to (9.37) can be reversed.] Consider the integral over the virtual energy, dk_0. Factoring the Feynman denominator,

$$\mu^2 - k^2 - i\epsilon = \omega_k^2 - k_0^2 - i\epsilon = (\omega_k - k_0 - i\epsilon)(\omega_k + k_0 - i\epsilon) \ , \tag{9.44}$$

shows that there are two poles in the complex k_0 plane, located at $\pm(\omega - i\epsilon)$ as shown in Fig. 9.4. To evaluate the integral, we must extend the path of integration into either the upper or lower half plane, making a closed contour so that the integral may be evaluated using the calculus of residues. If $t = t_1 - t_2 > 0$, the contour must be closed in the lower half plane where the integrand converges to zero at infinity. If $t = t_1 - t_2 < 0$, it must be inclosed in the upper half plane. This gives

$$\int \frac{dk_0}{2\pi i} e^{-ik_0(t_1-t_2)} \left(\frac{1}{\mu^2 - k^2 - i\epsilon} \right) = \frac{e^{-i\omega_k(t_1-t_2)}}{2\omega_k}\theta(t_1 - t_2) + \frac{e^{i\omega_k(t_1-t_2)}}{2\omega_k}\theta(t_2 - t_1) \ , \tag{9.45}$$

in agreement with our starting point. We call attention to the fact that this result leads *uniquely to the physical interpretation that positive energy states propagate forward in time and negative energy states propagate backward.* This interpretation is a *mathematical consequence of our placement of the poles in the complex plane* using the "$i\epsilon$" prescription we introduced in Eq. (9.36), and this in turn is a consequence of our use of the identity Eq. (4.77).

Now examine the role of the momentum space propagator, Rule 2, in the final result for the scattering matrix, Eq. (9.38). Factoring the propagator into two terms gives

$$\frac{1}{\mu^2 - (k - k')^2} = \frac{1}{\omega^2 - (E_1 - E_1')^2} = \frac{1}{2\omega} \Bigg[\underbrace{\frac{1}{\omega + E_1' - E_1}}_{\text{A}} + \underbrace{\frac{1}{\omega + E_1 - E_1'}}_{\text{B}} \Bigg] , \tag{9.46}$$

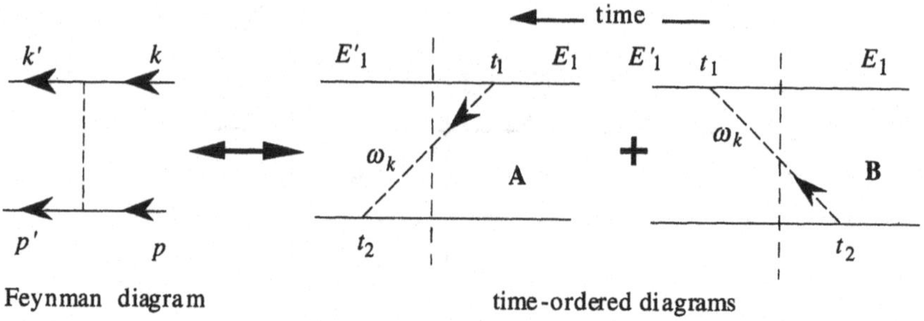

Fig. 9.5 The Feynman diagram on the left is the sum of the two time-ordered diagrams A and B.

where $\omega^2 = \mu^2 + (\boldsymbol{k} - \boldsymbol{k}')^2$ and $E_1 = \sqrt{m_1^2 + k^2}$, $E_1' = \sqrt{m_1^2 + k'^2}$. This decomposition is interpreted diagrammatically in Fig. 9.5. It illustrates a very general feature of Feynman diagrams: a single Feynman diagram is equal to the *sum of a number of time-ordered diagrams.* A *time-ordered diagram* is one in which the times at which particles are emitted or absorbed are in a definite well-defined order, which is *not* the case for Feynman diagrams, where times at which particles are emitted or absorbed can have *all possible orderings.* For a diagram with n vertices, there are $n!$ possible time-orderings and hence $n!$ time-ordered diagrams correspond to *each* Feynman diagram. The two time-ordered diagrams A and B shown in Fig. 9.5 are similar to the two diagrams we discussed in Fig. 4.3.

9.5 CALCULATION OF THE CROSS SECTION

In this section we return to our general discussion of scattering theory and work out the general rules for obtaining the relativistic cross section from the $\mathcal{M}$-matrix. We will work out the details for the two body scattering case labeled in Fig. 9.1.

As in the case of decays, we must calculate the transition rate *per unit volume,* since the scattering is now viewed as taking place over all space (corresponding to momentum conservation). However, our discussion following Eq. (3.46) is essentially unchanged, because the rate per unit volume occurs in both the numerator *and* the denominator. We have

$$\frac{d\sigma \times (\#\text{target particles})}{(\text{area of beam–target interaction})} = \frac{(\#\,\text{particles scattered into } \Delta\Omega/\text{sec-vol})}{(\#\,\text{particles incident/sec-vol})}\,. \tag{9.47}$$

Note that this way of writing the expression shows clearly that $d\sigma$ has dimensions of an area.

Now, the number of particles scattered into $\Delta\Omega$/sec-vol is proportional to [using Eq. (9.14) and the discussion following]

$$\frac{|S|^2}{TL^3} = \frac{(2\pi)^4\delta^4(p'+k'-p-k)|\mathcal{M}|^2}{L^{12}16E_1E_1'E_2E_2'} \,, \tag{9.48}$$

where the particle energies will be denoted $E_1(k) = E_1$, etc., for simplicity. The rest of the analysis parallels our development in Sec. 3.1, except we now have an extra factor of L^3,

$$d\sigma = \sum_{\Delta\Omega} \frac{(2\pi)^4\delta^4(p'+k'-p-k)|\mathcal{M}|^2}{L^{12}\, v\, 16E_1E_1'E_2E_2'} L^6 \,, \tag{9.49}$$

where the sum is over all momenta in the final state, consistent with the restriction to solid angle $\Delta\Omega$, and v is the flux factor introduced in our discussion in Sec. 3.1. Note that the L^6 term cancels, and

$$d\sigma = \sum_{p'\in\Delta\Omega}\sum_{k'\in\Delta\Omega} \frac{1}{L^6} \frac{(2\pi)^4\delta^4(p'+k'-p-k)|\mathcal{M}|^2}{v\, 16E_1E_1'E_2E_2'} \,.$$

Going to the continuum limit gives

$$\boxed{d\sigma = \int_{\Delta\Omega} \frac{d^3p'\, d^3k'}{(2\pi)^6} \frac{(2\pi)^4\delta^4(p'+k'-p-k)}{v\, 16E_1E_1'E_2E_2'} |\mathcal{M}|^2 \,.} \tag{9.50}$$

This is a special case of the general formula for the cross section, which builds $d\sigma$ from the following factors:

- a factor of $(2\pi)^4\delta^4(p_f - p_i)/v$, where v is the flux,
- a factor of
$$\frac{d^3p_i}{(2\pi)^3\, 2E_i}$$
for each particle in the final state,
- a factor of
$$\frac{1}{4EE'} \,,$$
where E and E' are the energies of the two particles in the initial state,
- the absolute square of the $\mathcal{M}$-matrix.

Now, reducing the cross section Eq. (9.50) in the center of mass (CM) frame gives

$$d\sigma = \int_{\Delta\Omega} \frac{d^3k'}{(2\pi)^2} \frac{\delta\left(E_1' + E_2' - E_1 - E_2\right)}{\left(\frac{k}{E_1} + \frac{k}{E_2}\right) 16 E_1 E_1' E_2 E_2'} |\mathcal{M}|^2 \,, \tag{9.51}$$

where all the energies can be expressed in terms of $E_i' = \sqrt{m_i^2 + k'^2}$ and $E_i = \sqrt{m_i^2 + k^2}$, because, in the CM frame, $\boldsymbol{p} = -\boldsymbol{k}$ (which implies that $p = k$), and the three-momentum delta function therefore gives $\boldsymbol{p}' = -\boldsymbol{k}'$ (implying $p' = k'$). The flux factor v, which is the sum of the approach velocities of the two particles, is

$$v = \left(\frac{k}{E_1} + \frac{k}{E_2}\right) = \frac{k}{E_1 E_2}(E_1 + E_2) \ . \tag{9.52}$$

To evaluate the energy conserving δ-function, note that

$$\frac{d(E_1' + E_2')}{dk'} = \frac{k'}{E_1'} + \frac{k'}{E_2'} = \frac{k'}{E_1' E_2'}(E_1' + E_2') \ . \tag{9.53}$$

Dividing by this factor enables us to evaluate the δ-function easily. If $\Delta\Omega \to d\Omega$, then

$$d\sigma = d\Omega \frac{k'^2}{(2\pi)^2} \frac{E_1' E_2' E_1 E_2 |\mathcal{M}|^2}{k'(E_1' + E_2') k (E_1 + E_2) 16 E_1 E_1' E_2 E_2'} \,,$$

where the energy conserving delta function gives $k' = k$ and $E' = E$. Hence, for elastic scattering of spinless particles in the CM system,

$$\boxed{\frac{d\sigma}{d\Omega} = \frac{|\mathcal{M}|^2}{(2\pi)^2 16 (E_1 + E_2)^2}} \ . \tag{9.54}$$

Evaluation of Cross Section in the Laboratory Frame

We return to Eq. (9.50) and show that $d\sigma$ is covariant, at least in all reference frames colinear with the velocity of the incident particle.

To show this, note that

$$\int \frac{d^3p'}{2E_2'} = \int d^4p' \delta_+(m_2^2 - p'^2) \,, \tag{9.55}$$

where $\delta_+(m^2 - p^2) = \delta(m^2 - p^2)\,\theta(p_0)$. This shows that $\int d^3p'/(2E_2')$ is covariant. Also, $\delta^4(p' + k' - p - k)$ is obviously covariant, and $\mathcal{M}$ is covariant. The only remaining factors in the cross section are $E_1 E_2 v$, which can be written

$$\begin{aligned} E_1 E_2 v &= E_1 E_2 \left(\frac{k}{E_1} + \frac{p}{E_2}\right) \\ &= E_2 k + E_1 p = \sqrt{(E_2 k + E_1 p)^2} \\ &= \sqrt{E_2^2 k^2 + E_1^2 p^2 + 2 E_1 E_2 k p} \ . \end{aligned}$$

Compare this to

$$\sqrt{(k \cdot p)^2 - m_1^2 m_2^2} = \sqrt{(E_1 E_2 + kp)^2 - m_1^2 m_2^2}$$
$$= \sqrt{E_1^2 p^2 + E_2^2 k^2 + 2E_1 E_2 kp} \ .$$

Hence, for any colinear frame,

$$E_1 E_2 v = \sqrt{(k \cdot p)^2 - m_1^2 m_2^2} \ , \tag{9.56}$$

which demonstrates the covariance of $E_1 E_2 v$, and hence $d\sigma$.

Now evaluate $d\sigma$ in the laboratory (LAB) frame, where m_2 is at rest. The initial expression is

$$d\sigma = d\Omega \int \frac{k'^2\, dk'\, \delta(E_1 + m_2 - E_1' - E_2')|\mathcal{M}|^2}{(2\pi)^2 \left(\frac{k}{E_1}\right) 16 E_1 m_2 E_1' E_2'} \ . \tag{9.57}$$

The δ-function is now far from trivial. Recalling that θ is the scattering angle in the LAB system and using

$$(E_1' + E_2') = \sqrt{m_1^2 + k'^2} + \sqrt{m_2^2 + p'^2} \ , \tag{9.58}$$

where, by three-momentum conservation,

$$p'^2 = (\boldsymbol{k} - \boldsymbol{k}')^2 = (k' \sin\theta)^2 + (k - k'\cos\theta)^2$$
$$= k'^2 + k^2 - 2kk'\cos\theta \ , \tag{9.59}$$

gives

$$E_1' E_2' \frac{d(E_1' + E_2')}{dk'}\bigg|_{\theta \text{ fixed}} = E_2' k' + E_1'(k' - k\cos\theta)$$
$$= (E_1 + m_2 - E_1')k' + E_1'(k' - k\cos\theta)$$
$$= (E_1 + m_2)k' - E_1' k \cos\theta \ . \tag{9.60}$$

It turns out to be convenient to express this in terms of the square of the four-momentum transfer, which can be written in many different ways and will be useful in later applications:

$$q^2 = (k' - k)^2 = 2m_1^2 - 2E_1 E_1' + 2kk'\cos\theta$$
$$= (p' - p)^2 = 2m_2^2 - 2m_2 E_2' = 2m_2(m_2 - E_2')$$
$$= 2m_2(E_1' - E_1) \ . \tag{9.61}$$

Hence:

$$\begin{aligned}
E_1'E_2'\frac{d(E_1'+E_2')}{dk'} &= (E_1+m_2)k' - \frac{E_1'}{2k'}\left(q^2 - 2m_1^2 + 2E_1E_1'\right) \\
&= (E_1+m_2)k' - \frac{q^2E_1'}{2k'} + \frac{E_1'm_1^2}{k'} - \frac{E_1}{k'}(m_1^2+k'^2) \\
&= m_2k' - \frac{q^2E_1'}{2k'} + \frac{m_1^2}{k'}(E_1'-E_1) \\
&= m_2k'\left[1 - \frac{q^2}{2m_2k'^2}\left(E_1' - \frac{m_1^2}{m_2}\right)\right] \quad . \qquad (9.62)
\end{aligned}$$

The *exact formula* for the elastic cross section in the LAB frame, in which m_2 is at rest, is therefore

$$\left.\frac{d\sigma}{d\Omega}\right|_{\text{LAB}} = \frac{1}{(2\pi)^2}\left(\frac{k'}{k}\right)\frac{|\mathcal{M}|^2}{16m_2^2\left[1 - \dfrac{q^2}{2m_2k'^2}\left(E_1' - \dfrac{m_1^2}{m_2}\right)\right]} \quad . \qquad (9.63)$$

This exact formula is rarely used, but two limits are extremely important. The first limit occurs when $m_1 \ll m_2 \to \infty$. In this case no energy is transmitted to m_2, $k' = k$, and the term involving q^2 can be neglected. The formula reduces to

$$\frac{d\sigma}{d\Omega} \underset{m_2\to\infty}{\longrightarrow} \frac{|\mathcal{M}|^2}{(2\pi)^2 16m_2^2} \quad . \qquad (9.64)$$

This agrees with the CM result in the same limit,

$$\left.\frac{d\sigma}{d\Omega}\right|_{CM} = \frac{|\mathcal{M}|^2}{(2\pi)^2 16(E_1+E_2)^2} \underset{m_2\to\infty}{\longrightarrow} \frac{|\mathcal{M}|^2}{(2\pi)^2 16m_2^2} \quad , \qquad (9.65)$$

as it must.

The other limit of interest occurs when $k \to \infty$, and therefore we may take $m_1 \to 0$. In this case set $k = E_1$ and $k' = E_1'$, and note that

$$\begin{aligned}
q^2 &= 2m_1^2 - 2E_1E_1' + 2kk'\cos\theta \\
&\to -4E_1E_1'\sin^2\frac{\theta}{2} \quad . \qquad (9.66)
\end{aligned}$$

Then

$$\frac{d\sigma}{d\Omega} \underset{k\to\infty}{\longrightarrow} \frac{1}{(2\pi)^2}\left(\frac{E_1'}{E_1}\right)\frac{|\mathcal{M}|^2}{16m_2^2\left[1 + \dfrac{2E_1}{m_2}\sin^2\dfrac{\theta}{2}\right]} \quad . \qquad (9.67)$$

Note that the recoil factor may now be significant. We will return to this formula later.

9.6 EFFECTIVE NONRELATIVISTIC POTENTIAL

We can use Eqs. (9.29) and (9.54) to relate $\mathcal{M}$ to the nonrelativistic potential (in momentum space). Recall that the nonrelativistic S-matrix [from Schiff (1968), for example] for non-forward scattering is

$$\begin{aligned} S &= -\frac{i}{\hbar} 2\pi\delta(E_f - E_i)\, T(q) \\ &\cong -\frac{i}{\hbar} 2\pi\delta(E_f - E_i)\, \tilde{V}(q) \ , \end{aligned} \tag{9.68}$$

where T is the reduced nonrelativistic scattering amplitude (which plays a role similar to $\mathcal{M}$) and the second equation expresses the result in the first Born approximation. The momentum transfer is $\boldsymbol{q} = (\boldsymbol{k} - \boldsymbol{k}')$, and $\tilde{V}(q)$ is the potential in momentum space

$$\tilde{V}(q) = \int d^3r\, V(r)\, e^{-i\boldsymbol{q}\cdot\boldsymbol{r}} \ . \tag{9.69}$$

In terms of these quantities, the differential cross section is

$$\begin{aligned} \frac{d\sigma}{d\Omega} &= \left(\frac{m}{2\pi\hbar^2}\right)^2 |T(q)|^2 \\ &\cong \left(\frac{m}{2\pi\hbar^2}\right)^2 |\tilde{V}(q)|^2 = \frac{m^2}{(2\pi)^2} |\tilde{V}(q)|^2 \ , \end{aligned} \tag{9.70}$$

where m is the reduced mass,

$$m = \frac{m_1 m_2}{m_1 + m_2} \ . \tag{9.71}$$

To extract the *effective* nonrelativistic potential from the relativistic theory, assume that V should be chosen to give the correct result for $\mathcal{M}$ and $d\sigma/d\Omega$ in the *first Born approximation*. Equating the relativistic formula (9.54) to the nonrelativistic formula (9.70) gives, in the nonrelativistic limit,

$$\tilde{V}(q) = \left(\frac{m_1 + m_2}{E_1 + E_2}\right) \frac{\mathcal{M}}{4m_1 m_2} \underset{\text{nonrel}}{\longrightarrow} \frac{1}{4m_1 m_2} \mathcal{M} \ , \tag{9.72}$$

where the *sign*, which cannot be determined from the cross sections (which involve the square of both quantities), is fixed by a comparison of the nonrelativistic scattering amplitude T, given in Eq. (9.68), and the relativistic scattering amplitude $\mathcal{M}$, defined in Eq. (9.29). In both cases the factor relating these quantities to the S-matrix is a factor of $-i$ (because of Rule 0), so they have the same sign.

Hence the effective coordinate space potential between the charged particles m_1 and m_2, which arises from the *exchange* of the neutral particle in the ϕ^3 theory investigated above, is obtained by Fourier transforming the result (9.38),

$$V(r) = \frac{1}{(2\pi)^3}\int d^3q\, e^{i\boldsymbol{q}\cdot\boldsymbol{r}}\tilde{V}(q)$$
$$= -\frac{\lambda_1\lambda_2}{4m_1m_2}\int \frac{d^3q}{(2\pi)^3} e^{i\boldsymbol{q}\cdot\boldsymbol{r}}\, \frac{1}{\mu^2+q^2}\,, \tag{9.73}$$

where $E_1 - E_1' = 0$ in the CM system. Recalling the Yukawa integral, Eq. (4.67),

$$\int \frac{d^3r}{(2\pi)^3} e^{-i\boldsymbol{q}\cdot\boldsymbol{r}}\, \frac{e^{-\mu r}}{r} = \frac{4\pi}{(2\pi)^3}\, \frac{1}{\mu^2+q^2}\,, \tag{9.74}$$

we obtain

$$\int \frac{d^3q}{(2\pi)^3} e^{i\boldsymbol{q}\cdot\boldsymbol{r}}\, \frac{1}{\mu^2+q^2} = \frac{1}{4\pi}\, \frac{e^{-\mu r}}{r}\,. \tag{9.75}$$

This gives finally

$$\boxed{V(r) = -\left(\frac{\lambda_1\lambda_2}{4\pi}\right)\frac{1}{4m_1m_2}\,\frac{e^{-\mu r}}{r}}\,. \tag{9.76}$$

The famous Yukawa potential therefore arises from the exchange of a particle in field theory and is a natural consequence of the simplest interaction.

There are three general features of this result which are of considerable significance and should be noted for future reference:

- $V(r)$ is attractive. This is a general feature of scalar particle exchange.
- The strength of $V(r)$ depends on the coupling constants $\lambda_1\lambda_2$, which in ϕ^3 theory have the dimensions of energy (mass).
- The range of $V(r)$ depends on the mass of the exchanged particle, μ.

The extraction of effective interactions from field theory is a major industry, and we will soon see how the famous one-pion exchange (OPE) potential is derived. The study of effective interactions is continued in Chapter 12.

9.7 IDENTICAL PARTICLES

In this section we discuss some of the additional complications which arise when identical particles are present in the initial or final state. The ϕ^3 theory introduced in the first section will be used to illustrate the discussion, but all of the results will be quite general.

For identical particles we must be careful how the states are normalized. For a two-particle state, we have

$$|p_1p_2\rangle = \mathcal{N}\, a_1^\dagger a_2^\dagger|0\rangle \quad , \tag{9.77}$$

where $\mathcal{N}$ is a normalization constant to be fixed shortly, and we will adopt the convention that the order of the variables in the "bra" is the same as the order of the operators, which matters only for fermions [recall Eq. (7.28)]. For "kets" we will use the same rule, so that

$$\langle p_2p_1| = \mathcal{N}\langle 0|\, a_2a_1 \quad . \tag{9.78}$$

In either case,

$$|p_1p_2\rangle = \pm|p_2p_1\rangle \qquad \begin{matrix} + \text{ for bosons} \\ - \text{ for fermions.} \end{matrix} \tag{9.79}$$

The normalization constant $\mathcal{N}$ is fixed by the completeness requirement

$$\sum_{p_1p_2} |p_1p_2\rangle\langle p_2p_1| = \mathbf{1} \; , \tag{9.80}$$

which implies

$$\sum_{p_1p_2} \langle k_2k_1|p_1p_2\rangle \, \langle p_2p_1|k_1'k_2'\rangle = \langle k_2k_1|k_1'k_2'\rangle \quad . \tag{9.81}$$

But with identical particles

$$\begin{aligned} \langle k_2k_1|k_1'k_2'\rangle &= \mathcal{N}^2 \left\langle 0|a_2a_1\, a_{1'}^\dagger a_{2'}^\dagger|0\right\rangle \\ &= \mathcal{N}^2 \left[\delta_{k_1k_1'}\delta_{k_2k_2'} \pm \delta_{k_1k_2'}\delta_{k_2k_1'}\right] \quad . \end{aligned} \tag{9.82}$$

Substituting (9.82) into the completeness relation (9.81) and doing the sums over p_1 and p_2 give

$$2\mathcal{N}^4 \left[\delta_{k_1k_1'}\delta_{k_2k_2'} \pm \delta_{k_1k_2'}\delta_{k_2k_1'}\right] = \mathcal{N}^2 \left[\delta_{k_1k_1'}\delta_{k_2k_2'} \pm \delta_{k_1k_2'}\delta_{k_2k_1'}\right] \; ,$$

which leads to the requirement

$$\mathcal{N} = \frac{1}{\sqrt{2}} \quad . \tag{9.83}$$

Now the factor $1/\sqrt{2}$ is inconvenient and can be *omitted* if we use the *convention*, for identical particles only, that $p_1 > p_2$, or in the center of mass system where both momenta always have precisely the same magnitude, $\boldsymbol{p}_1$ will be the momentum which has a polar angle $\theta_1 \le \pi/2$. In this case the second Kronecker delta $\delta_{k_1k_2'}\delta_{k_2k_1'} = 0$, and the above derivation would give $\mathcal{N} = 1$. *This is the convention we will adopt.* This convention can be extended to more that one identical particle, in which case the momenta are restricted by $p_1 > p_2 > \cdots p_n$, or $\theta_1 < \theta_2 \cdots \theta_n$; with this restriction the n-particle normalization also equals unity.

This convention requires care when integrating over final states in the calculation of cross sections or decay rates. If we choose to integrate over the *full* solid angle 4π then, for a two-body final state, we are counting both $p_1 \geq p_2$ *and* $p_2 \geq p_1$, so we *must divide by 2.* [Note that this factor would arise naturally if we had used $\mathcal{N} = 1/\sqrt{2}$ for final states (which is appropriate if we place no restriction on the final momenta), giving a factor of $\mathcal{N}^2 = \frac{1}{2}$ in cross sections and decay rates.] For an n-body final state, we can again integrate over all final momenta *if we divide by a factor of* $n!$. These factors are called *statistical factors.* In summary, our convention for the treatment of identical particles involves two new rules:

- for identical particles take $\mathcal{N} = 1$ and order the momenta

$$p_1 > p_2 > \cdots > p_n \ ,$$

- when integrating over final states, ignore the ordering and divide by the statistical factor $n!$.

With these rules in mind, we will now calculate the elastic scattering of two particles of type-1 (i.e., $1+1 \to 1+1$) to lowest order in ϕ^3 theory. Our beginning is very similar to Eq. (9.23) for the S-matrix,

$$S = \frac{(-i)^2}{2}\lambda_1^2 \left\langle p_2' \, p_1' \left| \int d^4x_1 \, d^4x_2 : \Phi_1^\dagger(x_1)\Phi_1(x_1) :: \Phi_1^\dagger(x_2)\Phi_1(x_2) : \right| p_1 \, p_2 \right\rangle \\ \times \ \langle 0|T\left(\phi(x_1)\phi(x_2)\right)|0\rangle \ , \qquad (9.84)$$

except that now there is only one term (the square of the λ_1 interaction term) which can account for the scattering, so the factor of $\frac{1}{2}$ remains. Also, since the initial state contains two creation operators, $a^\dagger_{1p_1} a^\dagger_{1p_2}|0\rangle$, and the final state two annihilation operators, $\langle 0|a_{1p_2'} a_{1p_1'}$, we will need both of the annihilation operators contained in $\Phi(x_1)$ and $\Phi(x_2)$ and both of the creation operators contained in $\Phi^\dagger(x_1)$ and $\Phi^\dagger(x_2)$ to "balance" the a's and $a^\dagger$'s in the initial and final states and give a non-zero result. Specifically, in place of (9.24) we need the following matrix element:

$$\left\langle p_2' \, p_1' \left| a^\dagger_{1n} a_{1m} a^\dagger_{1i} a_{1j} \right| p_1 \, p_2 \right\rangle = \left\langle 0 \left| a_{1p_2'} a_{1p_1'} a^\dagger_{1n} a_{1m} a^\dagger_{1i} a_{1j} a^\dagger_{1p_1} a^\dagger_{1p_2} \right| 0 \right\rangle \\ = \left(\delta_{p_1'n}\delta_{p_2'i} + \delta_{p_1'i}\delta_{p_2'n}\right)\left(\delta_{mp_1}\delta_{jp_2} + \delta_{mp_2}\delta_{jp_1}\right) \ , \qquad (9.85)$$

which displays all the pairings which are possible when the particles are identical. Note that as the creation operator a_{1m} is moved to the left, for example, a term δ_{mi} arising from the commutator $[a_{1m}, a^\dagger_{1i}]$ will appear, but all such terms can be ignored because they will eventually require the momenta in the initial and final states to be equal, and we are specifically excluding forward scattering from consideration. Substituting (9.85) into (9.84) and inserting the wave functions give four terms:

$$S = -\frac{1}{2}\lambda_1^2 \int \frac{d^4x_1 d^4x_2}{L^6\sqrt{16E_1E_2E_1'E_2'}} \langle 0|T\phi(x_1)\phi(x_2)|0\rangle \times \left(e^{i(p_1'-p_1)\cdot x_1 + i(p_2'-p_2)\cdot x_2} + e^{i(p_1'-p_1)\cdot x_2 + i(p_2'-p_2)\cdot x_1} + e^{i(p_1'-p_2)\cdot x_1 + i(p_2'-p_1)\cdot x_2} + e^{i(p_1'-p_2)\cdot x_2 + i(p_2'-p_1)\cdot x_1}\right). \quad (9.86)$$

Using the fact that the propagator depends only on $x_1 - x_2$ and introducing the sum and difference variables R and r as we did in Eq. (9.26) allow us to extract the $\mathcal{M}$-matrix as we did before. The four terms collapse in two different terms, each multiplied by 2, giving

$$\mathcal{M} = -\lambda_1^2 \int d^4r \, \langle 0|T(\phi(r)\phi(0))|0\rangle \left(e^{i(p_1'-p_1)\cdot r} + e^{i(p_2'-p_1)\cdot r}\right) = \frac{-\lambda_1^2}{\mu^2 + (p_1'-p_1)^2} + \frac{-\lambda_1^2}{\mu^2 + (p_2'-p_1)^2}. \quad (9.87)$$

The first term is the same as the result we obtained when the particles were not identical, and the second term is obtained from the first by *symmetrizing the final state* (or the initial state, but we will adopt the convention that the final state is to be symmetrized). This illustrates a new Feynman rule, Rule 4 (as they are numbered in Appendix B):

Rule 4: symmetrize between identical bosons in the final state.

Let's briefly review the calculation and see where the two terms given in (9.87) came from. For this purpose it is best to think of the interaction unfolding in space–time, where, according to (9.84), one interaction takes place at point x_1 and the other at x_2, where the points x_1 and x_2 will be *defined* to be the points of arrival of the incoming particles with momentum p_1 and p_2, respectively. The propagator describes the propagation of the neutral particle between these two interaction points. Now, because the interactions at x_1 and x_2 involve the *same particles*, the particle in the final state with momentum p_1' could emerge equally well from *either of these two points*, as illustrated in Fig. 9.6, and this explains why both of the diagrams shown in Fig. 9.6 must be present.

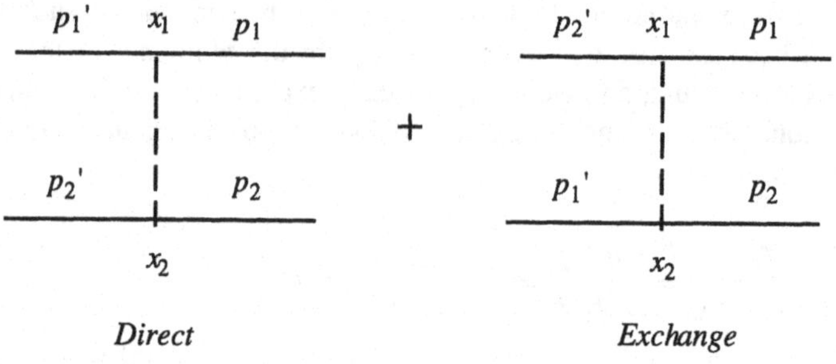

Fig. 9.6 If the particles are identical, the final particle with momentum p_1' can emerge from either of the interaction points x_1 or x_2, and hence there must be two diagrams as discussed in the text.

As a final example, consider the elastic scattering of two identical neutral particles, $\mu + \mu \to \mu + \mu$. This scattering is described by the third term in the interaction Hamiltonian (9.6), and the S-matrix is

$$S = \frac{(-i)^2}{2}\frac{\lambda^2}{(3!)^2}\left\langle p_2' \, p_1' \middle| \int d^4x_1 \, d^4x_2 \, T\left(:\phi^3(x_1)::\phi^3(x_2):\right) \middle| p_1 \, p_2 \right\rangle , \tag{9.88}$$

where now the initial and final states contain the creation and annihilation operators of the neutral ϕ field. As before, two of the fields in (9.88) must pair to produce a propagator which connects the space–time points at x_1 and x_2 (otherwise the matrix element will be zero), but now all the fields are identical, so *any field from either interaction term may be used to construct the propagator.* There are therefore $3 \times 3 = 9$ possible pairings to form $\langle 0|T\left(\phi(x_1)\phi(x_2)\right)|0\rangle$, and the S-matrix reduces to

$$S = \frac{(-i)^2}{2}\frac{\lambda^2}{4}\left\langle p_2' \, p_1' \middle| \int d^4x_1 \, d^4x_2 \, T\left(:\phi^2(x_1)::\phi^2(x_2):\right) \middle| p_1 \, p_2 \right\rangle \times \langle 0|T\left(\phi(x_1)\phi(x_2)\right)|0\rangle . \tag{9.89}$$

However, since all the fields are Hermitian, each contains the same operators, and therefore a completely new possibility emerges; it is now possible for the two annihilation operators in the $\phi^2(x_2)$ term (for example) to balance the two creation operators in the initial state and for the two creation operators in the $\phi^2(x_1)$ term to balance the two annihilation operators in the final state. [This was not possible before because the term $\Phi_1^\dagger(x_2)\Phi_1(x_2)$ contained only *one* term, an $a^\dagger a$ term,

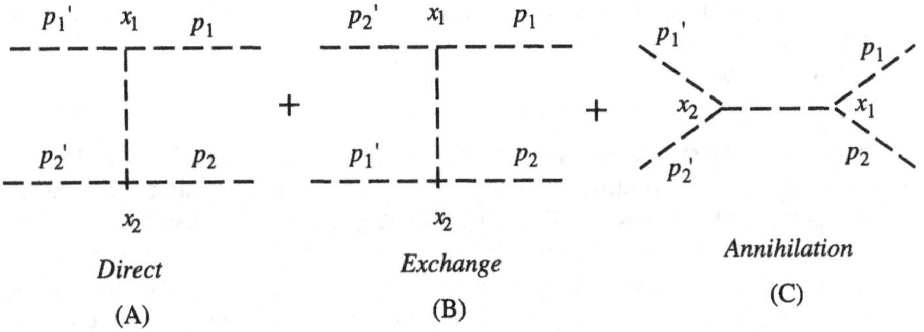

Fig. 9.7 For the symmetric ϕ^3 theory, these three diagrams contribute to second order scattering. Note the (new) annihilation diagram C.

which could balance the initial and final operators.] Furthermore, the $\phi^2(x_2)$ term can pair with the initial state in two ways, or with the final state in two ways, and counting up all of these possibilities gives $2 \times 2 + 2 \times 2 = 8$ possible terms of this kind, which cancels the remaining factor of 8 in Eq. (9.89). A similar factor of 8 emerges for each of the two possible diagrams shown in Fig. 9.6, so the final result for the $\mathcal{M}$-matrix for this example is

$$\mathcal{M} = \frac{-\lambda^2}{\mu^2 + (p_1' - p_1)^2} + \frac{-\lambda^2}{\mu^2 + (p_2' - p_1)^2} + \frac{-\lambda^2}{\mu^2 + (p_2 + p_1)^2} , \tag{9.90}$$

which corresponds to the three diagrams shown in Fig. 9.7.

The new process, Fig. 9.7C, describes the virtual combination of the two incoming particles into a single off-shell particle, followed by its disintegration into the original two particles. While this process is physically very different from the other two, its existence and mathematical form follows from the same Feynman rules we have already obtained, including the requirement that the momentum of the propagating particle be fixed by four-momentum conservation. This example shows that we must understand what processes are possible before we can draw all of the allowed Feynman diagrams.

The reason for inserting the additional factor of 1/3! into the interaction Hamiltonian for the pure ϕ^3 theory is now clear. This factor was eventually canceled by the many combinations of identical terms which led to the same final physical process. However, when these particles appear inside internal *loops* (see Chapter 11), this factor is not canceled completely, and loop diagrams involving such particles carry with them special suppression factors, called *symmetry factors*. These will be discussed in Chapters 11 and 15 and need not concern us now.

With the experience we have acquired, are now ready to study a more realistic problem.

9.8 PION–NUCLEON INTERACTIONS AND ISOSPIN

We now consider the case of two nucleons (neutron and proton) interacting with neutral *and* charged pseudoscalar fields (pions). The Lagrangian density is composed of four terms

$$\mathcal{L} = \mathcal{L}_N + \mathcal{L}_+ + \mathcal{L}_0 + \mathcal{L}_{\text{int}} \,, \tag{9.91}$$

where $\mathcal{L}_N$ is the sum of two non-interacting spin $\frac{1}{2}$ Lagrangian densities for the proton, ψ_p, and the neutron, ψ_n, with $m_n = m_p = m_N$ for simplicity [recall Eq. (7.46)]. The $\mathcal{L}_+$ is the Lagrangian density for a non-interacting *charged* pseudoscalar field, identical to the charged scalar field, Eq. (7.32) [or its operator form Eq. (7.62)]. (The difference between scalar and pseudoscalar fields does not show up until we discuss $\mathcal{L}_{\text{int}}$.) This field will be denoted ϕ_+ and describes charged π^+ pions. Next, $\mathcal{L}_0$ is the Lagrangian density for a non-interacting *neutral* pseudoscalar field, identical to $\mathcal{L}_0$ given in Eq. (9.3). This field is a self-conjugate (Hermitian) field, denoted by ϕ_0, and describes neutral pions, π_0 and $\bar{\pi}_0$, which are identical. For simplicity, we also assume the mass of the charged and neutral pions are equal to m_π.

Hence, the first three Lagrangian densities describe the following four non-interacting fields:

ψ_p :	proton p	mass $= m_N$	+ charge
ψ_n :	neutron n	mass $= m_N$	0 charge
ϕ_+ :	π^+; antiparticle $\bar{\pi}^+ = \pi^-$	mass $= m_\pi$	+ charge
ϕ_0 :	π^0; identical to its antiparticle	mass $= m_\pi$	0 charge.

These four fields interact through $\mathcal{L}_{\text{int}}$, which has a generic ϕ^3 structure. We take

$$\begin{aligned}\mathcal{L}_{\text{int}} = &- ig_p\!:\!\bar{\psi}_p\gamma^5\psi_p\!:\!\phi_0 - ig_n\!:\!\bar{\psi}_n\gamma^5\psi_n\!:\!\phi_0 \\ &- ig_+\!:\!\bar{\psi}_n\gamma^5\psi_p\!:\!\phi_+^\dagger - ig_+\!:\!\bar{\psi}_p\gamma^5\psi_n\!:\!\phi_+ \,,\end{aligned} \tag{9.92}$$

where g_p, g_n, and g_+ are three *real constants*. Note that:

(i) The interaction conserves charge. The neutral field ϕ_0, which describes π^0 mesons, couples to the proton and the neutron with independent coupling constants g_p and g_n. The $p \to n + \pi^+$ and $\pi^+ + n \to p$ charge conserving interactions are described by the two terms with coupling constant g_+.

(ii) $\mathcal{L}_{\text{int}}$ is Hermitian if g_p, g_n, and g_+ are all real. The g_p and g_n terms are individually Hermitian because $\phi_0 = \phi_0^\dagger$. The two g_+ terms must have the same coupling constant to preserve hermiticity. To prove this, recall that $\gamma_0^2 = 1$ and $\gamma^{5\dagger} = \gamma^5$, so that

$$\left(\bar{\psi}_n\gamma^5\psi_p\right)^\dagger = \psi_p^\dagger\gamma^5\bar{\psi}_n^\dagger = \underbrace{\psi_p^\dagger\gamma_0}_{\bar{\psi}_p}\underbrace{\gamma^0\gamma^5\gamma^0}_{-\gamma^5}\psi_n \,.$$

The factor of i therefore makes these terms Hermitian if g_+ is real.

Isospin and $SU(2)$ Symmetry

Now suppose that the neutron and proton are indistinguishable under πN interactions. This means that they can be transformed into each other without altering the interaction. Mathematically, it means that p and n are components of a two-vector

$$|p\rangle = \begin{pmatrix} 1 \\ 0 \end{pmatrix} \qquad |n\rangle = \begin{pmatrix} 0 \\ 1 \end{pmatrix} \tag{9.93}$$

in some abstract two-dimensional space and that the *interaction is invariant under all unitary transformations* in this space. The group of all unitary transformations separates into multiplication by a common phase, the $U(1)$ group, and the remaining group of unitary transformations with unit determinant (a condition which fixes the overall phase of the transformation), the familiar $SU(2)$ group. Each of these groups can be considered separately. The abstract two dimensional space is referred to as *isospin* space, and the $SU(2)$ transformations in this space are called *isospin* transformations. The mathematics of isospin transformations is identical to ordinary spin $\frac{1}{2}$ transformations, but the space is a different, abstract spin space; hence the name isospin.

Denoting the (now) two-component nucleon field by

$$\psi = \begin{pmatrix} \psi_p \\ \psi_n \end{pmatrix} , \tag{9.94}$$

the interaction Lagrangian density can be written

$$\mathcal{L}_{\text{int}} = -i\!:\bar{\psi}\gamma^5\Phi\psi\!: , \tag{9.95}$$

where Φ is now a 2×2 matrix with the form

$$\begin{aligned}\Phi &= g_p\tfrac{1}{2}(1+\tau_3)\phi_0 + g_n\tfrac{1}{2}(1-\tau_3)\phi_0 + g_+\tfrac{1}{2}(\tau_1+i\tau_2)\phi_+ + g_+\tfrac{1}{2}(\tau_1-i\tau_2)\phi_+^\dagger \\ &= \tfrac{1}{2}(g_p+g_n)\phi_0 + \tfrac{1}{2}(g_p-g_n)\tau_3\phi_0 + g_+\tau_1\tfrac{1}{2}(\phi_+ + \phi_+^\dagger) + ig_+\tau_2\tfrac{1}{2}(\phi_+ - \phi_+^\dagger) .\end{aligned} \tag{9.96}$$

It is convenient at this point to introduce two self-conjugate fields ϕ_1 and ϕ_2 defined by

$$\phi_1 = \frac{1}{\sqrt{2}}\left(\phi_+ + \phi_+^\dagger\right) \qquad \phi_2 = \frac{i}{\sqrt{2}}\left(\phi_+ - \phi_+^\dagger\right) . \tag{9.97}$$

Note that ϕ_1 and ϕ_2 commute,

$$[\phi_1(r,t),\phi_1(r',t)] = [\phi_2(r,t),\phi_2(r',t)] = [\phi_1(r,t),\phi_2(r',t)] = 0 , \tag{9.98}$$

and that the CCR's for ϕ_1 (and similarly for ϕ_2) become

$$\begin{aligned}[\pi_1(r,t),\phi_1(r',t)] &= \tfrac{1}{2}\left\{\left[\pi_+^\dagger(r,t),\phi_+^\dagger(r',t)\right] + [\pi_+(r,t),\phi_+(r',t)]\right\} \\ &= -i\delta^3(r-r') ,\end{aligned} \tag{9.99}$$

as expected. Hence, ϕ_1 and ϕ_2 are independent self-conjugate fields similar to $\phi_0 \equiv \phi_3$, and the field matrix Φ can be written

$$\Phi = \tfrac{1}{2}(g_p + g_n)\phi_3 + \tfrac{1}{2}(g_p - g_n)\tau_3\phi_3 + \frac{g_+}{\sqrt{2}}(\tau_1\phi_1 + \tau_2\phi_2) \ . \tag{9.100}$$

Note that the free pion Lagrangian

$$\mathcal{L}_\pi = \mathcal{L}_+ + \mathcal{L}_0 = \partial_\mu\phi_+^\dagger\partial^\mu\phi_+ - m_\pi^2\phi_+^\dagger\phi_+ + \tfrac{1}{2}\left\{\partial_\mu\phi_0\partial^\mu\phi_0 - m_\pi^2\phi_0^2\right\} \tag{9.101}$$

can be simplified if we use

$$\phi_+ = \frac{1}{\sqrt{2}}(\phi_1 - i\phi_2)$$

and

$$\phi_+^\dagger\phi_+ = \tfrac{1}{2}\left(\phi_1^2 + \phi_2^2\right) \tag{9.102}$$

so that

$$\begin{aligned}\mathcal{L}_\pi &= \tfrac{1}{2}\left\{\partial_\mu\phi_1\partial^\mu\phi_1 + \partial_\mu\phi_2\partial^\mu\phi_2 + \partial_\mu\phi_3\partial^\mu\phi_3 - m_\pi^2\left(\phi_1^2 + \phi_2^2 + \phi_3^2\right)\right\} \\ &= \tfrac{1}{2}\left[\partial_\mu\boldsymbol{\phi}\cdot\partial^\mu\boldsymbol{\phi} - m_\pi^2\boldsymbol{\phi}^2\right] \ ,\end{aligned} \tag{9.103}$$

where $\boldsymbol{\phi}$ is a vector in an abstract three-dimensional space (corresponding to a state with isospin 1),

$$\boldsymbol{\phi} = \begin{pmatrix}\phi_1\\ \phi_2\\ \phi_3\end{pmatrix} \qquad \boldsymbol{\phi}\cdot\boldsymbol{\phi} = \phi_1^2 + \phi_2^2 + \phi_3^2 \ . \tag{9.104}$$

Because $\mathcal{L}_\pi$ depends only on the square of the length of the vector $\partial_\mu\boldsymbol{\phi}$, it is clear that it is invariant under rotations of the $\boldsymbol{\phi}$ field in the isospin (one) space.

Now we demand that the full Lagrangian be invariant under isospin rotations, i.e., under transformations of

$$\psi = \begin{pmatrix}\psi_p\\ \psi_n\end{pmatrix} \qquad \text{and} \qquad \boldsymbol{\phi} = \begin{pmatrix}\phi_1\\ \phi_2\\ \phi_3\end{pmatrix} \ .$$

We have just seen that $\mathcal{L}_N + \mathcal{L}_\pi$ is invariant; it only remains to see what requirements must be imposed if $\mathcal{L}_{\rm int}$ is also to be invariant. Invariance of $\mathcal{L}_N$ and $\mathcal{L}_\pi$ requires that the transformation of ψ be unitary in two dimensions and the transformation of $\boldsymbol{\phi}$ be an orthogonal transformation (rotation) in three dimensions. The transformations correspond to different representations of the same rotation.

Using our experience with rotations and angular momentum, we therefore expect (recall Sec. 5.8)

$$\begin{aligned}\psi' &= e^{-i\theta_i \frac{1}{2}\tau_i}\psi \cong \left[1 - i\theta_i \tfrac{1}{2}\tau_i\right]\psi \\ \phi' &= e^{-i\theta_i L^i}\phi \cong \left[1 - i\theta_i L^i\right]\phi\ ,\end{aligned} \tag{9.105}$$

where θ_i are three continuous parameters describing the transformation and $\frac{1}{2}\tau_i$ and L^i are the generators, defined by

$$\begin{aligned}&\left[\tfrac{1}{2}\tau_i, \tfrac{1}{2}\tau_j\right] = i\epsilon_{ijk}\tfrac{1}{2}\tau_k \\ &\left[L_i, L_j\right] = i\epsilon_{ijk}L_k \qquad \left(L_i\right)_{jk} = -i\epsilon_{ijk}\ .\end{aligned} \tag{9.106}$$

These are the familiar algebraic properties of $SU(2)$ symmetry.

The requirement that the interaction Lagrangian be invariant to lowest order implies

$$\begin{aligned}\bar\psi'\gamma^5\Phi'\psi' &= \bar\psi\gamma^5\Phi\psi \\ &\cong \bar\psi\left[1 + i\theta_i\tfrac{1}{2}\tau_i\right]\gamma^5\left[1 - i\theta_j L^j\right]\Phi\left[1 - i\theta_k\tfrac{1}{2}\tau_k\right]\psi \\ &\cong \bar\psi\gamma^5\Phi\psi + i\tfrac{1}{2}\theta_i\bar\psi\gamma^5\left\{\left[\tau_i, \Phi\right] - 2L^i\Phi\right\}\psi\ .\end{aligned} \tag{9.107}$$

The term proportional to θ_i must be zero, which requires that

$$\begin{aligned}\left[\tau_i, \Phi\right] - 2L^i\Phi =&0 \\ =&\overbrace{\left[\tau_i, \tau_3\right]}^{2i\epsilon_{i3j}\tau_j}\frac{1}{2}(g_p - g_n)\phi + \frac{g_+}{\sqrt{2}}\Big\{\overbrace{\left[\tau_i, \tau_1\right]}^{2i\epsilon_{i1j}\tau_j}\phi_1 + \overbrace{\left[\tau_i, \tau_2\right]}^{2i\epsilon_{i2j}\tau_j}\phi_2\Big\} \\ &+ 2i\epsilon_{i3j}\frac{1}{2}(g_p + g_n)\phi_j + 2i\epsilon_{i3j}(g_p - g_n)\tau_3\phi_j \\ &+ \frac{g_+}{\sqrt{2}}2i\left[\epsilon_{i1j}\tau_1\phi_j + \epsilon_{i2j}\tau_2\phi_j\right]\ .\end{aligned} \tag{9.108}$$

We get three equations, one for each i:

$$\begin{aligned}i = 1 \qquad & -i(g_p - g_n)\tau_2\phi_3 + i\sqrt{2}\,g_+\tau_3\phi_2 + i(g_p + g_n)\phi_2 \\ & \quad - i(g_p - g_n)\tau_3\phi_2 + i\sqrt{2}\,g_+\tau_2\phi_3 = 0 \\ i = 2 \qquad & i(g_p - g_n)\tau_1\phi_3 - i\sqrt{2}\,g_+\tau_3\phi_1 + i(g_p + g_n)\phi_1 \\ & \quad + i(g_p - g_n)\tau_3\phi_1 - i\sqrt{2}\,g_+\tau_1\phi_3 = 0 \\ i = 3 \qquad & i\sqrt{2}\,g_+\left[\tau_2\phi_1 - \tau_1\phi_2\right] + i\sqrt{2}\,g_+\left[\tau_1\phi_2 - \tau_2\phi_1\right] = 0\ .\end{aligned} \tag{9.109}$$

The third equation is satisfied identically; the first two are satisfied for *any* ϕ_i if

$$g_p - g_n = \sqrt{2}\,g_+ \qquad g_p + g_n = 0\ . \tag{9.110}$$

Hence

$$g_p = -g_n = \frac{g_+}{\sqrt{2}} \equiv g \tag{9.111}$$

and the interaction can be written

$$\boxed{\mathcal{L}_{\rm int} = -ig\!:\bar{\psi}\gamma^5\boldsymbol{\tau}\psi\cdot\boldsymbol{\phi}: \ .} \tag{9.112}$$

This structure is obvious once we recall that $\bar{\psi}\boldsymbol{\tau}\psi$ must transform like a vector (by analogy with ordinary spin), so its contraction with $\boldsymbol{\phi}$ will be a scalar.

9.9 ONE-PION EXCHANGE

We are now ready to calculate the one-pion exchange (OPE) diagrams in perturbation theory. Since this is the first time we have dealt with spinor fields, we will work out all the details carefully.

The basic structure of the second order result for non-forward scattering is identical to the results we worked out for ϕ^3 theory. The S-matrix is

$$\begin{aligned} S =& \frac{(-i)^2}{2}(ig)^2\langle p_2's_2', p_1's_1'| \int d^4x_1\, d^4x_2 \\ &\times T\left\{:\bar{\psi}(x_1)\gamma^5\tau_i\psi(x_1)\phi_i(x_1)::\bar{\psi}(x_2)\gamma^5\tau_j\psi(x_2)\phi_j(x_2):\right\}|p_1s_1, p_2s_2\rangle\,, \end{aligned} \tag{9.113}$$

where the s_i are the spin quantum numbers of the nucleons, the two particle states are

$$\begin{aligned} |p_1s_1, p_2s_2\rangle &= b^\dagger_{p_1,s_1} b^\dagger_{p_2,s_2}|0\rangle \\ \langle p_2's_2', p_1's_1'| &= \langle 0| b_{p_2',s_2'} b_{p_1',s_1'}\ , \end{aligned}$$

and the field expansions are

$$\begin{aligned} \psi(x) &= \sum_{k,s}\sqrt{\frac{1}{2E_kL^3}}\left\{b_{k,s}u(\boldsymbol{k},s)\,e^{-ik\cdot x} + d^\dagger_{k,s}v(\boldsymbol{k},s)\,e^{ik\cdot x}\right\} \\ \bar{\psi}(x) &= \sum_{k,s}\sqrt{\frac{1}{2E_kL^3}}\left\{b^\dagger_{k,s}\bar{u}(\boldsymbol{k},s)\,e^{ik\cdot x} + d_{k,s}\bar{v}(\boldsymbol{k},s)\,e^{-ik\cdot x}\right\}\ . \end{aligned}$$

Throughout this discussion we will suppress the fact that the nucleon field ψ and the Dirac spinors u and v are really direct products of *two-component spinors in isospin space* and four-component spinors in Dirac space; the two-component isospin structure is implied but *not written explicitly*.

The evaluation of (9.113) parallels the evaluation of (9.84) quite closely, so the steps will be quite familiar. As in the earlier calculation, only the b and $b^\dagger$ terms from the field expansions will contribute; they are precisely what is needed to "balance" the two $b^\dagger$'s in the initial state and the two b's in the final state, which is required if the matrix element is to be non-zero. One new feature is the anticommutation relations satisfied by the annihilation and creation operators; in place of Eq. (9.85), we have the identity

$$\begin{aligned}
&\left\langle p_2's_2', p_1's_1'\right| b^\dagger_{k_1',r_1'} b_{k_1,r_1} b^\dagger_{k_2',r_2'} b_{k_2,r_2} \left| p_1s_1, p_2s_2 \right\rangle \\
&\quad = \left\langle 0 \right| b_{p_2',s_2'} b_{p_1',s_1'} b^\dagger_{k_1',r_1'} b_{k_1,r_1} b^\dagger_{k_2',r_2'} b_{k_2,r_2} b^\dagger_{p_1,s_1} b^\dagger_{p_2,s_2} \left| 0 \right\rangle \\
&\quad = \left[\delta_{p_2'k_1'}\delta_{s_2'r_1'}\delta_{p_1'k_2'}\delta_{s_1'r_2'} - \delta_{p_1'k_1'}\delta_{s_1'r_1'}\delta_{p_2'k_2'}\delta_{s_2'r_2'}\right] \\
&\qquad \times \left[-\delta_{k_1p_1}\delta_{r_1s_1}\delta_{k_2p_2}\delta_{r_2s_2} + \delta_{k_1p_2}\delta_{r_1s_2}\delta_{k_2p_1}\delta_{r_2s_1}\right] . \qquad (9.114)
\end{aligned}$$

To prove this, note that there are four possible pairings of b's with $b^\dagger$'s and that the signs are different because of the anticommutation relations. For example, two terms with different signs result when $b^\dagger_{k_1'r_1'}$ is moved to the left in the following expression:

$$\begin{aligned}
&\left\langle 0 \right| b_{p_2's_2'} b_{p_1's_1'} b^\dagger_{k_1'r_1'} b_{k_1r_1} b^\dagger_{k_2'r_2'} b_{k_2r_2} \cdots \left| 0 \right\rangle \\
&\quad = \delta_{p_1'k_1'}\delta_{s_1'r_1'} \left\langle 0 \right| b_{p_2's_2'} b_{k_1r_1} b^\dagger_{k_2'r_2'} b_{k_2r_2} \cdots \left| 0 \right\rangle \\
&\qquad - \left\langle 0 \right| b_{p_2's_2'} b^\dagger_{k_1'r_1'} b_{p_1's_1'} b_{k_1r_1} b^\dagger_{k_2'r_2'} b_{k_2r_2} \cdots \left| 0 \right\rangle \\
&\quad = \delta_{p_1'k_1'}\delta_{s_1'r_1'} \left\langle 0 \right| b_{p_2's_2'} b_{k_1r_1} b^\dagger_{k_2'r_2'} b_{k_2r_2} \cdots \left| 0 \right\rangle \\
&\qquad - \delta_{p_2'k_1'}\delta_{s_2'r_1'} \left\langle 0 \right| b_{p_1's_1'} b_{k_1r_1} b^\dagger_{k_2'r_2'} b_{k_2r_2} \cdots \left| 0 \right\rangle . \qquad (9.115)
\end{aligned}$$

Using this identity, the second order S-matrix element becomes

$$\begin{aligned}
S = \frac{1}{2} g^2 \int & \frac{d^4x_1 d^4x_2}{L^6\sqrt{16E_1E_2E_1'E_2'}} \left\langle 0 \right| T\phi_i(x_1)\phi_j(x_2) \left| 0 \right\rangle \\
& \times \Big\{ \left[\bar{u}(p_1',s_1')\gamma^5\tau_i u(p_1,s_1)\right] \left[\bar{u}(p_2',s_2')\gamma^5\tau_j u(p_2,s_2)\right] \\
& \qquad \times \left(e^{i(p_1'-p_1)\cdot x_1 + i(p_2'-p_2)\cdot x_2} + e^{i(p_1'-p_1)\cdot x_2 + i(p_2'-p_2)\cdot x_1} \right) \\
& \quad - \left[\bar{u}(p_1',s_1')\gamma^5\tau_i u(p_2,s_2)\right] \left[\bar{u}(p_2',s_2')\gamma^5\tau_j u(p_1,s_1)\right] \\
& \qquad \times \left(e^{i(p_1'-p_2)\cdot x_1 + i(p_2'-p_1)\cdot x_2} + e^{i(p_1'-p_2)\cdot x_2 + i(p_2'-p_1)\cdot x_1} \right) \Big\} , \qquad (9.116)
\end{aligned}$$

where it was assumed that $\langle 0|T\phi_i(x_1)\phi_j(x_2)|0\rangle$ depends on $x_1 - x_2$ only and is symmetric in i, j. But this is clearly the case, because ϕ_i and ϕ_j commute if $j \neq i$, so that

$$\begin{aligned}
\langle 0|T\left(\phi_i(x_1)\phi_j(x_2)\right)|0\rangle &= \delta_{ij} \langle 0|T\left(\phi_1(x)\phi_1(x_2)\right)|0\rangle \\
&= \int \frac{d^4k}{(2\pi)^4} e^{-ik\cdot(x_1-x_2)} \left[\frac{-i\delta_{ij}}{m_\pi^2 - k^2 - i\epsilon}\right] , \qquad (9.117)
\end{aligned}$$

where, in the first line, we used the fact that the time ordered product can be evaluated using *any one* of the fields ϕ_i (because they all give the same answer) and the last step follows from Eq. (9.37). Finally, we call attention to the sign difference between the first and second term in Eq. (9.116), which is due to the anticommutation relations.

We now can simplify the result (9.116) for S if we introduce

$$\begin{aligned} X &= \tfrac{1}{2}(x_1 + x_2) \qquad & x_1 &= X + \tfrac{1}{2}x \\ x &= x_1 - x_2 \qquad & x_2 &= X - \tfrac{1}{2}x \; . \end{aligned} \tag{9.118}$$

With this substitution we can integrate out the overall energy conserving δ-function, and separate the $\mathcal{M}$-matrix, defined in Eq. (9.12). This gives the following result:

$$\begin{aligned} \mathcal{M} =& i(-i)g^2 \frac{\left[\bar{u}(\boldsymbol{p}_1', s_1')\gamma^5 \tau_i u(\boldsymbol{p}_1, s_1)\right] \left[\bar{u}(\boldsymbol{p}_2', s_2')\gamma^5 \tau_i u(\boldsymbol{p}_2, s_2)\right]}{m_\pi^2 - (p_1 - p_1')^2 - i\epsilon} \\ & - \text{exchange term with } \{p_1' s_1'\} \leftrightarrow \{p_2' s_2'\} \; . \end{aligned} \tag{9.119}$$

This result leads to the two Feynman diagrams shown in Fig. 9.6, except that in this case the states are *antisymmetric* so the second diagram comes in with a *minus* sign. The following Feynman rules apply to this example:

Rule 0: a factor of i.

Rule 1: the operator $g\tau_i\gamma^5$ at each πNN vertex.

Rule 2: a factor of

$$\frac{-i\delta_{ij}}{m_\pi^2 - k^2 - i\epsilon}$$

for each pion propagator with four-momentum k and isospin indices i, j. (Fix k by momentum conservation.)

Rule 3: for fermions, assemble the incoming fermion spinors, vertex operators, and outgoing fermion spinors in *order* along each fermion line to make a well-formed matrix element. In particular:

- multiply from the *left* by $\bar{u}(\boldsymbol{p}, s)$ for each *outgoing fermion* with momentum $\boldsymbol{p}$ and spin s.
- multiply from the *right* by $u(\boldsymbol{p}, s)$ for each *incoming fermion* with momentum $\boldsymbol{p}$ and spin s.

Rule 4: antisymmetrize between identical fermions in the final state.

Except for Rule 3 we have encountered versions of all of these rules before. Note that Rule 1 assumes a different form because of the spin–isospin dependence of the πNN interaction, and Rule 2 includes the isospin of the exchanged pion. Rule 3 is new; in cases where the initial or final state particles have spin or other quantum numbers, there will always be a spinor or a vector carrying this information which will be a part of the $\mathcal{M}$-matrix.

The OPE Potential

We saw in Sec. 9.6 how the effective potential for two equal mass particles of mass m_N can be related to the nonrelativistic limit of the $\mathcal{M}$-matrix,

$$\tilde{V}(q) \simeq \frac{1}{4m_N^2}\mathcal{M} \ .$$

Hence the famous nuclear OPE potential can be obtained from Eq. (9.119) by taking the Fourier transform of its nonrelativistic limit. In the CM system, where $E_1 = E_1' = E$, etc., the Dirac matrix elements become

$$\begin{aligned}
\bar{u}(\boldsymbol{p}_1', s_1')\gamma^5 u(\boldsymbol{p}_1, s_1) &= (E+m_N)\ \chi_1'^{\dagger} \begin{pmatrix} 1 & \frac{-\boldsymbol{\sigma}\cdot\boldsymbol{p}_1'}{E+m_N} \end{pmatrix} \begin{pmatrix} 0 & 1 \\ 1 & 0 \end{pmatrix} \begin{pmatrix} 1 \\ \frac{\boldsymbol{\sigma}\cdot\boldsymbol{p}_1}{E+m_N} \end{pmatrix} \chi_1 \\
&= (E+m_N)\,\chi_1'^{\dagger}\left[\frac{\boldsymbol{\sigma}\cdot\boldsymbol{p}_1}{E+m_N} - \frac{\boldsymbol{\sigma}\cdot\boldsymbol{p}_1'}{E+m_N}\right]\chi_1 \\
&= \chi_1'^{\dagger}\boldsymbol{\sigma}\cdot\boldsymbol{q}\chi_1 = \boldsymbol{\sigma}_1\cdot\boldsymbol{q} \ ,
\end{aligned} \tag{9.120}$$

where we use the compact notation $\boldsymbol{\sigma}_1 = \chi_1'^{\dagger}\boldsymbol{\sigma}\chi_1$. In this notation,

$$\tilde{V}(q) = -\frac{g^2(\boldsymbol{\tau}_1\cdot\boldsymbol{\tau}_2)}{(2m_N)^2}\,\frac{\boldsymbol{\sigma}_1\cdot\boldsymbol{q}\,\boldsymbol{\sigma}_2\cdot\boldsymbol{q}}{m_\pi^2+\boldsymbol{q}^2} \quad - \quad \text{(exchange term)}, \tag{9.121}$$

where the minus sign in the first (direct) term comes from $\boldsymbol{\sigma}_2\cdot(\boldsymbol{p}_2 - \boldsymbol{p}_2') = -\boldsymbol{\sigma}\cdot\boldsymbol{q}$ and $\boldsymbol{\tau}_1$ is the matrix element of the isospin operators $\boldsymbol{\tau}$ for nucleon number 1. Then the direct term in the potential, V_D, is

$$\begin{aligned}
V_D(r) &= \int \frac{d^3q}{(2\pi)^3} e^{i\boldsymbol{q}\cdot\boldsymbol{r}}\,\tilde{V}_D(q) \\
&= \frac{g^2(\boldsymbol{\tau}_1\cdot\boldsymbol{\tau}_2)}{(2m_N)^2}\,\boldsymbol{\sigma}_1\cdot\nabla\,\boldsymbol{\sigma}_2\cdot\nabla \int \frac{d^3q}{(2\pi)^3} e^{i\boldsymbol{q}\cdot\boldsymbol{r}}\,\frac{1}{m_\pi^2+\boldsymbol{q}^2} \\
&= \frac{g^2(\boldsymbol{\tau}_1\cdot\boldsymbol{\tau}_2)}{4\pi(2m_N)^2}\,\boldsymbol{\sigma}_1\cdot\nabla\,\boldsymbol{\sigma}_2\cdot\nabla\left(\frac{e^{-m_\pi r}}{r}\right) \ .
\end{aligned}$$

Carrying out the differentiation gives

$$\begin{aligned}
V_D(r) =& \frac{g^2}{4\pi}\,\frac{m_\pi^2}{12m_N^2}(\boldsymbol{\tau}_1\cdot\boldsymbol{\tau}_2) \\
&\times\Bigg\{\underbrace{\boldsymbol{\sigma}_1\cdot\boldsymbol{\sigma}_2}_{\text{central}}\,\frac{e^{-m_\pi r}}{r} \\
&+ \underbrace{(3\,\boldsymbol{\sigma}_1\cdot\hat{\boldsymbol{r}}\,\boldsymbol{\sigma}_2\cdot\hat{\boldsymbol{r}} - \boldsymbol{\sigma}_1\cdot\boldsymbol{\sigma}_2)}_{\text{tensor}}\,\frac{e^{-m_\pi r}}{r}\left(1+\frac{3}{m_\pi r}+\frac{3}{(m_\pi r)^2}\right)\Bigg\} \ .
\end{aligned} \tag{9.122}$$

This is the famous OPE potential with central and tensor parts. The full potential is the direct term minus the exchange term, as indicated in Eq. (9.121).

Remarks and Phenomenology

(i) The angular dependent part is referred to as the tensor force. Its structure,

$$S_{12} = 3\boldsymbol{\sigma}_1 \cdot \hat{r}\, \boldsymbol{\sigma}_2 \cdot \hat{r} - \boldsymbol{\sigma}_1 \cdot \boldsymbol{\sigma}_2 \ , \tag{9.123}$$

clearly displays the $L = 2$ character of the tensor force and shows that it averages to zero when integrated over solid angles,

$$\int d\Omega\, S_{12} = 0 \ . \tag{9.124}$$

Note also that the tensor potential is highly singular at short distances ($\sim \frac{1}{r^3}$).

(ii) The properties of the central potential can be inferred from the values of $\boldsymbol{\sigma}_1 \cdot \boldsymbol{\sigma}_2$ and $\boldsymbol{\tau}_1 \cdot \boldsymbol{\tau}_2$. If the total spin of the state is S and the total isospin is I, then

$$\begin{aligned} \boldsymbol{\sigma}_1 \cdot \boldsymbol{\sigma}_2 &= 2\left[S(S+1) - \tfrac{3}{2}\right] = \begin{cases} 1 & S = 1 \\ -3 & S = 0 \end{cases} \\ \boldsymbol{\tau}_1 \cdot \boldsymbol{\tau}_2 &= \begin{cases} 1 & I = 1 \\ -3 & I = 0 \ . \end{cases} \end{aligned} \tag{9.125}$$

Hence the *central part* of the potential is attractive and has the same strength in the two S-states, but is repulsive in the P-states. Using the spectroscopic notation introduced in Sec. 8.6,

$$\left.\begin{array}{lll} {}^1S_0: & S = 0,\ I = 1 & (\boldsymbol{\sigma}_1 \cdot \boldsymbol{\sigma}_2)(\boldsymbol{\tau}_1 \cdot \boldsymbol{\tau}_2) = -3 \\ {}^3S_1: & S = 1,\ I = 0 & (\boldsymbol{\sigma}_1 \cdot \boldsymbol{\sigma}_2)(\boldsymbol{\tau}_1 \cdot \boldsymbol{\tau}_2) = -3 \end{array}\right\} \text{attractive}$$

$$\left.\begin{array}{lll} {}^1P_1: & S = 0,\ I = 0 & (\boldsymbol{\sigma}_1 \cdot \boldsymbol{\sigma}_2)(\boldsymbol{\tau}_1 \cdot \boldsymbol{\tau}_2) = \ \ 9 \\ {}^3P_0, {}^3P_1, {}^3P_2: & S = 1,\ I = 1 & (\boldsymbol{\sigma}_1 \cdot \boldsymbol{\sigma}_2)(\boldsymbol{\tau}_1 \cdot \boldsymbol{\tau}_2) = \ \ 1 \end{array}\right\} \text{repulsive.}$$

(iii) The central force obtained from the OPE is about 10 times smaller than the central force which is inferred from a phenomenological analysis of NN scattering data. Furthermore, the empirical central force is stronger (more attractive) in ${}^1S_0$ than it is in ${}^3S_1$, yet the only bound state, the deuteron, is a mixture of ${}^3S_1 - {}^3D_1$. The reason that the ${}^3S_1 - {}^3D_1$ channel is more tightly bound than the ${}^1S_0$ channel is due to the tensor force, which is attractive in the ${}^3S_1 - {}^3D_1$ channel (but zero in the ${}^1S_0$ channel) and provides the necessary binding. It turns out that the tensor force is well described by the OPE potential.

The study of the nuclear force continues to be a problem of current research. The approach discussed in this section can be extended by adding the exchange of other mesons; such a model is referred to as a one-boson exchange (OBE) model. It has been found that the force can be very well parameterized by an OBE model with pion, scalar, and vector meson exchanges.

9.10 ELECTROWEAK DECAYS

For a final example, we consider the decay of the *charged* pion. This decay takes place through the *weak charged current*, which couples the charged pion with leptons. First, we will discuss these charged currents, and then we will compute the decay of the π^+.

The weak interactions (which are unified with the electromagnetic interactions, as discussed in Sec. 15.4) are now known to include both charged and neutral currents, but in this section we will discuss the charged currents only. These are interactions of the generic ϕ^3 structure, and add the following interaction term to the Lagrangian:

$$\mathcal{L}_{\rm int}^{W\pm} = -g_{\rm eff}\left(J_\mu^W W^{\dagger\mu} + J_\mu^{W\dagger} W^\mu\right) , \tag{9.126}$$

where $g_{\rm eff}$ is the effective weak coupling constant,* W^μ is a complex (positively charged) vector field, and J_μ^W is the weak charged current. The field W_μ will be described by the free Lagrangian

$$\mathcal{L}_W = -\tfrac{1}{2}F_{\mu\nu}^\dagger F^{\mu\nu} - M_W^2 W_\mu^\dagger W^\mu \quad . \tag{9.127}$$

Note that this Lagrangian density is almost identical to the one introduced in Sec. 2.5, except that it is constructed from complex vector fields and, as in the charged ϕ^3 theory, must therefore be regarded as describing two real fields. Hence the Lagrangian density (9.127) is twice as large as the neutral density (2.39) (see Prob. 7.3). Other aspects of the massive spin one theory are the same; in particular, the W_μ field satisfies the Lorentz condition and the wave equation with mass term

$$\begin{aligned} \partial_\mu W^\mu &= 0 \\ \left(\Box + M_W^2\right) W^\mu &= 0 \ , \end{aligned} \tag{9.128}$$

and the fields are described by an expansion similar to (2.45), except that the charged field is not self-conjugate. We have

$$\begin{aligned} W^\mu &= \sum_{n,\alpha} \frac{1}{\sqrt{2E_n L^3}} \left\{\epsilon_n^{\alpha\mu} A_{n,\alpha}\, e^{-ik_n\cdot x} + \epsilon_n^{\alpha\mu\,*} C_{n,\alpha}^\dagger e^{ik_n\cdot x}\right\} \\ &= \sum_{n,\alpha} \left\{\epsilon_n^{\alpha\mu} A_{n,\alpha}\Phi_n^{(+)}(x) + \epsilon_n^{\alpha\mu\,*} C_{n,\alpha}^\dagger \Phi_{-n}^{(-)}(x)\right\} , \end{aligned} \tag{9.129}$$

where $\Phi_n^{(\pm)}(x)$ are the plane wave solutions for the W boson.

The weak current is a sum of *hadron* and *lepton* parts:

$$J_\mu^W = J_{H\mu}^W + J_{\ell\mu}^W \ . \tag{9.130}$$

---

*This effective coupling constant is $g_W \cos\theta_W$ in the notation of Sec. 15.4.

The *lepton* part has the famous *vector – axial vector* ($V - A$) structure,

$$J^W_{\ell\mu} = \left[\bar{\psi}_e \gamma_\mu (1-\gamma^5)\psi_{\nu_e}\right] + \left[\bar{\psi}_{\rm muon} \gamma_\mu (1-\gamma^5)\psi_{\nu_\mu}\right] + \cdots \,, \tag{9.131}$$

where there are believed now to be only three terms in the sum (9.131), one for each *generation* of leptons (see the table in Appendix D). In the examples discussed in this section, we will be concerned only with the first two generations, consisting of the electron and the electron neutrino, denoted ν_e, and the muon and muon neutrino, denoted ν_μ. (The muon is often denoted by μ but we will use the subscript "muon" in order to avoid confusion with the vector index μ.) Note that the coupling (9.131) only involves *left-handed neutrinos* (and *right-handed antineutrinos*) and therefore violates parity. There is no role for right-handed neutrinos (or left-handed antineutrinos) in the electroweak interactions, and whether or not they exist, along with the question of whether or not the mass of the neutrino is *exactly* zero, is currently not known.

The most fundamental definition of the hadronic current is in terms of quark fields. However, for phenomenological applications we can express these currents directly in terms of the composite hadrons which are observed in the laboratory. Among these will be contributions from the nucleon, the pion, and other hadrons, but in this section we will discuss the pion contribution only, which can be written

$$J^W_{H\mu} = f_\pi \partial_\mu \phi \,, \tag{9.132}$$

where ϕ is the positively charged pion field and f_π is the famous *pion decay constant*. Since the pion and W fields have dimensions of mass, the constant f_π must also have dimensions of mass, and as defined here its value turns out to be

$$f_\pi = 93.0 \text{ MeV} \ . \tag{9.133}$$

In Chapter 13 we will show how the current (9.132) can be obtained from a particular model, and more generally such a current can be justified directly from the quark structure of the pion, but now we will simply assume that a current of the form (9.132) exists, and use it to study pion decay.

Using the interaction Lagrangian (9.126) with the currents (9.131) and (9.132), the second order matrix element for π^- decay is

$$\begin{aligned} S = (-i)^2 g^2_{\rm eff} f_\pi \int d^4x_1\, d^4x_2 \,\langle p\,\ell\,|\bar{\psi}_e(x_1)\gamma_\mu(1-\gamma^5)\psi_{\nu_e}(x_1)\partial_\nu\phi^\dagger(x_2)|\,q\rangle \\ \times \langle 0\,|T\left(W^{\mu\dagger}(x_1)W^\nu(x_2)\right)|\,0\rangle \,, \end{aligned} \tag{9.134}$$

where $|q\rangle$ is the initial π^- state with four-momentum q and $\langle p\,\ell|$ is the final state of an outgoing electron with momentum p and an electron antineutrino with momentum ℓ. The decay is illustrated in Fig. 9.8. We are forced to consider the decay into leptons because energy–momentum conservation would not permit the decay of a π^- into a real W boson unless its mass were identical to the mass

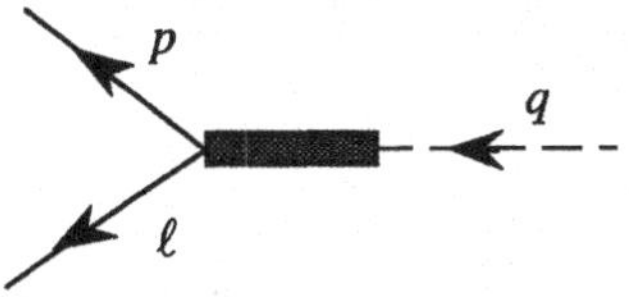

Fig. 9.8 Diagram describing pion decay. The heavy shaded line is the virtual W boson.

of the π^-, which is approximately 139 MeV. In fact, the mass of the W^- is 80.2 GeV, a *very* large mass compared to the mass of the π^-. Because of this, the decay must be second order, with the virtual W connecting the pion to the leptons. Charge conservation requires that the final state be either an electron and an electron–*antineutrino* or a muon and a muon–antineutrino; we will speak of the electron channel, but the calculation for both cases is the same.

The matrix element of the W fields is the propagator and can be reduced as we have done several times before,

$$
\begin{aligned}
&\left\langle 0 \left| T\left(W^{\mu\dagger}(x_1)W^{\nu}(x_2)\right)\right| 0\right\rangle \\
&= \left\langle 0 \left| \left(\sum_{n,\alpha} \epsilon_n^{\alpha\mu} C_{n,\alpha} \Phi_{-n}^{(-)*}(x_1)\right)\left(\sum_{n',\alpha'} \epsilon_{n'}^{\alpha'\nu *} C^{\dagger}_{n',\alpha'} \Phi_{-n'}^{(-)}(x_2)\right)\right| 0\right\rangle \theta(t_1-t_2) \\
&+ \left\langle 0 \left| \left(\sum_{n',\alpha'} \epsilon_{n'}^{\alpha'\nu} A_{n,\alpha} \Phi_{n'}^{(+)}(x_2)\right)\left(\sum_{n,\alpha} \epsilon_{n}^{\alpha\mu *} A^{\dagger}_{n,\alpha} \Phi_{n}^{(+)*}(x_1)\right)\right| 0\right\rangle \theta(t_2-t_1) \\
&= \sum_{n,\alpha} \left[\epsilon_{-n}^{\alpha\mu}\epsilon_{-n}^{\alpha\nu *}\Phi_n^{(-)*}(x_1)\Phi_n^{(-)}(x_2)\theta(t_1-t_2)\right. \\
&\qquad \left. + \, \epsilon_n^{\alpha\nu}\epsilon_n^{\alpha\mu *}\Phi_n^{(+)}(x_2)\Phi_n^{(+)*}(x_1)\theta(t_2-t_1)\right] \; . \qquad (9.135)
\end{aligned}
$$

The new feature is the sum over the polarization states, which is easily done if we remember that $k_n \cdot \epsilon_n = 0$. Hence, the polarization vectors for a boson with momentum k_n in the $\hat{z}$-direction are

$$
\begin{aligned}
\epsilon_n^{1\mu} &= (0,1,0,0) \\
\epsilon_n^{2\mu} &= (0,0,1,0) \\
\epsilon_n^{3\mu} &= \frac{1}{M_W}\left(k_n, 0, 0, E_n\right) \; .
\end{aligned} \qquad (9.136)
$$

These are uniquely defined by the requirements that they be normalized to -1 and that they reduce to the vectors $\hat{x}$, $\hat{y}$, and $\hat{z}$ when the boson is out at rest. Using these definitions, you can show by direct computation (Prob. 9.7) that

$$
\sum_{\alpha} \epsilon_n^{\alpha\mu}\epsilon_n^{\alpha\nu} = -g^{\mu\nu} + \frac{\tilde{k}_n^{\mu}\tilde{k}_n^{\nu}}{M_W^2} \; , \qquad (9.137)
$$

where M_W is the mass of the boson and $\tilde{k}^\mu = (E_n, \boldsymbol{k}_n)$, with $E_n = \sqrt{M_W^2 + \boldsymbol{k}_n^2}$. Hence, going to continuum normalization and using (4.77) to eliminate the θ-functions give immediately

$$\begin{aligned}
&\langle 0 \left| T\left(W^{\mu\dagger}(x_1) W^\nu(x_2)\right) \right| 0 \rangle \\
&= -\int \frac{d^3k}{(2\pi)^3\, 2E} \left(g^{\mu\nu} - \frac{\tilde{k}^\mu \tilde{k}^\nu}{M_W^2}\right) \Big[e^{-i\tilde{k}\cdot(x_1-x_2)}\, \theta(t_1 - t_2) \\
&\qquad\qquad + e^{-i\tilde{k}\cdot(x_1-x_1)}\, \theta(t_2 - t_1)\Big] \\
&= \int \frac{d^4k}{(2\pi)^4} e^{-ik\cdot(x_1-x_2)} \left[\frac{i\left(g^{\mu\nu} - k^\mu k^\nu / M_W^2\right)}{M_W^2 - k^2 - i\epsilon}\right] \quad . \qquad (9.138)
\end{aligned}$$

To obtain this result, we used (4.77), with $n = 0$ for the spatial components and with $n = 2$ for the $(0,0)$ part. This is the general form for the propagator of a massive vector meson.

Now we must reduce the matrix element

$$\langle p\,\ell \left| \theta_{\mu\nu} \right| q \rangle = \langle p\,\ell \left| \bar{\psi}_e(x_1) \gamma_\mu (1 - \gamma^5) \psi_{\nu_e}(x_1) \partial_\nu \phi^+(x_2) \right| q \rangle \quad . \qquad (9.139)$$

Since ψ_{ν_e} destroys electron neutrinos and creates electron antineutrinos, this matrix element describes the decay

$$\pi^- \to e + \bar{\nu}_e \quad . \qquad (9.140)$$

For the pion at rest, this matrix element becomes (letting $L \to 2\pi$)

$$\langle p\,\ell \left| \theta_{\mu\nu} \right| q \rangle = -i \frac{e^{i[(p+\ell)\cdot x_1 - q\cdot x_2]}}{\sqrt{(2\pi)^9\, 8E_e \omega_\nu m_\pi}}\, q_\nu \bar{u}_e(\boldsymbol{p}, \lambda_e) \gamma_\mu (1 - \gamma^5) v_{\lambda_\nu}(\boldsymbol{\ell}, \lambda_\nu) . \qquad (9.141)$$

Inserting (9.141) and (9.138) into (9.134), doing the integration over x_1, x_2, and k, and extracting the $\mathcal{M}$-matrix from the result give

$$\begin{aligned}
\mathcal{M} &= -i g_{\text{eff}}^2 f_\pi\, q_\nu\, \bar{u}_e(\boldsymbol{p}, \lambda_e) \gamma_\mu (1 - \gamma^5) v_{\nu_e}(\boldsymbol{\ell}, \lambda_\nu) \frac{\left[g^{\mu\nu} - q^\mu q^\nu / M_W^2\right]}{M_W^2 - q^2} \\
&= -i \frac{g_{\text{eff}}^2}{M_W^2} f_\pi \bar{u}_e(\boldsymbol{p}, \lambda_e)\, \not{q}\, (1 - \gamma^5) v_{\nu_e}(\boldsymbol{\ell}, \lambda_\nu) \,, \qquad (9.142)
\end{aligned}$$

where the second expression neglects terms of order $m_\pi^2 / M_W^2 \cong 10^{-6}$.

The Feynman diagram for this process has already been drawn in Fig. 9.8. This evaluation illustrates some new Feynman rules:

Rule 1: • a factor of

$$-i g_{\text{eff}} \gamma^\mu (1 - \gamma^5)$$

for each $W^- \to e + \nu_e$ weak vector, where μ is the polarization index carried by the W.

- a factor of

$$-g_{\text{eff}} f_\pi q^\mu$$

for each interaction in which a π^- of momentum q turns into a virtual W^- with polarization index μ.

Rule 2: a factor of

$$\frac{i\left(g^{\mu\nu} - k^\mu k^\nu/M^2\right)}{M^2 - k^2 - i\epsilon} \tag{9.143}$$

for each internal line describing the propagation of a spin one boson with mass M, momentum k, and polarization indices μ and ν.

Rule 3: multiply from the *right* by $v(\boldsymbol{p}, \lambda)$ for each *outgoing antifermion* with momentum $\boldsymbol{p}$ and spin (or helicity) λ.

Using these rules (and Rule 0) it is easy to reconstruct the result (9.142).

The quantity g_{eff}^2/M_W^2 is related to the Fermi coupling constant G,

$$\frac{G}{\sqrt{2}} = \frac{g_{\text{eff}}^2}{M_W^2} = 1.015 \times 10^{-5}\frac{1}{m_P^2} \cong 1.15 \times 10^{-5}\ \text{GeV}^{-2} \ . \tag{9.144}$$

The Fermi constant G was introduced when the weak interactions were first described as a current–current interaction of the form

$$\mathcal{L}_{\text{int}}^{\text{Fermi}} = \frac{G}{\sqrt{2}} J_\mu^{W\dagger} J^{W\mu} \ . \tag{9.145}$$

Since the boson mass M_W is very large, the lowest order results obtained from (9.126) and (9.145) are equivalent as long as the identification (9.144) is made. However, if one tries to calculate higher order corrections, one finds that (9.145) is *unrenormalizable* (see Chapter 16) while (9.126) can be renormalized if the theory is converted into a gauge theory. These issues will be discussed further in Chapter 15.

For now, we take the result (9.144) and compute the decay rate of the pion. Summing over the final spin states and integrating over all momenta, as in Eq. (9.19), give

$$W_\pi = \int \frac{d^3p\, d^3\ell}{(2\pi)^6}\, \frac{(2\pi)^4\delta^4(p+\ell-q)}{8E_p\omega_\ell m_\pi} \sum_{\text{spin}} |\mathcal{M}|^2 \ . \tag{9.146}$$

To evaluate the spin sum over $|\mathcal{M}|^2$, we first simplify $\mathcal{M}$ by using the Dirac equation and the fact that the neutrino mass is zero. Hence

$$\begin{aligned}
\bar{u}_e(\boldsymbol{p}, \lambda_e)\, \not{q}\, (1-\gamma^5) v_{\nu_e}(\boldsymbol{\ell}, \lambda_\nu) &= \bar{u}_e(\boldsymbol{p}, \lambda_e)(\not{p} + \not{\ell})(1-\gamma^5) v_{\nu_e}(\boldsymbol{\ell}, \lambda_\nu) \\
&= m_e \bar{u}_e(\boldsymbol{p}, \lambda_e)(1-\gamma^5) v_{\nu_e}(\boldsymbol{\ell}, \lambda_\nu) \\
&\quad + \bar{u}_e(\boldsymbol{p}, \lambda_e)(1+\gamma^5) \underbrace{\not{\ell}\, v_{\nu_e}(\boldsymbol{\ell}, \lambda_\nu)}_{=0} \\
&= m_e \bar{u}_e(\boldsymbol{p}, \lambda_e)(1-\gamma^5) v_{\nu_e}(\boldsymbol{\ell}, \lambda_\nu) \ .
\end{aligned} \tag{9.147}$$

Next we must square the $\mathcal{M}$-matrix and sum over the spins of the outgoing electron and antineutrino. A simple and elegant way of doing such spin sums will be presented in the next chapter, but in this case it is instructive to do the sum directly using the helicity representation for the spinors given in Sec. 5.11.

First, we must define the electron and antineutrino states correctly. In the center of mass of the pion, we will take the momentum of the electron to be in the $+\hat{z}$-direction and the momentum of the antineutrino to be in the $-\hat{z}$-direction. This means that the antineutrino spinor can be obtained from a spinor initially oriented along the $+\hat{z}$-axis by *rotating the state through an angle* π *about the* $\hat{y}$*-axis* which rotates $+\hat{z}$ into $-\hat{z}$ [other definitions are possible, but this is the conventional one* and is consistent with the definitions (8.50)]. Using the helicity spinor given in Eq. (5.148) and the Dirac rotation operators from (5.116) give

$$\begin{aligned} v_{\nu_e}(\boldsymbol{\ell}, \lambda_\nu) = S_y(R_\pi)\, v_{\nu_e}(\boldsymbol{\ell}_{\hat{z}}, \lambda_\nu) &= \sqrt{\omega_\ell} \begin{pmatrix} 0 & -i\sigma_2 \\ -i\sigma_2 & 0 \end{pmatrix} \begin{pmatrix} -2\lambda_\nu \\ 1 \end{pmatrix} (-i\sigma_2 \chi_{\lambda_\nu}) \\ &= \sqrt{\omega_\ell} \begin{pmatrix} 1 \\ -2\lambda_\nu \end{pmatrix} \chi_{\lambda_\nu} \; . \end{aligned} \tag{9.148}$$

Hence the matrix element is

$$\begin{aligned} &\bar{u}_e(\boldsymbol{p}, \lambda_e)(1-\gamma^5) v_{\nu_e}(\boldsymbol{\ell}, \lambda_\nu) \\ &\quad = 2\delta_{\lambda_\nu,1/2} \sqrt{\omega_\ell (E_p + m_e)}\; \chi^\dagger_{\lambda_e} \begin{pmatrix} 1 & \frac{-2\lambda_e p}{E_p + m_e} \end{pmatrix} \begin{pmatrix} 1 \\ -1 \end{pmatrix} \chi_{1/2} \\ &\quad = 2\delta_{\lambda_\nu,1/2}\, \delta_{\lambda_e,1/2} \sqrt{\frac{\omega_\ell}{E_p + m_e}}\, [E_p + m_e + p] \; . \end{aligned} \tag{9.149}$$

Note that the antineutrino must be right-handed, as we discussed in Sec. 5.11, and hence the $\chi^\dagger_{\lambda_e} \chi_{\frac{1}{2}}$ matrix element in Eq. (9.149) automatically restricts the electron to its right-handed state also (as required by angular momentum conservation). Hence the spin sum contains only one term and reduces to

$$\sum_{\text{spin}} |\mathcal{M}|^2 = \frac{G^2}{2} f_\pi^2 m_e^2 \frac{4\omega_\ell}{E_p + m_e} [E_p + m_e + p]^2 = 4G^2 f_\pi^2 m_e^2\, \omega_\ell\, [E_p + p] \; . \tag{9.150}$$

Hence (9.146) reduces to (using $\omega_\ell = \omega_p = p$)

$$\begin{aligned} W_\pi &= \frac{G^2 f_\pi^2 m_e^2}{2(2\pi)^2} \int \frac{d^3p\, [E_p + p]}{E_p m_\pi}\, \delta(E_p + \omega_p - m_\pi) \\ &= \frac{G^2 f_\pi^2 m_e^2\, p^2}{2\pi m_\pi} = \frac{G^2 f_\pi^2}{8\pi} m_e^2 m_\pi \left(1 - \frac{m_e^2}{m_\pi^2}\right)^2 \; . \end{aligned} \tag{9.151}$$

*For a clear discussion of the definition and properties of helicity states see the classic paper by Jacob and Wick [JW 59].

Note that the ratio of this decay channel to the muon decay channel is

$$\frac{R(\pi^- \to e + \bar{\nu}_e)}{R(\pi^- \to \mu + \bar{\nu}_\mu)} = \frac{m_e^2}{m_\mu^2}\left(\frac{m_\pi^2 - m_e^2}{m_\pi^2 - m_\mu^2}\right) \cong 1.28 \times 10^{-4} \ . \tag{9.152}$$

Hence the decay of the π^- into $\mu + \bar{\nu}_\mu$ is *10,000 times more probable, even though the phase space is much smaller.* [For a discussion of phase space, refer back to Eq. (9.21).] This surprising result is an example of a general feature of vector (and axial vector) interactions. Generalizing the discussion which led to Eq. (9.150), we can show that the coupling of a vector (or axial vector) current to *massless* fermions *conserves helicity*, which in this example means that helicity combinations $\lambda_e + \lambda_\nu$ which add to zero (the total helicity of the initial state) should be favored over those which add to unity. However, in this case the vector current couples to a spin zero state and angular momentum conservation *requires* $\lambda_e + \lambda_\nu = 1$ (i.e., both particles must be either right- or left-handed). The resulting matrix element will therefore be suppressed by the factor $1 - p/(E_p + m) \cong m/E_p$ (if $E_p \gg m$), which is proportional to the mass of the fermion and a measure of the extent to which helicity conservation can be violated by vector interactions. In our discussion, this suppression factor of the fermion mass appeared automatically when we used the Dirac equation to reduce Eq. (9.147).

Since W_π in Eq. (9.151) has dimension of mass, it can be converted to inverse seconds by dividing by $\hbar \cong 6.6 \times 10^{-22}$ MeV/sec. Substituting numbers into the formula gives

$$\tau = \frac{1}{W_{\text{muon}}} \cong \frac{4\pi\, 6.6 \times 10^{-15}}{(1.15)^2(0.093)^2(0.105)^2(0.139)\left(1 - \left(\frac{0.105}{0.139}\right)^2\right)^2}$$
$$= 2.57 \times 10^{-8} \text{ sec.} \ , \tag{9.153}$$

in good agreement with the measured decay rate of 2.60×10^{-8} sec.

PROBLEMS

9.1 Suppose the interaction Hamiltonian contains a term like

$$g\left\{:\phi^\dagger \phi\, \phi_K: + :\phi^\dagger \phi\, \phi_K^\dagger:\right\} \ ,$$

where ϕ is the π^+ meson field and ϕ_K is the field corresponding to a neutral K_0 meson of mass 498 MeV. (The mass of the charged pion is 140 MeV, and anti-K_0 meson, denoted $\bar{K}_0$, is not identical to K_0.) Suppose $g = 4 \times 10^{-4}$ MeV. Compute the lifetime of the K_0 meson assuming that its dominant decay mode is into charged π pairs. (The real K_0 decays are more complicated than described above.)

9.2 The neutral pion, π^0, usually decays into two photons but can also decay into an electron–positron pair. The effective Hamiltonian density for the latter decay is

$$\mathcal{H}_{\rm int} = ig\,\bar{\psi}(x)\gamma^5\psi(x)\,\phi(x) \quad ,$$

where ϕ is the π^0 field and ψ is the electron field.

(a) Show that $\mathcal{H}_I$ is Hermitian.

(b) Calculate the rate for the decay $\pi^0 \to e^- + e^+$. [Note that $m_{\pi^0} = 135$ MeV and $m_e = 0.511$ MeV, so you may approximate $m_e = 0$.]

9.3 [Taken from Sakurai (1967).] Consider the decay of the Λ^0 into a $n + \pi^0$ (which happens about 35% of the time). Represent the Λ^0, which is a neutral Dirac particle, by

$$\psi_{\Lambda_0}(x) = \sum_{\boldsymbol{k},s} \frac{1}{\sqrt{2E_\Lambda L^3}} \left\{ b^\Lambda_{\boldsymbol{k},s} u_\Lambda(\boldsymbol{k},s) e^{-ik\cdot x} + d^{\Lambda\dagger}_{\boldsymbol{k},s} v_\Lambda(\boldsymbol{k},s) e^{ik\cdot x} \right\} \quad ,$$

where u_Λ and v_Λ are Dirac spinors for the Λ and

$b^\Lambda_{\boldsymbol{k},s}$ destroys a Λ of momentum k, spin s

$d^\Lambda_{\boldsymbol{k},s}$ destroys a $\bar{\Lambda}$ of momentum k, spin s,

and $b^{\Lambda\dagger}$, b^Λ, d^Λ, and $d^{\Lambda\dagger}$ satisfy the usual anticommutation relations characteristic of a Dirac field. Represent the interaction describing the decay by

$$\mathcal{H}_{\rm int} =: \phi(x) \left[g_\Lambda \bar{\psi}_n(x)\gamma^5\psi_{\Lambda_0}(x) - g^*_\Lambda \bar{\psi}_{\Lambda_0}(x)\gamma^5\psi_n(x) \right] : \quad ,$$

where g_Λ is a constant and ϕ and ψ_n are the π^0 and neutron fields, respectively. Since π^0 is neutral, $\phi^\dagger(x) = \phi(x)$.

(a) Compute the transition rate for the decay $\Lambda^0 \to n + \pi^0$. Express your answer in terms of numbers times $|g_\Lambda|^2$.

(b) From the experimental lifetime of the Λ^0 and from the fact that 35% of all Λ^0's decay into $n + \pi^0$, compute $|g_\Lambda|$.

9.4 Compute the phase space integral $\rho(M; m_1, m_2)$ for the most general two-body decay. Be sure to express your results only in terms of numbers and the masses M, m_1, and m_2.

9.5 Suppose the two poles in the propagator (9.43) are placed in the upper half plane. Show that the resulting propagator differs from the Feynman propagator only by a homogeneous solution of the KG equation. Discuss the physical difference between this propagator and the Feynman propagator. [Compare with the discussion in Sec. 4.9.]

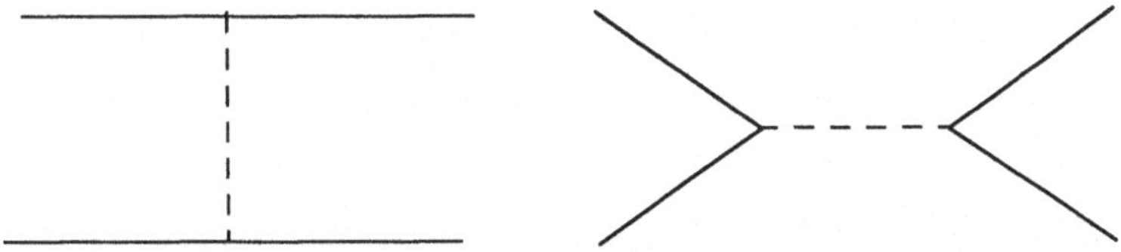

Fig. 9.9 Diagrams for $N\bar{N}$ scattering (Prob. 9.6).

9.6 Suppose the nucleon interacts with a neutral, *scalar*, meson according to

$$\mathcal{H}_{\text{int}} = g\bar{\psi}(x)\psi(x)\phi(x) \; ,$$

where ψ is the nucleon field and $\phi = \phi^\dagger$ is the neutral scalar meson field.

(a) Compute the scattering *amplitude* to order g^2, and using Eq. (9.72), find the precise form of the NN potential which arises from scalar meson exchange.

(b) Investigate nucleon–*antinucleon* scattering to order g^2 in the scattering amplitude. There are two terms corresponding to the diagrams shown in Fig. 9.9. By explicit calculation of the second order S-matrix, find the $\mathcal{M}$-matrix for each of these diagrams and extract the Feynman rules for the treatment of antiparticles.

9.7 Prove the relation (9.137) which is needed in the derivation of the propagator for a massive vector meson.

CHAPTER 10

QUANTUM ELECTRODYNAMICS

We now turn to what is perhaps the most important of all theories — Quantum Electrodynamics (referred to as QED). In addition to being important in its own right, it is also the prototype for Quantum Chromodynamics (QCD), which will be discussed in greater detail in Chapter 15.

Quantum Electrodynamics is the theory which describes how structureless (point-like) charged particles (usually with spin $\frac{1}{2}$) interact with the EM field. As such, it is the foundation of the subject of Atomic Physics and of fundamental importance to Condensed Matter, Nuclear, and Particle Physics. We already presented a preliminary discussion of some of these topics in Chapters 2 and 3. The quantization of the EM field follows the development given in Chapter 2, and the results obtained there will be carried over without further change. The new aspect of our discussion in this chapter is the treatment of the charged fermions, which will now be described by a fermion quantum field of the type introduced in the preceding three chapters.

10.1 THE HAMILTONIAN

We start with two species of charged fermions and the neutral electromagnetic field. For definiteness we take the fermions to be electrons and protons, so that our fields are

ψ_e destroys electrons

ψ_p destroys protons

A^μ destroys *and* creates photons.

Quantum Electrodynamics assumes that these fermions only interact through the EM field, which is an excellent approximation for electrons (and muons, which will be considered later in the chapter), but a poor approximation for quarks and protons, which also interact strongly. Nevertheless, in some situations the strong interactions can be taken into account without explicitly calculating their effects. For example, it turns out that a good estimate for the total cross section for the production of hadronic matter in $e^+ e^-$ annihilation at high energy can be calculated from QED alone, and we will discuss this in Sec. 10.4. And even though the proton is a bound state of quarks and gluons with a complex structure and a significant size, it is still very useful to calculate electron–proton scattering by first ignoring these effects, as we do in Sec. 10.2, and then include them by modifying the calculation later. With these applications in mind, we include a point-like hadron (which we call the proton) in our description of QED.

The Lagrangian density for this theory is

$$\mathcal{L} = \mathcal{L}_e + \mathcal{L}_p - \tfrac{1}{4}F^{\mu\nu}F_{\mu\nu} - J^\mu A_\mu \ , \tag{10.1}$$

where J^μ is the current, written

$$J^\mu = e\left[\bar{\psi}_p \gamma^\mu \psi_p - \bar{\psi}_e \gamma^\mu \psi_e\right] \ , \tag{10.2}$$

$\mathcal{L}_e$ and $\mathcal{L}_p$ are free Dirac Lagrangians (discussed in Sec. 7.4), and $F^{\mu\nu}$ and the electromagnetic Lagrangian with current were encountered before in Sec. 2.2. Normal ordering is understood. We leave it as an exercise to show that this Lagrangian gives the correct Dirac equation with minimal electromagnetic substitution for both the electron and the proton as well as the correct Maxwell equations with J^μ as current. Note also that the Dirac equation insures that the current is conserved.

Next, we impose the Coulomb gauge, solve for A^0, and eliminate it from the Lagrangian just as we did in Sec. 2.2. This gives us

$$\mathcal{L} = \mathcal{L}_p + \mathcal{L}_e + \frac{1}{2}\left(\boldsymbol{E}_\perp^2 - \boldsymbol{B}^2\right) + \boldsymbol{J}\cdot\boldsymbol{A} - \frac{1}{2}\int \frac{d^3r'}{4\pi}\,\frac{\rho(r,t)\cdot\rho(r',t)}{|r-r'|} \ , \tag{10.3}$$

where $\rho = J^0$, and

$$\boldsymbol{E}_\perp = -\frac{\partial \boldsymbol{A}}{\partial t} \qquad \boldsymbol{B} = \nabla\times\boldsymbol{A} \ . \tag{10.4}$$

Next calculate the corresponding Hamiltonian. The reduction is easier in this case than it was for the nonrelativistic case (Sec. 2.3) because the current does *not* depend on the generalized velocities. Neglecting the self-energy terms in the Coulomb interaction gives immediately

$$\begin{aligned} H =& H_e^0 + H_p^0 + H_{EM}^0 + \frac{1}{4\pi}\int \frac{d^3r\,d^3r'}{|r-r'|} J_p^0(r,t) J_e^0(r',t) \\ &- \int d^3r\,\left(\boldsymbol{J}_p(r,t) + \boldsymbol{J}_e(r,t)\right)\cdot\boldsymbol{A}(r,t) \ , \end{aligned} \tag{10.5}$$

where

$$\begin{aligned} H_e^0 &= \int d^3r\,\mathcal{H}_e^0 = \int d^3r\;\psi_e^\dagger(r,t)\left[-i\boldsymbol{\alpha}\cdot\nabla + m_e\beta\right]\psi_e(r,t) \\ H_{EM}^0 &= \frac{1}{2}\int d^3r\;\left\{\boldsymbol{E}_\perp^2(x) + \boldsymbol{B}^2(x)\right\} \ , \end{aligned} \tag{10.6}$$

and, as always, $x = (\boldsymbol{r}, t)$ is understood.

With this Hamiltonian we may treat both relativistic atoms and relativistic scattering problems, but the method is somewhat different in the two cases.

Relativistic Atoms — Here the instantaneous Coulomb term is treated to all orders by solving the Dirac equation with a Coulomb interaction exactly. Then the interaction with the radiation field is treated perturbatively. [In bound state problems, the weak binding potential must always be treated to all orders, since the bound state owes its existence to higher order effects of the potential—see Chapter 12.] Therefore, for problems of this type we separate the Hamiltonian into the free (unperturbed) and interacting parts as follows:

$$\begin{aligned} H_0 &= H_e^0 + H_p^0 + H_{EM}^0 + \frac{1}{4\pi}\int \frac{d^3r\, d^3r'}{|r - r'|} J_p^0(r,t) J_e^0(r',t) \\ H_{\text{int}} &= -\int d^3r\, (\boldsymbol{J}_p(r,t) + \boldsymbol{J}_e(r,t)) \cdot \boldsymbol{A}(r,t) \quad , \end{aligned} \tag{10.7}$$

where, as in Chapter 3, we omit the self-energies of the electron and nucleus.

Relativistic Scattering — Here the interaction takes place only over a short time, so that *both* the instantaneous Coulomb *and* the radiation interaction are treated perturbatively. Hence,

$$\begin{aligned} H_0 &= H_e^0 + H_p^0 + H_{EM}^0 \\ H_{\text{int}} &= \frac{1}{4\pi}\int \frac{d^3r\, d^3r'}{|r - r'|} J_p^0(r,t) J_e^0(r',t) - \int d^3r\, (\boldsymbol{J}_p(r,t) + \boldsymbol{J}_e(r,t)) \cdot \boldsymbol{A}(r,t) \quad . \end{aligned} \tag{10.8}$$

We shall show later that these two interaction terms combine to give an explicitly covariant result.

Before treating either of these cases, it is helpful to note that the structure of the unperturbed fermion fields depends on the structure of the unperturbed Hamiltonian H_0. We will now study this correspondence in general.

Suppose the unperturbed Hamiltonian can be written

$$H_0 = \int d^3r\, \psi^\dagger(r,t)\, \hat{\mathcal{O}}\, \psi(r,t) \; , \tag{10.9}$$

where $\hat{\mathcal{O}}$ is an operator which operates on the Dirac space, and the unperturbed fields satisfy the usual anticommutation relations:

$$\begin{aligned} &\{\psi_\alpha(r,t), \psi_\beta(r',t)\} = 0 \\ &\left\{\psi_\alpha(r,t), \psi_\beta^\dagger(r',t)\right\} = \delta_{\alpha\beta}\delta^3(r - r') \quad . \end{aligned} \tag{10.10}$$

Furthermore, we require that H_0 be the generator of time translations:

$$[H_0, \psi_\alpha(r,t)] = -i\frac{\partial}{\partial t}\psi_\alpha(r,t) \quad . \tag{10.11}$$

To find the structure of the field ψ implied by these conditions, note first that (10.11) and (10.10) imply

$$\begin{aligned}H_0\psi_\alpha(r,t)-\psi_\alpha(r,t)H_0 \\ &= -\int d^3r' \left[\psi^\dagger_\beta(r',t)\psi_\alpha(r,t)+\psi_\alpha(r,t)\psi^\dagger_\beta(r',t)\right]\hat{O}_{\beta\gamma}\psi_\gamma(r',t) \\ &= -\hat{O}_{\alpha\gamma}\psi_\gamma(r,t) = -i\frac{\partial}{\partial t}\psi_\alpha(r,t)\end{aligned}$$

or, in a more compact notation,

$$i\frac{\partial}{\partial t}\psi = \hat{O}\psi \ . \tag{10.12}$$

The operator $\mathcal{O}$ therefore specifies the wave equation which the unperturbed fields must satisfy.

Therefore, if the field is expanded in terms of annihilation and creation operators,

$$\psi_\alpha(x) = \sum_A \left\{b_A\psi^{(+)}_{A\alpha}(r)\,e^{-iE_At} + d^\dagger_A\psi^{(-)}_{A\alpha}(r)\,e^{iE_At}\right\} \ , \tag{10.13}$$

where

$$\left\{b_A, b^\dagger_{A'}\right\} = \delta_{AA'} = \left\{d_A, d^\dagger_{A'}\right\} \ ,$$

then (10.12) implies that the expansion functions $\psi^{(+)}_A(r)$ must satisfy the following equations:

$$\begin{aligned}E_A\psi^{(+)}_A &= \hat{O}\psi^{(+)}_A \\ -E_A\psi^{(-)}_A &= \hat{O}\psi^{(-)}_A \ .\end{aligned} \tag{10.14}$$

Therefore, they are eigenfunctions of the operator $\mathcal{O}$. Furthermore, since the fields satisfy the anticommutation relations (10.10), it follows that the expansion functions satisfy the following relation (the completeness condition):

$$\sum_A \left\{\psi^{(+)}_{A\alpha}(r)\psi^{(+)\dagger}_{A\beta}(r') + \psi^{(-)}_{A\alpha}(r)\psi^{(-)\dagger}_{A\beta}(r')\right\} = \delta_{\alpha\beta}\delta^3(r-r') \ , \tag{10.15}$$

from which it follows that any function (which satisfies the correct boundary conditions) can be expanded in terms of them. Hence *all* of the eigenfunctions of the operator $\mathcal{O}$ must be used in the field expansion (10.13).

Thus we see that the "free" (or more correctly, the unperturbed) field can be expanded in terms of the solutions of any Hamiltonian. In the study of relativistic atoms, where the unperturbed Hamiltonian includes the Coulomb interaction, this is equivalent to the requirement that ψ be expanded in terms of a complete set of normalized bound state wave functions (including negative energy states) and that

$$\begin{array}{ll} b_A & \text{annihilates positive energy states } A \\ d_A & \text{annihilates negative energy states } A \ , \end{array}$$

and

$$|A\rangle \equiv b_A^\dagger|0\rangle \qquad |\bar{A}\rangle \equiv d_A^\dagger|0\rangle \ . \tag{10.16}$$

Note also that

$$\begin{aligned} \langle 0|\psi(r,t)|A\rangle &= \psi_A^{(+)}(r)\, e^{-iE_At} \\ \langle 0|\psi(r,t)|\bar{A}\rangle &= \psi_A^{(-)\dagger}(r)\, e^{-iE_At} \ . \end{aligned} \tag{10.17}$$

In summary, for *relativistic atoms* the ψ_A's are solutions to the *wave equation with the instantaneous Coulomb interaction*, while for *relativistic scattering* the ψ_A's are *plane wave states*.

10.2 PHOTON PROPAGATOR: *ep* SCATTERING

In this section we calculate the differential cross section for electron–proton elastic scattering. This calculation is not only important in its own right, but it also illustrates how the photon propagator arises in QED.

Use the relativistic scattering form of the Hamiltonian, where

$$H_{\text{int}} = \underbrace{\frac{1}{4\pi}\int \frac{d^3r\, d^3r'}{|r-r'|} J_p^0(r,t) J_e^0(r',t)}_{H_{\text{INST}}} - \underbrace{\int d^3r\, (\boldsymbol{J}_p(r,t) + \boldsymbol{J}_e(r,t)) \cdot \boldsymbol{A}(r,t)}_{H_{\text{RAD}}} \ , \tag{10.18}$$

where H_{INST} is the contribution from the instantaneous Coulomb interaction and H_{RAD} is the contribution from the radiation field. We are interested in the lowest non-trivial scattering result, which is of order e^2,

$$\begin{aligned} &S(P_f, p_f; P_i, p_i) \\ &\quad = \left\langle P_f\, p_f \right| \left\{ -i \int_{-\infty}^{+\infty} H_{\text{INST}}(t) dt \right. \\ &\qquad \left. + \frac{(-i)^2}{2} \int d^4x_1\, d^4x_2\, T\left(\mathcal{H}_{\text{RAD}}(x_1) \mathcal{H}_{\text{RAD}}(x_2)\right) \right\} \left| P_i\, p_i \right\rangle \ . \end{aligned} \tag{10.19}$$

Note that the annihilation and creation operators for the proton (denoted by a subscript p, such as b_p) commute with the annihilation and creation operators for the electron (denoted by a subscript e, such as b_e), so the order of the momenta in $|P_i\, p_i\rangle$ does not matter.

The *first* term in (10.19) is particularly straightforward, giving

$$\begin{aligned}
&-i\left\langle P_f\, p_f \left| \int dt\, H_{\text{INST}}(t) \right| P_i\, p_i \right\rangle \\
&\quad = i\frac{e^2}{4\pi}\sqrt{\frac{1}{16\mathcal{E}_f\mathcal{E}_i E_f E_i}}\,\frac{1}{(2\pi)^6}\int_{-\infty}^{+\infty} dt\, e^{i(E_f+\mathcal{E}_f-E_i-\mathcal{E}_i)t} \\
&\quad \times \int \frac{d^3r'\, d^3r}{|\boldsymbol{r}'-\boldsymbol{r}|}\, e^{-i(\boldsymbol{p}_f-\boldsymbol{p}_i)\cdot\boldsymbol{r}'}\, e^{-i(\boldsymbol{P}_f-\boldsymbol{P}_i)\cdot\boldsymbol{r}} \left[\bar{u}(\boldsymbol{P}_f)\gamma^0 u(\boldsymbol{P}_i)\right]\left[\bar{u}(\boldsymbol{p}_f)\gamma^0 u(\boldsymbol{p}_i)\right],
\end{aligned} \tag{10.20}$$

where $\mathcal{E}_i = \sqrt{m_p^2 + P_i^2}$ and $E_i = \sqrt{m_e^2 + p_i^2}$ and the spins of the fermions have been suppressed for simplicity, i.e., $u(\boldsymbol{p}) = u(\boldsymbol{p}, s)$. To reduce this, follow the now standard procedure and introduce

$$\begin{aligned}
\boldsymbol{\rho} &= \boldsymbol{r} - \boldsymbol{r}' & \boldsymbol{r} &= \boldsymbol{R} + \tfrac{1}{2}\boldsymbol{\rho} \\
\boldsymbol{R} &= \tfrac{1}{2}(\boldsymbol{r} + \boldsymbol{r}') & \boldsymbol{r}' &= \boldsymbol{R} - \tfrac{1}{2}\boldsymbol{\rho}\ .
\end{aligned} \tag{10.21}$$

Then the integrals give δ-functions, and the reduced $\mathcal{M}$-matrix can be extracted,

$$\begin{aligned}
\mathcal{M}_{\text{INST}} &= -\frac{e^2}{4\pi}\int \frac{d^3\rho}{\rho}\, e^{-i(\overbrace{\boldsymbol{P}_f - \boldsymbol{P}_i}^{\boldsymbol{q}} + \overbrace{\boldsymbol{p}_i - \boldsymbol{p}_f}^{\boldsymbol{q}})\cdot\frac{1}{2}\boldsymbol{\rho}} \\
&\quad \times \left[\bar{u}(\boldsymbol{P}_f)\gamma^0 u(\boldsymbol{P}_i)\right]\left[\bar{u}(\boldsymbol{p}_f)\gamma^0 u(\boldsymbol{p}_i)\right] .
\end{aligned}$$

The integral over ρ is the familiar Yukawa integral, giving

$$\mathcal{M}_{\text{INST}} = -\frac{e^2}{\boldsymbol{q}^2}\left[\bar{u}(\boldsymbol{P}_f)\gamma^0 u(\boldsymbol{P}_i)\right]\left[\bar{u}(\boldsymbol{p}_f)\gamma^0 u(\boldsymbol{p}_i)\right] . \tag{10.22}$$

The *second* term in (10.19) involves a contraction of the A^i field. Most of the details are similar to the OPE calculation worked out in Sec. 9.8, except that the particles are no longer identical. Keeping *only* the *e–p* interaction term gives

$$\begin{aligned}
&-\tfrac{1}{2}\langle P_f\, p_f| \int d^4x_1\, d^4x_2\, \langle 0|T\left(A^i(x_1)A^j(x_2)\right)|0\rangle \\
&\qquad \times(-e^2)\left[:\bar{\psi}_p(x_1)\gamma^i\psi_p(x_1)::\bar{\psi}_e(x_2)\gamma^j\psi_e(x_2): + \begin{matrix} x_1 \leftrightarrow x_2 \\ i \leftrightarrow j \end{matrix}\right]|P_i\, p_i\rangle \\
&\quad = e^2\int \frac{d^4x_1\, d^4x_2}{\sqrt{16\mathcal{E}_f\mathcal{E}_i E_f E_i}(2\pi)^6}\, \langle 0|T\left(A^i(x_1)A^j(x_2)\right)|0\rangle \\
&\qquad \times e^{i(P_f-P_i)\cdot x_1 + i(p_f-p_i)\cdot x_2}\left[\bar{u}(\boldsymbol{P}_f)\gamma^i u(\boldsymbol{P}_i)\right]\left[\bar{u}(\boldsymbol{p}_f)\gamma^j u(\boldsymbol{p}_i)\right],
\end{aligned} \tag{10.23}$$

where we anticipated the fact that $\langle 0|T\left(A^i(x_1)A^j(x_2)\right)|0\rangle$ is symmetric in x_1,x_2 and i, j. Again, introduce

$$\begin{aligned} x &= x_1 - x_2 & x_1 &= X + \tfrac{1}{2}x \\ X &= \tfrac{1}{2}(x_1 + x_2) & x_2 &= X - \tfrac{1}{2}x \end{aligned} \tag{10.24}$$

and use the fact that $\langle 0|T\left(A^i(x_1)A^j(x_2)\right)|0\rangle$ depends only on x to obtain

$$\begin{aligned} \mathcal{M}_{\text{RAD}} = i\, e^2 \int d^4x\, \langle 0|T\left(A^i(x_1)A^j(x_2)\right)|0\rangle\, e^{iq\cdot x} \\ \times \left[\bar{u}(\boldsymbol{P}_f)\gamma^i u(\boldsymbol{P}_i)\right]\left[\bar{u}(\boldsymbol{p}_f)\gamma^j u(\boldsymbol{p}_i)\right] \quad . \end{aligned} \tag{10.25}$$

Next, compute the transverse photon propagator, iD_{tr}, which is defined to be

$$\begin{aligned} iD^{ij}_{\text{tr}}(x) &\equiv \langle 0|T\left(A^i(x_1)A^j(x_2)\right)|0\rangle \\ &= \left\langle 0|A^{i(+)}(x_1)A^{j(-)}(x_2)\theta(t_1 - t_2) + A^{j(+)}(x_2)A^{i(-)}(x_1)\theta(t_2 - t_1)|0\right\rangle \\ &= \sum_{k,\alpha} \frac{e^{-ik\cdot(x_1 - x_2)}}{2\omega L^3}\epsilon^{\alpha i}_k \epsilon^{\alpha j\,*}_k \theta(t_1 - t_2) \\ &\quad + \sum_{k,\alpha} \frac{e^{-ik\cdot(x_2 - x_1)}}{2\omega L^3}\epsilon^{\alpha j}_k \epsilon^{\alpha i\,*}_k \theta(t_2 - t_1)\,, \end{aligned} \tag{10.26}$$

where $A^{i(\pm)}$ are the positive and negative frequency parts of the vector field operator A^i and the ϵ's are the polarization vectors first introduced in Chapter 2. Recall that the ϵ's are transverse; i.e., $\boldsymbol{k}\cdot\boldsymbol{\epsilon} = 0$. Hence

$$\sum_\alpha \epsilon^{\alpha i}_k\, \epsilon^{\alpha j\,*}_k = -g^{ij} - \frac{k^i k^j}{\boldsymbol{k}^2} \quad . \tag{10.27}$$

Replacing the sums over k in (10.26) by integrals over k, using Eq. (4.77) to express the θ functions as integrals over $k_0 \neq \omega$, and recalling that $\omega^2 = \boldsymbol{k}^2$ give

$$iD^{ij}_{\text{tr}}(x) = \int \frac{d^4k}{(2\pi)^4}\, e^{-ik\cdot x} \left(\frac{-i}{\boldsymbol{k}^2 - k_0^2 - i\epsilon}\right)\left[-g^{ij} - \frac{k^i k^j}{\boldsymbol{k}^2}\right], \tag{10.28}$$

where the two poles in the denominator give contributions for $x_0 > 0$ and $x_0 < 0$ as described in Eq. (9.45). Equation (10.28) for the transverse photon propagator is very similar to the expressions for the other propagators we have obtained previously; the only difference is the factor in square brackets. To obtain a more useful form, let us introduce the reference vector $\eta^\mu = (1, \boldsymbol{0})$. Then

$$-g^{ij} - \frac{k^i k^j}{\boldsymbol{k}^2} \Longrightarrow -g^{\mu\nu} - \frac{k^\mu k^\nu}{\boldsymbol{k}^2} + \frac{\eta^\mu k^\nu k_0}{\boldsymbol{k}^2} + \frac{k^\mu \eta^\nu k_0}{\boldsymbol{k}^2} - \eta^\mu\eta^\nu\left(\frac{k_0^2}{\boldsymbol{k}^2} - 1\right) \tag{10.29}$$

and doing the d^4k integration gives

$$\mathcal{M}_{\text{RAD}} = i\,e^2\left[\bar{u}(\boldsymbol{P}_f)\gamma_\mu u(\boldsymbol{P}_i)\right]\left[\bar{u}(\boldsymbol{p}_f)\gamma_\nu u(\boldsymbol{p}_i)\right]\left(\frac{-i}{-q^2 - i\epsilon}\right)$$
$$\times\left\{-g^{\mu\nu} - \frac{q^\mu q^\nu}{\boldsymbol{q}^2} + \frac{\eta^\mu q^\nu q_0}{\boldsymbol{q}^2} + \frac{q^\mu \eta^\nu q_0}{\boldsymbol{q}^2} - \eta^\mu\eta^\nu\frac{q^2}{\boldsymbol{q}^2}\right\}. \tag{10.30}$$

Next, note that

$$\begin{aligned} q^\mu\,\bar{u}(\boldsymbol{P}_f)\gamma_\mu u(\boldsymbol{P}_i) &= (P_f - P_i)^\mu\,\bar{u}(\boldsymbol{P}_f)\gamma_\mu u(\boldsymbol{P}_i) \\ &= (m_p - m_p)\,\bar{u}(\boldsymbol{P}_i)\,u(\boldsymbol{P}_i) = 0 \end{aligned} \tag{10.31}$$

so that all terms with a q^μ can be dropped. (This is just current conservation.) Hence

$$\mathcal{M}_{\text{RAD}} = i\,e^2\left[\bar{u}(\boldsymbol{P}_f)\gamma_\mu u(\boldsymbol{P}_i)\right]\left[\bar{u}(\boldsymbol{p}_f)\gamma_\nu u(\boldsymbol{p}_i)\right]\left(\frac{i}{-q^2 - i\epsilon}\right)\left[g^{\mu\nu} + \eta^\mu\eta^\nu\frac{q^2}{\boldsymbol{q}^2}\right]. \tag{10.32}$$

Combining Eqs (10.22) and (10.32) gives the total result for the $\mathcal{M}$-matrix

$$\begin{aligned} \mathcal{M} &= \mathcal{M}_{\text{RAD}} + \mathcal{M}_{\text{INST}} \\ &= i\,e^2\left[\bar{u}(\boldsymbol{P}_f)\gamma_\mu u(\boldsymbol{P}_i)\right]\left[\bar{u}(\boldsymbol{p}_f)\gamma_\nu u(\boldsymbol{p}_i)\right] \\ &\quad\times\left(\frac{i}{-q^2 - i\epsilon}\right)\left[g^{\mu\nu} + \eta^\mu\eta^\nu\frac{q^2}{\boldsymbol{q}^2} - \eta^\mu\eta^\nu\frac{q^2}{\boldsymbol{q}^2}\right] \end{aligned}$$

and the non-covariant term cancels, giving, finally,

$$\boxed{\mathcal{M} = i\,e^2\left[\bar{u}(\boldsymbol{P}_f)\gamma_\mu u(\boldsymbol{P}_i)\right]\left[\bar{u}(\boldsymbol{p}_f)\gamma_\nu u(\boldsymbol{p}_i)\right]\left(\frac{ig^{\mu\nu}}{-q^2 - i\epsilon}\right).} \tag{10.33}$$

Manifest covariance has been restored in the final result.

This example illustrates how the Feynman Rules 1 and 2 are modified for EM interactions involving the exchange of a virtual photon:

> **Rule 1:** the operator $-ie\,\gamma_\mu$ at each vertex where a photon with polarization μ is emitted from or absorbed by a fermion with *positive* charge e.
>
> **Rule 2:** a factor of
> $$\frac{ig^{\mu\nu}}{-q^2 - i\epsilon}$$
> for each internal photon line carrying momentum q and polarization indices μ and ν. (Fix q by momentum conservation.)

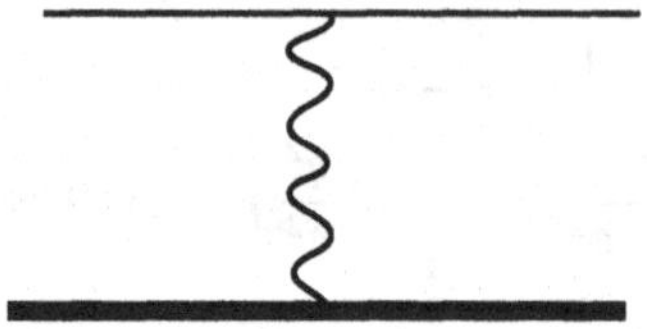

Fig. 10.1 Feynman diagram for *ep* scattering.

The final result is obtained by multiplying by the appropriate spinors u or $\bar{u}$ as before. Photon propagators are usually represented by wavy lines, as shown in Fig. 10.1, which shows the Feynman diagram corresponding to the result (10.33).

The Unpolarized *ep* Cross Section

The differential cross section for unpolarized elastic scattering of electrons from protons in the LAB frame can be found from our general result Eq. (9.63). With notational changes including $m_2 = m_p = M$; $m_1 = m_e = m$, we have

$$\frac{d\sigma}{d\Omega} = \frac{1}{(2\pi)^2\, 16M^2}\left(\frac{p'}{p}\right)\frac{1}{\left[1 - \dfrac{q^2}{2p'^2 M}\left(E' - \dfrac{m^2}{M}\right)\right]}\frac{1}{4}\sum_{\text{spins}} |\mathcal{M}|^2 \ , \qquad (10.34)$$

where the cross section is unpolarized because we have *summed* over all final spins of the proton and the electron and *averaged* over all initial spins. Since there are two spin states of the proton and the electron, the initial average is computed by summing over all initial spins and dividing by $2 \times 2 = 4$, which explains the origin of the factor of $\frac{1}{4}$ before the spin sum in (10.34).

The spin sum is

$$\begin{aligned}\sum_{\text{spins}} |\mathcal{M}|^2 = &\frac{e^4}{q^4} \sum_{S_f S_i} \bar{u}(\boldsymbol{P}_f, S_f)\gamma^\mu u(\boldsymbol{P}_i, S_i)\bar{u}(\boldsymbol{P}_i, S_i)\gamma^\nu u(\boldsymbol{P}_f, S_f) \\ &\times \sum_{s_i s_f} \bar{u}(\boldsymbol{p}_f, s_f)\gamma_\mu u(\boldsymbol{p}_i, s_i)\bar{u}(\boldsymbol{p}_i, s_i)\gamma_\nu u(\boldsymbol{p}_f, s_f)\,,\end{aligned} \qquad (10.35)$$

where the following identity has been used:

$$\begin{aligned}&\{\bar{u}(\boldsymbol{P}_f, S_f)\gamma^\mu\, u(\boldsymbol{P}_i, S_i)\}^* \\ &\quad = \left\{u^\dagger(\boldsymbol{P}_f, S_f)\gamma^0\gamma^\mu\, u(\boldsymbol{P}_i, S_i)\right\}^* = u^\dagger(\boldsymbol{P}_i, S_i)\gamma^{\mu\dagger}\gamma^0\, u(\boldsymbol{P}_f, S_f) \\ &\quad = u^\dagger(\boldsymbol{P}_i, S_i)\gamma^0\gamma^\mu\, u(\boldsymbol{P}_f, S_f) = \bar{u}(\boldsymbol{P}_i, S_i)\gamma^\mu\, u(\boldsymbol{P}_f, S_f)\end{aligned} \qquad (10.36)$$

which depends on $\gamma^{0\dagger} = \gamma^0$ and $\gamma^0\gamma^{\mu\dagger}\gamma^0 = \gamma^\mu$. Now recall from our discussion of Dirac projection operators that

$$\sum_s u_\alpha(\boldsymbol{p}, s)\bar{u}_\beta(\boldsymbol{p}, s) = (m + \not{p})_{\alpha\beta} \ . \qquad (10.37)$$

Hence the spin sum in Eq. (10.35) can be reduced to a trace

$$\begin{aligned}
\sum_{S_f S_i} \bar{u}(\boldsymbol{P}_f, S_f)\gamma^\mu \left[u(\boldsymbol{P}_i, S_i)\bar{u}(\boldsymbol{P}_i, S_i)\right] \gamma^\nu u(\boldsymbol{P}_f, S_f) \\
&= \sum_{S_f} \bar{u}_\alpha(\boldsymbol{P}_f, S_f) \left[\gamma^\mu (M + \not{P}_i) \gamma^\nu\right]_{\alpha\beta} u_\beta (\boldsymbol{P}_f, S_f) \\
&= \sum_{S_f} u_\beta (\boldsymbol{P}_f, S_f) \bar{u}_\alpha (\boldsymbol{P}_f, S_f) \left[\gamma^\mu (M + \not{P}_i) \gamma^\nu\right]_{\alpha\beta} \\
&= \left(M + \not{P}_f\right)_{\beta\alpha} \left[\gamma^\mu (M + \not{P}_i) \gamma^\nu\right]_{\alpha\beta} \\
&= \text{trace}\left\{\left(M + \not{P}_f\right) \gamma^\mu (M + \not{P}_i) \gamma^\nu\right\} \quad . \qquad (10.38)
\end{aligned}$$

This is a general technique we will employ frequently from now on. It gives spin sums as traces of products of γ-matrices. There are tricks for evaluating such traces which will now be discussed.

Theorems for Computing Traces of Products of γ-Matrices

Theorem 1: The trace of an odd number of γ-matrices is zero.

Proof:

$$\begin{aligned}
\text{tr}\,(\not{a}_1 \not{a}_2 \ldots \not{a}_n) = \text{tr}\left\{\overbrace{\gamma^5\gamma^5}^{1} \not{a}_1 \ldots \not{a}_n\right\} &= (-1)^n\, \text{tr}\,\left\{\gamma^5 \not{a}_1 \ldots \not{a}_n \gamma^5\right\} \\
&= (-1)^n\, \text{tr}\,\{\not{a}_1 \ldots \not{a}_n\} \quad . \qquad (10.39)
\end{aligned}$$

Theorem 2: The traces of zero, two, and four powers of γ-matrices are

$$\text{tr}\,(1) = 4 \qquad (10.40a)$$

$$\text{tr}\,(\not{a}\,\not{b}) = 4\, a \cdot b \qquad (10.40b)$$

$$\text{tr}\,(\not{a}\,\not{b}\,\not{c}\,\not{d}) = 4\,(a \cdot b\, c \cdot d - a \cdot c\, b \cdot d + a \cdot d\, b \cdot c) \quad . \qquad (10.40c)$$

Proof:

$$\text{tr}\,(\not{a}\,\not{b}) = \text{tr}\,(\not{b}\,\not{a}) = \tfrac{1}{2}\,\text{tr}\,(\not{a}\not{b} + \not{b}\not{a}) = a \cdot b\; \text{tr}\,(1) = 4\, a \cdot b$$

$$\begin{aligned}
\text{tr}\,(\not{a}\,\not{b}\,\not{c}\,\not{d}) &= 2\, a \cdot b\, \text{tr}\,(\not{c}\,\not{d}) - \text{tr}\,(\not{b}\,\not{a}\,\not{c}\,\not{d}) \\
&= 8\, a \cdot b\; c \cdot d - 2\, a \cdot c\, \text{tr}\,(\not{b}\,\not{d}) + \text{tr}\,(\not{b}\,\not{c}\,\not{a}\,\not{d}) \\
&= 8\, a \cdot b\; c \cdot d - 8\, a \cdot c\; b \cdot d + 2\, a \cdot d\, \text{tr}\,(\not{b}\,\not{c}) - \text{tr}\,(\not{a}\,\not{b}\,\not{c}\,\not{d}) \\
&= 8(a \cdot b\; c \cdot d - a \cdot c\; b \cdot d + a \cdot d\; b \cdot c) - \text{tr}\,(\not{a}\,\not{b}\,\not{c}\,\not{d}) \; .
\end{aligned}$$

Hence

$$2\,\mathrm{tr}\,(\not{a}\,\not{b}\,\not{c}\,\not{d}) = 8(a\cdot b\; c\cdot d - a\cdot c\; b\cdot d + a\cdot d\; b\cdot c)\;. \qquad \blacksquare$$

Using these theorems, the trace (10.38) becomes

$$\mathrm{tr}\Big\{\Big(M+\not{P}_f\Big)\gamma^\mu\,(M+\not{P}_i)\,\gamma^\nu\Big\} = 4\Big\{M^2 g^{\mu\nu} + P_f^\mu P_i^\nu - P_f\cdot P_i\, g^{\mu\nu} + P_f^\nu P_i^\mu\Big\} \tag{10.41}$$

and Eq. (10.35) becomes

$$\begin{aligned}\sum_{\text{spins}} |\mathcal{M}|^2 &= 16\,\frac{e^4}{q^4}\Big\{(M^2 - P_f\cdot P_i)g^{\mu\nu} + P_f^\mu P_i^\nu + P_f^\nu P_i^\mu\Big\} \\ &\qquad \times \big\{(m^2 - p_f\cdot p_i)g_{\mu\nu} + p_{f\,\mu}p_{i\,\nu} + p_{f\,\nu}p_{i\,\mu}\big\} \\ &= 16\,\frac{e^4}{q^4}\Big\{\overbrace{4(M^2 - P_f\cdot P_i)(m^2 - p_f\cdot p_i)}^{q^4} + \overbrace{2\,P_f\cdot P_i}^{2M^2-q^2}\,\overbrace{(m^2 - p_f\cdot p_i)}^{\frac{1}{2}q^2} \\ &\qquad + \overbrace{2\,p_f\cdot p_i}^{2m^2-q^2}\,\overbrace{(M^2 - p_f\cdot p_i)}^{\frac{1}{2}q^2} \\ &\qquad + 2\,P_f\cdot p_f\,P_i\cdot p_i + 2\,P_f\cdot p_i\,P_i\cdot p_f\Big\}.\end{aligned} \tag{10.42}$$

In simplifying these formula, it is convenient to recall that in the LAB system q^2 can be expressed in many ways:

$$\begin{aligned} q^2 &= (P_f - P_i)^2 = 2M^2 - 2P_f\cdot P_i = 2\left(m^2 - p_f\cdot p_i\right) \\ &= 2M(M - \mathcal{E}_f) = 2M(E' - E)\;, \end{aligned} \tag{10.43}$$

where $\mathcal{E}_f$ is the energy of the final proton [recall the discussion following Eq. (10.20)]. The q^4 terms in the { } cancel, and evaluating the other terms in the LAB system permits us to use

$$P_i\cdot p_f = ME' \qquad P_i\cdot p_i = ME \tag{10.44}$$

so that we get

$$\begin{aligned}\sum_{\text{spins}} |\mathcal{M}|^2 &= 16\,\frac{e^4}{q^4}\Big\{q^2(M^2+m^2) + 2M^2EE' + 2ME\overbrace{(p_i\cdot p_f - m^2)}^{-\frac{1}{2}q^2} \\ &\qquad + 2M^2EE' + 2ME'\underbrace{(m^2 - p_f\cdot p_i)}_{\frac{1}{2}q^2}\Big\} \\ &= 16\,\frac{e^4}{q^4}\Big\{4M^2EE' + q^2(M^2+m^2) + q^2\,\underbrace{M(E'-E)}_{\frac{1}{2}q^2}\Big\}\;.\end{aligned} \tag{10.45}$$

Hence, combining the other factors from Eq. (10.34) gives

$$\frac{d\sigma}{d\Omega} = \frac{e^4}{(4\pi)^2}\left(\frac{p'}{p}\right)\frac{1}{q^4}\frac{\left[4EE' + q^2\left(1 + \frac{m^2}{M^2}\right) + \frac{q^4}{2M^2}\right]}{1 - \frac{q^2}{2p'^2 M}\left(E' - \frac{m^2}{M}\right)} \quad . \tag{10.46}$$

This is an exact formula rarely found in the literature. Two limiting cases are of particular interest.

Limiting Cases

The first limiting case of interest is the nonrelativistic limit, which can be realized by letting $M \to \infty$. In this limit, proton recoil is unimportant, and

$$p \simeq p' \qquad E \simeq E' \ ,$$

and the differential cross section reduces to

$$\frac{d\sigma}{d\Omega} = \frac{\alpha^2}{q^4}[4E^2 + q^2] \ , \tag{10.47}$$

where, as always, $\alpha = e^2/4\pi$. Using

$$\begin{aligned} q^2 &= (p_f - p_i)^2 = 2m^2 - 2EE' + 2pp'\cos\theta \\ &\cong -2p^2(1 - \cos\theta) = -4p^2\sin^2\theta/2 \ , \end{aligned} \tag{10.48}$$

the formula can be written

$$\frac{d\sigma}{d\Omega} = \frac{\alpha^2}{q^4}\left[4m^2 + 4p^2\cos^2\frac{\theta}{2}\right] \quad . \tag{10.49}$$

If $m \to 0$, which implies $p \to E$, we have the familiar Mott cross section

$$\sigma_M \equiv \frac{d\sigma}{d\Omega} = \left(\frac{2\alpha E\cos\frac{\theta}{2}}{q^2}\right)^2 = \left(\frac{\alpha\cos\frac{\theta}{2}}{2E\sin^2\frac{\theta}{2}}\right)^2 \qquad \text{Mott cross section.} \tag{10.50}$$

The second interesting case is the ultra-relativistic limit when proton recoil becomes important. Here $p' \sim E'$, $p \sim E$, $m \ll E$, but $E \neq E'$. We have

$$\frac{d\sigma}{d\Omega} = \frac{\alpha^2}{q^4}\left(\frac{E'}{E}\right)\frac{\left[4EE' + q^2 + \frac{q^4}{2M^2}\right]}{1 - \frac{q^2}{2E'M}} \quad . \tag{10.51}$$

In this limit, $q^2 = -4EE' \sin^2 \frac{\theta}{2}$, so

$$\frac{d\sigma}{d\Omega} = \left(\frac{\alpha^2}{16E^2E'^2 \sin^4 \frac{\theta}{2}}\right)\left(\frac{E'}{E}\right)\frac{\left[4EE'\overbrace{\left(1-\sin^2\frac{\theta}{2}\right)}^{\cos^2\frac{\theta}{2}} - \frac{q^2}{2M^2}4EE'\sin^2\frac{\theta}{2}\right]}{1+\frac{2E}{M}\sin^2\frac{\theta}{2}} \tag{10.52}$$

or simplifying,

$$\frac{d\sigma}{d\Omega} = \underbrace{\left(\frac{\alpha\cos\frac{\theta}{2}}{2E\sin^2\frac{\theta}{2}}\right)^2}_{\text{Mott cross section}} \underbrace{\frac{\left[1-\overbrace{\frac{q^2}{2M^2}\tan^2\frac{\theta}{2}}^{\text{magnetic moment}}\right]}{1+\frac{2E}{M}\sin^2\frac{\theta}{2}}}_{\text{recoil factor}} . \tag{10.53}$$

This is the famous *Rosenbluth* cross section for a *point Dirac* proton. It is built up from three factors:

- the Mott cross section,
- the recoil factor $\left[1+\frac{2E}{M}\sin^2\left(\frac{\theta}{2}\right)\right]^{-1}$,
- an extra term due to scattering from the Dirac magnetic moment of the proton.

An alternative expression for this cross section often found in the literature follows from the observation

$$\frac{E'}{E} = \frac{1}{1+\frac{2E}{M}\sin^2\frac{\theta}{2}} . \tag{10.54}$$

Hence

$$\boxed{\frac{d\sigma}{d\Omega} = \sigma_M\left(\frac{E'}{E}\right)\left[1-\frac{q^2}{2M^2}\tan^2\frac{\theta}{2}\right]} \qquad \begin{array}{c}\text{Rosenbluth}\\ \text{cross section}\\ \textit{point}\text{ proton.}\end{array} \tag{10.55}$$

Form Factors and the Structure of the Proton

The strong interactions modify the proton current by introducing a significant proton structure. This structure leads to the appearance of *form factors* in the proton electromagnetic current. A complete description of the structure of a physical proton requires only two form factors, denoted by F_1 (the Dirac form factor) and F_2 (the Pauli form factor), and these modify the proton current as follows:

$$e\,\bar{u}(\boldsymbol{P}_f)\gamma^\mu u(\boldsymbol{P}_i) \to e\,\bar{u}(\boldsymbol{P}_f)\left[F_1(q^2)\gamma^\mu + F_2(q^2)\frac{i\sigma^{\mu\nu}}{2M}q^\nu\right]u(\boldsymbol{P}_i) . \tag{10.56}$$

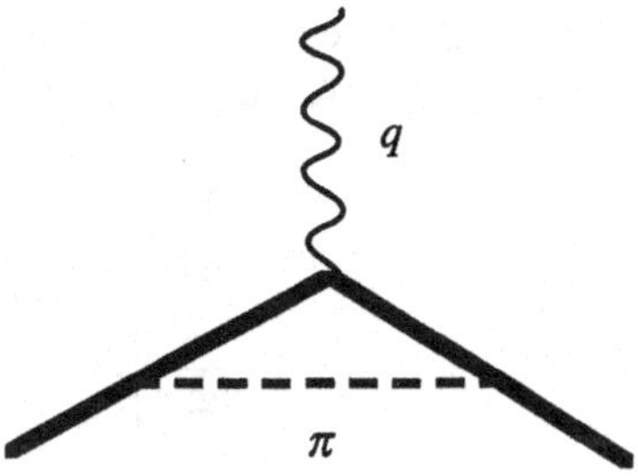

Fig. 10.2 One of the many Feynman diagrams which contributes to the proton form factor.

For physical protons ($P_f^2 = M^2 = P_i^2$), the form factors can depend only on q^2. Furthermore, at $q^2 = 0$ the above expression must reduce to the static expression for the generalized current discussed in Prob. 5.6 Hence

$$F_1(0) = 1 \quad \text{(charge)}$$
$$F_2(0) = \kappa_p \quad \text{(anomalous magnetic moment = 1.79)}.$$

These form factors are thought of as arising from the quark structure of the proton, or in the older meson theory would arise from loop diagrams like that shown in Fig. 10.2, which we will discuss in detail in the next chapter.

The cross section will be modified by the form factors. For this purpose it is customary to introduce new form factors,

$$\begin{aligned} G_E(q^2) &= F_1(q^2) + \tau F_2(q^2) \\ G_M(q^2) &= F_1(q^2) + F_2(q^2) \quad , \end{aligned} \tag{10.57}$$

where $\tau = q^2/4M^2$. Then

$$G_E(0) = 1 \qquad \text{(charge)}$$
$$G_M(0) = \mu_p = 2.79 \quad \text{(full magnetic moment)}.$$

In terms of these, the *ep* scattering cross section becomes

$$\boxed{\left.\frac{d\sigma}{d\Omega'}\right|_{\text{LAB}} = \left.\frac{d\sigma}{d\Omega'}\right|_{\text{NS}} \left[\frac{G_E^2(q^2) - \tau G_M^2(q^2)}{1-\tau} - 2\tau G_M^2(q^2)\tan^2\frac{\theta}{2}\right]} \; , \tag{10.58}$$

where

$$\left.\frac{d\sigma}{d\Omega'}\right|_{\text{NS}} = \sigma_M \left(\frac{E'}{E}\right) = \sigma_M \frac{1}{1 + \frac{2E}{M}\sin^2\frac{\theta}{2}} \tag{10.59}$$

is the cross section for scattering from a target with no structure (NS). If the proton is structureless, with no anomalous moment, then

$$G_E = G_M = 1 \tag{10.60}$$

and the general formula reduces to the simpler expression Eq. (10.55).

Remarks

Electron–proton scattering is used to measure G_E and G_M. These can be separated by measuring the differential cross section at two different scattering angles θ for the *same* q^2 (referred to as a Rosenbluth separation). Specifically, one measures

$$\frac{\left.\dfrac{d\sigma}{d\Omega'}\right|_{\text{LAB}}}{\left.\dfrac{d\sigma}{d\Omega'}\right|_{\text{NS}}} = A(q^2) + B(q^2)\tan^2\frac{\theta}{2} \tag{10.61}$$

and separates A and B by plotting the ratio as a function of $\tan^2\frac{\theta}{2}$. The structure functions are related to the form factors by

$$\begin{aligned} A(q^2) &= \frac{G_E^2(q^2) - \tau G_M^2(q^2)}{1-\tau} \\ B(q^2) &= -2\tau G_M^2(q^2) \ . \end{aligned} \tag{10.62}$$

Forward scattering (scattering at small electron angles θ) is dominated by G_E and backward scattering (where θ is near 180 deg) by G_M. At very high q^2, G_E is hard to measure because $\tau >> 1$. At very small q^2, $\tau << 1$ and then G_M is hard to measure.

10.3 ANTIPARTICLES: $e^+e^- \rightarrow \mu^+\mu^-$

We now turn to another important illustration of the power of field theory. In addition to electron scattering, the same expressions also describe the production of $p\bar{p}$ pairs from e^+e^- annihilation. Instead of describing $p\bar{p}$ production, we describe $\mu^+\mu^-$ production, because the μ meson does *not* interact strongly, and therefore the lowest order EM result is more accurate. While the same mechanism works for $p\bar{p}$ production, it is modified by subsequent strong interactions in the final state, so that the QED result is not very reliable.

The relevant Feynman diagram is shown in Fig. 10.3. It must arise from

$$\begin{aligned} &S(p_-p_+, k_-k_+) \\ &\quad = \langle p_-p_+| \Big\{ -i\int_{-\infty}^{+\infty} H_{\text{INST}}(t)dt \\ &\qquad\qquad + (-i)^2\frac{1}{2}\int d^4x_1\, d^4x_2\, T\left(\mathcal{H}_{\text{RAD}}(x_1)\mathcal{H}_{\text{RAD}}(x_2)\right)\Big\} |k_-k_+\rangle\,, \end{aligned} \tag{10.63}$$

where $\psi_p \rightarrow \psi_{\mu^-}$ (with a corresponding change in the sign of the charge), but

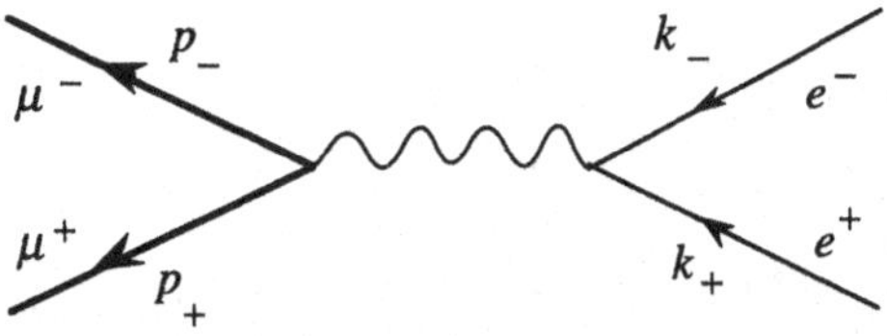

Fig. 10.3 Feynman diagram for production of $\mu^+ \mu^-$ pairs.

otherwise the expressions are identical to those treated before. The momenta of the particles (e^- and μ^-) are k_- and p_-, and those of the antiparticles (e^+ and μ^+) are k_+ and p_+, as shown in the figure.

The first term in (10.63) gives

$$
\begin{aligned}
&-i\langle p_- p_+| \int_{-\infty}^{+\infty} dt\, H_{\rm INST}(t)\, |k_-\, k_+\rangle \\
&= -\frac{ie^2}{4\pi} \sum_{q_1 q_2 q_3 q_4} \sqrt{\frac{1}{16\mathcal{E}_{q_3}\mathcal{E}_{q_4}E_{q_1}E_{q_2}}}\frac{1}{L^6} \int dt\, e^{i(\mathcal{E}_{q_3}+\mathcal{E}_{q_4}-E_{q_1}-E_{q_2})t} \\
&\times \int \frac{d^3r\, d^3r'}{|r-r'|} \langle 0|D_{p_+} B_{p_-} \Big\{ \underbrace{B^\dagger_{q_3}\, e^{-i\boldsymbol{q}_3\cdot \boldsymbol{r}} \bar{u}(\boldsymbol{q}_3)}_{\bar{\psi}_\mu^{(+)}} \gamma^0\, \underbrace{v(\boldsymbol{q}_4) D^\dagger_{q_4}\, e^{-i\boldsymbol{q}_4\cdot \boldsymbol{r}}}_{\psi_\mu^{(-)}} \Big\} \\
&\times \Big\{ \underbrace{d_{q_2}\, e^{i\boldsymbol{q}_2\cdot \boldsymbol{r}'} \bar{v}(\boldsymbol{q}_2)}_{\bar{\psi}_e^{(-)}} \gamma^0\, \underbrace{u(\boldsymbol{q}_1) b_{q_1}\, e^{i\boldsymbol{q}_1\cdot \boldsymbol{r}'}}_{\psi_e^{(+)}} \Big\} b^\dagger_{k_-}\, d^\dagger_{k_+}\, |0\rangle\,, \qquad (10.64)
\end{aligned}
$$

where b and d are the creation and annihilation operators for the electron, B and D are the creation and annihilation operators for the muon, and the muon and electron spinors will be distinguished by their argument (with spins suppressed). For example, the familiar field expansion for the muon is

$$
\begin{aligned}
\psi_\mu(r,t) &= \sum_{q_4} \sqrt{\frac{1}{2\mathcal{E}_{q_4}L^3}} \left\{ B_{q_4}\, u(\boldsymbol{q}_4) e^{i\boldsymbol{q}_4\cdot \boldsymbol{r}}\, e^{-i\mathcal{E}_{q_4}t} + D^\dagger_{q_4}\, v(\boldsymbol{q}_4) e^{-i\boldsymbol{q}_4\cdot \boldsymbol{r}}\, e^{i\mathcal{E}_{q_4}t} \right\} \\
&= \psi_\mu^{(+)}(r,t) + \psi_\mu^{(-)}(r,t)\ . \qquad (10.65)
\end{aligned}
$$

Note that (10.64) will be zero unless the interaction term contains precisely one of each of the operators b, d, $B^\dagger$, and $D^\dagger$ and that this requires the unique combination of positive and negative frequency parts of the fields given in (10.64). Now, from the anticommutation relations of the b and d operators belonging to the same particle,

$$
\Big\langle 0|D_{p_+}\ B_{p_-}\ B^\dagger_{q_3}\ D^\dagger_{q_4}\ \ d_{q_2}\ b_{q_1}\ b^\dagger_{k_-}\ \ d^\dagger_{k_+}\ |0\Big\rangle = \delta_{q_3,p_-}\delta_{q_4,p_+}\delta_{q_1,k_-}\delta_{q_2,k_+} \qquad (10.66)
$$

and the sums over $q_1 \dots q_4$ collapse, giving

$$
\begin{aligned}
S_{\text{INST}}(p_-p_+, k_-k_+) \\
&= -\frac{ie^2}{4\pi}\frac{2\pi\delta(\mathcal{E}_+ + \mathcal{E}_- - E_+ - E_-)}{(2\pi)^6\sqrt{16\mathcal{E}_+\mathcal{E}_-E_+E_-}} \\
&\times \int \frac{d^3r\, d^3r'}{|r-r'|} e^{i(\boldsymbol{k}_+ + \boldsymbol{k}_-)\cdot \boldsymbol{r}' - i(\boldsymbol{p}_+ + \boldsymbol{p}_-)\cdot \boldsymbol{r}} \left[\bar{u}(\boldsymbol{p}_-)\gamma^0 v(\boldsymbol{p}_+)\right]\left[\bar{v}(\boldsymbol{k}_+)\gamma^0 u(\boldsymbol{k}_-)\right] .
\end{aligned}
\tag{10.67}
$$

The r, r' integrals can be done using the familiar substitution (10.21), and pulling out the factors which relate the S- to the $\mathcal{M}$-matrix gives

$$
\mathcal{M}_{\text{INST}} = \frac{e^2}{4\pi}\left[\bar{u}(p_-)\gamma^0 v(p_+)\right]\left[\bar{v}(k_+)\gamma^0 u(k_-)\right]\int \frac{d^3\rho}{\rho} e^{-i(\boldsymbol{k}_+ + \boldsymbol{k}_-)\cdot \boldsymbol{\rho}} . \tag{10.68}
$$

Defining

$$
K = k_+ + k_- \tag{10.69}
$$

and using $\eta^\mu = (1, 0)$ permit us to write the instantaneous $\mathcal{M}$-matrix in the following way:

$$
\mathcal{M}_{\text{INST}} = \frac{e^2}{\boldsymbol{K}^2}\left[\bar{u}(\boldsymbol{p}_-)\gamma^\mu v(\boldsymbol{p}_+)\right]\left[\bar{v}(\boldsymbol{k}_+)\gamma^\nu u(\boldsymbol{k}_-)\right]\eta_\mu\eta_\nu . \tag{10.70}
$$

Note that this is similar to Eq. (10.22) except for the sign, which is now + because both the μ^- and e^- have negative charge, and the appearance of v spinors, which will be discussed below.

The second term, due to H_{RAD}, becomes

$$
\begin{aligned}
S_{\text{RAD}} = -2\times\frac{e^2}{2}\sum_{q_1q_2q_3q_4}\int \frac{d^4x_1\, d^4x_2}{\sqrt{16\mathcal{E}_{q_3}\mathcal{E}_{q_4}E_{q_1}E_{q_2}}\; L^6}\langle 0|T\left(A^i(x_1)A^j(x_2)\right)|0\rangle \\
\times \langle 0|D_{p_+}B_{p_-}\Big\{\underbrace{B^\dagger_{q_3}\, e^{iq_3\cdot x_1}\bar{u}(\boldsymbol{q}_3)}_{\bar{\psi}_\mu(x_1)}\gamma^i\,\underbrace{v(\boldsymbol{q}_4)D^\dagger_{q_4}\, e^{iq_4\cdot x_1}}_{\psi_\mu(x_1)}\Big\} \\
\times \Big\{\underbrace{d_{q_2}\, e^{-iq_2\cdot x_2}\bar{v}(\boldsymbol{q}_2)}_{\bar{\psi}_e(x_2)}\gamma^j\,\underbrace{\bar{u}(\boldsymbol{q}_1)b_{q_1}\, e^{-iq_1\cdot x_2}}_{\psi_e(x_2)}\Big\} b^\dagger_{k_-}d^\dagger_{k_+}|0\rangle,
\end{aligned}
\tag{10.71}
$$

where use was made of the fact that the two terms with $x_1 \leftrightarrow x_2$ are identical and can be accounted for by multiplying by 2. Using the same pairings we worked out before gives us

$$
\begin{aligned}
S_{\text{RAD}} = -e^2\int \frac{d^4x_1\, d^4x_2\; e^{i(p_+ + p_-)\cdot x_1 - i(k_+ + k_-)\cdot x_2}}{(2\pi)^6\sqrt{16E_+E_-\mathcal{E}_+\mathcal{E}_-}}\langle 0|T\left(A^i(x_1)A^j(x_2)\right)|0\rangle \\
\times \left[\bar{u}(\boldsymbol{p}_-)\gamma^i v(\boldsymbol{p}_+)\right]\left[\bar{v}(\boldsymbol{k}_+)\gamma^j u(\boldsymbol{k}_-)\right] .
\end{aligned}
\tag{10.72}
$$

The transverse photon propagator was evaluated in Eq. (10.28),

$$\langle 0|T\left(A^i(x_1)A^j(x_2)\right)|0\rangle = i\int \frac{d^4k}{(2\pi)^4}\,\frac{e^{-ik\cdot x}}{-k^2-i\epsilon}\underbrace{\left[g^{ij}+\frac{k^ik^j}{\boldsymbol{k}^2}\right]}_{g^{\mu\nu}+\eta^\mu\eta^\nu\frac{k^2}{\boldsymbol{k}^2}}\ .$$

Substituting $X = \frac{1}{2}(x_1+x_2)$ and $x = x_1 - x_2$ makes it a simple matter to extract $\mathcal{M}_{\rm RAD}$ from Eq. (10.72):

$$\mathcal{M}_{\rm RAD} = -ie^2\left[\bar{u}(\boldsymbol{p}_-)\gamma^\mu v(\boldsymbol{p}_+)\right]\left[\bar{v}(\boldsymbol{k}_+)\gamma^\nu u(\boldsymbol{k}_-)\right]\frac{i}{-K^2-i\epsilon}\left[g_{\mu\nu}+\eta_\mu\eta_\nu\frac{K^2}{\boldsymbol{K}^2}\right]\ . \tag{10.73}$$

When added to the instantaneous result (10.70), the $\eta_\mu\eta_\nu$ term cancels, giving the final result:

$$\boxed{\mathcal{M} = e^2\left[\bar{u}(\boldsymbol{p}_-)\gamma^\mu v(\boldsymbol{p}_+)\right]\left[\bar{v}(\boldsymbol{k}_+)\gamma^\nu u(\boldsymbol{k}_-)\right]\left(\frac{g_{\mu\nu}}{-K^2-i\epsilon}\right)\ .} \tag{10.74}$$

Discussion

(i) The derivation made use of the fact that the terms like $\eta^\mu K^\nu$ and $K^\mu\eta^\nu$ are again zero. This follows from current conservation in the form

$$\begin{aligned}\bar{v}(\boldsymbol{k}_+)\gamma^\nu u(\boldsymbol{k}_-)K_\nu &= \bar{v}(\boldsymbol{k}_+)(\not{k}_+ + \not{k}_-)u(\boldsymbol{k}_-)\\ &= \bar{v}(\boldsymbol{k}_+)(-m_e+m_e)u(\boldsymbol{k}_-) = 0\ .\end{aligned} \tag{10.75}$$

(ii) **Feynman rules:** The Feynman diagram corresponding to this process was given in Fig. 10.3 but is redrawn in Fig. 10.4 with the antiparticles (e^+ and μ^+) labeled as if they are moving in the opposite directions and with their momentum given an opposite sign so that the new diagram is identical to the old one. This labeling is often done to help with the construction of matrix elements involving antiparticles. The Feynman rules illustrated by this calculation are:

Rule 0: a factor of i.

Rule 1: the operator $-ie\,\gamma_\mu$ at each vertex where a photon with polarization μ is emitted from or absorbed by a fermion with *positive* charge e.

Rule 2: a factor of

$$\frac{ig^{\mu\nu}}{-q^2-i\epsilon}$$

for each internal photon line carrying momentum q and polarization indices μ and ν. (Fix q by momentum conservation.)

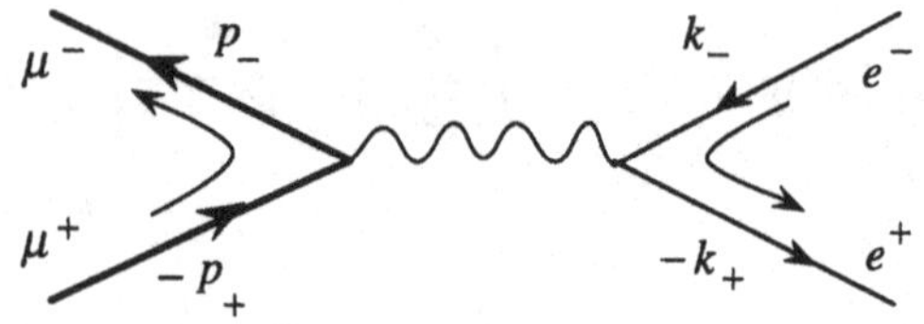

Fig. 10.4 Feynman diagram for the production process showing the flow of charge.

Rule 3: for fermions, assemble the incoming fermion spinors, vertex operators, and outgoing fermion spinors in *order* along each fermion line to make a well-formed matrix element. In particular:

- multiply from the *left* by $\bar{u}(\boldsymbol{p}_-, s^-)$ for each *outgoing fermion* with momentum $\boldsymbol{p}_-$ and spin s_-.
- multiply from the *right* by $u(\boldsymbol{k}_-, s_-)$ for each *incoming fermion* with momentum $\boldsymbol{k}_-$ and spin s_-.
- multiply from the *right* by $v(\boldsymbol{p}_+, s_+)$ for each *outgoing antifermion* with momentum $\boldsymbol{p}_+$ and spin s_+.
- multiply from the *left* by $\bar{v}(\boldsymbol{k}_+, s_+)$ for each *incoming antifermion* with momentum $\boldsymbol{k}_+$ and spin s_+.

Note the peculiar fact that $\bar{v}$ *is associated with incoming antiparticles*, yet must be on the left to make a Lorentz invariant matrix element. Similarly, v *is associated with outgoing antiparticles* but must be on the right. The labeling of antiparticles given in Fig. 10.4 helps suggest this ordering. On the electron side of the diagram, the direction of the momentum of the incoming positron is reversed, suggesting that the incoming positive charge is to be regarded as equivalent to an outgoing negative charge. The negative electron charge flows into the vertex along the electron line with momentum $\boldsymbol{k}_-$ and "out" of the vertex along the positron line with momentum $-\boldsymbol{k}_+$. This flow of negative charge is the same as the ordering of the Dirac indices:

$$\bar{v}(\boldsymbol{k}_+)(+ie\gamma^\mu)u(\boldsymbol{k}_-) \quad \Longleftrightarrow \quad \begin{array}{c}\text{order of Dirac indices}\\ \Longleftarrow \\ \text{flow of negative charge.}\end{array} \tag{10.76}$$

Similarly, the order in which the μ^- matrix element is constructed follows the flow of negative muon charge. In this case the incoming negative charge is carried by the positive muon flowing backward and into the vertex with momentum $-\boldsymbol{p}_+$ and by the negative muon flowing out of the vertex with momentum $\boldsymbol{p}_-$:

$$\bar{u}(\boldsymbol{p}_-)(+ie\gamma^\mu)v(\boldsymbol{p}_+) \quad \Longleftrightarrow \quad \begin{array}{c}\text{order of Dirac indices}\\ \Longleftarrow \\ \text{flow of negative charge.}\end{array} \tag{10.77}$$

Warning: The momentum in the v or $\bar{v}$ spinors is the actual physical momentum of the antiparticle.

(iii) **Sign ambiguity:** If we choose the μ^+ to be the particle and the μ^- to be the antiparticle, the current would be of the opposite sign, and we would have for the muon

$$\bar{u}(\boldsymbol{p}_+)(-ie\gamma^\mu)v(\boldsymbol{p}_-) \ .$$

Taking the transpose of this expression and using Eq. (5.38), from which the relations

$$v(\boldsymbol{p}) = C\bar{u}^{\mathsf{T}}(\boldsymbol{p}) \qquad \bar{u}(\boldsymbol{p}) = v^{\mathsf{T}}(\boldsymbol{p})\, C$$

follow, give

$$\left[\bar{u}(\boldsymbol{p}_+)(-ie\gamma^\mu)v(\boldsymbol{p}_-)\right]^{\mathsf{T}} = \bar{u}(\boldsymbol{p}_-)C^{\mathsf{T}}\left(-ie\gamma^{\mu\mathsf{T}}\right)C^{\mathsf{T}}v(\boldsymbol{p}_+) \ . \tag{10.78}$$

But, $C^{\mathsf{T}} = -C = C^{-1}$, and $C\gamma^{\mu\mathsf{T}}C^{-1} = -\gamma^\mu$, and therefore

$$\left[\bar{u}(\boldsymbol{p}_+)(-ie\gamma^\mu)v(\boldsymbol{p}_-)\right]^{\mathsf{T}} = \bar{u}(\boldsymbol{p}_-)(-ie\gamma^\mu)v(\boldsymbol{p}_+) \tag{10.79}$$

which gives the *opposite sign* to (10.77). However, this sign ambiguity is completely unphysical, because the $\mu^+\mu^-$ sector is completely independent of all other sectors. This means that amplitudes with *one* $\mu^+\mu^-$ pair can interfere *only* with other amplitudes with *one* $\mu^+\mu^-$ pair and cannot interfere with amplitudes with a different number of $\mu^+\mu^-$ pairs. Hence the overall sign of such amplitudes cannot be measured and may be separately fixed by an arbitrary sign convention.

(iv) **Application to $e^+e^- \to e^+e^-$:** Our calculation also applies to $e^+e^- \to e^+e^-$ scattering. However, in this case there is another diagram. The two diagrams are shown in Fig. 10.5, and the $\mathcal{M}$-matrix corresponding to them is

$$\begin{aligned}
\mathcal{M} = i&\left(\frac{i}{-K^2 - i\epsilon}\right)\left[\bar{u}(\boldsymbol{p}_-)ie\gamma^\mu v(\boldsymbol{p}_+)\right]\left[\bar{v}(\boldsymbol{k}_+)ie\gamma_\mu u(\boldsymbol{k}_-)\right] \\
&\quad + i\left(\frac{i}{-q^2 - i\epsilon}\right)\left[\bar{u}(\boldsymbol{p}_-)ie\gamma^\mu u(\boldsymbol{k}_-)\right]\left[\bar{v}(\boldsymbol{k}_+)(-ie\gamma_\mu)v(\boldsymbol{p}_+)\right] \\
= &\left(\frac{e^2}{-K^2 - i\epsilon}\right)\left[\bar{u}(\boldsymbol{p}_-)\gamma^\mu v(\boldsymbol{p}_+)\right]\left[\bar{v}(\boldsymbol{k}_+)\gamma_\mu u(\boldsymbol{k}_-)\right] \\
&\quad - \left(\frac{e^2}{-q^2 - i\epsilon}\right)\left[\bar{u}(\boldsymbol{p}_-)\gamma^\mu u(\boldsymbol{k}_-)\right]\left[\bar{v}(\boldsymbol{k}_+)\gamma_\mu v(\boldsymbol{p}_+)\right] ,
\end{aligned} \tag{10.80}$$

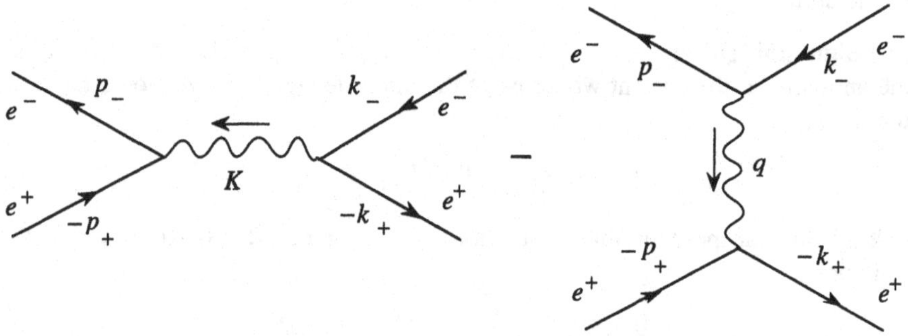

Fig. 10.5 The two Feynman diagrams which describe e^+e^- scattering to second order.

where $K^2 = (k_- + k_+)^2$ and $q^2 = (k_- - p_-)^2$. Note the *relative minus sign between the two terms* and the fact that *the second term can be obtained from the first by the interchange of*

$$-k_+ \longleftrightarrow p_- \ , \tag{10.81}$$

provided we also interpret $\bar{u}(-\boldsymbol{k}_+) = \bar{v}(\boldsymbol{k}_+)$ and $\bar{v}(-\boldsymbol{p}_-) = \bar{u}(\boldsymbol{p}_-)$. This can be understood as a generalization of the Pauli principle which holds when identical particles exist in the initial and final states. The interchange is shown diagrammatically in Fig. 10.6. To use this symmetry, it is best to regard an incoming positron with four-momentum k_+ as equivalent to an outgoing electron with four-momentum $-k_+$. From this point of view there are two outgoing electrons in this problem — one with four-momentum p_- and one with four-momentum $-k_+$. Hence, it is expected that the amplitude should be antisymmetric under interchange of the two.

This sign arises naturally from the field theory reduction. For identical particles, the matrix element (10.66) must be replaced by two matrix elements:

$$\langle 0|D_+B_-B_3^\dagger D_4^\dagger d_2\, b_1\, b_-^\dagger d_+^\dagger|0\rangle \quad \text{is replaced by}$$
$$\langle 0|d_{p+} b_{p-} b_3^\dagger\, d_4^\dagger\, d_2\, b_1\, b_{k-}^\dagger\, d_{k+}^\dagger|0\rangle - \langle 0|d_{p+} b_{p-} b_3^\dagger\, b_4\, d_1^\dagger\, d_2\, b_{k-}^\dagger\, d_{k+}^\dagger|0\rangle \ . \tag{10.82}$$

The two terms arise because there are now two qualitatively different ways the matrix element can be non-zero. To see this, recall that the full normal ordered expansion of the fermion product in the Hamiltonian density is

$$:\bar{\psi}_e\gamma^\mu\psi_e: \longrightarrow \underbrace{b_2^\dagger\, b_1\, [\bar{u}(\boldsymbol{q}_2)\gamma^\mu u(\boldsymbol{q}_1)]}_{\text{term 1}} + \underbrace{b_2^\dagger\, d_1^\dagger\, [\bar{u}(\boldsymbol{q}_2)\gamma^\mu v(\boldsymbol{q}_1)]}_{\text{term 2}}$$
$$+ \underbrace{d_2\, b_1\, [\bar{v}(\boldsymbol{q}_2)\gamma^\mu u(\boldsymbol{q}_1)]}_{\text{term 3}} - \underbrace{d_1^\dagger\, d_2\, [\bar{v}(\boldsymbol{q}_2)\gamma^\mu v(\boldsymbol{q}_1)]}_{\text{term 4}} \ , \tag{10.83}$$

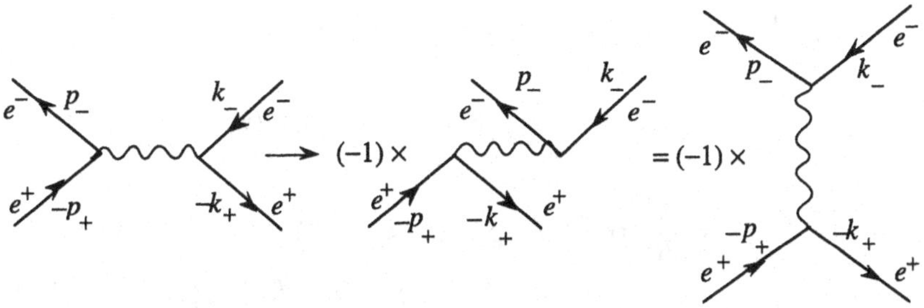

Fig. 10.6 Illustration showing how the exchange diagram can be obtained from the annihilation diagram by antisymmetrization of the two "outgoing" fermions.

where the minus sign in front of term 4 comes from the interchange of the d and $d^\dagger$ operators required by the normal ordering. For $e^+e^- \to \mu^+\mu^-$, only term 3 occurred (paired with term 2 from the muon matrix element), but there were two terms which give rise to it. In $e^+e^- \to e^+e^-$, the requirement that we have precisely one of each of the b, $b^\dagger$, d, $d^\dagger$ can occur four ways, by the following pairings:

$$\text{term 2} \times \text{term 3} + \text{term 3} \times \text{term 2} = 2 \times [\text{term 2} \times \text{term 3}]$$
$$\text{term 1} \times \text{term 4} + \text{term 4} \times \text{term 1} = 2 \times [\text{term 1} \times \text{term 4}] \quad .$$

However, as the brackets in Eq. (10.83) show, these lead to the two diagrams given in Fig. 10.5 with a *relative minus sign*. Furthermore, the two terms differ only in the exchange of $-k_+ \leftrightarrow p_-$, as has already been discussed.

Cross Section

The next task is to calculate the total cross section for the production of $\mu^+\mu^-$ pairs in e^+e^- collisions. These experiments are normally carried out in colliding beam accelerators, where the LAB system is the same as the CM system. Hence the calculation will be carried out in the CM system. In this system, the energies of the initial and final particles are the same, but the magnitude of the momenta are different. Introduce

$$\begin{aligned} s = 4E^2 &= (k_+ + k_-)^2 = 2m^2 + 2k_+ \cdot k_- \\ &= (p_+ + p_-)^2 = 2M^2 + 2p_+ \cdot p_- \,, \end{aligned} \tag{10.84}$$

where $m = m_e$ is the electron mass and $M = m_\mu$ is the muon mass.

The unpolarized cross section from Eq. (9.54), is

$$\frac{d\sigma}{d\Omega} = \frac{1}{(2\pi)^2 16s} \frac{1}{4} \sum_{\text{spins}} |\mathcal{M}|^2 \,, \tag{10.85}$$

where [recall the discussion following Eq. (10.35)]

$$\sum_{\text{spin}} |\mathcal{M}|^2 = \frac{e^4}{K^4} \sum_{\substack{\mu \\ \text{spins}}} [\bar{u}(p_-)\gamma^\mu v(p_+)] \, [\bar{v}(p_+)\gamma^\nu u(p_-)]$$
$$\times \sum_{\substack{\text{electron} \\ \text{spins}}} [\bar{v}(k_+)\gamma_\mu u(k_-)] \, [\bar{u}(k_-)\gamma_\nu v(k_+)]$$
$$= \frac{e^4}{K^4} \text{tr} \left\{ (M + \not{p}_-) \, \gamma^\mu \, (\not{p}_+ - M) \, \gamma^\nu \right\}$$
$$\times \text{tr} \left\{ (\not{k}_+ - m) \, \gamma_\mu \, (m + \not{k}_-) \, \gamma_\nu \right\} , \qquad (10.86)$$

where the projection operators for u and v spinors, Eqs. (5.137) and (5.141), have been used.

It is left as an exercise (Prob. 10.2) to complete the calculation of the cross section. If the energy $E \gg M$ or m, the total cross section reduces to

$$\sigma = \frac{4\pi\alpha^2}{3s} , \qquad (10.87)$$

where, as always, $\alpha = \frac{e^2}{4\pi}$ is the fine structure constant. This is an interesting result; it leads to a discussion of the total cross section for the production of strongly interacting particles (hadrons) and to a discussion of the evidence for quarks.

10.4 e^+e^- ANNIHILATION

As an application of the ideas developed so far, consider e^+e^- annihilation into hadrons at very high energy. We can compute the total cross section for this process if we borrow two facts from high energy physics:

- All hadrons are made of spin $\frac{1}{2}$ quarks, which are charged, and spin 1 gluons, which are neutral.
- The strong coupling constant, g_s, is small at very high energies. In particular, the strong fine structure constant, $\alpha_s = g_s^2/4\pi$, is a function of q^2, the square of the momentum transfer, and $\alpha_s(q^2)$ decreases as q^2 increases. For high q^2 in the range of 10's to 100's of GeV^2, $\alpha_s(q^2) \lesssim 0.3$, and we can use perturbation theory.

The first of these ideas was already used in Sec. 6.2.

Using these facts, the production of hadrons from a virtual photon (created by the e^+e^- annihilation) must proceed first through the creation of a $q\bar{q}$ pair. The diagram is identical in structure to the production of a $\mu\bar{\mu}$ pair and is shown in Fig. 10.7. Since gluons are neutral, the first correction to this diagram is single gluon emission by a quark, which is described by the two Feynman diagrams shown in Fig. 10.8. In Feynman diagrams, it is customary to represent

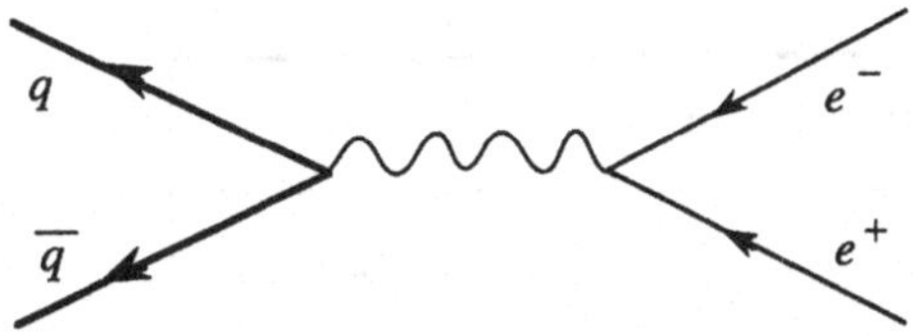

Fig. 10.7 Diagram for the production of hadronic matter from electron–positron annihilation.

the gluon by a curly, corkscrew shaped line. Since the qqg and $\bar{q}\bar{q}g$ vertices in these diagrams are proportional to the strong coupling constant g_s, they are smaller than the leading diagram (10.7) at high energy. Corrections from such processes will be calculated in Chapter 17; their effect can be expressed as a multiplicative enhancement of the lowest order cross section

$$\begin{pmatrix} \text{full} \\ \text{cross section} \end{pmatrix} = \begin{pmatrix} \text{lowest order} \\ \text{cross section} \end{pmatrix} \times \left[1 + \left(\frac{\alpha_s}{\pi}\right) + \ldots\right] .$$

The first correction is about 10% of the lowest order result, and it is natural to assume the higher order terms are negligible.

Comment: If the hadrons are expressed in terms of q, $\bar{q}$, *and gluon degrees of freedom*, the calculation is as simple as described above. However, the $q\bar{q}$ pair is not seen in the final state. Somehow, the $q\bar{q}$ pair converts itself into a variety of hadrons: N, $\bar{N}$, π^+, π^-, K's, etc. If we were to expand the hadrons in terms of these degrees of freedom, the calculation would be very complicated, and it would be impossible to predict definite results. Hence, using q, $\bar{q}$, and g degrees of freedom, we are able to predict *total cross sections*, but we cannot predict the production of $N\bar{N}$ pairs, say, without understanding how $q\bar{q} \to N\bar{N}$, a very complex process referred to as *hadronization*.

The prediction of the total hadronic cross section is usually expressed in terms of the ratio R, where

$$R = \frac{\sigma_{\text{had}}}{\sigma_{\mu^+\mu^-}} , \tag{10.88}$$

where $\sigma_{\mu^+\mu^-}$ is the $\mu^+\mu^-$ production cross section given in Eq. (10.87). It sets

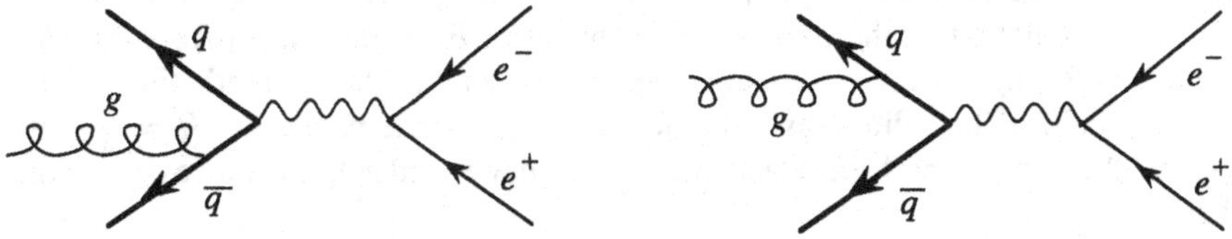

Fig. 10.8 The leading corrections to R come from these one-gluon emission diagrams.

Table 10.1 The six flavors and three generations of quarks.

	flavor	charge	mass	name
first	u	$\frac{2}{3}$	$\simeq 5-7$ MeV	*up*
	d	$-\frac{1}{3}$		*down*
second	c	$\frac{2}{3}$	~ 1.5 GeV	*charmed*
	s	$-\frac{1}{3}$	~ 150 MeV	*strange*
third	t	$\frac{2}{3}$	not seen	*top*
	b	$-\frac{1}{3}$	~ 4.5 GeV	*bottom*

the scale of the hadronic cross section.

Since the $q\bar{q}$ production diagram (10.7) differs from the $\mu^-\mu^+$ production diagram (10.3) only by the charge of the quark, we see immediately that

$$R = \frac{\sigma_{\rm had}}{\sigma_{\mu^+\mu^-}} = \sum_i Q_i^2 \,, \tag{10.89}$$

where Q_i *is the charge of the* ith *quark* (in units of e) and the *sum is over all quarks which can be produced* at the energy $\sqrt{s}$. This is because the cross section is independent of mass, etc., and depends only on $(Q_i e)^2$. Thus, only Q_i^2 enters the ratio.

The known and conjectured, quarks are grouped into three families (or generations), as shown in Table 10.1. (See Appendix D for more discussion.) According to QCD, each *flavor* of quark comes in three colors (an internal quantum number analogous to spin), so that we have the following predictions:

$$\sum_i Q_i^2 = \begin{cases} 3\left(\frac{4}{9}+\frac{1}{9}+\frac{1}{9}\right) = 2 & \text{for } W = \sqrt{s} \lesssim 3 \text{ GeV} \\ 2+3\left(\frac{4}{9}\right) = \frac{10}{3} & \text{for } 3 \lesssim W \lesssim 9 \text{ GeV} \\ \frac{10}{3}+3\left(\frac{1}{9}\right) = \frac{11}{3} & \text{for } 9 \lesssim W \lesssim 2m_{\rm top}\,. \end{cases} \tag{10.90}$$

The corrections to these simple predictions are about +10%, as described above.

A compilation of the data for R is shown in Fig. 10.9 [taken from RP 92]. Added to the upper figure are solid lines corresponding to the predictions given in Eq. (10.90). The solid lines in the lower figure are the prediction $R = \frac{11}{3}$ with theoretical corrections from electroweak and higher order QCD processes. Note that:

- The ratio R does increase at the c and b thresholds and is more or less constant between thresholds.

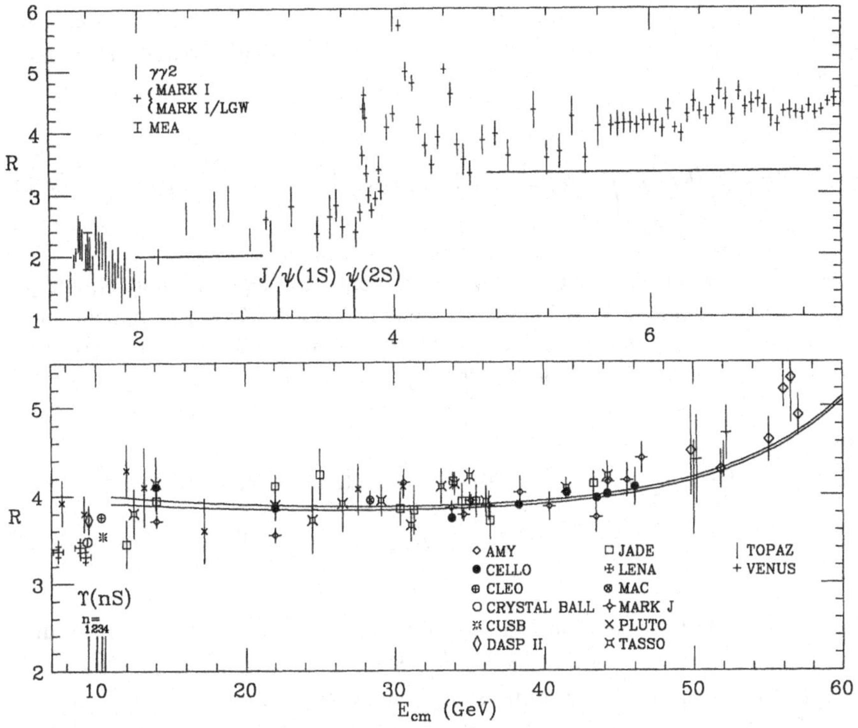

Fig. 10.9 The ratio R as a function of center of mass energy $E_{cm} = W$ (taken from [RP 92]).

- At the thresholds, $c\bar{c}$ bound states (the J/ψ-states), and $b\bar{b}$ bound states (the Υ-states) exist. These produce strong final state enhancements which modify R, as expected. Note that above the thresholds, R "settles down."
- The theoretical curves at high energy [which include corrections to the simple prediction (10.90)] are in good agreement with the data.

In all, these predictions are a beautiful confirmation of the correctness of QCD and the quark model. In particular, they give strong *support for the choice of three colors.*

10.5 FERMION PROPAGATOR: COMPTON SCATTERING

As a final example, we treat Compton scattering, which is the scattering of photons from electrons $\gamma + e \rightarrow \gamma + e$. This introduces two new Feynman rules: (i) the treatment of (real) γ's in the initial and final state and (ii) the use of a fermion propagator (in this case an electron) describing the propagation of a virtual, off-mass shell spin $\frac{1}{2}$ particle.

The first (lowest order) non-zero contribution to non-forward Compton scattering comes from the second order contribution of the radiation part of the Hamiltonian:

$$S = (-i)^2 \tfrac{1}{2} \langle k_f \, p_f | \int d^4x_1 \, d^4x_2 \, T\left(\mathcal{H}_{\mathrm{RAD}}(x_1)\mathcal{H}_{\mathrm{RAD}}(x_2)\right) | k_i \, p_i \rangle \; , \tag{10.91}$$

where k_i and k_f are the momenta of the incident and outgoing photons, and p_i and p_f are the momenta of the electron. (Why doesn't the H_{INST} term contribute in this case?) Interchanging x_1 and x_2 permits us to eliminate T in favor of a $\theta(t_1 - t_2)$ function and a factor of 2, giving

$$\begin{aligned} S = -e^2 \langle 0 | b_{p_f} \, a_{k_f} \int d^4x_1 \, d^4x_2 \, \theta(t_1 - t_2) : \bar{\psi}_e(x_1) \boldsymbol{\gamma} \cdot \boldsymbol{A}(x_1) \psi_e(x_1) : \\ \times : \bar{\psi}_e(x_2) \boldsymbol{\gamma} \cdot \boldsymbol{A}(x_2) \psi_e(x_2) : \, a^\dagger_{k_i} \, b^\dagger_{p_i} \, |0\rangle \; , \end{aligned} \tag{10.92}$$

where the a's are the annihilation operators for the A field. Next, recall that $\psi_e(x) = \psi_e^{(+)}(x) + \psi_e^{(-)}(x)$, where $\psi_e^{(+)} \sim b$ and $\psi_e^{(-)} \sim d$. Hence, to "balance" the b and $b^\dagger$ of the final and initial states, we need precisely one $\psi^{(+)}$ and one $\bar{\psi}^{(+)}$. *The other ψ and $\bar{\psi}$ operators must be left to balance each other*; i.e., their b, $b^\dagger$, d, $d^\dagger$ must pair off so that they give a non-zero vacuum expectation value. *Such internal pairing is referred to as contraction*, and we have seen it before whenever a propagator arose. This case is different only because the contraction is between the same field (electron) which also occurs in the initial and final state. *Since contractions cannot occur between fields in a single $\mathcal{H}_{\mathrm{RAD}}$ (because they are normal ordered and all vacuum expectation values give zero), there are precisely two terms which contribute*, corresponding to the two different possible contractions:

$$\begin{aligned} S = & - e^2 \langle 0 | b_{p_f} \, a_{k_f} \int d^4x_1 \, d^4x_2 \, \theta(t_1 - t_2) \\ & \times \Big\{ : \bar{\psi}_e^{(+)}(x_1) \boldsymbol{\gamma} \cdot \boldsymbol{A}(x_1) \underline{\psi_e(x_1) :: \bar{\psi}_e}(x_2) \boldsymbol{\gamma} \cdot \boldsymbol{A}(x_2) \psi_e^{(+)}(x_2) : \\ & + : \underline{\bar{\psi}_e(x_1) \boldsymbol{\gamma} \cdot \boldsymbol{A}(x_1) \psi_e^{(+)}(x_1) :: \bar{\psi}_e^{(+)}(x_2) \boldsymbol{\gamma} \cdot \boldsymbol{A}(x_2) \psi_e}(x_2) : \Big\} a^\dagger_{k_i} \, b^\dagger_{p_i} \, |0\rangle \; , \end{aligned} \tag{10.93}$$

where the contractions are shown by the horizontal brackets. Each of these terms generates two more terms corresponding to different pairings of each $\boldsymbol{A}$ with a_{k_f}

or $a^\dagger_{k_i}$. Hence we obtain

$$S = -\, e^2 \int \frac{d^4x_1\, d^4x_2\, \theta(t_1 - t_2)}{(2\pi)^6 \sqrt{16\omega_i\, \omega_f E_i E_f}}$$
$$\times \left[\epsilon_f^{i\,*} \epsilon_i^j \, e^{i(k_f\cdot x_1 - k_i \cdot x_2)} + \epsilon_f^{j\,*} \epsilon_i^i \, e^{i(k_f\cdot x_2 - k_i\cdot x_1)} \right]$$
$$\times \left\{ \bar{u}_\alpha(\boldsymbol{p}_f)\gamma^i_{\alpha\beta} \, \langle 0|\psi_\beta(x_1)\bar{\psi}_\gamma(x_2)|0\rangle \, \gamma^j_{\gamma\delta} u_\delta(\boldsymbol{p}_i) e^{i(p_f\cdot x_1 - p_i\cdot x_2)} \right.$$
$$\left. - \, \bar{u}_\alpha(\boldsymbol{p}_f)\gamma^j_{\alpha\beta} \, \langle 0|\bar{\psi}_\gamma(x_1)\psi_\beta(x_2)|0\rangle \, \gamma^i_{\gamma\delta} u_\delta(\boldsymbol{p}_i) e^{i(p_f\cdot x_2 - p_i\cdot x_1)} \right\}, \tag{10.94}$$

where the minus sign in front of the second fermion term comes from the fact that this term requires an odd number of interchanges of anticommuting Fermi fields to get $\psi^{(+)}(x_1)$ to the right and $\bar{\psi}^{(+)}(x_2)$ to the left. Both terms in { } can be combined if we interchange $i \leftrightarrow j$ and $x_1 \leftrightarrow x_2$ in the second term. The combined result can be written

$$S = -e^2 \int \frac{d^4x_1\, d^4x_2}{(2\pi)^6 \sqrt{16\omega_i\, \omega_f E_i E_f}} \left[\epsilon_f^{i\,*} \epsilon_i^j \, e^{i(k_f\cdot x_1 - k_i\cdot x_2)} + \begin{pmatrix} i \leftrightarrow j \\ x_1 \leftrightarrow x_2 \end{pmatrix} \right]$$
$$\times e^{i(p_f\cdot x_1 - p_i\cdot x_2)} \, \bar{u}_\alpha(\boldsymbol{p}_f)\gamma^i_{\alpha\beta} \, iS_{\beta\gamma}(x_1, x_2) \, \gamma^j_{\gamma\delta} u_\delta(\boldsymbol{p}_i) \,, \tag{10.95}$$

where

$$iS_{\beta\gamma}(x_1, x_2) = \theta(t_1 - t_2) \, \langle 0|\psi_\beta(x_1)\bar{\psi}_\gamma(x_2)|0\rangle - \theta(t_2 - t_1) \, \langle 0|\bar{\psi}_\gamma(x_2)\psi_\beta(x_1)|0\rangle$$
$$= \langle 0| T \left(\psi_\beta(x_1)\bar{\psi}_\gamma(x_2)\right) |0\rangle \quad . \tag{10.96}$$

Note the extra minus sign in the *vacuum expectation value of the time-ordered product of Fermi field operators*, which arises from the change in sign which must accompany any interchange of Fermi operators. This is the Fermi propagator.

To evaluate the Fermi propagator, first note that only some of the ($\pm$) parts of the fields will contribute,

$$iS_{\beta\gamma} = \theta(t_1 - t_2) \left\langle 0|\psi^{(+)}_\beta(x_1)\bar{\psi}^{(+)}_\gamma(x_2)|0\right\rangle - \theta(t_2 - t_1) \left\langle 0|\bar{\psi}^{(-)}_\gamma(x_2)\psi^{(-)}_\beta(x_1)|0\right\rangle$$
$$= \sum_{k,s} \left(\frac{1}{2E_k L^3}\right) \left\{ u_\beta(\boldsymbol{k}, s)\bar{u}_\gamma(\boldsymbol{k}, s) \, e^{-i\hat{k}\cdot(x_1 - x_2)} \, \theta(t_1 - t_2) \right.$$
$$\left. - \, v_\beta(\boldsymbol{k}, s)\bar{v}_\gamma(\boldsymbol{k}, s) \, e^{i\hat{k}\cdot(x_1 - x_2)} \, \theta(t_2 - t_1) \right\} , \tag{10.97}$$

where $\hat{k} = (E_k, \boldsymbol{k})$. Recall the matrix form of the Dirac spin sums,

$$\sum_s u_\beta(\boldsymbol{k}, s)\bar{u}_\gamma(\boldsymbol{k}, s) = \left(m + \hat{\not{k}}\right)_{\beta\gamma}$$
$$-\sum_s v_\beta(\boldsymbol{k}, s)\bar{v}_\gamma(\boldsymbol{k}, s) = \left(m - \hat{\not{k}}\right)_{\beta\gamma} \quad .$$

Replacing the spin sums by these matrices gives

$$iS_{\beta\gamma} = \int \frac{d^3k}{(2\pi)^3 2E_k} \Big\{ (m + \hat{\not k})_{\beta\gamma}\, e^{i\boldsymbol{k}\cdot(\boldsymbol{r}_1-\boldsymbol{r}_2)}\, e^{-iE_k(t_1-t_2)}\, \theta(t_1 - t_2)$$
$$+ (m - \hat{\not k})_{\beta\gamma}\, e^{-i\boldsymbol{k}\cdot(\boldsymbol{r}_1-\boldsymbol{r}_2)}\, e^{iE_k(t_1-t_2)}\, \theta(t_2 - t_1) \Big\}$$
$$= \int \frac{d^3k}{(2\pi)^3 2E_k} \Big\{ \left(m - \boldsymbol{\gamma}\cdot\boldsymbol{k} + E_k\gamma^0\right)\, e^{i\boldsymbol{k}\cdot(\boldsymbol{r}_1-\boldsymbol{r}_2)}\, e^{-iE_k(t_1-t_2)}\, \theta(t_1 - t_2)$$
$$+ \left(m - \boldsymbol{\gamma}\cdot\boldsymbol{k} - E_k\gamma^0\right)\, e^{i\boldsymbol{k}\cdot(\boldsymbol{r}_1-\boldsymbol{r}_2)}\, e^{iE_k(t_1-t_2)}\, \theta(t_2 - t_1) \Big\}, \tag{10.98}$$

where we have changed $\boldsymbol{k} \to -\boldsymbol{k}$ in the second term. Now, using the identities (4.77), we may re-express these integrals as

$$iS_{\beta\gamma} = \int \frac{d^3k}{(2\pi)^3 2E_k}\, e^{i\boldsymbol{k}\cdot(\boldsymbol{r}_1-\boldsymbol{r}_2)}$$
$$\Big\{ \int \frac{dk_0}{2\pi i}\, e^{-ik_0(t_1-t_2)} \frac{\left(m - \boldsymbol{\gamma}\cdot\boldsymbol{k} + E_k\gamma^0\right)_{\beta\gamma}}{E_k - k_0 - i\epsilon}$$
$$+ \int \frac{dk_0}{2\pi i}\, e^{-ik_0(t_1-t_2)} \frac{\left(m - \boldsymbol{\gamma}\cdot\boldsymbol{k} - E_k\gamma^0\right)_{\beta\gamma}}{E_k + k_0 - i\epsilon} \Big\}, \tag{10.99}$$

Note that

$$\frac{1}{2E_k}\left\{ \frac{m - \boldsymbol{\gamma}\cdot\boldsymbol{k} + E_k\gamma^0}{E_k - k_0 - i\epsilon} + \frac{m - \boldsymbol{\gamma}\cdot\boldsymbol{k} - E_k\gamma^0}{E_k + k_0 - i\epsilon} \right\}$$
$$= \frac{m - \boldsymbol{\gamma}\cdot\boldsymbol{k}}{E_k^2 - k_0^2 - i\epsilon} + \frac{k_0\gamma^0}{E_k^2 - k_0^2 - i\epsilon}$$
$$= \frac{m + \not k}{m^2 - k^2 - i\epsilon}\,, \tag{10.100}$$

where $k = (k_0, \boldsymbol{k})$. We have just proved a very useful identity, which is

$$\boxed{\frac{m + \not k}{m^2 - k^2 - i\epsilon} = \left(\frac{1}{2E_k}\right)\left\{ \sum_s \frac{u(\boldsymbol{k},s)\bar{u}(\boldsymbol{k},s)}{E_k - k_0 - i\epsilon} - \sum_s \frac{v(-\boldsymbol{k},s)\bar{v}(-\boldsymbol{k},s)}{E_k + k_0 - i\epsilon} \right\}.} \tag{10.101}$$

This is an important tool for relating relativistic and nonrelativistic calculations.

Our final result for the fermion propagator is

$$\boxed{\langle 0|T\left(\psi_\beta(x_1)\bar{\psi}_\gamma(x_2)\right)|0\rangle = -i \int \frac{d^4k}{(2\pi)^4} \frac{e^{-ik\cdot(x_1-x_2)}}{m^2 - k^2 - i\epsilon} \underbrace{[m + \not k]_{\beta\gamma}}_{\text{new factor}}\,.} \tag{10.102}$$

Inserting this result into the original expression (10.95) for S and separating the x_1, x_2 integrations into $X = \frac{1}{2}(x_1 + x_2)$ and $x = x_1 - x_2$ give the following result for the $\mathcal{M}$-matrix:

$$\begin{aligned}\mathcal{M} = -ie^2 \int d^4x \int \frac{d^4k}{(2\pi)^4} e^{-ik\cdot x} \\ \times \bar{u}(\boldsymbol{p}_f)\Big\{ e^{i(k_f+p_f)\cdot x}\, \not{\epsilon}_f^* \left(\frac{-i(m+\not{k})}{m^2 - k^2 - i\epsilon}\right) \not{\epsilon}_i \\ + e^{i(p_f-k_i)\cdot x}\, \not{\epsilon}_i \left(\frac{-i(m+\not{k})}{m^2 - k^2 - i\epsilon}\right) \not{\epsilon}_f^* \Big\} u(\boldsymbol{p}_i)\,,\end{aligned} \tag{10.103}$$

where we used the fact that ϵ_i and ϵ_f have no time components to write $\gamma^i \epsilon_f^i = -\gamma^\mu \epsilon_{\mu f} = -\not{\epsilon}_f$. Carrying out the x and k integrals gives

$$\boxed{\begin{aligned}\mathcal{M} = -ie^2\, \bar{u}(\boldsymbol{p}_f)\Big\{ \not{\epsilon}_f^* \left(\frac{-i(m+\not{k}_f + \not{p}_f)}{m^2 - (k_f + p_f)^2 - i\epsilon}\right) \not{\epsilon}_i \\ + \not{\epsilon}_i \left(\frac{-i(m+\not{p}_f - \not{k}_i)}{m^2 - (p_f - k_i)^2 - i\epsilon}\right) \not{\epsilon}_f^* \Big\} u(\boldsymbol{p}_i)\,.\end{aligned}} \tag{10.104}$$

This result corresponds to the two Feynman diagrams shown in Fig. 10.10 and gives some new Feynman rules. In addition to those previously encountered, we have:

Rule 2: a factor of

$$\frac{-i(m+\not{k})_{\alpha\beta}}{m^2 - k^2 - i\epsilon}$$

for each internal fermion line carrying momentum k and Dirac indices α and β. (Fix k by momentum conservation.)

Rule 3: take matrix elements along each fermion line by assembling the incoming fermion spinor, vertex operators, propagators, and outgoing fermion spinors in *order* along the fermion line to make a well-formed matrix element.

for photons, construct well-formed vector products by saturating any free vector polarization indices μ on current operators γ^μ by:

- multiplying by ϵ_μ^* for each *outgoing photon* with polarization index μ.
- multiplying by ϵ_μ for each *incoming photon* with polarization index μ.

Rule 4: symmetrize between identical bosons in the initial and final states. (This rule is analogous to the need to anti-symmetrize when identical fermions are in the initial and final states.)

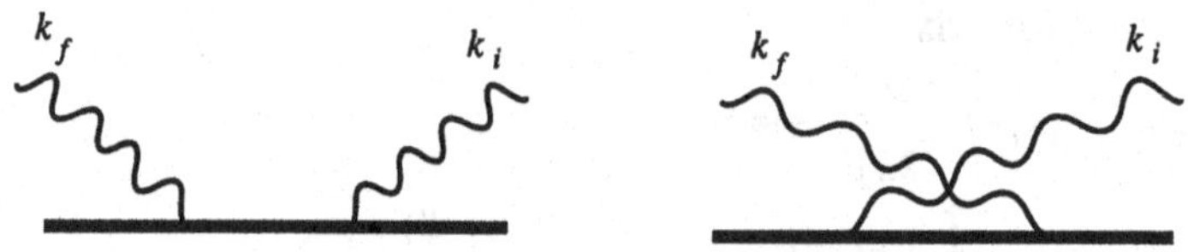

Fig. 10.10 Feynman diagrams for Compton scattering.

With these new additions to Rules 2–4, we have most of the basic rules needed for elementary calculations. We are still *missing the rules associated with renormalization* (so far, all renormalization constants have been set to unity), closed loops, and some of the rules associated with isospin. We have also not discussed the electrodynamics of spin 0 and 1 bosons. We will cover some of these topics in the next chapter and later in Chapters 13 and 15.

Cross Section

We now calculate the unpolarized Compton scattering cross section in the LAB frame. To simplify the expression, we use

$$\epsilon_k \cdot k_i = \epsilon_f \cdot k_f = \epsilon_i \cdot p_i = \epsilon_f \cdot p_i = 0 \,. \tag{10.105}$$

The last two conditions hold in the LAB frame, where $p_i = (m, \mathbf{0})$ and $\epsilon = (0, \boldsymbol{\epsilon})$, but would not hold in an arbitrary frame. (This is a feature of the Coulomb gauge, which requires that $\epsilon^0 = 0$, a condition which is not frame independent. Nevertheless, the total result is frame independent.) Also, $(p_i + k_i)^2 = m^2 + 2m\omega_i$ because $k_i^2 = k_f^2 = 0$. Taking ϵ_f to be real for simplicity,

$$\begin{aligned}
\mathcal{M} &= -e^2 \bar{u}(\boldsymbol{p}_f) \left\{ \not{\epsilon}_f \, \frac{(m + \not{p}_i + \not{k}_i)}{-2m\omega_i} \, \not{\epsilon}_i + \not{\epsilon}_i \, \frac{(m + \not{p}_i - \not{k}_f)}{2m\omega_f} \, \not{\epsilon}_f \right\} u(\boldsymbol{p}_i) \\
&= e^2 \bar{u}(\boldsymbol{p}_f) \left[\frac{\not{\epsilon}_f \not{k}_i \not{\epsilon}_i}{2m\omega_i} + \frac{\not{\epsilon}_i \not{k}_f \not{\epsilon}_f}{2m\omega_f} \right] u(\boldsymbol{p}_i) \,,
\end{aligned} \tag{10.106}$$

we have reduced $\mathcal{M}$ using $\not{p}_i \not{\epsilon}_i = 2p_i \cdot \epsilon_i - \not{\epsilon}_i \not{p}_i = -\not{\epsilon}_i \not{p}_i$, and $\not{p}_i u(\boldsymbol{p}_i) = m u(\boldsymbol{p}_i)$, etc.

Next, calculate the sum over spins of $|\mathcal{M}|^2$. Using the technique for carrying out spin sums which we worked out in the previous sections gives

$$\begin{aligned}
\sum_{\text{spins}} |\mathcal{M}|^2 = \frac{e^4}{(2m)^2} \operatorname{tr} \Bigg\{ & (m + \not{p}_f) \left[\frac{\not{\epsilon}_f \not{k}_i \not{\epsilon}_i}{\omega_i} + \frac{\not{\epsilon}_i \not{k}_f \not{\epsilon}_f}{\omega_f} \right] \\
& \times (m + \not{p}_i) \left[\frac{\not{\epsilon}_i \not{k}_i \not{\epsilon}_f}{\omega_i} + \frac{\not{\epsilon}_f \not{k}_f \not{\epsilon}_i}{\omega_f} \right] \Bigg\} \,.
\end{aligned}$$

To reduce this expression (essential because we have products of eight γ-matrices, which are too complicated to work with), expand out the product and use results like $\not{k}_i^2 = \not{k}_f^2 = 0$ and $\not{\epsilon}_i^2 = -1 = \not{\epsilon}_f^2$. We have

$$
\begin{aligned}
\sum_{\text{spins}} |\mathcal{M}|^2 \\
= \frac{e^4}{(2m)^2} \operatorname{tr} \Bigg\{ & \overbrace{(m+\not{p}_f)}^{\overbrace{m+\not{p}_i}^{=0}+\not{k}_i-\not{k}_f} \Bigg[\overbrace{\frac{\not{\epsilon}_f \not{k}_i \not{\epsilon}_i \not{\epsilon}_i \not{k}_i \not{\epsilon}_f}{\omega_i^2}}^{=0} + \frac{\not{\epsilon}_f \not{k}_i \not{\epsilon}_i \not{\epsilon}_f \not{k}_f \not{\epsilon}_i}{\omega_i \omega_f} \\
& + \frac{\not{\epsilon}_i \not{k}_f \not{\epsilon}_f \not{\epsilon}_i \not{k}_i \not{\epsilon}_f}{\omega_i \omega_f} + \overbrace{\frac{\not{\epsilon}_i \not{k}_f \not{\epsilon}_f \not{\epsilon}_f \not{k}_f \not{\epsilon}_i}{\omega_f^2}}^{=0} \Bigg] (m - \not{p}_i) \\
& - 2(m+\not{p}_f) \left[\frac{\not{\epsilon}_f \not{k}_i \not{\epsilon}_i}{\omega_i} + \frac{\not{\epsilon}_i \not{k}_f \not{\epsilon}_f}{\omega_f} \right] \overbrace{\left(\not{\epsilon}_i \not{\epsilon}_f + \not{\epsilon}_f \not{\epsilon}_i \right)}^{2\epsilon_i \cdot \epsilon_f} m \Bigg\} \\
= \frac{e^4}{(2m)^2} \operatorname{tr} \Bigg\{ & \frac{(\not{k}_i - \not{k}_f)}{\omega_i \omega_f} \left[\not{\epsilon}_i \not{k}_f \not{\epsilon}_f \not{\epsilon}_i \not{k}_i \not{\epsilon}_f + \not{\epsilon}_f \not{k}_i \not{\epsilon}_i \not{\epsilon}_f \not{k}_f \not{\epsilon}_i \right] (\overbrace{m}^{=0} - \not{p}_i) \\
& - 4m(\epsilon_i \cdot \epsilon_f)(\overbrace{m}^{=0} + \not{p}_f) \left[\frac{\not{\epsilon}_f \not{k}_i \not{\epsilon}_i}{\omega_i} + \frac{\not{\epsilon}_i \not{k}_f \not{\epsilon}_f}{\omega_f} \right] \Bigg\} \\
= \frac{e^4}{4m^2} \operatorname{tr} \Bigg\{ & \frac{2(k_f \cdot \epsilon_i)}{\omega_i \omega_f} \not{k}_f \not{\epsilon}_f \not{\epsilon}_i \not{k}_i \not{\epsilon}_f \not{p}_i + \frac{2m}{\omega_i} \not{\epsilon}_f \not{k}_i \not{\epsilon}_i \not{\epsilon}_f \not{k}_f \not{\epsilon}_i \\
& - \frac{2(k_f \cdot \epsilon_i)}{\omega_i \omega_f} \not{\epsilon}_f \not{k}_i \not{\epsilon}_i \not{\epsilon}_f \not{k}_f \not{p}_i - \frac{2(k_i \cdot \epsilon_f)}{\omega_i \omega_f} \not{k}_i \not{\epsilon}_i \not{\epsilon}_f \not{k}_f \not{\epsilon}_i \not{p}_i \\
& - \frac{2m}{\omega_f} \not{\epsilon}_i \not{k}_f \not{\epsilon}_f \not{\epsilon}_i \not{k}_i \not{\epsilon}_f + \frac{2(k_i \cdot \epsilon_f)}{\omega_i \omega_f} \not{\epsilon}_i \not{k}_f \not{\epsilon}_f \not{\epsilon}_i \not{k}_i \not{p}_i \\
& - 4m(\epsilon_i \cdot \epsilon_f) \not{p}_f \left[\frac{\not{\epsilon}_f \not{k}_i \not{\epsilon}_i}{\omega_i} + \frac{\not{\epsilon}_i \not{k}_f \not{\epsilon}_f}{\omega_f} \right] \Bigg\} ,
\end{aligned}
$$

where the last expression was obtained by moving the $\not{k}$'s in the factor $(\not{k}_i - \not{k}_f)$ to the right (or left by cyclically permuting the trace) and using $\not{k}^2 = 0$. Next, note that

$$
\operatorname{tr} \{ \not{a} \not{b} \not{c} \not{d} \not{e} \not{f} \} = \operatorname{tr} \{ \not{f} \not{e} \not{d} \not{c} \not{b} \not{a} \} . \tag{10.107}
$$

Proof:

$$
\begin{aligned}
\operatorname{tr} \{ \not{a} \not{b} \not{c} \not{d} \not{e} \not{f} \} &= \operatorname{tr} \left\{ \not{f}^{\mathsf{T}} \not{e}^{\mathsf{T}} \not{d}^{\mathsf{T}} \not{c}^{\mathsf{T}} \not{b}^{\mathsf{T}} \not{a}^{\mathsf{T}} \right\} \\
&= \operatorname{tr} \left\{ C^{-1} C \not{f}^{\mathsf{T}} \not{e}^{\mathsf{T}} \not{d}^{\mathsf{T}} \not{c}^{\mathsf{T}} \not{b}^{\mathsf{T}} \not{a}^{\mathsf{T}} \right\} \\
&= \operatorname{tr} \left\{ C^{-1} \not{f} \not{e} \not{d} \not{c} \not{b} \not{a} \, C \right\} = \operatorname{tr} \{ \not{f} \not{e} \not{d} \not{c} \not{b} \not{a} \} .
\end{aligned}
$$

∎

Then the two $k_f \cdot \epsilon_i/(\omega_i\omega_f)$ terms cancel, the two $k_i \cdot \epsilon_f/(\omega_i\omega_f)$ terms cancel, and the coefficients of the $1/\omega_i$ term and the $1/\omega_f$ term are equal. Finally, we get

$$\sum_{\text{spins}} |\mathcal{M}|^2 = \frac{e^4}{4m^2}\,\mathrm{tr}\left\{\left(\frac{1}{\omega_i}-\frac{1}{\omega_f}\right)2m\, \overbrace{\not\epsilon_f \not k_i \not\epsilon_i \not\epsilon_f \not k_f \not\epsilon_i}^{-2(\epsilon_f\cdot\epsilon_i)\not\epsilon_f \not k_i \not\epsilon_i \not k_f - \not k_i \not\epsilon_i \not k_f \not\epsilon_i}\right.$$
$$- 4m(\epsilon_i\cdot\epsilon_f)\left[(\not p_i + \not k_i - \not k_f)\frac{\not\epsilon_f \not k_i \not\epsilon_i}{\omega_i}\right.$$
$$\left.\left. + (\not p_i + \not k_i - \not k_f)\frac{\not\epsilon_i \not k_f \not\epsilon_f}{\omega_f}\right]\right\}. \tag{10.108}$$

After noting the cancellations and using $\not k^2 = 0$, we obtain

$$\sum_{\text{spins}} |\mathcal{M}|^2 = \frac{e^4}{4m^2}\,\mathrm{tr}\left\{2m\left(\frac{1}{\omega_i}-\frac{1}{\omega_f}\right)\overbrace{(-\not k_i \not\epsilon_i \not k_f \not\epsilon_i)}^{-\not k_i \not k_f}\right.$$
$$\left. - 4m(\epsilon_i\cdot\epsilon_f)\,\not p_i\left(\frac{\not\epsilon_f \not k_i \not\epsilon_i}{\omega_i} + \frac{\not\epsilon_i \not k_f \not\epsilon_f}{\omega_f}\right)\right\}$$
$$= e^4\left[8(\epsilon_i\cdot\epsilon_f)^2 + 2\frac{(k_i\cdot k_f)}{m}\left(\frac{1}{\omega_f}-\frac{1}{\omega_i}\right)\right]. \tag{10.109}$$

Now return to the cross section. Using the general formula (9.63) gives (let $k_f = k'$, $k_i = k$)

$$\left.\frac{d\sigma}{d\Omega}\right|_{\text{LAB}} = \frac{1}{16(2\pi)^2 m^2}\left(\frac{k'}{k}\right)\frac{1}{1-\frac{q^2}{2mk'}}\frac{1}{2}\sum_{\text{spins}}|\mathcal{M}|^2\,, \tag{10.110}$$

where the $\frac{1}{2}$ is for the *average* over initial electron spins. Here

$$\begin{aligned} q^2 &= (k_f - k_i)^2 = -2k\cdot k' = -2kk'(1-\cos\theta) \\ &= (p_f - p_i)^2 = 2m^2 - 2mE_f = 2m(k'-k)\ . \end{aligned} \tag{10.111}$$

Hence, as we found before, the recoil factor is

$$\frac{1}{1-\frac{q^2}{2mk'}} = \frac{1}{1+\frac{k}{m}(1-\cos\theta)} = \frac{k'}{k}\ . \tag{10.112}$$

Thus

$$\left.\frac{d\sigma}{d\Omega}\right|_{\text{LAB}} = \left(\frac{k'}{k}\right)^2\frac{\alpha^2}{4m^2}\left[4(\epsilon_i\cdot\epsilon_f)^2 + \frac{k_i\cdot k_f}{m}\left(\frac{1}{\omega_f}-\frac{1}{\omega_i}\right)\right] \tag{10.113}$$

or, in terms of $k' = \omega_f$, $k = \omega_i$, and θ, using $k_i\cdot k_f = kk'(1-\cos\theta) = m(k-k')$, we have

$$\boxed{\left.\frac{d\sigma}{d\Omega}\right|_{\text{LAB}} = \frac{\alpha^2}{4m^2}\left(\frac{k'}{k}\right)^2\left[4\left(\epsilon_i\cdot\epsilon_f\right)^2 + \frac{k}{k'} + \frac{k'}{k} - 2\right]} \quad . \tag{10.114}$$

This is the famous *Klein–Nishina* formula.

Thomson Cross Section and the Classical Limit

We now assume that k and $k' \ll m$, so there is no recoil and $k \cong k'$ (long wavelength limit). Then

$$\frac{d\sigma}{d\Omega} = \frac{\alpha^2}{m^2}\left(\epsilon\cdot\epsilon'\right)^2 \quad . \tag{10.115}$$

If we average over initial polarization states and sum over final states, using

$$\sum_\alpha \epsilon^{\alpha i}\epsilon^{\alpha j} = \delta_{ij} - \frac{k^i k^j}{k^2} \quad ,$$

we obtain

$$\begin{aligned}\frac{1}{2}\sum_{\text{spins}}\left(\epsilon\cdot\epsilon'\right)^2 &= \frac{1}{2}\left(\delta_{ij} - \hat{k}^i\hat{k}^j\right)\left(\delta_{ij} - \hat{k}'^i\hat{k}'^j\right)\\ &= \frac{1}{2}\left(1+\cos^2\theta\right) \quad .\end{aligned} \tag{10.116}$$

Hence

$$\left.\frac{d\sigma}{d\Omega}\right|_{\text{unpolarized}} = \frac{\alpha^2}{m^2}\,\frac{1}{2}\left(1+\cos^2\theta\right) \quad , \tag{10.117}$$

and integrating over the solid angle gives the famous Thomson cross section

$$\boxed{\sigma_{\text{total}} = \frac{8\pi}{3}\,\frac{\alpha^2}{m^2}} \quad . \tag{10.118}$$

This *classical result* can be obtained quite easily directly from the relativistic $\mathcal{M}$-matrix by letting $k \simeq k' \to 0$ right from the start. A remarkable fact emerges from this calculation. If we use the identity (10.101)

$$\frac{m+\not{p}}{m^2-p^2} = \left(\frac{1}{2E_p}\right)\sum_s\left\{\frac{u(\boldsymbol{p},s)\bar{u}(\boldsymbol{p},s)}{E_p - p_0} - \frac{v(-\boldsymbol{p},s)\bar{v}(-\boldsymbol{p},s)}{E_p + p_0}\right\} \quad ,$$

the $\mathcal{M}$-matrix can be written in the form

$$\mathcal{M} = \sum_s \frac{[\bar{u}(\boldsymbol{p}_f, s_f)\, \mathcal{O}'\, u(\boldsymbol{p}_i + \boldsymbol{k}_i, s)]\, [\bar{u}(\boldsymbol{p}_i + \boldsymbol{k}_i, s)\, \mathcal{O}\, u(\boldsymbol{p}_i, s_i)]}{2E_{p_i+k_i}\, [E_{p_i+k_i} - (E_{p_i} + \omega_i)]}$$
$$- \sum_s \frac{[\bar{u}(\boldsymbol{p}_f, s_f)\, \mathcal{O}'\, v(-\boldsymbol{p}_i - \boldsymbol{k}_i, s)]\, [\bar{v}(-\boldsymbol{p}_i - \boldsymbol{k}_i, s)\, \mathcal{O}\, u(\boldsymbol{p}_i, s_i)]}{2E_{p_i+k_i}\, [E_{p_i+k_i} + (E_{p_i} + \omega_i)]}$$
$$+ \text{crossed (exchange) term} \ . \tag{10.119}$$

This is a convenient form for taking the classical limit, and the Thomson cross section can be easily computed in this way. However, it turns out that the classical limit comes *entirely* from the $v\,\bar{v}$ terms (the negative energy contribution). It is left as an exercise (Prob. 10.3) to work this out and discuss the results.

PROBLEMS

10.1 Calculate the differential cross section, in the one-photon exchange approximation, for the scattering of electrons from pions (pseudoscalar particles) initially at rest. First, write down the correct $\mathcal{M}$-matrix using the Feynman rules (the form of the vertex for a spin zero boson is given in the Appendix). Then, square and calculate the unpolarized cross section. Finally, show that when the energy E of the incoming electron becomes very large,

$$\frac{d\sigma}{d\Omega} = \left(\frac{\alpha}{2E}\right)^2 \frac{\cos^2\frac{\theta}{2}}{\sin^4\frac{\theta}{2}\left[1 + \frac{2E}{m_\pi}\sin^2\frac{\theta}{2}\right]} \ .$$

10.2 Calculate the total cross section for the annihilation of electrons and positrons into muons and antimuons. (The muon is just like a heavy electron.) Do the calculation in the center of mass system, which is also the LAB system in a colliding beam accelerator where this experiment would usually be performed. When the energy E of the electron becomes very large, show that the unpolarized *total* cross section becomes

$$\sigma = \frac{4\pi\alpha^2}{3s} \qquad s \equiv 4E^2 \ .$$

[You may use the results from Eq. (10.85) and (10.86).]

10.3 Show that the Thomson cross section can be obtained from the relativistic Feynman diagrams for Compton scattering using only the negative energy part of the virtual electron propagator. That is, using the decomposition

$$\frac{m + \not{p}}{m^2 - p^2} = \left(\frac{1}{2E_p}\right) \sum_s \left\{ \frac{u(\boldsymbol{p}, s)\bar{u}(\boldsymbol{p}, s)}{E_p - p_0} - \frac{v(-\boldsymbol{p}, -s)\bar{v}(-\boldsymbol{p}, -s)}{E_p + p_0} \right\} ,$$

show that the full result for the Thomson cross section comes from the *second term* in this decomposition, the first term giving a vanishingly small contribution. To make the calculation simple, carry out the calculation in the limit when all momenta are $\ll m$ right from the start.

10.4 Suppose the muon could decay onto an electron and a photon through an electromagnetic-like term of the form

$$\mathcal{H}_{\text{int}} = -g\left[\bar{\psi}_{\text{muon}}(x)\gamma^{\alpha}\psi_{\text{e}}(x) + \bar{\psi}_{\text{e}}(x)\gamma^{\alpha}\psi_{\text{muon}}(x)\right] A_{\alpha}(x) \ ,$$

where g is an unknown constant. (As far as we know, this process does not occur.) Calculate the total rate for this decay in terms of the unknown constant g.

10.5 Consider the annihilation of electron–positron pairs into two photons, i.e., $e^- + e^+ \to 2\gamma$.

(a) Draw all of the Feynman diagrams which contribute to this process to order e^2 in the electric charge. Let the momenta of the incoming electron be p_-, of the incoming positron p_+, and of the outgoing photons k_1 and k_2. Label each diagram with these momenta and the momenta of any internal lines.

(b) Write down the correct Feynman amplitude for each diagram.

10.6 Assume two protons scatter by exchanging either a photon or a neutral π^0 meson.

(a) Draw clearly labeled Feynman diagrams showing the interactions to lowest order in the electric charge e or the πNN coupling constant g. Give the mathematical expression for the $\mathcal{M}$-matrix corresponding to each diagram.

(b) Give a *rough estimate* of the comparative size of the different diagrams when the scattering takes place at high energy and at non-forward angles ($\theta > 10°$, for example). Which process is more important?

10.7 Consider electron–proton (ep) and positron–proton ($\bar{e}p$) scattering in the framework of QED.

(a) Draw all the Feynman diagrams which contribute to ep and $\bar{e}p$ scattering in *lowest* order perturbation theory. Write the $\mathcal{M}$-matrix corresponding to each diagram.

(b) Calculate the *difference* in the cross sections for ep and $\bar{e}p$ scattering. It is sufficient to find

$$\sum_{\text{spins}} |\mathcal{M}|^2_{ep} - \sum_{\text{spins}} |\mathcal{M}|^2_{\bar{e}p} \ .$$

Alternatively, you may be able to see the answer by examining $\mathcal{M}_{ep}$ and $\mathcal{M}_{\bar{e}p}$ directly.

(c) Using the insight gained in part (b), roughly how accurate an experiment would be required to distinguish between ep and $\bar{e}p$ scattering?

Fig. 10.11 Diagrams for πN scattering (Prob. 10.8).

10.8 Pion nucleon scattering.

(a) Calculate the *total* cross section for $\pi^+ p$ and $\pi^- p$ scattering near threshold using only the two Feynman diagrams shown in Fig. 10.11. Assume the πNN coupling is $-\sqrt{2}\, g\gamma^5$ for positively charged pions, $g\,\gamma^5$ for $\pi^0 pp$, and $-g\gamma^5$ for $\pi^0 nn$, where $g^2/4\pi = 14.0$. Carry out the following steps:

(i) Write down the exact Feynman amplitudes, from the two diagrams in Fig. 10.11, for the following processes:

$$\pi^+ p \to \pi^+ p$$
$$\pi^- p \to \pi^- p$$
$$\pi^- p \to \pi^0 n \ .$$

(ii) Evaluate these amplitudes in the center of mass system in the limit when the momenta $|\boldsymbol{p}|$ of both the proton and pion is zero.

(iii) Calculate the total cross section for $\pi^+ + p \to$ anything and $\pi^- + p \to$ anything and compare with the experimental results, which are

$$\sigma_{\pi^+ p} = \frac{4\pi}{m_\pi^2}\{0.0114 \pm 0.0006\}$$
$$\sigma_{\pi^- p} = \frac{4\pi}{m_\pi^2}\{0.0249 \pm 0.0014\} \ ,$$

where m_π is the pion mass.

(b) Redo the calculation of part (a) with the πNN coupling replaced by

$$-\sqrt{2}\, g \frac{\left(\not p_f - \not p_i\right)}{2m_N}\gamma^5 \quad \text{for positively charged pions}$$
$$\pm g \frac{\left(\not p_f - \not p_i\right)}{2m_N}\gamma^5 \quad \text{for neutral pions} \begin{cases} + & \text{for } \pi^0 pp \\ - & \text{for } \pi^0 nn\,, \end{cases}$$

where p_f is the outgoing nucleon four-momentum, p_i is the incoming nucleon four-momentum, and m_N is the nucleon mass. Do all three steps of this calculation just as you did for part (a) above. (If you are very careful, you will discover an important result in πN physics.)

CHAPTER 11

LOOPS AND INTRODUCTION TO RENORMALIZATION

We now turn to the general question of how to calculate the higher order terms which arise in the perturbation expansion for the S-matrix. We have chosen to introduce this study by examining all the terms which arise in second order QED. The discussion therefore serves two purposes: the specific examples we study are of practical importance, and they also are rich enough to illustrate most of the issues which will arise in a general study.

A survey of all Feynman diagrams generated by second order processes shows that they are of two kinds. In some, the momenta of the internal (or virtual) particles is fixed by energy–momentum conservation. Such diagrams are referred to as "tree diagrams," and their "computation" involves no more that writing them down and evaluating all internal momenta using energy–momentum conservation. Examples of tree diagrams are scattering in the one-photon exchange approximation, annihilation and pair production through a single intermediate photon, and Compton scattering, all of which were studied in the last chapter. The second kind of diagram has closed *loops* in which the momenta of all of the internal particles are *not* fixed by the four-momenta of the external particles. For *each* loop there is *one* four-momentum completely unspecified by energy–momentum conservation. Each different value of this internal four-momentum will give a different amplitude connecting the *same* initial and final states, and these amplitudes are therefore *indistinguishable*, and the rules of quantum mechanics tell us that these must be added together (by integrating over all possible values of the internal four-momentum) *before* we square the result to obtain predictions for physical observables. The evaluation of loop diagrams therefore requires that loop integrals be carried out, and this is much more difficult. Remarkably, there is a very powerful, standard method for evaluating loops. This method, referred to as *dimensional regularization*, will be introduced in Sec. 11.6. We will see that these integrations will sometimes produce infinities, which must be systematically removed if we are to obtain meaningful answers from the higher order terms. The infinities are first isolated through a process called *regularization* and then removed from the theory through a process referred to as *renormalization*. This involves systematically redefining the coupling constants and masses of the theory so that the infinities are systematically absorbed into these parameters, which are then taken from experiment.

This chapter prepares the way for the more complete study of renormalization presented in Chapters 16 and 17. It also is a prerequisite for the study of bound states and unitarity presented in Chapter 12. However, a reading of this chapter in not necessary for the additional study of symmetries presented in Chapter 13, nor for a large part of Chapters 14 and 15, and the reader may prefer to turn to these topics first.

11.1 WICK'S THEOREM

We begin this chapter with a brief study of Wick's theorem, a standard tool for the study of higher order processes when quantum fields are treated as operators on a Fock space. Later, after we have introduced the path integral formalism in Chapter 14, we will be able to obtain the same results we obtain here using a completely different method.

The first problem we encounter in computation of higher order terms is the computation of the matrix elements of products of field operators, and Wick's theorem provides a systematic way to reduce these products. There are two theorems: one tells how to reduce a product of field operators to a sum of terms, *each* of which is a *normal-ordered* product, and the second tells how to reduce the product to a sum of terms, *each* of which is a *time-ordered* product. In the following discussion, we will use the symbol ϕ to denote any quantum field: scalar, vector, or spinor.

Wick's Theorem for Normal-ordered Products

A *contraction* of two fields will be defined to be their vacuum expectation value and will be denoted by a square bracket connecting the two fields,

$$\underbracket{\phi(x)\phi^\dagger}(y) \equiv \langle 0|\phi(x)\phi^\dagger(y)|0\rangle \quad . \tag{11.1}$$

This contraction occurs naturally when normal ordering a product of two fields. Recall that fields can be broken into positive and negative frequency parts,

$$\begin{aligned} \phi^{(+)}(x) &= \sum_k \frac{1}{\sqrt{2E_k L^3}}\, a_k\, e^{-ik\cdot x} \\ \phi^{(-)}(x) &= \sum_k \frac{1}{\sqrt{2E_k L^3}}\, c_k^\dagger\, e^{ik\cdot x} , \end{aligned} \tag{11.2}$$

and since the $(-)$ parts are always associated with creation operators,

$$\begin{aligned} \phi^{\dagger(-)}(x) &= \sum_k \frac{1}{\sqrt{2E_k L^3}}\, a_k^\dagger\, e^{ik\cdot x} \neq \left[\phi^{(-)}(x)\right]^\dagger \\ \phi^{\dagger(+)}(x) &= \sum_k \frac{1}{\sqrt{2E_k L^3}}\, c_k\, e^{-ik\cdot x} \neq \left[\phi^{(+)}(x)\right]^\dagger \quad . \end{aligned} \tag{11.3}$$

In order to treat commuting and anticommuting fields at the same time, we will use the notation

$$[a, b]_{\pm} \equiv ab \pm ba \ . \tag{11.4}$$

Then, it follows that, for both Bose and Fermi fields,

$$\phi(x)\phi^{\dagger}(y) =: \phi(x)\phi^{\dagger}(y): + \left[\phi^{(+)}(x), \phi^{\dagger(-)}(y)\right]_{\pm} \ . \tag{11.5}$$

Also, since $\phi^{(+)}(x)|0\rangle = 0$ and $\phi^{\dagger(+)}(x)|0\rangle = 0$, it follows that

$$\langle 0|\phi(x)\phi^{\dagger}(y)|0\rangle = \left[\phi^{(+)}(x), \phi^{\dagger(-)}(y)\right]_{\pm} \ , \tag{11.6}$$

and therefore, for both Bose and Fermi fields,

$$\boxed{\phi(x)\phi^{\dagger}(y) =: \phi(x)\phi^{\dagger}(y): + \underbrace{\phi(x)\phi^{\dagger}}(y) \ .} \tag{11.7}$$

This observation was previously encountered in Eq. (7.43), but now will be generalized to products of more than two fields.

When more than two fields are present, we will frequently encounter the *product* of a contraction (a c-number) *multiplied by* a normal-ordered product of fields (a q-number). For this purpose it is convenient to *define* a *rearrangement* of this product in the following way*:

$$\begin{aligned}(-1)^{p}\underbrace{\phi_j\phi_k} : \phi_1 \cdots \phi_{j-1}\phi_{j+1} \cdots \phi_{k-1}\phi_{k+1} \cdots \phi_n: \\ =: \phi_1\phi_2 \cdots \underbrace{\phi_j \cdots \phi_k} \cdots \phi_n: \ ,\end{aligned} \tag{11.8}$$

where $\phi_i = \phi_i(x_i)$ and p is the number of interchanges of *Fermi* fields (even or odd) required to move ϕ_j and ϕ_k from their position on the RHS of the equation to their position on the LHS of the equation where they are in front of the product. This notation is convenient, but don't forget that *no contractions are possible within a normal ordered product because the normal ordering insures that any vacuum expectation value of any two fields is zero.* The product on the RHS of (11.8) should *not* be thought of as a contraction of fields within a normal-ordered product, but only as a shorthand for the LHS of the equation. Finally, note that the phase $(-1)^p$ is always $+1$ unless both ϕ_j and ϕ_k are Fermi fields, and there are other Fermi fields "in the way" which must be passed in pulling ϕ_j and ϕ_k to the front.

With these definitions, the Wick theorem for normal-ordered products can be stated in a deceptively simple way:

> **Theorem:** The *ordinary* product of field operators is equal to the sum of *normal* products with *all possible contractions*, including the normal product with no contractions.

*The discussion here follows Bogoliubov and Shirkov (1959).

Formally, we can write

$$\begin{aligned}
\phi_1\cdots\phi_n =&: \phi_1\cdots\phi_n : + : \wick{\c1\phi_1\c1\phi_2}\cdots\phi_n : + : \wick{\c1\phi_1\phi_2\c1\phi_3}\cdots\phi_n : + \cdots \\
&+ : \phi_1\wick{\c1\phi_2\c1\phi_3}\cdots\phi_n : + \cdots \\
&+ : \wick{\c1\phi_1\c1\phi_2\c2\phi_3\c2\phi_4}\phi_5\cdots\phi_n : + \cdots \\
&+ : \wick{\c1\phi_1\c1\phi_2\c2\phi_3\c2\phi_4\c3\phi_5\c3\phi_6}\phi_7\cdots\phi_n + \cdots \quad .
\end{aligned} \tag{11.9}$$

Note that if all pairs of fields have non-zero contractions (which is not generally the case, of course), then for even n the above sum contains

1	uncontracted term
$\frac{n(n-1)}{2}$	terms with one contraction
$\frac{1}{2}\,\frac{n(n-1)}{2}\,\frac{(n-2)(n-3)}{2}$	terms with two contractions
$\vdots$	
$\frac{n!}{2^{n/2}(n/2)!}$	fully contracted terms.

However, only fields with non-zero commutation or anticommutation relations can give a non-zero contraction, and thus in practice most of the fields will not contract. Hence, for example,

$$\wick{\c1\psi\c1\psi} = 0 \qquad \wick{\c1\phi\c1\psi} = 0 \qquad \wick{\c1\psi\c1{\bar\psi}} \neq 0 \qquad \wick{\c1\phi\c1\phi^\dagger} \neq 0$$

$$\wick{\c1\phi\c1\phi} \begin{cases} = 0 & \text{if } \phi \text{ is a charged field} \\ \neq 0 & \text{if } \phi \text{ is self-conjugate} \end{cases} \quad .$$

The general proof of Wick's theorem (which is not difficult) can be found in many texts. Rather than present a general proof, we will work it out for second order QED. We limit ourselves to spinor QED, where the radiation part of the interaction Hamiltonian has the form $H =: \bar\psi(x)\psi(x)a(x):$, where, in the interests of simplicity, the Dirac matrix and vector indices have been suppressed. The second order product of two radiation terms (time-ordered products will come later) which we will encounter is

$$H_2 =: \bar\psi(x)\psi(x)A(x): \; : \bar\psi(y)\psi(y)A(y): \; . \tag{11.10}$$

Fields within a normal-ordered product must not be contracted, because they are already in normal order. Applied to the above product, Wick's theorem gives eight terms:

$$\begin{aligned}
H_2 =&: \bar\psi(x)\psi(x)A(x)\bar\psi(y)\psi(y)A(y): + : \bar\psi(x)\psi(x)\wick{\c1 A(x)\bar\psi(y)\psi(y)\c1 A}(y): \\
&+ : \wick{\c1{\bar\psi}(x)\psi(x)A(x)\bar\psi(y)\c1\psi}(y)A(y): + : \bar\psi(x)\wick{\c1\psi(x)A(x)\c1{\bar\psi}}(y)\psi(y)A(y): \\
&+ : \wick{\c1{\bar\psi}(x)\psi(x)\c2 A(x)\bar\psi(y)\c1\psi(y)\c2 A}(y): + : \bar\psi(x)\wick{\c1\psi(x)\c2 A(x)\c1{\bar\psi}(y)\psi(y)\c2 A}(y): \\
&+ : \wick{\c1{\bar\psi}(x)\c2\psi(x)A(x)\c2{\bar\psi}(y)\c1\psi}(y)A(y): + : \wick{\c1{\bar\psi}(x)\c2\psi(x)\c3 A(x)\c2{\bar\psi}(y)\c1\psi(y)\c3 A}(y): \; .
\end{aligned} \tag{11.11}$$

Bringing all of the contracted fields to the front gives

$$
\begin{aligned}
H_2 =&: \bar\psi(x)\psi(x)\bar\psi(y)\psi(y)A(y): + \underbrace{A(x)A(y)} : \bar\psi(x)\psi(x)\bar\psi(y)\psi(y): \\
&+ \underbrace{\bar\psi(x)\psi(y)} : A(x)A(y)\psi(x)\bar\psi(y): + \underbrace{\psi(x)\bar\psi(y)} : A(x)A(y)\bar\psi(x)\psi(y): \\
&+ \underbrace{\bar\psi(x)\psi(y)}\ \underbrace{A(x)A(y)} : \psi(x)\bar\psi(y): + \underbrace{\psi(x)\bar\psi(y)}\ \underbrace{A(x)A(y)} : \bar\psi(x)\psi(y) : \\
&+ \underbrace{\bar\psi(x)\psi(y)}\ \underbrace{\psi(x)\bar\psi(y)} : A(x)A(y): + \underbrace{\psi(x)\bar\psi(y)}\ \underbrace{\bar\psi(x)\psi(y)}\ \underbrace{A(x)A(y)} \ .
\end{aligned}
\tag{11.12}
$$

In every case in this example the p of Eq. (11.8) is even.

We will now prove that Eq. (11.12) is the correct result by putting all of the fields into normal order, being careful to keep any commutators or anticommutators which may arise in the process. Since the A's commute with the ψ's, we can place the A's in normal order independent of the ψ's. The A's therefore give

$$
A(x)A(y) =: A(x)A(y): + \underbrace{A(x)A(y)} \ . \tag{11.13}
$$

Now, look at the ψ terms. These are

$$
\begin{aligned}
&: \bar\psi(x)\psi(x): \ : \bar\psi(y)\psi(y): \\
&\quad = \left(\bar\psi^{(-)}(x)\left[\psi^{(+)}(x) + \psi^{(-)}(x)\right] + \bar\psi^{(+)}(x)\psi^{(+)}(x) - \psi^{(-)}(x)\bar\psi^{(+)}(x)\right) \\
&\quad\quad \times \left(\bar\psi^{(-)}(y)\left[\psi^{(+)}(y) + \psi^{(-)}(y)\right] + \bar\psi^{(+)}(y)\psi^{(+)}(y) - \psi^{(-)}(y)\bar\psi^{(+)}(y)\right) .
\end{aligned}
\tag{11.14}
$$

To put in normal order, all $\psi^{(-)}$'s to the right of any $\psi^{(+)}$ must be moved to the left, and all $\psi^{(+)}$'s to the left of any $\psi^{(-)}$ must be moved to the right. The terms not already in normal order are

$$
\begin{aligned}
\text{Terms} = \Big[&\bar\psi^{(-)}(x)\psi^{(+)}(x)\bar\psi^{(-)}(y)\psi(y) + \bar\psi^{(+)}(x)\psi^{(+)}(x)\bar\psi^{(-)}(y)\psi(y) \\
&- \bar\psi^{(+)}(x)\psi^{(+)}(x)\psi^{(-)}(y)\bar\psi^{(+)}(y) - \psi^{(-)}(x)\bar\psi^{(+)}(x)\bar\psi^{(-)}(y)\psi^{(-)}(y) \\
&+ \psi^{(-)}(x)\bar\psi^{(+)}(x)\psi^{(-)}(y)\bar\psi^{(+)}(y)\Big] .
\end{aligned}
$$

These reduce to a sum of normal-ordered terms plus additional terms as follows:

$$
\begin{aligned}
&\text{Terms} \\
&= \begin{pmatrix}\text{normal-ordered} \\ \text{terms}\end{pmatrix} + \left[\psi^{(+)}(x), \bar\psi^{(-)}(y)\right]_+ \left[\bar\psi^{(-)}(x)\psi(y) + \bar\psi^{(+)}(x)\psi(y)\right] \\
&\quad + \left[\bar\psi^{(+)}(x), \psi^{(-)}(y)\right]_+ \Big[\psi^{(+)}(x)\bar\psi^{(+)}(y) + \psi^{(-)}(x)\bar\psi^{(-)}(y) \\
&\qquad\qquad + \psi^{(-)}(x)\bar\psi^{(+)}(y) - \bar\psi^{(-)}(y)\psi^{(+)}(x)\Big]
\end{aligned}
$$

$$
\begin{aligned}
&= \begin{pmatrix}\text{normal-ordered} \\ \text{terms}\end{pmatrix} + \left[\psi^{(+)}(x), \bar{\psi}^{(-)}(y)\right]_+ :\bar{\psi}(x)\psi(y): \\
&\qquad + \left[\psi^{(+)}(x), \bar{\psi}^{(-)}(y)\right]_+ \left[\bar{\psi}^{(+)}(x), \psi^{(-)}(y)\right]_+ \\
&\qquad + \left[\bar{\psi}^{(+)}(x), \psi^{(-)}(y)\right]_+ :\psi(x)\bar{\psi}(y): \; .
\end{aligned}
$$

Now, combining the ψ terms with the A terms gives

$$
\begin{aligned}
H_2 = &\left\{ :A(x)A(y): + \underbracket{A(x)A}(y) \right\} \left\{ :\bar{\psi}(x)\psi(x)\bar{\psi}(y)\psi(y): \right. \\
&+ \left[\psi^{(+)}(x), \bar{\psi}^{(-)}(y)\right]_+ :\bar{\psi}(x)\psi(y): + \left[\bar{\psi}^{(+)}(x), \psi^{(-)}(y)\right]_+ :\psi(x)\bar{\psi}(y): \\
&\left. + \left[\psi^{(+)}(x), \bar{\psi}^{(-)}(y)\right]_+ \left[\bar{\psi}^{(+)}(x), \psi^{(-)}(y)\right]_+ \right\} \; .
\end{aligned}
$$

Multiplying these out generates the eight terms given in Eq. (11.12) if we recall that $\left[\psi^{(+)}(x), \bar{\psi}^{(-)}(y)\right]_+ = \underbracket{\psi(x)\bar{\psi}}(y)$, etc.

Wick's theorem for normal-ordered products is a useful result, but what is really needed is an analogous theorem for the time-ordered products which occur in the perturbation expansion for U_I. We turn to this now.

Wick's Theorem for Time-ordered Products

First, consider the time-ordered product of a single pair of field operators. This is defined to be

$$
T\left(\phi(x)\phi^\dagger(y)\right) = \begin{cases} \phi(x)\phi^\dagger(y) & x_0 > y_0 \\ \eta\phi^\dagger(y)\phi(x) & x_0 < y_0 \end{cases} , \tag{11.15}
$$

where η is -1 for the Fermi fields and $+1$ for the Bose fields. However, following our previous discussion this is just

$$
T\left(\phi(x)\phi^\dagger(y)\right) = \begin{cases} :\phi(x)\phi^\dagger(y): + \underbracket{\phi(x)\phi}^\dagger(y) & x_0 > y_0 \\ \eta :\phi^\dagger(y)\phi(x): + \eta\underbracket{\phi^\dagger(y)\phi}(x) & x_0 < y_0 \end{cases} . \tag{11.16}
$$

However, the order of terms in a normal-ordered product can be changed (if we respect the anticommutation relations) and hence we can define a *time-ordered contraction* by

$$
\boxed{T\left(\phi(x)\phi^\dagger(y)\right) = :\phi(x)\phi^\dagger(y): + \overbracket{\phi(x)\phi}^\dagger(y) ,} \tag{11.17}
$$

where the *time-ordered contraction* is distinguished from the normal-ordered contraction by placing the square brackets *above* the fields, and from Eqs. (11.16) and (11.17)

$$
\overbracket{\phi(x)\phi}^\dagger(y) = \begin{cases} \underbracket{\phi(x)\phi}^\dagger(y) & x_0 > y_0 \\ \eta\underbracket{\phi^\dagger(y)\phi}(x) & y_0 > x_0 \end{cases} . \tag{11.18}
$$

From the definition of the normal-ordered contraction, (11.18) can also be written

$$\overline{\phi(x)\phi}^\dagger(y) = \begin{cases} \langle 0|\phi(x)\phi^\dagger(y)|0\rangle & x_0 > y_0 \\ \eta\,\langle 0|\phi^\dagger(y)\phi(x)|0\rangle & x_0 < y_0 \end{cases} \tag{11.19}$$

or

$$\boxed{\overline{\phi(x)\phi}^\dagger(y) = \langle 0|T\left(\phi(x)\phi^\dagger(y)\right)|0\rangle \quad .} \tag{11.20}$$

We see that the time-ordered contractions are just the propagators we have already computed. Recall Eq. (9.37) for self-conjugate fields,

$$\begin{aligned}\langle 0|T\left(\phi(x)\phi(y)\right)|0\rangle &= i\Delta(x-y) \\ &= \int \frac{d^4k}{(2\pi)^4} e^{-ik\cdot(x-y)} \left[\frac{-i}{\mu^2 - k^2 - i\epsilon}\right] \quad ;\end{aligned} \tag{11.21a}$$

Eq. (10.102) for Dirac fields,

$$\begin{aligned}\langle 0|T\left(\psi_\alpha(x)\bar{\psi}_\beta(y)\right)|0\rangle &= i\,S_{\alpha\beta}(x-y) \\ &= \int \frac{d^4k}{(2\pi)^4}\, e^{-ik\cdot(x-y)} \frac{-i[m+\not{k}]_{\alpha\beta}}{m^2 - k^2 - i\epsilon} \quad ;\end{aligned} \tag{11.21b}$$

and Eq. (10.28) for the radiation field,

$$\begin{aligned}\langle 0|T\left(A^i(x)A^j(y)\right)|0\rangle &= i\,D^{ij}_{\text{tr}}(x-y) \\ &= \int \frac{d^4k}{(2\pi)^4}\, e^{-ik\cdot(x-y)} \left(\frac{-i}{-k^2 - i\epsilon}\right)\left[-g^{ij} - \frac{k^i k^j}{\mathbf{k}^2}\right] \quad .\end{aligned}$$

We now are ready to state and prove Wick's theorem for time-ordered products:

> **Theorem:** The T product of a system of linear operators is the sum of their normal products with all possible time-ordered contractions, including the term with no contractions.

The formal statement of the theorem is almost identical to the previous theorem:

$$\begin{aligned}T(\phi_1\cdots\phi_n) =&: \phi_1\cdots\phi_n : + : \overline{\phi_1\phi}_2\cdots\phi_n : + : \overline{\phi_1\phi_2\phi}_3\cdots\phi_n : + \cdots \\ &+ : \phi_1\overline{\phi_2\phi}_3\cdots\phi_n : + \cdots + : \overline{\phi_1\phi}_2\overline{\phi_3\phi}_4\phi_5\cdots\phi_n : + \cdots \\ &+ : \overline{\phi_1\phi}_2\overline{\phi_3\phi}_4\overline{\phi_5\phi}_6\cdots\phi_n : + \cdots \quad .\end{aligned} \tag{11.22}$$

Proof: For each particular time ordering $i_1 i_2 i_3 \cdots i_n$, the time-ordered product is

$$T(\phi_1\cdots\phi_n) = \eta\,\phi_{i_1}\phi_{i_2}\cdots\phi_{i_n} \ ,$$

where η is ± 1 depending on whether there are an even or odd number of interchanges of Dirac fields required to put the ϕ's in the desired order. Now apply the Wick normal order theorem to this product. The normal-ordered contractions will all be present with the correct time ordering. *Next*, using the fact that the orderings of fields within a normal-ordered term can be interchanged (with the usual phase for Fermi interchanging), we can restore the normal-ordered terms to their standard order, $\phi_1\phi_2\cdots\phi_n$. The resulting phase which will remain will be ± 1, depending only on whether the *normal-ordered* contractions *have an even or odd number of Fermi interchanges*. *Finally*, for all terms with the normal contraction

$$\underbrace{\phi_1\phi_2}\,\chi$$

corresponding to $t_1 > t_2$, there exists identical terms corresponding to $t_1 < t_2$ which give

$$\eta\,\underbrace{\phi_2\phi_1}\,\chi\ .$$

These terms may be combined, giving

$$\left(\underbrace{\phi_1\phi_2} + \eta\,\underbrace{\phi_2\phi_1}\right)\,\chi = \overbrace{\phi_1\phi_2}\chi\ ,$$

which proves the theorem. ∎

We are now ready to apply these ideas to a systematic study of QED in second order.

11.2 QED TO SECOND ORDER

We use QED to illustrate these ideas because it is the simplest, most successful quantum field theory for which perturbation theory works. (Recall $\alpha \simeq 1/137$). We will consider electron interactions only, so the Hamiltonian is a version of the one given in Eq.(10.8) and has the form

$$\begin{aligned} H_0 &= H_e^0 + H_{EM}^0 \\ H_I &= \frac{1}{8\pi}\int \frac{d^3r\,d^3r'}{|r-r'|} J_e^0(r,t)J_e^0(r',t) - \int d^3r\,\boldsymbol{J}_e(r,t)\cdot\boldsymbol{A}(r,t) \\ &= H_{\rm INST} + H_{\rm RAD}\ , \end{aligned} \tag{11.23}$$

where

$$J_e^\mu = -e\!:\left[\bar{\psi}\gamma^\mu\psi\right]:\ . \tag{11.24}$$

Note that there are no protons (there is only one fermion, the electron) and that we have restored the electron instantaneous self-energy term ignored in our previous discussions. Also, our treatment of $H_{\rm INST}$ will depart from the normal procedure.

Instead of normal ordering the entire term, we will only normal order each J_e^0 in this term. That is, we use

$$J_e^0(r,t)J_e^0(r',t) = e^2 \colon \left[\bar{\psi}(r,t)\gamma^0\psi(r,t)\right] \colon \colon \left[\bar{\psi}(r',t)\gamma^0\psi(r',t)\right] \colon \tag{11.25}$$

instead of

$$e^2 \colon \left[\bar{\psi}(r,t)\gamma^0\psi(r,t)\right] \left[\bar{\psi}(r',t)\gamma^0\psi(r',t)\right] \colon \ . \tag{11.26}$$

These give Hamiltonian densities differing from each other only by a constant, so the physics is the same, but the first choice is more convenient, as we will see below. Using this Hamiltonian, the time translation operator to second order was given in Eq. (9.7),

$$U_I = 1 - i\int_{-\infty}^{\infty} dt\, H_{\mathrm{INST}}(t) + \frac{(-i)^2}{2}\int dt_1\, dt_2\, T\left[H_{\mathrm{RAD}}(t_1)H_{\mathrm{RAD}}(t_2)\right] .$$

We now reduce this expression for the time translation operator using Wick's theorem. First, note that the instantaneous term can be written

$$\begin{aligned}
-i\int dt\, H_{\mathrm{INST}}(t) &= -i\int \frac{dt}{8\pi}\frac{d^3r_1\, d^3r_2}{|r_1 - r_2|} J^0(r_1,t)J^0(r_2,t) \\
&= -i\int \frac{d^4x_1\, d^4x_2}{8\pi}\frac{\delta(t_1-t_2)}{|r_1-r_2|}\, T\left[J^0(x_1)J^0(x_2)\right] \\
&= -\frac{i}{2}\int d^4x_1 d^4x_2 \underbrace{\int dk_0 \frac{e^{-ik_0(t_1-t_2)}}{2\pi}}_{\delta(t_1-t_2)} \underbrace{\int \frac{d^3k}{(2\pi)^3}\frac{e^{i\boldsymbol{k}\cdot(\boldsymbol{r}_1-\boldsymbol{r}_2)}}{\boldsymbol{k}^2}}_{\frac{1}{4\pi|r_1-r_2|}} T\left[J^0(x_1)J^0(x_2)\right] \\
&= -\frac{i}{2}\int d^4x_1 d^4x_2\, T\left[J^0(x_1)J^0(x_2)\right] \int \frac{d^4k}{(2\pi)^4}\frac{e^{-ik\cdot(x_1-x_2)}}{\boldsymbol{k}^2} ,
\end{aligned} \tag{11.27}$$

where the insertion of the T product in the second line has no effect but produces a formula more easily compared with the radiation part. The radiation part of the time translation operator, to second order, includes terms where the photon field is contracted. As we have just seen, these reduce to

$$\begin{aligned}
-\frac{1}{2}\int d^4x_1\, d^4x_2\, T\left[J^i(x_1)J^j(x_2)\right] \langle 0|T\left(A^i(x_1)A^j(x_2)\right)|0\rangle \\
= -\frac{i}{2}\int d^4x_1\, d^4x_2\, T\left[J^i(x_1)J^j(x_2)\right] \int \frac{d^4k}{(2\pi)^4}\frac{e^{-ik\cdot(x_1-x_2)}}{-k^2-i\epsilon}\left[g^{ij} + \frac{k^ik^j}{\boldsymbol{k}^2}\right].
\end{aligned}$$

Now, *if either the incoming or outgoing electron is virtual*, it is no longer true that $k_\mu J^\mu = 0$, but it can be proved, to any given order in the electric charge e, that *all* $k_\mu J^\mu$ *terms cancel*, so that we may make the replacement

$$k^iJ^i \Rightarrow k^0J^0 \ , \tag{11.28}$$

just as if $k_\mu J^\mu = 0$. This is a consequence of the gauge invariance of the theory. With this replacement, the instantaneous and radiation term can be added together as we did in Sec. 10.2, giving

$$\overline{A(x)A}(y) \text{ terms} = -\frac{i}{2}\int d^4x_1\, d^4x_2\, T\left[J^\mu(x_1)J^\nu(x_2)\right] \times \int \frac{d^4k}{(2\pi)^4}\, \frac{e^{-ik\cdot(x_1-x_2)}}{-k^2-i\epsilon}\left[g_{\mu\nu} - \eta_\mu\eta_\nu \underbrace{\left(1-\frac{k_0^2}{\boldsymbol{k}^2}+\frac{k^2}{\boldsymbol{k}^2}\right)}_{=0}\right]. \tag{11.29}$$

Hence, whenever $\overline{AA}$ contractions arise, we may use the four-current J^μ and the relativistically invariant photon propagator, defined to be

$$\langle 0|T\left(A^\mu(x)A^\nu(y)\right)|0\rangle = \int \frac{d^4k}{(2\pi)^4}\, e^{-ik\cdot(x-y)}\left[\frac{i\,g^{\mu\nu}}{-k^2-i\epsilon}\right]. \tag{11.21c}$$

In conclusion, the effective second order interaction time translation operator for QED consists of six terms which contribute to the following physical processes:

$$\begin{aligned}
-i\int dt\, H^{(2)}_{\text{eff}}(t) &= -\frac{e^2}{2}\int d^4x_1\, d^4x_2 \Big\{ \underbrace{:\bar\psi(x_1)\gamma^i\psi(x_1)A^i(x_1)\bar\psi(x_2)\gamma^j\psi(x_2)A^j(x_2):}_{\text{disconnected term equal to 0}} \\
&+ \underbrace{\overline{A_\mu(x_1)A}_\nu(x_2):\bar\psi(x_1)\gamma^\mu\psi(x_1)\bar\psi(x_2)\gamma^\nu\psi(x_2):}_{\text{annihilation and exchange terms}} \\
&+ \underbrace{2:\bar\psi(x_1)\gamma^i A^i(x_1)\overline{\psi(x_1)\bar\psi}(x_2)\gamma^j A^j(x_2)\psi(x_2):}_{\text{Compton scattering terms}} \\
&+ \underbrace{2\overline{A_\mu(x_1)A}_\nu(x_2):\bar\psi(x_1)\gamma^\mu\overline{\psi(x_1)\bar\psi}(x_2)\gamma^\nu\psi(x_2):}_{\text{electron self-energy term}} \\
&- \underbrace{\gamma^j_{\beta'\alpha'}\overline{\psi_{\alpha'}(x_1)\bar\psi}_\alpha(x_2)\gamma^i_{\alpha\beta}\overline{\psi_\beta(x_2)\bar\psi}_{\beta'}(x_1) :A^j(x_1)A^i(x_2):}_{\text{vacuum polarization term}} \\
&- \underbrace{\gamma^\mu_{\beta'\alpha'}\overline{\psi_{\alpha'}(x_1)\bar\psi}_\alpha(x_2)\gamma^\nu_{\alpha\beta}\overline{\psi_\beta(x_2)\bar\psi}_{\beta'}(x_1)\overline{A_\mu(x_1)A}_\nu(x_2)}_{\text{vacuum bubble term}}\Big\}.
\end{aligned} \tag{11.30}$$

These six terms arise because the original time-ordered product has been replaced by normal-ordered products, each of which can be identified with particular physical processes, as we will discuss. The contractions are the propagators given in Eqs. (11.21 a–c), and they describe the propagation of virtual particles from an interaction point at x_1 to one at x_2.

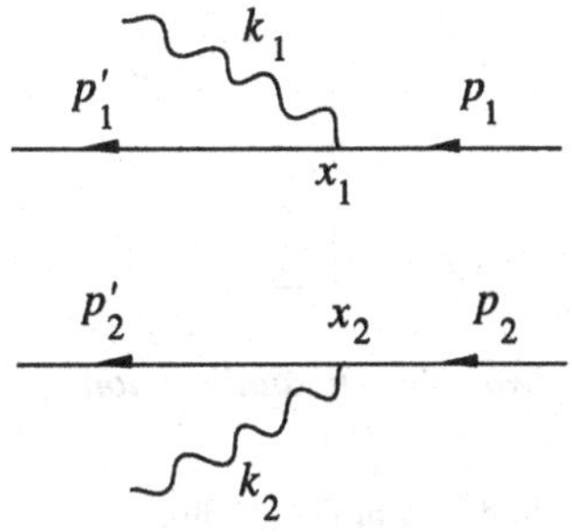

Fig. 11.1 Example of disconnected Feynman diagrams generated by the first term in Eq. (11.30).

The first term, the fully normal-ordered term, has no contractions, and hence the integration over d^4x_1 and d^4x_2 *can be carried out independently*. This generates two independent energy–momentum conserving delta functions, the form of which depends on the process under consideration. Since the term is fully normal ordered, only matrix elements involving a combined presence of six particles in both the initial and final states can contribute. An example of such a process, shown in Fig. 11.1, is $e^- + e^- \to e^- + e^- + 2\gamma$. Since there are no contractions, the interactions at x_1 and x_2 are independent of each other, and the delta functions in this case are

$$\delta^4\left(p_1' + k_1 - p_1\right)\delta^4\left(p_2' + k_2 - p_2\right) = 0 \ . \tag{11.31}$$

They are zero because a physical electron cannot decay into a real photon and another physical electron. Examination of other processes generated by this interaction shows that the above analysis holds in every case. The first term makes no contribution to any process.

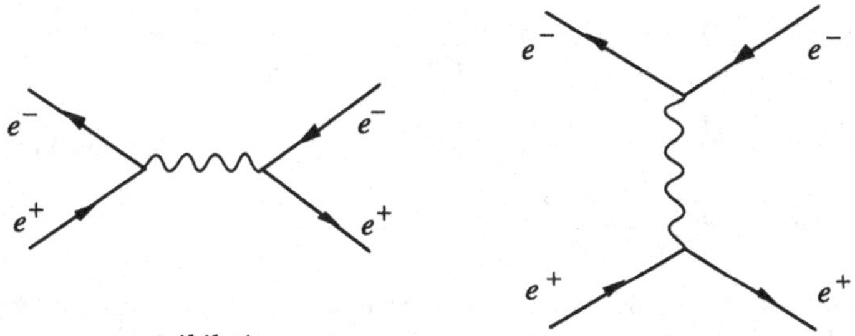

Fig. 11.2 Annihilation and exchange diagrams generated by the second term in Eq. (11.30).

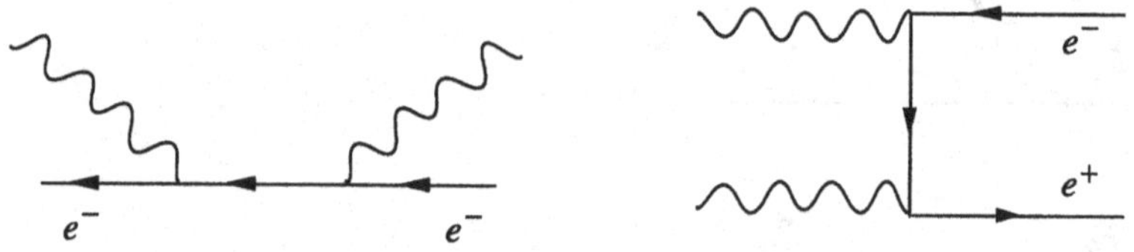

Compton scattering *two -photon annihilation*

Fig. 11.3 Examples of diagrams generated by the third term in Eq. (11.30).

The second term gives rise to annihilation or photon exchange diagrams like those shown in Fig. 11.2. These were discussed in Secs. 10.2 and 10.3. The third term gives rise to Compton scattering and *two-photon processes*, such as those given in Fig. 11.3. Compton scattering was discussed in Sec. 10.5, and the annihilation of an $e^- \, e^+$ pair into two γ's is a process which contributes to the decay of positronium, as discussed in Sec. 8.5. The reader should be able to calculate this diagram with the Feynman rules we have already obtained. All the diagrams involving a *single* contraction in *second order* are examples of *tree diagrams*, which are diagrams with no *loops*.

The first loop diagram which we will discuss now in some detail comes from the fourth term in Eq. (11.30). It gives rise to the electron self-energy and also introduces us to the infinities which can arise in field theory.

11.3 ELECTRON SELF-ENERGY

The fourth term will contribute to matrix elements of a single free electron:

$$
\begin{aligned}
S_{fi} = & -e^2 \int d^4x_1\, d^4x_2\, \langle p's' | \overline{A_\mu(x_1) A}_\nu(x_2) : \bar{\psi}(x_1)\gamma^\mu \overline{\psi(x_1)\bar{\psi}}(x_2)\gamma^\nu \psi(x_2) \colon |ps\rangle \\
= & -\frac{(-i)^2 e^2}{\sqrt{4E'EL^6}} \int d^4x_1\, d^4x_1 \int \frac{d^4k}{(2\pi)^4} \frac{e^{-ik\cdot(x_1-x_2)}}{-k^2 - i\epsilon} (-g_{\mu\nu}) \\
& \times \langle 0 | b_{p',s'}\, b^\dagger_{p',s'}\, e^{ip'\cdot x_1} \int \frac{d^4\ell}{(2\pi)^4} e^{-i\ell\cdot(x_1-x_2)}\, e^{-ip\cdot x_2}\, b_{p,s}\, b^\dagger_{p,s} | 0 \rangle \\
& \times \bar{u}(\boldsymbol{p}', s')\, \gamma^\mu \left[\frac{m + \not{\ell}}{m^2 - \ell^2 - i\epsilon} \right] \gamma^\nu u(\boldsymbol{p}, s) \,,
\end{aligned}
\tag{11.32}
$$

where the sums in the $\bar{\psi}$ and ψ fields have been reduced to one term anticipating the result $\left\{ b_{q,t}, b^\dagger_{p,s} \right\} = \delta_{q,p}\, \delta_{t,s}$. Making this reduction and doing the integrals over x_1 and x_2 give

$$
S_{fi} = -i\, \frac{(2\pi)^4\, \delta^4(p' - p)}{\sqrt{4EE'L^6}}\, \mathcal{M}_\Sigma \,, \tag{11.33}
$$

where

$$\mathcal{M}_\Sigma = i\,(-i)^2 e^2 \int \frac{d^4k}{(2\pi)^4}\, d^4\ell\, \delta^4(p' - k - \ell)\, \frac{\bar{u}(\boldsymbol{p}', s')\,\gamma^\mu (m + \not{\ell})\gamma_\mu\, u(\boldsymbol{p}, s)}{(-k^2 - i\epsilon)(m^2 - \ell^2 - i\epsilon)}$$
$$= -i\,e^2 \int \frac{d^4k}{(2\pi)^4}\, \frac{\bar{u}(\boldsymbol{p}', s')\,\gamma^\mu (m + \not{p}' - \not{k})\,\gamma_\mu\, u(\boldsymbol{p}, s)}{(-k^2 - i\epsilon)\,(m^2 - (p' - k)^2 - i\epsilon)} \quad . \tag{11.34}$$

Note that the $\mathcal{M}$ now involves a non-trivial integral over the internal four-momentum of the virtual photon, k, and corresponds to the Feynman diagram shown in Fig 11.4. In this process, the four-momentum of the virtual photon is not constrained, even though energy–momentum is conserved at every vertex. The process illustrates a new Feynman rule:

Rule 5: integrate over each internal four-momentum k not fixed by energy–momentum conservation with a weight

$$\int \frac{d^4k}{(2\pi)^4} \quad .$$

The *electron self-energy*, denoted by $\Sigma(p)$, is related to $\mathcal{M}_\Sigma$ by

$$\mathcal{M}_\Sigma = \bar{u}(\boldsymbol{p}', s')\; \Sigma(p)\; u(\boldsymbol{p}, s) \quad . \tag{11.35}$$

Equation (11.34) therefore gives, to second order in the electron charge, the following result for the electron self-energy:

$$\Sigma(p) = -ie^2 \int \frac{d^4k}{(2\pi)^4}\, \frac{\gamma^\mu\,(m + \not{p} - \not{k})\,\gamma_\mu}{(-k^2 - i\epsilon)\,(m^2 - (p - k)^2 - i\epsilon)} \quad . \tag{11.36}$$

We will postpone discussion of how to evaluate such integrals until Sec. 11.6. Now we will discuss the physical significance of the self-energy.

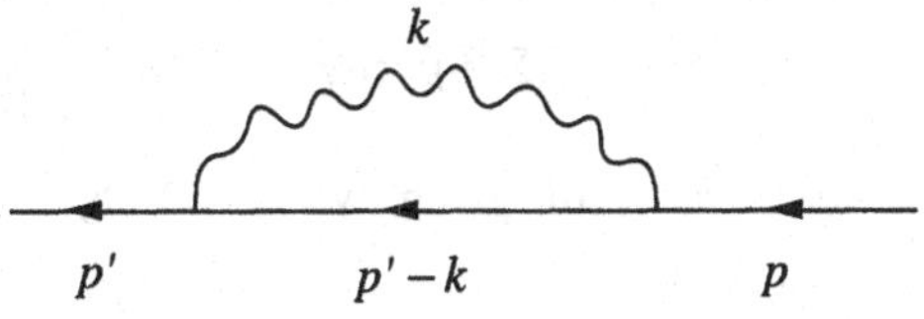

Fig. 11.4 Feynman diagram for the electron self-energy. Note that the loop momentum k cannot be constrained by momentum conservation.

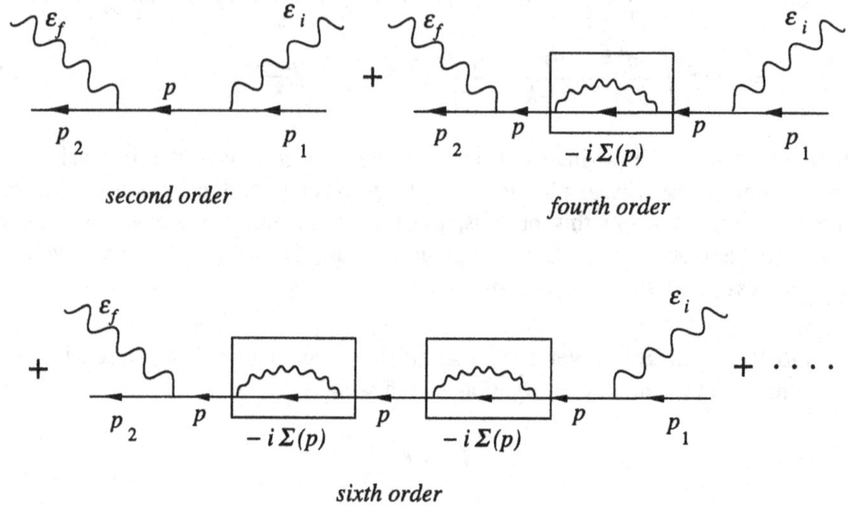

Fig. 11.5 First three of an infinite class of Feynman diagrams which defines the dressed electron propagator.

Mass and Wave Function Renormalization

To understand the physical significance of the electron self-energy, consider Compton scattering to higher order in e. There are an infinity of diagrams, but they can be organized into an infinite number of classes, with each class itself containing an infinite number of diagrams. The first three diagrams in one class are shown in Fig. 11.5. Recalling the definition of the electron propagator, Eq. (11.21b), these first three diagrams can be written

$$\begin{aligned}
\mathcal{M}_2 &= -ie^2\bar{u}(\boldsymbol{p}_2)\ \not{\epsilon}_f\ iS(p)\ \not{\epsilon}_i\ u(\boldsymbol{p}_1) \\
\mathcal{M}_4 &= -ie^2\bar{u}(\boldsymbol{p}_2)\ \not{\epsilon}_f\ iS(p)\,[-i\Sigma(p)]\,iS(p)\ \not{\epsilon}_i\ u(\boldsymbol{p}_1) \\
\mathcal{M}_6 &= -ie^2\bar{u}(\boldsymbol{p}_2)\ \not{\epsilon}_f\ iS(p)\,[-i\Sigma(p)]\,iS(p)\,[-i\Sigma(p)]\,iS(p)\ \not{\epsilon}_i\ u(\boldsymbol{p}_1)\ ,
\end{aligned} \tag{11.37}$$

where in these diagrams $\Sigma(p)$ is still given by Eq. (11.36) even though p^2 is no longer equal to m^2, and

$$iS(p) = \frac{-i(m+\not{p})}{m^2 - p^2 - i\epsilon} = \frac{-i}{m-\not{p}-i\epsilon}\ . \tag{11.38}$$

This infinite class of diagrams can be summed using the geometric series

$$\begin{aligned}\mathcal{M} &= \sum_{n=1}^{\infty} \mathcal{M}_{2n} \\ &= -ie^2\bar{u}(\mathbf{p}_2)\ \not{\epsilon}_f\ iS(p)\left[1+\Sigma(p)S(p)+[\Sigma(p)S(p)]^2+\cdots\right]\not{\epsilon}_i\ u(\mathbf{p}_1) \\ &= -ie^2\bar{u}(\mathbf{p}_2)\ \not{\epsilon}_f\ iS'(p)\ \not{\epsilon}_i\ u(\mathbf{p}_1)\ ,\end{aligned} \tag{11.39}$$

where the *dressed propagator*, $S'(p)$, is

$$\begin{aligned}iS'(p) &= i\,S(p)\left[1+\Sigma(p)S(p)+[\Sigma(p)S(p)]^2+\cdots\right] \\ &= \frac{i\,S(p)}{1-\Sigma(p)S(p)}\ .\end{aligned} \tag{11.40}$$

Because $\Sigma(p)$ is a matrix in Dirac space which transforms like a scalar and because p^μ is the only four-vector on which it can depend, $\Sigma(p)$ must have the form

$$\Sigma(p) = mA(p^2) + \not{p}\,B(p^2)\ , \tag{11.41}$$

where $A(p^2)$ and $B(p^2)$ are scalar functions of p^2. Hence $\Sigma(p)$ commutes with $iS(p)$, and

$$\boxed{iS'(p) = \frac{-i}{m-\not{p}+\Sigma(p)-i\epsilon}\ .} \tag{11.42}$$

The effect of the self-energy $\Sigma(p)$ is to modify both the mass and the normalization of the propagator. To see how this comes about, we regard $\Sigma(p)$ as a function of $\not{p}$ [which is consistent with Eq. (11.41) because $\not{p}^2 = p^2$], and expand $\Sigma(\not{p})$ in a power series in the quantity $(\not{p}-\overline{m})$,

$$\begin{aligned}\Sigma(\not{p}) &= \Sigma(\overline{m}) + (\not{p}-\overline{m})\Sigma'(\overline{m}) + \tfrac{1}{2}(\not{p}-\overline{m})^2\Sigma''(\overline{m}) + \cdots \\ &= \Sigma(\overline{m}) + (\not{p}-\overline{m})\Sigma'(\overline{m}) + \Sigma_R(\not{p})\ ,\end{aligned} \tag{11.43}$$

where $\overline{m}$ is a constant to be chosen shortly and the coefficient of the second term can be found in the usual way,

$$\Sigma'(\overline{m}) = \left.\frac{d\Sigma}{d\not{p}}\right|_{\not{p}=\overline{m}}\ ,$$

even though the matrix $\not{p}$, which is constructed from the γ-matrices, can never equal $\overline{m}$, which is a multiple of the identity. Note that the second line of (11.43) is exact because Σ_R is simply the sum of all the higher order terms in the expansion, and therefore, by construction,

$$\Sigma_R(\not{p}) = (\not{p}-\overline{m})^2\,R\ , \tag{11.44}$$

where R is some function of $\not p$ which may not be zero at $\not p = \overline{m}$.

Now, we substitute the expansion (11.43) into the dressed propagator (11.42) and *choose* $\overline{m}$ so that the propagator will have the form

$$\boxed{i\,S'(\not p) = \frac{-i\,Z_2}{\overline{m} - \not p + Z_2\Sigma_R(\not p) - i\epsilon}\ .} \tag{11.45}$$

We will show that this form insures that the dressed propagator has a pole at $\overline{m}$, with residue Z_2, and relate these two quantities to the self-energy. To show that the dressed propagator has a pole at $\overline{m}$, use (11.44) and multiply numerator and denominator of (11.45) by $\not p + \overline{m}$,

$$iS'(\not p) = \frac{-i(\overline{m} + \not p)\,Z_2}{(\overline{m}^2 - p^2)\,[1 + (\overline{m} - \not p)\,Z_2 R]}\ . \tag{11.46}$$

This displays the pole at $p^2 = \overline{m}^2$, justifying our interpretation of $\overline{m}$ as the physical mass. Furthermore, since

$$\not p = \overline{m} + \frac{p^2 - \overline{m}^2}{\not p + \overline{m}} = \overline{m} + \mathcal{O}(p^2 - \overline{m}^2)\ , \tag{11.47}$$

S' assumes a form identical to S *near* the pole,

$$iS'(\not p) \underset{p^2\to\overline{m}^2}{\longrightarrow} \frac{-i(\overline{m} + \not p)\,Z_2}{\overline{m}^2 - p^2} = \frac{-i\,Z_2}{\overline{m} - \not p}\ , \tag{11.48}$$

which shows that the residual self-energy Σ_R is negligible near the pole and that Z_2 is the residue of the dressed propagator at the pole.

Furthermore, from Eqs. (11.42), (11.43), and (11.45) it is easy to see that

$$\begin{aligned} \overline{m} - m &= \Sigma(\overline{m}) \\ Z_2^{-1} &= 1 - \Sigma'(\overline{m})\ . \end{aligned} \tag{11.49}$$

In terms of A and B defined in Eq. (11.41),

$$\Sigma(\not p) = mA(\not p^2) + \overline{m}B(\not p^2) + (\not p - \overline{m})B(\not p^2)\ , \tag{11.50}$$

so that

$$\begin{aligned} \Sigma(\overline{m}) &= mA(\overline{m}^2) + \overline{m}B(\overline{m}^2) \\ \left.\frac{d\Sigma}{d\not p}\right|_{\not p=\overline{m}} &= 2\overline{m}\,(mA_0' + \overline{m}B_0') + B(\overline{m}^2)\ , \end{aligned} \tag{11.51}$$

where

$$A_0' = \left.\frac{dA(p^2)}{dp^2}\right|_{p^2=\overline{m}^2} \qquad B_0' = \left.\frac{dB(p^2)}{dp^2}\right|_{p^2=\overline{m}^2}\ . \tag{11.52}$$

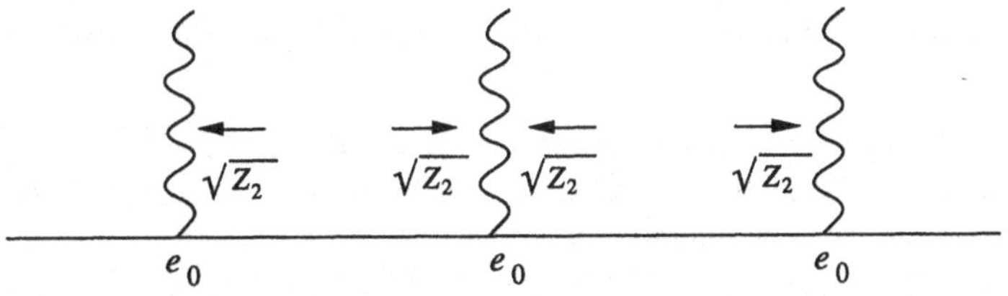

Fig. 11.6 The factor Z_2 is removed by absorbing it into charges which occur at the end of each internal fermion line.

In *principle*, these equations can be solved for $\overline{m}$ and Z_2, but in *practice*, Z_2 is removed from the theory (discussed below) and $\overline{m}$ is fixed at the physical electron mass. An exception to this general rule occurs in the special case when the unrenormalized mass is *zero*. In this case, the renormalized mass will also be zero. This follows from Eq. (11.49) for the renormalized mass, which is

$$\overline{m} = \overline{m}B(\overline{m}^2) \quad , \tag{11.53}$$

and hence, in the absence of special conditions, Eq. (11.53) tells us that $\overline{m} = 0$. In anticipation of Chapter 13, we point out now that in theories with *spontaneous symmetry breaking* (not QED), special conditions are established so that mass is *spontaneously generated* by the interaction. In this case the self-energy functions A and B are calculated from Feynman diagrams using the anticipated $\overline{m}$ in place of m (which is zero, by assumption). Then the contribution from A is no longer zero, and Eq. (11.53) is replaced by a transcendental equation for $\overline{m}$, known as the *gap* equation. We defer any further discussion of these ideas to Chapter 13.

The process by which Z_2 is removed from the theory is referred to as *wave function renormalization*, and the change of m to $\overline{m}$ is referred to as *mass renormalization*. It is important that this can be carried out, because $A(\overline{m}^2)$ and $B(\overline{m}^2)$ are infinite, and if these infinities could not be removed, we could not obtain predictions from QED. After the renormalization is carried out, the remaining expressions are finite, and the theory makes meaningful predictions. In the end, the only thing lost in the renormalization process is the ability to calculate the mass shift of the electron and the change in its electric charge, and since the theory does not tell us how to calculate the electron mass and charge anyway, this does not reduce the predictive power of the original theory.

There are two steps which must be taken to remove the renormalization constant Z_2. First, for internal electron lines, break $Z_2 = \sqrt{Z_2}\sqrt{Z_2}$ and absorb one factor of $\sqrt{Z_2}$ into each charge at each end of the electron line, as shown in Fig. 11.6. Hence the "bare" charge, e_0, must be renormalized as follows:

$$e_0 \to \sqrt{Z_2}\, e_0 \sqrt{Z_2} = Z_2 e_0 \quad . \tag{11.54}$$

Next, we must also multiply the wave function of each external electron by $\sqrt{Z_2}$ so that the charge operator connected to each incoming and outgoing electron is similarly renormalized. The Feynman rule which incorporates this step is:

Rule 8: for each external fermion, a factor of $\sqrt{Z_2}\,u$, $\sqrt{Z_2}\,v$, $\sqrt{Z_2}\,\bar{u}$, or $\sqrt{Z_2}\,\bar{v}$, depending on whether or not the fermion is a particle or an antiparticle, and incoming or outgoing.

The origin of the external $\sqrt{Z_2}$ factor is associated with renormalization of the free field functions in the presence of interactions.

The reader is warned that *the charge will undergo further renormalization, so this is* not *what is called "charge renormalization."*

Note that the dressed propagator (11.45) has been defined so that the Z_2 multiplying Σ_R can also be absorbed into the renormalization of the charges in Σ_R, so the *renormalized* dressed propagator, denoted by $\overline{S}$, is

$$i\overline{S}(\not{p}) = \frac{-i}{\overline{m} - \not{p} + \overline{\Sigma}_R(\not{p})} \underset{p^2 \to \overline{m}^2}{\longrightarrow} \frac{-i}{\overline{m} - \not{p}} \ . \tag{11.55}$$

Near the particle pole, the renormalization insures that the dressed propagator has the same form as the original undressed propagator.

11.4 VACUUM BUBBLES

Turn now to the last term in Eq. (11.30). Since it is fully contracted, it is a c-number and has a non-zero vacuum expectation value. It describes (see Fig. 11.7) a vacuum fluctuation in which an electron of four-momentum p, a photon of four-momentum k, and a positron of four-momentum $-p-k$ spontaneously materialize from the vacuum at space–time point x_1 and propagate to space–time point x_2, where they annihilate. It is not zero because the particles are off-shell, so energy and momentum can be conserved.

However, it is not necessary to calculate such vacuum fluctuations; they can be shown to disappear from the theory. To prove this, note that the second order contribution from vacuum bubbles can be written

$$-i\,\langle 0| \int dt\, H^{(2)}_{\text{eff}}(t)|0\rangle = -ic_2 \ , \tag{11.56}$$

where it can be shown from (11.30) that c_2 is real. This same matrix element also occurs (squared) in fourth order, where the four space–time points x_1, x_2, x_3,

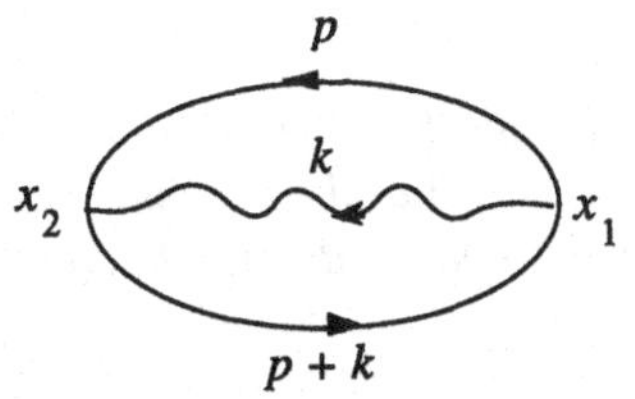

Fig. 11.7 The vacuum bubble in second order QED.

and x_4 can be connected pairwise in three different combinations: $[x_1\,x_2]\,[x_3\,x_4]$, $[x_1\,x_3]\,[x_2\,x_4]$, and $[x_1\,x_4]\,[x_2\,x_3]$. In fact, in $(2n)$th order, there are $(2n)!/(2^n\,n!)$ combinations, so that the sum of these contributions to the interaction time translation operator to all orders is

$$\begin{aligned}U_I^{B_2} &= 1 + \frac{1}{2}(-ic_2) + \frac{1}{4!}3(-ic_2)^2 + \cdots + \frac{1}{(2n)!}\left(\frac{(2n)!}{2^n\,n!}\right)(-ic_2)^n + \cdots \\ &= 1 + \left(-\frac{ic_2}{2}\right) + \frac{1}{2}\left(-\frac{ic_2}{2}\right)^2 + \cdots + \frac{1}{n!}\left(-\frac{ic_2}{2}\right)^n + \cdots \\ &= e^{-ic_2/2} \ . \end{aligned} \tag{11.57}$$

This result can be generalized to bubble diagrams of all orders, in which case $c_2/2$ is replaced by c, the sum of all bubbles.

However, bubbles can also be present in the background while other processes, such as scattering or annihilation, are occurring. By an extension of the above argument, we may show that the bubble diagrams modify these processes by the *same multiplicative factor* we worked out above. The full time translation operator can therefore be written

$$U_I = e^{-ic}\,U_I' \ , \tag{11.58}$$

where U_I' is the time translation operator *without* the bubble diagrams. Recalling the exponential structure of the perturbation expansion for the time translation operator, Eq. (3.25), and using the fact that c is a c-number, we can write

$$\begin{aligned}U_I &= e^{-ic}\,T\exp\left(-i\left[\int_{-\infty}^{\infty} dt\,H_I(t) - c\right]\right) \\ &= e^{-ic}\,U_I' \ , \end{aligned} \tag{11.59}$$

which shows that extraction of the bubble diagrams from U_I is equivalent to redefining the Hamiltonian by *subtracting all bubble contributions*. Furthermore, because of our definition of the S-matrix in Eq. (3.28),

$$S_{fi} = \frac{\langle f|U_I|i\rangle}{\langle 0|U_I|0\rangle} \ ,$$

the phase from bubble diagrams cancels,

$$S_{fi} = \frac{e^{-ic}\langle f|U_I'|i\rangle}{e^{-ic}\langle 0|U_I'|0\rangle} = \frac{\langle f|U_I'|i\rangle}{\langle 0|U_I'|0\rangle} \ . \tag{11.60}$$

Hence, the S-matrix may be calculated from U_I', which has no vacuum bubbles. The removal of the vacuum bubbles does not change any physics because vacuum bubbles contribute only to an overall phase, which is unobservable.

11.5 VACUUM POLARIZATION

Now consider the fifth term in Eq. (11.30). This contributes to the self-energy of a photon and is referred to as the *vacuum polarization*. The relevant Feynman diagram is given in Fig. 11.8. From our experience so far, we *expect* the Feynman diagram to give the following integral:

$$\begin{aligned}\mathcal{M}_{\Pi} \underset{\text{expected}}{=} & \; i\,(ie)^2(-i)^2 \int \frac{d^4p}{(2\pi)^4} \frac{\gamma^j_{\beta'\alpha'}\epsilon_f^{j\,*}\,[m+\not p]_{\alpha'\alpha}\,\gamma^i_{\alpha\beta}\epsilon^i_i\,[m+\not p-\not q]_{\beta\beta'}}{[m-p^2-i\epsilon]\,[m^2-(p-q)^2-i\epsilon]} \\ = & \; i\,e^2 \int \frac{d^4p}{(2\pi)^4} \frac{\mathrm{tr}\left\{\not\epsilon_f^*\,[m+\not p]\,\not\epsilon_i\,[m+\not p-\not q]\right\}}{[m^2-p^2-i\epsilon]\,[m^2-(p-q)^2-i\epsilon]} \ .\end{aligned} \tag{11.61}$$

This result is almost correct. The correct result includes an extra minus sign which is associated with every closed fermion loop and gives us another Feynman rule:

Rule 6: for each closed fermion loop, a minus sign.

To derive this result, with the correct sign, compute the matrix element of the fifth term in Eq. (11.30):

$$\begin{aligned}S_{fi} = & \frac{e^2}{2} \int d^4x_1\, d^4x_2\, \gamma^j_{\beta'\alpha'} \int \frac{d^4p}{(2\pi)^4} e^{-ip\cdot(x_1-x_2)} \frac{-i\,(m+\not p)_{\alpha'\alpha}}{m^2-p^2-i\epsilon}\,\gamma^i_{\alpha\beta} \\ & \times \int \frac{d^4p'}{(2\pi)^4}\, e^{-ip'\cdot(x_2-x_1)}\, \frac{-i\left(m+\not p'\right)_{\beta\beta'}}{m^2-p'^2-i\epsilon}\, \langle q'\epsilon_f|: A^j(x_1)A^i(x_2): |q\epsilon_i\rangle \, .\end{aligned} \tag{11.62}$$

The photon matrix element gives

$$\begin{aligned}\langle q'\epsilon_f|: A^j(x_1)A^i(x_2): |q\,\epsilon_i\rangle = & \frac{1}{\sqrt{2\omega 2\omega' L^3}} \\ & \times \left\{e^{i\left(q'\cdot x_1 - q\cdot x_2\right)}\epsilon_f^{j\,*}\epsilon_i^i + e^{i\left(q'\cdot x_2 - q\cdot x_1\right)}\epsilon_f^{i\,*}\epsilon_i^j\right\} .\end{aligned} \tag{11.63}$$

The two terms come from the two possible pairings of annihilation and creation operators in $:A\,A:$. The rest of the integrand is symmetric under $i \leftrightarrow j$, $x_1 \leftrightarrow x_2$, and $p \leftrightarrow p'$, because the trace can be cyclically rearranged. Hence the two terms

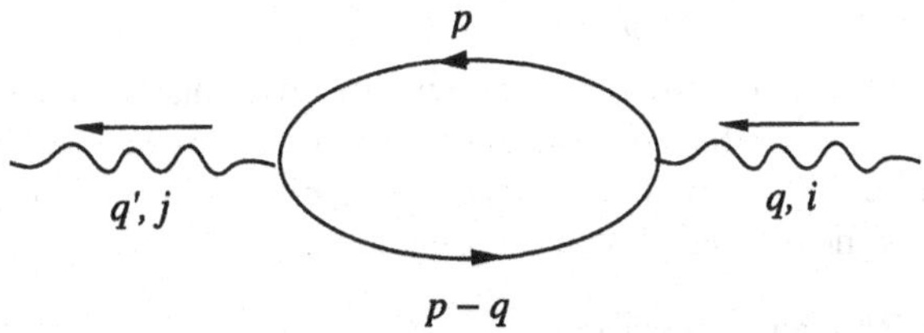

Fig. 11.8 Feynman diagram which gives the vacuum polarization to second order.

add, and the combined result is

$$\begin{aligned} S_{fi} &= -e^2 \int \frac{d^4x_1\, d^4x_2}{\sqrt{2\omega 2\omega'}\, L^3} \int \frac{d^4p}{(2\pi)^4} \int \frac{d^4p'}{(2\pi)^4}\, e^{i(q'-p+p')\cdot x_1} e^{-i(q-p+p')\cdot x_2} \\ &\qquad \times \frac{\mathrm{tr}\left\{\not\epsilon_f^*\,(m+\not p)\,\not\epsilon_i\,(m+\not p')\right\}}{(m^2-p^2-i\epsilon)(m^2-p'^2-i\epsilon)} \\ &= -i\frac{(2\pi)^4\delta^4(q'-q)}{\sqrt{4\omega\omega'}\,(2\pi)^3}\,\mathcal{M}_\Pi\,, \end{aligned} \tag{11.64}$$

where

$$\boxed{\mathcal{M}_\Pi = -ie^2 \int \frac{d^4p}{(2\pi)^4}\, \frac{\mathrm{tr}\left\{\not\epsilon_f^*\,(m+\not p)\,\not\epsilon_i\,(m+\not p-\not q)\right\}}{[m^2-p^2-i\epsilon]\,[m^2-(p-q)^2-i\epsilon]}\,.} \tag{11.65}$$

We see that this differs from the answer we guessed by an overall sign. The minus sign can be traced to the interchange of Fermi field operators required to obtain this term in Eq. (11.30).

As in the electron case, we generalize our discussion to off-shell photons, and introduce

$$\mathcal{M}_\Pi = \epsilon^*_{\mu f}\,\epsilon_{\nu i}\,\Pi^{\mu\nu}(q)\,, \tag{11.66}$$

where

$$\begin{aligned} \Pi^{\mu\nu}(q) &= -ie^2 \int \frac{d^4p}{(2\pi)^4}\, \frac{\mathrm{tr}\left\{\gamma^\mu(m+\not p)\gamma^\nu(m+\not p-\not q)\right\}}{(m^2-p^2-i\epsilon)(m^2-(p-q)^2-i\epsilon)} \\ &= -4ie^2 \int \frac{d^4p}{(2\pi)^4}\, \frac{\left[m^2 g^{\mu\nu} + p^\mu(p-q)^\nu + p^\nu(p-q)^\mu - g^{\mu\nu}p\cdot(p-q)\right]}{(m^2-p^2-i\epsilon)(m^2-(p-q)^2-i\epsilon)}. \end{aligned} \tag{11.67}$$

Note that this integral has terms which go, at large p, like

$$\int \frac{d^4p}{p^4}\, p^2 \sim \int^N d^2p \sim N^2 \tag{11.68}$$

and hence is quadratically divergent! Since the vacuum polarization is an especially important quantity, we will study how these divergences are handled in detail in the next section (11.6). For now we anticipate the results of that section and complete our discussion of the dressed photon propagator.

The Dressed Photon Propagator

In the next section, we will show that the vacuum polarization has the form

$$\boxed{\Pi^{\mu\nu}(q) = \left(g^{\mu\nu}q^2 - q^\mu q^\nu\right)\Pi(q^2)\,,} \tag{11.69}$$

where $\Pi(q^2)$ is a scalar function of q^2, so that (11.69) displays the dependence of the vacuum polarization on the initial and final photon spin indices. Also, note that $\Pi^{\mu\nu}$ satisfies the gauge invariant constraints

$$q_\mu \Pi^{\mu\nu} = 0 = \Pi^{\mu\nu} q_\nu \ . \tag{11.70}$$

Following the discussion of the electron self-energy we consider the infinite sum of photon self-energy terms, such as might occur in $e^+e^- \to \mu^+\mu^-$. These are shown in Fig. 11.9. The infinite sum of these terms gives the dressed propagator, which becomes

$$\begin{aligned} iD'^{\mu\nu}(q) &= \frac{ig^{\mu\nu}}{-q^2 - i\epsilon} + \frac{ig^{\mu\nu}}{-q^2 - i\epsilon}\left(\frac{q^2\Pi(q^2)}{-q^2}\right) + \frac{ig^{\mu\nu}}{-q^2 - i\epsilon}\left(\frac{q^2\Pi(q^2)}{-q^2}\right)^2 + \cdots \\ &= \frac{ig^{\mu\nu}}{-q^2 - i\epsilon}\left(\frac{1}{1+\Pi(q^2)}\right), \end{aligned} \tag{11.71}$$

where we used the relation $g^{\mu\lambda}\left[g_{\lambda\lambda'}q^2 - q_\lambda q_{\lambda'}\right]g^{\lambda'\nu} = q^2 g^{\mu\nu} - q^\mu q^\nu \to q^2 g^{\mu\nu}$ because the $q^\mu q^\nu$ terms will give zero when they are contracted into the free final $\mu^+\mu^-$ or initial e^+e^- currents. Note that *the photon pole remains at $q^2 = 0$, a consequence of gauge invariance*. Hence there can be no mass shift for the photon, and we introduce only *one renormalization constant*, called Z_3, as follows:

$$\boxed{iD'^{\mu\nu}(q) = \frac{ig^{\mu\nu}}{-q^2 - i\epsilon}\,\frac{Z_3}{1 + Z_3\overline{\Pi}(q^2)}\,,} \tag{11.72}$$

where $\overline{\Pi}(q^2) = \Pi(q^2) - \Pi(0)$ and the equivalence of this result with Eq. (11.71) requires

$$\boxed{\frac{1}{Z_3} = 1 + \Pi(0)\ .} \tag{11.73}$$

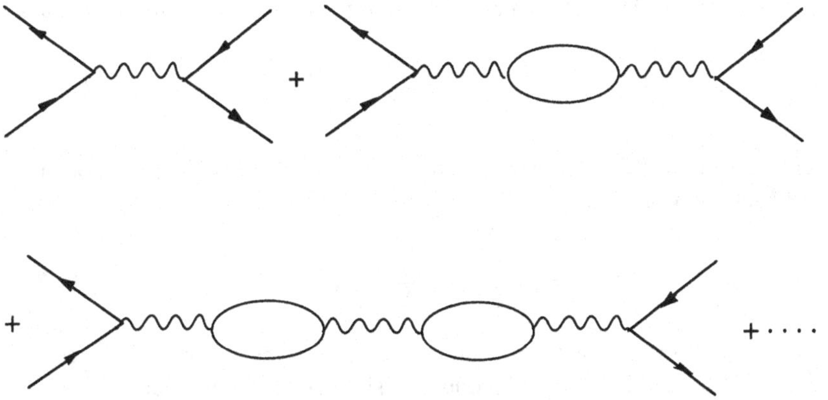

Fig. 11.9 Three diagrams which contribute to the dressed photon propagator.

Hence Z_3 removes the infinity contained in $\Pi_d(0)$, and its removal from the theory will eliminate the infinities associated with vacuum polarization.

The constant Z_3 is absorbed into the charge, just as was done with Z_2 for the electron. There must be one charge at the end of each photon line, and hence $\sqrt{Z_3}$ is absorbed into each charge (in this case it is only $\sqrt{Z_3}$ because only one photon is connected to each charge). Therefore the result [Eq. (11.54)] for charge renormalization gets extended to

$$e_0 \longrightarrow Z_2\sqrt{Z_3}\, e_0 \quad . \tag{11.74}$$

However, we have still not finished with charge renormalization!

Finally, external photons must be renormalized in the same way as electrons, giving us an addition to Rule 8:

Rule 8: for each external photon, a factor $\sqrt{Z_3}\,\epsilon^\mu$ or $\sqrt{Z_3}\,\epsilon^{\mu\,*}$ depending on whether or not the photon is incoming or outgoing.

We close this discussion with a final observation. For $|q^2|$ small, after removal of Z_3, the dressed photon propagator becomes (see the next section)

$$iD'^{\mu\nu}(q) \underset{q^2\ \mathrm{small}}{\simeq} -\frac{ig^{\mu\nu}}{q^2+i\epsilon}\left(\frac{1}{1+\frac{\alpha}{15\pi}\frac{q^2}{m^2}}\right) \quad . \tag{11.75}$$

For electron scattering, the momentum transfers q^2 are negative, and hence we see that the *effective force between charged particles* which are scattering *increases with higher energy* (momentum transfer) corresponding to an increase in the force at shorter distances. This can be restated by saying that *the effective charge at short distances (high momentum transfer) grows.* To get a quantitative estimate of the importance of this effect on atomic systems, expand (11.75) for $q^2 = -\boldsymbol{q}^2$,

$$D'^{\mu\nu}(q) \underset{q^2 \text{ small}}{\simeq} g^{\mu\nu}\left(\frac{1}{\boldsymbol{q}^2} + \frac{\alpha}{15\pi m^2}\right) . \tag{11.76}$$

Fourier transforming this to momentum space gives the familiar Coulomb potential plus the Uehling term,

$$V(r) = \frac{-Ze^2}{4\pi r} - \frac{Ze^4}{60\pi^2 m^2}\delta^3(r) . \tag{11.77}$$

Note that this affects S-states only, and contributes to the Lamb shift which we estimated in Chapter 3. It is of the opposite sign, contributing about -27 MHz to the overall shift of about 1058 MHz. Since the Lamb shift is known to about 0.01 MHz, this effect makes a small but important contribution to the total, and the overall agreement between theory and experiment confirms the correctness of this estimate.

In QCD, other terms due to gluon self-interactions contribute to the gluon self-energy. These terms change the sign of the corresponding Π-function, giving the result that *the effective coupling constant decreases at short distances (high momentum transfer).* This leads to the remarkable property of QCD known as *asymptotic freedom*, in which the forces go to zero at high energy. It also suggests that the forces will increase at high distances (low energy) and hence suggests *confinement.* These very interesting subjects will be taken up in Chapter 17.

We now discuss the evaluation of the loop integral (11.67).

11.6 LOOP INTEGRALS AND DIMENSIONAL REGULARIZATION

In this section we develop a general method which can be used to evaluate any one-loop Feynman integral. The method will be extended in Sec. 16.2 to the evaluation of Feynman integrals with more than one loop, and with these techniques we will be able to evaluate all Feynman diagrams. All of the formulae needed are summarized in Appendix C.

A general one-loop Feynman integral (in four space–time dimensions) is of the following form:

$$I = \int \frac{d^4k}{(2\pi)^4} \frac{N}{A_1 A_2 \cdots A_n} , \tag{11.78}$$

where the A_i are the denominators of Feynman propagators [cf. Eq. (11.36) for the electron self-energy and Eq. (11.67) for the vacuum polarization] and N is a numerator function which is a polynomial in the loop momentum k^μ. The calculation of this Feynman integral is carried out in two steps.

- The different denominators are combined into a single denominator, and the combined denominator is reduced to standard form by translating, or shifting, the internal loop momentum.
- The integral is then evaluated using an integral identity.

The first step makes use of identities of the form

$$\begin{aligned}\frac{1}{A_1A_2} &= \int_0^1 dz\frac{1}{[A_1z+A_2(1-z)]^2} = \int_0^1 dz\frac{1}{[D(z)]^2}\\ \frac{1}{A_1A_2A_3} &= 2\int_0^1 dz_1\int_0^{1-z_1} dz_2\frac{1}{[A_1z_1+A_2z_2+A_3(1-z_1-z_2)]^3}\\ &= 2\int_0^1 dz_1\int_0^{1-z_1} dz_2\frac{1}{[D(z_1,z_2)]^3}\ ,\end{aligned} \tag{11.79}$$

which are easily proved by direct integration. The integration variables z_i are referred to as *Feynman parameters*. The two identities (11.79) are the only two we will need in this chapter, but a completely general identity which covers any case which might be encountered is proven in Sec. 16.2, and given in Appendix C.

To complete the reduction to standard form (the first step), it is necessary to observe that the combined denominator D always has the form

$$\begin{aligned}D &= A_1z_1+A_2z_2+A_3(1-z_1-z_2)\\ &= B^2+2k\cdot Q-k^2\ ,\end{aligned} \tag{11.80}$$

where k is the internal loop momentum and Q is a vector function of the external momenta and the Feynman parameters. This form follows from the observation that each of the individual propagators in the loop is itself of the form $A_i = m_i^2 + k\cdot q_i - k^2$, so that when they are combined as in Eq. (11.80), the k^2 term has the coefficient $z_1 + z_2 + (1 - z_1 - z_2) = 1$. This holds for a loop with any number of propagators. Thus the square of the denominator can always be completed by shifting $k = k' + Q$, which gives

$$D \to D' = B^2 + Q^2 - k'^2\ . \tag{11.81}$$

This shift must also be carried out in the numerator N, which assumes the general form

$$N = N_0 + k'_\mu N_1^\mu + k'_\mu k'_\nu N_2^{\mu\nu} + k'_\mu k'_\nu k'_\sigma N_3^{\mu\nu\sigma} + k'_\mu k'_\nu k'_\sigma k'_\lambda N_4^{\mu\nu\sigma\lambda} + \cdots\ , \tag{11.82}$$

where the N^i are tensors which do not depend on k'. Since the denominator is even in k' (in fact, it depends on k'^2 only), all of the odd terms reduce to zero and the even ones can be simplified using identities we will introduce shortly.

After step one has been completed, we are confronted with an integral of the following form:

$$F = \int \frac{d^4k}{(2\pi)^4} \frac{N}{(C^2 - k^2 - i\epsilon)^n} \, , \tag{11.83}$$

where n is an integer and we will assume for now that the numerator N is independent of k. We will assume that there are no zeros in the denominator for finite k^2, and consider the convergence of the integral at large k. If $n > 2$, the integral will converge, but we will encounter many cases when $n \leq 2$, and the integral is divergent (the vacuum polarization and electron self-energies are examples where $n = 2$). The general method for treating these divergent integrals is to imagine that we are evaluating them in a number of space–time dimensions $d < 4$. In this case, the volume integration goes like $d^dk \sim k^d$, but the denominator still goes like k^{2n}, so the integral will converge as long as $d < 2n$. As $d \to 4$, the singularity returns, but it is easily identified and isolated into a renormalization constant, as we have already discussed briefly, and the finite part of the integral is then clearly defined. The process of separating the integral into its finite and infinite parts is referred to as *regularization* and must be done before the infinity can be removed by absorbing it into the coupling constants of the theory, a process referred to as *renormalization.* The general procedure for renormalizing theories is discussed in some detail in Chapter 16; in this chapter we introduce these ideas using second order QED as an example.

The integral (11.83) can be evaluated using the following identity:

$$\int \frac{d^dk}{(2\pi)^d} \frac{1}{(C^2 - k^2 - i\epsilon)^n} = \frac{i}{(4\pi)^{d/2}} \frac{\Gamma(n - d/2)}{\Gamma(n)} \left(\frac{1}{C^2}\right)^{n-d/2} , \tag{11.84}$$

where d is the number of dimensions (as discussed above) and $\Gamma(\alpha)$ is the familiar generalization of the factorial function with $\Gamma(\alpha) = (\alpha - 1)\Gamma(\alpha - 1)$ and $\Gamma(1) = 1$. For $\alpha = n$, an integer, $\Gamma(n) = (n - 1)!$, but $\Gamma(\alpha)$ is also defined for noninteger values of α. A convenient integral representation for Γ which we will use frequently is*

$$\Gamma(\alpha) = \int_0^\infty dt \, t^{\alpha-1} \, e^{-t} \; . \tag{11.85}$$

We will use this representation to prove (11.84).

Proof: We begin with the observation that

$$\frac{-i}{C^2 - k^2 - i\epsilon} = \int_0^\infty dz \, e^{-iz[C^2 - k^2 - i\epsilon]} \; . \tag{11.86}$$

*A good reference for special functions is Abramowitz and Stegun (1964).

This identity is easily proved by direct integration:

$$\int_0^\infty dz\, e^{-iz[D-i\epsilon]} = \int_0^\infty dz\, e^{-z(iD+\epsilon)}$$
$$= \frac{e^{-z(iD+\epsilon)}}{-(iD+\epsilon)}\bigg|_0^\infty = \frac{1}{iD+\epsilon} = \frac{-i}{D-i\epsilon} \ .$$

Note the crucial role played by the "$i\epsilon$" prescription; it defines the integral in (11.86) by providing convergence for large z and plays a similar role by defining the function at any singular points $D^2 = 0$.

The identity (11.86) is now generalized by differentiating both sides $n-1$ times with respect to C^2:

$$\frac{-i}{(C^2-k^2-i\epsilon)^n} = \frac{i^{n-1}}{\Gamma(n)}\int_0^\infty dz\, z^{n-1}\, e^{-iz[C^2-k^2-i\epsilon]} \ . \tag{11.87}$$

Next, we integrate (11.87) over k using the following identities, which hold because $z > 0$,

$$\begin{aligned}\int_{-\infty}^\infty \frac{dk_0}{2\pi}\, e^{ik_0^2 z} &= \frac{i^{1/2}}{(4\pi z)^{1/2}} \\ \int_{-\infty}^\infty \frac{dk_1}{2\pi} e^{-ik_1^2 z} &= \frac{(-i)^{1/2}}{(4\pi z)^{1/2}} \ .\end{aligned} \tag{11.88}$$

These integrals may be evaluated using well-known methods for integrating functions in the complex plane. Initially, the integrals are along the real axis in the complex k_0 (or k_1) plane. To evaluate the first integral, rotate the k_0 contour through a positive angle ϕ. Then $k_0 \to r\,e^{i\phi}$, and $k_0^2 = r^2\,e^{2i\phi} = r^2(\cos 2\phi + i\sin 2\phi)$, so that the integral converges as long as $\pi/2 > \phi > 0$ (and the contribution from the arc at $k_0 = \infty$ is zero). At $\phi = \pi/4 = 45°$ we have optimal convergence:

$$\begin{aligned}\int_{-\infty}^\infty \frac{dk_0}{2\pi} e^{ik_0^2 z} &= e^{i\pi/4}\int_{-\infty}^\infty \frac{dr}{2\pi} e^{-zr^2} \\ &= \frac{e^{i\pi/4}}{2\pi\sqrt{z}}\int_{-\infty}^\infty dy\, e^{-y^2} = \frac{i^{1/2}}{(4\pi z)^{1/2}} \ .\end{aligned} \tag{11.89}$$

For the dk_1 integral, convergence requires rotating by $\phi = -\pi/4$, giving the opposite sign for i.

For an integral with one time dimension and $d-1$ space dimensions, the combined effect of the identities (11.88) is stated in the following identity:

$$\int \frac{d^d k}{(2\pi)^d}\, e^{ik^2 z} = \frac{(-i)^{\frac{d}{2}-1}}{(4\pi z)^{d/2}} \ . \tag{11.90}$$

To prove this identity, note that in d dimensions, $k^2 = k_0^2 - \sum_{i=1}^{d-1} k_i^2$ and $d^dk = dk_0 \prod_{i=1}^{d-1} dk_i$. Hence the integral factors into d terms,

$$\int \frac{d^dk}{(2\pi)^d}\, e^{ik^2 z} = \left(\int_{-\infty}^{\infty} \frac{dk_0}{2\pi}\, e^{ik_0^2 z}\right)\left(\int_{-\infty}^{\infty} \frac{dk_1}{2\pi} e^{-ik_1^2 z}\right)^{d-1}$$
$$= \frac{i^{\frac{1}{2}}}{(4\pi z)^{1/2}}\left(\frac{-i}{4\pi z}\right)^{\frac{d}{2}-\frac{1}{2}} = \frac{(-i)^{\frac{d}{2}-1}}{(4\pi z)^{d/2}}\;,$$

which proves (11.90).

Finally, combining the results (11.87) and (11.90) gives the result

$$\int \frac{d^dk}{(2\pi)^d}\, \frac{1}{(C^2-k^2-i\epsilon)^n} = \frac{(-i)^{\frac{d}{2}-n-1}}{(4\pi)^{d/2}}\, \frac{1}{\Gamma(n)} \int_0^\infty dz z^{n-\frac{d}{2}-1}\, e^{-iz[C^2-i\epsilon]}\;. \tag{11.91}$$

Scaling this integral by substituting $t = iz(C^2 - i\epsilon)$ gives

$$\int \frac{d^dk}{(2\pi)^d}\, \frac{1}{(C^2-k^2-i\epsilon)^n} = \frac{i}{(4\pi)^{d/2}}\, \frac{1}{\Gamma(n)} \left(\frac{1}{C^2}\right)^{n-d/2} \int_0^\infty dt\, t^{n-\frac{d}{2}-1}\, e^{-t}\;. \tag{11.92}$$

However, the integral over t is just the integral representation for $\Gamma(n - d/2)$, Eq. (11.85), and hence the identity (11.84) has been proved. ∎

Before returning to our discussion of vacuum polarization, observe that the integral over the vector components of k can be quickly reduced using the results we have previously obtained. We will show that

$$\int \frac{d^dk}{(2\pi)^d}\, \frac{k^\mu k^\nu}{(C^2-k^2-i\epsilon)^n} = \frac{g^{\mu\nu}}{d} \int \frac{d^dk}{(2\pi)^d}\, \frac{k^2}{(C^2-k^2-i\epsilon)^n}$$
$$= -\frac{i\, g^{\mu\nu}}{2(4\pi)^{d/2}}\, \frac{\Gamma(n-1-d/2)}{\Gamma(n)} \left(\frac{1}{C^2}\right)^{n-1-d/2}. \tag{11.93}$$

To prove this, first note that terms with $\mu \neq \nu$ are zero because they are odd under changing $k^\mu \to -k^\mu$ (or $k^\nu \to -k^\nu$). For the $\mu = \nu$ terms, assume that C^2 is real and positive, and note that the singularities in the k_0 complex plane are therefore in the second and fourth quadrants:

$$k_0 = \pm\left(\sqrt{C^2 + \mathbf{k}^2} - i\epsilon\right)\;.$$

Hence we may rotate the k_0 integration contour as we did above by letting $k_0 = r\, e^{i\phi}$ and changing ϕ continuously from 0 to $\pi/2$. This changes $k_0^2 \to -k_0^2$, and the resulting integral is transformed from a d-dimensional Minkowski space* to a

---

*A *Minkowski space* is one with an indefinite metric (in our example the diagonal elements of the metric are $+1, -1, -1, -1$ in $d = 4$ dimensions). The rotation of the k_0 contour has the effect of changing the metric to a Euclidean form: $-1, -1, -1, -1$.

d-dimensional Euclidean space, where the integrand is completely symmetric in all components of k, and from this symmetry we can conclude that $k^\mu k^\nu \to \delta^{\mu\nu}\, k^2/d$. Rotating back to Minkowski space changes the sign of the k_0^2 term on both sides, giving the first line of the identity (11.93).

To get the second, we use the properties of the Γ-function, as follows

$$\begin{aligned}
\frac{1}{d}\int \frac{d^dk}{(2\pi)^d}\,&\frac{k^2}{(C^2-k^2-i\epsilon)^n} \\
&= \frac{C^2}{d}\int \frac{d^dk}{(2\pi)^d}\,\frac{1}{(C^2-k^2-i\epsilon)^n} - \frac{1}{d}\int \frac{d^dk}{(2\pi)^d}\,\frac{1}{(C^2-k^2-i\epsilon)^{n-1}} \\
&= \frac{i}{(4\pi)^{d/2}\,d}\left(\frac{1}{C^2}\right)^{n-1-d/2}\left[\frac{\Gamma(n-d/2)}{\Gamma(n)} - \frac{\Gamma(n-1-d/2)}{\Gamma(n-1)}\right] \\
&= \frac{i}{(4\pi)^{d/2}\,d}\left(\frac{1}{C^2}\right)^{n-1-d/2}\frac{\Gamma(n-1-d/2)}{\Gamma(n)}\left[n-1-\frac{d}{2}-(n-1)\right] \\
&= -\frac{i}{2\,(4\pi)^{d/2}}\left(\frac{1}{C^2}\right)^{n-1-d/2}\frac{\Gamma(n-1-d/2)}{\Gamma(n)}\,, \qquad (11.94)
\end{aligned}$$

which completes the proof of (11.93).

While all of these identities have been derived for integral d, the final results are expressed as functions of d which can be analytically continued into the complex d plane. Hence, from now on, we will think of d *as a continuous variable.*

Finally, we are ready to return to the integral (11.67). We will evaluate it following the steps we have just discussed. First, we write the integral in d dimensions, and start off by assuming that $d < 2$, so that the integral is convergent and everything is well defined. Then we combine the two propagators using the first of the identities (11.79),

$$\begin{aligned}
\Pi_d^{\mu\nu}(q) &= -4ie^2\int \frac{d^dp}{(2\pi)^d}\,\frac{\left[m^2g^{\mu\nu}+2p^\mu p^\nu - p^\mu q^\nu - q^\mu p^\nu - g^{\mu\nu}p\cdot(p-q)\right]}{(m^2-p^2-i\epsilon)\underbrace{\left(m^2-(p-q)^2-i\epsilon\right)}_{x}} \\
&= -4ie^2\int_0^1 dx\int \frac{d^dp}{(2\pi)^d}\,\frac{N^{\mu\nu}}{\left[m^2-p^2+2p\cdot qx-q^2x-i\epsilon\right]^2}\,. \qquad (11.95)
\end{aligned}$$

Next we complete the square in the denominator by introducing $p = k + qx$. Because the integration is over all of space–time, adding or subtracting a fixed four-vector to p does not change the volume of integration, and the new integral is

$$\Pi_d^{\mu\nu}(q) = -4ie^2\int_0^1 dx\int \frac{d^dk}{(2\pi)^d}\,\frac{N'^{\mu\nu}}{\left[m^2-k^2-q^2x(1-x)-i\epsilon\right]^2}\,,$$

where the transformed numerator is

$$\begin{aligned}
N'^{\mu\nu} =& m^2g^{\mu\nu} + 2k^\mu k^\nu + (q^\mu k^\nu + k^\mu q^\nu)(2x-1) - 2q^\mu q^\nu x(1-x) \\
&- g^{\mu\nu}(k+qx)\cdot(k-q(1-x))\,.
\end{aligned}$$

Next, drop terms odd in k, and use the identity (11.93), $k^\mu k^\nu \to g^{\mu\nu}k^2/d$, to reduce $N'^{\mu\nu}$ to

$$N'^{\mu\nu} = g^{\mu\nu}\left(m^2 + \left(\frac{2}{d} - 1\right)k^2 + q^2x(1-x)\right) - 2q^\mu q^\nu x(1-x) \ .$$

Now we may evaluate the integrals over k using the identity (11.84) and the identity (11.93) for the term proportional to k^2 in the numerator,

$$\begin{aligned}\Pi_d^{\mu\nu}(q) = &\frac{4e^2}{(4\pi)^{d/2}}\frac{\Gamma(2-d/2)}{\Gamma(2)}\int_0^1 dx\left(\frac{1}{m^2 - q^2x(1-x)}\right)^{2-d/2} \\ &\times\left[g^{\mu\nu}\left(m^2 - q^2x(1-x)\right)\left\{1 - \frac{d}{2}\left(\frac{2}{d} - 1\right)\frac{\Gamma(1-d/2)}{\Gamma(2-d/2)}\right\}\right. \\ &\left. + \left(g^{\mu\nu}q^2 - q^\mu q^\nu\right)2x(1-x)\right] \ .\end{aligned}$$

Using the properties of the Γ-function, we see immediately that the coefficient of the $g^{\mu\nu}$ term is zero! This is a nice feature of dimensional regularization; it respects the gauge invariance of the theory. Other regularization methods give the same result, but only after considerable labor. The remaining term is gauge invariant and has the form we anticipated in Eq. (11.69). Extracting the scalar part, $\Pi_d(q^2)$, gives

$$\Pi_d(q^2) = \frac{\alpha}{\pi}\frac{\Gamma(3-d/2)}{(2-d/2)(4\pi)^{d/2-2}}\int_0^1 dx\frac{2x(1-x)}{\left[m^2 - q^2x(1-x)\right]^{2-d/2}} \ . \qquad (11.96)$$

In this expression, the singularity which exists for $d = 4$ dimensions appears as a pole in Π_d. This pole corresponds to a logarithmic singularity in the original integral; the quadratic divergence has disappeared because it was contained in the gauge violating $g^{\mu\nu}$ term, which integrated to zero. This means that the vacuum polarization is now well defined for all $d < 4$, and the physical result can be obtained from the limit $d \to 4$. The scalar vacuum polarization will now be written as the sum of two terms, an (infinite) constant corresponding to its value at $q^2 = 0$ and a (finite) term obtained by subtracting the integral (11.96) at $q^2 = 0$. This gives

$$\begin{aligned}\Pi_d(q^2) &= \Pi_d(0) + \left(\Pi_d(q^2) - \Pi_d(0)\right) \\ &= \Pi_d(0) + \overline{\Pi}_d(q^2) \ ,\end{aligned} \qquad (11.97)$$

where $\Pi_d(0)$ is singular,

$$\Pi_d(0) = \frac{\alpha}{\pi}\frac{2\Gamma(1+\epsilon/2)}{\epsilon\,(4\pi)^{-\epsilon/2}}\int_0^1 dx\frac{2x(1-x)}{m^\epsilon} \ , \qquad (11.98)$$

with $\epsilon = 4 - d$ now regarded as a small quantity which approaches zero as $d \to 4$. The subtracted self-energy is

$$\overline{\Pi}_d(q^2) = \frac{\alpha}{\pi} \frac{4\Gamma(1+\epsilon/2)}{\epsilon\,(4\pi)^{-\epsilon/2}} \int_0^1 x(1-x) \left[\frac{1}{[m^2 - q^2 x(1-x)]^{\epsilon/2}} - \frac{1}{m^\epsilon} \right] . \tag{11.99}$$

This part is finite and can be evaluated by taking the limit $\epsilon \to 0$. The only term which survives in this limit comes from the expansion of the fractional powers involving m^2. Using

$$\lim_{\epsilon\to 0} A^\epsilon = 1 + \epsilon \,\log A + \mathcal{O}(\epsilon^2) ,$$

we obtain

$$\overline{\Pi}_d(q^2) \to \overline{\Pi}(q^2) = -\frac{2\alpha}{\pi} \int_0^1 dx\, x(1-x) \log\left[1 - \frac{q^2}{m^2} x(1-x)\right] . \tag{11.100}$$

Note that this is zero at $q^2 = 0$, as expected. For small $|q|^2 \ll m^2$, $\overline{\Pi}(q^2)$ can be approximated by expanding the logarithm

$$\overline{\Pi}(q^2) \underset{|q^2| \ll m^2}{\longrightarrow} \frac{2\alpha}{\pi} \frac{q^2}{m^2} \int_0^1 dx\, x^2(1-x)^2 = \frac{\alpha}{15\pi} \frac{q^2}{m^2} , \tag{11.101}$$

which is the result we anticipated in (11.77).

Note that the subtracted self-energy $\overline{\Pi}(q^2)$ is complex if $q^2 \geq 4m^2$. To see this, note that the maximum value of $x(1-x)$ in the interval $[0,1]$ is $\frac{1}{4}$, and hence when $q^2/4m^2 > 1$, the argument of the log in Eq. (11.100) becomes negative at some point in the region of integration, and the log becomes complex. As this is an example of a general property of Feynman diagrams which is of great importance, we will discuss it in more detail in the next section.

11.7 DISPERSION RELATIONS

We begin our discussion of dispersion relations by considering the self-energy of a neutral particle in the symmetric ϕ^3 theory. The self-energy in second order comes from the matrix element

$$S_{fi} = -\frac{\lambda^2}{2(3!)^2} \langle p'| \int d^4x_1\, d^4x_2\, T\left(:\phi^3(x_1):\,:\phi^3(x_2):\right) |p\rangle . \tag{11.102}$$

Using Wick's theorem, four of these fields must be contracted into propagators, leaving the remaining two to balance to annihilation and creation operators contained in the final and initial states. However, since all of the fields are identical, the two contractions can be made in many ways. There are $3 \times 3 = 9$ possible ways to make the first pairing, and $2 \times 2 = 4$ ways to make the second, but the correct number of different choices is $(9\times 4)/2 = 18$ because it does not matter in which order the pairings are made (they are all identical). Finally, the remaining normal product $:\phi(x_1)\phi(x_2):$ can balance against the creation operator from the initial state and the annihilation operator from the final state in two ways. The resulting factor of $18 \times 2 = 36$ cancels the factor of $(3!)^2$ in (11.102), leaving an extra factor of $\frac{1}{2}$, giving the following result for the self-energy:

$$\Sigma(q^2) = i\frac{\lambda^2}{2}\int \frac{d^4p}{(2\pi)^4}\frac{1}{(\mu^2 - p^2 - i\epsilon)\,(\mu^2 - (p-q)^2 - i\epsilon)} \,. \tag{11.103}$$

The extra factor of $\frac{1}{2}$ is referred to as a *symmetry factor* and is a new Feynman rule:

> **Rule 7:** for each bubble diagram involving identical bosons, a symmetry factor of $\frac{1}{2}$.

Except for the symmetry factor, the self-energy (11.103) has the same structure as the vacuum polarization (without the numerator). It diverges logarithmically in $d = 4$ dimensions. If we evaluate it for $d < 4$, the steps leading to the evaluation of the vacuum polarization can be retraced and give (we will return to this computation in Chapter 16)

$$\Sigma(q^2) = -\frac{\lambda^2}{2(4\pi)^{d/2}}\Gamma\left(2-\frac{d}{2}\right)\int_0^1 dx\frac{1}{\left[\mu^2 - q^2x(1-x)\right]^{2-d/2}} \,. \tag{11.104}$$

This diagram has the same singularities as the vacuum polarization diagram and will be discussed first.

Before we discuss dispersion relations in general, we will cast the integral (11.104) into a *dispersion form.* To this end, integrate (11.104) by parts to obtain a more convenient form:

$$\Sigma(q^2) = -\frac{\lambda^2}{2(4\pi)^{d/2}}\Gamma\left(2-\frac{d}{2}\right) \times\left\{\frac{1}{\mu^{4-d}} - q^2\,(2-d/2)\int_0^1 dx\frac{x(1-2x)}{\left[\mu^2 - q^2x(1-x)\right]^{3-d/2}}\right\}. \tag{11.105}$$

The first term is simply $\Sigma(0)$ and is singular. We consider the finite part, defined by

$$\begin{aligned}\frac{\Sigma(q^2)-\Sigma(0)}{q^2} &= \frac{\lambda^2}{2(4\pi)^{d/2}}\left(2-\frac{d}{2}\right)\Gamma\left(2-\frac{d}{2}\right)\int_0^1 dx\frac{x(1-2x)}{\left[\mu^2 - q^2x(1-x)\right]^{3-d/2}} \\ &\xrightarrow[d\to 4]{} \frac{\lambda^2}{2(4\pi)^2}\int_0^1 dx\frac{x(1-2x)}{\left[\mu^2 - q^2x(1-x)\right]} \\ &= -\frac{\lambda^2}{2(4\pi)^2}\int_{\frac{1}{2}}^1 dx\frac{(1-2x)^2}{\left[\mu^2 - q^2x(1-x)\right]} \,.\end{aligned} \tag{11.106}$$

Next, change integration variables from x to $s = \mu^2/x(1-x)$. This gives finally

$$\begin{aligned} A(q^2) = \frac{\Sigma(q^2) - \Sigma(0)}{q^2} &= -\frac{\lambda^2}{2(4\pi)^2}\int_{4\mu^2}^{\infty} ds\sqrt{1-\frac{4\mu^2}{s}}\,\frac{1}{s(s-q^2-i\epsilon)} \\ &= -\frac{\lambda^2}{4\pi}\int_{4\mu^2}^{\infty} ds\frac{\rho(\sqrt{s};\mu,\mu)}{s(s-q^2-i\epsilon)} \\ &= \frac{1}{\pi}\int_{4\mu^2}^{\infty} ds\frac{Im\,A(s)}{(s-q^2-i\epsilon)}\ , \end{aligned} \tag{11.107}$$

where $\rho(\sqrt{s};\mu,\mu)$ is the two-body phase space factor defined in Chapter 9, Eq. (9.21).

Equation (11.107) is an example of a *dispersion integral.* It expresses the amplitude as an integral over the region where it is singular. In addition to displaying the singularities of the Feynman amplitude explicitly, this integral representation defines the amplitude as an analytic function, so that we may study it using the powerful mathematics of complex analysis. If the location of its singularities is known, the mere knowledge that a dispersion relation exists can sometimes be used to estimate the behavior of an amplitude. But the real power of dispersion theory rests in three facts:

- All Feynman diagrams satisfy dispersion relations, and hence the exact amplitudes probably do also.
- There will be a singularity, or a *cut*, in an amplitude *whenever the external variables have values for which it is possible for all the particles in an intermediate state to be on-mass-shell, i.e., to be physical.*
- The imaginary part of the amplitude (referred to as its *absorptive part*) along any of its cuts can be determined from unitarity.

The last two observations give dispersion theory an element of predictive power, and the first means that it is a very general technique for the study of relativistic interactions. In the 1960's, before the advent of gauge theories, it was believed by some that dispersion theory might be the best method for the study of the strong interactions. This did not turn out to be true, but these methods still belong in the arsenal of the well-equipped physicist.*

Our task here is to use the self-energy (11.107) and the vacuum polarization to illustrate the last two of the above general facts about dispersion theory [the first is already illustrated by (11.107)]. The second is illustrated in Fig. 11.10. For the ϕ^3 self-energy, the two intermediate particles can be physical whenever the total energy in their center of mass is greater that 2μ, and since q^2 is the square of this energy, the cut runs from $4\mu^2 \to \infty$. The same is true of the vacuum polarization diagram; the production of physical $e^+\,e^-$ pairs is possible whenever the energy of a virtual photon at rest is greater than $2m$, or when $q^2 > 4m^2$.

---

*For a review of dispersion methods see, for example, Barton (1965). For details about the singularities of Feynman amplitudes, see Todorov (1971).

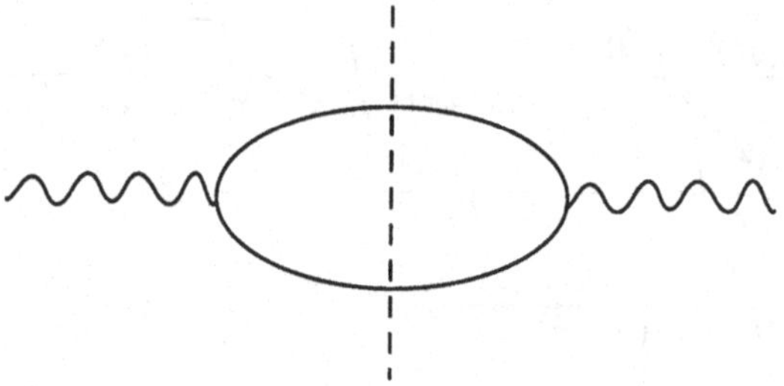

Fig. 11.10 Self-energy diagram showing the cut where the two intermediate particles can be physical if the energy is sufficiently great.

The third principle says that the imaginary part is given by unitarity, which follows from the unitarity of the time translation operator. We will derive this relation in the next chapter (see Sec. 12.8); for now we merely note that the unitarity statement is

$$Im\, \mathcal{M}_{fi}(s) = -\frac{1}{2}\sum_{\alpha} \rho(\sqrt{s}, m_1, m_2, \cdots, m_n)\, \mathcal{M}^{\dagger}_{\alpha f}(s)\, \mathcal{M}_{\alpha i}(s) \quad , \tag{11.108}$$

where ρ is the phase space operator defined in Eq. (9.22) and the sum over α and integration (contained in the phase space operator) are over all momenta and spins in the intermediate state. For the symmetric ϕ^3 theory, the decay amplitude is given simply by $\mathcal{M} = \lambda$, so the unitarity statement becomes

$$Im\, \mathcal{M} = -\tfrac{1}{4}\, \rho(\sqrt{s}; \mu, \mu)\, \lambda^2 \quad , \tag{11.109}$$

where the extra factor of $\frac{1}{2}$ is the statistical factor for identical particle decays which we discussed in Sec. 9.7. Using the fact that $\Sigma(0)$ is real, and remembering that $A = \mathcal{M}/q^2$, we obtain precisely the result for $Im\, A$ implied by (11.107).

We conclude this discussion by returning to the vacuum polarization amplitude. Turning the discussion around, we first calculate the imaginary part of the polarization diagram from

$$\begin{aligned} Im\, \mathcal{M}^{\mu\nu} & \\ &= -\tfrac{1}{2}\rho(\sqrt{s}; m, m)\, e^2 \sum_{s,s'} \Big(\bar{u}(\boldsymbol{q}-\boldsymbol{p}, s')\gamma^{\mu} v(\boldsymbol{p}, s)\Big)^{\dagger} \Big(\bar{u}(\boldsymbol{q}-\boldsymbol{p}, s')\gamma^{\nu} v(\boldsymbol{p}, s)\Big) \\ &= -\tfrac{1}{2}\rho(\sqrt{s}; m, m)\, e^2 \sum_{s,s'} \Big(\bar{v}(\boldsymbol{p}, s)\gamma^{\mu} u(\boldsymbol{q}-\boldsymbol{p}, s')\Big) \Big(\bar{u}(\boldsymbol{q}-\boldsymbol{p}, s')\gamma^{\nu} v(\boldsymbol{p}, s)\Big) \\ &= -\tfrac{1}{2}\rho(\sqrt{s}; m, m)\, e^2 \;\mathrm{tr}\, \{\gamma^{\mu}\, (m + \not{q} - \not{p})\, \gamma^{\nu}\, (\not{p} - m)\} \\ &= -2\rho(\sqrt{s}; m, m)\, e^2\, (-g^{\mu\nu} p\cdot q + q^{\mu} p^{\nu} + q^{\nu} p^{\mu} - 2p^{\mu} p^{\nu}) \,, \end{aligned} \tag{11.110}$$

where both intermediate particles are on-shell, so that $p^2 = (q-p)^2 = m^2$. In the center of mass, where $q^\mu = (\sqrt{s}, \mathbf{0})$, one can readily see that the $(0,0)$ component of (11.110) is zero, as it must be, and averaging over the direction of $\boldsymbol{p}$ gives

$$Im\,\mathcal{M}^{ij} = -2\rho(\sqrt{s}; m, m)\, e^2\, \delta_{ij} \left(\frac{s}{2} - \frac{2}{3}\left(\frac{s}{4} - m^2\right)\right) . \tag{11.111}$$

Extracting the vacuum polarization scalar and dividing by q^2 give the following dispersion integral:

$$B(q^2) = \frac{\Pi(q^2) - \Pi(0)}{q^2} = \frac{2e^2}{3\pi} \int_{4m^2}^{\infty} ds\, \frac{\rho(\sqrt{s}; m, m)\,[s + 2m^2]}{s^2(s - q^2 - i\epsilon)} . \tag{11.112}$$

We will now derive this same expression directly from Eq. (11.100).

First, get a usable expression by integrating (11.100) by parts, which gives

$$\begin{aligned} B(q^2) = \frac{\overline{\Pi}(q^2)}{q^2} &= -\frac{e^2}{12\pi^2} \int_0^1 dx\, \frac{x^2(3-2x)(1-2x)}{[m^2 - q^2 x(1-x) - i\epsilon]} \\ &= -\frac{e^2}{12\pi^2} \int_{\frac{1}{2}}^1 dx\, \frac{(6x^2 - 4x^3 - 1)(1-2x)}{[m^2 - q^2 x(1-x) - i\epsilon]} . \end{aligned} \tag{11.113}$$

This integral is very similar to (11.106), and we reduce it using the same transformation, which gives the correspondences

$$s = \frac{m^2}{x(1-x)} \qquad x = \frac{1}{2}\left(1 + \sqrt{1 - \frac{4m^2}{s}}\right)$$

$$6x^2 - 4x^3 - 1 = \sqrt{1 - \frac{4m^2}{s}}\left(1 + \frac{2m^2}{s}\right) .$$

Substituting these into (11.113) gives (11.112), and the equivalence is established.

We will return to a discussion involving dispersion relations several times in the remainder of this book.

11.8 VERTEX CORRECTIONS

We have completed our discussion of the six second order terms in Eq. (11.30), and now turn to a discussion of the second order correction to the electromagnetic *vertex*. This is an extremely important special example, leading to a calculation of the anomalous magnetic moment of the electron and additional contributions needed to conclude our discussion of charge renormalization.

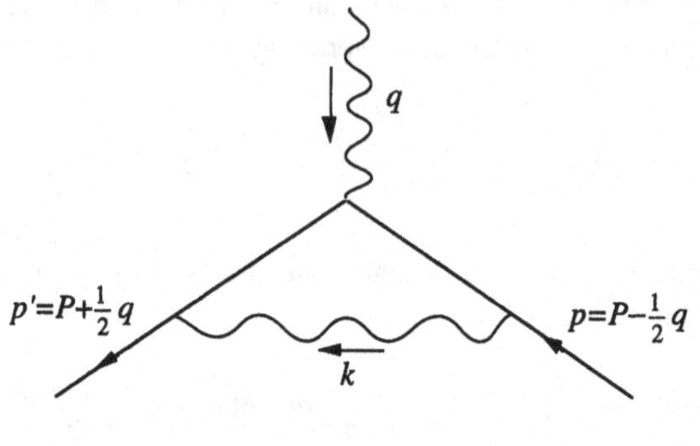

Fig. 11.11 The lowest order "correction" to the electromagnetic vertex.

The diagram we will calculate is shown in Fig. 11.11. Since the electron charge is negative, the bare vertex is $ie\gamma^\mu$, and the lowest order term shown in Fig. 11.11 adds a "correction" factor of the form $ie\Lambda^\mu(P,q)$, where $P=\frac{1}{2}(p'+p)$ and $q=p'-p$, as shown in the figure. Adding Rule 0 and external electron states, the Feynman amplitude corresponding to this diagram is

$$
\begin{aligned}
&-e\bar{u}(\mathbf{p}')\Lambda^\mu(P,q)u(\mathbf{p}) \\
&\quad = i(ie)^3(-i)^3\int\frac{d^4k}{(2\pi)^4} \\
&\qquad \times\frac{\bar{u}(\mathbf{p}')\gamma^\nu[m+\not{p}'-\not{k}]\gamma^\mu[m+\not{p}-\not{k}]\gamma^{\nu'}(-g_{\nu\nu'})u(\mathbf{p})}{[m^2-(p'-k)^2-i\epsilon][m^2-(p-k)^2-i\epsilon][\lambda^2-k^2-i\epsilon]}\,. \qquad (11.114)
\end{aligned}
$$

where the fictitious photon mass λ will be taken to zero after the integrals have been done.

With the understanding that we may later want to reduce Λ^μ by using the Dirac equation, so that

$$
\begin{aligned}
\not{p}' &= m \quad \text{(when operating towards the left)} \\
\not{p} &= m \quad \text{(when operating towards the right,)}
\end{aligned} \qquad (11.115)
$$

we will drop the initial u and final $\bar{u}$ spinors, and separate out Λ

$$
\begin{aligned}
&\Lambda^\mu(P,q) \\
&\quad = ie^2\int\frac{d^4k}{(2\pi)^4}\,\frac{\gamma^\nu[m+\not{p}'-\not{k}]\gamma^\mu[m+\not{p}-\not{k}]\gamma_\nu}{[m^2-(p'-k)^2-i\epsilon]\,[m^2-(p-k)^2-i\epsilon]\,[\lambda^2-k^2-i\epsilon]}\,.
\end{aligned} \qquad (11.116)
$$

This integral is evaluated following the steps described in Sec. 11.6. We first combine the three Feynman denominators using the identity (11.79):

$$
\frac{1}{A_1A_2A_3}=2\int_0^1 dz_1\int_0^{1-z_1}dz_2\,\frac{1}{\underbrace{[A_1z_1+A_2z_2+A_3(1-z_1-z_2)]^3}_{D}}\;, \qquad (11.79)
$$

where

$$A_1 = m^2 - (p' - k)^2 - i\epsilon = 2p' \cdot k - k^2 - i\epsilon$$
$$A_2 = m^2 - (p - k)^2 - i\epsilon = 2p \cdot k - k^2 - i\epsilon$$
$$A_3 = \lambda^2 - k^2 - i\epsilon \ .$$

The combined denominator becomes

$$D = 2\,(z_1 p' + z_2 p) \cdot k + \lambda^2 (1 - z_1 - z_2) - k^2 - i\epsilon \ .$$

Next, shift k so as to complete the square of the denominator:

$$k = k' + z_1 p' + z_2 p \ . \tag{11.117}$$

The shifted denominator reduces to

$$\begin{aligned} D &= (z_1 p' + z_2 p)^2 + \lambda^2 (1 - z_1 - z_2) - k'^2 - i\epsilon \\ &= \left(z_1^2 + z_2^2\right) m^2 + 2 z_1 z_2\, p' \cdot p + \lambda^2 (1 - z_1 - z_2) - k'^2 - i\epsilon \\ &= (z_1 + z_2)^2\, m^2 - z_1 z_2\, q^2 + \lambda^2\,(1 - z_1 - z_2) - k'^2 - i\epsilon \,, \end{aligned}$$

which shows that D is also *symmetric* in z_1 and z_2.

This shift in $k \to k'$ must also be carried out in the numerator, where it gives

$$\begin{aligned} N^\mu =& \gamma^\nu \left[m + \not{p}' \,(1 - z_1) - z_2 \not{p} - \not{k}'\right] \gamma^\mu \left[m + \not{p}\,(1 - z_2) - \not{p}'\, z_1 - \not{k}'\right] \gamma_\nu \\ =& \gamma^\nu \left[m + \not{p}' \,(1 - z_1) - z_2 \not{p}\right] \gamma^\mu \left[m + \not{p}\,(1 - z_2) - z_1 \not{p}'\right] \gamma_\nu \\ &+ \gamma^\nu\, \not{k}'\, \gamma^\mu\, \not{k}'\, \gamma_\nu \end{aligned}$$

where all terms linear in k' have been dropped (because they will integrate to zero). Using the identities

$$\begin{aligned} \gamma^\mu \not{a}\, \gamma_\mu &= -2\not{a} \\ \gamma^\mu \not{a}\not{b}\, \gamma_\mu &= 4\, a \cdot b \\ \gamma^\mu \not{a}\not{b}\not{c}\, \gamma_\mu &= -2\not{c}\not{b}\not{a} \end{aligned} \tag{11.118}$$

enables us to further reduce the numerator:

$$\begin{aligned} N^\mu =& -2m^2\gamma^\mu + 4m\{p'^\mu(1 - z_1) - z_2 p^\mu + p^\mu(1 - z_2) - z_1 p'^\mu\} \\ & -2(\underbrace{\not{p}}_{\not{p}' - \not{q}}(1 - z_2) - z_1 \not{p}')\gamma^\mu(\underbrace{\not{p}'}_{\not{p} + \not{q}}(1 - z_1) - z_2 \not{p}) - 2\not{k}'\,\gamma^\mu\,\not{k}' \\ =& -2m^2\gamma^\mu + 4m\left\{(p' + p)^\mu (1 - z_1 - z_2) + (p' - p)^\mu (z_2 - z_1)\right\} \\ & -2m(1 - z_1 - z_2)\gamma^\mu m(1 - z_1 - z_2) \\ & -2m(1 - z_1 - z_2)[\gamma^\mu \not{q}\,(1 - z_1) - \not{q}\,\gamma^\mu (1 - z_2)] \\ & +2(1 - z_2)(1 - z_1)\underbrace{\not{q}\,\gamma^\mu\,\not{q}}_{2q^\mu\not{q} - \gamma^\mu q^2} - 4k'^\mu\,\not{k}' + 2\gamma^\mu k'^2 \\ =& \gamma^\mu\Big(-2m^2 - 2m^2(1 - z_1 - z_2)^2 - 2(1 - z_2)(1 - z_1)q^2 + 2k'^2\Big) - 4k'^\mu\,\not{k}' \\ & + 4m(1 - z_1 - z_2)(p' + p)^\mu - 2m(1 - z_1 - z_2)\left(1 - \tfrac{1}{2}(z_1 + z_2)\right)[\gamma^\mu, \not{q}], \end{aligned}$$

where Eq. (11.115) was used in the second step, $z_2 - z_1$ terms have been dropped because they integrate to zero, and we used the Dirac equation to reduce $\not{q} \rightarrow 0$. Next, use the identity (11.93) to replace $k^\mu k^\nu$ by $g^{\mu\nu}k^2/4$ in the numerator:

$$\begin{aligned} N^\mu =& \gamma^\mu \left(-2m^2\left[1+(1-z_1-z_2)^2\right] - 2q^2(1-z_2)(1-z_1) + k'^2\right) \\ &+ 4m(1-z_1-z_2)(p'+p)^\mu + 4m(1-z_1-z_2)\left(1-\tfrac{1}{2}(z_1+z_2)\right) i\sigma^{\mu\nu}q_\nu , \end{aligned}$$

where $\frac{i}{2}[\gamma^\mu, \not{q}] = \sigma^{\mu\nu}q_\nu$. Finally, use the Gordon decomposition (see Prob. 11.1)

$$(p'+p)^\mu = -i\sigma^{\mu\nu}q_\nu + 2m\gamma^\mu \tag{11.119}$$

to get

$$\begin{aligned} N^\mu =& \gamma^\mu \Big\{-2m^2\left[1-4(1-z_1-z_2)+(1-z_1-z_2)^2\right] \\ &\qquad - 2q^2(1-z_1)(1-z_2) + k'^2\Big\} \\ &- i2m\,\sigma^{\mu\nu}q_\nu\,(1-z_1-z_2)(z_1+z_2) \ . \end{aligned}$$

Our calculation has shown that the first correction to the "bare" electromagnetic coupling generates a correction to the γ^μ term and a new term of the form $i\sigma^{\mu\nu}q_\nu$. Specifically, the form of the electromagnetic vertex function is

$$\Lambda^\mu(P,q) = F_1(q^2)\gamma^\mu + F_2(q^2)\frac{i\sigma^{\mu\nu}q_\nu}{2m} , \tag{11.120}$$

where the functions F_1 and F_2 are scalar functions of q^2, the γ^μ term is the familiar Dirac current, and the $\sigma^{\mu\nu}q^\nu$ term is the induced anomalous (or Pauli) current discussed in Prob. 5.6 and Sec. 10.2. While we have not shown it, (11.120) is the most general form which Λ^μ can take. Higher order corrections will not contribute any new operators; they will only add to the scalar functions F_1 and F_2. To second order, these functions are

$$\begin{aligned} F_1(q^2) =& \ 2ie^2 \int \frac{d^4k'}{(2\pi)^4}\int_0^1 dz_1 \int_0^{1-z_1} dz_2 \,\frac{N_1}{[D(k'^2)]^3} \\ F_2(q^2) =& -2ie^2 \int \frac{d^4k'}{(2\pi)^4}\int_0^1 dz_1 \int_0^{1-z_1} dz_2 \,\frac{4m^2(z_1+z_2)[1-z_1-z_2]}{[D(k'^2)]^3} , \end{aligned} \tag{11.121}$$

where the numerator N_1 is

$$N_1 = k'^2 - 2q^2(1-z_1)(1-z_2) - 2m^2\left[1-4(1-z_1-z_2)+(1-z_1-z_2)^2\right] .$$

Note that F_1 diverges (because of the k'^2 term in the numerator), but F_2 is finite. We will return to a discussion of F_1 later. For now, we evaluate F_2 at $q^2 = 0$. The value of F_2 at this point is the *anomalous magnetic moment of the electron*, which we denote by κ.

At $q^2 = 0$, the integral for $F_2(0) = \kappa$ becomes

$$\begin{aligned}\kappa = &- 2ie^2 \int_0^1 dz_1 \int_0^{1-z_1} dz_2 \\ &\times \int \frac{d^4k'}{(2\pi)^4} \frac{4m^2(z_1+z_2)(1-z_1-z_2)}{[m^2(z_1+z_2)^2 + \lambda^2(1-z_1-x_2) - k'^2 - i\epsilon]^3} \, . \end{aligned} \tag{11.122}$$

We first do the d^4k' integration using the identity (11.84) with $d = 4$ and $n = 3$,

$$i \int \frac{d^4k}{(2\pi)^4} \frac{1}{(B^2 - k^2 - i\epsilon)^3} = -\frac{1}{32\pi^2 B^2} \, .$$

This reduces the integral to

$$\kappa = \frac{2e^2}{32\pi^2} 4m^2 \int_0^1 dz_1 \int_0^{1-z_1} dz_2 \frac{(z_1+z_2)(1-z_1-z_2)}{[m^2(z_1+z_2)^2 + \lambda^2(1-z_1-z_2)]} \, . \tag{11.123}$$

The singularity at $z_1 + z_2 = 0$ is only a point in a two-dimensional space and hence is integrable. Let $\lambda^2 \to 0$ and change variables from z_1 and z_2 to ξ and η, where

$$\begin{aligned} \xi &= z_1 + z_2 \\ \eta &= \frac{1}{2\xi}(z_1 - z_2) \, . \end{aligned} \tag{11.124}$$

Then the volume element transforms to

$$\int_0^1 dz_1 \int_0^{1-z_1} dz_2 = \int_0^1 \xi \, d\xi \int_{-\frac{1}{2}}^{\frac{1}{2}} d\eta \tag{11.125}$$

and the anomalous moment is quickly calculated, giving the famous result

$$\kappa = \frac{e^2}{4\pi^2} \int_0^1 d\xi \int_{-\frac{1}{2}}^{\frac{1}{2}} d\eta \, (1-\xi) = \frac{\alpha}{\pi} \int_0^1 d\xi(1-\xi) = \frac{\alpha}{2\pi} \, . \tag{11.126}$$

This value was first calculated by Schwinger in 1948 [Sc 48].

The current agreement between theory and experiment represents an impressive confirmation of the correctness of QED.* The magnetic moment is often expressed in terms of the *gyromagnetic ratio* g related to the magnetic moment by

$$\mu = g\frac{e}{2m} s = g\frac{e}{2m}\frac{\sigma}{2} \, ,$$

where, as we saw in Chapter 5, the value predicted by the Dirac equation is $g = 2$. It turns out that the departure from this value, usually expressed in terms of the

*For a recent account, see [KL 90].

anomalous moment $\kappa = g/2 - 1$, can be measured directly and has been measured recently with very high accuracy [VS 87]:

$$\kappa_{\text{expt}} = .001\,159\,652\,1884(43) \ ,$$

where the numbers in parentheses are an estimate of the error (in the last digits given). The theoretical value has been calculated to eighth order [KL 90]:

$$\begin{aligned}\kappa_{\text{th}} &= \frac{\alpha}{2\pi} - 0.328\,478\,966 \left(\frac{\alpha}{\pi}\right)^2 + 1.17611(42) \left(\frac{\alpha}{\pi}\right)^3 - 1.434(138) \left(\frac{\alpha}{\pi}\right)^4 \\ &= .001\,159\,652\,140(28) \ ,\end{aligned}$$

where the theoretical error is mainly do to the uncertainty in α (which is currently determined from the quantized Hall effect) but also includes the uncertainty in the numerical evaluation of the sixth and eighth order integrals. Hence the difference (experiment − theory) is 0.000 000 000 048(28), giving agreement (within 1.7 standard deviations) for the value of g to a part in 10^{12}!

11.9 CHARGE RENORMALIZATION

As we saw, the vertex correction to the electron current diverges as $k \to \infty$. However, the divergence is localized entirely in the F_1 term which multiplies γ^μ, and hence only affects the charge. We can renormalize it by subtracting the value of the vertex at $q^2 = 0$, which guarantees that the remainder term is zero at $q^2 = 0$, and hence does not affect the charge. We write

$$\Lambda^\mu(P, q) = \left[F_1(q^2) - F_1(0)\right] \gamma^\mu + F_2(q^2) \frac{i\sigma^{\mu\nu} q_\nu}{2m} + F_1(0)\gamma^\mu \ . \tag{11.127}$$

The infinite constant $F_1(0)$ will be a new renormalization constant which we will define to be

$$F_1(0) = \frac{1}{Z_1} - 1 \ . \tag{11.128}$$

We can now fully discuss the renormalization of the charge.

First, note that the sum of *all* the Feynman diagrams which describe interactions "near" a single charge can be organized into four classes as shown in Fig. 11.12. The central circle represents "proper" vertex corrections, illustrated by the diagram (1) in the upper right corner of the figure. Proper vertex corrections are those which cannot be separated into two disconnected pieces by cutting one electron or one photon line. One says that they are *one particle irreducible.* The other three diagrams shown in the figure are one particle reducible, or "improper" contributions to the vertex function; they can all be separated into two parts by cutting a *single* line which connects them to the vertex. The vacuum polarization contribution (2) can be separated from the vertex by cutting the photon line (the dark dashed line shown in the figure is the "cut"), and both of the electron self-energy contributions, (3) and (4), can be separated by cutting an electron line, as shown.

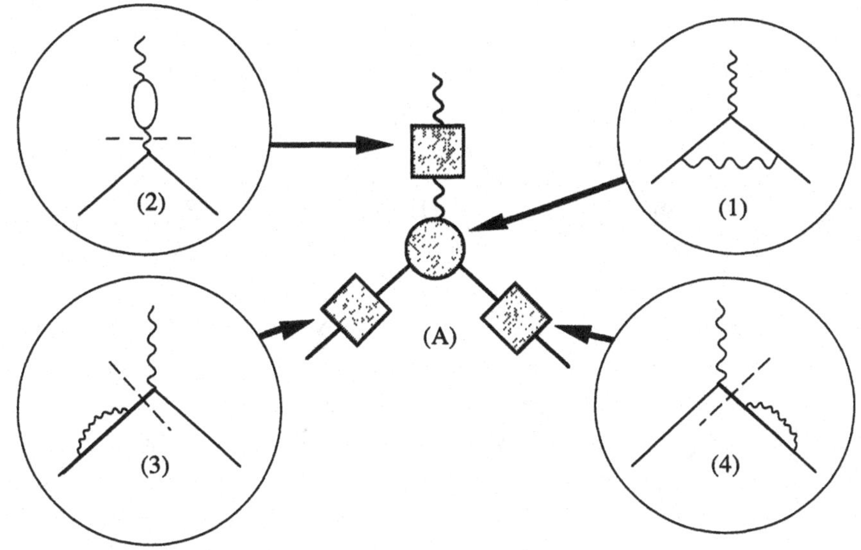

Fig. 11.12 The full, improper vertex (A) is a product of the proper vertex corrections, symbolized by figure (1), and contributions to the dressed propagators, arising in lowest order from the self-energy diagrams (2), (3), and (4).

The significance of this analysis is that *the sum of all Feynman diagrams which contribute to the vertex (both proper and improper) is the product of all of the diagrams in each of the four separate classes.* This is the justification for considering dressed propagators and (proper) vertex corrections separately. The full vertex, Γ'^{μ}, can therefore be expressed in terms of the proper vertex, $\Gamma^{\mu} = \gamma^{\mu} + \Lambda^{\mu}$, through the following relation:

$$S(\not{p}')\Gamma'^{\mu}(p',p)S(\not{p})\Delta(p'-p) = S'(\not{p}')\Gamma^{\mu}(p',p)S'(\not{p})\Delta'(p'-p) \quad , \qquad (11.129)$$

where S' and Δ' are dressed propagators and S and Δ are bare, undressed propagators.

Remembering that the renormalization of the propagator must be shared equally between the two charges at either end, we can use (11.129) to obtain the following final result for the *renormalization of the electric charge*:

$$\boxed{e_R = e_0\sqrt{Z_3}\left(\frac{Z_2}{Z_1}\right) .} \qquad (11.130)$$

If the charge is renormalized in the above fashion, the three renormalization constants Z_1, Z_2, and Z_3 will all be removed from the theory. In Chapter 16 we will discuss how it can be shown that *this procedure works to all orders.*

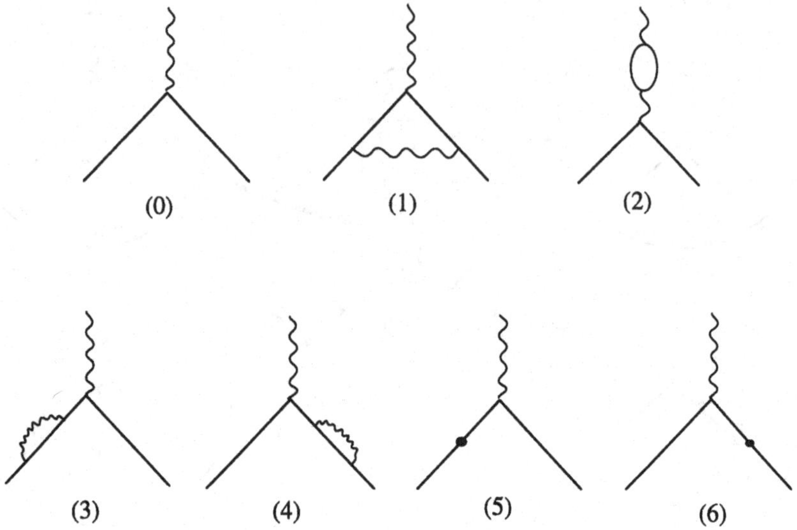

Fig. 11.13 The Feynman diagrams which contribute to the renormalization of the electric charge to order e^3.

To gain a better understanding of how this works out in perturbation theory, it is amusing to look at the diagrams which can contribute to the electric charge to order e^3. There are seven such diagrams as shown in Fig. 11.13. These contribute the following terms:

$$-e_R\gamma^\mu = -e_0\Big\{\underbrace{\gamma^\mu}_{(0)} + \underbrace{\Lambda^\mu(P,q)}_{(1)} - \tfrac{1}{2}\gamma^\mu\underbrace{\Pi(q^2)}_{(2)} - \tfrac{1}{2}\big[\underbrace{\Sigma(p')}_{(3)} - \underbrace{\delta m}_{(5)}\big]\left(\frac{1}{m - \not{p}' - i\epsilon}\right)\gamma^\mu$$
$$- \gamma^\mu\left(\frac{1}{m - \not{p} - i\epsilon}\right)\tfrac{1}{2}\big[\underbrace{\Sigma(p)}_{(4)} - \underbrace{\delta m}_{(6)}\big]\Big\}\ . \qquad (11.131)$$

In this expression e_R is the renormalized charge and e_0 the bare charge, and the following remarks apply:

- The electron self-energy, $-i\Sigma(p)$, the photon self-energy, $-q^2\Pi(q^2)$, and vertex correction, $\Lambda^\mu(P,q)$, have all been included.

- Mass counter terms ($i\delta m$), not previously discussed, have been introduced. These terms are represented by diagrams (5) and (6) and are denoted by a small black circle. They will be discussed shortly.

- The contributions from diagrams (2) – (6) are multiplied by $\frac{1}{2}$ because *only $\frac{1}{2}$ of these contributions are identified with the charge under study* (the other half go with other charges not under consideration).

These factors of $\frac{1}{2}$ are the perturbative equivalent of the square root encountered in Secs. 11.2 and 11.5 and in Eq. (11.130) above. In the discussion of self-energies presented in these sections, we summed contributions to *all orders* in perturbation theory. For the vacuum polarization discussed in Sec. 11.5, the self-energy, as $q^2 \to 0$ where the renormalization is defined, had the form $\Pi \to q^2(Z_3^{-1}-1)$, so the infinite sum of powers of the self-energy multiplied by the propagator given in Eq. (11.71) becomes

$$\sum_{n=0}^{\infty}\left(1-\frac{1}{Z_3}\right)^n = \frac{1}{1-\left(1-\frac{1}{Z_3}\right)} = Z_3 \ . \tag{11.132}$$

Since $\sqrt{Z_3}$ is associated with each charge, the relevant expansion for *each* charge is

$$\sqrt{Z_3} = \frac{1}{\sqrt{1-\left(1-\frac{1}{Z_3}\right)}} = 1+\tfrac{1}{2}\left(1-Z_3\right)+\cdots \ , \tag{11.133}$$

which explains the factor of $\frac{1}{2}$ for the second order term. Note that these arguments make use of the fact that $Z_3^{-1}-1$ is of leading order e^2 and is *considered small, even though the integral which defines it is divergent.* A similar argument holds for the electron self-energies.

However, a significant difference between the electron self-energy terms and the vacuum polarization is that the electron mass is shifted by the self-energy. The infinite electron sum analogous to (11.132), as $p^2 \to \overline{m}^2$, becomes

$$\sum_{n=0}^{\infty}\left(\frac{-\Sigma(\not p)}{m-\not p}\right)^n \to Z_2\,\frac{m-\not p}{\overline{m}-\not p} \ . \tag{11.134}$$

Since $m \neq \overline{m}$, the pole at $p^2 = \overline{m}^2$ is not canceled as $p^2 \to \overline{m}^2$. However, this change in mass is clearly unphysical, because, *to each order in perturbation theory, the mass is to be fixed to the observed electron mass.* Thus we must add a counterterm to cancel this mass shift; this is easily done by subtracting the term $\delta m = \overline{m} - m$ from Σ. Then the sum (11.134) is changed to

$$\sum_{n=0}^{\infty}\left(\frac{-\Sigma(\not p)+\delta m}{m-\not p}\right)^n \to Z_2\,\frac{m-\not p}{m-\not p} = Z_2 \ , \tag{11.135}$$

giving only the renormalization factor Z_2. The mass counterterm therefore keeps track of the mass shift and cancels out, order-by-order, any shift which the calculation produces. This explains our last Feynman rule:

Rule 9: for each particle with a mass which could be shifted by self-interactions, a mass counterterm $i\delta m$ is added to remove the mass shift.

Making these substitutions, recalling the definition of Z_1, Eq. (11.128), and canceling the mass shifts as above give the following reduction of Eq. (11.131)

$$\begin{aligned}
-e_R\gamma^\mu =& -e_0\Big\{\gamma^\mu + \tfrac{1}{2}\left(1 - Z_2^{-1}\right)\gamma^\mu + \gamma^\mu\tfrac{1}{2}\left(1 - Z_2^{-1}\right) \\
& - \tfrac{1}{2}\gamma^\mu\left(Z_3^{-1} - 1\right) + \gamma^\mu\left(Z_1^{-1} - 1\right)\Big\} \\
=& -e_0\gamma^\mu\left\{1 + \left(1 - Z_2^{-1}\right) + \tfrac{1}{2}\left(1 - Z_3^{-1}\right) + \left(Z_1^{-1} - 1\right)\right\} \\
\Rightarrow& -e_0\gamma^\mu\left\{\frac{1 + \left(Z_1^{-1} - 1\right)}{\left[1 - \left(1 - Z_2^{-1}\right)\right]\left[1 - \left(1 - Z_3^{-1}\right)\right]^{1/2}}\right\} \\
=& -\left(e_0\sqrt{Z_3}\frac{Z_2}{Z_1}\right)\gamma^\mu ,
\end{aligned} \tag{11.136}$$

where the next to the last step is suggested by Eq. (11.130), which indicates what happens if Z_2 and Z_3 terms are treated to higher order. Our argument shows the role of the mass counterterms and how the renormalization works to second order.

We now ask a crucial question: if the bare charge is universal, does it remain so after renormalization? Specifically, *does the renormalized charge depend on the fermion mass*? It *could* depend on it through the factors Z_2 and Z_1, which *do* depend on the fermion mass. We shall now show that gauge invariance insures that $Z_1 = Z_2$, so that there can be *no* mass dependence.

The Ward–Takahashi Identity

We will first prove that the vertex correction, Λ, and the electron self-energy, Σ, satisfy the following relationship:

$$\boxed{\Lambda^\mu(p, 0) = -\frac{\partial}{\partial p_\mu}\Sigma(p)\ .} \tag{11.137}$$

This is the infinitesimal form of the *Ward–Takahashi identity* [Wa 50, Ta 57]. Recall

$$\Lambda^\mu(P, q) = ie^2\int\frac{d^4k}{(2\pi)^4}\,\gamma^\nu\,\frac{1}{m - \not{p}' + \not{k} - i\epsilon}\,\gamma^\mu\,\frac{1}{m - \not{p} + \not{k} - i\epsilon}\,\gamma_\nu\,\frac{1}{\lambda^2 - k^2 - i\epsilon}$$

and Eq. (11.36),

$$\Sigma(p) = -ie^2\int\frac{d^4k}{(2\pi)^4}\gamma^\nu\,\frac{1}{m - \not{p} + \not{k} - i\epsilon}\,\gamma_\nu\,\frac{1}{\lambda^2 - k^2 - i\epsilon}\ ,$$

where a photon mass term λ^2 has been added to $\Sigma(p)$. Next, note that

$$\frac{1}{m-\not{p}'+\not{k}} - \frac{1}{m-\not{p}+\not{k}} = \frac{1}{m-\not{p}'+\not{k}}\left(\not{p}'-\not{p}\right)\frac{1}{(m-\not{p}+\not{k})} .$$

Hence we see immediately that

$$\boxed{q_\mu \Lambda^\mu\left[\tfrac{1}{2}(p'+p), q\right] = \Sigma(p) - \Sigma(p') .} \tag{11.138}$$

This is the finite difference form of the Ward–Takahashi identity and it turns out that this relation between Λ^μ in Σ holds *to all orders in e*. As $p' \to p$, $q^\mu \to 0$, and expanding

$$\Sigma(p') \cong \Sigma(p) + \frac{\partial \Sigma(p)}{\partial p_\mu}(p'-p)_\mu$$

gives the relation (11.137).

Now, near $p^2 = m^2$, Λ^μ and Σ can be expressed in terms of renormalization constants,

$$\Lambda^\mu(p,0) = \left(Z_1^{-1} - 1\right)\gamma^\mu$$
$$\Sigma(p) = \delta m + (m - \not{p})\left(Z_2^{-1} - 1\right) .$$

Hence

$$-\frac{\partial \Sigma(p)}{\partial p_\mu} = \gamma^\mu\left(Z_2^{-1} - 1\right) = \Lambda^\mu(p,0) = \gamma^\mu\left(Z_1^{-1} - 1\right)$$

and therefore

$$\boxed{Z_1 = Z_2} \tag{11.139}$$

which was to be proved. We conclude that the full charge renormalization reduces to

$$\boxed{e_R = \sqrt{Z_3}\, e_0} \tag{11.140}$$

and this is the reason why only the constant Z_3 need be studied in order to draw the conclusions we did from Eq. (11.75).

11.10 BREMSSTRAHLUNG AND RADIATIVE CORRECTIONS

As a final example of the treatment of loops and renormalization, we calculate the cross section for the radiation of soft photons from external particles in any physical scattering process. For definiteness, think of soft photons radiating from the electron in electron–proton scattering. The two bremsstrahlung diagrams are shown in Fig. 11.14.

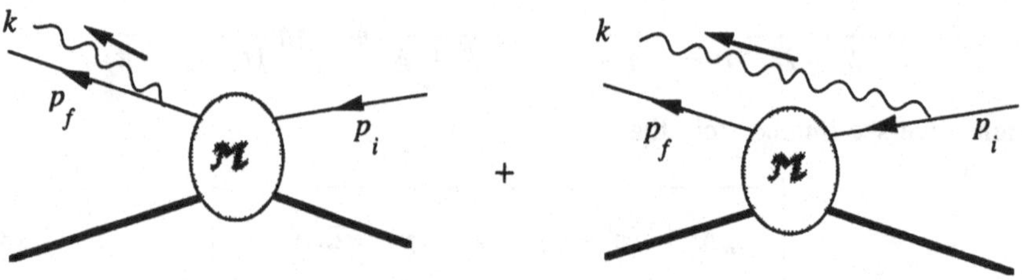

Fig. 11.14 Feynman diagrams for the bremsstrahlung process.

Denoting the on-shell *ep* scattering amplitude by $\bar{u}(\boldsymbol{p}_f)\mathcal{M}u(\boldsymbol{p}_i)$, the Feynman diagrams for these two processes are

$$\mathcal{M}_B = e\bar{u}(\boldsymbol{p}_f)\, \not{\epsilon}^*\, \frac{(m+\not{p}_f+\not{k})}{m^2-(p_f+k)^2}\mathcal{M}(p_f+k, p_i)u(\boldsymbol{p}_i)$$
$$+ e\bar{u}(\boldsymbol{p}_f)\mathcal{M}(p_f, p_i-k)\frac{(m+\not{p}_i-\not{k})}{m^2-(p_i-k)^2}\,\not{\epsilon}\, u(\boldsymbol{p}_i) \quad . \tag{11.141}$$

We are most interested in this process when $|\boldsymbol{k}| = k_0 \to 0$, the *soft photon limit.* In this limit we may ignore the $\not{k}$ in the numerator of the propagators, and using

$$\bar{u}(\boldsymbol{p}_f)\,\not{\epsilon}^*\,(m+\not{p}_f) = \bar{u}(\boldsymbol{p}_f)\left[2p_f\cdot\epsilon^* + (m-\not{p}_f)\,\not{\epsilon}^*\right]$$
$$= \bar{u}(\boldsymbol{p}_f)\, 2p_f\cdot\epsilon^* \tag{11.142}$$

obtain the result

$$\mathcal{M}_B = e\bar{u}(\boldsymbol{p}_f)\left[\mathcal{M}(p_f, p_i-k)\,\frac{p_i\cdot\epsilon^*}{p_i\cdot k} - \frac{p_f\cdot\epsilon^*}{p_f\cdot k}\mathcal{M}(p_f+k, p_i)\right]u(\boldsymbol{p}_i) \quad . \tag{11.143}$$

Note that this diverges in the limit as $k \to 0$. This poses no problems if we measure the outgoing photon, because then its energy is known and is not zero. However, the application we have in mind is *elastic scattering, where the soft photons are not observed.* In this case, in the limit as $k \to 0$, the bremsstrahlung amplitude gives a divergent contribution which multiplies the elastic amplitude:

$$\mathcal{M}_B\Big|_{k\to 0} = \underbrace{\bar{u}(\boldsymbol{p}_f)\mathcal{M}(p_f, p_i)u(\boldsymbol{p}_i)}_{\text{elastic}}\;\underbrace{e\left[\frac{p_i\cdot\epsilon^*}{p_i\cdot k} - \frac{p_f\cdot\epsilon^*}{p_f\cdot k}\right]}_{\substack{\text{multiplicative factor}\\ \text{which is singular}}}$$
$$+ \text{(small terms finite as } k\to 0) \quad . \tag{11.144}$$

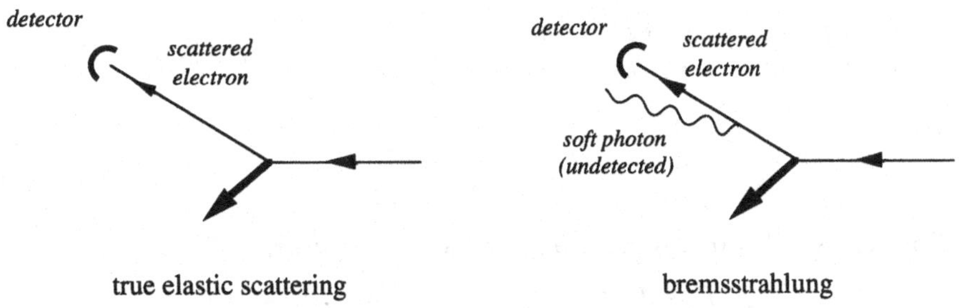

Fig. 11.15 Placement of detectors in a measurement of elastic scattering will also "measure" bremsstrahlung.

Ignoring the small terms which are finite as $k \to 0$, the bremsstrahlung amplitude appears to be infinite, and it does not seem possible to distinguish it from elastic scattering. Every time we measure elastic scattering, we also measure bremsstrahlung (see Fig. 11.15), and the latter appears to be larger. How can we measure elastic scattering?

We will now see how *the infinity arising from the bremsstrahlung amplitude is actually canceled in QED.* This cancellation leaves *finite corrections to elastic scattering which arise from soft photon processes* and which cannot be distinguished from elastic scattering. These corrections are referred to as *radiative corrections* and must be removed before true elastic data can be extracted from any experimental measurement. It is interesting and important to see how they can be calculated from QED.

Since the bremsstrahlung process is a final state distinct from elastic scattering, the cross sections add incoherently. The bremsstrahlung part is therefore just

$$d\sigma_B = \int d\sigma_{\rm elastic} \frac{d^3k}{2k} \frac{e^2}{(2\pi)^3} \left| \frac{p_i \cdot \epsilon^*}{p_i \cdot k} - \frac{p_f \cdot \epsilon}{p_f \cdot k} \right|^2 \delta^4(p_f + k + p_f - p_i - P_i) \ , \tag{11.145}$$

where the overall δ^4-function is removed from $d\sigma_{\rm elastic}$ and is replaced with the bremsstrahlung one. Again, if $k \to 0$, we can drop the k in the δ^4-function and get

$$\begin{aligned} d\sigma_B &= d\sigma_{\rm elastic} \int \frac{d^3k}{2k} \frac{e^2}{(2\pi)^3} \left| \frac{p_i \cdot \epsilon^*}{p_i \cdot k} - \frac{p_f \cdot \epsilon^*}{p_f \cdot k} \right|^2 \\ &= d\sigma_{\rm elastic} |C|^2 \ , \end{aligned} \tag{11.146}$$

where the range of the d^3k integral in the multiplicative factor $|C|^2$ is determined by the experimental conditions, to be discussed shortly.

To evaluate $|C|^2$, first sum over the polarization states of the photon. This is done by noting that

$$\sum_{\alpha=\pm 1}\left(\frac{p_i\cdot\epsilon_k^{\alpha\,*}}{p_i\cdot k}-\frac{p_f\cdot\epsilon_k^{\alpha\,*}}{p_f\cdot k}\right)\left(\frac{p_i\cdot\epsilon_k^{\alpha}}{p_i\cdot k}-\frac{p_f\cdot\epsilon_k^{\alpha}}{p_f\cdot k}\right)=J_x\,J_x+J_y\,J_y\,, \quad (11.147)$$

where

$$J^\mu=\frac{p_i^\mu}{p_i\cdot k}-\frac{p_f^\mu}{p_f\cdot k}\ . \quad (11.148)$$

However, $k_\mu J^\mu=0$, and hence $J_0=J_z$, and we may write

$$\begin{aligned} J_x\,J_x+J_y\,J_y &= J_x\,J_x+J_y\,J_y+J_z\,J_z-J_0\,J_0 \\ &= -J_\mu J^\mu\ . \end{aligned} \quad (11.149)$$

Hence

$$\begin{aligned} |C|^2 &= \int\frac{d^3k}{2k}\,\frac{e^2}{(2\pi)^3}\left[-\frac{m^2}{(p_i\cdot k)^2}-\frac{m^2}{(p_f\cdot k)^2}+\frac{2\,p_i\cdot p_f}{(p_i\cdot k)(p_f\cdot k)}\right] \\ &= \frac{2\alpha}{\pi}\chi\int_R\frac{dk}{k}\,, \end{aligned} \quad (11.150)$$

where

$$\begin{aligned} \chi=\int\frac{d\Omega_k}{4\pi}&\left[\frac{p_i\cdot p_k}{\left(E_i-\boldsymbol{p}_i\cdot\hat{k}\right)\left(E_f-\boldsymbol{p}_f\cdot\hat{k}\right)}\right. \\ &\left.-\frac{m^2}{2\left(E_i-\boldsymbol{p}_i\cdot\hat{k}\right)^2}-\frac{m^2}{2\left(E_f-\boldsymbol{p}_f\cdot\hat{k}\right)^2}\right] \end{aligned} \quad (11.151)$$

does not depend on $|\boldsymbol{k}|=k$, and the range of integration, R, is to be determined. Note that χ *depends strongly on the direction of* $\boldsymbol{k}$, particularly if the electrons are ultrarelativistic so that $p_i\sim E_i$ and the denominators are *sharply peaked in the forward direction.*

Radiative Corrections

Now we focus on the $\int_R dk/k$, which diverges both at short wavelengths (ultraviolet) and long wavelengths (infrared).

What sets limits on the value of k? The upper limit is fixed by the angular and energy resolution of the experimental equipment and the energy resolution of the incident beam, which together define the energy resolution of the detection system. If this energy resolution is ΔE, and the angular resolution is $\Delta\theta$, then the detector will count all electrons between energies $E+\frac{1}{2}\Delta E$ and $E-\frac{1}{2}\Delta E$ and with scattering angles between $\theta+\frac{1}{2}\Delta\theta$ and $\theta-\frac{1}{2}\Delta\theta$, where the kinematics of elastic scattering fixes the relationship between E and θ that is expected (see Fig. 11.16). Now, for purposes of this discussion, we may assume forward peaking,

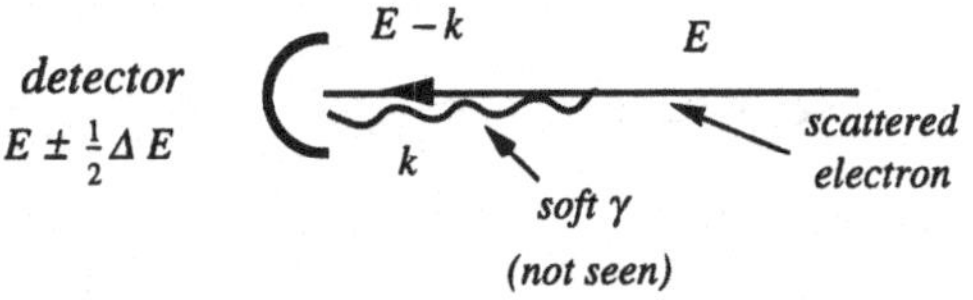

Fig. 11.16 Soft, forward going photons will always be present in the detector.

in which case $\boldsymbol{k} \parallel \boldsymbol{p}$, since this is where the cross section is largest. Hence, k *cannot be bigger than* ΔE; if it were, it would change the energy of the electron so much that it would either not be seen by the detector at all or it would be recognized as not coming from elastic scattering. (For example, if k were $2\Delta E$, and the detector were set to measure scattered electrons of energy E and angle θ, the scattered electron would have to have an energy, *before emitting the photon,* of at least $E + \frac{3}{2}\Delta E$, and such an electron would be traveling at the wrong angle to be confused with an elastically scattered electron.)

One of the central problems in the computation of radiative corrections is that there is *no lower limit on the bremsstrahlung photon energy*, k, so that the integral *diverges* at the lower limit. If we choose an arbitrary lower limit, k_{min}, the measured cross section becomes

$$\left.\frac{d\sigma}{d\Omega}\right|_{\text{measured}} = \left.\frac{d\sigma}{d\Omega}\right|_{\text{actual}} \underbrace{\left\{1 + \frac{2\alpha}{\pi}\chi(q)\ \ln\frac{\Delta E}{k_{\text{min}}}\right\}}_{\text{radiative correction factor}} . \tag{11.152}$$

As $k_{\text{min}} \to 0$, the correction factor becomes infinite. How does QED control this effect and give *finite* radiative corrections?

To get finite results, we need to treat corrections to elastic electron scattering of order e^4. It turns out that the *interference* of such corrections with the lowest order process (of order e^2) is of order e^6 (the same order as bremsstrahlung) and *give infinities which precisely cancel those which arise from bremsstrahlung.* Diagrammatically, the situation is shown in Fig. 11.17.

There are three types of terms: A (order e^2) and B (order e^4) are contributions to elastic ep scattering which can interfere and C (order e^3) are bremsstrahlung contributions which are added incoherently. Hence, the differential cross section has the form

$$\begin{aligned}\frac{d\sigma}{d\Omega} &\cong e^4|A|^2 + 2e^6\, Re\,(AB^*) + e^6|C|^2 \\ &= \underbrace{e^4|A|^2}_{\substack{\text{lowest order}\\ \text{elastic}}} + e^6\left\{\underbrace{2\, Re\,(AB^*)}_{\substack{\text{interference}\\ \text{elastic}}} + \underbrace{|C|^2}_{\text{bremsstrahlung}}\right\} .\end{aligned} \tag{11.153}$$

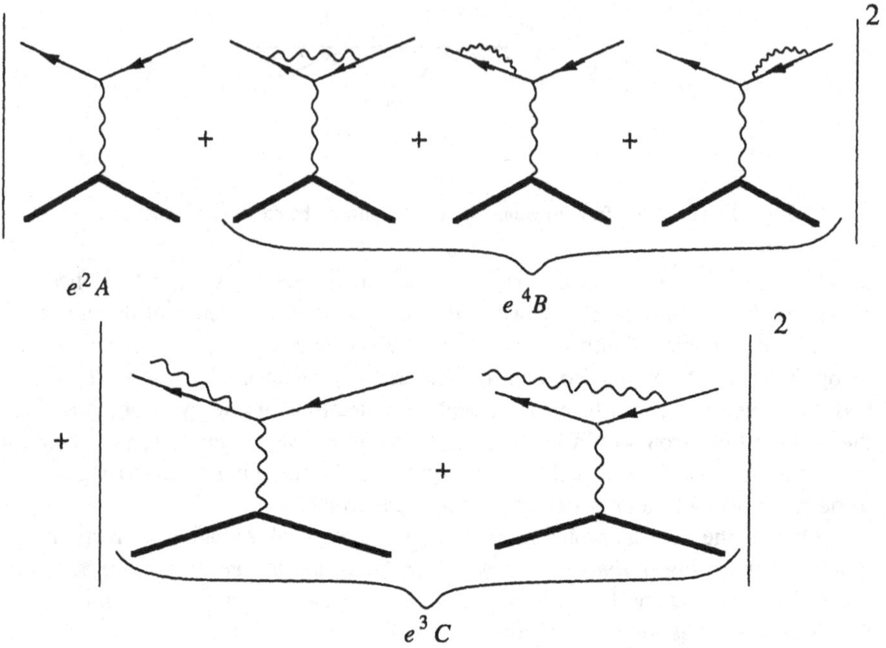

Fig. 11.17 Radiative corrections arise both from internal self-energy corrections and from external, bremsstrahlung processes.

We will show that the infinities in the $Re\,(AB^*)$ and $|C|^2$ terms cancel.

Since large k^2 contributions are finite, we will track only those terms which diverge as $k \to 0$. Since we integrate over k in the bremsstrahlung contributions, it does not matter that we also integrate over k in the vertex and self-energy corrections.

The infinite terms at small k arise from the photon pole. Recall Eq. (11.116) for the vertex correction, $\Lambda^\mu[\frac{1}{2}(p_f + p_i), q]$,

$$\Lambda^\mu\left[\tfrac{1}{2}(p_f + p_i), q\right] = ie^2 \int \frac{d^4k}{(2\pi)^4}\left(\frac{1}{D}\right)\gamma^\nu\left(m + \not{p}_f - \not{k}\right)\gamma^\mu\left(m + \not{p}_i - \not{k}\right)\gamma_\nu\,,$$

where, with $\lambda^2 = 0$, the demoninator D can be written in a factored form which displays its dependence on the virtual photon energy, k_0,

$$D = \underbrace{(k - k_0 - i\epsilon)}_{(1)}\underbrace{(k + k_0 - i\epsilon)}_{(2)}\underbrace{(E_{p_f-k} - E_f + k_0 - i\epsilon)}_{(3)}$$
$$\times \underbrace{(E_{p_f-k} + E_f - k_0 - i\epsilon)}_{(4)}\underbrace{(E_{p_i-k} - E_i + k_0 - i\epsilon)}_{(5)}\underbrace{(E_{p_i-k} + E_i - k_0 - i\epsilon)}_{(6)}. \tag{11.154}$$

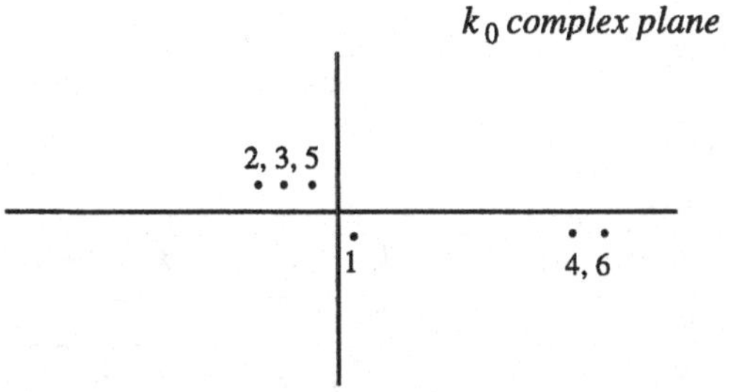

Fig. 11.18 Location of the singularities arising from the zeros of the denominator (11.154).

When we do the k_0 integration in the complex k_0 plane, the demoninator has six poles, located as shown schematically in Fig. 11.18.

If we close the contour in the lower half plane, only pole (1) will be singular as $k \to 0$. [Poles (4) and (6) always remain at least a distance $2m$ from the singularities in the upper half plane, and hence are finite.] This term gives

$$\Lambda^\mu\left[\tfrac{1}{2}(p_f+p_i),q\right] \underset{\substack{\text{photon}\\\text{pole}}}{\longrightarrow} -e^2\int\frac{d^3k}{(2\pi)^3}\,\frac{1}{2k}\,\frac{\gamma^\nu(m+\not{p}_f-\not{k})\gamma^\mu(m+\not{p}_i-\not{k})\gamma_\nu}{(2p_f\cdot k)(2p_i\cdot k)}$$

$$\underset{\substack{\text{divergent}\\\text{part}}}{\longrightarrow} -\frac{e^2}{4\pi^2}\int_{k_{\rm min}}^{k_c}\frac{dk}{k}\int\frac{d\Omega}{4\pi}\,\frac{\gamma^\mu(m+\not{p}_f)\gamma^\mu(m+\not{p}_i)\gamma_\nu}{4(E_f-\mathbf{p}_f\cdot\hat{\mathbf{k}})(E_i-\mathbf{p}_i\cdot\hat{\mathbf{k}})}\,, \tag{11.155}$$

where we have used the fact that the square of the photon four-momentum is zero under the integral (because we are at the photon pole), and in the second step we introduced a cutoff k_c with a value $k_{\rm min} << k_c \ll p_f$ or p_i, so that we can regard k as small everywhere under the integral. Thus, only the terms which are singular as $k \to 0$ need be retained in the second step. The cutoff k_c can eventually be eliminated by calculating the large k terms correctly (including renormalization). Next, exploiting the fact that Λ will eventually be sandwiched between mass shell electron spinors permits us to simplify the numerator,

$$\gamma^\nu(m+\not{p}_f)\gamma^\mu(m+\not{p}_i)\gamma_\nu = 4\,p_f\cdot p_i\,\gamma^\mu\quad . \tag{11.156}$$

The interference of this term with the leading term therefore contributes a correction factor of

$$\begin{pmatrix}\text{Vertex interference}\\\text{term}\end{pmatrix} = -2\frac{\alpha}{\pi}\int_{k_{\rm min}}^{k_C}\frac{dk}{k}\int\frac{d\Omega}{4\pi}\,\frac{p_i\cdot p_f}{(E_f-\mathbf{p}_f\cdot\hat{\mathbf{k}})(E_i-\mathbf{p}_i\cdot\hat{\mathbf{k}})}\quad . \tag{11.157}$$

Note that this has the same structure as part of the bremsstrahlung result.

Next, turn to the contributions from the electron self-energy terms. Recall that the self-energy is

$$\Sigma(\not{p}) = -ie^2 \int \frac{d^4k}{(2\pi)^4}\, \gamma^\mu \,\frac{1}{m-\not{p}+\not{k}-i\epsilon}\, \gamma_\mu \,\frac{1}{-k^2-i\epsilon} \;.$$

Hence

$$\begin{aligned}\frac{\partial \Sigma}{\partial \not{p}} &= -ie^2 \int \frac{d^4k}{(2\pi)^4}\, \gamma^\mu \,\frac{1}{(m-\not{p}+\not{k}-i\epsilon)^2}\, \gamma^\nu \,\frac{1}{-k^2-i\epsilon} \\ &= -ie^2 \int \frac{d^4k}{(2\pi)^4}\, \gamma^\mu \,\frac{(m+\not{p}-\not{k})^2}{\left[m^2-(p-k)^2-i\epsilon\right]^2}\, \gamma_\mu \,\frac{1}{-k^2-i\epsilon} \;. \end{aligned} \tag{11.158}$$

From the discussion of Eq. (11.131), we recall that the contribution of the self-energy of *each* electron is only $\frac{1}{2}$ of the full result, and hence

$$\begin{aligned} e^4B &= \frac{1}{2}\left[\Sigma(\not{p}_f)-\delta m\right]\frac{1}{m-\not{p}_f-i\epsilon}\mathcal{M} - \frac{1}{2}\mathcal{M}\frac{1}{(m-\not{p}_i-i\epsilon)}\left[\Sigma(\not{p}_i)-\delta m\right] \\ &= \frac{1}{2}\left\{\left.\frac{\partial\Sigma(\not{p}_f)}{\partial \not{p}_f}\right|_{\not{p}_f=m} + \mathcal{O}(m-\not{p}_f)\right\}\mathcal{M} + \frac{1}{2}\mathcal{M}\left\{\frac{\partial\Sigma(\not{p}_i)}{\partial\not{p}_i} + \mathcal{O}(m-\not{p}_i)\right\} \\ &\underset{\not{p}_i,\not{p}_f\to m}{\longrightarrow} \left.\frac{1}{2}\frac{\partial\Sigma(\not{p}_f)}{\partial\not{p}_f}\right|_{\not{p}_f=m}\mathcal{M} + \frac{1}{2}\mathcal{M}\left.\frac{\partial\Sigma(\not{p}_i)}{\partial\not{p}_i}\right|_{\not{p}_i=m} \;. \end{aligned} \tag{11.159}$$

But near $\not{p}_f = m$,

$$\left.\frac{\partial\Sigma(\not{p})}{\partial\not{p}}\right|_{\not{p}=m} = (1-Z_2^{-1}) + \text{infrared singularities} \;. \tag{11.160}$$

The $(1-Z_2^{-1})$ term is associated with the ultra-violet divergence of the self-energy integral and is absorbed in the charge renormalization, as we have previously discussed. The infrared divergences, which we have not yet discussed, are radiative corrections. These infrared singularities can be obtained from the photon pole contributions to $\partial\Sigma/\partial\not{p}$, just as they were for Λ^μ. Assuming k is small under the integral, as we did in Eq. (11.155) for Λ^μ, we reduce $\partial\Sigma/\partial\not{p}$ at $\not{p}= m$ as follows:

$$\begin{aligned}\left.\frac{\partial\Sigma(\not{p})}{\partial\not{p}}\right|_{\not{p}=m} &\underset{\substack{\text{photon}\\ \text{pole}}}{=} e^2\int\frac{d^3k}{(2\pi)^3 2k}\,\frac{\gamma^\mu(m+\not{p})^2\gamma_\mu}{(2p\cdot k)^2} = e^2\int\frac{d^3k}{(2\pi)^3 2k}\,\frac{4m^2}{(2p\cdot k)^2} \\ &= \frac{\alpha}{\pi}\int_{k_{\min}}^{k_c}\frac{dk}{k}\int\frac{d\Omega}{4\pi}\,\frac{m^2}{(E_p-\boldsymbol{p}\cdot\hat{k})^2} \;. \end{aligned} \tag{11.161}$$

Hence the two self-energy interference terms contribute:

$$\begin{pmatrix}\text{Self-energy}\\ \text{interference}\\ \text{terms}\end{pmatrix} = \frac{2\alpha}{\pi}\int_{k_{\min}}^{k_c}\frac{dk}{k}\int\frac{d\Omega}{4\pi}\,\frac{1}{2}\left[\frac{m^2}{(E_f-p_f\cdot\hat{k})^2}+\frac{m^2}{(E_i-p_i\cdot\hat{k})^2}\right] . \tag{11.162}$$

Combining (11.162) and (11.157), we get

$$e^4|A|^2 + 2e^6\, Re\,(AB^*) = \left.\frac{d\sigma}{d\Omega}\right|_{\substack{\text{lowest}\\ \text{order}}} \left\{1 - \frac{2\alpha}{\pi}\chi(q)\ln\left(\frac{k_c}{k_{\min}}\right)\right\}, \tag{11.163}$$

where χ was defined in Eq. (11.151).

Adding the interference corrections to the bremsstrahlung cross section gives

$$\begin{aligned} \left.\frac{d\sigma}{d\Omega}\right|_{\text{measured}} &= \left.\frac{d\sigma}{d\Omega}\right|_{\substack{\text{lowest}\\ \text{order}}} \left\{1 + \frac{2\alpha}{\pi}\chi(q)\left[\ln\frac{\Delta E}{k_{\min}} - \ln\frac{k_c}{k_{\min}}\right]\right\} \\ &= \left.\frac{d\sigma}{d\Omega}\right|_{\substack{\text{lowest}\\ \text{order}}} \Bigg\{1 + \overbrace{\frac{2\alpha}{\pi}\chi(q)\,\ln\frac{\Delta E}{k_c}}^{\text{finite correction}}\Bigg\}\,. \end{aligned} \tag{11.164}$$

We see that the unknown cutoff $k_{\min}$ is canceled (so that we may let $k_{\min} \to 0$ now), and the result of combining bremsstrahlung and radiative corrections is finite and well defined. The upper limit ΔE is experimentally determined, while the lower limit k_c is theoretically determined. The precise treatment of k_c requires calculation of the finite differences between Z_1 and Z_2 when they are first calculated with a finite photon mass and then with a zero photon mass.

PROBLEMS

11.1 Prove the Gordon decomposition, Eq. (11.119). Specifically, if $p^2 = p'^2 = m^2$, show that

$$-\bar{u}(\boldsymbol{p}')\left(\frac{i\sigma^{\mu\nu}q_\nu}{2m}\right)u(\boldsymbol{p}) = \bar{u}(\boldsymbol{p}')\left[\frac{(p'+p)^\mu}{2m} - \gamma^\mu\right]u(\boldsymbol{p})\,,$$

where $q = p' - p$.

11.2 Using the techniques developed in Sec. 11.6, explain why the fermion self-energy term A does not enter into Eq. (11.53) if the undressed fermion mass is zero. Discuss the significance of this result. Under what circumstances can an interaction produce a mass even when there would be none without the interaction (this is referred to as *spontaneous* generation of mass)? What is the correct equation in this case? If mass is spontaneously generated, is it still correct to ignore the self-energy term A?

11.3 Photon–photon scattering in QED.

(a) Write down the amplitude for the Feynman electron box diagram shown in Fig. 11.19, which contributes to photon–photon scattering. (Let p_i and ϵ_i be the momentum and polarization of photon i.)

(b) Speaking naively (i.e., just counting powers of momentum in the numerator and denominator), is the above diagram finite or infinite?

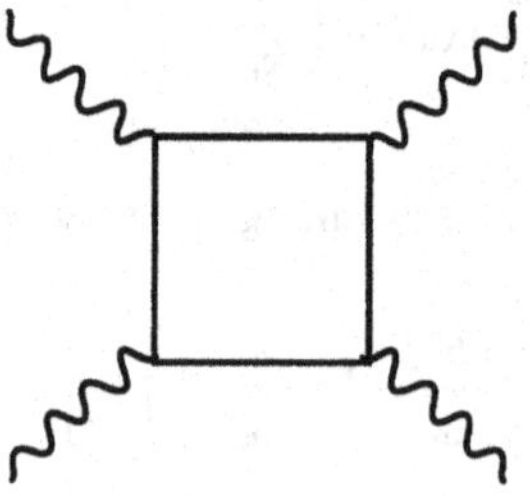

Fig. 11.19 Feynman diagram for photon–photon scattering (Prob. 11.3).

(c) Draw all the distinct diagrams that can contribute to $\gamma + \gamma \to \gamma + \gamma$ in the lowest non-vanishing order of perturbation theory.

CHAPTER 12

BOUND STATES AND UNITARITY

In the previous chapters it has been implicitly assumed that perturbation theory is adequate and that we can obtain a reasonable estimate of the scattering amplitude by calculating a few Feynman diagrams of lowest order. However, there are many problems for which the calculation of a few Feynman diagrams is inadequate. The study of *bound states* is one of these problems. A bound state produces a pole in the scattering matrix in the channel in which it appears. If the bound state is truly composite, no such pole exists in any Feynman diagram (or any finite sum); a pole can only be generated by an infinite sum. The same observations apply to the description of low energy elastic scattering; an exact treatment of *unitary* requires an infinite number of diagrams.

Ideally, we would like to sum all Feynman diagrams which describe the reaction; if we could do this, we assume we would have the correct answer. However, this is not possible in general, and we must settle for an infinite sum of a particular class of diagrams we believe to be particularly important physically. This is done by finding an integral equation, the solution of which can be interpreted as the sum of the class of diagrams under consideration. The equation used depends on the physics of the problem.

The only problems which will be discussed in this chapter are those in which long range peripheral interactions are expected to be important. We consider systems of two heavy particles interacting through the exchange of light mesons, and further assume that self-energy diagrams and vertex corrections can be ignored or treated phenomenologically. Systems which may be approximated in such a way include atomic, nuclear, and heavy quark bound states and low energy elastic scattering (below particle production thresholds).

We will first consider the *ladder* and *crossed ladder* sums of Feynman diagrams, which leads to a discussion of the Bethe–Salpeter equation [SB 51], and then to other relativistic two-body equations. All of these equations can be shown to produce bound states and to satisfy elastic unitary. We conclude this chapter with a brief discussion of the application of dispersion theory to bound states, which requires an understanding of *anomalous thresholds*.

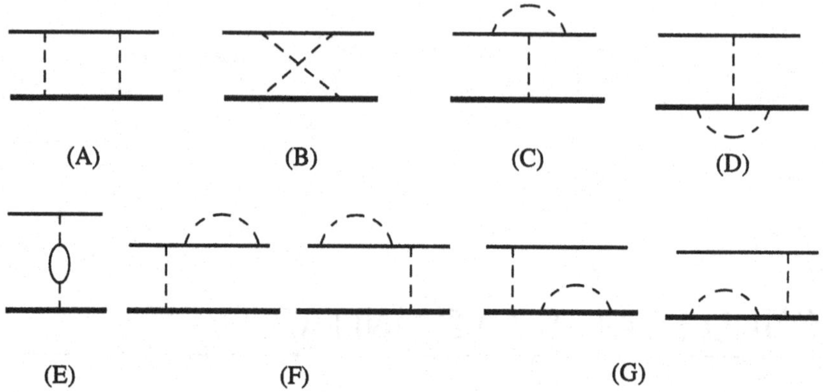

Fig. 12.1 The fourth order diagrams in ϕ^3 theory. The light solid line is the particle with mass m_1, the heavy line has mass m_2, and the dashed line is the light boson with mass μ.

12.1 THE LADDER DIAGRAMS

For definiteness, we return to the ϕ^3 theory first introduced in Chapter 9 and consider the scattering of two heavy particles (1 and 2) through the exchange of a light neutral particle. This problem was already solved to lowest order in perturbation theory in Sec. 9.3; the only Feynman diagram which contributes to the $\mathcal{M}$ matrix in lowest (second) order in the OBE (one-boson exchange) diagram shown in Fig. 9.3.

Now consider the diagrams which contribute to fourth order (there are no third order diagrams which contribute to elastic scattering; why?). Using the experience obtained from the study of loops and higher order processes in Chapter 11, we see that there will be nine diagrams, shown in Fig. 12.1. The first of these, 12.1A, is referred to as the *box* diagram, and it is the only diagram which contributes to the ladder sum. In higher order, the ladder diagrams are those which replicate the structure of the OBE and box diagram (like the rungs of a ladder) as shown in Fig. 12.2. Our task in this section is

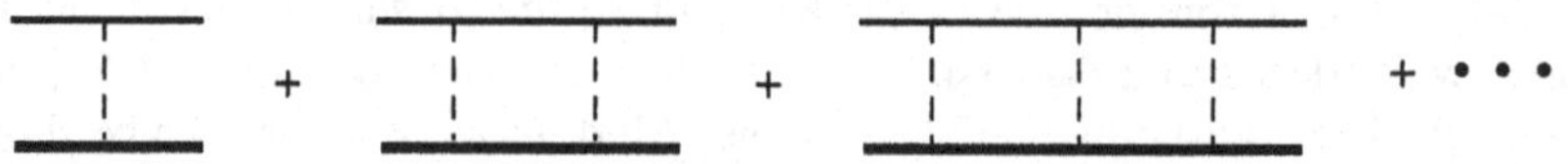

Fig. 12.2 The ladder diagrams to sixth order.

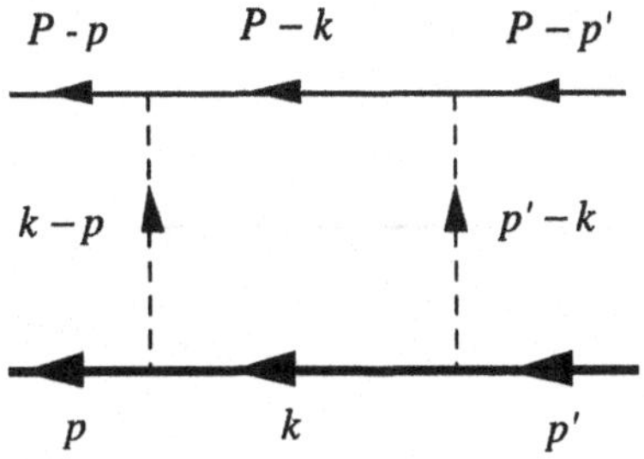

Fig. 12.3 The box diagram showing how the momenta are labeled.

to show that when the physics is controlled by *long range peripheral interactions*, these diagrams give the largest contribution in each order, and it is therefore reasonable to assume that the summation of all the diagrams in this class will give a good description of such problems.

First, assume the coupling constants (λ_1 and λ_2) are very small, so that all of the higher order diagrams can be assumed, *a priori*, to be quite small. Then if the scattering takes place very near threshold (at low energies), we will show that the ladder diagrams are exceptionally large, and hence it is justified to single them out for special consideration.

Using the Feynman rules for ϕ^3 theory, the box diagram (labeled in Fig. 12.3) is

$$\mathcal{M}_{\rm box} = i\lambda_1^2\lambda_2^2 \int \frac{d^4k}{(2\pi)^4}\,\frac{1}{D_1 D_2\, D_0 D_0'}\,, \tag{12.1}$$

where the denominators are

$$\begin{aligned}
D_1 &= m_1^2 - (P-k)^2 - i\epsilon &&= E_1^2 - (W-k_0)^2 - i\epsilon \\
D_2 &= m_2^2 - k^2 - i\epsilon &&= E_2^2 - k_0^2 - i\epsilon \\
D_0 &= \mu^2 - (k-p)^2 - i\epsilon &&= \omega^2 - \left(k_0 - E_2(p)\right)^2 - i\epsilon \\
D_0' &= \mu^2 - (k-p')^2 - i\epsilon &&= \omega'^2 - \left(k_0 - E_2(p')\right)^2 - i\epsilon\,,
\end{aligned} \tag{12.2}$$

where $m_1 < m_2$ are the masses of the two heavy particles being scattered, $\mu \ll m_1$ is the mass of the light meson being exchanged between them, and

$$E_i = \sqrt{m_i^2 + \boldsymbol{k}^2} \qquad\qquad \omega = \sqrt{\mu^2 + (\boldsymbol{k}-\boldsymbol{p})^2}$$
$$E_i(p) = \sqrt{m_i^2 + \boldsymbol{p}^2} \qquad\qquad \omega' = \sqrt{\mu^2 + (\boldsymbol{k}-\boldsymbol{p}')^2}\ .$$

Here we assume the external particles are on-shell, so that if the diagram is evaluated in the center of mass frame, $W = E_1(p) + E_2(p) = E_1(p') + E_2(p')$.

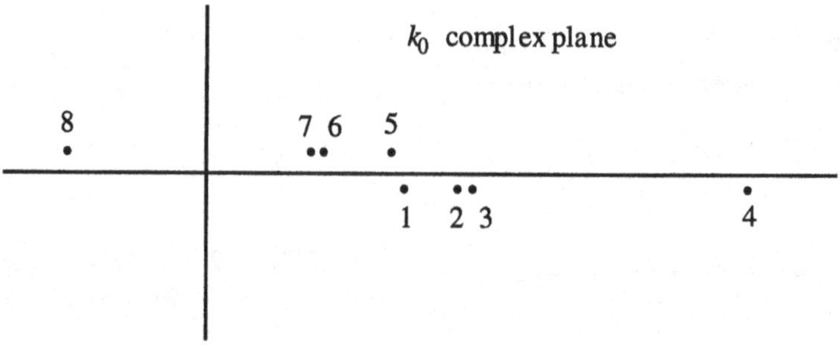

Fig. 12.4 The location of the singularities of the box diagram in the complex k_0 plane, when $|\boldsymbol{k}|$ is small. As $|\boldsymbol{k}|$ increases, the singularities in the lower half plane move to the right and those in the upper half plane move to the left.

To estimate this diagram near threshold (where $\boldsymbol{p}$ and $\boldsymbol{p}'$ are small) it is helpful to examine its singularity structure in the complex k_0 plane, as we did in Sec. 11.7 when discussing the structure of the vertex function. There are eight poles, as shown in Fig. 12.4. These are found from the zeros of the denominators in (12.2) which can be factored into eight factors,

$$\begin{aligned}
D_1 &= \underbrace{\left(E_1 - W + k_0 - i\epsilon\right)}_{5}\underbrace{\left(E_1 + W - k_0 - i\epsilon\right)}_{4} \\
D_2 &= \underbrace{\left(E_2 + k_0 - i\epsilon\right)}_{8}\underbrace{\left(E_2 - k_0 - i\epsilon\right)}_{1} \\
D_0 &= \underbrace{\left(\omega - E_2(p) + k_0 - i\epsilon\right)}_{6}\underbrace{\left(\omega + E_2(p) - k_0 - i\epsilon\right)}_{2} \\
D_0' &= \underbrace{\left(\omega' - E_2(p') + k_0 - i\epsilon\right)}_{7}\underbrace{\left(\omega' + E_2(p') - k_0 - i\epsilon\right)}_{3} ,
\end{aligned} \tag{12.3}$$

where the numbering of the factors in (12.3) corresponds to the numbering of the corresponding poles in Fig. 12.4. If we evaluate the box diagram by closing the contour in the lower half k_0 complex plane, we see that the pole at E_2 will dominate, because it is very close to the singularity at $k_0 = W - E_1$ in the upper half plane. Keeping this term only, the box diagram reduces to

$$\begin{aligned}
\mathcal{M}_{\rm box} \simeq &- \lambda_1^2\lambda_2^2 \int \frac{d^3k}{(2\pi)^3} \frac{1}{2E_2\left[E_1^2 - (W - E_2)^2 - i\epsilon\right]} \\
&\times \frac{1}{\left[\omega^2 - (E_2 - E_2(p))^2\right]\left[\omega'^2 - (E_2 - E_2(p'))^2\right]} \\
\simeq &- \frac{\lambda_1^2\lambda_2^2}{4m_1m_2} \int \frac{d^3k}{(2\pi)^3} \frac{1}{\left[E_1 + E_2 - W - i\epsilon\right]\omega^2\omega'^2} ,
\end{aligned} \tag{12.4}$$

where we have approximated $E_2 \sim m_2$ and $E_1 + W - E_2 \sim 2m_1$ in terms where the weak k-dependence is not critical. The only term where the k-dependence of E_1 and E_2 is critical is the factor $E_1 + E_2 - W$, which has a zero in the region of integration, giving the scattering amplitude (12.4) an imaginary part associated with the elastic scattering. Later we will study this singularity in much greater detail (see Sec. 12.8). For now we complete our estimate by considering the case when m_1 and m_2 are both very large and $\boldsymbol{p} = \boldsymbol{p}'$ (scattering in the forward direction). Then the integral in (12.4) is cut off by the meson energies $\omega = \omega'$, and $k = |\boldsymbol{k}| \simeq \mu$. Hence it is permissible to expand the energies in the integrand, so that $E_1 \simeq m_1 + \frac{k^2}{2m_1}$, *etc.*, and the integral may be approximated by

$$\begin{aligned}
\mathcal{M}_{\text{box}}\Big|_{\substack{\text{large} \\ m}} &\simeq -\frac{\lambda_1^2\lambda_2^2}{4m_1m_2}\int \frac{d^3k}{(2\pi)^3}\,\frac{2m}{(k^2-p^2-i\epsilon)\left(\mu^2+(\boldsymbol{k}-\boldsymbol{p})^2\right)^2} \\
&= -\frac{\lambda_1^2\lambda_2^2}{4m_1m_2}\int_0^\infty \frac{k^2dk}{2\pi^2}\,\frac{2m}{(k^2-p^2-i\epsilon)\left[(k^2+p^2+\mu^2)^2-4k^2p^2\right]} \\
&= -\frac{\lambda_1^2\lambda_2^2}{16\pi}\,\frac{1}{(m_1+m_2)\,\mu^2}\left(\frac{1}{\mu-2ip}\right), \qquad (12.5)
\end{aligned}$$

where m is the reduced mass and the integral was evaluated in the last step by extending the k integration to $-\infty$ and using the calculus of residues.

Comparing (12.5) with the OBE amplitude, also evaluated for forward scattering, shows that they are comparable when

$$\frac{\lambda_1\lambda_2}{\mu^2} \simeq \frac{\lambda_1^2\lambda_2^2}{16\pi}\,\frac{1}{(m_1+m_2)\,\mu^2}\left(\frac{1}{\mu-2ip}\right). \qquad (12.6)$$

Recalling that the *effective dimensionless* coupling strength for the ϕ^3 Yukawa interaction is [from Eq. (9.76)]

$$g_{\text{eff}}^2 = \frac{\lambda_1\lambda_2}{4m_1m_2}, \qquad (12.7)$$

the condition (12.6) becomes

$$\frac{g_{\text{eff}}^2}{4\pi}\left(\frac{m}{\mu-2ip}\right) \simeq 1\,. \qquad (12.8)$$

When this condition is satisfied, the fourth order box diagram is comparable to the second order OBE term.

The condition (12.8) may also be analytically continued below threshold by letting $p \to i\delta$, where, if the binding energy is $\epsilon_B = m_1 + m_2 - W$, then $\delta = \sqrt{2m\epsilon_B}$. This leads to the following conjecture.

Conjecture: Even if the effective ϕ^3 coupling constant $g^2_{\rm eff}$ is much less than unity, so that the use of perturbation theory would normally be justified, there may still exist a bound state. This can occur if the exchange meson mass μ and the wave number δ are small enough to guarantee that *all* of the diagrams in the ladder sum are of comparable magnitude, so that the sum of an infinite number of ladder diagrams will diverge, reflecting the appearance of a *pole* in the scattering matrix $\mathcal{M}$. A sufficient condition for this to occur is that

$$\frac{g^2_{\rm eff}}{4\pi}\,\frac{m}{\mu+2\delta} \simeq 1 \ . \tag{12.9}$$

Our argument is not sufficiently polished or complete to constitute a "proof" of the above conjecture; in particular, we have not demonstrated that (12.9) is sufficient to insure that the sixth and higher order ladder diagrams are of comparable size to the fourth order diagram we just estimated [Gr 69]. But we will see below that relativistic bound state equations have solutions when condition (12.9) is satisfied, and our main purpose here is to provide a physical understanding of why this is so.

Note that (12.9) tells us that a potential with a *finite* range ($\mu \neq 0$) will have a bound state ($\delta \geq 0$) only when

$$\frac{g^2_{\rm eff}}{4\pi}\,\frac{m}{\mu} \gtrsim 1 \ . \tag{12.10}$$

It also tells us that a potential with an *infinite* range ($\mu = 0$, as in the Coulomb potential) will *always* have a bound state. In this case, (12.9) tells us that the ground state energy, which we can estimate from $-\delta^2/2m$, will be of the order of

$$E_0 \simeq -\frac{\delta^2}{2m} \simeq -\frac{m}{8}\left(\frac{g^2_{\rm eff}}{4\pi}\right)^2 \ . \tag{12.11}$$

Recalling that the ground state of a Coulomb potential has a binding energy of $E_0 = -m\,\alpha^2/2$, we see that this is consistent with (12.11).

The condition (12.9) for a finite range potential can also be understood nonrelativistically. Consider a particle of mass m bound by a Hulthén potential (introduced in Sec. 3.5)

$$V(r) = -\frac{g^2_{\rm eff}}{4\pi}\,\mu\,\frac{e^{-\mu r}}{1-e^{-\mu r}} \ . \tag{12.12}$$

Then the (exact) solution of the S-state Schrödinger equation has the form

$$\psi(r) = N\left(1-e^{-\mu r}\right)\frac{e^{-\delta r}}{r} \ , \tag{12.13}$$

where the eigenvalue condition for the parameter δ is

$$\delta = \frac{1}{2}\left(m\frac{g_{\text{eff}}^2}{4\pi} - \mu\right) . \tag{12.14}$$

Rewriting this condition gives

$$\frac{g_{\text{eff}}^2}{4\pi} = \frac{\mu + 2\delta}{m} , \tag{12.15}$$

in precise agreement with (12.9).

Relativistic Corrections

We have been led to the conclusion that the leading contribution from the sum of ladder diagrams can produce a bound state, but so far our discussion has been limited to the nonrelativistic limit. While it is extremely important and gratifying to see how a two-body Schrödinger equation emerges from field theory, it is even more interesting to find a relativistic generalization of the two-body Schrödinger equation. To prepare the way for this, we look at the relativistic corrections to the leading terms we have just discussed.

There are relativistic corrections to the approximations we made leading up to Eq. (12.4), and beyond, and these will be discussed later. For now we focus on the contribution from the meson pole terms, corresponding to poles 2 and 3 shown in Fig. 12.4 (we focus on poles 2 and 3 instead of 6 and 7 because we have decided to close the k_0 contour in the lower half plane). These contributions will be estimated by assuming, as before, that m_1 and m_2 are much greater than μ and assuming that p^2/m_1^2 or p^2/m_2^2 are much less than unity. In this case it is convenient to rewrite the box diagram in the following form:

$$\mathcal{M}_{\text{box}} \simeq -\frac{i\lambda_1^2\lambda_2^2}{4m_1m_2}\int\frac{d^4k'}{(2\pi)^4}\,\frac{1}{D}\,\frac{1}{\left(\omega^2 - k_0'^2 - i\epsilon\right)\left(\omega'^2 - k_0'^2 - i\epsilon\right)} , \tag{12.16}$$

where k_0 of Eq. (12.3) has been replaced by $k_0' + E_2(p)$, and

$$D = \left(\frac{p^2 - k^2}{2m_2} + k_0'\right)\left(\frac{k^2 - p^2}{2m_1} + k_0'\right) . \tag{12.17}$$

The meson pole contributions come from the two poles at $k_0' = \omega - i\epsilon$ and $k_0' = \omega' - i\epsilon$. A convenient way to obtain these contributions and reduce (12.16) quickly is to use Eq. (C.2) to combine the two denominators in (12.16) as follows:

$$\begin{aligned}\mathcal{M}_{\text{box}} \simeq& \frac{-i\lambda_1^2\lambda_2^2}{4m_1m_2}\int\frac{d^4k'}{(2\pi)^4}\,\frac{1}{D}\\ &\times\int_0^1 dx\frac{1}{\left[\mu^2 + k^2 + p^2 - k_0'^2 - 2\mathbf{k}\cdot(\mathbf{p}x + \mathbf{p}'(1-x))\right]^2} .\end{aligned} \tag{12.18}$$

Anticipating the fact that the value of k_0' fixed by the double pole in (12.18) is much larger than the $(p^2 - k^2)/2m_i$ terms in (12.17), we may approximate $D \simeq k_0'^2$ and complete the square in (12.18) by shifting $\boldsymbol{k} \to \boldsymbol{k} + \boldsymbol{p}x + \boldsymbol{p}'(1-x)$, obtaining

$$\begin{aligned}
\mathcal{M}_{\text{box}}\Big|_{\substack{\text{meson}\\ \text{poles}}} &\simeq -\frac{i\lambda_1^2\lambda_2^2}{4m_1m_2}\int\frac{d^4k'}{(2\pi)^4}\,\frac{1}{k_0'^2}\int_0^1 dx\frac{1}{\left[\mu^2+k^2+q^2x(1-x)-k_0'^2\right]^2} \\
&= \frac{3}{16}\,\frac{\lambda_1^2\lambda_2^2}{m_1m_2}\int\frac{d^3k}{(2\pi)^3}\int_0^1 dx\frac{1}{\left[\mu^2+k^2+q^2x(1-x)\right]^{5/2}} \\
&= \frac{\lambda_1^2\lambda_2^2}{32\pi^2m_1m_2}\int_0^1 dx\,\frac{1}{\mu^2+q^2x(1-x)}\ ,
\end{aligned} \tag{12.19}$$

where $q^2 = (\boldsymbol{p} - \boldsymbol{p}')^2$ is the three-momentum transferred by the scattering. Do not forget that the "pole" at $k_0' = 0$ is to be ignored, so that only the double pole at

$$k_0' = \sqrt{\mu^2 + k^2 + q^2x(1-x)}$$

is evaluated in going from the first to second line of Eq. (12.19).

If $q^2 = 0$ (forward scattering) we see that the ratio of the meson pole contributions to the leading pole contribution (12.5), with $p = i\delta$, is

$$\frac{\mathcal{M}_{\text{meson}}}{\mathcal{M}_{\text{leading}}} \cong -\frac{1}{2\pi}\left(\frac{\mu+2\delta}{m}\right) . \tag{12.20}$$

If the meson pole terms are regarded as a correction to the leading contribution, and if the mass of the exchanged meson is very small (or zero), then the correction is of order

$$\frac{\delta}{m} \cong \frac{v}{c}\ , \tag{12.21}$$

where v is the typical velocity of the bound constituents. This is a significant relativistic correction, and we conclude that the *meson poles terms cannot be ignored, unless they are canceled by some other contribution.*

It turns out that the contribution from the *crossed ladder diagram*, Fig. 12.1B, is of the *same size* as the meson pole contribution (12.19), and hence *the crossed ladder diagram must be included in order to obtain an accurate relativistic description of bound states.* In fact, for the ϕ^3 example under discussion, and for a large class of other theories, the contribution from the crossed ladder diagram *cancels* the meson pole contribution. Before we show this, it is instructive to recast (12.19) in a *dispersion*, or *spectral*, form.

Since the meson pole contributions to $\mathcal{M}$ are real and local (i.e., depend on q^2 only), it is appropriate to regard them as a candidate for a new contribution to the meson exchange *potential* between particles 1 and 2. Recall that the potential in the ϕ^3 theory example we are considering is obtained from the $\mathcal{M}$-matrix by dividing by $4m_1m_2$, as in Eq. (9.72), so that

$$\tilde{V}_{2\mu}(q) = \frac{1}{4m_1m_2}\mathcal{M}_{\rm box}\Big|_{\substack{\rm meson\\ \rm poles}} = \frac{1}{2}\left(\frac{g^2_{\rm eff}}{4\pi}\right)^2 \int_0^1 \frac{dy}{1-y^2}\left(\frac{1}{\dfrac{4\mu^2}{1-y^2}+q^2}\right), \tag{12.22}$$

where the integral in (12.19) was transformed by the substitution $x = \frac{1}{2}(1+y)$. Introducing $z^2 = 4\mu^2/(1-y^2)$ gives

$$\tilde{V}_{2\mu}(q) = \frac{1}{2}\left(\frac{g^2_{\rm eff}}{4\pi}\right)^2 \int_{2\mu}^{\infty} \frac{dz}{\sqrt{z^2-4\mu^2}}\left(\frac{1}{z^2+q^2}\right) . \tag{12.23}$$

In coordinate space, this representation shows that the potential $V_{2\mu}$ is a superposition of Yukawa shapes, with masses $z \geq 2\mu$,

$$V_{2\mu}(r) = \frac{1}{8\pi}\left(\frac{g^2_{\rm eff}}{4\pi}\right)^2 \int_{2\mu}^{\infty} \frac{dz}{\sqrt{z^2-4\mu^2}}\left(\frac{e^{-zr}}{r}\right) . \tag{12.24}$$

The potential has a maximum range of $(2\mu)^{-1}$, which identifies it as a force associated with the exchange of two quanta of mass μ, and it may be referred to as a two-boson exchange (TBE) potential. The form (12.24) is sometimes called a *spectral* form because it displays the "spectrum" of mass exchanges associated with the potential.

The relatively long range of this TBE contribution is not unrelated to its importance as a relativistic correction. To see the connection, it is sufficient to consider a weakly bound state in the nonrelativistic limit. In this case, using the asymptotic bound state wave function, the expectation value of an exponential potential of range σ^{-1} is

$$< e^{-\sigma r} > \sim \frac{1}{\sigma + 2\delta} .$$

The contribution of any potential derived from fourth order graphs compared to the OBE potential is therefore

$$\frac{< V_4 >}{< V_{\rm OBE} >} \sim \left(\frac{g^2_{\rm eff}}{4\pi}\right)\left(\frac{\mu+2\delta}{\sigma+2\delta}\right) \sim \frac{(\mu+2\delta)^2}{m(\sigma+2\delta)} , \tag{12.25}$$

where the bound state condition (12.9) has been used to relate the effective coupling constant to the masses. We see that the longer the range, the more significant the contribution; Eq. (12.25) is consistent with the estimate (12.20) only because the effective range of the TBE potential is $\sim (2\mu)^{-1} \sim \mu^{-1}$.

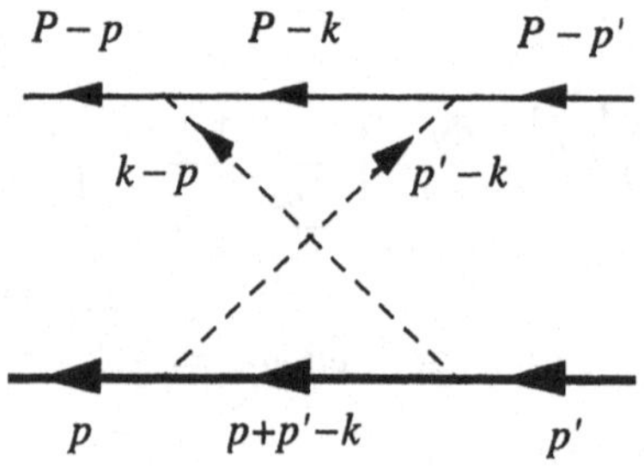

Fig. 12.5 The crossed box diagram showing how the momenta are labeled.

12.2 THE ROLE OF CROSSED LADDERS

We have just seen that the important corrections to weakly bound systems will come from the long range peripheral interactions. The only other fourth order diagram which describes such a long range (two-boson exchange) interaction is the crossed box, and it is for this reason that it is singled out for discussion. In this section we will show that the leading contribution from the crossed box is comparable to the TBE contribution which arises from the box and that, for a class of theories including the ϕ^3 theory we have been discussing, the two contributions cancel. Then we will show that these terms cancel to *all orders in perturbation theory*. This will prepare us for the subsequent discussion of relativistic two-body equations.

The crossed box diagram is labeled in Fig. 12.5 in such a way that only the internal propagator for particle 2 has a different momentum, so that

$$\mathcal{M}_{\times\text{box}} = i\lambda_1^2\lambda_2^2 \int \frac{d^4k}{(2\pi)^4} \frac{1}{D_1 D_2^\times D_0 D_0'} . \tag{12.26}$$

where D_1, D_0, and D_0' are identical to (12.2), but

$$\begin{aligned} D_2^\times &= m_2^2 - (p + p' - k)^2 - i\epsilon \\ &= \underbrace{\left(E_2^\times + 2E_2(p) - k_0 - i\epsilon\right)}_{8_\times}\underbrace{\left(E_2^\times - 2E_2(p) + k_0 - i\epsilon\right)}_{1_\times} , \end{aligned} \tag{12.27}$$

where

$$E_2^\times = \sqrt{m_2^2 + (\mathbf{p} + \mathbf{p}' - \mathbf{k})^2} \; .$$

There are still eight poles in the complex k_0 plane, but two of the poles, $1_\times$ and $8_\times$, are in different locations, as shown in Fig. 12.6. Ignoring poles 8 and $8_\times$, which

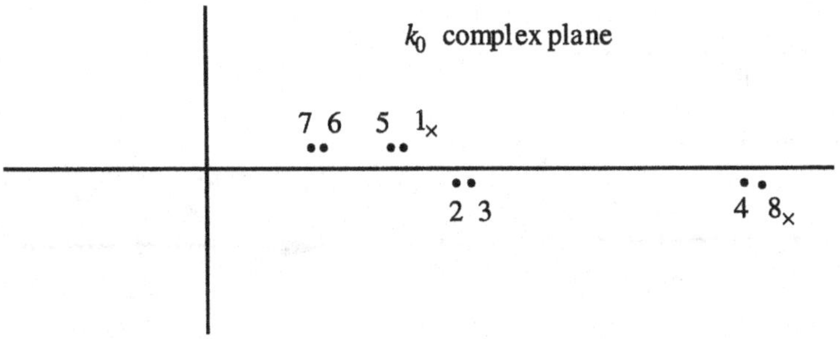

Fig. 12.6 The location of the singularities of the crossed box diagram in the complex k_0 plane. Compare with Fig. 12.4 and note that the only difference is that ploe 1 is replaced by $1_\times$ and pole 8 is replaced by $8_\times$.

are negligible, the major difference between the box and crossed box is that pole 1, *which dominated the box,* has moved from the *lower half plane to the upper half plane.* These two poles are located at:

$$\begin{aligned} &\text{pole } 1: && k_0 = E_2 - i\epsilon \cong m_2 + \frac{k^2}{2m_2} - i\epsilon \\ &\text{pole } 1_\times: && k_0 = 2E_2(p) - E_2^\times \cong m_2 + \frac{p^2}{m_2} - \frac{(p+p'-k)^2}{2m_2} + i\epsilon \ . \end{aligned} \tag{12.28}$$

Evaluating the crossed box by closing the contour in the lower half plane (as we did before) leads us to the following observations:

- The contribution which dominated the box (pole 1), is no longer present in the lower half plane, and hence this *leading contribution is missing from the crossed box.* (The curious reader may wonder how this argument would be affected if we were to close the contour in the upper half plane. In this case the two poles 5 and $1_\times$ would cancel, giving a similar result.) We conclude immediately that the *crossed box is smaller than the box* by the ratio (12.20).
- The meson poles dominate the crossed box, and the only difference between their contribution to the crossed box and the box is the denominator D_2.

Introducing $k_0 = k_0' + E_2(p)$, as we did in our discussion of the meson pole contribution to the box, these two denominators become

$$\begin{aligned} &\text{box:} && \frac{1}{D_2} \cong \frac{1}{2m_2\left(\frac{k^2-p^2}{2m_2} - k_0'\right)} \\ &\text{crossed box:} && \frac{1}{D_2^\times} \cong \frac{1}{2m_2\left(\frac{(p+p'-k)^2}{2m_2} - \frac{p^2}{2m_2} + k_0'\right)} \ . \end{aligned} \tag{12.29}$$

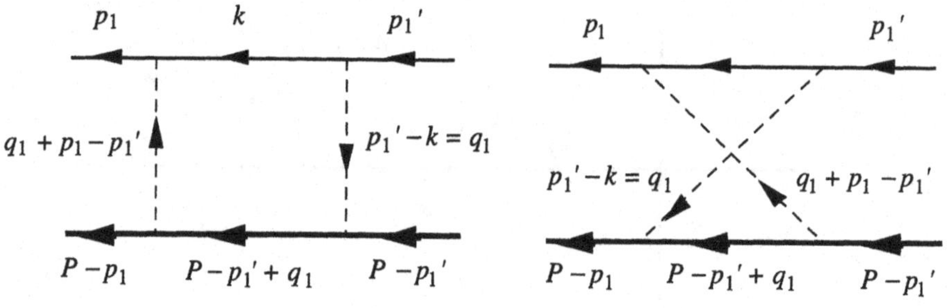

Fig. 12.7 The box and crossed box diagrams with a new labeling of momenta used to prove the cancellation theorem.

If m_2 is very large, the terms in (12.29) proportional to m_2^{-1} may be neglected compared to k_0' (which is equal to ω or ω' at the meson poles), and we see that

$$\frac{1}{D_2} \cong -\frac{1}{D_2^{\times}} \quad . \tag{12.30}$$

Hence, in this approximation the dominant contributions from the crossed box are equal to the meson pole contributions from the box but have the opposite sign, so that their sum (box plus crossed box) cancels. The *role of the crossed box is to cancel the meson pole contribution from the box.*

This cancellation is quite general, and we will now prove the following theorem:

> **Cancellation theorem:** In a theory in which a spin zero particle of mass m_1 interacts with a heavy particle of mass m_2 (which has no charge states) by exchanging a spin zero meson of mass μ, the meson pole contributions from the ladder diagram are canceled by meson pole contributions from crossed ladder diagrams, and this cancellation is *exact* in the *limit* as $m_2 \to \infty$.

We will prove the theorem by mathematical induction. First we will prove it in fourth order, and then we will show that if it is true in $(2n)$th order, it is also true in $(2n+2)$th order.

The proof in fourth order is essentially as given above, but it is instructive to repeat it using the notation we will use for the general proof. We will neglect the negative energy poles of the heavy particle (which can be proved to be small by a different argument) and assume that all integrals over internal three-momenta converge, so that the $m_2 \to \infty$ limit can be taken under the integral. We label

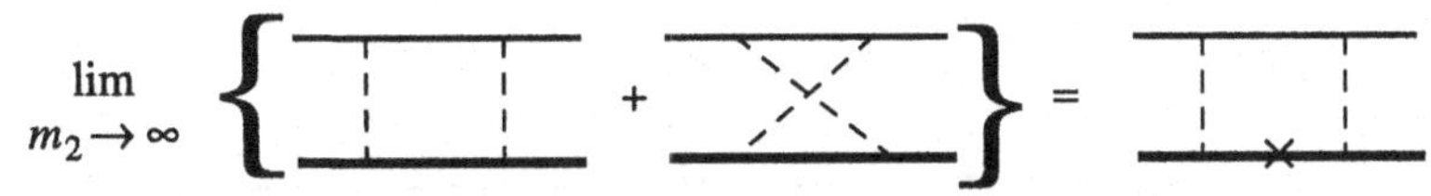

Fig. 12.8 Diagrammatic representation of the cancellation between the box and crossed box in the large m_2 limit. The cross means that the particle is on its positive energy mass shell.

the momenta for the box and crossed box as shown in Fig. 12.7. In this notation, with the above approximations, the sum of these two diagrams is

$$\mathcal{M}^{(4)} \cong -iK \int \frac{d^4q_1}{(2\pi)^4} \frac{I(q_1)}{2m_2} \left[\frac{1}{-q_{10} - i\epsilon} + \frac{1}{q_{10} - i\epsilon} \right] , \tag{12.31}$$

where $I(q_1)$ is the product of the two meson propagators and the propagator for particle 1, identical for the two diagrams, and K is a constant, also identical for the two diagrams. The approximation for the two heavy particle propagators in (12.31) is essentially the same as (12.29), with the $k^2/2m_2$ terms discarded. Using the familiar relation

$$\frac{1}{\pm q_0 - i\epsilon} = \pm \mathbb{P} \left(\frac{1}{q_0} \right) + i\,\pi\,\delta(q_0) , \tag{12.32}$$

$\mathcal{M}^{(4)}$ becomes

$$\mathcal{M}^{(4)} = K \int \frac{d^4q_1}{(2\pi)^3} I(q_1)\,\delta(q_{10}) = K \int \frac{d^3q_1}{(2\pi)^3} I(\hat{q}_1) \ . \tag{12.33}$$

Equation (12.33) tells us that the full fourth order result comes only from the pole at q_{10}, which corresponds to the contribution from the positive energy pole of particle 2. This result is represented symbolically in Fig. 12.8, where the *cross on an internal line* will mean the contribution from the positive energy pole. Operationally, one closes the contour in whichever half plane has the positive energy pole in question and evaluates the contribution from that pole using the calculus of residues, discarding all other terms [as we did in our calculation of the dominant contribution (12.4)].

Now we consider a typical $(2n)$th order diagram, shown in Fig. 12.9. To this diagram we add another exchange, fixed to particle 1, but with all possible "corrections" to particle 2, as shown in Fig. 12.10. The diagrams in 12.10 are

$$\begin{aligned} \mathcal{M}^{(2n+2)} = \int_n \Bigg\{ & \frac{1}{(q_{10} + q_{20} + \cdots + q_{n0} - i\epsilon) \ldots (q_{10} + q_{n0} - i\epsilon)\,(q_{n0} - i\epsilon)} \\ & + \frac{1}{(q_{10} + q_{20} + \ldots + q_{n0} - i\epsilon) \ldots (q_{10} + q_{n0} - i\epsilon)\,(q_{10} - i\epsilon)} \\ & + \ldots + \frac{1}{(-q_{n0} - i\epsilon) \ldots (q_{20} + q_{10} - i\epsilon)\,(q_{10} - i\epsilon)} \Bigg\} , \end{aligned} \tag{12.34}$$

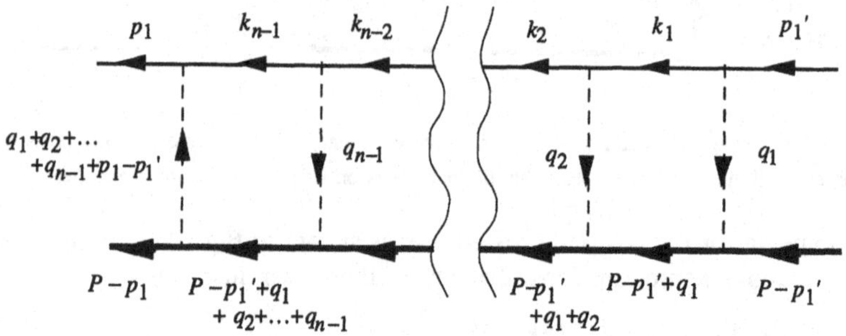

Fig. 12.9 A typical $(2n)$th order diagram with the momenta labeled.

where $\int_n$ is a shorthand for all of the n loop integrals and all of the propagators, excluding the heavy particle propagators. Now the terms in (12.34) can be added together successively. For example, the first two terms are combined as follows:

$$\frac{1}{(q_{10}+q_{n0}-i\epsilon)}\left[\frac{1}{q_{n0}-i\epsilon}+\frac{1}{q_{10}+i\epsilon}\right]=\frac{1}{(q_{n0}-i\epsilon)\,(q_{10}-i\epsilon)}\ . \tag{12.35}$$

In this way the denominators with sums containing q_{n0} are eliminated, giving finally

$$\begin{aligned}\mathcal{M}^{(2n+1)} &= \int_n \frac{1}{\left(q_{10}+q_{20}+\cdots+q_{(n-1)\,0}-i\epsilon\right)\ldots\left(q_{20}+q_{10}-i\epsilon\right)\left(q_{10}-i\epsilon\right)} \\ &\qquad\times\left[\frac{1}{q_{n0}-i\epsilon}+\frac{1}{-q_{n0}-i\epsilon}\right] \\ &= \int_n \frac{2\pi i\ \delta\left(q_{n0}\right)}{\left(q_{10}+q_{20}+\cdots+q_{(n-1)\,0}-i\epsilon\right)\ldots\left(q_{20}+q_{10}-i\epsilon\right)\left(q_{10}-i\epsilon\right)}.\end{aligned} \tag{12.36}$$

Since this argument is independent of how the original n mesons were ordered along the heavy particle line, we conclude that the addition of a new meson *in all possible ways* to any $(2n)$th order diagram is equivalent to adding a new "rung" with the heavy particle on-shell, as illustrated diagrammatically in Fig. 12.11. Hence, if the theorem were true for the $(2n)$th order diagrams, we have shown how it may be extended to $(2n+2)$th order diagrams, and the proof by mathematical induction is complete.

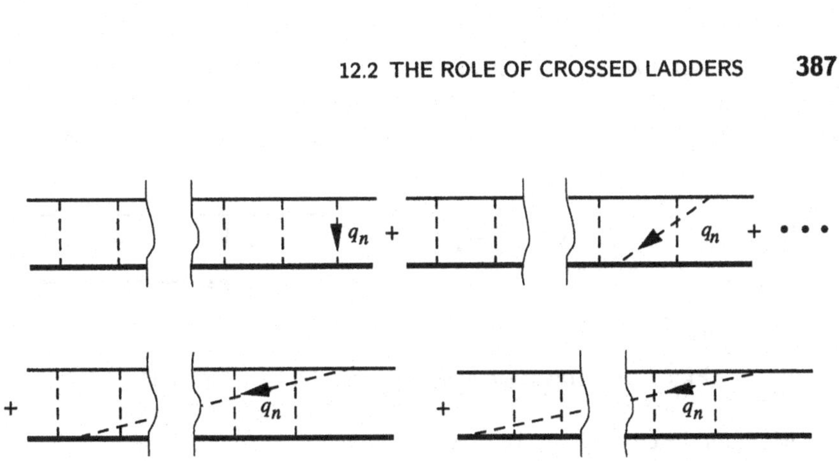

Fig. 12.10 The first and last two of the series of $n+1$ diagrams which give all the contributions to the ladders and crossed ladders which arise for the addition of another meson exchange to the particular $(2n)$th order diagram shown in Fig. 12.9.

The final result is that the sum of all ladder and crossed ladders is obtained from the single *ladder diagram with the heavy particle on-shell in each rung*:

$$\sum \begin{pmatrix} \text{all ladders} \\ \text{and crossed ladders} \end{pmatrix} = \int_n (2\pi i)^n \, \delta(q_{10})\delta(q_{20})\ldots\delta(q_{n0}) \ . \qquad (12.37)$$

This result is illustrated symbolically in Fig. 12.12.

A word of caution is in order, and the following remarks should be noted:

• The simple result (12.36) holds *only* in the limit as m_2 approaches infinity. For finite m_2, there are corrections of order m_2^{-1}.

• The result does not hold when the exchanged meson carries some "charge" (not necessarily electric) which affects the weight of the ladder and crossed ladder diagrams differently. For example, in the exchange of pions between nucleons, the coupling has an isospin factor τ_i and the factor for the box is

$$\tau_{1i}\tau_{2i}\tau_{1j}\tau_{2j} = 3 - 2\boldsymbol{\tau}_1 \cdot \boldsymbol{\tau}_2 = 9 - 4I(I+1) \qquad (12.38)$$

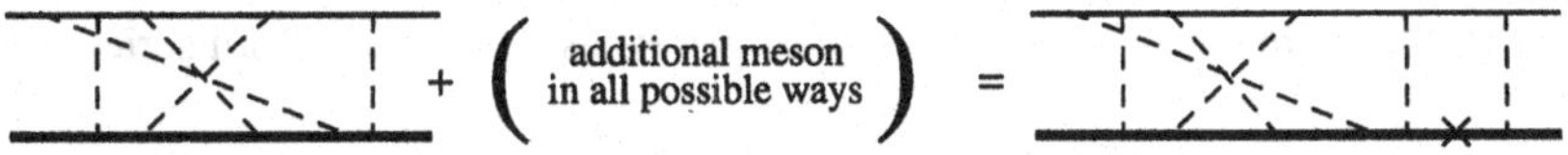

Fig. 12.11 Diagrammatic representation of the derivation of Eq. (12.36).

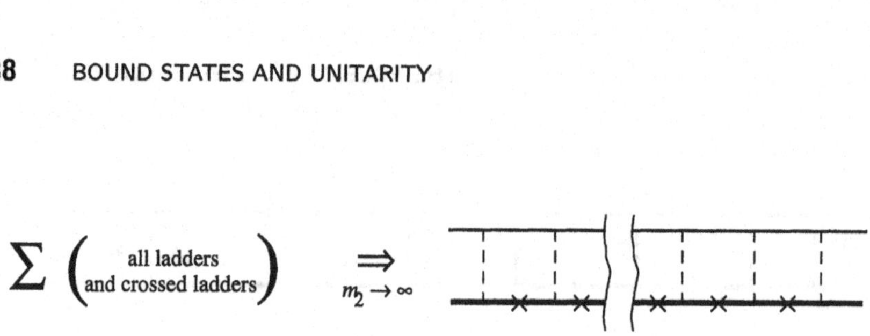

Fig. 12.12 Diagrammatic representation of the result Eq. (12.37).

while for the crossed box we have

$$\tau_{1i}\tau_{2j}\tau_{1j}\tau_{2i} = 3 + 2\boldsymbol{\tau}_1 \cdot \boldsymbol{\tau}_2 = -3 + 4I(I+1) \ . \qquad (12.39)$$

In these expressions I is the total isospin. Hence, for $I = 0$ states, the box and crossed box have additional factors of 9 and -3, while for $I = 1$ states the factors are 1 and 5.

- If the heavy particle has no charge states, the factors must be identical, regardless of the charges of the other particles. This is the case for which the theorem holds.

We are now ready to discuss relativistic two-body equations.

12.3 RELATIVISTIC TWO-BODY EQUATIONS

Using the previous discussion as motivation, we seek integral equations with a solution which can be interpreted as the sum of all ladder (and perhaps crossed ladder) diagrams. In this section we will discuss the general features of such equations and postpone the discussion of specific equations until the next section.

The equations we will study have the general form

$$\boxed{\mathcal{M}(p,p';P) = \mathcal{V}(p,p';P) + \int_k \mathcal{V}(p,k;P)\, G(k,P)\mathcal{M}(k,p';P) \ ,} \qquad (12.40)$$

where $\mathcal{M}$ is the scattering amplitude, G is the two-body propagator, $\mathcal{V}$ is the *kernel* of the equation, and $\int_k$ is the integration over the internal momenta. The specific forms of $\int_k$ and G will be given later when we discuss specific equations. Equation (12.40) is illustrated diagrammatically in Fig. 12.13.

If the kernel is "small," so that perturbation theory converges, the solution of (12.40) can be obtained by iteration. This generates a two-body Born series of the form

$$\mathcal{M} = \mathcal{V} + \int \mathcal{V}G\mathcal{V} + \int\int \mathcal{V}G\mathcal{V}G\mathcal{V} + \ldots + \left(\int \mathcal{V}G\right)^n \mathcal{V} + \ldots \ . \qquad (12.41)$$

Fig. 12.13 Diagrammatic representation of the integral equation (12.40).

This series is shown diagrammatically in Fig. 12.14. Typically, each of the terms in the series is identified with a Feynman diagram (or a part of a Feynman diagram), so that the sum (12.41) is indeed a sum of Feynman diagrams (or parts of Feynman diagrams). Note that the successive terms represent products of the kernel $\mathcal{V}$ connected by the propagator G. Any Feynman diagram which can be written in such a form, i.e.,

$$\mathcal{M}_R = \int \mathcal{M}_1 G \mathcal{M}_2 \,, \tag{12.42}$$

is said to be *reducible with respect to the propagator G*, and clearly such a diagram cannot be part of the kernel. The kernel must be built up *only from irreducible diagrams*. We will return to a discussion of how $\mathcal{V}$ is chosen later.

If we replace the integrals in (12.41) by sums over a finite set of discrete points in momentum space, so that $\mathcal{V}$ and $\mathcal{M}$ are matrices and G is a diagonal matrix, then the series (12.41) is a geometric series which can be formally summed, giving

$$\begin{aligned}\mathcal{M} &= \mathcal{V} + \mathcal{V}G\mathcal{V} + \mathcal{V}G\mathcal{V}G\mathcal{V} + \ldots + (\mathcal{V}G)^n\,\mathcal{V} + \ldots \\ &= (1-\mathcal{V}G)^{-1}\,\mathcal{V} \ .\end{aligned} \tag{12.43}$$

For cases when the Born series (12.43) does not converge, the solution of Eq. (12.40) may still exist and can be regarded as the analytic continuation of the sum (12.43) from a region where it converges to a region where it does not converge. This situation is familiar from the theory of complex functions; for example, the complex function

$$f(z) = \frac{z}{1-z} \tag{12.44}$$

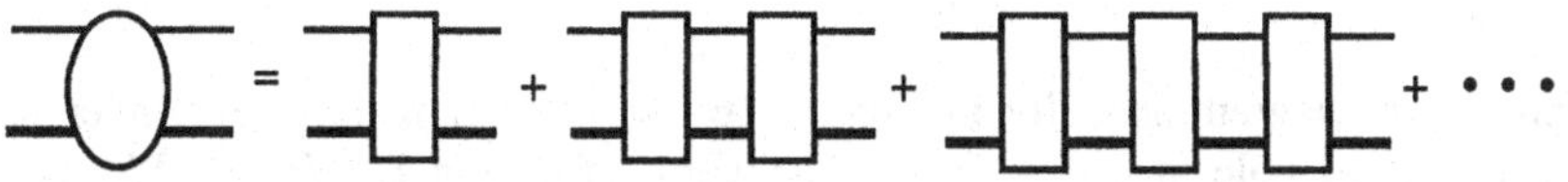

Fig. 12.14 Diagrammatic representation of the Born series generated by the integral equation (12.40).

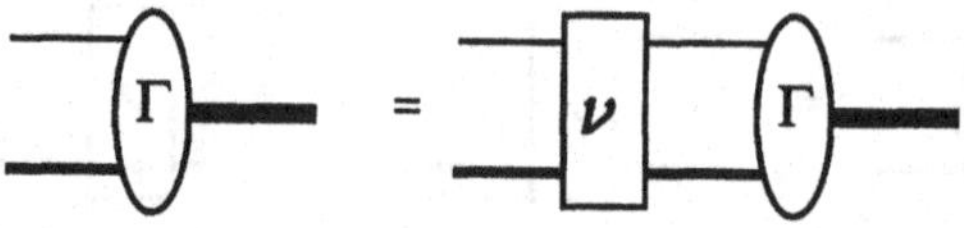

Fig. 12.15 Diagrammatic representation of the bound state integral equation.

is the unique analytic continuation of the series

$$f(z) = \sum_n z^n \tag{12.45}$$

from the region inside the unit circle $|z| < 1$ to the region outside, $|z| \geq 1$. Note that there is a pole at $z = 1$. If z is a matrix, the generalization of the condition $z = 1$ is that z has an eigenvalue equal to 1, so that if the corresponding eigenvector is $\boldsymbol{a}$, then the condition for a pole can be written

$$\boldsymbol{a} = z\boldsymbol{a} \ . \tag{12.46}$$

The corresponding condition for the existence of a pole in $\mathcal{M}$ is

$$\boxed{\Gamma\,(p,P) = \int_k \mathcal{V}\,(p,k;P)\,G(k,P)\Gamma(k,P) \ .} \tag{12.47}$$

This is the integral equation for a *bound state*, and the function Γ is referred to as the *vertex function*. The equation is illustrated in Fig. 12.15.

We have shown that the bound state Eq. (12.47) is a sufficient condition for a bound state. It is also a necessary condition. To see this, assume that a bound state exists, and study the consequences. The presence of a bound state is associated with a pole in the $\mathcal{M}$-matrix below threshold, so the $\mathcal{M}$-matrix would have the form

$$\mathcal{M}\,(p,p';P) = -\Gamma(p,P)\frac{1}{M_B^2 - P^2 - i\epsilon}\overline{\Gamma}\,(p',P) + \mathcal{R}\,(p,p';P) \ , \tag{12.48}$$

represented diagrammatically in Fig. 12.16. In (12.48) $\mathcal{R}$ is a remainder term which has no pole at $P^2 = M_B^2$. The vertex function for $P^2 \neq M_B^2$ is not uniquely defined, because the separation into a pole term and a non-pole term is not unique away from the pole. However, this will not be a problem because we will need the vertex function only at the pole.

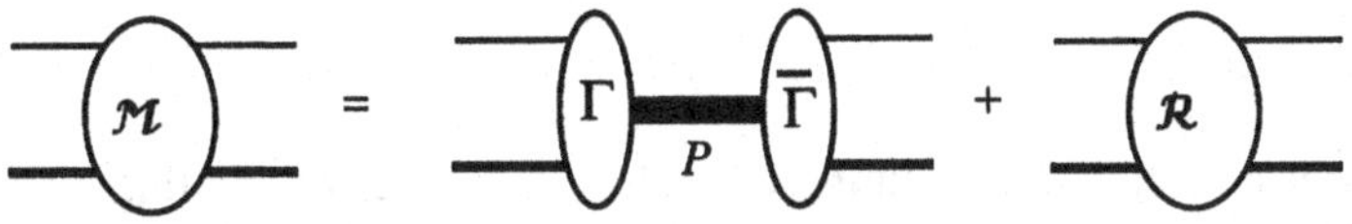

Fig. 12.16 Diagrammatic representation of the scattering matrix with a bound state pole, Eq. (12.48).

An equation for Γ can be derived by assuming that Eq. (12.40) holds everywhere, even at the pole. Substituting Eq. (12.48) into Eq. (12.40), multiplying by $M_B^2 - P^2$, and then taking the limit as $P^2 \to M_B^2$ eliminate all terms not singular at $P^2 = M_B^2$. Dropping the term $\overline{\Gamma}(p', P)$ from both sides gives Eq. (12.47). Note that, strictly speaking, $\Gamma(p, P)$ is uniquely defined only at the bound state pole, where $P^2 = M_B^2$. Alternatively, we may say that Eq. (12.47) does not hold except when $P^2 = M_B^2$, and hence it is an eigenvalue equation.

The relativistic bound state wave function is defined to be

$$\psi(p, P) = \mathcal{N} G(p, P) \Gamma(p, P) , \tag{12.49}$$

where $\mathcal{N}$ is a normalization constant, to be defined later. [Note that the normalization of Γ is defined by (12.48).]

12.4 NORMALIZATION OF BOUND STATES

The normalization condition for the bound state wave function can be obtained directly from Eq. (12.40) and the assumed form of the $\mathcal{M}$-matrix, Eq. (12.48). To this end, note that (12.40) can also be written

$$\mathcal{M} = \mathcal{V} + \int \mathcal{M} G \mathcal{V} , \tag{12.50}$$

where, for compactness, we will suppress all arguments of $\mathcal{M}$, G, and $\mathcal{V}$. The equivalent of (12.40) and (12.50) follows from the fact that they generate the same Born series. In general, $\mathcal{V}$ will be real but G will be complex because of the singularities associated with the zeros in its denominator. Hence (12.50) may also be written

$$\overline{\mathcal{M}} = \mathcal{V} + \int \overline{\mathcal{M} G} \mathcal{V} , \tag{12.51}$$

where the bar represents the adjoint, which includes complex conjugation and any additional operations (such as multiplication by γ_0 as in the Dirac theory). Writing (12.51) as

$$\mathcal{V} = \overline{\mathcal{M}} - \int \overline{\mathcal{M} G} \mathcal{V} \tag{12.52}$$

and substituting this expression for $\mathcal{V}$ under the $\int$ in (12.40) give the following equation:

$$\mathcal{M} = \mathcal{V} + \int \overline{\mathcal{M}} G \mathcal{M} - \int\int \overline{\mathcal{M}} G \mathcal{V} G \mathcal{M} \ . \tag{12.53}$$

Note, for later use, that substituting $\mathcal{V}$ obtained from (12.40) into Eq. (12.51) gives a similar equation for $\overline{\mathcal{M}}$,

$$\overline{\mathcal{M}} = \mathcal{V} + \int \overline{\mathcal{M}} G \mathcal{M} - \int\int \overline{\mathcal{M}} G \mathcal{V} G \mathcal{M} \ . \tag{12.54}$$

Only one of these equations is needed now, and it will be used below threshold (in the neighborhood of the bound state pole) where $\mathcal{M}$ and G are real. Substituting Eq. (12.48), written in shorthand as

$$\mathcal{M} = -\Gamma \frac{1}{M^2 - P^2} \overline{\Gamma} + \mathcal{R} \ ,$$

into (12.53) or (12.54) gives terms with a double pole at $M_B^2 = P^2$, a single pole, and no pole. The double pole terms occur only on the right-hand side (RHS) of the equation and are

$$\text{double poles} = \left(\frac{1}{M_B^2 - P^2} \right)^2 \left\{ \Gamma \int (\overline{\Gamma} G \Gamma) \overline{\Gamma} - \Gamma \int\int (\overline{\Gamma} G \mathcal{V} G \Gamma) \overline{\Gamma} \right\} . \tag{12.55}$$

The coefficient of the double pole term must be zero at $P^2 = M_B^2$. Dropping the initial factor of Γ and the final factor $\overline{\Gamma}$ gives

$$\begin{aligned} \int \overline{\Gamma} G \Gamma - \int\int \overline{\Gamma} \mathcal{V} G \Gamma &= \int \overline{\Gamma} G \left[\Gamma - \int \mathcal{V} G \Gamma \right] \\ &= \int \left[\overline{\Gamma} - \int \overline{\Gamma} G \mathcal{V} \right] G \Gamma = 0 \end{aligned} \tag{12.56}$$

because of the bound state Eq. (12.47). [Alternatively, Eq. (12.56) is another way to obtain the bound state Eq. (12.47).]

Next, look at the single poles. This is more complicated. There are terms from the single poles and terms from the expansion of the coefficient of the double poles near $P^2 = M_B^2$, the residue of the double poles.

First look at the terms involving $\mathcal{R}$. These do not contribute because

$$\begin{aligned} \mathcal{R}\,\text{terms} &= -\frac{1}{M_B^2 - P^2} \left\{ \Gamma \int \left[\overline{\Gamma} - \int \overline{\Gamma} \mathcal{V} \right] G \mathcal{R} + \int \mathcal{R} G \left[\Gamma - \int \mathcal{V} G \Gamma \right] \overline{\Gamma} \right\} \\ &= 0 \ . \end{aligned} \tag{12.57}$$

The expansion of the coefficient of the double pole terms near $P^2 = M_B^2$ will generate terms proportional to $\partial\Gamma/\partial P^2$ and $\partial\overline{\Gamma}/\partial P^2$. By an argument similar to the one above, the bound state wave equation guarantees that these are also zero. Finally, the only new result comes from the balancing of the single pole on the left-hand side with derivatives of G and $G\mathcal{V}G$ on the right-hand side. To find this result, introduce the expansion

$$G = G\bigg|_0 - \frac{\partial G}{\partial P^2}\bigg|_0 (M_B^2 - P^2) + \cdots \,, \tag{12.58}$$

where $|_0$ means that the quantity to the left of the vertical bar (usually a derivative) is evaluated at $P^2 = M_B^2$, and we obtain

$$\Gamma\overline{\Gamma} = \Gamma \int \left(\overline{\Gamma} \frac{\partial G}{\partial P^2}\bigg|_0 \Gamma \right) \overline{\Gamma} - \Gamma \int\!\!\int \left(\overline{\Gamma} \frac{\partial}{\partial P^2} (G\mathcal{V}G)\bigg|_0 \Gamma \right) \overline{\Gamma} \,. \tag{12.59}$$

This can be simplified. Dropping the common factor of $\Gamma\overline{\Gamma}$ and using the bound state equation when possible give

$$\boxed{1 = -\int \overline{\Gamma} \frac{\partial G}{\partial P^2}\bigg|_0 \Gamma - \int\!\!\int \overline{\Gamma} G \frac{\partial \mathcal{V}}{\partial P^2}\bigg|_0 G\Gamma \,.} \tag{12.60}$$

The derivation of this formula did not depend on any of the details, but only on the structure of the equation. It can be used to obtain the normalization condition for any relativistic bound state wave function. For cases when $\mathcal{V}$ is independent of energy, the condition reduces to

$$1 = -\int \overline{\Gamma} \frac{\partial G}{\partial P^2}\bigg|_0 \Gamma \,. \tag{12.61}$$

12.5 THE BETHE–SALPETER EQUATION

We are now fully prepared to discuss two-body relativistic wave equations. the Bethe–Salpeter (BS) equation was the first relativistic two-body equation, introduced in 1951 [SB 51]. For the ϕ^3 example we have been discussing, this equation is defined by*

$$\begin{aligned} \int_k &= i \int \frac{d^4k}{(2\pi)^4} \\ G(k,P) &= \frac{1}{\left[m_1^2 - (P-k)^2 - i\epsilon\right]\left[m_2^2 - k^2 - i\epsilon\right]} \,. \end{aligned} \tag{12.62}$$

*The use of *free* propagators in Eq. (12.62) is equivalent to ignoring all self energy contributions to the propagation of particles 1 and 2, sufficient for our purposes. Self energy contributions are included by *dressing* the single particle propagators used in G. (Thanks to David Owen for calling attention to this omission).

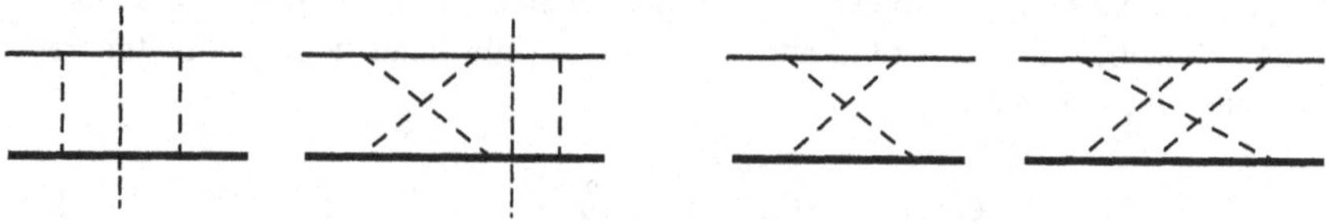

Fig. 12.17 Examples of reducible and irreducible diagrams. Reducible diagrams can be separated into two parts by a line which "cuts" only the two heavy particles.

Note that essential features of the equation are that the integration is over all four components of the internal momentum (and hence it is sometimes referred to as a "four-dimensional" equation in the literature) and that both of the particles are off-shell. Any equation of the general form (12.40) with the choices (12.62) is properly referred to as a Bethe–Salpeter equation.

The choice of the kernel $\mathcal{V}$ defines the approximation in which the BS equation is being employed. In principle, $\mathcal{V}$ can include any Feynman diagram which is two-particle irreducible [recall the discussion surrounding Eq. (12.42)]. Examples of *reducible* and *irreducible* diagrams for the BS equation are shown in Fig. 12.17. If the kernel is the sum of all two-particle irreducible diagrams, then the conventional view is that the solution of the BS equation should give the exact result for the scattering amplitude. In this case the equation can be viewed as producing and summing all diagrams which have a two-particle cut (are two-particle reducible) by combining diagrams which have no such cut. However, because the infinite sum of two-particle irreducible graphs is probably as difficult to calculate as the amplitude $\mathcal{M}$ itself, and since the kernel $\mathcal{V}$ exists order-by-order in perturbation theory, the kernel is usually approximated by the first few terms of its perturbation expansion. In theories where boson exchange is believed to describe the important physics, such as photon exchange in atomic physics, gluon exchange in perturbative quantum chromodynamics (QCD), and meson (in particular pion) exchange in low energy nuclear physics, $\mathcal{V}$ is often approximated by the lowest order one-boson exchange diagram. In this approximation, the solution to the BS equation can be regarded as the *exact* sum of the ladder diagrams. For the ϕ^3 example we have been discussing, this gives

$$\mathcal{V}(p, p'; P) = -\frac{\lambda_1 \lambda_2}{\mu^2 - (p - p')^2 - i\epsilon} . \tag{12.63}$$

Note that this kernel is independent of P^2, and hence the bound state normalization condition for the BS wave function assumes the simpler form (12.61).

If it is desired to sum the ladder *and* crossed ladder diagrams, then the kernel $\mathcal{V}$ must include all irreducible crossed ladder diagrams. These diagrams to sixth order are shown in Fig. 12.18. Since the number of irreducible crossed ladder

Fig. 12.18 Irreducible ladder and crossed ladder diagrams to sixth order.

diagrams grows rapidly with the order n, it is clear that the BS equation is not an efficient way to sum all ladders and crossed ladders.

Solutions to the BS equation can be obtained by rotating the k_0 contour to the imaginary axis (referred to as a *Wick rotation* [Wi 54]). This converts the equation to a Euclidean form and avoids the singularities always associated with Minkowski space. A disadvantage of this method is that it gives the solution for $\mathcal{M}$ (or Γ) along the k_0 imaginary axis, where it is unphysical. Nevertheless, using this technique, exact solutions for spinless particles interacting through the exchange of a massless scalar particle (in ladder approximation) have been obtained [Wi 54, Cu 54] by exploiting the $SU(4)$ symmetry of such a system. One finds additional bound states which do not exist in the nonrelativistic limit; these states may be associated with the inadequacy of the ladder description. The BS equation has also been applied to the description of nucleon–nucleon scattering, where it has been solved numerically [FT 75].

12.6 THE SPECTATOR EQUATION

One alternative to the BS equation is the *spectator equation* (sometimes referred to as the Gross equation [Gr 69]). When applied to the ϕ^3 theory we have been discussing, it is defined by

$$\begin{aligned} \int_k &= -\int \frac{d^4k}{(2\pi)^3}\delta_+\left(m_2^2 - k^2\right) \to -\int \frac{d^3k}{(2\pi)^3}\left(\frac{1}{2E_2}\right) \\ G\left(\hat{k}, P\right) &= \frac{1}{m_1^2 - (P-\hat{k})^2 - i\epsilon} = \frac{1}{E_1^2 - (W - E_2)^2 - i\epsilon}\,, \end{aligned} \tag{12.64}$$

where $\hat{k}$ is a four-vector which satisfies the mass-shell constraint $\hat{k}^2 = m_2^2$. Note that this equation is covariant, even though one of the components of the four-momentum (the energy k_0) is given in terms of the other three. This is because the constraint itself is covariant. Since only the three spatial components of the momentum are independent, the equation is referred to as a *three-dimensional* equation, or as a *quasipotential* equation. In all amplitudes and kernels, $\hat{p} = (E_2(p), \boldsymbol{p})$.

Once again, the physical content of this equation depends on the approximation one makes for the kernel $\mathcal{V}(\hat{p}, \hat{p}'; P)$. In the OBE approximation, the kernel is

$$\mathcal{V}_S(\hat{p}, \hat{p}'; P) = -\frac{\lambda_1\lambda_2}{\left(\mu^2 - (\hat{p} - \hat{p}')^2 - i\epsilon\right)}$$
$$= -\frac{\lambda_1\lambda_2}{\mu^2 + (\mathbf{p} - \mathbf{p}')^2 - (E_2(p) - E_2(p'))^2} \,. \tag{12.65}$$

This is the same as (12.63), but with the important difference that $\hat{p}^2 = \hat{p}'^2 = m_2^2$. In the large m_2 limit, the kernel (12.65) reduces to a form in coordinate space which is an instantaneous, local potential. Specifically, using the definition (12.49) for the bound state wave function, which for this equation is

$$\psi(\hat{p}, P) = \frac{\mathcal{N}}{m_1^2 - (P - \hat{p})^2}\Gamma(\hat{p}, P) \,, \tag{12.66}$$

the bound state equation becomes

$$\left[m_1^2 - (P - \hat{p})^2\right]\psi(\hat{p}, P) = -\int \frac{d^3k}{(2\pi)^3}\frac{1}{2m_2}\mathcal{V}_S\left(\hat{p}, \hat{k}; P\right)\psi\left(\hat{k}, P\right). \tag{12.67}$$

In the $m_2 \to 0$ limit, and taking $P = (W, \mathbf{0})$ with $W = m_2 + E$, this equation reduces to

$$\left(m_1^2 + \mathbf{p}^2 - E^2\right)\psi(\hat{p}, P) = -2m_1 \int \frac{d^3k}{(2\pi)^3} V_S(\mathbf{p} - \mathbf{k})\,\psi\left(\hat{k}, P\right) \,, \tag{12.68}$$

where the effective potential is

$$V_S(\mathbf{q}) = -\frac{g_{\text{eff}}^2}{\mu^2 + q^2} \tag{12.69}$$

with g_{eff} defined as in Eq. (12.7). Equation (12.68) is a Klein–Gordon equation for a particle of mass m_1 and energy E in an instantaneous scalar potential. In coordinate space it is simply

$$\left(m_1^2 - \nabla_r^2 + 2m_1 V_S(r)\right)\psi(r) = E^2\psi(r) \tag{12.70}$$

with

$$V_S(r) = -\frac{g_{\text{eff}}^2}{4\pi^2}\frac{e^{-\mu r}}{r} \,, \tag{12.71}$$

which is precisely Eq. (4.7) with $U(r) = 2m_1 V_S(r)$ and $\psi(x) = \psi(r)\, e^{-iEt}$. We see that the spectator equation has the property that it *reduces to a one-body equation in the limit when the mass of the on-shell particle approaches infinity.* We will refer to this property as a *one-body limit* [Gr 82].

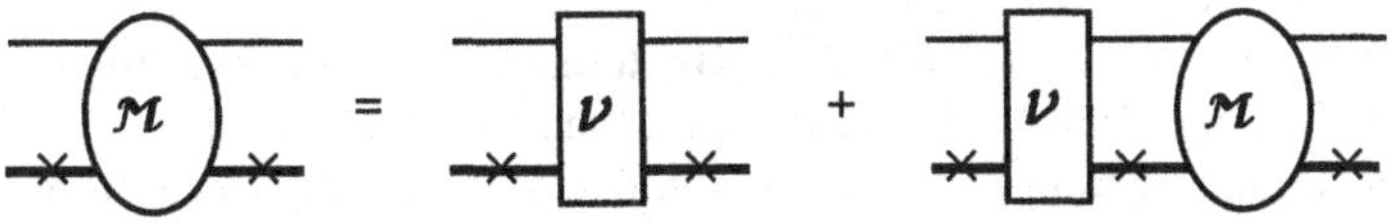

Fig. 12.19 Diagrammatic representation of the spectator equation. Recall that the cross on a meson line means that the particle is on its positive energy mass shell.

The reason why the spectator equation has a one-body limit and the BS equation does not can be understood from our discussion of the role of crossed ladders in Sec. 15.2. In the OBE approximation, the spectator equation *sums the leading terms from all ladders and crossed ladders*, as illustrated in Fig. 12.12. That this is the case is obvious from the diagrammatic representation of the equation given in Fig. 12.19 and from its definition. Another way to understand this result is to examine the contributions which ladders and crossed ladders make to the kernel of the spectator equation. These are shown, to sixth order, in Fig. 12.20. The open circle on the heavy particle line refers to all contributions from the loop in which the circle is found *except* the contribution from the positive energy pole of the particle with the circle. The reason the positive energy pole is excluded is that this part of the diagram is *reducible*, in the sense of the spectator equation. Now, from

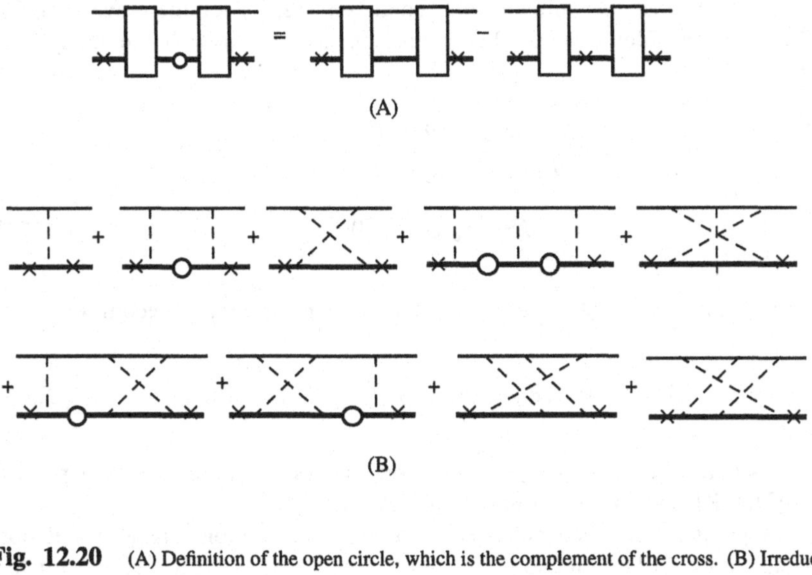

Fig. 12.20 (A) Definition of the open circle, which is the complement of the cross. (B) Irreducible (in the sense of the spectator equation) ladder and crossed ladder diagrams to sixth order.

the discussion in Sec. 15.2, we know that the meson pole contributions from ladders and crossed ladders cancel as $m_2 \to \infty$, and since these contributions dominate all of the higher order kernels shown in Fig. 12.20, they will all approach zero as $m_2 \to \infty$, leaving only the OBE term, which therefore gives the exact result for the sum of all ladders and crossed ladders in this limit.

In a similar fashion, it may be shown that the spectator equation for a Dirac particle of mass m_1 and a scalar particle of mass m_2 exchanging a scalar meson of mass μ reduces, in the $m_2 \to \infty$ limit, to a Dirac equation. This is left as an exercise (see Prob. 15.1).

In the $m_2 \to \infty$ limit, the relativistic wave function (12.49) for a bound state in the spectator formalism is

$$\psi\,(p,P) = \frac{\mathcal{N}}{m_1^2 + \mathbf{p}^2 - E_0^2}\,\Gamma\,(p,P)\;, \tag{12.72}$$

where E_0 is the bound state energy of particle 1. If the vertex function Γ is a constant, the coordinate space form of this wave function is

$$\psi(\mathbf{r}) = \int \frac{d^3p}{(2\pi)^{3/2}}\, e^{i\mathbf{p}\cdot\mathbf{r}}\psi(p,P) = \mathcal{N}\Gamma\,\sqrt{\frac{2}{\pi}}\,\frac{e^{-\sqrt{m_1^2-E_0^2}\,r}}{r}\;, \tag{12.73}$$

which is the familiar asymptotic S-state wave function. Hence the propagator factor in (12.72) gives the asymptotic part of the wave function, while the vertex function Γ contains all of the dynamical information contained in the intermediate and short-range part of the wave function.

Next, note that the normalization condition (12.61) for the wave function (12.72) becomes [don't forget the minus sign associated with the integral in (12.64)]

$$\begin{aligned} 1 &= \int \frac{d^3p}{(2\pi)^3}\,\frac{2E_0\,\Gamma^2(p,P)}{\left(m_1^2 + \mathbf{p}^2 - E_0^2\right)^2} \\ &= \int d^3p\, 2E_0\,\psi^2(p,P)\;, \end{aligned} \tag{12.74}$$

where we have chosen $\mathcal{N} = (2\pi)^{-3/2}$. In coordinate space this becomes

$$1 = \int d^3r\, 2E_0\,\psi^2(\mathbf{r}) = \int d^3r\,\psi^*(x)\,i\frac{\overleftrightarrow{\partial}}{\partial t}\psi(x)\;, \tag{12.75}$$

where, as before, $\psi(x) = \psi(\mathbf{r})\,e^{-iE_0t}$. Note that we have recovered the precise form of the Klein–Gordon normalization, Eq. (4.14).

The spectator equation has been used as the foundation for the calculation of higher order QED corrections in simple atomic systems [EK 91] and for the relativistic treatment of nucleon–nucleon scattering [GV 92].

12.7 EQUIVALENCE OF TWO-BODY EQUATIONS

The two different relativistic equations we have discussed so far correspond to two different ways of calculating the scattering matrix, and it is usually assumed that the exact answer could be obtained from either equation if the kernel included *all* irreducible diagrams. However, since the kernel is always approximated by a few irreducible diagrams, an approximate calculation of $\mathcal{M}$ using one equation will differ from an approximate calculation using another, and it is important to know how to compare the two approximations. Alternatively, by carefully choosing the kernels, it is possible to obtain the *same* solution for $\mathcal{M}$ from two *different* equations. In this sense different equations are equivalent. We will discuss this now.

We assume that the *same* solution for $\mathcal{M}$ has emerged from two different equations, and ask how their kernels must be related by this fact. Specifically, assume that

$$\begin{aligned}\mathcal{M} &= \mathcal{V}_1 + \int \mathcal{V}_1 G_1 \mathcal{M} = \mathcal{V}_1 + \int \mathcal{M} G_1 \mathcal{V}_1 \\ \mathcal{M} &= \mathcal{V}_2 + \int \mathcal{V}_2 G_2 \mathcal{M} = \mathcal{V}_2 + \int \mathcal{M} G_2 \mathcal{V}_2 \quad .\end{aligned} \tag{12.76}$$

Discretizing the integrals, so that $\mathcal{V}$ G and $\mathcal{M}$ become matrices, these equations imply

$$\mathcal{M} = (1 - \mathcal{V}_1 G_1)^{-1} \mathcal{V}_1 = \mathcal{V}_2 (1 - G_2 \mathcal{V}_2)^{-1} \tag{12.77}$$

and hence

$$\mathcal{V}_2 = \mathcal{V}_1 + \mathcal{V}_1 (G_1 - G_2) \mathcal{V}_2 \quad . \tag{12.78}$$

As an illustration of the content of this equation, suppose that equation 1 is the BS equation and $\mathcal{V}_1$ is the OBE approximation. Then Eq. (12.78) tells us that the kernel $\mathcal{V}_2$ of the spectator equation which *exactly* sums the *ladder diagrams* (since $\mathcal{V}_1$ does this) is given by the solution of the equation

$$\begin{aligned}\mathcal{V}_2 (p, \hat{p}'; P) &= \mathcal{V}_{\mathrm{OBE}} (p, \hat{p}'; P) \\ &\quad + i \int \frac{d^4k}{(2\pi)^4} \frac{\mathcal{V}_{\mathrm{OBE}}(p, k; P)}{[m_1^2 - (P-k)^2 - i\epsilon]} \\ &\quad \times \left[\frac{1}{(m_2^2 - k^2 - i\epsilon)} - 2\pi i \, \delta_+ \left(m_2^2 - k^2 \right) \right] \mathcal{V}_2 (k, p'; P) .\end{aligned} \tag{12.79}$$

Iterating this equation and then setting $p = \hat{p}$ generate the infinite series of diagrams shown in Fig. 12.21. The difference of the propagators $G_1 - G_2$ is equivalent diagrammatically to the open circle on the heavy particle line, as shown in Fig. 12.20. We conclude that the use of the spectator equation to sum the ladder diagrams is extremely inefficient. The kernel for this operation is an infinite series of terms, and evaluating it by solving (12.79) is just as difficult as solving the original BS equation in ladder approximation.

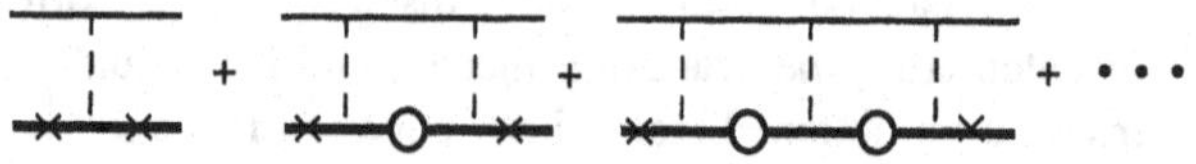

Fig. 12.21 The infinite series of diagrams which defines a kernel for the spectator equation which will give the exact ladder sum.

Alternatively, suppose equation 1 is the spectator equation and $\mathcal{V}_1$ is the OBE approximation to it. The spectator kernel can be extrapolated off-shell using (12.79) and the Feynman rules. The BS kernel equivalent to the spectator OBE kernel is shown diagrammatically in Fig. 12.22. Note that the new $\mathcal{V}_2$ is an infinite series of terms, very similar to those in Fig. 12.21, except that each term in this series has all external particles off-shell, and the even terms in the series have the opposite signs. We conclude that the use of the BS equation to sum the series of terms shown in Fig. 12.12 is inefficient.

The analysis we have just completed can be used to compare any other relativistic equations we might wish to consider, including the Blankenbecler–Sugar [BS 66] equation to be discussed below. It shows that each equation is efficient in summing a particular class of diagrams and inefficient in summing others. The choice of equation depends on which physical processes we wish to sum and how efficiently they can be summed by that equation (see Prob. 15.2).

12.8 UNITARITY

We now return to the box diagram and look at its structure from a different point of view. Recall our discussion of the singularities of the box diagram in the complex k_0 plane, shown in Fig. 12.4. Now we want to prove that the imaginary part of the box comes *only* from a "pinching" of the poles 1 and 5 and that all the other pole contributions [2, 3 and 4] *never* come opposite any of the poles

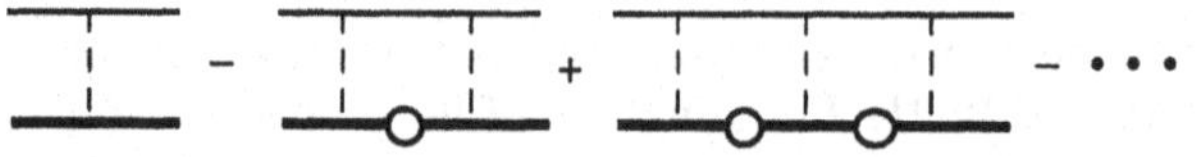

Fig. 12.22 The infinite series of diagrams which defines a kernel for the BS equation which will give the leading contributions to the sum of all ladder and crossed ladder diagrams.

in the upper half plane, and hence can never contribute to the imaginary part of $\mathcal{M}$. The proof will be carried out for physical values of W only. These are real values of $W \geq m_1 + m_2$.

To prove these statements, we will prove that the poles are ordered in the following sequence along the real k_0 axis:

$$8 < (7,6) < (5,1) < (2,3) < 4 , \tag{12.80}$$

where the ordering of the pairs (7,6), (5,1), and (2,3) is indeterminate. Using $W = E_1(p) + E_2(p)$, these inequalities become

$$\begin{aligned}
8 < (7,6) &\Longrightarrow -E_1 < \ E_2(p) - \omega \\
(7,6) < 5 &\Longrightarrow -E_1(p) - \omega < -E_1 \\
(7,6) < 1 &\Longrightarrow \ E_2(p) - \omega < \ E_2 \\
5 < (2,3) &\Longrightarrow \ E_1(p) - E_1 < \ \omega \\
1 < (2,3) &\Longrightarrow \ E_2 < \ \omega + E_2(p) \\
(2,3) < 4 &\Longrightarrow \ \omega < \ E_1(p) + E_1 ,
\end{aligned} \tag{12.81}$$

where inequalities involving p' are similar to those involving p and need not be considered explicitly. The only inequality which requires any demonstration is

$$|E_1(p) - E_1| < \omega \tag{12.82}$$

and a similar one for E_2. To prove this we square both sides giving the requirement

$$2m_1^2 + p^2 + k^2 - 2\sqrt{m_1^2 + p^2}\,\sqrt{m_1^2 + k^2} < \mu^2 + p^2 + k^2 - 2\boldsymbol{p}\cdot\boldsymbol{k} \ .$$

The minimum value of ω^2 occurs when $\boldsymbol{p}\cdot\boldsymbol{k} = pk$, and rearranging terms and squaring again give the requirement

$$\left(2m_1^2 + 2pk - \mu^2\right) < 4\left(m_1^4 + m_1^2\left(p^2 + k^2\right) + p^2k^2\right) \ .$$

Expanding out these terms shows that this equality is always satisfied, even if μ is very small, because $p^2 + k^2 \geq 2pk$.

Hence the *exact* result for the imaginary part of $\mathcal{M}_{\text{box}}$ can be obtained from the pinching of poles 1 and 5, which from Eq. (12.4) gives

$$\begin{aligned}
&Im\,\mathcal{M}_{\text{box}} \\
&= -\pi\lambda_1^2\lambda_2^2 \int \frac{d^3k}{(2\pi)^3 2E_2}\, \frac{\delta_+\left[E_1^2 - (W - E_2)^2\right]}{\left(\omega^2 - [E_2 - E_2(p)]^2\right)\left(\omega'^2 - [E_2 - E_2(p')]^2\right)} \\
&= -\pi\rho_2\,(W; m_1 m_2) \int \frac{d\Omega_k}{4\pi}\left(\frac{\lambda_1\lambda_2}{\omega^2}\right)\left(\frac{\lambda_1\lambda_2}{\omega'^2}\right) ,
\end{aligned} \tag{12.83}$$

where

$$\rho_2\left(W;m_1m_2\right) = \frac{p}{8\pi^2 W} = \frac{1}{8\pi^2 W}\sqrt{\frac{\left[W^2-(m_1+m_2)^2\right]\left[W^2-(m_1-m_2)^2\right]}{4W^2}} \tag{12.84}$$

is the two-body phase space factor introduced in Sec. 9.2. A diagrammatic interpretation of (12.84) is given in Fig. 12.23. The imaginary part of the box in the physical region is the product of two OBE amplitudes, with all of their external particles on-shell, integrated over all directions of the intermediate three-momentum $\boldsymbol{k}$ and multiplied by the two-body phase space factor.

This result anticipates the *unitarity relation* satisfied by a physical scattering amplitude. Recall, from a study of nonrelativistic scattering, that the scattering amplitude for the ℓth partial wave has the following general form:

$$f_\ell = K\frac{e^{i\delta_\ell}\sin\delta_\ell}{p} \,, \tag{12.85}$$

where δ is the *phase shift* and K is a constant, usually equal to unity in nonrelativistic theory. The scattering amplitude (12.85) satisfies the unitarity relation, which for (12.85) is

$$Im\, f_\ell = \frac{p}{K}\left|f_\ell\right|^2 \ . \tag{12.86}$$

The relativistic counterpart of this relation can be obtained from Eq. (12.53) and (12.54). Subtracting these two equations gives

$$\boxed{\mathcal{M}-\overline{\mathcal{M}} = \int \overline{\mathcal{M}}\left(G-\overline{G}\right)\mathcal{M} \ .} \tag{12.87}$$

This is the most general form of the unitary relation.

For the spectator equation in ϕ^3 theory, $\overline{\mathcal{M}} = \mathcal{M}^*$ because it is a complex number (and not a matrix) and hence the left-hand side of (12.87) is the imaginary part of $\mathcal{M}$. Recalling the defining relations (12.64) and (12.87) renders

$$\begin{aligned} Im\,\mathcal{M}\left(\hat{p},\hat{p}';P\right) &= \pi\int\frac{d^3k}{E_2}\,\delta_+\left[E_1^2-(W-E_2)^2\right]\mathcal{M}^*\left(\hat{p},\hat{k};P\right)\mathcal{M}\left(\hat{k},\hat{p}';P\right) \\ &= -\pi\rho_2\left(W;m_1m_2\right)\int\frac{d\Omega_k}{4\pi}\mathcal{M}^*\left(\hat{p},\hat{k};P\right)\mathcal{M}\left(\hat{k},\hat{p}';P\right) \ . \end{aligned} \tag{12.88}$$

If we expand $\mathcal{M}$ in partial waves,

$$\mathcal{M}\left(\hat{p},\hat{p}';P\right) = \sum_\ell (2\ell+1)\mathcal{M}_\ell(W)P_\ell(z) \,, \tag{12.89}$$

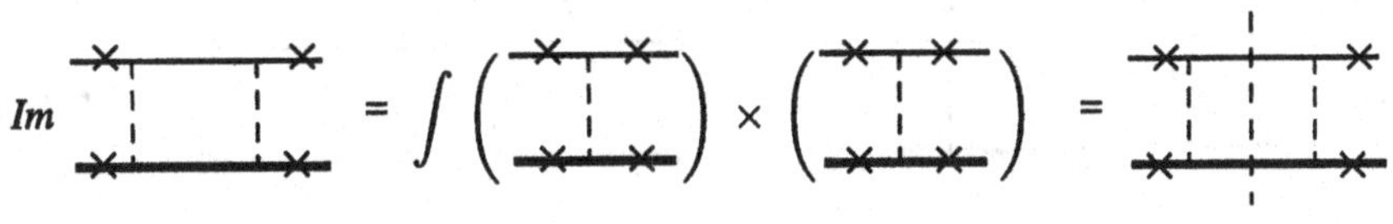

Fig. 12.23 Diagrammatic representation of the imaginary part, sometimes referred to as the unitarity cut, for the box diagram.

where $z = \hat{p} \cdot \hat{p}'$, and use the orthogonality and addition theorem for Legendre polynomials,

$$P_\ell(\hat{p} \cdot \hat{p}') = \frac{4\pi}{2\ell+1} \sum_{m=-\ell}^{m=\ell} Y^*_{\ell m}(\hat{p}) \, Y_{\ell m}(\hat{p}')$$
$$\int d\Omega_k Y^*_{\ell m}\left(\hat{k}\right) Y_{\ell' m'}\left(\hat{k}\right) = \delta_{\ell\ell'}\delta_{mm'} \quad , \tag{12.90}$$

then (12.88) reduces to

$$Im\, \mathcal{M}_\ell(W) = -\pi\rho_2\,(W; m_1 m_2)\, |\mathcal{M}_\ell(W)|^2 \quad . \tag{12.91}$$

This is the relativistic generalization of (12.86), and from it we have the identification

$$\mathcal{M}_\ell(W) = -8\pi W \frac{e^{i\delta_\ell}\sin\delta_\ell}{p} \quad , \tag{12.92}$$

which shows how the phase shift is related to the relativistic scattering amplitude. Note that, as a consequence of this relation,

$$\lim_{W\to\infty} |\mathcal{M}_\ell(W)| \leq 16\pi \quad . \tag{12.93}$$

The amplitude for the ℓth partial wave is bounded as the total energy W approaches infinity, and and by an extension of this argument it can be shown that the total amplitude is also bounded by a less that linear growth with energy. This limit is referred to as the *unitarity bound.* If a calculation (or theory) produces an amplitude which violates this limit, we know that the calculation (or theory) is incorrect. Early models of the weak interactions suffered from this disease, which is cured by the Standard Model (see Sec. 15.5).

The unitarity relation is a requirement which the exact scattering amplitude must satisfy. Yet no *finite* sum of Feynman diagrams can satisfy the relation. To

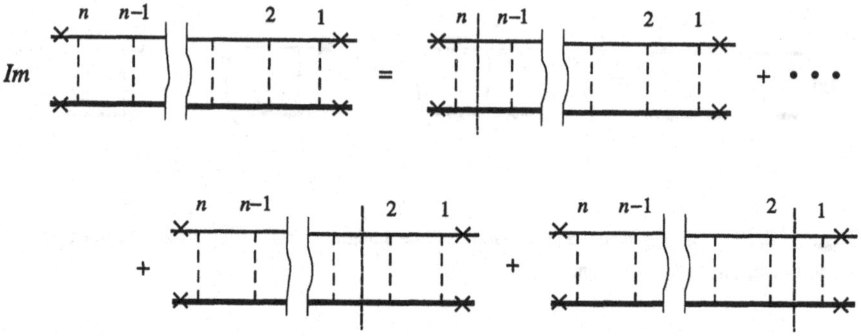

Fig. 12.24 Representation of the $n-1$ elastic unitarity cuts contained in the $(2n)$th order ladder diagram.

see why this is so, consider the $(2n)$th order diagram, and use Fig. 12.23 to see that the imaginary part of this diagram is composed of $n-1$ terms, as shown in Fig. 12.24. Using this figure, we see that the imaginary part of the $(2n)$th order diagram is built up of products of diagrams *less than* $(2n)$th order, and hence consistency can be achieved only if $n \to \infty$. The integral equation gives a unitary amplitude precisely because it sums an infinite number of diagrams. When the constraints imposed by unitarity are important, the use of integral equations is required.

12.9 THE BLANKENBECLER–SUGAR EQUATION

A relativistic two-body equation motivated by the unitarity relation was introduced by Blankenbecler and Sugar (BBS) in 1966 [BS 66]. In the ϕ^3 theory, this equation is defined by

$$\int_k = \int \frac{d^4k}{(2\pi)^4}$$
$$G_{\rm BBS} = \pi \int ds \frac{\delta_+\left(m_1^2 - (\bar{P}-k)^2\right)\delta_+\left(m_2^2 - k^2\right)}{s - P^2 - i\epsilon}, \tag{12.94}$$

where $\bar{P} = \left(\sqrt{s + \boldsymbol{P}^2}, \boldsymbol{P}\right)$ is the total four-momentum of the two particles if they are both on their mass shell. The imaginary part of $G_{\rm BBS}$, which can be obtained directly from the dispersion integral representation (12.94), is equal to

$$Im\, G_{\rm BBS} = \pi^2 \delta_+\left(m_1^2 - (P-k)^2\right)\delta_+\left(m_2^2 - k^2\right) \ . \tag{12.95}$$

The imaginary part of this propagator therefore restricts the intermediate particles to their mass shell, and in this way generates the correct two-body unitary cuts shown in Fig. 12.24. It is also a three-dimensional equation. Carrying out the integration over s gives, in the rest frame,

$$G_{\rm BBS} = \pi \frac{(E_1+E_2)}{E_1 E_2} \frac{\delta(E_2 - k_0)}{\left[(E_1+E_2)^2 - W^2 - i\epsilon\right]} , \tag{12.96}$$

showing that the relative energy is no longer an independent variable.

The two-body BBS equation was designed to preserve two-body unitarity, but, in fact, this is a feature it shares in common with the other two equations we have discussed previously. To compare the BBS propagator with the spectator propagator we must first remove the factor of π/E_2 contained in the spectator integral operator (12.64) and factor the denominator,

$$\tilde{G}_{\rm BBS} = \frac{E_2}{\pi} G_{\rm BBS} = \frac{(E_1+E_2)}{E_1} \frac{\delta(E_2-k_0)}{[E_1+E_2-W-i\epsilon]\,[E_1+E_2+W]} .$$

In the same form, the spectator propagator is

$$G_S \delta(E_2 - k_0) = \frac{\delta(E_2-k_0)}{(E_1+E_2-W-i\epsilon)(E_1+W-E_2-i\epsilon)} .$$

Note that these two propagators are identical along the unitarity cut (when $W = E_1 + E_2$) and differ only in how they describe the physics away from the unitarity cut.

One of the significant features of the BBS equation is that it treats the two particles symmetrically [this is most easily seen from the original form (12.94)], and hence it is easy to use the BBS equation for the description identical particles. Furthermore, the *only* singularities of the BBS propagator are those associated with the unitarity cut. The spectator equation shares neither of these features. It has additional singularities and can only be used to describe identical particles if it is explicitly symmetrized (or anti-symmetrized) by including channels in which either particle 1 or particle 2 is on-shell [GV 92].

The construction of relativistic two-body equations and the comparison between different methods are active areas of current research, and we will leave the subject at this point.

12.10 DISPERSION RELATIONS AND ANOMALOUS THRESHOLDS

We saw in Sec. 11.7 that the vacuum polarization diagram 11.10 satisfied a *dispersion relation*

$$\Pi(q^2) = \Pi(0) + \frac{q^2}{\pi} \int_{4m^2}^{\infty} \frac{ds\, Im\,\Pi(s)}{s(s-q^2)} . \tag{12.97}$$

This relation follows from the observation that $\Pi(q^2)$ is a *real analytic function* of its argument [i.e., $\Pi^*(q^2) = \Pi(q^{2*})$], with a cut which lies along the real $q^2 > 4m^2$

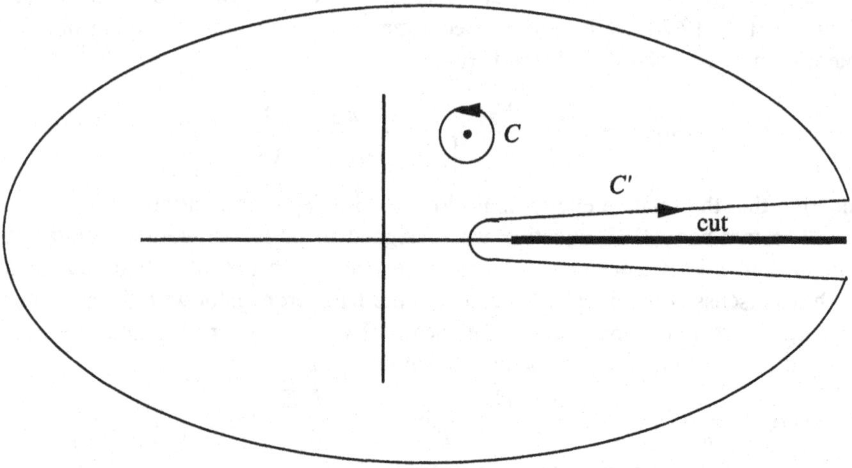

Fig. 12.25 Contours in the complex plane. C can be deformed into C'.

axis extending over the values of q^2 for which real production of electron–positron pairs is possible (m is the electron mass). Actually, the dispersion relation (12.97) is *subtracted once* in order to improve the convergence at $q^2 \to \infty$, which goes like $\log q^2$ for the simple diagram 11.10. It can be derived, as we did in Sec. 11.7, by considering the function

$$f(q^2) = \frac{\Pi(q^2) - \Pi(0)}{q^2} , \tag{12.98}$$

which approaches zero as $q^2 \to \infty$, and which, because of the subtraction, has no pole at $q^2 = 0$. Then, by Cauchy's theorem,

$$f(q^2) = \frac{1}{2\pi i} \int_C \frac{ds\, f(s)}{(s - q^2)} , \tag{12.99}$$

where q^2 is some reference point far from the cut, and C is a small contour circling q^2 in a *clockwise sense*. By opening up the contour to a larger contour C', as shown in Fig. 12.25, and using the fact that $f(q^2) \to 0$ at infinity, we obtain

$$\begin{aligned} f(q^2) &= \frac{1}{2\pi i} \int_{4m^2}^{\infty} \frac{ds\,[f(s+i\epsilon) - f(s-i\epsilon)]}{(s - q^2)} \\ &= \frac{1}{\pi} \int_{4m^2}^{\infty} ds \frac{Im\, f(s)}{(s - q^2)} \ . \end{aligned} \tag{12.100}$$

Substituting (12.98) into (12.100) gives the result (12.97).

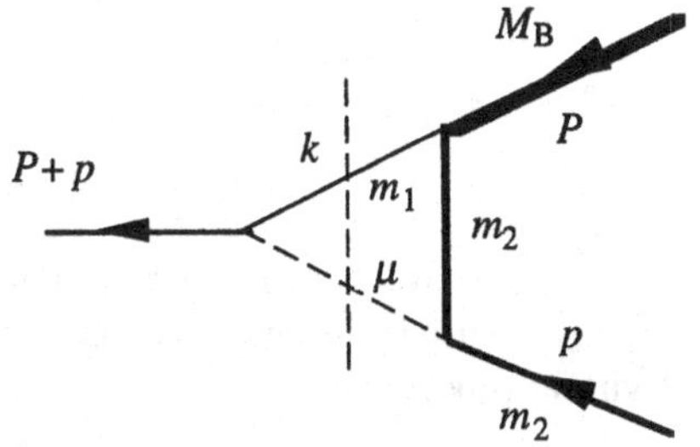

Fig. 12.26 Feynman diagram for estimating the relativistic wave function. The vertical dashed line is the two-body cut used in the dispersion relation.

As this argument shows, scattering amplitudes and vertex functions generally satisfy dispersion relations. Even though they are initially defined only along the real axis, they can usually be analytically continued to the complex plane, and since they are real over some interval of the real axis, the analytic continuation must satisfy the real analytic property

$$f(x^*) = f^*(x)$$

everywhere. When dispersion relations are combined with generalized unitarity, which specifies the imaginary part as the product of the initial and final state scattering amplitudes (or vertex functions), as discussed in Sec. 12.8 above, they can be an important tool for understanding the properties of matrix elements.

It is our purpose in this section to show how these ideas can be used to gain insight into the structure of bound states. In so doing, we point out that care must be exercised in applying dispersion theory to amplitudes involving bound states, because of the existence of *anomalous thresholds.*

As an example, consider the spectator vertex function for a bound state in the ϕ^3 theory, introduced in Eq. (12.66). As we have previously indicated, this vertex function can depend on only two independent four-vectors (the third is fixed by four-momentum conservation) and must be a scalar function of these two four-vectors. Only three scalar variables can be formed from the two four-vectors (they are p^2, P^2, and $p \cdot P$), and if we take the heavy particle to be on-shell (along with the bound state), there is only one scalar variable remaining, which we will choose to be the mass (squared) of the light off-mass-shell particle. In the notation of Fig. 12.26, this variable is $u = (P + \hat{p})^2$, and the vertex function will be regarded as a complex function of this variable,

$$\Gamma(\hat{p}, P) \equiv f(u) \quad . \tag{12.101}$$

To use dispersion theory to study this amplitude, we must know the location of its singularities in the complex u plane, and these are found (typically) by examining one of the simplest Feynman diagrams which contributes to the vertex. Such a diagram is shown in Fig. 12.26.

This diagram gives the following amplitude:

$$f(u) = \int \frac{d^4k}{(2\pi)^4} \frac{-i\, f_0\, \lambda_1 \lambda_2}{(m_1^2 - k^2 - i\epsilon)\,(m_2^2 - (P-k)^2 - i\epsilon)\left(\mu^2 - (P + \hat{p} - k)^2\right)}\,, \tag{12.102}$$

where $\hat{p}$ and P are the four-momenta of the initial heavy particle and bound state, respectively, and the masses are the same as the ones we introduced earlier in the chapter. Note that we have "turned the heavy particle around," so that the amplitude we are considering is actually the virtual process

$$\bar{m}_2 + M_B \rightarrow m_1\;, \tag{12.103}$$

where $\bar{m}_2$ is the antiparticle associated with particle 2, instead of the original process

$$M_B \rightarrow m_1 + m_2\;. \tag{12.104}$$

This new amplitude is an analytic continuation of the original amplitude and is more convenient to use with the dispersion relation.

The utility of the dispersion relation is evident if we use it to evaluate (12.102). As suggested by Fig. 12.26, we would expect the unitarity cut to begin at $u = (m_1 + \mu)^2$, the threshold for the production of a real intermediate state consisting of two particles of mass m_1 and μ, and hence the dispersion integral should have the form

$$f(u) = \frac{1}{\pi} \int_{(m_1+\mu)^2}^{\infty} \frac{du'\, Im\, f(u')}{u' - u}\,, \tag{12.105}$$

where $Im\, f(u')$ is evaluated from the generalized unitarity relation. Taking the general form of the unitarity relation from Eq. (12.83), we have

$$Im\, f(u) = -\pi f_0 \lambda_1 \lambda_2\, \rho_2\left(\sqrt{u}\,; m_1, \mu\right) \int \frac{d\Omega_k}{4\pi} \frac{1}{m_2^2 - \left(P - \hat{k}\right)^2}\,, \tag{12.106}$$

where ρ_2 is the two-body phase space factor and $\hat{k}$ is the four-momentum of the intermediate on-shell particle 1. A quick estimate of the function for negative values of u can be obtained by approximating $f(u)$ by a single pole at mass u_0 which would give

$$f(u) \cong \frac{N_0}{u_0 - u}\;. \tag{12.107}$$

Since the cut in (12.105) begins at $u = (m_1 + \mu)^2$, *we expect* $u_0 \geq (m_1 + \mu)^2$. Now, evaluating u in the frame in which the bound state is at rest and assuming the state is weakly bound so that $M_B \simeq m_1 + m_2$ and $\epsilon_B = m_1 + m_2 - M_B$ is small, give

$$u = (P - \hat{p}) = M_B^2 + m_2^2 - 2M_B E_2(p)$$
$$\cong (M_B - m_2)^2 - \frac{M_B}{m_2} p^2 \cong (M_B - m_2)^2 - p^2 \ , \qquad (12.108)$$

where p^2 is the square of the three-momentum of either of the bound particles in the rest system of the bound state. Inserting this into (12.107) gives

$$f(u) = \Gamma(\hat{p}, P) \simeq \frac{N_0}{u_0 - m_1^2 + \delta^2 + p^2} \ , \qquad (12.109)$$

where $\delta^2 = 2m_1\epsilon_B = 2m_1(m_1 + m_2 - M_B)$ and all terms of order p^4 have been discarded. If this estimate is to agree with the Hulthén model we discussed in Sec. 12.1, then we require

$$u_0 - m_1^2 + \delta^2 \cong (\mu + \delta)^2 \ . \qquad (12.110)$$

However, if $u_0 \cong (m_1 + \mu)^2$, this would imply that $m_1 \simeq \delta$, which is clearly not satisfied for a loosely bound state where $\delta \ll m_1$. *Our estimate does not work!*

The resolution of this problem lies in the fact that the dispersion integral has an *anomalous threshold* which lies far below the normal threshold at $(m_1 + \mu)^2$. It can be shown (Prob. 12.3) that if $M_B^2 > m_1^2 + m_2^2 + \mu m_1$, which is certainly the case for the loosely bound system we are considering here, then the exact location of the threshold is not at $u_0 = (m_1 + \mu)^2$ but is instead at

$$u_0 = m_1^2 + \frac{\mu}{2m_2}\sqrt{1 - \frac{\mu^2}{4m_2^2}}\ \sqrt{\Delta(M_B, m_1, m_2)} + \frac{\mu^2}{2m_2^2}\left[M_B^2 + m_2^2 - m_1^2\right] , \qquad (12.111)$$

where $\Delta(a, b, c) = 2a^2b^2 + 2a^2c^2 + 2b^2c^2 - a^4 - b^4 - c^4$. In the limit $m_2 \to \infty$, the above expression becomes

$$u_0 = m_1^2 + \mu^2 + 2\mu\delta \ , \qquad (12.112)$$

which agrees precisely with the requirement (12.110). The dispersion estimate, *including the anomalous threshold*, is in beautiful agreement with the estimate derived from the wave function.

To understand the origin of the anomalous threshold [Cu 61], it is helpful to regard the dispersion integral (12.105) as a function of the complex variable M_B^2 and to *analytically continue* this function from small values of M_B^2 (where there are only normal thresholds) to large values (where the anomalous thresholds will appear). To carry out the analytic continuation it is convenient to write the imaginary part, Eq. (12.106), in the following form:

$$\int d\Omega_k \frac{1}{m_2^2 - \left(P - \hat{k}\right)^2} = \int_a^{u_0} du'' \frac{P(u'')}{u'' - u} \ . \qquad (12.113)$$

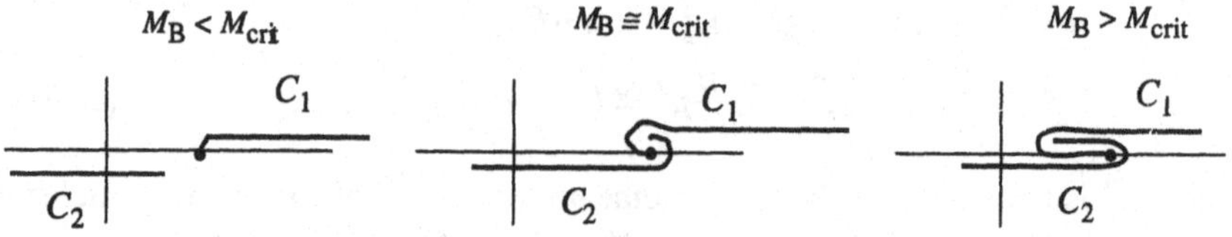

Fig. 12.27 The complex u plane showing motion of u_0, the end point of the contour C_2, as M_B increases. The contour C_1 begins at the fixed point $(m_1 + \mu)^2$ and must be deformed in order to avoid the moving singularity.

This displays the imaginary part as a dispersion integral with singularities along the real axis from a to u_0. The upper limit, u_0, will turn out to be the same u_0 which appears in (12.105), but the lower limit (which, in this application, is actually three numbers describing two disconnected line segments) will play no role in the subsequent discussion. The locus of singularities, and hence the value of u_0, can be found from the zeros of the denominator, which is a function of both u and M_B^2

$$m_2^2 - \left(P - \hat{k}\right)^2 = m_2^2 - \left(M_B^2 + m_1^2 - 2E_BE_1 + 2pkz\right) , \qquad (12.114)$$

where z is the cosine of the scattering angle, E_B and E_1 are the energies of the bound state and the on-shell intermediate particle 1 in the *rest system of the final virtual particle* 1, and p and k are the magnitudes of the respective three-momenta. Expressing these momenta and energies in terms of the energy $\sqrt{u}$, one can find u_0, which is the largest value of u at which the denominator (12.114) is zero (which occurrs at the end point $z = 1$ of the angular integration). (This is a straightforward but tedious calculation; see Prob. 12.3.)

Now examine (Prob. 12.3) the behavior of this upper limit u_0 as a *function of the bound state mass* M_B^2. Observe that $u_0 < (m_1 + \mu)^2$ for small M_B^2, but that as M_B^2 increases, u_0 increases to a maximum value of $(m_1 + \mu)^2$ and then decreases. The critical value of M_B^2 at which u_0 is equal to this maximum is easily found by differentiating u_0 with respect to M_B^2 and is

$$M_{\rm crit}^2 = m_1^2 + m_2^2 + \mu m_1 \ . \qquad (12.115)$$

Furthermore, if we give M_B^2 a small imaginary part (in order to define the singular denominators) we can show that u_0 moves in such a way that it *circles above* the point $(m_1 + \mu)^2$. Specifically, when $M_B^2 = M_{\rm crit}^2 \pm i\epsilon$,

$$Re\, u_{\rm crit} = (m_1 + \mu)^2 + \mathcal{O}(\epsilon^2) \ . \qquad (12.116)$$

The significant fact here is that $Re\, u_{\rm crit} > (m_1 + \mu)^2$, even if only by an infinitesimal amount. Therefore, the moving upper limit of the integral (12.113)

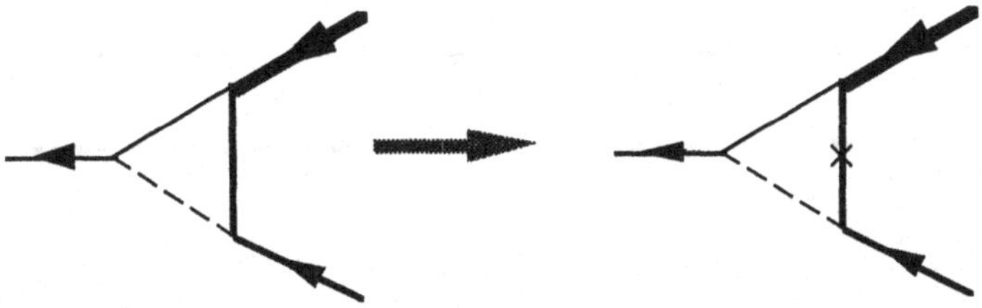

Fig. 12.28 The discontinuity in the anomalous region has the exchanged particle on-mass-shell, as in the spectator equation.

follows the path C_2 shown in Fig. 12.27. If C_1 is the cut from $(m_1 + \mu)^2$ to ∞ which defines the original dispersion integral (12.105), and if the overall dispersion integral (12.105) is to be a single analytic function for all values of M_B^2, then this cut must be deformed into the complex plane in order to avoid the moving singularity at the end of the contour C_2 as M_B^2 increases beyond M_{crit}^2. The contour C_1 then surrounds the path of integration C_2. This deformation is illustrated in Fig. 12.27. As the bound state mass increases, the protruding branch cut continues to move to the left toward smaller values of u_0, moving the anomalous threshold further and further toward m_1^2, as suggested in the figure. This is the mathematical origin of the anomalous threshold; the physical origin has already been discussed.

Finally, observe that the integrand of the new dispersion integral obtained from Eq. (12.105) in the anomalous region is the discontinuity (or imaginary part) of the dispersion integral for the exchanged particle pole, Eq. (12.104). But this integral is only singular when the exchanged denominator is singular, which means that the exchanged particle is *on-shell. We see that the contribution in the anomalous region, which is closest to the physical region when $u \sim m_1^2$, arises from the condition that the internal particle 2 be on-shell.* In this way we recover the spectator equation, as illustrated in Fig. 12.28. Another way to describe the spectator equation is to observe that it sums up the anomalous contributions exactly.

We now turn to the study of gauge symmetries and gauge field theories, which will occupy our attention for the remainder of this book.

PROBLEMS

12.1 Write down the spectator equation for a Dirac particle of mass m and a scalar particle of mass m_2, exchanging a scalar meson of mass μ, and show that it

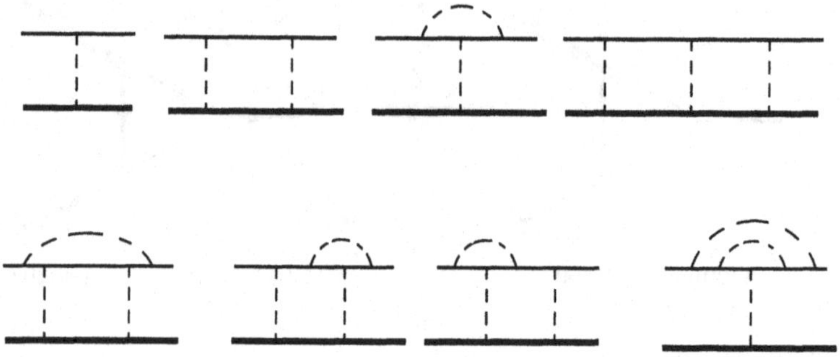

Fig. 12.29 Ladder and ladder-vertex diagrams up to sixth order (see Prob. 12.2).

reduces to a Dirac equation as $m_2 \to \infty$. Find the precise form of the Dirac potential corresponding to the OBE approximation for the kernel.

12.2 Construct an equation which sums the ladder and ladder vertex correction diagrams shown in Fig. 12.29. Prove that your equation works by iterating it to eighth order and showing that all diagrams to this order are included, with the correct weight.

12.3 Study of the anomalous threshold.

(a) Show that the upper limit u_0 of the dispersion integral Eq. (12.113) is given correctly by Eq. (12.111).

(b) Show that u_0 increases as M_B^2 increases if $M_B^2 < m_1^2 + m_2^2 + \mu m_1$, but if $M_B^2 > m_1^2 + m_2^2 + \mu m_1$, u_0 decreases as M_B^2 increases. Give M_B^2 a small negative imaginary part and show that the movement of u_0 in the complex plane is as described in Sec. 12.10.

(c) Using the techniques developed in Sec. 11.6, combine the denominators of the amplitude (12.102) into a single term, and do the integration over d^4k. Study the zeros of the denominator (which locate the singularities of the amplitude) as a function of the Feynman parameters x_i. Show that it has zeros for $u < (m_1 + \mu)^2$ only if $M_B^2 > m_1^2 + m_2^2 + \mu m_1$. Locate the branch point and confirm that it is the same u_0 given in Eq. (12.111).

PART IV

SYMMETRIES AND GAUGE THEORIES

CHAPTER 13

SYMMETRIES II

The remainder of this book is devoted to the study of dynamical symmetries and gauge field theories. We use the term "dynamical symmetry" to refer to any symmetry which is so restrictive that it completely determines the structure of the Lagrangian. The discovery that such symmetries exist, and that they actually seem to correctly describe the physical world, is one of the most remarkable, exciting, and successful of the recent developments in modern physics.

The most striking example of such symmetries are the *local, non-Abelian gauge symmetries*. Quantum Chromodynamics, or QCD, is a theory based on the $SU(3)$ gauge group. This is the modern theory of the strong interactions, which has the remarkable property that the coupling constant approaches zero as the momentum flowing through the interaction vertex approaches infinity, a property referred to as *asymptotic freedom* (which will be discussed in Chapter 17). Largely because of this, QCD has led to a plethora of successful predictions at high energies, which give us great faith in the theory even though predictions at low energies are hard to make. Another local gauge symmetry, based on the product of two groups, $SU(2) \times U(1)$, leads to the unification of the electromagnetic and weak interactions into a single *electroweak* theory, referred to as the *Standard Model.* Here a new feature is added; the gauge symmetry is "hidden" because the ground state of the theory, the vacuum, *spontaneously breaks the symmetry*. The breaking is said to be "spontaneous" because the mathematical form of the Lagrangian forces the vacuum to break the symmetry; no additional assumption is needed. In spite of its modest name, the Standard Model has also been a major success, but because there are variations of this model, based on larger gauge groups with more parameters, which are also consistent with the data, it is less clear that the Standard Model will survive into the next century without changes.

Both QCD and the Standard Model are formulated directly in terms of the elementary constituents of nature: the quarks and leptons. This makes it hard to extract predictions for the observed hadrons, which are complex composites of quarks. In this respect a third symmetry, *chiral symmetry*, which is only an approximate symmetry, has been very successful because it gives effective Lagrangians which can be expressed directly in terms of the observed hadrons.

The most successful version of chiral symmetry is also *spontaneously broken*.

In this chapter we will discuss gauge theories, chiral symmetry, and spontaneous symmetry breaking. The quantization of gauge fields will be the topic of Chapters 14 and 15. Discussion of the standard electroweak theory is also postponed until Chapter 15. Finally, we return to a more detailed discussion of renormalization in Chapter 16 and a derivation and discussion of asymptotic freedom in Chapter 17.

We begin this chapter by showing how the familiar gauge invariance of QED can be understood to be a consequence of a local $U(1)$ gauge symmetry. The $U(1)$ symmetry group is *Abelian*, and as a result QED has a rather simple structure, and the incredible power of the idea of local gauge invariance is not clearly illustrated by QED alone. Only when we consider *non-Abelian* gauge groups does the rich structure of such theories become apparent, and we discuss these theories in the middle sections of this chapter. The chapter concludes with a study of chiral symmetry and how this symmetry can be spontaneously broken.

13.1 ABELIAN GAUGE INVARIANCE

We introduced gauge transformations very briefly in Eq. (8.10). There are two types of Abelian gauge transformations depending on whether or not θ is a function of x:

$$\begin{aligned} &\theta = \text{constant} \qquad &&\text{(global gauge transformation)} \\ &\theta = \theta(x) \qquad &&\text{(local gauge transformation)}\ . \end{aligned} \tag{13.1}$$

Consider global gauge transformations first.

If the Lagrangian is invariant under the gauge transformation, Noether's theorem tells us that there is a conserved quantity associated with this invariance. Gauge transformations always leave the space–time coordinates unchanged, and hence the $\lambda^{\mu}{}_{i}$ of Eq. (8.20) is always zero, and the conserved quantity associated with the gauge transformation, always referred to as a *current*, has the form

$$\boxed{-\mathcal{O}^{\mu}{}_{i} \equiv J^{\mu}_{i} = \frac{\partial \mathcal{L}}{\partial\left(\dfrac{\partial \psi_{\alpha}}{\partial x^{\mu}}\right)}\,\Omega_{\alpha i}\ .} \tag{13.2}$$

Note that, by convention, the sign of the current is opposite to the sign of $\mathcal{O}^{\mu}$ in Eq. (8.20).

Now, a global Abelian gauge transformation is defined by

$$\begin{aligned} &\left.\begin{aligned} \psi'_{\alpha}(x) &= e^{-iq\theta}\psi_{\alpha}(x) \\ \bar{\psi}'_{\alpha}(x) &= e^{iq\theta}\,\bar{\psi}_{\alpha}(x) \end{aligned}\right\} \quad \text{complex fields} \\ &A'^{\mu}(x) = A^{\mu}(x) \qquad\qquad \text{real fields,} \end{aligned} \tag{13.3}$$

where q can be a different number for each complex field (later, q will be associated with the charge). Note that there is only *one* infinitesimal parameter and that the group consists of multiplying complex fields by a complex number with unit modulus, which is the unitary group of one dimension, $U(1)$. The infinitesimal form of the transformations are

$$\psi'_\alpha(x) = (1 - iq\theta)\,\psi_\alpha(x) \ , \tag{13.4}$$

and hence $\Omega_\alpha = -iq\psi_\alpha(x)$, $\bar{\Omega}_\alpha = iq\bar{\psi}_\alpha(x)$, and the conserved quantity associated with this $U(1)$ symmetry is

$$J^\mu = -iq\left\{\frac{\partial\mathcal{L}}{\partial\left(\dfrac{\partial\psi_\alpha}{\partial x^\mu}\right)}\psi_\alpha - \bar{\psi}_\alpha\frac{\partial\mathcal{L}}{\partial\left(\dfrac{\partial\bar{\psi}_\alpha}{\partial x^\mu}\right)}\right\} \ , \tag{13.5}$$

In order for $\mathcal{L}$ to be invariant under the global gauge transformation, it is sufficient that it be bilinear in $\bar{\psi}$ and ψ, or for scalar fields, $\phi^\dagger$ and ϕ.

For a free spinor theory, where

$$\mathcal{L} = \bar{\psi}\left[\frac{i}{2}\gamma^\mu\frac{\overleftrightarrow{\partial}}{\partial x^\mu} - m\right]\psi \ ,$$

the conserved quantity is

$$\boxed{J^\mu = q\bar{\psi}\gamma^\mu\psi \ ,} \tag{13.6}$$

which we recognize as the EM current, Eq. (10.2), provided we identify

$$q = e \ . \tag{13.7}$$

Hence we see that *conservation of charge can be "understood" as a consequence of a global gauge symmetry of the theory.*

Local Gauge Invariance

We now generalize the gauge transformation, permitting the phase θ to depend on the *local* space–time point, i.e., $\theta = \theta(x)$. This means that a gauge transformation can be carried out in one region of space–time without "knowing" what is taking place elsewhere. If it were the case that the value of the gauge phase angle had any physical significance, it would be essential that the gauge transformation be local, in order to allow time for information about any changes in the phase angle to propagate from one locality to another. However, the phase angle probably contains no information, in which case the requirement of local gauge invariance is merely the (very powerful) requirement that this phase angle can be completely arbitrary from point to point, except for the requirement that it be a smooth, differentiable function.

The Lagrangian will no longer be invariant under the local gauge transformation unless it has a particular form. Consider the free fermion part first. We have

$$\begin{aligned}\mathcal{L}'_F &= \frac{1}{2}\bar{\psi}'\left[i\gamma^\mu \frac{\overrightarrow{\partial}}{\partial x^\mu} - m\right]\psi' + \frac{1}{2}\bar{\psi}'\left[-i\gamma^\mu \frac{\overleftarrow{\partial}}{\partial x^\mu} - m\right]\psi' \\ &= \mathcal{L}_F + q\,\bar{\psi}\gamma^\mu\psi\;\partial_\mu\theta = \mathcal{L}_F + J^\mu\partial_\mu\theta \ , \end{aligned} \tag{13.8}$$

where $\partial_\mu\theta = \partial\theta(x)/\partial x^\mu$. To eliminate the extra term and make $\mathcal{L}$ invariant, we need a vector field A^μ which interacts with the current and transforms in a special way. To find this special transformation law for the vector field, add an interaction of the form $J_\mu A^\mu$:

$$\mathcal{L} = \mathcal{L}_F - \frac{e}{q} J^\mu A_\mu \ . \tag{13.9}$$

Then, since $J' = J$,

$$\mathcal{L}' = \mathcal{L}'_F - \frac{e}{q} J^\mu A'_\mu = \mathcal{L}_F + J^\mu\partial_\mu\theta - \frac{e}{q} J^\mu A'_\mu \ . \tag{13.10}$$

Hence $\mathcal{L}' = \mathcal{L}$ if

$$\boxed{A'_\mu(x) = A_\mu(x) + \frac{q}{e}\,\partial_\mu\theta(x) \ .} \tag{13.11}$$

From now on we will take $q/e = 1$, so that the gauge transformation of the vector field is precisely what we wrote down for electromagnetism in Sec. 2.2 [Eq. (2.15) with $\theta = -\Lambda_c$].

Next, consider the Lagrangian for the fields A^μ. A general form for the Lagrangian is

$$\mathcal{L}_{\text{field}} = \lambda_1 F_{\mu\nu}F^{\mu\nu} + \lambda_2\, G_{\mu\nu}G^{\mu\nu} + m_\gamma^2 A^\mu A_\mu \ , \tag{13.12}$$

where m_γ is a possible mass term for the gauge fields and $G_{\mu\nu}$ is a possible *symmetric* combination of fields and derivatives,

$$G_{\mu\nu} = \partial_\mu A_\nu + \partial_\nu A_\mu \ , \tag{13.13}$$

which, together with the antisymmetric combination $F_{\mu\nu}$, insures that the hypothetical Lagrangian (13.12) contains any combination of the independent terms $\partial_\mu A_\nu \partial^\mu A^\nu$ and $\partial_\mu A_\nu \partial^\nu A^\mu$. Now the gauge transformation leaves $F_{\mu\nu}$ invariant, but

$$G'_{\mu\nu} = G_{\mu\nu} + 2\partial_\mu\partial_\nu\theta \tag{13.14}$$

and

$$A'^{\mu}A'_{\mu} = A^{\mu}A_{\mu} + 2A^{\mu}\partial_{\mu}\theta + [\partial_{\mu}\theta]\,[\partial^{\mu}\theta] \quad . \tag{13.15}$$

Hence neither of these terms is gauge invariant, and $\mathcal{L}'_{\text{field}} = \mathcal{L}_{\text{field}}$ requires that both $m_{\gamma}^2 = 0$ and $\lambda_2 = 0$. The vector field must be massless and have a free Lagrangian of the familiar form $\lambda_1 F_{\mu\nu}F^{\mu\nu}$, where $\lambda_1 = -\frac{1}{4}$ corresponds to the conventional normalization used for the EM field. We conclude that *the requirement of local gauge invariance dictates the form of QED.*

13.2 NON-ABELIAN GAUGE INVARIANCE

Next, we discuss how the concept of gauge invariance can be extended to non-Abelian groups. Assume the "charged" fields have several components, such as isospin or color, describing some internal degree of freedom, and consider a unitary transformation which transforms them into each other. This gauge transformation can be written

$$\psi'(x) = \mathbf{U}\psi(x) \ , \tag{13.16}$$

where, for n degrees of freedom, $\mathbf{U}$ is an $n \times n$ matrix which is unitary. The group of such matrices can always be written as a product of the $U(1)$ group (which in n dimensions is the product of a complex number with unit modulus and the $n \times n$ unit matrix) and the $SU(n)$ group of unitary matrices with unit determinant (a condition which fixes the phase). In this section we will limit discussion to the $SU(n)$ group, so that we may assume $\det \mathbf{U} = 1$, and when a detailed example is needed, we will use the familiar $SU(2)$ group. For $SU(2)$ the $\mathbf{U}$ matrix is

$$\mathbf{U} = e^{-ig\frac{1}{2}\tau_i\epsilon_i(x)} \ , \tag{13.17}$$

where $\frac{1}{2}\tau_i$ are the familiar Pauli matrices ($\times\frac{1}{2}$), the generators of $SU(2)$, g is the coupling constant, and $\epsilon_i(x)$ are three independent rotation "angles." To simplify the formulas, we will use the notation

$$\boldsymbol{\varepsilon}(x) = \tfrac{1}{2}\tau_i\epsilon_i(x) \ , \tag{13.18}$$

where $\boldsymbol{\varepsilon}(x)$ is now a 2×2 matrix.

If the Lagrangian is independent of the internal degree of freedom (which, for example, could be isospin or color), the free Lagrangian will be a sum over the Lagrangians for each component of the internal degree of freedom (denoted by the subscript ℓ), and

$$\mathcal{L}_F = \bar{\psi}_{\ell}\left[\frac{1}{2}i\gamma^{\mu}\frac{\overleftrightarrow{\partial}}{\partial x^{\mu}} - m\right]\psi_{\ell} = \bar{\psi}\left[\frac{1}{2}i\gamma^{\mu}\frac{\overleftrightarrow{\partial}}{\partial x^{\mu}} - m\right]\psi \ , \tag{13.19}$$

where, in the second expression, a unit matrix in n-dimensional space is implied. If the gauge transformation is *global* ($\epsilon_i = \text{const}$), the free Fermi Lagrangian is invariant under the gauge transformation as it stands (because $\mathbf{U}$ is unitary), and the conserved current follows from the observation that the λ and Ω of Noether's theorem are

$$\begin{aligned} \lambda^{\mu}{}_{i} &= 0 \\ \Omega_{\ell i} &= -ig\tfrac{1}{2}\left[\tau_i \psi(x)\right]_\ell \end{aligned} \tag{13.20}$$

so that the conserved current (13.2) is

$$J_i^\mu = -ig\left\{\frac{\partial \mathcal{L}}{\partial\left(\dfrac{\partial \psi_\ell}{\partial x^\mu}\right)}\frac{1}{2}\left[\tau_i\psi(x)\right]_\ell - \frac{1}{2}\left[\bar{\psi}(x)\tau_i\right]_\ell \frac{\partial \mathcal{L}}{\partial\left(\dfrac{\partial \bar{\psi}_\ell}{\partial x^\mu}\right)}\right\}$$

or

$$\boxed{J_i^\mu = g\,\bar{\psi}\gamma^\mu \tfrac{1}{2}\tau_i\psi\ .} \tag{13.21}$$

If the $SU(2)$ space is isospin, this is the conserved *isospin current.*

If the symmetry is to be a *local* one, $\mathcal{L}_F$ is no longer invariant. By analogy with the the discussion of local Abelian symmetry, we expect to have to add some vector gauge fields to make the Lagrangian invariant. First note that

$$\begin{aligned} \mathcal{L}_F' &= \bar{\psi}'\left[\frac{1}{2}i\gamma^\mu \frac{\overleftrightarrow{\partial}}{\partial x^\mu} - m\right]\psi' = \bar{\psi}\,\mathbf{U}^\dagger\left[\frac{1}{2}i\gamma^\mu\frac{\overleftrightarrow{\partial}}{\partial x^\mu} - m\right]\mathbf{U}\psi \\ &= \mathcal{L}_F + \bar{\psi}\,\frac{1}{2}i\gamma^\mu\left[\mathbf{U}^\dagger\left(\frac{\partial}{\partial x^\mu}\mathbf{U}\right) - \left(\frac{\partial}{\partial x^\mu}\mathbf{U}^\dagger\right)\mathbf{U}\right]\psi\ . \end{aligned} \tag{13.22}$$

But, $\mathbf{U}^\dagger\mathbf{U} = 1$ implies that

$$\left(\partial_\mu \mathbf{U}^\dagger\right)\mathbf{U} + \mathbf{U}^\dagger\left(\partial_\mu \mathbf{U}\right) = 0 \tag{13.23}$$

and hence the two terms can be combined to yield

$$\mathcal{L}_F' = \mathcal{L}_F + \bar{\psi}\,i\gamma^\mu\,\mathbf{U}^\dagger\left(\partial_\mu\mathbf{U}\right)\psi\ . \tag{13.24}$$

Since $\mathbf{U}^\dagger\left(\partial_\mu \mathbf{U}\right)$ is a 2×2 matrix, the generalized gauge fields added to the Lagrangian must also be a 2×2 matrix, and there must be one gauge field for each generator of the group:

$$\boxed{\mathbf{A}_\mu(x) = \tfrac{1}{2}\tau_i A_\mu^i(x)} \tag{13.25}$$

so that for $SU(n)$, there are $n^2 - 1$ gauge fields. Adding such a gauge field interaction to the Lagrangian and demanding gauge invariance give

$$\begin{aligned} \mathcal{L}_F' - g\,\bar{\psi}'\gamma^\mu\mathbf{A}_\mu'\psi' &= \mathcal{L}_F + \bar{\psi}\gamma^\mu\,i\mathbf{U}^\dagger\left(\partial_\mu\mathbf{U}\right)\psi - g\,\bar{\psi}\gamma^\mu\,\mathbf{U}^\dagger\mathbf{A}_\mu'\mathbf{U}\psi \\ &= \mathcal{L}_F - g\,\bar{\psi}\gamma^\mu\mathbf{A}_\mu\psi\ . \end{aligned} \tag{13.26}$$

Hence, the required transformation law for the gauge field is

$$\mathbf{A}_\mu(x) = \mathbf{U}^\dagger \mathbf{A}'_\mu(x)\mathbf{U} - \frac{i}{g}\mathbf{U}^\dagger\left(\partial_\mu \mathbf{U}\right) ,$$

which can be rewritten

$$\boxed{\mathbf{A}'_\mu(x) = \mathbf{U}\mathbf{A}_\mu(x)\mathbf{U}^\dagger + \frac{i}{g}\left(\partial_\mu \mathbf{U}\right)\mathbf{U}^\dagger \ .} \tag{13.27}$$

This is the transformation of $\mathbf{A}_\mu$ under a *finite* gauge transformation. For an *infinitesimal* transformation, we expand $\mathbf{U}$,

$$\mathbf{U} = 1 - ig\,\boldsymbol{\varepsilon}(x) , \tag{13.28}$$

and retain only first order terms, giving

$$\mathbf{A}'_\mu(x) = \mathbf{A}_\mu(x) + \partial_\mu \boldsymbol{\varepsilon}(x) + ig\left[\mathbf{A}_\mu(x), \boldsymbol{\varepsilon}(x)\right] \ . \tag{13.29}$$

Using the operator form of $\mathbf{A}_\mu$ and the group commutation relations

$$\left[\tfrac{1}{2}\tau_i, \tfrac{1}{2}\tau_j\right] = i\epsilon_{ijk}\tfrac{1}{2}\tau_k , \tag{13.30}$$

where the ϵ_{ijk} are structure constants of the $SU(2)$ group [and would be replaced by the appropriate *structure constants* in the $SU(n)$ case], Eq. (13.29) can be written

$$\boxed{A'^i_\mu(x) = A^i_\mu(x) + \partial_\mu \epsilon_i(x) - g\,\epsilon_{ijk} A^j_\mu(x)\epsilon_k(x) \ .} \tag{13.31}$$

Note the presence of the new term in this gauge transformation which arises because the structure constants of the group are not zero. Also, note that A^i_μ must transform non-trivially even if $\epsilon_i(x) = $ constant. This is associated with the fact that A^i_μ now transforms as a vector under the gauge group as well as a Lorentz four-vector.

It is convenient at this point to introduce covariant derivatives which have nice properties under the gauge transformation. We define

$$\begin{aligned} \overrightarrow{D}_\mu\, \psi(x) &= \left[\overrightarrow{\partial_\mu} + ig\,\mathbf{A}_\mu(x)\right]\psi(x) \\ \bar{\psi}(x)\,\overleftarrow{D}_\mu &= \bar{\psi}(x)\left[\overleftarrow{\partial_\mu} - ig\,\mathbf{A}_\mu(x)\right] \ . \end{aligned} \tag{13.32}$$

Then, under a local non-Abelian gauge transformation,

$$\begin{aligned} \overrightarrow{D}'_\mu \psi'(x) &= \mathbf{U}\, D_\mu \psi(x) \\ \bar{\psi}'(x)\overleftarrow{D}' &= \bar{\psi}(x)\overleftarrow{D}_\mu\, \mathbf{U}^\dagger \quad . \end{aligned} \tag{13.33}$$

These relations follow from a straightforward calculation:

$$\begin{aligned} \overrightarrow{D}'_\mu \psi'(x) &= \left[\partial_\mu + ig\, \mathbf{A}'_\mu(x)\right] \mathbf{U}\psi(x) \\ &= \left[\partial_\mu + ig\left(\mathbf{U}\mathbf{A}_\mu\mathbf{U}^\dagger + \frac{i}{g}\left(\partial_\mu \mathbf{U}\right)\mathbf{U}^\dagger\right)\right]\mathbf{U}\psi(x) \\ &= \mathbf{U}D_\mu\psi(x) + \left[(\partial_\mu \mathbf{U}) - (\partial_\mu \mathbf{U})\right]\psi(x) = \mathbf{U}D_\mu\psi(x) \quad . \end{aligned} \tag{13.34}$$

Using this notation it is almost trivial to see that

$$\mathcal{L}_0 = \bar{\psi}\left[\frac{i}{2}\gamma^\mu\left(\overleftrightarrow{D}_\mu\right) - m\right]\psi \tag{13.35}$$

is gauge invariant.

To complete the construction of the full gauge invariant Lagrangian, we must find the Lagrangian of the gauge fields themselves. This must be separately gauge invariant under the gauge transformations (13.27) of the gauge fields. One idea is to construct this Lagrangian by simply summing electromagnetic type Lagrangians over the $n^2 - 1$ gauge fields. This would give

$$\mathcal{L}_{\text{field}} = -\tfrac{1}{4}F^i_{\mu\nu}F^{i\mu\nu} \quad . \tag{13.36}$$

Equation (13.36) will eventually work, but only if the definition of $F^i_{\mu\nu}$ is suitably generalized. Before finding the correct definition, note that if we define an $\mathbf{F}_{\mu\nu}$ matrix by

$$\mathbf{F}_{\mu\nu}(x) = \tfrac{1}{2}\tau_i F^i_{\mu\nu}(x) \tag{13.37}$$

and use

$$\text{tr}\,(\tau_i\,\tau_j) = 2\delta_{ij} \quad , \tag{13.38}$$

we can rewrite the field Lagrangian in a very convenient form,

$$\boxed{\mathcal{L}_{\text{field}} = -\tfrac{1}{2}\,\text{tr}\,(\mathbf{F}_{\mu\nu}\mathbf{F}^{\mu\nu}) \quad .} \tag{13.39}$$

This form shows that a sufficient condition for the gauge invariance of $\mathcal{L}_{\text{field}}$ is that

$$\boxed{\mathbf{F}'_{\mu\nu} = \mathbf{U}\mathbf{F}_{\mu\nu}\mathbf{U}^\dagger} \tag{13.40}$$

since the trace is invariant under unitary transformations.

It can be readily seen that the simple definition

$$\mathbf{F}_{\mu\nu} = \partial_\nu \mathbf{A}_\mu - \partial_\nu \mathbf{A}_\mu \tag{13.41}$$

does not work. It turns out that the generalized definition with covariant derivatives,

$$\boxed{\mathbf{F}_{\mu\nu} = D_\mu \mathbf{A}_\nu - D_\nu \mathbf{A}_\mu ,} \tag{13.42}$$

does work, and this will be shown now. Consider

$$\begin{aligned}\mathbf{F}'_{\mu\nu} &= D'_\mu \mathbf{A}'_\nu - D'_\nu \mathbf{A}'_\mu \\ &= D'_\mu \left(\mathbf{U}\mathbf{A}_\nu \mathbf{U}^\dagger\right) - D'_\nu \left(\mathbf{U}\mathbf{A}_\mu \mathbf{U}^\dagger\right) + \frac{i}{g} D'_\mu \left(\partial_\nu \mathbf{U}\right) \mathbf{U}^\dagger - \frac{i}{g} D'_\nu \left(\partial_\mu \mathbf{U}\right) \mathbf{U}^\dagger \ .\end{aligned} \tag{13.43}$$

The original discussion of the covariant derivative assures us that

$$D'_\mu \mathbf{U}\mathbf{A}_\nu = \mathbf{U} D_\mu \mathbf{A}_\nu \tag{13.44}$$

so that the first terms can be quickly reduced, giving

$$\mathbf{F}'_{\mu\nu} = \mathbf{U}\mathbf{F}_{\mu\nu}\mathbf{U}^\dagger + R , \tag{13.45}$$

where the remainder term, R, consists of the action of the partial derivatives in the first two terms of (13.43) on $\mathbf{U}^\dagger$ [all that remain after using (13.44)] plus the last two terms of (13.43). Simplifying R gives

$$\begin{aligned}R &= \mathbf{U}\mathbf{A}_\nu \left(\partial_\mu \mathbf{U}^\dagger\right) - \mathbf{U}\mathbf{A}_\mu \left(\partial_\nu \mathbf{U}^\dagger\right) + \frac{i}{g} D'_\mu \left(\partial_\nu \mathbf{U}\right) \mathbf{U}^\dagger - \frac{i}{g} D'_\nu \left(\partial_\mu \mathbf{U}\right) \mathbf{U}^\dagger \\ &= \mathbf{U}\mathbf{A}_\nu \left(\partial_\mu \mathbf{U}^\dagger\right) - \mathbf{U}\mathbf{A}_\mu \left(\partial_\nu \mathbf{U}^\dagger\right) + \frac{i}{g}\partial_\mu \left[\left(\partial_\nu \mathbf{U}\right)\mathbf{U}^\dagger\right] - \frac{i}{g}\partial_\nu \left[\left(\partial_\mu \mathbf{U}\right)\mathbf{U}^\dagger\right] \\ &\quad -\left(\mathbf{U}\mathbf{A}_\mu\mathbf{U}^\dagger + \frac{i}{g}\left(\partial_\mu \mathbf{U}\right)\mathbf{U}^\dagger\right)\left(\partial_\nu \mathbf{U}\right)\mathbf{U}^\dagger + \left(\mathbf{U}\mathbf{A}_\nu\mathbf{U}^\dagger + \frac{i}{g}\left(\partial_\nu \mathbf{U}\right)\mathbf{U}^\dagger\right)\left(\partial_\mu \mathbf{U}\right)\mathbf{U}^\dagger \\ &= \mathbf{U}\mathbf{A}_\nu \left(\partial_\mu \mathbf{U}^\dagger\right) - \mathbf{U}\mathbf{A}_\mu \left(\partial_\nu \mathbf{U}^\dagger\right) \\ &\quad + \frac{i}{g}\left(\left(\partial_\mu\partial_\nu \mathbf{U}\right)\mathbf{U}^\dagger + \left(\partial_\nu \mathbf{U}\right)\left(\partial_\mu \mathbf{U}^\dagger\right) - \left(\partial_\nu\partial_\mu \mathbf{U}\right)\mathbf{U}^\dagger - \left(\partial_\mu \mathbf{U}\right)\left(\partial_\nu \mathbf{U}^\dagger\right)\right) \\ &\quad - \mathbf{U}\mathbf{A}_\mu\mathbf{U}^\dagger \underbrace{\left(\partial_\nu \mathbf{U}\right)\mathbf{U}^\dagger}_{-\mathbf{U}\partial_\nu\mathbf{U}^\dagger} + \mathbf{U}\mathbf{A}_\nu\mathbf{U}^\dagger \underbrace{\left(\partial_\mu \mathbf{U}\right)\mathbf{U}^\dagger}_{-\mathbf{U}\partial_\mu\mathbf{U}^\dagger} \\ &\quad - \frac{i}{g}\left(\left(\partial_\mu \mathbf{U}\right)\mathbf{U}^\dagger \underbrace{\left(\partial_\nu \mathbf{U}\right)\mathbf{U}^\dagger}_{-\mathbf{U}\partial_\nu\mathbf{U}^\dagger} - \left(\partial_\nu \mathbf{U}\right)\mathbf{U}^\dagger \underbrace{\left(\partial_\mu \mathbf{U}\right)\mathbf{U}^\dagger}_{-\mathbf{U}\partial_\mu\mathbf{U}^\dagger}\right) = 0 \ .\end{aligned} \tag{13.46}$$

This proves that $\mathbf{F}_{\mu\nu}$ defined with covariant derivatives has the desired transformation properties. Writing this out gives

$$\boxed{\mathbf{F}_{\mu\nu} = \partial_\mu \mathbf{A}_\nu - \partial_\nu \mathbf{A}_\mu + ig\left[\mathbf{A}_\mu, \mathbf{A}_\nu\right] \;.} \tag{13.47}$$

Using the commutation relations, the individual field components become

$$\boxed{F^i_{\mu\nu} = \partial_\mu A^i_\nu - \partial_\nu A^i_\mu - g\,\epsilon_{ijk} A^j_\mu A^k_\nu \;.} \tag{13.48}$$

Note the presence of the extra term which depends on g. This gives a theory with very different properties from electromagnetism. Such a theory is called a *Yang–Mills* theory.

13.3 YANG–MILLS THEORIES

The discussion of invariance under the local gauge group $SU(2)$ can be extended to larger groups $SU(n)$. A theory which is *invariant under the local gauge group $SU(n)$ is referred to as a Yang–Mills theory.* Quantum Chromodynamics (QCD) is a Yang–Mills theory with the gauge group $SU(3)$.

To describe the Yang–Mills theory corresponding to the gauge group $SU(n)$, introduce the $SU(n)$ gauge transformation $\mathbf{U}$,

$$\mathbf{U} = e^{-ig\frac{1}{2}\lambda_a \epsilon_a(x)} \;, \tag{13.49}$$

where $\frac{1}{2}\lambda_a$ are the generators of the group, with a running from 1 to $n^2 - 1$. If the generators are defined by

$$\boxed{\begin{aligned} &\mathrm{tr}\;(\lambda_a\,\lambda_b) = 2\delta_{ab} \\ &[\tfrac{1}{2}\lambda_a, \tfrac{1}{2}\lambda_b] = i f_{abc}\,\tfrac{1}{2}\lambda_c \,, \end{aligned}} \tag{13.50}$$

where f_{abc} are the *structure constants* of the group (they are antisymmetric in all three indices), then all the discussion given in the previous section can be carried over with minor modifications. If we define

$$\begin{aligned} \boldsymbol{\epsilon}(x) &= \tfrac{1}{2}\lambda_a \epsilon_a(x) \\ \mathbf{A}_\mu(x) &= \tfrac{1}{2}\lambda_a A^a_\mu(x) \\ \mathbf{F}_{\mu\nu}(x) &= \tfrac{1}{2}\lambda_a F^a_{\mu\nu}(x) \;, \end{aligned} \tag{13.51}$$

the full Yang–Mills Lagrangian can be written

$$\boxed{\mathcal{L} = \bar{\psi}(x)\left[\frac{i}{2}\gamma^\mu \overleftrightarrow{D}_\mu - m\right]\psi(x) - \frac{1}{2}\,\mathrm{tr}\,\left[\mathbf{F}_{\mu\nu}(x)\mathbf{F}^{\mu\nu}(x)\right]\ ,} \tag{13.52}$$

where the fermion part of the Lagrangian is explicitly

$$\begin{aligned}\bar{\psi}(x)\left[\frac{i}{2}\gamma^\mu \overleftrightarrow{D}_\mu - m\right]\psi(x) &= \bar{\psi}(x)\left[\frac{i}{2}\gamma^\mu \overleftrightarrow{\partial}_\mu - g\gamma^\mu \mathbf{A}_\mu(x) - m\right]\psi(x) \\ &= \bar{\psi}\left[\frac{i}{2}\gamma^\mu \overleftrightarrow{\partial}_\mu - g\gamma^\mu \tfrac{1}{2}\lambda_a A^a_\mu(x) - m\right]\psi(x)\end{aligned} \tag{13.53}$$

and the gauge field tensor has components

$$F^a_{\mu\nu} = \partial_\mu A^a_\nu - \partial_\nu A^a_\mu - g f_{abc} A^b_\mu A^c_\nu\ . \tag{13.54}$$

These expressions are generalizations of Eqs. (13.35), (13.39), and (13.48).

Yang–Mills theories have a number of special properties:

• For the $SU(n)$ gauge group, each fermion (f) has n internal degrees of freedom, and is coupled to $n^2 - 1$ vector gauge fields (g). From (13.53) the coupling of the gauge fields to the fermions has the form

$$\boxed{\textit{ffg coupling}\quad \mathcal{L}_{ffg} = -g\,\bar{\psi}\gamma^\mu \tfrac{1}{2}\lambda_a\,\psi\, A^a_\mu\ .} \tag{13.55}$$

• The part of the Lagrangian which describes the gauge fields (the $F^i_{\mu\nu}F^{i\mu\nu}$ piece) includes self-couplings of these fields, and, by itself, is a non-trivial theory (sometimes referred to as a *pure* Yang–Mills theory). Expanding out the gauge field part gives

$$\mathcal{L}_{\text{field}} = -\tfrac{1}{2}\partial^\mu A^{a\nu}\left(\partial_\mu A^a_\nu - \partial_\nu A^a_\mu\right) + \mathcal{L}_3 + \mathcal{L}_4\ , \tag{13.56}$$

where the two types of self-couplings are

$$\boxed{\begin{aligned} &\textit{3g coupling}\quad \mathcal{L}_3 = \ \ \tfrac{1}{2} g\, f_{abc}\, A^{b\mu} A^{c\nu}\left(\partial_\mu A^a_\nu - \partial_\nu A^a_\mu\right) \\ &\textit{4g coupling}\quad \mathcal{L}_4 = -\tfrac{1}{4} g^2\, f_{abe}\, f_{cde}\, A^a_\mu A^b_\nu A^{c\mu} A^{d\nu}\ .\end{aligned}} \tag{13.57}$$

QED is an Abelian theory, and hence there are no self-couplings. *The existence of the gauge self-couplings is a unique consequence of non-Abelian gauge invariance.*

• the operator

$$-ig\,\gamma^{\mu}\,\tfrac{1}{2}\lambda_a$$

at each vertex where a gluon with polarization μ and color a is emitted from or absorbed by a quark.

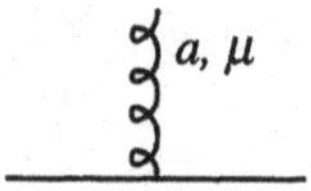

Fig. 13.1a

• the operator

$$g\,f_{abc}\left[g_{\mu\nu}\,(q-k)_{\sigma} + g_{\nu\sigma}\,(r-q)_{\mu} + g_{\sigma\mu}\,(k-r)_{\nu}\right]$$

at each $3g$ vertex, where *all* momenta flow into the vertex and the momenta, spin polarizations, and colors are as shown in Fig. 13.1b.

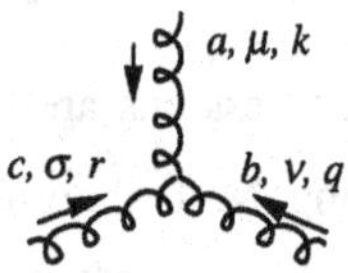

Fig. 13.1b

• the operator

$$-ig^2\Big[f_{abe}f_{cde}\,(g_{\mu\sigma}g_{\nu\rho}-g_{\mu\rho}g_{\nu\sigma}) + f_{ace}f_{bde}\,(g_{\mu\nu}g_{\sigma\rho}-g_{\mu\rho}g_{\nu\sigma}) + f_{ade}f_{cbe}\,(g_{\mu\sigma}g_{\nu\rho}-g_{\mu\nu}g_{\rho\sigma})\Big]$$

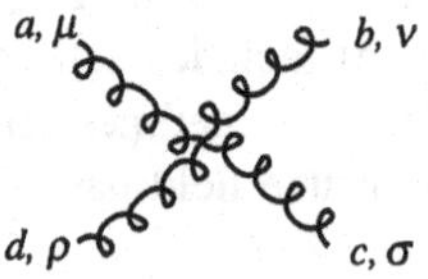

Fig. 13.1c

for each $4g$ vertex, where the spin polarizations and colors are as shown in Fig. 13.1c.

Fig. 13.1 Three of the Feynman rules for QCD. These, and the others, are summarized in Appendix B.

• There is only one coupling constant, which enters into the ffg coupling, the $3g$ coupling, and the $4g$ coupling. These theories will not be gauge invariant if these three couplings take on arbitrary values.

Quantum Chromodynamics

As previously noted, QCD can be defined as the Yang–Mills theory with $SU(3)$ local gauge invariance. The fermions are called quarks, and each *flavor* or type of quark has three internal degrees of freedom called color. There are $n^2 - 1 = 8$ vector gauge fields called gluons.

The interactions between the quarks and gluons in QCD are a special example of those already given in Eqs. (13.55) and (13.57) and suggest the additions to Feynman Rule 1 shown in Fig. 13.1. While these rules are some of the interactions which arise in QCD, it turns out that a complete discussion of the quantization of QCD will lead to several additional Feynman rules, which we are not equipped to introduce now. We will return to this topic in Chapter 15, where a systematic discussion and derivation of the Feynman rules for QCD will be given.

13.4 CHIRAL SYMMETRY

The last of the continuous symmetries which will be discussed in this chapter is chiral symmetry. Unlike the gauge symmetries discussed so far, chiral symmetry is believed to be only an *approximate* symmetry of the strong interactions (but a very good approximate symmetry). A simple model which displays this symmetry, the *sigma model*, will be discussed in the next section. Using this model, we will also discuss *spontaneous symmetry breaking*, a mechanism of great importance in physics. It turns out that a combination of gauge symmetry with spontaneous symmetry breaking is the basis for our present understanding of the electroweak forces.

We begin by considering the following transformations (referred to as chiral transformations) of a fermion field with two internal degrees of freedom (different from color):

$$\psi' = e^{-ig_5\gamma^5 \frac{1}{2}\tau_i\epsilon_i}\psi \; . \tag{13.58}$$

This looks like an $SU(2)$ gauge transformation but is quite different because of the presence of the γ^5. The γ^5 operates on the Dirac components of ψ, while the τ_i operates on the additional two-dimensional space corresponding to two internal degrees of freedom of the fermion, which could be isospin but more generally is referred to as a *flavor* space. In applications, the flavor space is usually isospin, which can be

$$\begin{pmatrix} p \\ n \end{pmatrix} \quad \text{for nucleons} \quad \text{or} \quad \begin{pmatrix} u \\ d \end{pmatrix} \quad \text{for quarks} \; .$$

We will denote the chiral transformation matrix by

$$\mathbf{U}_5 = e^{-ig_5\gamma^5 \frac{1}{2}\tau_i\epsilon_i} \quad . \tag{13.59}$$

Note that $\mathbf{U}_5$ is unitary, but because $\{\gamma^\mu, \gamma^5\} = 0$,

$$\gamma^\mu \mathbf{U}_5 = \mathbf{U}_5^\dagger \gamma^\mu \quad . \tag{13.60}$$

[This can be proved by expanding $\mathbf{U}_5$ in a power series, and noting that $(\gamma^5)^n \gamma^\mu = (-1)^n \gamma^\mu (\gamma^5)^n$, and resuming the power series.] Hence ψ and $\bar{\psi}$ transform in the same way,

$$\begin{aligned} \psi' &= \mathbf{U}_5 \psi \\ \bar{\psi}' &= \psi^\dagger \mathbf{U}_5^\dagger \gamma^0 = \psi^\dagger \gamma^0 \mathbf{U}_5 = \bar{\psi}\mathbf{U}_5 \ , \end{aligned} \tag{13.61}$$

so that under a global transformation, the kinetic energy term is invariant under a chiral transformation

$$\begin{aligned} i\bar{\psi}'\gamma^\mu \frac{\overleftrightarrow{\partial}}{\partial x^\mu}\psi' &= i\bar{\psi}\,\mathbf{U}_5 \gamma^\mu \frac{\overleftrightarrow{\partial}}{\partial x^\mu}\mathbf{U}_5 \psi \\ &= i\bar{\psi}\gamma^\mu \frac{\overleftrightarrow{\partial}}{\partial x^\mu}\mathbf{U}_5^\dagger \mathbf{U}_5 \psi = i\bar{\psi}\gamma^\mu \frac{\overleftrightarrow{\partial}}{\partial x^\mu}\psi \ , \end{aligned} \tag{13.62}$$

but the mass term is not,

$$m\,\bar{\psi}'\psi' = m\,\bar{\psi}\mathbf{U}_5\mathbf{U}_5\psi \neq m\,\bar{\psi}\psi \quad . \tag{13.63}$$

Therefore, in constructing a Lagrangian which is invariant under chiral transformations, one usually begins by assuming that the fermion masses are zero. The presence of a small fermion mass term provides a mechanism for breaking chiral symmetry.

In discussing chiral symmetry, it is customary to also include the usual (global) $SU(2)$ gauge symmetry, which is also a symmetry once we have ψ fields with two internal degrees of freedom (two flavors). Hence the overall symmetry is designated $SU(2) \times SU(2)$; one $SU(2)$ with a γ^5 in the generator and the other without a γ^5. Alternatively, these combined transformations are equivalent to the transformations

$$U_L U_R = e^{-ig_L \mathcal{P}_- \frac{1}{2}\tau_i \epsilon_{iL}} \times e^{-ig_R \mathcal{P}_+ \frac{1}{2}\tau_i \epsilon_{iR}} \ , \tag{13.64}$$

where the operators $\mathcal{P}_\pm$ in the exponents are projection operators,

$$\mathcal{P}_\pm = \tfrac{1}{2}\left(1 \pm \gamma^5\right) \ , \tag{13.65}$$

and the infinitesimal parameters ϵ_L and ϵ_R are independent. This group is designated $SU(2)_L \times SU(2)_R$, with one $SU(2)$ group containing the "left-handed" projection operator $\mathcal{P}_-$ and the other the "right-handed" operator $\mathcal{P}_+$ (recall the discussion of right- and left-handed spinors in Sec. 5.11).

An understanding of the physical meaning of chiral symmetry follows from the combined form (13.64). Chiral invariance says that an $SU(2)$ gauge symmetry can be *independently* realized on the two spaces projected out by the $\mathcal{P}_\pm$ operators; i.e., the gauge transformations on these two subspaces can have different parameters ϵ_R and ϵ_L as written in Eq. (13.64). For this to be true, the helicity of any particle must be conserved by all interactions, and *sufficient conditions* for this to be true are (1) the fermions are massless, so that their helicity cannot be changed by bringing them to rest and reversing their direction of motion, and (2) the interactions do not explicitly flip the spin. We see immediately that helicity conservation places strong restrictions on the interactions. However, in Sec. 13.6 we will show that interactions can be constructed in which the *mass of the fermions is non-zero.*

The conserved current associated with chiral symmetry is found from the infinitesimal transformations, for which

$$\begin{aligned}
\lambda^{\mu}{}_{i} &= 0 \\
\Omega_{\alpha i} &= -ig_5 \tfrac{1}{2}\tau_i \left(\gamma^5\psi\right)_\alpha \\
\bar{\Omega}_{\alpha i} &= -ig_5 \left(\bar{\psi}\gamma^5\right)_\alpha \tfrac{1}{2}\tau_i \quad .
\end{aligned} \tag{13.66}$$

Hence the conserved current is

$$J_5^{i\mu} = -ig_5 \left\{ \frac{\partial \mathcal{L}}{\partial\left(\dfrac{\partial\psi}{\partial x^\mu}\right)} \tfrac{1}{2}\tau_i\gamma^5\psi + \bar{\psi}\tfrac{1}{2}\tau_i\gamma^5 \frac{\partial \mathcal{L}}{\partial\left(\dfrac{\partial\bar{\psi}}{\partial x^\mu}\right)} \right\} \quad . \tag{13.67}$$

For massless fields the fermion Lagrangian is

$$\mathcal{L}_F = \frac{i}{2}\bar{\psi}\gamma^\mu \frac{\overleftrightarrow{\partial}}{\partial x^\mu}\psi$$

and the conserved current is an *axial* current:

$$\boxed{J_5^{i\mu} = g_5\, \bar{\psi}\gamma^\mu\gamma^5 \tfrac{1}{2}\tau_i\psi \quad .} \tag{13.68}$$

The ordinary $SU(2)$ gauge symmetry gave a conserved vector current [recall Eq. (13.21)]

$$J^{i\mu} = g\, \bar{\psi}\gamma^\mu \tfrac{1}{2}\tau_i\psi$$

so that together we have conserved vector and axial vector currents.

13.5 THE LINEAR SIGMA MODEL

We saw in the last section that chiral symmetry seems to imply that the fermions of the theory must be massless. However, this restriction can be eliminated by the construction of a theory in which the fermion mass arises as part of the interaction. There are many ways to do this. One way is to construct a theory in which the fermions start off as massless and then to generate fermion mass through *spontaneous symmetry breaking*. This is the route we will follow in the next two sections. In this section we will construct the *linear sigma model*, which is a theory of massless "nucleons" interacting with mesons, and then, in the next section, discuss the mechanism of spontaneous symmetry breaking and show how this mechanism can generate nucleon mass. Another way to build a chirally invariant theory with massive fermions is to construct the interaction in such a way that mass is included from the start. An example of such a model is the *non-linear sigma model*, which will be discussed in Sec. 13.7. Both the linear and the non-linear sigma models are very useful in understanding the interactions of nucleons with light mesons and in understanding the pion, which plays a special role in the description of the strong interactions at low energies.

We therefore begin by considering the interactions of the Fermi fields with mesons, particularly the pion. To be slightly more general, assume the mesons are self-conjugate scalar or pseudoscalar mesons. The interaction is therefore of the form

$$\mathcal{L}_{\text{int}} = -\bar{\psi}\mathbf{M}\psi \,, \tag{13.69}$$

where $\mathbf{M}$ is a superposition of scalar and pseudoscalar meson submatrices [GL 60]:

$$\mathbf{M} = g_\sigma \boldsymbol{\phi}_\sigma + i g_\pi \gamma^5 \boldsymbol{\phi}_\pi \,, \tag{13.70}$$

where g_σ and g_π are real constants, $\boldsymbol{\phi}_\sigma$ and $\boldsymbol{\phi}_\pi$ are 2×2 Hermitian matrix fields in the flavor space, and the i multiplying the $\boldsymbol{\phi}_\pi$ term comes from the fact that hermiticity implies

$$\mathcal{L}_{\text{int}}^\dagger = -\psi^\dagger \mathbf{M}^\dagger \gamma^0 \psi = -\bar{\psi}\mathbf{M}\psi \tag{13.71}$$

so that

$$\gamma^0 \mathbf{M}^\dagger \gamma^0 = \mathbf{M} \ . \tag{13.72}$$

It is the fact that $\mathbf{M}$ depends linearly on the meson fields which is the origin of the term *linear* sigma model.

Next, we determine the transformation laws of the meson field matrix from the requirement that $SU(2) \times SU(2)$ be a good symmetry of the interaction. The transformations in the $SU(2)$ gauge group will be denoted $\mathbf{U}$ and can be written [recall Eq. (13.17)]

$$\mathbf{U} = e^{-ig\frac{1}{2}\tau_i \epsilon_i} = e^{-ig\frac{1}{2}\boldsymbol{\mathcal{E}}} \,,$$

where in this section we adopt a *new convention* $\boldsymbol{\mathcal{E}} = \tau_i \epsilon_i$. Symmetry under the gauge group implies that

$$\mathbf{M}' = \mathbf{U}\mathbf{M}\mathbf{U}^\dagger \,, \tag{13.73}$$

which gives

$$\boldsymbol{\phi}'_\sigma = \mathbf{U}\boldsymbol{\phi}_\sigma\mathbf{U}^\dagger \qquad \boldsymbol{\phi}'_\pi = \mathbf{U}\boldsymbol{\phi}_\pi\mathbf{U}^\dagger \ . \tag{13.74}$$

Hence $\boldsymbol{\phi}_\sigma$ and $\boldsymbol{\phi}_\pi$ can be expanded in terms of $\mathbf{1}$ and the $SU(2)$ generators. It turns out to be sufficient to choose $\boldsymbol{\phi}_\sigma$ *to be pure isoscalar and* $\boldsymbol{\phi}_\pi$ *to be pure isovector*, so that

$$\boldsymbol{\phi}_\sigma = 1\ \phi_\sigma \ , \qquad \boldsymbol{\phi}_\pi = \tau_i\phi^i_\pi \ . \tag{13.75}$$

[This is the reason for the names σ and π. It is also possible to make the Dirac scalar interaction pure isovector and the Dirac pseudoscalar interaction pure isoscalar, but this will not be discussed here.]

Now, examine the implications of chiral symmetry. Require

$$\mathcal{L}'_{\rm int} = -\bar{\psi}'\mathbf{M}'\psi' = -\bar{\psi}\,\mathbf{U}_5\mathbf{M}'\mathbf{U}_5\psi = -\bar{\psi}\mathbf{M}\psi = \mathcal{L}_{\rm int} \ . \tag{13.76}$$

Hence, under the chiral transformation $\mathbf{U}_5$ we require

$$\mathbf{M}' = \mathbf{U}_5^\dagger\mathbf{M}\mathbf{U}_5^\dagger \ . \tag{13.77}$$

Using $\left(\gamma^5\right)^2 = 1$, $\boldsymbol{\epsilon}^2 = \epsilon_i\epsilon_i$, and denoting $|\epsilon| = \sqrt{\epsilon_i\epsilon_i}$, $\hat{\boldsymbol{\epsilon}} = \boldsymbol{\epsilon}/|\epsilon|$, and $\theta = \frac{1}{2}g_5|\epsilon|$, we can obtain a compact form for $\mathbf{U}_5$,

$$\begin{aligned}\mathbf{U}_5 &= 1 - i\gamma^5\,\tfrac{1}{2}g_5\boldsymbol{\epsilon} - \tfrac{1}{2}\left(\tfrac{1}{2}g_5\boldsymbol{\epsilon}\right)^2 + \tfrac{1}{6}i\gamma^5\left(\tfrac{1}{2}g_5\boldsymbol{\epsilon}\right)^3 + \ldots \\ &= \cos\theta - i\gamma^5\hat{\boldsymbol{\epsilon}}\sin\theta \ .\end{aligned} \tag{13.78}$$

Hence, the transformation law (13.77) can be written

$$\begin{aligned}g_\sigma\phi'_\sigma + ig_\pi\gamma^5\boldsymbol{\phi}'_\pi &= \left(\cos\theta + i\gamma^5\hat{\boldsymbol{\epsilon}}\,\sin\theta\right)\left(g_\sigma\phi_\sigma + ig_\pi\gamma^5\boldsymbol{\phi}_\pi\right)\left(\cos\theta + i\gamma^5\hat{\boldsymbol{\epsilon}}\,\sin\theta\right) \\ &= g_\sigma\phi_\sigma\left[\cos 2\theta + i\gamma^5\hat{\boldsymbol{\epsilon}}\,\sin 2\theta\right] \\ &\quad + ig_\pi\gamma^5\left[\boldsymbol{\phi}_\pi\cos^2\theta - \hat{\boldsymbol{\epsilon}}\boldsymbol{\phi}_\pi\hat{\boldsymbol{\epsilon}}\sin^2\theta\right] - g_\pi\tfrac{1}{2}\sin 2\theta\left\{\hat{\boldsymbol{\epsilon}},\boldsymbol{\phi}_\pi\right\} \ .\end{aligned} \tag{13.79}$$

Using

$$\hat{\boldsymbol{\epsilon}}\boldsymbol{\phi}_\pi\hat{\boldsymbol{\epsilon}} = 2\left(\boldsymbol{\phi}_\pi\cdot\hat{\epsilon}\right)\hat{\boldsymbol{\epsilon}} - \boldsymbol{\phi}_\pi \ ,$$

where $\boldsymbol{\phi}_\pi = (\phi^1_\pi, \phi^2_\pi, \phi^3_\pi)$ is the three-vector in isospin space (not to be confused with the matrix $\boldsymbol{\phi}$), gives

$$\begin{aligned}g_\sigma\phi'_\sigma + ig_\pi\gamma^5\boldsymbol{\phi}'_\pi &= g_\sigma\phi_\sigma\cos 2\theta - g_\pi\left(\boldsymbol{\phi}_\pi\cdot\hat{\epsilon}\right)\sin 2\theta \\ &\quad + i\gamma^5\left[g_\sigma\phi_\sigma\hat{\boldsymbol{\epsilon}}\,\sin 2\theta + g_\pi\boldsymbol{\phi}_\pi - 2g_\pi\left(\boldsymbol{\phi}_\pi\cdot\hat{\epsilon}\right)\hat{\boldsymbol{\epsilon}}\,\sin^2\theta\right] .\end{aligned} \tag{13.80}$$

Substituting $2\sin^2\theta = 1 - \cos 2\theta$ into the last term gives finally

$$\begin{aligned}\phi'_\sigma &= \phi_\sigma\cos 2\theta - \left(\frac{g_\pi}{g_\sigma}\right)\left(\boldsymbol{\phi}_\pi\cdot\hat{\epsilon}\right)\sin 2\theta \\ \phi'^i_\pi &= \left[\phi^i_\pi - \left(\boldsymbol{\phi}_\pi\cdot\hat{\epsilon}\right)\,\hat{\epsilon}_i\right] \\ &\quad + \hat{\epsilon}_i\left[\left(\frac{g_\sigma}{g_\pi}\right)\phi_\sigma\sin 2\theta + \left(\boldsymbol{\phi}_\pi\cdot\hat{\epsilon}\right)\cos 2\theta\right] \ .\end{aligned} \tag{13.81}$$

The transformation is therefore a rotation in four-dimensional space through angle 2θ in the plane defined by ϕ_σ and $\hat{\epsilon}$. The components of ϕ_π perpendicular to $\hat{\epsilon}$ are unchanged.

The transformation above looks very complicated but is much simpler when written out to first order in ϵ. It becomes

$$\begin{aligned} \phi'_\sigma &= \phi_\sigma - g_5\left(\frac{g_\pi}{g_\sigma}\right)(\phi^i_\pi\epsilon_i) \\ \phi'^i_\pi &= \phi^i_\pi + g_5\left(\frac{g_\sigma}{g_\pi}\right)\phi_\sigma\,\epsilon_i \end{aligned} \qquad \textit{infinitesimal.} \tag{13.82}$$

Note that the chiral rotation mixes the σ and one of the π components of the four-dimensional field vector $(\phi_\sigma, \phi^i_\pi)$, and hence, in the linear sigma model, both fields must be present in order to have chiral invariance.

Now, examine the kinetic energy part of the Lagrangian for the π and σ fields. Since these fields are self-conjugate, the KE term must have the form

$$\mathcal{L}_{\mathrm{KE}} = \tfrac{1}{2}\partial_\mu\phi_\sigma\partial^\mu\phi_\sigma + \tfrac{1}{2}\partial_\mu\phi^i_\pi\partial^\mu\phi^i_\pi \ . \tag{13.83}$$

This is clearly invariant under the gauge group $SU(2)$. Invariance under an infinitesimal chiral transformation requires

$$\begin{aligned} \mathcal{L}'_{\mathrm{KE}} &= \tfrac{1}{2}\left[\partial_\mu\phi'_\sigma\partial^\mu\phi'_\sigma + \partial_\mu\phi'^i_\pi\partial^\mu\phi'^i_\pi\right] \\ &= \mathcal{L}_{\mathrm{KE}} + g_5\left[\frac{g_\sigma}{g_\pi}\left(\partial_\mu\phi^i_\pi\right)\partial^\mu\left(\phi_\sigma\,\epsilon_i\right) - \frac{g_\pi}{g_\sigma}\left(\partial_\mu\phi_\sigma\right)\partial^\mu\left(\phi^i_\pi\,\epsilon_i\right)\right] \ . \end{aligned} \tag{13.84}$$

Since ϵ_i is a constant, the terms proportional to g_5 have the same structure and

$$\Delta\mathcal{L}_{\mathrm{KE}} = g_5\left(\frac{g_\sigma}{g_\pi} - \frac{g_\pi}{g_\sigma}\right)\left[\partial_\mu\phi^i_\pi\,\partial^\mu\phi_\sigma\right]\epsilon_i \ . \tag{13.85}$$

Hence these will cancel, and $\mathcal{L}_{\mathrm{KE}}$ will be invariant if $g_\sigma = g_\pi$. Furthermore, as we did in our discussion of gauge invariance, we will take $g_5 = g_\pi$, so that there is *only one coupling*,

$$g_\sigma = g_5 = g_\pi \ . \tag{13.86}$$

The meson matrix can therefore be written $\mathbf{M} = g_\pi\,\boldsymbol{\phi}$, where the new, simplified meson matrix is

$$\boxed{\boldsymbol{\phi} = \phi_\sigma + i\gamma_5\tau_i\phi^i_\pi \ .} \tag{13.87}$$

The trace of $\boldsymbol{\phi}\boldsymbol{\phi}^\dagger$ in flavor space can be used to define a chirally invariant length $|\boldsymbol{\phi}|$,

$$\begin{aligned} \tfrac{1}{2}\,\mathrm{tr}\,(\boldsymbol{\phi}\,\boldsymbol{\phi}^\dagger) &= \tfrac{1}{2}\,\mathrm{tr}\left(\left[\phi_\sigma + i\gamma^5\tau_i\phi^i_\pi\right]\left[\phi_\sigma - i\gamma^5\tau_i\phi^i_\pi\right]\right) \\ &= \phi^2_\sigma + \phi^i_\pi\phi^i_\pi = |\boldsymbol{\phi}|^2 \ , \end{aligned} \tag{13.88}$$

and also leads to a compact notation for the KE term,

$$\mathcal{L}_{\rm KE} = \tfrac{1}{4}\,{\rm tr}\left(\partial_\mu \boldsymbol{\phi}\, \partial^\mu \boldsymbol{\phi}^\dagger\right) \,, \tag{13.89}$$

which makes it obvious that this term is invariant under chiral transformations

$$\begin{aligned}\mathcal{L}'_{\rm KE} &= \tfrac{1}{4}\,{\rm tr}\left(\partial_\mu \boldsymbol{\phi}'\, \partial^\mu \boldsymbol{\phi}'^\dagger\right) \\ &= \tfrac{1}{4}\,{\rm tr}\left(\partial_\mu \left(\mathbf{U}_5^\dagger \boldsymbol{\phi} \mathbf{U}_5^\dagger\right) \partial^\mu \left(\mathbf{U}_5 \boldsymbol{\phi}^\dagger \mathbf{U}_5\right)\right) \\ &= \tfrac{1}{4}\,{\rm tr}\left(\mathbf{U}_5^\dagger\, \partial_\mu \boldsymbol{\phi}\, \partial^\mu \boldsymbol{\phi}^\dagger\, \mathbf{U}_5\right) = \mathcal{L}_{\rm KE}\end{aligned} \tag{13.90}$$

because $\mathbf{U}_5$ is independent of x. Using this notation, we can write the chirally invariant Lagrangian for the linear sigma model in a very compact form:

$$\boxed{\begin{aligned}\mathcal{L} = &\tfrac{1}{2}\bar{\psi}\, i\gamma^\mu \frac{\overleftrightarrow{\partial}}{\partial x^\mu}\psi - g_\pi \bar{\psi}\, \boldsymbol{\phi}\, \psi \\ &+ \tfrac{1}{4}\,{\rm tr}\left(\partial_\mu \boldsymbol{\phi}\, \partial^\mu \boldsymbol{\phi}^\dagger\right) - \tfrac{1}{2} m^2 |\boldsymbol{\phi}|^2 + V\left(|\boldsymbol{\phi}|^2\right) \,,\end{aligned}} \tag{13.91}$$

where m is the meson mass and V is some "potential" function which describes the meson–meson self-interactions in a chirally invariant way. A simple choice for the self-interaction potential is

$$V\left(|\boldsymbol{\phi}|^2\right) = -\tfrac{1}{4}\lambda^2 |\boldsymbol{\phi}|^4 \,, \tag{13.92}$$

where $\lambda^2 > 0$ in order to have a stable vacuum (see the next section).

Using the simplification (13.86), we summarize the results. We have found that $SU(2) \times SU(2)$ symmetry leads to a Lagrangian of the form Eq. (13.91) with the meson matrix Eq. (13.87). This Lagrangian contains three parameters: one fermion–meson coupling constant g_π, a meson mass m, and a ϕ^4 type meson–meson interaction strength λ^2. The meson matrix $\boldsymbol{\phi}$ can be expanded in terms of four real fields and satisfies the following global transformation laws:

$$\boxed{\begin{aligned}\boldsymbol{\phi}' &= \mathbf{U}\, \boldsymbol{\phi}\, \mathbf{U}^\dagger \\ \boldsymbol{\phi}' &= \mathbf{U}_5^\dagger\, \boldsymbol{\phi}\, \mathbf{U}_5^\dagger \,.\end{aligned}} \tag{13.93}$$

In terms of the four real fields, the explicit form of the finite chiral $\mathbf{U}_5$ transformation is

$$\boxed{\begin{aligned}\phi'_\sigma &= \phi_\sigma \cos 2\theta - (\boldsymbol{\phi}_\pi \cdot \hat{\epsilon}) \sin 2\theta \\ \phi'^{i}_\pi &= \left[\phi_\sigma \sin 2\theta + (\boldsymbol{\phi}_\pi \cdot \hat{\epsilon}) \cos 2\theta\right] \hat{\epsilon}_i + \left[\phi^i_\pi - (\boldsymbol{\phi}_\pi \cdot \hat{\epsilon})\, \hat{\epsilon}_i\right] \,,\end{aligned}} \tag{13.94}$$

and of the infinitesimal chiral transformation is

$$\boxed{\begin{aligned} \phi'_\sigma &= \phi_\sigma - g_\pi\, \boldsymbol{\phi}_\pi \cdot \boldsymbol{\epsilon} \\ \phi'^i_\pi &= \phi^i_\pi + g_\pi\, \phi_\sigma\, \epsilon_i \end{aligned} \qquad \textit{infinitesimal.}} \tag{13.95}$$

The conserved axial vector current associated with this Lagrangian now includes two additional terms from the meson fields:

$$\boxed{J_5^{i\mu} = g_\pi\, \bar{\psi}\gamma^\mu\gamma^5 \tfrac{1}{2}\tau_i\psi - g_\pi\left(\partial^\mu\phi_\sigma\right)\phi^i_\pi + g_\pi\left(\partial^\mu\phi^i_\pi\right)\phi_\sigma \ ,} \tag{13.96}$$

where the last two terms come from

$$\begin{aligned} \Omega_{\sigma i} &= -g_\pi \phi^i_\pi && \text{for} \quad \phi_\sigma \\ \Omega_{ji} &= +g_\pi \phi_\sigma \delta_{ij} && \text{for} \quad \phi^j_\pi \ . \end{aligned} \tag{13.97}$$

We now turn to a discussion of spontaneous symmetry breaking.

13.6 SPONTANEOUS SYMMETRY BREAKING

If the linear sigma model is to be applied to the study of nucleons, as is the intention, it is necessary to find some way to produce a significant fermion mass. This can be done by a mechanism called spontaneous symmetry breaking [Go 61, NJ 61], in which the symmetry is broken by the vacuum and *not* by the Lagrangian. This will produce a fermion mass. This mechanism is of great importance in modern physics; in particular, it is the mechanism by which gauge bosons $W^\pm$ and Z_0 acquire mass, and this discussion therefore has quite wide validity.

First, look at the classical limit of the meson Hamiltonian obtained from the linear sigma model. In the long wave length limit, when $\partial_\mu \phi \simeq 0$, the minimum energy for the mesons will occur when

$$H_\phi = \tfrac{1}{2} m^2 |\boldsymbol{\phi}|^2 + \tfrac{1}{4}\lambda^2 |\boldsymbol{\phi}|^4 \tag{13.98}$$

is a minimum. (Note that λ^2 must be positive in order that H_ϕ have a well defined minimum, and this is the reason for the requirement $\lambda^2 > 0$ alluded to in the last section.) The H_ϕ surface is a quartic with a minimum at

$$\frac{dH_\phi}{d|\boldsymbol{\phi}|} = m^2 |\boldsymbol{\phi}| + \lambda^2 |\boldsymbol{\phi}|^3 = 0 \ . \tag{13.99}$$

If the meson masses are physical, so that $m^2 \geq 0$, this equation has only one solution, and the minimum is at $|\boldsymbol{\phi}| = 0$ [see Fig. 13.2A]. Recalling that

$$|\boldsymbol{\phi}| = \sqrt{\phi_\sigma^2 + \boldsymbol{\phi}_\pi \cdot \boldsymbol{\phi}_\pi} \ ,$$

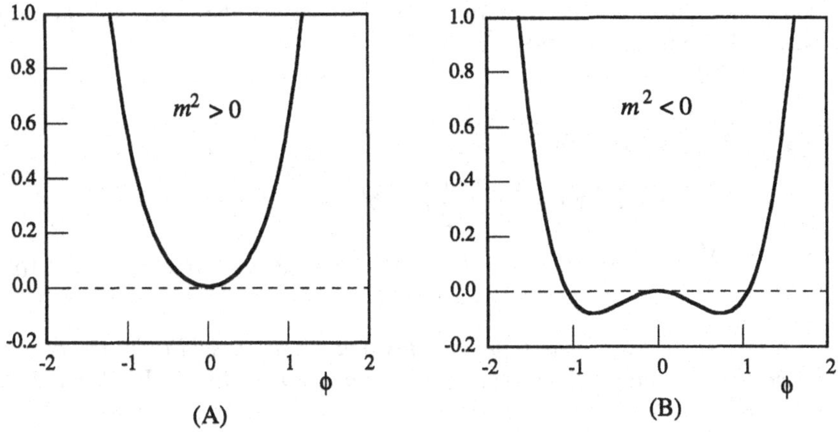

Fig. 13.2 The Hamiltonian H_ϕ as a function of ϕ for two signs of the mass term.

the requirement that $|\boldsymbol{\phi}| = 0$ automatically implies that the value of *all* the classical fields in the minimum energy state (the vacuum) is zero:

$$\phi_\sigma = \phi_\pi = 0 \qquad \text{(for the vacuum)} \ . \tag{13.100}$$

This is in agreement with our intuition that the vacuum should have no σ or π particles present.

Consider now the interesting case when $m^2 < 0$, so that $-m^2 > 0$. This appears to correspond to choosing an unphysical meson mass, but as we shall soon see, this is not really the case. In this case, the H_ϕ curve now has a minimum for a finite value of $|\boldsymbol{\phi}| = v$, determined from the solution to Eq. (13.99):

$$|\boldsymbol{\phi}| = \sqrt{-\frac{m^2}{\lambda^2}} = v \ . \tag{13.101}$$

This case is illustrated in Fig. 13.2B. The Lagrangian is still mathematically invariant under chiral transformations, but the vacuum no longer is. This is because $|\boldsymbol{\phi}| = v$ implies that at least one of the four fields $(\phi_\sigma, \phi_\pi^i)$ must be non-zero, and hence, in general, we cannot treat all four components of $\boldsymbol{\phi}$ in a symmetric way. The choice $m^2 < 0$ forces the vacuum to break the chiral symmetry *spontaneously*, and this is the origin of the term *spontaneous symmetry breaking*.

We choose ϕ_σ to be non-zero and introduce a new field which is

$$s = \phi_\sigma - v \ . \tag{13.102}$$

Note that $s = 0$ for the vacuum state, corresponding to the intuitive notion that the vacuum contains no s particles. Hence s, and not ϕ_σ, is the appropriate field to use if we want the Lagrangian to describe the interactions of physical particles, which are not present in the vacuum. Since v is constant, the Lagrangian (13.91), expressed in terms of the field s, becomes

$$\begin{aligned}\mathcal{L}_s = & \frac{1}{2}\bar{\psi} i\gamma^\mu \frac{\overleftrightarrow{\partial}}{\partial x^\mu}\psi - g_\pi v \bar{\psi}\psi \\ & - g_\pi \bar{\psi}\left[s + i\gamma^5 \boldsymbol{\tau}\cdot\boldsymbol{\phi}_\pi\right]\psi + \frac{1}{2}\left[\partial_\mu s \partial^\mu s + \partial_\mu \phi_\pi^i \partial^\mu \phi_\pi^i\right] \\ & - \frac{\lambda^2}{4}\left[s^2 + \phi_\pi^2\right]^2 - \lambda^2 v s\left[s^2 + \phi_\pi^2\right] - \lambda^2 v^2 s^2 + \text{constant} \ . \end{aligned} \tag{13.103}$$

Note that *mass terms for the fermion fields and for the s field (with the correct sign) have been generated but that the pion mass term vanishes.* Introducing the parameters

$$M = g_\pi v \qquad m_\sigma^2 = 2\lambda^2 v^2 \tag{13.104}$$

and dropping the constant term, which contains no physics, give a new Lagrangian with physically interesting parameters:

$$\boxed{\begin{aligned}\mathcal{L}_s = & \frac{1}{2}\bar{\psi} i\gamma^\mu \frac{\overleftrightarrow{\partial}}{\partial x^\mu}\psi - M\bar{\psi}\psi - g_\pi \bar{\psi}\left[s + i\gamma^5 \boldsymbol{\tau}\cdot\boldsymbol{\phi}_\pi\right]\psi \\ & + \frac{1}{2}\left[\partial_\mu s \partial^\mu s - m_\sigma^2 s^2\right] + \frac{1}{2}\left[\partial_\mu \phi_\pi^i \partial^\mu \phi_\pi^i\right] \\ & - \frac{g_\pi^2 m_\sigma^2}{8M^2}\left(s^2 + \phi_\pi^2\right)^2 - \frac{g_\pi m_\sigma^2}{2M} s\left[s^2 + \phi_\pi^2\right] \ . \end{aligned}} \tag{13.105}$$

Remarks

(i) This Lagrangian is still chirally invariant (only the vacuum breaks the symmetry). Note that the mass of the pion is zero. There turns out to be a deep reason for this. Whenever a symmetry is spontaneously broken, a massless boson is always left behind. These are referred to as *Goldstone bosons* [Go 61]. The fact that the pion mass is so small is often believed to reflect the fact that chiral symmetry is "almost" an exact symmetry.

(ii) The above Lagrangian has many desirable features for nuclear physics. It has a nucleon mass and a reasonable πNN coupling. The pion mass can be made non-zero by breaking the symmetry slightly. The principal problem is that it has a sigma meson, which is not observed in nature. But this problem can also be solved by defining a new pion field. We turn to these issues now.

13.7 THE NON-LINEAR SIGMA MODEL

The linear sigma model requires both a pion and a scalar (sigma) field in order to maintain chiral symmetry. The presence of the scalar field makes the model less than satisfactory for low energy applications, because, at low energy, scalar particles exist only as very broad resonances with masses in the neighborhood of 1 GeV, very much larger that the pion mass. It does not seem natural for such particles to play the fundamental role suggested by the linear sigma model, and in this section we will discuss how the sigma model can be reformulated so that it contains *no* sigma mesons.

The most obvious idea on how to eliminate the sigma meson is to let the σ mass become infinite (since it is a free parameter); we would then hope that all contributions from σ exchange forces would be vanishingly small. For example, the Lagrangian (13.105) predicts a σ exchange force in NN scattering which goes like

$$\mathcal{M}_{NN} \sim -\frac{g_\pi^2}{m_\sigma^2 - q^2} \Longrightarrow 0 \qquad (\text{as } m_\sigma^2 \to \infty) \ . \tag{13.106}$$

However, this idea does not work because the sigma does not decouple from the pion. This is because the $\pi\pi s$ (or σ) coupling also is proportional to m_σ^2, so the σ exchange force between pions cannot be ignored:

$$\mathcal{M}_{\pi\pi} \sim \left(\frac{g_\pi m_\sigma^2}{2M}\right)^2 \frac{1}{m_\sigma^2 - q^2} \Longrightarrow \frac{g_\pi^2 m_\sigma^2}{4M^2} \Longrightarrow \infty \qquad (\text{as } m_\sigma^2 \to \infty) \ . \tag{13.107}$$

Since the sigma couples to the pion, and the pion to the nucleon, it cannot be eliminated from the Lagrangian (13.105).

We can eliminate the sigma, however, if we are willing to consider a meson matrix $\boldsymbol{\phi}$ which is *non-linear* in the pion fields. This approach was developed by Weinberg in the late 1960's [We 68]. The idea is to exploit the fact that the length $|\boldsymbol{\phi}|$ is invariant under chiral rotations, and hence if we set it equal to a *constant* (which is denoted f_π), we still preserve chiral symmetry, but ϕ_σ is no longer an independent field but is related to ϕ_π by

$$f_\pi^2 = |\boldsymbol{\phi}|^2 = \phi_\sigma^2 + \phi_\pi^2 \ . \tag{13.108}$$

With this constraint, we can construct a Lagrangian which depends on the pion field only, but the replacement of ϕ_σ by a function of ϕ_π will give a meson matrix *non-linear* in the pion field. The non-linearities are quite severe; the Lagrangian will include *all powers* of the pion field.

Theories based on a non-linear Lagrangian of the type discussed in this section cannot be quantized using the methods we have discussed in the preceding chapters and require the techniques we will discuss in Chapter 14. Generally, these Lagrangians give simple reliable results when used in tree approximation, but the calculation of loops with such a Lagrangian leads to many infinities which can only be removed by introducing many renormalization constants. Recently a systematic method for absorbing these renormalization constants into undetermined parameters (known as *chiral perturbation theory* [DW 69]) has been developed. It can be used to calculate the interactions of very low energy pions and nucleons with considerable reliability.

Once the idea of a non-linear Lagrangian is accepted, we have great freedom, and a new pion field $\boldsymbol{\pi}$ can be chosen so as to give a convenient form for the Lagrangian. Using the notation

$$\mathbf{y} = \frac{\boldsymbol{\tau}\cdot\boldsymbol{\pi}}{2f_\pi} = \boldsymbol{\tau}\cdot\boldsymbol{y} \ , \tag{13.109}$$

two forms which appear frequently in the literature are

$$\begin{aligned} \Phi_1 &= f_\pi \, e^{2i\gamma^5 \mathbf{y}} \\ \Phi_2 &= f_\pi \left(\frac{1 - y^2 + 2i\gamma^5 \mathbf{y}}{1+y^2}\right) \ , \end{aligned} \tag{13.110}$$

where $\mathbf{y}^2 = \boldsymbol{y}\cdot\boldsymbol{y} = y^2$. Both of these forms satisfy the constraint (13.108) and, up to *second order* in the pion field, are equal to

$$\Phi = f_\pi + i\gamma^5\boldsymbol{\tau}\cdot\boldsymbol{\pi} - \frac{\pi^2}{2f_\pi} \ , \tag{13.111}$$

in agreement with Eq. (13.87) and the constraint (13.108) if we identify $\phi_\pi \cong \pi$. Hence the two fields ϕ_π and π differ only in higher order.

Starting from the linear sigma model Lagrangian (13.91), we consider the Lagrangian

$$\mathcal{L}_2 = \tfrac{1}{2}\bar{\psi}\, i\gamma^\mu \frac{\overleftrightarrow{\partial}}{\partial x^\mu}\psi - g_\pi \bar{\psi}\,\Phi_2\,\psi + \tfrac{1}{4}\,\mathrm{tr}\left(\partial_\mu\Phi_2\,\partial^\mu\Phi_2^\dagger\right) \ . \tag{13.112}$$

Note that this Lagrangian is identical to (13.91) except that terms involving $|\Phi_2|^2$ have been dropped *because* $|\Phi_2|^2 = f_\pi^2$ *is a constant* and is no longer a dynamical variable and cannot be used to construct the Lagrangian.

Next we simplify this Lagrangian by reducing the meson kinetic energy term. Note that

$$\Phi_2 = f_\pi + 2f_\pi\left(\frac{-y^2 + i\gamma^5\mathbf{y}}{1+y^2}\right)$$

and, using $\partial_\mu y^2 = 2\boldsymbol{y}\cdot\partial_\mu\boldsymbol{y}$, we have

$$\begin{aligned} \partial_\mu\Phi_2 &= 2f_\pi\left(\frac{-2\boldsymbol{y}\cdot\partial_\mu\boldsymbol{y} + i\gamma^5\,\partial_\mu\mathbf{y}}{1+y^2}\right) - 2f_\pi\,\frac{(-y^2 + i\gamma^5\mathbf{y})}{(1+y^2)^2}2\boldsymbol{y}\cdot\partial_\mu\boldsymbol{y} \\ &= -2f_\pi\,\frac{(1+i\gamma^5\mathbf{y})}{(1+y^2)^2}2\boldsymbol{y}\cdot\partial_\mu\boldsymbol{y} + 2f_\pi\left(\frac{i\gamma^5\,\partial_\mu\mathbf{y}}{1+y^2}\right) \ . \end{aligned} \tag{13.113}$$

Hence, recalling that $\text{tr}(\mathbf{y}\,\partial_\mu \mathbf{y}) = 2\mathbf{y}\cdot\partial_\mu \mathbf{y}$, we have

$$\begin{aligned}\tfrac{1}{2}\,\text{tr}\,(\partial_\mu \Phi_2\, \partial^\mu \Phi_2) =& 4f_\pi^2 \frac{(2\mathbf{y}\cdot\partial_\mu \mathbf{y})(2\mathbf{y}\cdot\partial^\mu \mathbf{y})}{(1+y^2)^3} - 4f_\pi^2 \frac{(2\mathbf{y}\cdot\partial_\mu \mathbf{y})(2\mathbf{y}\cdot\partial^\mu \mathbf{y})}{(1+y^2)^3} \\ &+ 4f_\pi^2 \frac{\partial_\mu \mathbf{y}\cdot\partial^\mu \mathbf{y}}{(1+y^2)^2} \\ =& \frac{\partial_\mu \boldsymbol{\pi}\cdot\partial^\mu \boldsymbol{\pi}}{(1+y^2)^2} \quad . \end{aligned} \tag{13.114}$$

With this simplification, the Lagrangian (13.112) can be written

$$\boxed{\begin{aligned}\mathcal{L}_2 = \bar{\psi}\left[\frac{i}{2}\gamma^\mu \frac{\overleftrightarrow{\partial}}{\partial x^\mu} - M\right]\psi + \frac{1}{1+y^2}\bar{\psi}\left(\frac{g_\pi^2 \pi^2}{2M} - ig_\pi \gamma^5 \boldsymbol{\tau}\cdot\boldsymbol{\pi}\right)\psi \\ + \frac{1}{2}\left(\frac{1}{1+y^2}\right)^2 \partial_\mu \boldsymbol{\pi}\cdot\partial^\mu \boldsymbol{\pi} \ ,\end{aligned}} \tag{13.115}$$

where, as for the linear sigma model with spontaneous symmetry breaking,

$$M = g_\pi f_\pi \quad . \tag{13.116}$$

The chirally symmetric Lagrangian (13.115) has massive fermions right from the start. However, the pion mass is zero. In some respects this is a reasonable approximation for low energy interactions, because the mass of the pion is so much smaller that the masses of any other *strongly* interacting particle. In many other respects, however, the pion mass should not be neglected, and we will see in the next section how a pion mass term may be added and discuss the consequences of adding such a term. It turns out that, because $|\Phi|^2$ = constant, we cannot add a pion mass term without breaking chiral symmetry.

At low energies it is a good approximation to neglect the higher order terms in the Lagrangian. Keeping terms up to order $y^2 \approx \pi^2$ only, the Lagrangian becomes

$$\mathcal{L}_2' = \bar{\psi}\left[\frac{i}{2}\gamma^\mu \frac{\overleftrightarrow{\partial}}{\partial x^\mu} - M + \frac{g_\pi^2 \pi^2}{2M} - ig_\pi \gamma^5 \boldsymbol{\tau}\cdot\boldsymbol{\pi}\right]\psi + \frac{1}{2}\left(\partial_\mu \boldsymbol{\pi}\cdot\partial^\mu \boldsymbol{\pi}\right) \ . \tag{13.117}$$

The inclusion of the $g_\pi^2 \pi^2 \bar{\psi}\psi/2M$ term in this Lagrangian corrects some of the bad features of the elementary γ^5-type πNN coupling (see Prob. 13.3). It is what remains of the contributions from the original σ meson, and gives rise to a new "contact" interaction which contributes to π-N scattering (see Fig. 13.3). When this new contact interaction is added to the direct and exchange nucleon pole terms (Prob. 10.8) chiral symmetry is restored and a good description of low energy π-N scattering obtained.

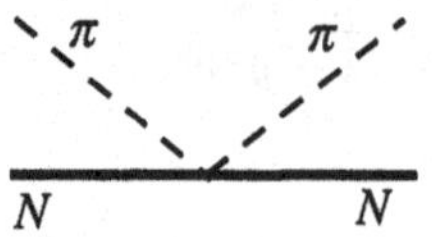

Fig. 13.3 Additional Feynman diagram corresponding to the $\pi\pi NN$ contact term.

We conclude this section by finding the form taken by the infinitesimal chiral transformation, Eq. (13.95), when applied to the non-linear sigma model, and the form of the conserved axial vector current. To find the infinitesimal chiral transformation, apply (13.95) to the meson matrix ϕ_2, giving

$$\begin{aligned} \frac{1-y'^2}{1+y'^2} &= \frac{1-y^2}{1+y^2} - g_\pi\left(\frac{2\mathbf{y}\cdot\boldsymbol{\epsilon}}{1+y^2}\right) \\ \frac{2y'^i}{1+y'^2} &= \frac{2y^i}{1+y^2} + g_\pi\left(\frac{1-y^2}{1+y^2}\right)\epsilon_i \ . \end{aligned} \tag{13.118}$$

These two relations are consistent with each other. To find the transformation of y implied by them, note that the first relation can be simplified,

$$\frac{1}{1+y'^2} = \frac{1}{1+y^2} - g_\pi\left(\frac{\mathbf{y}\cdot\boldsymbol{\epsilon}}{1+y^2}\right) ,$$

and substituting this into the second gives

$$y'^i\left[\frac{1}{1+y^2} - g_\pi\left(\frac{\mathbf{y}\cdot\boldsymbol{\epsilon}}{1+y^2}\right)\right] = \frac{y^i}{1+y^2} + \frac{g_\pi}{2}\left(\frac{1-y^2}{1+y^2}\right)\epsilon_i \ .$$

Replacing y' by y in the term of order ϵ gives the following infinitesimal transformation of π:

$$\boxed{\pi'^i = \pi^i\,(1+g_\pi\mathbf{y}\cdot\boldsymbol{\epsilon}) + g_\pi f_\pi(1-y^2)\epsilon_i \ .} \tag{13.119}$$

It is an easy matter to recover both of the Eqs. (13.118) from this transformation law.

The conserved axial vector current implied by the Lagrangian (13.115) can be determined from the infinitesimal transformation (13.119) of the jth component of the pion field,

$$\Omega_{ji} = g_\pi f_\pi\left[\delta_{ji}(1-y^2) + 2y^j y^i\right] , \tag{13.120}$$

and is composed of the same fermion term as before together with a new meson term,

$$\boxed{J_5^{i\mu} = g_\pi\,\bar\psi\gamma^\mu\gamma^5\tfrac{1}{2}\tau_i\psi + g_\pi f_\pi\left[\frac{2\,\partial^\mu\pi^i}{(1+y^2)^2} - \partial^\mu\left(\frac{\pi^i}{1+y^2}\right)\right] \ .} \tag{13.121}$$

If we expand the new meson current in powers of π^i, keep the lowest order term only, and compare it with the hadronic axial vector isospin current $\bar{\psi}\gamma^\mu\gamma^5\frac{1}{2}\tau_i\psi$ (i.e., remove the overall factor of g_π, which gives a strength appropriate for the strong interactions only), we obtain

$$J^i_{\pi\mu} = f_\pi\, \partial_\mu \pi^i \ ,$$

which is precisely the current used for the calculation of the weak decay of the pion in Sec. 9.10 (the field ϕ in Sec. 9.10 is the same as π). Our discussion here provides the justification needed for the use of the current (9.132).

We turn now to a discussion of the explicit breaking of chiral symmetry and the origin of the pion mass.

13.8 CHIRAL SYMMETRY BREAKING AND PCAC

Chiral symmetry can be broken in two ways: (i) *spontaneously*, through the emergence of a vacuum state which is not chirally invariant, and (ii) *explicitly*, through the presence of a small term in the Lagrangian which is not chirally invariant. Spontaneous symmetry breaking was discussed in detail in Sec. 13.6, and in the preceding section we saw that the consequences of spontaneous symmetry breaking, namely the generation of fermion mass and the emergence of a massless Goldstone boson (the pion), could also be obtained directly from a non-linear Lagrangian which is exactly symmetric and for which the vacuum is also symmetric. So spontaneous symmetry breaking does not really remove the symmetry from the theory. In particular, there still exists a conserved axial vector current. In this section we discuss *explicit* chiral symmetry breaking. When a symmetry is explicitly broken, the current is no longer conserved. Another consequence of breaking chiral symmetry explicitly is the possibility of having *both* fermions *and* a pion with finite mass.

Since the breaking of chiral symmetry is "small" (in a sense to be precisely defined shortly), the axial vector current is "almost" conserved, or "partially" conserved. This is referred to as PCAC, for *partially conserved axial-vector current*, and the precise statement of PCAC amounts to a specific statement about the extent to which the conservation of the axial-vector current is broken. This relation takes the form

$$\partial_\mu J_5^{i\mu} = m_\pi^2 \phi^i \ , \tag{13.122}$$

where ϕ_i is proportional to the pion field and m_π is the pion mass. Note that the current is conserved if the pion mass is zero, and that if the pion mass can be regarded as a small quantity, the current can be thought of as "almost" conserved. The consequences of the PCAC relation (13.122), initially put forward as a hypothesis, were among the earliest observations which led to an interest in chiral symmetry.

In this section we will see how the PCAC relation emerges from both the linear and non-linear sigma models.

PCAC in the Linear Sigma Model

To prepare for a discussion of explicit symmetry breaking in the linear sigma model, we first prove, from the equations of motion for the fields, that the axial vector current (13.96) is conserved. The equations of motion for the fields in the linear sigma model can be quickly found from the Euler–Lagrange equations and are

$$\begin{aligned} i\,\partial\!\!\!/\,\psi &= \;g_\pi\,\Phi\,\psi \\ i\bar\psi\,\overleftarrow{\partial\!\!\!/} &= -\,g_\pi\bar\psi\Phi \\ \Box\phi_\sigma &= -\,m^2\phi_\sigma + 2V'\phi_\sigma - g_\pi\bar\psi\psi \\ \Box\phi_\pi^i &= -\,m^2\phi_\pi^i + 2V'\phi_\pi^i - ig_\pi\bar\psi\gamma^5\tau_i\psi \;, \end{aligned} \tag{13.123}$$

where $V' = dV/d|\phi|$. Using these equations,

$$\begin{aligned} \partial_\mu J_5^{i\mu} &= -g_\pi\bar\psi\gamma^5\tfrac{1}{2}\tau_i\,\overrightarrow{\partial\!\!\!/}\,\psi + g_\pi\bar\psi\,\overleftarrow{\partial\!\!\!/}\,\gamma^5\tfrac{1}{2}\tau_i\psi - g_\pi\left(\Box\phi_\sigma\right)\phi_\pi^i + g_\pi\left(\Box\phi_\pi^i\right)\phi_\sigma \\ &= ig_\pi^2\bar\psi\,\gamma^5\tfrac{1}{2}\tau_i\Phi\,\psi + ig_\pi^2\bar\psi\,\Phi\gamma^5\tfrac{1}{2}\tau_i\psi \\ &\qquad + g_\pi\left[m^2\phi_\sigma\phi_\pi^i - 2V'\phi_\sigma\phi_\pi^i + g_\pi\bar\psi\psi\phi_\pi^i\right] \\ &\qquad - g_\pi\left[m^2\phi_\pi^i\phi_\sigma - 2V'\phi_\pi^i\phi_\sigma + ig_\pi\bar\psi\gamma^5\tau_i\psi\phi_\sigma\right] \\ &= ig_\pi^2\bar\psi\left[\gamma^5\tau_i\phi_\sigma + i\phi_\pi^i\right]\psi + g_\pi\bar\psi\left[g_\pi\phi_\pi^i - ig_\pi\gamma^5\tau_i\phi_\sigma\right]\psi = 0\,. \end{aligned} \tag{13.124}$$

Now consider the effect of adding an explicit chiral symmetry breaking term of the form

$$\mathcal{L}_{\text{breaking}} = -c\phi_\sigma \tag{13.125}$$

to the linear Lagrangian. Then the axial vector current will no longer be conserved. The new term only affects the equation for ϕ_σ, which becomes

$$\Box\phi_\sigma + m^2\phi_\sigma - 2V'\phi_\sigma + g_\pi\bar\psi\psi = -c \;, \tag{13.126}$$

and this new equation, when used in the calculation of Eq. (13.124), adds an extra term to $\partial_\mu J_5^{i\mu}$, giving

$$\partial_\mu J_5^{i\mu} = g_\pi c\phi_\pi^i \;. \tag{13.127}$$

The non-conservation is seen to be proportional to the pion field.

To determine c (in particular, to show that it is proportional to m_π^2), consider the effect of this new term on the spontaneous symmetry breaking described in Sec. 13.6. Break the symmetry of the vacuum by choosing $\phi_\pi^i = 0$ and $\phi_\sigma \neq 0$ as before. Then the new vacuum value of the sigma field is determined from the minimum of the sigma part of the meson Hamiltonian:

$$H_{\phi_\sigma} = \tfrac{1}{2}m^2\phi_\sigma^2 + \tfrac{1}{4}\lambda^2\phi_\sigma^4 + c\phi_\sigma \tag{13.128}$$

with a minimum at

$$\frac{dH}{d\phi_\sigma} = m^2\phi_\sigma + \lambda^2\phi_\sigma^3 + c = 0 \;. \tag{13.129}$$

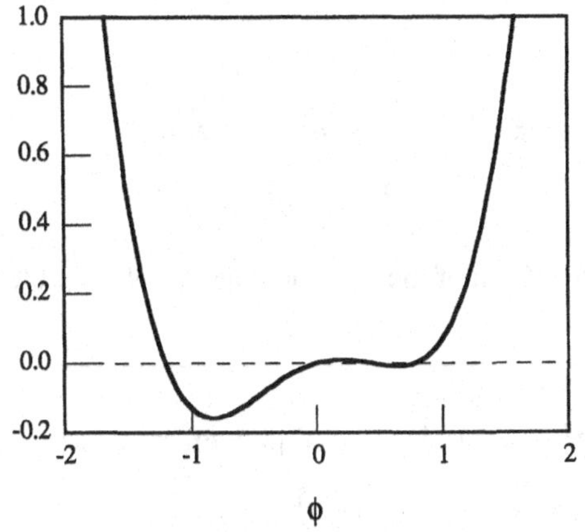

Fig. 13.4 The Hamiltonian with a chiral symmetry breaking term. Now there is only one minimum [compare with Fig. 13.2].

The energy curve is now asymmetric, as shown in Fig. 13.4. Assuming c is small, we can solve for the minimum perturbatively by dividing (13.129) by ϕ_σ and, in the small term, setting $\phi_\sigma = v$, its unperturbed value. We then obtain the new vacuum value of ϕ_σ, which we will denote by v', from the equation

$$m^2 + \lambda^2\phi_\sigma^2 + \frac{c}{v} = 0 \ ,$$

which has the solution

$$\phi_\sigma = v' = \sqrt{-\frac{m^2}{\lambda^2} - \frac{c}{\lambda^2 v}} \simeq v\left(1 + \frac{c}{2m^2 v}\right) \ . \tag{13.130}$$

This solution is correct to first order in c. As Fig. 13.4 shows, c and v must have opposite signs.

The physical Lagrangian is now obtained by subtracting v' (instead of v) from ϕ_σ,

$$s = \phi_\sigma - v' = \phi_\sigma - v - \frac{c}{2m^2} \ , \tag{13.131}$$

and the new Lagrangian, expressed in terms of s instead of ϕ_σ, becomes

$$\begin{aligned}
\mathcal{L}'_s = & \frac{1}{2}\bar{\psi} i\gamma^\mu \frac{\overleftrightarrow{\partial}}{\partial x^\mu}\psi - g_\pi v' \bar{\psi}\psi - g_\pi \bar{\psi}\left[s + i\gamma^5 \boldsymbol{\tau}\cdot\boldsymbol{\phi}_\pi\right]\psi \\
& + \frac{1}{2}\left[\partial_\mu s \partial^\mu s + \partial_\mu \phi_\pi^i \partial^\mu \phi_\pi^i\right] - \frac{1}{2}(m^2 + \lambda^2 v'^2)\left[s^2 + \phi_\pi^2\right] \\
& - \frac{\lambda^2}{4}\left[s^2 + \phi_\pi^2\right]^2 - \lambda^2 v' s\left[s^2 + \phi_\pi^2\right] - \lambda^2 v'^2 s^2 + \text{constant}\,.
\end{aligned} \tag{13.132}$$

Note that the pion mass is now non-zero (because of the symmetry breaking) and c is proportional to m_π^2,

$$m_\pi^2 = m^2 + \lambda^2 v'^2 \cong \frac{\lambda^2 v c}{m^2} = -\frac{c}{v} > 0 \ . \tag{13.133}$$

The new σ mass is $m_\sigma^2 = 2\lambda^2 v'^2 + m_\pi^2$, so that now

$$\lambda^2 v'^2 = \tfrac{1}{2}\left(m_\sigma^2 - m_\pi^2\right) \ . \tag{13.134}$$

Generalizing the definition of the nucleon mass to $M = g_\pi v'$, the new Lagrangian becomes

$$\begin{aligned}
\mathcal{L}_B =& \bar{\psi}\left(\frac{i}{2}\gamma^\mu \overleftrightarrow{\partial_\mu} - M\right)\psi - g_\pi \bar{\psi}\left[s + i\gamma^5 \boldsymbol{\tau}\cdot\boldsymbol{\phi}_\pi\right]\psi \\
&+ \frac{1}{2}\left[\partial_\mu s \partial^\mu s - m_\sigma^2 s^2\right] + \frac{1}{2}\left[\partial_\mu \phi_\pi^i \partial^\mu \phi_\pi^i - m_\pi^2 \phi_\pi^2\right] \\
&- \frac{g_\pi^2\left(m_\sigma^2 - m_\pi^2\right)}{4M^2}\left(s^2 + \phi_\pi^2\right)^2 - \frac{g_\pi\left(m_\sigma^2 - m_\pi^2\right)}{2M} s\left[s^2 + \phi_\pi^2\right] \ .
\end{aligned} \tag{13.135}$$

This Lagrangian is correct to first order in m_π^2.

The statement of PCAC given in Eq. (13.127) can now be made explicit by substituting for c:

$$\partial_\mu J_5^{i\mu} = -M\, m_\pi^2 \phi_\pi^i \ . \tag{13.136}$$

This displays the fact that the current is conserved if $m_\pi^2 = 0$.

PCAC in the Non-Linear Sigma Model

Now look at the PCAC relation in the context of the non-linear sigma model. The equations of motion for the non-linear sigma model fields are

$$\begin{aligned}
(i\not\partial - M)\psi =& -\frac{1}{1+y^2}\left(\frac{g_\pi^2 \pi^2}{2M} - i g_\pi \gamma^5 \boldsymbol{\tau}\cdot\boldsymbol{\pi}\right)\psi \\
-\bar{\psi}(i\overleftarrow{\not\partial} + M) =& -\frac{1}{1+y^2}\bar{\psi}\left(\frac{g_\pi^2 \pi^2}{2M} - i g_\pi \gamma^5 \boldsymbol{\tau}\cdot\boldsymbol{\pi}\right) \\
\partial_\mu\left[\frac{\partial^\mu \pi^i}{(1+y^2)^2}\right] =& -\frac{2\pi^i(\partial_\mu \boldsymbol{y}\cdot\partial^\mu \boldsymbol{y})}{(1+y^2)^3} \\
&+ \frac{g_\pi}{(1+y^2)^2}\bar{\psi}\left(2y^i + i\gamma^5\left[2y^i \boldsymbol{\tau}\cdot\boldsymbol{y} - \tau_i(1+y^2)\right]\right)\psi .
\end{aligned} \tag{13.137}$$

From these equations we can demonstrate that the axial vector current (13.121) is conserved, but in this case the demonstration requires some work.

First note that the divergence of the axial vector current is

$$
\begin{aligned}
\partial_\mu J_5^{i\mu} =& -g_\pi \bar{\psi}\gamma^5 \tfrac{1}{2}\tau_i \overrightarrow{\partial\!\!\!/}\, \psi + g_\pi \bar{\psi}\, \overleftarrow{\partial\!\!\!/}\, \gamma^5 \tfrac{1}{2}\tau_i \psi \\
& + M\left(2\,\partial_\mu\left[\frac{\partial\pi^i}{(1+y^2)^2}\right] - \Box\left[\frac{\pi^i}{1+y^2}\right]\right) \\
=& -g_\pi M\,\bar{\psi}\left[\frac{2\,y^i}{1+y^2} - i\tau_i\gamma^5\left(\frac{1-y^2}{1+y^2}\right)\right]\psi \\
& + M\left(2\,\partial_\mu\left[\frac{\partial\pi^i}{(1+y^2)^2}\right] - \Box\left[\frac{\pi^i}{1+y^2}\right]\right) \;.
\end{aligned}
\tag{13.138}
$$

This will be zero if

$$
2\,\partial_\mu\left[\frac{\partial\pi^i}{(1+y^2)^2}\right] - \Box\left[\frac{\pi^i}{1+y^2}\right] = g_\pi\,\bar{\psi}\left[\frac{2\,y^i}{1+y^2} - i\tau_i\gamma^5\left(\frac{1-y^2}{1+y^2}\right)\right]\psi \;. \tag{13.139}
$$

At first glance, this equation does not appear to be consistent with the field equation for the π field, the last of equations (13.137). However, it is possible to transform the field equation into the relation (13.139). To do this, consider an operator of the form $\mathcal{O}_{ij} = a\delta_{ij} + b\,y^i y^j$ and require that

$$
\frac{1}{1+y^2}\,\mathcal{O}_{ij}\,\bar{\psi}\left(2y^j + i\gamma^5\left[2y^j\boldsymbol{\tau}\cdot\mathbf{y} - \tau_j(1+y^2)\right]\right)\psi = \bar{\psi}\left[2\,y^i - i\tau_i\gamma^5(1-y^2)\right]\psi. \tag{13.140}
$$

This relation will be satisfied if

$$
\begin{aligned}
\mathcal{O}_{ij}\,y^i &= y^i\,(1+y^2) \\
\mathcal{O}_{ij}\left[\tau_i - \frac{2\,y^i\boldsymbol{\tau}\cdot\mathbf{y}}{1+y^2}\right] &= \tau_i\,(1-y^2) \;\;,
\end{aligned}
\tag{13.141}
$$

which can be satisfied if $a = 1 - y^2$ and $b = 2$. With this observation the current is conserved *if* we can show that

$$
\mathcal{O}_{ij}\left(\partial_\mu\left[\frac{\partial^\mu\pi^j}{(1+y^2)^2}\right] + \frac{2\,\pi^j(\partial_\mu\mathbf{y}\cdot\partial^\mu\mathbf{y})}{(1+y^2)^3}\right) = 2\,\partial_\mu\left[\frac{\partial\pi^i}{(1+y^2)^2}\right] - \Box\left[\frac{\pi^i}{1+y^2}\right] \;. \tag{13.142}
$$

The proof of (13.142) is straightforward and is left as an exercise.

Now, we consider the effect of adding a pion mass term to the non-linear Lagrangian (13.115). The new Lagrangian will be

$$
\mathcal{L}_{2B} = \mathcal{L}_2 - \tfrac{1}{2}m_\pi^2\pi^2 f(y^2) \;\;, \tag{13.143}
$$

where f is any smooth function of y^2, for example $f(y^2) = 1/(1+y^2)$. This pion mass term changes the equation of the pion field to

$$
\begin{aligned}
\partial_\mu&\left[\frac{\partial^\mu\pi^i}{(1+y^2)^2}\right] + \frac{2\,\pi^i(\partial_\mu\mathbf{y}\cdot\partial^\mu\mathbf{y})}{(1+y^2)^3} + m_\pi^2\pi^i\left(f(y^2) + y^2\,\frac{df(y^2)}{dy^2}\right) \\
&= \frac{g_\pi}{(1+y^2)^2}\,\bar{\psi}\left(2y^i + i\gamma^5\left[2y^i\boldsymbol{\tau}\cdot\mathbf{y} - \tau_i(1+y^2)\right]\right)\psi \;,
\end{aligned}
\tag{13.144}
$$

and gives the following specific form for the PCAC relation:

$$\partial_\mu J_5^{i\mu} = -M\, m_\pi^2 \pi^i (1+y^2)\left(f(y^2) + y^2\,\frac{df(y^2)}{dy^2}\right) . \qquad (13.145)$$

Note that, once again, the non-zero value of the pion mass is responsible for breaking chiral symmetry.

This concludes our discussion of symmetries. In the next chapter we discuss the path integral formulation of quantum mechanics and prepare the way for a discussion of the quantization of QCD.

PROBLEMS

13.1 Charged scalar theory.

(a) Find the current which is conserved as a consequence of *global* gauge invariance for a *charged scalar* field theory.
(b) If J^μ is the current found in part (a), show that the interaction $-J_\mu A^\mu$ is not sufficient to insure that the total Lagrangian is invariant under *local* $U(1)$ gauge transformations. An extra interaction term of the form

$$\lambda\phi^\dagger(x)\phi(x)A_\mu(x)A^\mu(x)$$

is needed. Show that with this extra term the Lagrangian is *locally* gauge invariant *provided* the constant λ takes a particular value. Find λ.

13.2 Consider the scattering of photons from a π^+ meson, $\gamma + \pi^+ \to \gamma + \pi^+$ (this is Compton scattering with the pion replacing the electron).

(a) Using the Feynman rules for "scalar QED" given in Appendix B, draw all Feynman graphs which contribute to order e^2. Label all momenta and write the correct $\mathcal{M}$-matrix corresponding to each diagram. Do not simplify your results at this stage.

(b) If $\mathcal{M} = \mathcal{M}_{\mu\nu}\epsilon_f^{\mu *}\epsilon_i^\nu$, where ϵ_f and ϵ_i are the four-polarization vectors of the final and initial photons, show that $\mathcal{M}_{\mu\nu}$ conserves current. In particular, prove that

$$k_f^\mu \mathcal{M}_{\mu\nu} = 0 = k_i^\nu \mathcal{M}_{\mu\nu} ,$$

where k_f and k_i are the four-momentum vectors of the final and initial photons, respectively.

13.3 Redo Prob. 10.8(a) including the new diagram drawn in Fig. 13.3 (consult Appendix B for the Feynman rules for this case, if you need to). How do the results change? Discuss the significance of this calculation. What important physical principle does this problem illustrate?

13.4 Consider a classical field theory of massive neutral scalar mesons, ϕ, and massless fermions, ψ, described by the Lagrangian density

$$\mathcal{L} = \tfrac{1}{2}\left[\partial_\mu\phi(x)\partial^\mu\phi(x)\right] - \tfrac{1}{2}m^2\phi^2(x) - \tfrac{\lambda}{4!}\phi^4(x) + \tfrac{1}{2}\bar{\psi}(x)\left[i\gamma^\mu \overleftrightarrow{\partial_\mu}\right]\psi(x) - g\bar{\psi}(x)\psi(x)\phi(x) ,$$

where $\lambda > 0$.

(a) Find the Hamiltonian density for this theory.

(b) Consider the classical, near static case when $\partial_\mu\phi(x) \simeq 0$ and $\psi(x) \simeq 0$. Find the value of the field ϕ which minimizes the energy for $m^2 > 0$ and for $m^2 < 0$.

(c) Define a new scalar field $s(x) = \phi(x) - \langle\phi\rangle$, where $\langle\phi\rangle$ is the value at the minimum found in part (b), and rewrite the Lagrangian in terms of this field. What are the masses of the scalar and Fermi particles? Discuss and interpret your result.

CHAPTER 14

PATH INTEGRALS

In this chapter an alternative formulation of quantum mechanics and field theory, based on path integrals, is presented. The great advantage of this formulation is that it allows us to quantize a theory using only c-number fields, without the need to turn the fields into operators. Other advantages of this approach to quantum mechanics are:

- It provides the simplest, most direct way to obtain the Feynman rules for any field theory. In particular, we can obtain the Feynman rules for QCD (some of which were introduced in Chapter 13) using this method.

- It provides a method for obtaining exact, numerical solutions of strongly interacting field theories (where the perturbation expansion does not work). These methods, referred to as lattice gauge calculations, will not be discussed in this book.

- It provides a connection between field theory and statistical mechanics, which gives insight into the nature of both subjects.

- It provides a general theoretical framework of a systematic discussion of a number of advanced topics in field theory. Among these is the study of renormalization and the appearance of *anomalies*. Using path integrals one can prove general theorems about renormalization in a comparatively easy way, and anomalies, not discussed in this book, are most easily understood using these techniques.

Because of these many advantages, the path integral approach to quantum mechanics is an essential part of a modern study of field theory.*

In this chapter we introduce the idea of a path integral from a consideration of the propagator and then show how the S-matrix can be expressed as a path integral. To show the power of this approach and to acquire needed experience,

*The use of path integrals for practical calculations was first proposed by Feynman [Fe 48]. For further reading, see Feynman and Hibbs (1965), Negele and Orland (1988), and the "new" books listed in the References.

we then obtain the Feynman rules for ϕ^3 theory. In the following chapter we use this approach to obtain the Feynman rules for QED and QCD. Obtaining the rules for QED serves as another example of how the method works, but the quantization of QCD requires the power of path integrals. Several of the topics developed in subsequent chapters will depend on path integrals for their development.

14.1 THE WAVE FUNCTION AND THE PROPAGATOR

In this chapter we will work in the Heisenberg representation. In this representation the states are independent of time, while the operators depend on time. The operator which will be the center of attention is the *generalized* coordinate operator, denoted by $Q = Q(t)$. In this section this operator represents the position of a particle, but in subsequent sections the discussion will be extended to fields, which are also generalized coordinates. Since this operator depends on time, its eigenfunctions, denoted by $|q, t\rangle$, must also depend on time. Hence the coordinate space wave function for the state n is

$$\psi_n(q,t) = \langle q, t|n\rangle_H \quad , \tag{14.1}$$

where the subscript H reminds us that the matrix element is in the Heisenberg representation. The corresponding Schrödinger states are equal to the Heisenberg states at some time $t = t_0$ [recall Eq. (1.31)], so that

$$|n, t\rangle_S = U(t, t_0)|n\rangle_H \quad , \tag{14.2}$$

where U is the familiar time translation operator equal to unity at $t = t_0$. Demanding that the wave function be the same for either picture,

$$\psi_n(q,t) = \langle q, t|n\rangle_H = \langle q|n, t\rangle_S \quad , \tag{14.3}$$

gives us the following relation between the eigenfunctions of the operator Q in the two pictures:

$$|q\rangle_S = U(t, t_0)|q, t\rangle_H = U^\dagger(t_0, t)|q, t\rangle_H \quad . \tag{14.4}$$

This relation, which will be used shortly, is identical to Eq. (14.2), but it looks different because the time dependence of the eigenfunctions of the coordinate operator is *opposite* to the usual rule that Heisenberg states are independent of time and Schrödinger states depend on time. From now on the subscript H will be implied but not written as we work in the Heisenberg representation unless explicitly stated to the contrary.

Using the completeness of the states, the wave function at a different position and later time can be written

$$\begin{aligned}
\psi_n(q_f, t_f) = \langle q_f, t_f|n\rangle &= \int dq_i \, \langle q_f, t_f|q_i, t_i\rangle \, \langle q_i, t_i|n\rangle \\
&= \int dq_i \, \langle q_f, t_f|q_i, t_i\rangle \, \psi_n(q_i, t_i) \\
&= \int dq_i \, K(q_f, t_f; q_i, t_i) \, \psi_n(q_i, t_i) \quad .
\end{aligned} \tag{14.5}$$

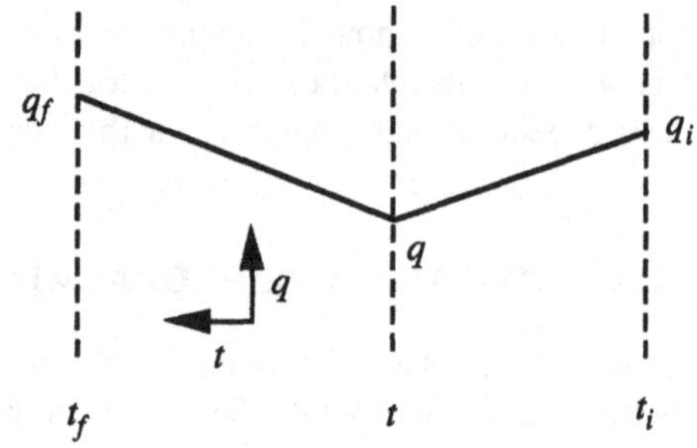

Fig. 14.1 Diagram illustrating the superposition property of the propagator. The path of the particle from the initial to the final space–time point is a superposition of paths through *any* intermediate space–time point.

The quantity $\langle q_f, t_f | q_i, t_i \rangle = K(q_f, t_f; q_i, t_i)$ is referred to as the propagator and is the basis of the path integral formulation of quantum mechanics.

From the preceding discussion, it follows that the propagator has the following superposition property:

$$\begin{aligned} K(q_f, t_f; q_i, t_i) &= \int dq \, \langle q_f, t_f | q, t \rangle \, \langle q, t | q_i, t_i \rangle \\ &= \int dq \, K(q_f, t_f; q, t) K(q, t; q_i, t_i) \quad . \end{aligned} \tag{14.6}$$

If the propagator represents the motion of a particle from some initial point q_i at time t_i to a fixed point q_f at time t_f, then the above equation can be represented diagrammatically as shown in Fig. 14.1. Equation (14.6) and the figure show that the total propagator is obtained by *summing* over *all possible* paths from q_i through *any* point q at time t to the final point q_f at t_f. Only *one* possible path is shown in the figure. The treatment of a particle passing through a double slit is a practical example of this method. In that case there are only two possible paths, one through each slit. If the slit is removed, there are an infinite number of such paths.

The superposition principle can be generalized. Divide the interval into n time intervals and label the initial time t_0 and the final time t_n. Then

$$K(q_n, t_n; q_0, t_0) = \int \prod_{i=1}^{n-1} dq_i \prod_{i=0}^{n-1} K\left(q_{i+1}, t_{i+1}; q_i, t_i\right) \quad . \tag{14.7}$$

This integral is represented in Fig. 14.2, which displays the total propagator as the superposition of all paths through all possible points q_i at times t_i where $1 \leq i \leq n-1$. Note that, as required by causality, we do not allow the paths to

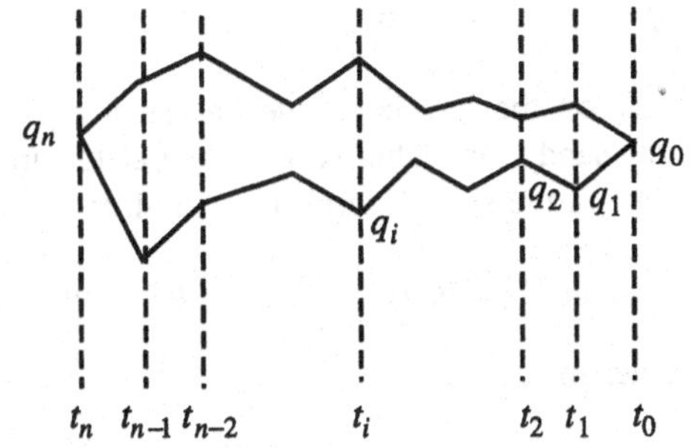

Fig. 14.2 Repeated application of the superposition principle means that the propagator can be written as an integral over all possible paths through intermediate points in space–time, as illustrated above.

propagate backward in time. Now let the number $n \to \infty$, so that the time interval between successive slabs $t_{i+1} - t_i = \epsilon \to 0$. Then the matrix element connecting neighboring slabs can be estimated by assuming the Hamiltonian is constant over the small time interval ϵ, and using Eq. (14.4) which connects Heisenberg states $|q,t\rangle$ with Schrödinger states $|q\rangle$ gives

$$\begin{aligned} \langle q_{i+1}, t_{i+1} | q_i, t_i\rangle_H &= \langle q_{i+1}|U(t_{i+1}, t_0)U^\dagger(t_i, t_0)|q_i\rangle_S \\ &= \langle q_{i+1}|e^{-iH\epsilon}|q_i\rangle_S \quad . \end{aligned} \tag{14.8}$$

Furthermore, since ϵ is infinitesimal, we can estimate the exponential by computing only the first term in its expansion,

$$\begin{aligned} \langle q_{i+1}, t_{i+1}|q_i, t_i\rangle_H &\cong \langle q_{i+1}|1 - iH\epsilon|q_i\rangle_S \\ &= \delta(q_{i+1} - q_i) - i\epsilon\,\langle q_{i+1}|H|q_i\rangle_S \\ &= \int \frac{dp}{2\pi} e^{ip(q_{i+1}-q_i)} - i\epsilon \left\langle q_{i+1} \left| \frac{P^2}{2m} + V(Q) \right| q_i \right\rangle_S \, , \end{aligned} \tag{14.9}$$

where P and Q are the momentum and position *operators* and (for now) we use the H appropriate to the nonrelativistic motion of a single particle. If p and q are eigenfunctions of P and Q, the first factor in the second term becomes

$$\left\langle q_{i+1} \left| \frac{P^2}{2m} \right| q_i \right\rangle_S = \int dp'dp\, \langle q_{i+1}|p'\rangle_S \left\langle p' \left| \frac{P^2}{2m} \right| p \right\rangle_S \langle p|q_i\rangle_S \quad . \tag{14.10}$$

In one dimension

$$\begin{aligned} \langle q|p\rangle_S &= \frac{e^{ipq}}{(2\pi)^{1/2}} \\ \left\langle p' \left| \frac{P^2}{2m} \right| p \right\rangle_S &= \frac{p^2}{2m}\,\delta(p' - p) \end{aligned} \tag{14.11}$$

and hence

$$\left\langle q_{i+1}\left|\frac{P^2}{2m}\right|q_i\right\rangle_S = \int \frac{dp}{2\pi}\frac{p^2}{2m}\, e^{ip(q_{i+1}-q_i)} \quad . \tag{14.12}$$

Note that the *operator* on the left-hand side has been replaced by an *ordinary c-number* on the right-hand side. This is an essential feature of this formalism. Next, the second factor in the second term of Eq. (14.9) is

$$\langle q_{i+1}|V(Q)|q_i\rangle_S = V(q_i)\,\delta(q_{i+1}-q_i) \tag{14.13}$$

and hence, to first order in ϵ,

$$\begin{aligned}\langle q_{i+1}, t_{i+1}|q_i, t_i\rangle_H &\cong \int \frac{dp}{2\pi}\left[1 - i\epsilon\overbrace{\left(\frac{p^2}{2m} + V(q_i)\right)}^{H(p,q_i)}\right] e^{ip(q_{i+1}-q_i)} \\ &\cong \int \frac{dp}{2\pi}\, e^{i[p(q_{i+1}-q_i)-\epsilon H(p,q_i)]} \quad .\end{aligned} \tag{14.14}$$

If ϵ is small,

$$\frac{q_{i+1}-q_i}{\epsilon} \cong \dot{q}_i \tag{14.15}$$

and substituting $p \rightarrow p_i$ gives

$$\langle q_{i+1}, t_{i+1}|q_i, t_i\rangle_H \cong \int \frac{dp_i}{2\pi}\, e^{i\epsilon[p_i\dot{q}_i - H(p_i,q_i)]} \quad . \tag{14.16}$$

The quantity p_i is the independent momentum associated with each q_i. Then, the original expression Eq. (14.7) can be cast into the following form:

$$K(q_n, t_n; q_0, t_0) = \int \underbrace{\prod_{i=1}^{n-1} dq_i \prod_{i=0}^{n-1}\frac{dp_i}{2\pi}}_{\left[\frac{dq\,dp}{2\pi}\right]}\ \underbrace{\prod_{i=0}^{n-1} e^{i\epsilon[p_i\dot{q}_i - H(p_i,q_i)]}}_{e^{i\int_{t_0}^{t_n} dt[p\dot{q}-H(p,q)]}} \tag{14.17}$$

or, in a shorthand notation,

$$\boxed{K(q_n, t_n; q_0, t_0) = \int \left[\frac{dq\,dp}{2\pi}\right] e^{i\int_{t_0}^{t_n} dt\,[p\dot{q}-H(p,q)]}} \quad . \tag{14.18}$$

This is the path integral expression for the propagator.

The path integral is an integral over *all smooth paths in phase space which connect the initial and final points* $q_0(t_0)$ and $q_n(t_n)$, each path expressible in the parametric form $\{q(t), p(t)\}$ and weighted by the phase factor

$$e^{i[p\dot{q}-H(p,q)]} \quad .$$

Note that the original operator $H(P,Q)$ has been replaced by its equivalent c-number $H(p,q)$. Because the path integral is an *integral over all functions* $q(t)$ [and $p(t)$], it is referred to as a *functional integral*, and this whole approach is sometimes referred to as a functional method for handling quantum mechanics and field theory.

In deriving the path integral for the propagator, we used the Schrödinger equation when we took the infinitesimal time translation operator to be $\exp(-iH\epsilon)$. However, it is informative to prove directly that the wave function (14.5) satisfies the Schrödinger equation. Substituting our final answer (14.18) into (14.5) and differentiating with respect to the last time, t_n, gives

$$\begin{aligned} i\frac{\partial}{\partial t_n}\psi(q_n,t_n) &= i\frac{\partial}{\partial t_n}\int dq_0 K(q_n,t_n;q_0,t_0)\psi(q_0,t_0) \\ &= \int \prod_{i=0}^{n-1} dq_i \prod_{i=0}^{n-1}\frac{dp_i}{2\pi}\, H(p_{n-1},q_{n-1}) \prod_{i=0}^{n-1} e^{i\epsilon[p_i\dot q_i - H(p_i,q_i)]}\psi(q_0,t_0) \\ &= \int dq_{n-1}\frac{dp_{n-1}}{2\pi}H(p_{n-1},q_{n-1})e^{i[p_{n-1}(q_n-q_{n-1})-\epsilon H(p_{n-1},q_{n-1})]} \\ &\qquad \times \int dq_0 K(q_{n-1},t_{n-1};q_0,t_0)\psi(q_0,t_0)\ , \end{aligned} \tag{14.19}$$

where the derivative with respect to t_n does not bring down the $p_{n-1}\dot q_{n-1}$ term from the exponential because $\epsilon\dot q_{n-1}$ is independent of time. The integral over p_{n-1} can be replaced by reversing the steps which led up to (14.14):

$$\begin{aligned} &\int \frac{dp_{n-1}}{2\pi}H(p_{n-1}q_{n-1})e^{i[p_{n-1}(q_n-q_{n-1})-\epsilon H(p_{n-1}q_{n-1})]} \\ &\qquad = H(-i\nabla_n,q_n)\langle q_n,t_n|q_{n-1},t_{n-1}\rangle_H\ , \end{aligned} \tag{14.20}$$

and hence we obtain

$$\begin{aligned} i\frac{\partial}{\partial t_n}\psi(q_n,t_n) =& H(-i\nabla_n,q_n)\int dq_{n-1}\langle q_n,t_n|q_{n-1},t_{n-1}\rangle_H \\ & \times \int dq_0 K(q_{n-1},t_{n-1};q_0,t_0)\psi(q_0,t_0) \\ =& H(-i\nabla_n,q_n)\int dq_{n-1}\langle q_n,t_n|q_{n-1},t_{n-1}\rangle_H \\ & \times \int dq_0\,\langle q_{n-1},t_{n-1}|q_0,t_0\rangle_H\,\psi(q_0,t_0) \\ =& H(-i\nabla_n,q_n)\int dq_0 K(q_n,t_n;q_0,t_0)\psi(q_0,t_0) \\ =& H(-i\nabla_n,q_n)\,\psi(q_n,t_n)\ , \end{aligned} \tag{14.21}$$

where the completeness relation was used in the third step. Hence the path integral (14.18) is fully equivalent to the Schrödinger equation, and we have an alternative way to describe quantum mechanics.

If the Hamiltonian is quadratic in the generalized momentum, the path integral (14.18) can be further reduced by doing the p integrations explicitly. To this end, complete the p^2 square, giving

$$\begin{aligned}\int_{-\infty}^{\infty} \frac{dp}{2\pi} e^{i\left[p\dot{q}-\frac{p^2}{2m}\right]\epsilon} &= \int_{-\infty}^{\infty} \frac{dp}{2\pi} e^{-\frac{i\epsilon}{2m}(p-m\dot{q})^2}\, e^{i\epsilon\frac{1}{2}m\dot{q}^2} \\ &= \frac{1}{2\pi}\sqrt{\frac{2m}{\epsilon}}\, e^{i\epsilon\frac{m}{2}\dot{q}^2} \int_{-\infty}^{\infty} dy\, e^{-iy^2} . \end{aligned} \tag{14.22}$$

The Gaussian integral and constants can be lumped together into an overall multiplicative factor, which is

$$\frac{1}{2\pi}\sqrt{\frac{2m}{\epsilon}} \int_{-\infty}^{\infty} dy\, e^{-iy^2} = \frac{1}{2\pi}\sqrt{\frac{2m}{\epsilon}}\sqrt{\frac{\pi}{i}} = \sqrt{\frac{m}{2\pi i\epsilon}} = \frac{1}{N} . \tag{14.23}$$

Hence

$$K(q_n, t_n; q_0, t_0) = \int \left[\frac{dq}{N}\right] \frac{1}{N}\, e^{i\int_{t_0}^{t_n} dt\left[\frac{1}{2}m\dot{q}^2 - V(q)\right]}$$

or, in terms of the Lagrangian,

$$\boxed{K(q_n, t_n; q_0, t_0) = \int \left[\frac{dq}{N}\right] \frac{1}{N}\, e^{i\int_{t_0}^{t_n} dt\, L(q,\dot{q})}} . \tag{14.24}$$

This is the form originally introduced by Feynman and is strictly equivalent to our starting expression (14.18) only for systems quadratic in the generalized momentum, i.e., with a Hamiltonian of the form $H = cp^2 + V(q)$.

14.2 THE S-MATRIX

In this section we show how the S-matrix for a particle moving under the influence of a potential is described in this formalism. (The extensions to field theory will be developed in subsequent sections.) We saw in Sec. 3.1, Eq. (3.28), that the S-matrix was

$$\begin{aligned} S_{\beta\alpha} &= \frac{\langle\beta|U_I(\infty, -\infty)|\alpha\rangle}{\langle 0|U_I(\infty, -\infty)|0\rangle} \\ &= \lim_{\substack{t_n\to\infty \\ t_0\to-\infty}} \mathcal{N}\,\langle\beta|U_I(t_n, t_0)|\alpha\rangle \quad , \end{aligned} \tag{14.25}$$

where $\mathcal{N}$ is a constant. The key quantity in this expression is $\langle\beta|U_I(t_n, t_0)|\alpha\rangle$, the matrix element of the interaction time translation operator for the finite time interval $[t_n, t_0]$. Because this differs from the S-matrix only by the limiting process and by a constant, as shown in Eq. (14.25), we will study the quantity

$$\langle\beta|U_I(t_n, t_0)|\alpha\rangle = \bar{S}_{\beta\alpha} \ . \tag{14.26}$$

Substituting the perturbation expansion for U_I, Eq. (3.24), gives

$$\bar{S}_{\beta\alpha} = (\bar{S}_{\beta\alpha})_0 + (\bar{S}_{\beta\alpha})_1 + (\bar{S}_{\beta\alpha})_2 + \cdots, , \tag{14.27}$$

where

$$\begin{aligned}
(\bar{S}_{\beta\alpha})_0 &= \langle\beta|\alpha\rangle = \delta_{\beta\alpha} \\
(\bar{S}_{\beta\alpha})_1 &= -i\int_{t_0}^{t_n} dt\, \langle\beta|H_I(t)|\alpha\rangle \\
(\bar{S}_{\beta\alpha})_2 &= \frac{(-i)^2}{2}\int_{t_0}^{t_n} dt_1\, dt_2\, \langle\beta|T\left(H_I(t_2)H_I(t_1)\right)|\alpha\rangle \ .
\end{aligned} \tag{14.28}$$

We will now show that the $\bar{S}$-matrix can be obtained from the path integral by using the relation

$$\boxed{\bar{S}_{\beta\alpha} = \mathcal{N}'\int dq_n\, dq_0\, \phi^*_\beta(q_n, t_n)K(q_n, t_n; q_0, t_0)\phi_\alpha(q_0, t_0)\,,} \tag{14.29}$$

where $\mathcal{N}'$ is a normalization constant to be chosen shortly, K is the propagator for which we obtained a path integral expression in the last section, and

$$\phi_\alpha(q, t) = {}_0\langle q, t|\alpha\rangle$$

is the *free* coordinate space wave function for the state α (to be defined shortly, and not to be confused with ψ_α). As in Eq. (14.25), the S-matrix is then obtained from (14.29) by taking the limits $t_n \to \infty$ and $t_0 \to -\infty$. The significance of this result is that the *path integral gives a closed form for the exact S-matrix*, which can, in principle at least, be *numerically evaluated.*

To prove Eq. (14.29), return to path integral (14.18), and think of the Hamiltonian as decomposed of a free term, H_0, and an interacting term, H_I, which we assume (for now) is a function of q only. Then the full propagator, K, and the corresponding free propagator, K_0, are

$$\begin{aligned}
K(q_n, t_n; q_0, t_0) &= \int \prod_{k=1}^{n-1} dq_k \prod_{i=0}^{n-1}\left(\frac{dp_i}{2\pi}\, e^{i\epsilon[p_i\dot{q}_i - H_0(p_i,q_i) - H_I(q_i)]}\right) \\
K_0(q_n, t_n; q_0, t_0) &= \int \prod_{k=1}^{n-1} dq_k \prod_{i=0}^{n-1}\left(\frac{dp_i}{2\pi}\, e^{i\epsilon[p_i\dot{q}_i - H_0(p_i,q_i)]}\right) \ .
\end{aligned} \tag{14.30}$$

The free wave function which enters into Eq. (14.29) is the evolution of the state α under the free Hamiltonian H_0, given by

$$\phi_\alpha(q_n, t_n) = \int dq_0\, K_0(q_n, t_n; q_0 t_0)\phi_\alpha(q_0, t_0) \quad , \tag{14.31}$$

where the free propagator is $K_0(q_2, t_2; q_1, t_1) = {}_0\langle q_2, t_2|q_1, t_1\rangle_0$. These states can be regarded as plane wave states (although they could be atomic states, or some other basis of exact solutions of H_0).

We will now prove Eq. (14.29) by demonstrating that it gives the same perturbation expansion for $\bar{S}$ as the previously derived expansion given in Eqs. (14.27) and (14.28). We will be satisfied to show this for the first three terms in the expansion. Using perturbation theory, the exact propagator (14.30) can be expanded in a power series in H_I,

$$K = K_0 + K_1 + K_2 + \cdots , \tag{14.32}$$

where K_n is proportional to $(H_I)^n$, and K_0 is just the free propagator. Hence, we must prove

$$\left(\bar{S}_{\beta\alpha}\right)_n = \mathcal{N}' \int dq_n\, dq_0\, \phi^*_\beta(q_n, t_n) K_n(q_n, t_n; q_0, t_0)\phi_\alpha(q_0, t_0) . \tag{14.33}$$

The first term in this series is

$$\begin{aligned}\left(\bar{S}_{\beta\alpha}\right)_0 &= \mathcal{N}' \int dq_n\, dq_0\, \phi^*_\beta(q_n, t_n) K_0(q_n, t_n; q_0, t_0)\phi_\alpha(q_0, t_0) \\ &= \mathcal{N}' \int dq_n\, \phi^*_\beta(q_n, t_n)\phi_\alpha(q_n t_n) = \mathcal{N}'\delta_{\beta\alpha} \quad ,\end{aligned} \tag{14.34}$$

where the free propagation of the wave function, as described in Eq. (14.31), was used in the last step. This will agree with the first term in Eq. (14.28) if we choose $\mathcal{N}' = 1$.

To evaluate the second term, we must first find K_1. Expanding the exact propagator to first order in H_I proceeds in two steps:

$$\begin{aligned}K_1 &= \sum_{j=1}^{n-1} \int dq_j \left[\int \prod_{k=j+1}^{n-1} dq_k \prod_{i=j}^{n-1} \frac{dp_i}{2\pi}\, e^{i\epsilon[p_i \dot{q}_i - H_0(p_i, q_i)]}\right] \\ &\quad \times e^{-i\epsilon H_I(q_j)} \left[\prod_{k'=1}^{j-1} dq_{k'} \prod_{i'=0}^{j-1} \frac{dp_{i'}}{2\pi}\, e^{i\epsilon[p_{i'} \dot{q}_{i'} - H_0(p_{i'}, q_{i'})]}\right] \\ &= -i \sum_{j=1}^{n-1} \int dq_j\, \epsilon\, K_0(q_n, t_n; q_j, t_j) H_I(q_j) K_0(q_j, t_j; q_0, t_0) \\ &= -i \int dq \int_0^n dt\, K_0(q_n, t_n; q, t) H_I(q) K_0(q, t; q_0, t_0) \quad .\end{aligned} \tag{14.35}$$

Note that the first order term is obtained by retaining the *first order* contribution from $\exp(-i\epsilon H_I)$ to *only one* time "slice" (if H_I contributes to two time slices, the result is already second order) and summing (integrating) over all possible times at which H_I can contribute. The time summation is converted to a time integral using $\epsilon \to dt$. Now, inserting this into the expression for $\bar{S}_1$ and using $\mathcal{N}' = 1$ give

$$\begin{aligned}(\bar{S}_{\beta\alpha})_1 &= -i \int dq_n\, dq_0\, dq \int_{t_0}^{t_n} dt\, \phi_\beta^*(q_n, t_n) K_0(q_n, t_n; q, t) H_I(q) \\ &\qquad\qquad \times K_0(q, t; q_0, t_0)\phi_\alpha(q_0, t_0) \\ &= -i \int_{t_0}^{t_n} dt \int dq\, \phi_\beta^*(q, t) H_I(q) \phi_\alpha(q, t) \ . \end{aligned} \tag{14.36}$$

However, since $H_I = H_I(q) = {}_0\langle q, t|H_I(Q(t))|q', t\rangle_0 \delta(q - q')$, with the operator Q in the interaction representation (in agreement with the formalism of Chapter 3), this equation can be written

$$\begin{aligned}(\bar{S}_{\beta\alpha})_1 &= -i \int_{t_0}^{t_n} dt \int dq\, dq'\, \langle\beta|q, t\rangle_{00}\langle q, t|H_I(Q(t))|q', t\rangle_{00}\langle q', t|\alpha\rangle \\ &= -i \int dt\, \langle\beta|H_I(Q(t))|\alpha\rangle \ , \end{aligned} \tag{14.37}$$

where, in the last step, the completeness of the position eigenfunctions at any time

$$\int dq\, |q, t\rangle_0\, {}_0\langle q, t| = \mathbf{1} \tag{14.38}$$

has been used to remove the integrations over q and q'. Note that this result agrees with (14.28), proving the equality to first order in H_I. Note also that in this form H_I is now an operator, depending on the generalized coordinates Q.

Similarly, K_2 can be obtained by using the above expansion at two different times,

$$\begin{aligned} K_2 &= \sum_{j=2}^{n-1} \sum_{j'=1}^{j-1} \int dq_j\, dq_{j'} \left[\int \prod_{k=j+1}^{n-1} dq_k \prod_{i=j}^{n-1} \frac{dp_i}{2\pi} e^{i\epsilon[p_i \dot{q}_i - H_0(p_i, q_i)]} \right] \\ &\times (-i\epsilon) H_I(q_j) \left[\int \prod_{k'=j'+1}^{j-1} dq_{k'} \prod_{i'=j'}^{j-1} \frac{dp_i}{2\pi} e^{i\epsilon[p_{i'} \dot{q}_{i'} - H_0(p_{i'} q_{i'})]} \right] \\ &\times (-i\epsilon) H_I(q_{j'}) \left[\int \prod_{k''=1}^{j'-1} dq_{k''} \prod_{i''=1}^{j'-1} \frac{dp_i}{2\pi} e^{i\epsilon[p_{i''} \cdot \dot{q}_{i''} - H_0(p_{i''} q_{i''})]} \right] \\ &= (-i)^2 \sum_{j=2}^{n-1} \sum_{j'=1}^{j-1} \int dq_j\, dq_{j'}\, \epsilon^2\, K_0(q_n t_n; q_j, t_j) H_I(q_j) K_0(q_j, t_j; q_{j'}, t_{j'}) \\ &\qquad \times H_I(q_{j'}) K_0(q_{j'}, t_{j'}; q_0, t_0) \end{aligned}$$

and finally, because $j > j'$,

$$K_2 = (-i)^2 \int dq\, dq' \int_{t_0}^{t_n} dt \int_{t_0}^{t} dt'\, K_0(q_n, t_n; q, t) H_I(q) K_0(q, t; q', t') \times H_I(q') K_0(q', t'; q_0, t_0) \ , \qquad (14.39)$$

where t is associated with the slab at time t_j and t' at time $t_{j'}$. The second order contributions from a *single time slab* are ignored because they are negligibly small compared to the contributions from two different times. To see this, note that there are $n-1$ contributions from a single slab, but $(n-1)(n-2)/2$ contributions from two slabs, so that the error in neglecting the single slab contributions goes like $\sim 2/n \to 0$ as $n \to \infty$. Note that the time-ordered structure of (14.39) emerges automatically in an almost trivial fashion; it comes from the fact that one of the H_I's must necessarily follow the other. Now to find the second order $\bar{S}$-matrix element implied by K_2, exploit the fact that K_0 is the free propagator to carry out the following reduction:

$$\begin{aligned}
(S_{\beta\alpha})_2 =& (-i)^2 \int dq_n\, dq_0\, dq_1\, dq_2 \int_{t_0}^{t_n} dt_2\, \phi^*_\beta(q_n, t_n)\, K_0(q_n, t_n; q_2, t_2)\, H_I(q_2) \\
& \times \int_{t_0}^{t_2} dt_1\, K_0(q_2, t_2; q_1, t_1) H_I(q_1) K_0(q_1, t_1; q_0, t_0) \phi_\alpha(q_0, t_0) \\
=& (-i)^2 \int dq_1\, dq_2 \int_{t_0}^{t_n} dt_2\, \phi^*_\beta(q_2, t_2)\, H_I(q_2) \\
& \times \int_{t_0}^{t_2} dt_1\, K_0(q_2, t_2; q_1, t_1) H_I(q_1)\, \phi_\alpha(q_1, t_1) \ . \qquad (14.40)
\end{aligned}$$

Using the definition of K_0 and the completeness of the position eigenstates gives

$$\begin{aligned}
(S_{\beta\alpha})_2 =& (-i)^2 \int dq_1\, dq_2\, dq'_1\, dq'_2 \int_{t_0}^{t_n} dt_2 \langle\beta|q_2, t_2\rangle_{0\,0}\langle q_2, t_2|H_I(Q(t_2))|q'_2, t_2\rangle_0 \\
& \times \int_{t_0}^{t_2} dt_1\, {}_0\langle q'_2, t_2|q'_1, t_1\rangle_{0\,0}\langle q'_1, t_1|H_I(Q(t_1))|q_1, t_1\rangle_{0\,0}\langle q_1, t_1|\alpha\rangle \\
=& (-i)^2 \int_{t_0}^{t_n} dt_2 \int_{t_0}^{t_2} dt_1\, \langle\beta|H_I(Q(t_2)) H_I(Q(t_1))|\alpha\rangle \\
=& \frac{(-i)^2}{2} \int_{t_i}^{t_f} dt_1\, dt_2\, \langle\beta|T\left(H_I(t_2) H_I(t_1)\right)|\alpha\rangle \ . \qquad (14.41)
\end{aligned}$$

Again recognize the familiar second order result, Eq. (3.25). The structure of the other terms in the series is now apparent and leads to the suggestive diagrammatic representation shown in Fig. 14.3.

In preparation for application of these ideas to field theory, we next discuss time-ordered products and the generating function.

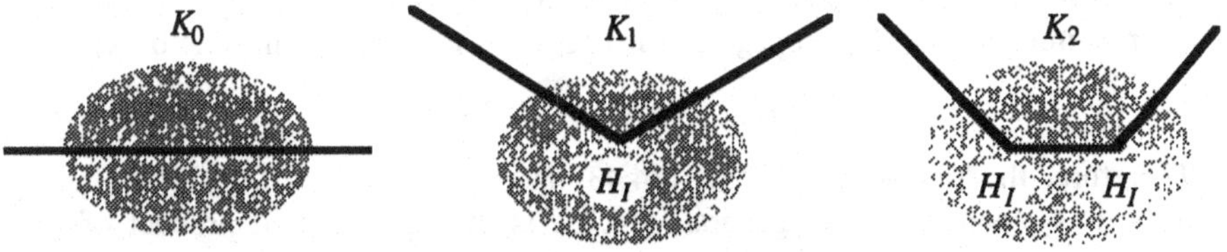

Fig. 14.3 Diagrammatic representation of the perturbation series for the *S-matrix.*

14.3 TIME-ORDERED PRODUCTS

In applications to field theory, one is interested in the expectation values of time-ordered products of fields. Since the fields play the role of generalized coordinates, the analogous quantities in nonrelativistic quantum mechanics are quantities such as

$$\langle q_f, t_f | T\left(Q(t_1)Q(t_2)\ldots Q(t_n)\right) | q_i, t_i\rangle \ ,$$

where, from now on, the initial coordinate and time will be denoted by q_i and t_i (instead of q_0 and t_0) and the final coordinate and time by q_f and t_f (instead of q_n and t_n), and the times t_j, $j = 1$ to n, all lie between t_i and t_f, the initial and final times.

These matrix elements are readily obtained by generalizing the discussion at the end of the previous section. For example, note that

$$\begin{aligned}\langle q_f, t_f | Q(t_1) | q_i, t_i\rangle &= \int dq_1\, dq_1'\, \langle q_f, t_f | q_1, t_1\rangle \overbrace{\langle q_1, t_1 | Q(t_1) | q_1', t_1\rangle}^{q_1(t_1)\delta(q_1-q_1')} \langle q_1', t_1 | q_i, t_i\rangle \\ &= \int dq_1\, q_1(t_1)\, \langle q_f, t_f | q_1, t_1\rangle \langle q_1, t_1 | q_i, t_i\rangle \\ &= \int \left[\frac{dq\, dp}{2\pi}\right] q(t_1) e^{i\int_{t_i}^{t_f} dt[p\dot q - H(p,q)]} \ .\end{aligned} \tag{14.42}$$

Similarly, if $t_1 > t_2$ and both are in the interval $[t_f, t_i]$, then

$$\begin{aligned}&\langle q_f, t_f | Q(t_1)Q(t_2) | q_i, t_i\rangle \\ &\quad = \int dq_1\, dq_1'\, dq_2\, dq_2'\, \langle q_f, t_f | q_1, t_1\rangle \langle q_1, t_1 | Q(t_1) | q_1', t_1\rangle \\ &\qquad\qquad \times \langle q_1', t_1 | q_2', t_2\rangle \langle q_2', t_2 | Q(t_2) | q_2, t_2\rangle \langle q_2, t_2 | q_i, t_i\rangle \\ &\quad = \int dq_1\, dq_2\, q_1(t_1) q_2(t_2) \langle q_f, t_1 | q_1, t_1\rangle \langle q_1, t_1 | q_2, t_2\rangle \langle q_2, t_2 | q_i, t_i\rangle \\ &\quad = \int \left[\frac{dq\, dp}{2\pi}\right] q(t_1) q(t_2) e^{i\int_{t_i}^{t_f} dt[p\dot q - H(p,q)]} \ .\end{aligned} \tag{14.43}$$

The condition that the paths do not double back on themselves was implicit in our construction of the path integral, and this constraint is implemented through the requirement that the propagator for backward propagation in time be zero,

$$\langle q, t|q', t'\rangle = 0 \quad \text{if } t < t' \ . \tag{14.44}$$

Therefore, the right-hand side of (14.43) is equal to $\langle q_f, t_f|Q(t_2)Q(t_1)|q_i, t_i\rangle$ if $t_2 > t_1$, and hence, in general, (14.43) is the time-ordered product of two operators,

$$\langle q_f, t_f|T\left(Q(t_1)Q(t_2)\right)|q_i, t_i\rangle = \int \left[\frac{dq\,dp}{2\pi}\right] q(t_1)\,q(t_2)\,e^{i\int_{t_i}^{t_f} dt[p\dot{q}-H(p,q)]} \ . \tag{14.45}$$

This result clearly generalizes to

$$\langle q_f, t_f|T\left(Q(t_1)Q(t_2)\ldots Q(t_n)\right)|q_i, t_i\rangle = \int \left[\frac{dq\,dp}{2\pi}\right] q(t_1)\,q(t_2)\ldots q(t_n)\,e^{i\int_{t_i}^{t_f} dt[p\dot{q}-H(p,q)]} \ . \tag{14.46}$$

It is convenient to introduce a generating function from which an arbitrary time-ordered product can be determined. To this end, introduce the function

$$z[J] = \int \left[\frac{dq\,dp}{2\pi}\right] e^{i\int_{t_i}^{t_f} dt[p\dot{q}-H(p,q)+J(t)q(t)]} \equiv \langle q_f, t_f|q_i, t_i\rangle^J \ . \tag{14.47}$$

Be careful not to confuse $\langle q_f, t_f|q_i, t_i\rangle^J$ with $\langle q_f, t_f|q_i, t_i\rangle$; they are very different objects, but

$$z[0] = \langle q_f, t_f|q_i, t_i\rangle^0 = \langle q_f, t_f|q_i, t_i\rangle \tag{14.48}$$

is the propagator over the finite time interval $[t_f, t_i]$. If the *functional derivative* is defined by the relation

$$\frac{\delta F[f(x)]}{\delta f(y)} = \lim_{\epsilon\to 0}\left(\frac{1}{\epsilon}\right)\Big(F[f(x)+\epsilon\delta(x-y)] - F[f(x)]\Big) = \frac{\delta F[f]}{\delta f}, \tag{14.49}$$

then

$$\begin{aligned}\frac{\delta z[J(t)]}{\delta J(t_0)} &= \lim_{\epsilon\to 0}\int \left[\frac{dq\,dp}{2\pi}\right] e^{i\int_{t_i}^{t_f} dt[p\dot{q}-H(p,q)+J(t)q(t)]}\left\{\frac{e^{i\epsilon q(t_0)}-1}{\epsilon}\right\}\\ &= i\int \left[\frac{dq\,dp}{2\pi}\right] q(t_0)\,e^{i\int_{t_i}^{t_f} dt[p\dot{q}-H(p,q)+J(t)q(t)]} \ .\end{aligned} \tag{14.50}$$

It follows immediately that

$$\langle q_f, t_f|T\left(Q(t_1)Q(t_2)Q(t_3)\ldots Q(t_n)\right)|q_i, t_i\rangle = (-i)^n \left.\frac{\delta^n z[J(t)]}{\delta J(t_1)\delta J(t_2)\ldots\delta J(t_n)}\right|_{J=0} \ . \tag{14.51}$$

Any time-ordered product can be obtained from a functional derivative of z with respect to J.

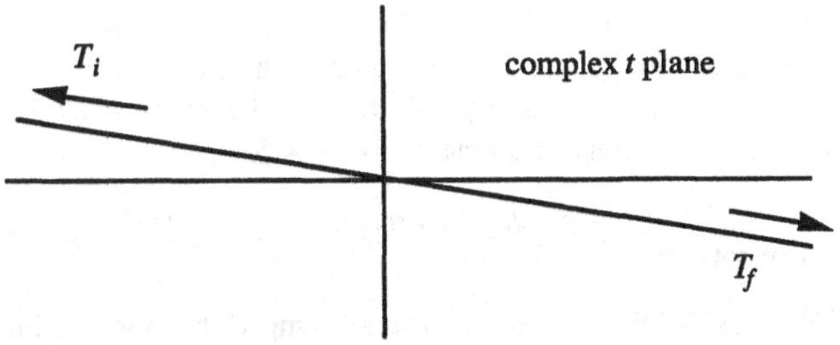

Fig. 14.4 The complex t plane showing the rotated time integration path.

Ground State Expectation Values

We conclude this preparatory discussion by considering the expectation values of the time-ordered products of the Q's in the *ground state*. The ground state is the one with minimum energy $E_0 < E_n$, where E_n are the energies of the excited states enumerated by their quantum numbers n. For free systems, the ground state is the vacuum, and $E_0 = 0$. This ground state is described by the wave function [recall Eq. (14.1)]

$$\psi_0(q,t) = \langle q, t|0\rangle = \psi_0(q)\, e^{-iE_0 t} \; . \tag{14.52}$$

Assume that the $J(t)q(t)$ term [which can be interpreted physically as a source (or sink) for the dynamical variables q (later to be identified as fields)] is switched on at the time t_i in the past and switched off at time t_f in the future. This is not a serious limitation, because t_i and t_f can be made as large as we like, but it does allow us to define all integrals precisely. Then, in the time intervals $T_f > t_f$ and $t_i > T_i$, the states propagate as if $J = 0$, and the matrix elements are

$$\begin{aligned}\langle q_i, t_i|q, T_i\rangle_H &= \left\langle q_i \left| e^{-iH(t_i-T_i)} \right| q\right\rangle_S \\ &= \sum_{n=0}^{\infty} \psi_n(q_i)\psi_n^*(q) e^{-iE_n(t_i-T)} = \sum_{n=0}^{\infty} \psi_n(q_i,t_i)\psi_n^*(q) e^{iE_n T_i} \; .\end{aligned} \tag{14.53}$$

Now, the ground state contribution to this sum is the one which oscillates less rapidly as $T_i \to -\infty$. This behavior can be converted into a practical method for extracting the ground state if we rotate the time axis into the complex plane, as shown in Fig. 14.4. Then the time varies along the line $t(1 - i\epsilon)$, so that it has an imaginary part which approaches $-\infty$ as $t \to \infty$ and $+\infty$ as $t \to -\infty$. While it is sufficient to rotate the time axis through a small angle as suggested by the figure, for numerical calculations it is best to rotate the time axis all the way to the imaginary axis, where the most rapid convergence is achieved. In this case Minkowski space–time is transformed to Euclidean space–time. Or, instead of rotating the time axis, the identical effect is obtained if all *energies* are multiplied by the factor $1 - i\epsilon$. In either case the oscillating factors are converted to exponentials, which approach zero as $T_i \to -\infty$, with the ground state approaching zero least rapidly. Multiplying by $\exp(-iE_0T_i)$ enables us to extract the ground state matrix element from (14.53):

$$\lim_{T_i\to-\infty(1-i\epsilon)} e^{-iE_0T_i} \langle q_i, t_i|q, T_i\rangle = \psi_0(q_i, t_i)\psi_0^*(q) \ , \qquad (14.54)$$

because all other terms in the sum are damped by the exponential factor $e^{\epsilon(E_n-E_0)T_i}$, which approaches zero for $T_i < 0$ and $E_n > E_0$.

This, then, provides a way to project out the ground state matrix element, even if we do not know the ground state wave function $\psi_0(q)$ explicitly. Start with the matrix element $\langle q', T_f|q, T_i\rangle^J$, and insert a complete set of states,

$$\begin{aligned}\langle q', T_f|q, T_i\rangle^J &= \int dq_f\, dq_i\, \langle q', T_f|q_f, t_f\rangle^0 \langle q_f, t_f|q_i, t_i\rangle^J \langle q_i, t_i|q, T_i\rangle^0 \\ &= \sum_{n'n} \psi_{n'}(q')\psi_n^*(q) e^{-iE_{n'}T_f + iE_nT_i} \\ &\quad \times \int dq_f\, dq_i\, \psi_{n'}^*(q_f, t_f) \langle q_f, t_f|q_i, t_i\rangle^J \psi_n(q_i, t_i) . \end{aligned} \qquad (14.55)$$

Then, using Eq. (14.54) and the analogous relation for the final state gives

$$\begin{aligned}&\lim_{\substack{T_i\to-\infty(1-i\epsilon)\\ T_f\to+\infty(1-i\epsilon)}} e^{iE_0(T_f-T_i)} \langle q', T_f|q, T_i\rangle^J \\ &= \psi_0(q')\psi_0^*(q) \int dq_f\, dq_i\, \psi_0^*(q_f, t_f) \langle q_f, t_f|q_i, t_i\rangle^J \psi_0(q_i, t_i) , \end{aligned} \qquad (14.56)$$

and dividing by the various factors, we see that the ground state expectation value is proportional to $\langle q', \infty|q, -\infty\rangle$,

$$\begin{aligned}&\int dq_f\, dq_i\, \psi_0^*(q_f, t_f) \langle q_f, t_f|q_i, t_i\rangle^J \psi_0(q_i, t_i) \\ &= \lim_{\substack{T_i\to i\infty\\ T_f\to-i\infty}} \frac{e^{iE_0(T_f-T_i)} \langle q', T_f|q, T_i\rangle^J}{\psi_0(q')\psi_0^*(q)} = \mathcal{N}_Z(q', q)\ \langle q', \infty|q, -\infty\rangle^J . \end{aligned} \qquad (14.57)$$

where $\mathcal{N}_Z(q', q)$ is a function of proportionality which insures that the product on the right-hand side is independent of q and q'. The *ground state expectation value is therefore proportional to the original functional* $z[J]$, provided we let T_f and $-T_i \to \infty$ along a line which passes from the second to the fourth quadrant, as shown in Fig. 14.4.

As discussed above, an *alternative* way to insure convergence is to give the *energy a small negative imaginary part*, $E \to E(1-i\epsilon)$. With applications to field theory in mind, where the non-interacting Hamiltonian has the form of a simple harmonic oscillator, it is sufficient to approximate the energy by $\frac{1}{2}q^2$, where q^2 is the square of the coordinate. This is a simple positive definite quantity which is a minimum for the ground state. This is the procedure which we will follow, and hence the ground state expectation values will all be obtained from

$$\boxed{Z[J] = \int \left[\frac{dq\,dp}{2\pi}\right] e^{i\int_{-\infty}^{\infty} dt\left[p\dot{q} - H(p,q) + i\epsilon\frac{1}{2}q^2 + J(t)q(t)\right]}} , \tag{14.58}$$

where Z differs from z only in the limits of the time integration and in the small negative real part (proportional to q^2 to insure convergence in q as well as t and coming from the negative imaginary part given to the energy).

The ground state expectation values are *proportional* to derivatives of the generating functional $Z[J]$. How are we to determine the unknown constant $\mathcal{N}_Z$? From Eq. (14.57) this appears to be a very difficult task, and indeed it would be if it was necessary. However, all the physics can be extracted from $Z[J]$ *without knowing the proportionality constant.* This is because the vacuum state must be normalized to unity, and if $J = 0$, Eq. (14.57) tells us that

$$\langle 0|0\rangle = 1 = \mathcal{N}_Z\, Z[0] \ . \tag{14.59}$$

Hence the vacuum expectation values are obtained from the normalized generating function $\overline{Z}$,

$$\overline{Z}[J] = \frac{Z[J]}{Z[0]} \ . \tag{14.60}$$

This gives $\overline{Z}[0] = 1$, as required. From now on we will freely ignore any multiplicative constants which may emerge in the computation of $Z[J]$, anticipating the fact that the physics comes finally from $\overline{Z}[J]$, where all constants cancel.

14.4 PATH INTEGRALS FOR SCALAR FIELD THEORIES *

We now apply the previous ideas to field theory. For simplicity we first treat the symmetric ϕ^3 theory introduced in Sec. 9.1. The results will be extended to spinor fields in Sec. 14.6 and to gauge theories in the next chapter.

The central idea in extending the path integral formalism to field theory is to replace the generalized coordinates Q of the previous discussion by the fields ϕ, which will be the new generalized coordinates. However, two problems prevent us from carrying this over directly. The first problem is that the ϕ's themselves have an uncountable number of degrees of freedom (the values of ϕ at each space–time

---

*I thank Michael Frank for helpful conversations on the definition of path integrals in field theory. See [Fr 91].

point), but this is easily handled by dividing up space into N^3 non-overlapping cells, centered at the points $\alpha = (x_\alpha, y_\alpha, z_\alpha)$. We then average the fields over each small cell (with volume V_α) centered at α:

$$\phi_\alpha(t) = \frac{1}{\sqrt{V_\alpha}} \int_{V_\alpha} d^3r \, \phi(r,t) \ . \tag{14.61}$$

The averaged quantities $\phi_\alpha(t)$ are now a countable number of independent coordinates which are also better behaved than the original fields (recall Prob. 1.5).

The second problem is more serious and requires some discussion. The key to the development of the path integral is the introduction of *eigenstates* of the operators which correspond to the generalized coordinates, but the *quantum field operators have no eigenstates.* However, the *coherent states,* which we introduced briefly in Sec. 1.7, are eigenstates of the annihilation operators A_i (in this chapter, these operators will be denoted by capital letters in order to distinguish them from their eigenvalues, a_i). These states will be denoted $|a\rangle$, where

$$|a\rangle = \prod_i \sum_{n_i} \frac{1}{n_i!} \left(a_i A_i^\dagger\right)^{n_i} |0\rangle \ , \tag{14.62}$$

where the product is over all frequencies i, n_i is the number of quanta with frequency i, and the state is described by the numbers $a = \{a_i\}$. These states are not normalized; their norm is

$$\langle a|a\rangle = \prod_i \sum_{n_i} \frac{(a_i^* a_i)^{n_i}}{n_i!} = e^{\sum_i a_i^* a_i} = e^{a^* \cdot a} \ . \tag{14.63}$$

As we saw in Sec. 1.7,

$$\begin{aligned} A_i|a\rangle &= a_i|a\rangle \qquad A_i^\dagger|a\rangle = \frac{\partial}{\partial a_i}|a\rangle \\ \langle a|A_i^\dagger &= \langle a|a_i^* \qquad \langle a|A_i = \frac{\partial}{\partial a_i^*}\langle a| \ . \end{aligned} \tag{14.64}$$

Note that these relations are consistent with the commutation relations $[A_i, A_j^\dagger] = \delta_{ij}$, and if we make the identification

$$\begin{aligned} A_i^\dagger &\to Q_i \\ -iA_i &\to P_i \ , \end{aligned} \tag{14.65}$$

the operators Q_i and P_i *can be regarded as canonical coordinates,* because they fully describe the degrees of freedom of the field and satisfy the required commutation relations $[Q_i, P_j] = i\delta_{ij}$.

Because these coordinates have eigenfunctions (the coherent states) with eigenvalues a and a^*, we will first express the path integral in terms of the c-numbers a and a^* and then replace the integrations over a and a^* with integrations over the c-number fields ϕ_α and π_α.

To prepare the way for the construction of the path integral, first observe that the matrix element of an operator built from *normal-ordered* products of the annihilation and creation operators is

$$\left\langle a \middle| \mathcal{O}\left(A_i^\dagger, A_j\right) \middle| a' \right\rangle = \mathcal{O}(a_i^*, a_j') \langle a|a'\rangle$$
$$= \mathcal{O}(a_i^*, a_j')\, e^{a^* \cdot a'} \quad , \tag{14.66}$$

where it is understood that $a = \{a_i\}$ and $a' = \{a_i'\}$. Next, we show that the completeness relation for the coherent states takes the following form:

$$\mathbf{1} = \int \prod_i \frac{da_i\, da_i^*}{2\pi i}\, e^{-\sum_i a_i^* a_i} |a\rangle\langle a| \ . \tag{14.67}$$

The presence of the exponential factor is related to the norm (14.63) of the states.

Proof: We will first show that (14.67) commutes with all creation and annihilation operators, which establishes that it must be a multiple of the identity. Then we will show that the constant of proportionality is unity.

Using Eq. (14.64), the commutator is

$$\left[A_i, \int \prod_i \frac{da_i\, da_i^*}{2\pi i}\, e^{-\sum_i a_i^* a_i} |a\rangle\langle a|\right]$$
$$= \int \prod_i \frac{da_i\, da_i^*}{2\pi i}\, e^{-\sum_i a_i^* a_i} \left(a_i - \frac{\partial}{\partial a_i^*}\right) |a\rangle\langle a| = 0 \ , \tag{14.68}$$

where we integrated by parts in the last step, flipping the a_i^* derivative over onto the exponential where it gives a factor which cancels a_i. Taking the Hermitian conjugate shows that the right-hand side of (14.67) also commutes with the creation operators, and since the annihilation and creation operators are a complete set, (14.67) must be a multiple of the identity. To evaluate this multiple, compute its vacuum expectation value

$$\int \prod_i \frac{da_i\, da_i^*}{2\pi i}\, e^{-\sum_i a_i^* a_i} \langle 0|a\rangle\langle a|0\rangle = \int \prod_i \frac{da_i\, da_i^*}{2\pi i}\, e^{-\sum_i a_i^* a_i} \ . \tag{14.69}$$

If we introduce $a_i = r\, e^{i\theta}$, then $da_i\, da_i^* = 2ir\, dr\, d\theta$, and

$$\int \frac{da_i\, da_i^*}{2\pi i}\, e^{-a_i^* a_i} = \frac{1}{\pi} \int_0^{2\pi} d\theta \int_0^\infty r\, dr\, e^{-r^2} = 1 \ ,$$

and hence the quantity (14.69) is unity. ∎

We are now ready to find the path integral for the "coherent state representation" of the time translation operator in field theory. We begin by introducing the field theory equivalent to the wave function (14.1). Since the annihilation operators depend on time, the coherent states will also depend on time, and by analogy with (14.1), the coherent state wave function for the state $|s\rangle$ is therefore

$$\psi_s(a^*, t) = \langle a, t|s\rangle \quad . \tag{14.70}$$

Note that the wave function depends on a^*, and not a. Using the completeness relation (14.67), the time evolution of the wave function for the state $|s\rangle$ can be written

$$\begin{aligned}\psi_s(a_n^*, t_n) = \langle a_n, t_n|s\rangle &= \int \prod_i \frac{da_{i,0}\, da_{i,0}^*}{2\pi i}\, e^{-\sum_i a_{i,0}^* a_{i,0}} \\ &\qquad \times \langle a_n, t_n|a_0, t_0\rangle\langle a_0, t_0|s\rangle \\ &= \int [da_0^*] U(a_n^*, t_n; a_0^*, t_0)\psi_s(a_0^*, t_0) \quad ,\end{aligned} \tag{14.71}$$

where $a_{i,0}$ are the quantities a_i at the time t_0 and the time translation operator is

$$\begin{aligned}U(a_n^*, t_n; a_0^*, t_0) &= \int \prod_i \frac{da_{i,0}}{2\pi i}\, e^{-\sum_i a_{i,0}^* a_{i,0}}\, \langle a_n, t_n|a_0, t_0\rangle \\ &= \int [da_0]\, e^{-a_0^* \cdot a_0}\, \langle a_n, t_n|a_0, t_0\rangle \quad ,\end{aligned} \tag{14.72}$$

and the integration volumes are

$$\begin{aligned}\int [da_0^*] &= \int \prod_i da_{i,0}^* \\ \int [da_0] &= \int \prod_i \frac{da_{i,0}}{2\pi i} \quad .\end{aligned} \tag{14.73}$$

Note the similarity between the definition of the time translation operator in field theory, Eq. (14.71), and the single particle time translation operator, which is the propagator given in Eq. (14.5). The principal difference is that now we have an infinite number of coordinates $\{a_i^*\}$ instead of a single coordinate q. There will be a close similarity between most of the following steps and the discussion in Sec. 14.1.

Next we divide the interval $[t_n, t_0]$ into n intervals and use the completeness of the coherent states to obtain the following general formula for the time translation operator:

$$U(a_n^*, t_n; a_0^*, t_0) = \int \prod_{j=1}^{n-1} [da_j^*] \prod_{j=0}^{n-1} [da_j]\, e^{-a_j^* \cdot a_j}\, \langle a_{j+1}, t_{j+1}|a_j, t_j\rangle \quad . \tag{14.74}$$

If the number of intervals $n \to \infty$, so that $t_{j+1} - t_j = \epsilon \to 0$, then we may estimate the overlap between the coherent states at neighboring times t_{j+1} and t_j by expressing this in terms of coherent states in the Schrödinger representation (as we did in Sec. 14.1), and using Eq. (14.66)

$$\begin{aligned}\langle a_{j+1}, t_{j+1} | a_j, t_j\rangle &= \langle a_{j+1} \left| U(t_{j+1}, t_0) U^\dagger(t_j, t_0) \right| a_j\rangle \\ &\cong \langle a_{j+1} \left| \exp\{-i\epsilon H\left(A^\dagger, A\right) \right| a_j\rangle \\ &\cong \langle a_{j+1} | a_j\rangle - i\epsilon \langle a_{j+1} | H\left(A^\dagger, A\right) | a_j\rangle \\ &\cong e^{a^*_{j+1}\cdot a_j}\left[1 - i\epsilon H\left(a^*_{j+1}, a_j\right)\right] \\ &\cong e^{\left[a^*_{j+1}\cdot a_j - i\epsilon H\left(a^*_{j+1}, a_j\right)\right]} \quad .\end{aligned} \tag{14.75}$$

Inserting this into (14.74) gives the following expression for the time translation operator:

$$U(a^*_n, t_n; a^*_0, t_0) = \int \prod_{j=1}^{n-1} [da^*_j] \prod_{j=0}^{n-1} [da_j]\, e^{\sum_{j=0}^{n-1} \left\{[a^*_{j+1} - a^*_j]\cdot a_j - i\epsilon H\left(a^*_{j+1}, a_j\right)\right\}} \quad . \tag{14.76}$$

Replacing a by ip and a^* by q, as suggested by Eq. (14.65), gives

$$U(q_n, t_n; q_0, t_0) = \int \prod_{j=1}^{n-1} [dq_j] \prod_{j=0}^{n-1} [dp_j]\, e^{\sum_{j=0}^{n-1} i\epsilon\{\dot q_j \cdot p_j - H(q_{j+1}, p_j)\}} \quad , \tag{14.77}$$

where now

$$[dp_j] = \prod_i \frac{dp_{i,j}}{2\pi} \quad .$$

The expression (14.77) is identical to its counterpart (14.17) if *each* q and p in (14.17) is replaced by the set $q = \{q_i\}$ and $p = \{p_i\}$.

To complete the derivation, we must replace the q_i and p_i in (14.77) by the field functions ϕ_α and π_α, the c-number equivalents of the operators Φ and Π. The connection between these averaged fields and the q_i and p_i is

$$\begin{aligned}\phi_\alpha(t) &= \sum_i \frac{1}{\sqrt{2\omega_i}} \left\{ip_i(t)\, f_{\alpha,i} + q_i(t)\, f^*_{\alpha,i}\right\} \\ \pi_\alpha(t) &= \sum_i \sqrt{\frac{\omega_i}{2}} \left\{p_i(t)\, f_{\alpha,i} + iq_i(t)\, f^*_{\alpha,i}\right\} \quad ,\end{aligned} \tag{14.78}$$

where the sum is over all energies i and $f_{\alpha,i}$ is a plane wave averaged over the small volume centered at α,

$$f_{\alpha,i} = \frac{1}{\sqrt{V_\alpha}} \int_{V_\alpha} d^3r \frac{e^{i\boldsymbol{k}_i\cdot \boldsymbol{r}}}{\sqrt{L^3}} \quad . \tag{14.79}$$

The f's satisfy the following completeness and orthogonality relations (see Prob. 14.1):

$$\sum_\alpha f_{\alpha,i} f^*_{\alpha,i'} = \delta_{i\,i'}$$
$$\sum_i f_{\alpha,i} f^*_{\alpha',i} = \delta_{\alpha\alpha'} \quad , \tag{14.80}$$

The relations (14.78) can be regarded as a canonical transformation of the coordinates $\{q_i, p_i\} \to \{\phi_\alpha, \pi_\alpha\}$. The details are saved for Prob. 14.1. We obtain

$$U(\phi_n, t_n; \phi_0, t_0) = \int \prod_\alpha [d\phi_\alpha d\pi_\alpha]\, e^{i\int_{-\infty}^{\infty} dt \sum_\alpha \{\dot\phi_\alpha \pi_\alpha - H(\phi_\alpha, \pi_\alpha)\}} \, . \tag{14.81}$$

Note that, as in the one-particle case, *all operators have been replaced by c-numbers.*

For the ϕ^3 theory under discussion, the interaction terms do not involve any derivatives, and hence the π_α may be integrated out, and the $\pi_\alpha(\partial\phi_\alpha/\partial t) - \mathcal{H}(\phi_\alpha, \pi_\alpha)$ may be replaced immediately by the Lagrangian. Adding a convergence factor and a source term, as we did before, gives the following generating function:

$$Z[J] \cong \int \prod_\alpha [d\phi_\alpha]\, e^{i\int_{-\infty}^{\infty} dt \sum\limits_\alpha \{L(\phi_\alpha) + i\epsilon\frac{1}{2}\phi_\alpha^2 + J_\alpha(t)\phi_\alpha(t)\}} \, , \tag{14.82}$$

where the normalization constants (or factors) N which are encountered when integrating over the ϕ_α [recall Eq. (14.23)] are dropped because the overall normalization is not important in determining the vacuum expectation values (as we showed in the last section). It is somewhat more elegant (but perhaps less precise) to replace the sum over α by an integral over d^3r, giving the following generating function, which is taken as our starting point:

$$\boxed{Z[J] = \int \mathcal{D}[\phi]\, e^{i\int d^4x \{\mathcal{L}(\phi) + i\epsilon\frac{1}{2}\phi^2(x) + J(x)\phi(x)\}} \, ,} \tag{14.83}$$

where, for the symmetric theory introduced in Chapter 9,

$$\mathcal{L} = \mathcal{L}_0 + \mathcal{L}_{\text{int}}$$
$$\mathcal{L}_0 = \tfrac{1}{2}\left(\partial_\mu\phi(x)\partial^\mu\phi(x) - \mu^2\phi^2(x)\right) \tag{14.84}$$
$$\mathcal{L}_{\text{int}} = -\frac{\lambda}{3!}\phi^3(x) \; .$$

We will now discuss the computation of propagators and scattering amplitudes from this generating function.

Generating Function for Free Fields

Begin by ignoring the interaction Lagrangian $\mathcal{L}_{\text{int}}$. Then the generating function is

$$Z_0[J] = \int \mathcal{D}[\phi]\, e^{i\int d^4x\{\mathcal{L}_0(\phi)+i\epsilon\frac{1}{2}\phi^2(x)+J(x)\phi(x)\}} \quad . \tag{14.85}$$

It is convenient to express the $\int d^4x$ integral in terms of momentum space fields using

$$\phi(p) = \int \frac{d^4x}{(2\pi)^2} e^{ip\cdot x}\phi(x) \quad . \tag{14.86}$$

We will reduce the integral for the general case of a complex field, where

$$\begin{aligned}\int d^4x\, &\{\partial_\mu\phi^*(x)\partial^\mu\phi(x) - (\mu^2 - i\epsilon)\phi^*(x)\phi(x)\} \\ &= \int d^4x \int \frac{d^4p_1\, d^4p_2}{(2\pi)^4}\, e^{i(p_1-p_2)\cdot x}\, \left[p_1\cdot p_2 - \mu^2 + i\epsilon\right]\phi^*(p_1)\phi(p_2) \\ &= \int d^4p\, \phi^*(p)\left[p^2 - \mu^2 + i\epsilon\right]\phi(p) \quad .\end{aligned} \tag{14.87}$$

If the field is real (which is the case in the symmetric ϕ^3 theory), $\phi^*(p) = \phi(-p)$. Also, if the field is complex,

$$\begin{aligned}\int d^4x\, &(J(x)^*\phi(x) + \phi^*(x)J(x)) \\ &= \int d^4x \int \frac{d^4p_1\, d^4p_2}{(2\pi)^4} e^{-i(p_1+p_2)\cdot x}\left(J^*(p_1)\phi(p_2) + \phi^*(p_1)J(p_2)\right) \\ &= \int d^4p\left[J^*(p)\phi(p) + \phi^*(p)J(p)\right] \quad .\end{aligned} \tag{14.88}$$

If the field is real, then $J^*(p) = J(-p)$ and only one of these terms is present, or alternatively, the two terms in (14.88) must be multiplied by $\frac{1}{2}$. Hence, for the symmetric ϕ^3 theory, with real field ϕ, the phase θ of the exponential in $Z_0[J]$ is

$$\begin{aligned}\theta &= \int d^4x\left\{\mathcal{L}_0 + i\epsilon\tfrac{1}{2}\phi^2 + J(x)\phi(x)\right\} \\ &= \frac{1}{2}\int d^4p\left\{\phi^*(p)\left[p^2 - \mu^2 + i\epsilon\right]\phi(p) + J^*(p)\phi(p) + \phi^*(p)J(p)\right\} \quad .\end{aligned} \tag{14.89}$$

This can be diagonalized by introducing

$$\begin{aligned}\phi(p) &= \phi_0(p) - \frac{J(p)}{p^2 - \mu^2 + i\epsilon} \\ \phi^*(p) &= \bar{\phi}_0(p) - \frac{J^*(p)}{p^2 - \mu^2 + i\epsilon} \quad .\end{aligned} \tag{14.90}$$

Then

$$\theta = \frac{1}{2}\int d^4p \left\{ \bar{\phi}_0(p)\left[p^2 - \mu^2 + i\epsilon\right]\phi_0(p) - \frac{J^*(p)J(p)}{p^2 - \mu^2 + i\epsilon} \right\} . \tag{14.91}$$

The exponential is now a product of a term depending only on J and one depending only on ϕ. Since we are only interested in derivatives with respect to J, normalized to the amplitude when $J = 0$, the first term, which does not involve J, will play no role, and the normalized generating function $\overline{Z}$ from which the physics emerges is

$$\begin{aligned} \overline{Z}_0[J] = \frac{Z_0[J]}{Z_0[0]} &= \frac{\int \mathcal{D}[\phi]F(\phi)e^{-i\frac{1}{2}\int d^4p \frac{J^*(p)J(p)}{p^2-\mu^2+i\epsilon}}}{\int \mathcal{D}[\phi]F(\phi)} \\ &= \exp\left\{ -i\,\frac{1}{2}\int d^4p\, \frac{J^*(p)J(p)}{p^2 - \mu^2 + i\epsilon} \right\} . \end{aligned} \tag{14.92}$$

In using this with the symmetric theory, remember that $J^*(p) = J(-p)$, so that J^* and J are *not* independent, as they would be for a charged theory.

Calculation of Tree Diagrams

As our first example of how to use this generating function, we find the free propagator. This is the vacuum expectation value of the time-ordered product of field operators, and hence

$$\langle 0|T\left(\Phi(x)\Phi(y)\right)|0\rangle = -\left.\frac{\delta^2 \overline{Z}_0[J]}{\delta J(x)\delta J(y)}\right|_{J=0} . \tag{14.93}$$

To find the derivative, re-express $\overline{Z}_0$,

$$\overline{Z}_0[J] = \exp\left\{ -\frac{i}{2}\int d^4x'\, d^4y'\, J(x')J(y') \int \frac{d^4p}{(2\pi)^4}\, \frac{e^{-ip\cdot(x'-y')}}{p^2 - \mu^2 + i\epsilon} \right\} , \tag{14.94}$$

and therefore (remembering that $J(x)$ and $J(y)$ should be treated as identical)

$$\langle 0|T\left(\Phi(x)\Phi(y)\right)|0\rangle = -i\int \frac{d^4p}{(2\pi)^4}\, \frac{e^{-ip\cdot(x-y)}}{\mu^2 - p^2 - i\epsilon} = i\Delta(x-y) \tag{14.95}$$

which is precisely the result obtained in Eq. (8.31). Note that the matrix element of field *operators* has been obtained from an expression involving *c-number* fields only. This is the familiar feature of the path integral formalism.

Now we include interactions. The functional derivative method may be used to express $\mathcal{L}_{\rm int}$ as an operator. For example,

$$\mathcal{L}_{\rm int}(\phi) = -\frac{\lambda}{3!}\phi^3(x) \to -(-i)^3\frac{\lambda}{3!}\left(\frac{\delta}{\delta J}\right)^3 . \tag{14.96}$$

Putting this in the exponential gives

$$Z[J] = e^{i\int d^4x\, \mathcal{L}_{\rm int}\left(-i\frac{\delta}{\delta J}\right)}\, \overline{Z}_0[J] \ . \tag{14.97}$$

To obtain results in momentum space, we will re-express $\mathcal{L}_{\rm int}$ using the following reduction:

$$\begin{aligned}\int d^4x\, \phi^3(x) &= \int d^4x \frac{d^4p_1\, d^4p_2\, d^4p_3}{(2\pi)^6} e^{-i(p_1+p_2+p_3)\cdot x}\phi(p_1)\phi(p_2)\phi(p_3)\\ &= \int \frac{d^4p_1\, d^4p_2\, d^4p_3}{(2\pi)^2}\, \delta^4(p_1+p_2+p_3)\phi(p_1)\phi(p_2)\phi(p_3)\\ \Rightarrow (-i)^3 &\int \frac{d^4p_1\, d^4p_2\, d^4p_3}{(2\pi)^2}\, \delta^4(p_1+p_2+p_3)\frac{\delta}{\delta J(p_1)}\frac{\delta}{\delta J(p_2)}\frac{\delta}{\delta J(p_3)} \ .\end{aligned} \tag{14.98}$$

Now, as a second example of the use of path integrals, we calculate the $\mathcal{M}$-matrix for the elastic scattering of two particles. We will only obtain the result to second order in the couplings, but many of the steps, including the method for extracting the $\mathcal{M}$-matrix from the path integral, are quite general. Recall that the second order result for this case was already obtained in Sec. 9.7, Eq. (9.90).

From the discussion in Sec. 14.2, we know that the S-matrix is proportional to the path integral, but in field theory a single Lagrangian describes many different interactions, so we must develop a way to project out the particular channel in which we are interested. Recall that the initial and final states are constructed from the vacuum by the action of creation operators and that these creation operators can be obtained from the field operators by projecting out the coefficient of the positive frequency part of the field. The form of this projection depends on the properties of the field. For scalar fields we use the orthogonality of the Klein–Gordon wave functions, and continuum normalization, to obtain

$$a^\dagger(p) = -i\int \frac{d^3r}{\sqrt{2\omega(p)(2\pi)^3}}\, e^{-ip\cdot x}\frac{\overleftrightarrow{\partial}}{\partial t}\Phi(x) = \mathcal{P}(t,\boldsymbol{p})\Phi(x) \ . \tag{14.99}$$

We will also need to express this projection operator in terms of momentum space variables, which can obtained by substituting the momentum space form of Φ into the above expression, giving

$$\begin{aligned}a^\dagger(p) &= -i\int \frac{d^3r}{\sqrt{2\omega(p)(2\pi)^3}}\, e^{-ip\cdot x}\frac{\overleftrightarrow{\partial}}{\partial t}\int\frac{d^4p'}{(2\pi)^2}\, e^{ip'\cdot x}\Phi(p')\\ &= \int \frac{dp_0}{\sqrt{2\omega(p)(2\pi)}}\, e^{-i(\omega(p)-p_0)t}(\omega(p)+p_0)\Phi(p_0,\boldsymbol{p}) = \mathcal{P}(t,\boldsymbol{p})\Phi(p) \ ,\end{aligned} \tag{14.100}$$

where, in the last expression, it is understood that the projection operator $\mathcal{P}(t,\boldsymbol{p})$ involves an integration over the virtual energy p_0 on which the operator $\Phi(p)$

depends and somehow projects in onto the physical energy of the particle, which is $\omega(p)$ (the details of how this works out will be given shortly). Note also that the projection is carried out at a particular time t.

Using the position space projection operator (14.99), the S-matrix, in the notation of Fig. 9.7, is

$$\begin{aligned} S =& \langle 0|a(p_2')a(p_1')\, U_I\, a^\dagger(p_1)a^\dagger(p_2)|0\rangle \\ =&\mathcal{P}^\dagger(\infty,\boldsymbol{p}_2')\mathcal{P}^\dagger(\infty,\boldsymbol{p}_1')\mathcal{P}(-\infty,\boldsymbol{p}_1)\mathcal{P}(-\infty,\boldsymbol{p}_2) \\ &\times \langle 0|\Phi(x_2')\Phi(x_1')U_I\Phi(x_1)\Phi(x_2)|0\rangle \\ =&\mathcal{P}_{\rm total}\,\langle 0|T\left(\Phi(x_2')\Phi(x_1')U_I\Phi(x_1)\Phi(x_2)\right)|0\rangle \quad , \end{aligned} \tag{14.101}$$

where the particles have momenta $p_1 + p_2 \to p_1' + p_2'$, and $\mathcal{P}_{\rm total}$ is the product of the four individual projection operators which project the S-matrix from the vacuum expectation value in (14.101). The time ordering symbol is added for free because all of the fields internal to U_I are already time ordered and are at times between $\pm\infty$, and the *projections* of the initial and final fields commute, so their time order does not matter. Now, as we discussed in Sec. 14.3, this *vacuum expectation value of a time-ordered product* can be calculated from the normalized generating function $\overline{Z}[J]$. Hence, casting this time-ordered product into momentum space and using the momentum form of the projection operators (14.100), we obtain the following expression for the S-matrix which describes elastic scattering:

$$S = (-i)^4\mathcal{P}_{\rm total}\left\{\frac{\delta^4\overline{Z}[J]}{\delta J^*(p_2')\delta J^*(p_1')\delta J(p_1)\delta J(p_2)}\bigg|_{J=0}\right\} \quad . \tag{14.102}$$

The projection operator $\mathcal{P}_{\rm total}$ is now the product of four *momentum space* projection operators with the form given in (14.100) and is not the same as the operator given in Eq. (14.101). Also, in the process of going from the initial expression (14.101) to our final expression (14.102), all operators are replaced by their c-number equivalents, as has been discussed in detail in the previous sections.

Equation (14.102) is a specific example of the general formula for the scattering matrix in the path integral formalism. In the general case, there is one derivative with respect to J for each particle in the initial state and one with respect to J^* for each particle in the final state and projection operators for each external particle. We will discuss the form of the projection operator for fermions later.

We specialize to *non-forward scattering*, where $p_1' \neq p_1$ and $p_2' \neq p_2$. This means that, when computing the above derivatives, we *must involve the interaction terms*, because the free generating function only contains products of the form $J^*(p)J(p)$, in which both initial and final momenta are necessarily equal. The second order generating function, which we denote by $Z^{(2)}$, is

$$\begin{aligned} Z^{(2)}[J] =&(-i)^2 \frac{\lambda^2}{2(3!)^2} \int \frac{d^4k_1\, d^4k_2\, d^4k_3\, d^4k_1'\, d^4k_2'\, d^4k_3'}{(2\pi)^4} \\ &\times \delta^4(k_1+k_2+k_3)\delta^4(k_1'+k_2'+k_3') \\ &\times (-i)^6 \frac{\delta}{\delta J(k_1)} \frac{\delta}{\delta J(k_2)} \frac{\delta}{\delta J(k_3)} \frac{\delta}{\delta J(k_1')} \frac{\delta}{\delta J(k_2')} \frac{\delta}{\delta J(k_3')} \overline{Z}_0[J] \ , \end{aligned} \tag{14.103}$$

where $\overline{Z}_0$ is the free generating function (14.92), which contains the product $J^*(p)J(p) = J(-p)J(p)$. The second order elastic scattering amplitude therefore results from 6 "internal" differentiations from Eq. (14.103) and 4 "external" differentiations from Eq. (14.102), for a total of $6+4=10$ differentiations of J on $\overline{Z}_0$, after which all J's are set equal to zero. Because $\overline{Z}_0 \sim \exp(J^2)$, each differentiation by J brings down a factor of J, which must be eliminated eventually if the final result is to be non-zero when $J \to 0$. Therefore, 5 of the 10 derivatives must act directly on $\overline{Z}_0$, bringing down 5 powers of J, which are then eliminated by the 5 remaining derivatives. Hence, all the derivatives must be "paired" so that the factor of J brought down by one is eliminated by the other. When two derivatives are "paired," their momenta must sum to zero because they act on a single $J(-p)J(p)$ term. Therefore the external derivatives cannot be paired with each other because the scattering is in the non-forward direction, requiring $p_1 \neq p_1'$, etc. Hence each external derivative must pair with one internal derivative, leaving two internal derivatives to pair with each other. However, only internal derivatives from *different* interaction terms can pair; any terms which might arise from a pairing of derivatives within the *same* interaction are zero. To see this, note that if the derivatives with respect to $J(k_1)$ and $J(k_2)$ act on the same $J(-p)J(p)$ term, for example, they will force $k_1+k_2=0$ and the delta function will then force $k_3=0$. Since one of the external four-momenta must pair with k_3, it will be therefore also be zero, which is impossible. Now, since all the derivatives are identical (i.e., the J's are identical even if their arguments are not), there are many ways to obtain the final answer. As we have just shown, the only restriction on how the derivatives are evaluated is that one derivative from each interaction must pair, and there are therefore $3\times 3 = 9$ identical possibilities. Therefore the action of the 6 internal derivatives gives

$$\begin{aligned} &Z^{(2)}[J] \\ &= \frac{\lambda^2}{8} \int \frac{d^4k_1\, d^4k_2\, d^4k_3\, d^4k_1'\, d^4k_2'\, d^4k_3'}{(2\pi)^4} \delta^4(k_1+k_2+k_3)\delta^4(k_1'+k_2'+k_3') \\ &\times \frac{(-i)^5\, J(-k_1)\, J(-k_2)\, J(-k_1')\, J(-k_2')\, \delta^4(k_3'+k_3)\overline{Z}_0[J]}{[k_1^2-\mu^2+i\epsilon][k_2^2-\mu^2+i\epsilon][k_1'^2-\mu^2+i\epsilon][k_2'^2-\mu^2+i\epsilon][k_3^2-\mu^2+i\epsilon]} \\ &= i\frac{\lambda^2}{8} \int \frac{d^4k_1\, d^4k_2\, d^4k_1'\, d^4k_2'}{(2\pi)^4} \frac{\delta^4(k_1+k_2+k_1'+k_2')}{\mu^2 - \frac{1}{4}(k_1+k_2-k_1'-k_2')^2 - i\epsilon} \\ &\times \left(\frac{J(-k_1)}{k_1^2-\mu^2+i\epsilon}\right)\left(\frac{J(-k_2)}{k_2^2-\mu^2+i\epsilon}\right)\left(\frac{J(-k_1')}{k_1'^2-\mu^2+i\epsilon}\right)\left(\frac{J(-k_2')}{k_2'^2-\mu^2+i\epsilon}\right), \end{aligned}$$

where, by convention, the 9 identical terms have been expressed as a factor of 9 times the term with k_3 and k'_3 paired and the factor $\overline{Z}_0[J]$ has been set to unity in the last step, anticipating the fact that none of the 4 external derivatives will act on it, and it will become unity when $J \to 0$ in the final step. Note also that $Z^{(2)}[0] = 0$, showing that this factor does not contribute to the overall normalization factor $Z[0]$, justifying the use of $\overline{Z}_0[0]$ (instead of $Z_0[J]/Z[0]$) in the above equation. Next, differentiating this four times, as required by Eq. (14.102), gives 4!=24 terms, which can be organized into three different terms, each multiplied by 8. Because the two final momenta are fixed by differentiations with respect to $J(-p'_1)$ and $J(-p'_2)$, and the initial momenta by differentiations with respect to $J(p_1)$ and $J(p_2)$, the argument of the delta function becomes $p_1+p_2-p'_1-p'_2$ *regardless of how the derivatives act*, and the different terms are distinguished *only* by different values of the quantity $\Delta = \frac{1}{4}(k_1+k_2-k'_1-k'_2)^2$. Only three different values of Δ are possible, arising as follows:

$$\Delta = (p_1+p_2)^2 \qquad \begin{cases} -p_1,\ -p_2 \to k_1,\ k_2 & p'_1,\ \ p'_2 \to k'_1,\ k'_2 \\ -p_1,\ -p_2 \to k'_1,\ k'_2 & p'_1,\ \ p'_2 \to k_1,\ k_2 \end{cases}$$

$$\Delta = (p_1-p'_1)^2 \qquad \begin{cases} -p_1,\ \ p'_1 \to k_1,\ k_2 & -p_2,\ \ p'_2 \to k'_1,\ k'_2 \\ -p_1,\ \ p'_1 \to k'_1,\ k'_2 & -p_2,\ \ p'_2 \to k_1,\ k_2\,. \end{cases}$$

$$\Delta = (p_1-p'_2)^2 \qquad \begin{cases} -p_1,\ \ p'_2 \to k_1,\ k_2 & p'_1,\ -p_2 \to k'_1,\ k'_2 \\ -p_1,\ \ p'_2 \to k'_1,\ k'_2 & p'_1,\ -p_2 \to k_1,\ k_2 \end{cases}$$

Hence, after the four external derivatives are computed, *and the remaining J's are set to zero*, we have

$$\frac{\delta^4 Z^{(2)}[J]}{\delta J^*(p'_2)\delta J^*(p'_1)\delta J(p_1)\delta J(p_2)}\bigg|_{J=0}$$
$$= i\,\frac{\lambda^2}{(2\pi)^4}\,\delta^4\left(p_1+p_2-p'_1-p'_2\right)$$
$$\times\left(\frac{1}{\mu^2-p'^2_2-i\epsilon}\right)\left(\frac{1}{\mu^2-p'^2_1-i\epsilon}\right)\left(\frac{1}{\mu^2-p^2_1-i\epsilon}\right)\left(\frac{1}{\mu^2-p^2_2+i\epsilon}\right)$$
$$\times\left[\frac{1}{\mu^2-(p_1-p'_1)^2-i\epsilon}+\frac{1}{\mu^2-(p_1-p'_2)^2-i\epsilon}+\frac{1}{\mu^2-(p_1+p_2)^2-i\epsilon}\right].$$

Finally, the S-matrix is obtained by application of the projection operator $P_{\rm total}$. First, replace the energy conserving δ function by its integral representation

$$\delta(p_{10}+p_{20}-p'_{10}-p'_{20}) = \lim_{T\to\infty}\frac{1}{2\pi}\int_{-T}^{T} d\tau\; e^{i(p_{10}+p_{20}-p'_{10}-p'_{20})\,\tau}\,,$$

and, holding T fixed, consider a typical initial time t much earlier than $-T$, so that $-T > t \to -\infty$. The integral over p_{10} (for example) then becomes

$$\int\frac{dp_{10}\,e^{-i[\omega(p_1)t-p_{10}(t+\tau)]}}{\sqrt{2\omega(p_1)}\,2\pi}\frac{[\omega(p_1)+p_{10}]}{[\omega^2(p_1)-p^2_{10}-i\epsilon]}\,f(\{p_i\}) = i\,e^{i\omega(p_1)\tau}\sqrt{\frac{2\pi}{2\omega(p_1)}}\,f(\{p_i\})\,, \tag{14.104}$$

where, anticipating four-momentum conservation, any dependence of the function f (the remaining factors in the integral) on the momentum p_1 has been expressed in terms of the other momenta, $\{p_i\}$. Note that the contour must be closed in the lower half plane because $t+\tau$ is always negative, and the integrand is damped only when p_{10} has a negative imaginary part. Similarly, for the final state projections chose an initial time t' later than T, so that $T < t' \to +\infty$ and

$$\int \frac{dp'_{10}\, e^{i[\omega(p'_1)t'-p'_{10}(t'+\tau)]}}{\sqrt{2\omega(p'_1)\,2\pi}} \frac{[\omega(p'_1)+p'_{10}]}{[\omega^2(p'_1)-p'^2_{10}-i\epsilon]} f(\{p_i\}) = i\, e^{-i\omega(p'_1)\tau}\sqrt{\frac{2\pi}{2\omega(p_1)}}\, f(\{p_i\})$$

because $t' + \tau$ will always be positive, again forcing the contour to be closed in the lower half plane. Hence, *the action of each projection operator puts an external particle on-mass-shell and turns the corresponding external propagator into a factor of* $i\sqrt{\pi/\omega}$. The emergence of these mass shell poles, which are the lowest energy states for the external particles, is a practical application of the discussion leading up to Fig. 14.4. Now, let $T \to \infty$ and reconstruct the delta function, giving

$$S = -i\,\frac{\delta\,(p_1+p_2-p'_1-p'_2)}{(2\pi)^2\sqrt{16\omega_1\omega'_1\omega_2\omega'_2}}\,\mathcal{M}\ , \tag{14.105}$$

where

$$\mathcal{M} = \left[\frac{-\lambda^2}{\mu^2-(p_1-p'_1)^2-i\epsilon} + \frac{-\lambda^2}{\mu^2-(p_1-p'_2)^2-i\epsilon} + \frac{-\lambda^2}{\mu^2-(p_1+p_2)^2-i\epsilon}\right]. \tag{14.106}$$

In (14.106), *each of the external four-momenta is on-mass-shell* (as we just discussed), and hence Eq. (14.106) is identical to Eq. (9.90), illustrating that the path integral formalism and the operator formalism are equivalent. We have recovered the Feynman rules for ϕ^3 "tree" diagrams.

Next, we will see how the Feynman rules for "loops" emerge from the path integral. Again, we confine our discussion to the symmetric ϕ^3 theory. In the next section we calculate the self-energy of the scalar particle, which was previously obtained using the operator formalism in Sec. 11.7. A study of this simple case also leads naturally to a discussion of disconnected diagrams and vacuum bubbles. The discussion will illustrate the similarities and differences between the method of path integrals and the operator formalism.

14.5 LOOP DIAGRAMS IN ϕ^3 THEORY

In this section we calculate the propagator for a neutral, self-conjugate particle to order λ^2. The result for the free propagator was already given in Eq. (14.95), and now we will obtain the first loop contribution to the self-energy.

As before, the propagator is

$$i\Delta(x-y) = \langle 0|T\left(\Phi(x)\Phi(y)\right)|0\rangle = -\left.\frac{\delta^2\overline{Z}[J]}{\delta J(x)\delta J(y)}\right|_{J=0}, \tag{14.107}$$

but now the *generating function includes interactions* in contrast to the free generating function $\overline{Z}_0$ used in Eq. (14.93).

The first non-zero contribution which depends on the coupling strength λ occurs in second order and can be computed from Eq. (14.103), except that in this case it will be necessary to normalize the generator by dividing by $Z[0]$ (it was sufficient to divide by $Z_0[0]$ for the scattering problem we treated above but is no longer sufficient here). The term which is first order in λ does not contribute to the propagator, because when it is inserted into Eq. (14.107), it contains an odd number of J derivatives, leaving at least one factor of J after differentiation and hence insuring that it is zero when $J = 0$.

Now we carry out the six "internal" J derivatives in (14.103). Our analysis is similar to the steps following Eq. (14.103), except that only terms with *two* factors of J brought down from the exponential in Z_0 or *no* factors of J will survive the final action of the two "external" J derivatives in (14.107). Since there are six internal derivatives to be evaluated, the terms with two factors of J require that four of the internal derivatives be paired [as defined in the discussion following Eq. (14.103)], and those with no factors of J require that all of the internal derivatives be paired.

Consider the terms with no factors of J first. These require three pairings, and since all of the derivatives are identical, there are only two distinct ways in which these pairings can be made. We may pair momenta from one interaction with momenta from the other, for example

$$k_1 \leftrightarrow k_1' \qquad k_2 \leftrightarrow k_2' \qquad k_3 \leftrightarrow k_3' \ ,$$

or we may pair momenta within a single interaction, for example

$$k_1 \leftrightarrow k_2 \qquad k_1' \leftrightarrow k_2' \qquad k_3 \leftrightarrow k_3' \ .$$

There are $3! = 6$ ways to make the first pairing and $3 \times 3 = 9$ ways to make the second, giving the following result:

$$\begin{aligned}
\overline{Z}_A^{(2)}[J] &= \frac{-i\lambda^2}{Z[0]} \int \frac{d^4k_1\, d^4k_2\, d^4k_3\, d^4k_1'\, d^4k_2'\, d^4k_3'}{(2\pi)^4}\, \delta^4(k_1+k_2+k_3)\delta^4(k_1'+k_2'+k_3') \\
&\quad \times \left\{ \frac{1}{12} \frac{\delta^4(k_1+k_1')\delta^4(k_2+k_2')\delta^4(k_3+k_3')}{[\mu^2-k_1^2-i\epsilon]\,[\mu^2-k_2^2-i\epsilon]\,[\mu^2-k_3^2-i\epsilon]} \right. \\
&\qquad \left. + \frac{1}{8} \frac{\delta^4(k_1+k_2)\delta^4(k_1'+k_2')\delta^4(k_3+k_3')}{[\mu^2-k_1^2-i\epsilon]\,[\mu^2-k_1'^2-i\epsilon]\,[\mu^2-k_3^2-i\epsilon]} \right\} Z_0[J] \\
&= \frac{-i\lambda^2}{Z[0]}\, \delta^4(0)\, Z_0[J] \int \frac{d^4k_1\, d^4k_2\, d^4k_3\; \delta^4(k_1+k_2+k_3)}{12(2\pi)^4\, [\mu^2-k_1^2-i\epsilon]\,[\mu^2-k_2^2-i\epsilon]\,[\mu^2-k_3^2-i\epsilon]} \\
&\quad + \frac{-i\lambda^2}{Z[0]}\, \delta^4(0)\, Z_0[J] \int \frac{d^4k_1\, d^4k_1'}{8(2\pi)^4\, \mu^2} \frac{1}{[\mu^2-k_1^2-i\epsilon]\,[\mu^2-k_1'^2-i\epsilon]} \\
&= -\frac{i}{2} \frac{Z_0[J]}{Z[0]} (2\pi)^4 \delta^4(0) \left[B + \frac{\lambda^2}{4\mu^2} \Delta^2(0) \right] \ ,
\end{aligned} \tag{14.108}$$

where $\Delta(0)$ is an (infinite) constant obtained from the $\Delta(x-y)$ given in Eq. (14.95) by setting its argument to zero,

$$\Delta(0) = -\int \frac{d^4k}{(2\pi)^4} \frac{1}{\mu^2 - k^2 - i\epsilon} , \tag{14.109}$$

and B is the integral

$$B = \frac{\lambda^2}{6} \int \frac{d^4k_1\, d^4k_2\, d^4k_3\, \delta^4(k_1+k_2+k_3)}{(2\pi)^8 [\mu^2 - k_1^2 - i\epsilon][\mu^2 - k_2^2 - i\epsilon][\mu^2 - k_3^2 - i\epsilon]} . \tag{14.110}$$

We will discuss these quantities after we have completed the calculation.

Now we compute the terms with two factors of J. Now there are three distinct types of terms which occur. There are $6 \times 3 = 18$ terms corresponding to the first pairing given above (with the J^2 term associated with any one of the pairings), $9 \times 1 = 9$ terms corresponding to the second pairing with the J^2 term associated with momenta in different interactions, and $9 \times 2 = 18$ terms corresponding to the second paring with the J^2 term associated with the momenta in the same interaction. Thus we have

$$\begin{aligned}
&\overline{Z}_B^{(2)}[J] \\
&= \frac{\lambda^2}{Z[0]} \int \frac{d^4k_1\, d^4k_2\, d^4k_3\, d^4k_1'\, d^4k_2'\, d^4k_3'}{(2\pi)^4} \delta^4(k_1+k_2+k_3)\delta^4(k_1'+k_2'+k_3') \\
&\quad \times \left\{ \frac{1}{4} \frac{\delta^4(k_1+k_1')\delta^4(k_2+k_2')\, J(-k_3)J(-k_3')}{[\mu^2-k_1^2-i\epsilon][\mu^2-k_2^2-i\epsilon][\mu^2-k_3^2-i\epsilon][\mu^2-k_3'^2-i\epsilon]} \right. \\
&\quad + \frac{1}{8} \frac{\delta^4(k_1+k_2)\delta(k_1'+k_2')\, J(-k_3)J(-k_3')}{[\mu^2-k_1^2-i\epsilon][\mu^2-k_1'^2-i\epsilon][\mu^2-k_3^2-i\epsilon][\mu^2-k_3'^2-i\epsilon]} \\
&\quad \left. + \frac{1}{4} \frac{J(-k_1)J(-k_2)\, \delta(k_1'+k_2')\delta(k_3+k_3')}{[\mu^2-k_1^2-i\epsilon][\mu^2-k_2^2-i\epsilon][\mu^2-k_1'^2-i\epsilon][\mu^2-k_3^2-i\epsilon]} \right\} Z_0[J] \\
&= \frac{\lambda^2}{Z[0]} Z_0[J] \int \frac{d^4k_1\, d^4k_2\, d^4k_3\, \delta^4(k_1+k_2+k_3)\, J(-k_3)J(k_3)}{4(2\pi)^4\, [\mu^2-k_1^2-i\epsilon][\mu^2-k_2^2-i\epsilon][\mu^2-k_3^2-i\epsilon]^2} \\
&\quad + \frac{\lambda^2}{Z[0]} \frac{\Delta(0)\, Z_0[J]}{4\mu^2} \left[\frac{(2\pi)^4\Delta(0)\, J^2(0)}{2\mu^2} - \int \frac{d^4k_1\, J(-k_1)J(k_1)}{[\mu^2-k_1^2-i\epsilon]^2} \right] \\
&= \frac{Z_0[J]}{2Z[0]} \left\{ \int \frac{d^4k\, J(-k)J(k)}{[\mu^2-k^2-i\epsilon]^2} \left[-i\Sigma(k^2) - \frac{\lambda^2\Delta(0)}{2\mu^2} \right] + \frac{\lambda^2(2\pi)^4\Delta^2(0)\, J^2(0)}{4\mu^4} \right\},
\end{aligned} \tag{14.111}$$

where

$$\Sigma(k^2) = \frac{i\lambda^2}{2} \int \frac{d^4k_1}{(2\pi)^4} \frac{1}{[\mu^2-k_1^2-i\epsilon][\mu^2-(k_1+k)^2-i\epsilon]} . \tag{14.112}$$

Note that this quantity is the second order self-energy previously encountered in Sec. 11.7. Combining these terms with those with no factors of J computed

Fig. 14.5 Vacuum bubble diagrams which contribute to $Z^{(2)}$. (A) The term proportional to $\Delta^2(0)/\mu^2$. (B) The term proportional to B.

in Eq. (14.110) above gives the result

$$\overline{Z}^{(2)}[J] = \frac{Z_0[J]}{2Z[0]}\left\{\int \frac{d^4k\, J(-k)J(k)}{[\mu^2 - k^2 - i\epsilon]^2}\left[-i\Sigma(k^2) - \frac{\lambda^2\Delta(0)}{2\mu^2}\right] + \frac{\lambda^2\,(2\pi)^4\Delta^2(0)\,J^2(0)}{4\mu^4} - (2\pi)^4 i\delta^4(0)\left[B + \frac{\lambda^2}{4\mu^2}\Delta^2(0)\right]\right\} . \tag{14.113}$$

We now examine the various terms.

Cancellation of Vacuum Bubbles

Note that the terms proportional to $\delta^4(0)$ on the right-hand side of the above expression are those which do *not* depend on J. They have the general structure

$$\overline{Z}^{(2)}[J]\Big|_{\text{no J}} = C\,\frac{Z_0[J]}{Z[0]} , \tag{14.114}$$

where C is a c-number constant. Hence these terms are also present in $Z[0]$, and when added to the lowest order term, $Z_0[0]$, we get

$$\overline{Z}[J] = \frac{Z_0[J] + Z^{(2)}[J]\Big|_{\text{no J}}}{Z[0]} = \frac{(1+C)Z_0[J]}{(1+C)Z_0[0]} = \overline{Z}_0[J] . \tag{14.115}$$

Hence they are *canceled* by the renormalization of $Z[J]$, and the generality of this argument shows that such terms are canceled to *all orders* and thus need never to be considered in any calculation. These terms are examples of vacuum bubbles, first discussed in Sec. 11.4. They describe the excitation, or fluctuation, of the vacuum. Using the experience acquired in Chapter 11, the interested student can be convinced that the term proportional to $\Delta^2(0)/\mu^2$ corresponds to the Feynman diagram shown in Fig. 14.5A, while the term proportional to B is shown in Fig. 14.5B (see Prob. 14.2). Diagrams of the type shown in Fig. 14.5A are referred to as "tadpole" diagrams; they consist of closed loops with only one interaction.

Such diagrams do *not* occur if the Hamiltonian is normal ordered, as was the case in Chapter 11, but they do occur in the path integral formalism, and this is one of the interesting technical differences between the two formalisms. In this case they cancel, but in later applications they will make important contributions.

With this cancellation, we may replace the constant $Z[0]$ with $Z_0[0]$, drop the J-independent terms, and consider

$$\overline{Z}^{(2)}[J] = \frac{1}{2}\,\overline{Z}_0[J]\left\{\int \frac{d^4k\, J(-k)J(k)}{\left[\mu^2-k^2-i\epsilon\right]^2}\left[-i\Sigma(k^2)-\frac{\lambda^2\Delta(0)}{2\mu^2}\right] + \frac{\lambda^2\,(2\pi)^4\Delta^2(0)\,J^2(0)}{4\mu^4}\right\} . \tag{14.116}$$

Now compute the propagator. Anticipating the fact that the $\overline{Z}_0[J]$ factor will eventually approach unity [after the action of the final two J derivatives in (14.107) followed by the $J\to 0$ limit], we drop this factor and replace the J's by their configuration space representations:

$$\overline{Z}^{(2)}[J] = \frac{1}{2}\int \frac{d^4x\, d^4y\, J(x)J(y)}{(2\pi)^4}\int d^4k\, \frac{e^{-ik\cdot(x-y)}}{[\mu^2-k^2-i\epsilon]^2}\left[-i\Sigma(k^2)-\frac{\lambda^2\Delta(0)}{2\mu^2}\right] + \lambda^2\frac{(2\pi)^4\Delta^2(0)}{8\mu^4}\int\frac{d^4xJ(x)}{(2\pi)^2}\int\frac{d^4yJ(y)}{(2\pi)^2} . \tag{14.117}$$

The second order contribution to the propagator is then

$$i\Delta^{(2)}(x-y) = -\frac{\lambda^2\Delta^2(0)}{4\mu^4} + \int\frac{d^4k\, e^{-ik\cdot(x-y)}}{(2\pi)^4\,[\mu^2-k^2-i\epsilon]^2}\left[i\Sigma(k^2)+\frac{\lambda^2\Delta[0]}{2\mu^2}\right] . \tag{14.118}$$

In momentum space this becomes

$$i\Delta^{(2)}(p) = -\frac{\lambda^2\Delta^2(0)}{4\mu^4}(2\pi)^4\delta^4(p) + \frac{1}{\left[\mu^2-p^2-i\epsilon\right]^2}\left[i\Sigma(p^2)+\frac{\lambda^2\Delta[0]}{2\mu^2}\right] . \tag{14.119}$$

The first term is non-zero only when $p=0$ and will be dropped for now (it will be discussed shortly). Introducing the quantity

$$i\Sigma_T(p) = i\Sigma(p^2) + \frac{\lambda^2\Delta[0]}{2\mu^2} , \tag{14.120}$$

the combined effect of the free propagator plus the second order contributions

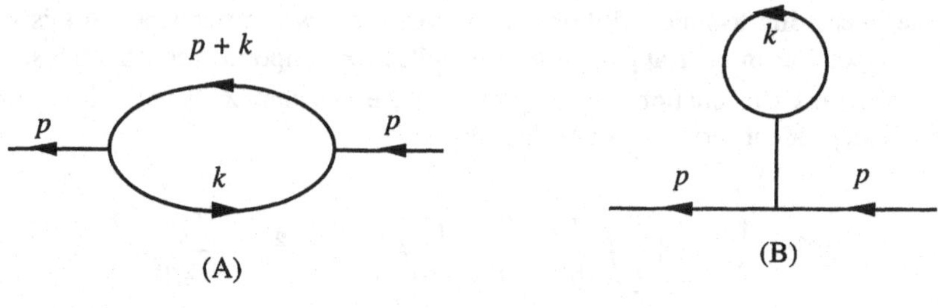

Fig. 14.6 Feynman diagrams for the self-energy. (A) The pair production diagram discussed in Chapter 11 and (B) the tadpole diagram.

become

$$
\begin{aligned}
i\Delta(p) &= \frac{-i}{\mu^2 - p^2 - i\epsilon} + \left(\frac{-i}{\mu^2 - p^2 - i\epsilon}\right)[-i\Sigma_T(p)]\left(\frac{-i}{\mu^2 - p^2 - i\epsilon}\right) \\
&\cong \frac{-i}{[\mu^2 - p^2 - i\epsilon]\left[1 + i\Sigma_T(p)\left(\dfrac{-i}{\mu^2 - p^2 - i\epsilon}\right)\right]} \\
&= \frac{-i}{\mu^2 - p^2 + \Sigma_T(p) - i\epsilon}\,,
\end{aligned}
\tag{14.121}
$$

where we assumed that the first two terms are the beginning of a geometric series which can be summed as we did previously for the electron propagator in Sec. 11.3 and the photon propagator in Sec. 11.5. Note that the self-energy $\Sigma_T(p)$ will change the location of the pole of the propagator, and hence the mass μ of the particle.

Now examine the self-energy $\Sigma_T(p)$. The two contributions can be constructed from the two Feynman diagrams shown in Fig. 14.6:

$$
\begin{aligned}
\Sigma_T(p) = {} & i\frac{(-i\lambda)^2}{2}\int \frac{d^4k}{(2\pi)^4}\left[\frac{-i}{(\mu^2 - k^2 - i\epsilon)}\right]\left[\frac{-i}{\mu^2 - (p+k)^2 - i\epsilon}\right] \\
& + i\frac{(-i\lambda)^2}{2}\left[\frac{-i}{\mu^2}\right]\int \frac{d^4k}{(2\pi)^4}\left[\frac{-i}{(\mu^2 - k^2 - i\epsilon)}\right]\,.
\end{aligned}
\tag{14.122}
$$

Note the symmetry factors of $\frac{1}{2}$ associated with each term. The first term is the $\Sigma(p^2)$ of Eq. (14.112) which was previously encountered in Sec. 11.7, but the second term, shown in Fig. 14.6B, is new. Both diagrams are loop diagrams, where the four-momentum, k, is *unfixed* by the constraints of energy–momentum conservation. We see that the Feynman rules for these loops are identical to those encountered in Chapter 11.

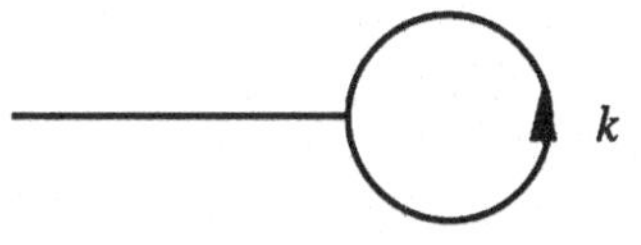

Fig. 14.7 The elementary tadpole diagram

The new diagram, Fig. 14.6B, is another example of a "tadpole" diagram. The simplest example of such a diagram, and the one which originally suggested the name "tadpole," is the diagram shown in Fig. 14.7, which is proportional to

$$i(-i\lambda)\frac{1}{2}\left(\frac{-i}{\mu^2}\right)\int\frac{d^4k}{(2\pi)^4}\frac{-i}{\mu^2-k^2-i\epsilon}=\frac{\lambda}{2\mu^2}\Delta(0)\ . \qquad (14.123)$$

In general, a diagram is referred to as a tadpole diagram if it contains a factor of $\Delta(0)$, which corresponds to a loop which couples to an external particle in only one place. Such a loop arises from a factor of

$$\langle 0|T\left[\phi(x)\phi(y)\right]|0\rangle\ ,$$

where *both ϕ's come from the same interaction term in the Hamiltonian.* In a theory in which the *Hamiltonian is normal ordered, as in the operator formalism developed in Chapter 11, such terms cannot appear,* but in the path integral formalism they arise naturally; this is one difference between the two formalisms. In some sense, the path integral formalism does not permit normal ordering of the fields, which seems intuitive once one realizes that fields behave like c-numbers in the path integral formalism, and as such their order must be the same as multiplication by c-numbers.

We see that the operator formalism of Chapter 11 and the path integral formalism give *different* results for the self-energy of a neutral particle; the difference is the tadpole term shown in Fig. 14.6B. However, this lack of uniqueness has no physical consequences, *because the tadpole 14.6B is a constant* and will therefore be absorbed into the renormalization constants which are ultimately fixed by the physical charge and mass of the particles. The finite parts which remain after the renormalization is completed are identical.

Now, we return to the first term in $i\Delta^{(2)}(p)$ given in Eq. (14.119) and previously ignored. This term corresponds to a *disconnected diagram* for the propagator; specifically, it gives the diagram shown in Fig. 14.8. This explains why it contributes only at zero momenta.

As we noted previously in Sec. 11.2, diagrams which are disconnected are a product of two (or more) *independent* lower order processes. In some cases these processes are unphysical and therefore do not contribute to the S-matrix. The

Fig. 14.8 The disconnected tadpole contribution to the propagator.

diagrams in Figs. 11.1 and 14.8 are examples of such processes. The disconnected tadpole, Fig. 14.8, when considered in isolation, corresponds to the simultaneous absorption and emission of a scalar particle by the vacuum, each of which is an unphysical process. In cases where each of the separate parts of a disconnected diagram are physical, they are more properly regarded as the simultaneous occurrence of more that one process, and not a proper contribution to a description of a single physical process. For all of these reasons, disconnected diagrams should be removed from the theory, and they can be removed systematically by working with the generating function $W[J]$ instead of $Z[J]$, where

$$Z[J] = e^{W[J]} \tag{14.124}$$

or

$$W[J] = \ln Z[J] \ . \tag{14.125}$$

We leave it as an exercise (Prob. 14.3) to show that the disconnected term in $\Delta^{(2)}(p)$ is canceled when $\Delta^{(2)}(p)$ is calculated from W, but that the connected term is unchanged.

14.6 FERMIONS

The extension of path integral techniques to fermions poses a special problem. An essential aspect to the description of Dirac fields is their anticommuting nature, and in the path integral formalism fields are c-numbers! What we need is a formalism or mathematics of anticommuting c-numbers. As it turns out, such a mathematics was developed by Grassmann in the latter half of the 19th century; the algebra of such anticommuting numbers is referred to as a Grassmann algebra.

The Grassmann numbers are constructed from real (or complex) numbers and generators. The generators are denoted C_i, where $i = 1$ to N and N may be infinite. The generators satisfy

$$\{C_i, C_j\} = 0 \ . \tag{14.126}$$

Note that $C_i^2 = 0$. In applications, we are interested in functions of products of η and $\bar{\eta}$, where

$$\begin{aligned} \eta &= \psi C_i \\ \bar{\eta} &= \bar{\psi} C_j \ , \end{aligned} \tag{14.127}$$

and ψ and $\bar{\psi} = \psi^\dagger \gamma_0$ are Dirac c-number spinors and conjugate spinors (respectively) and C_i and C_j are independent Grassmann generators. If there is only one η and one $\bar{\eta}$, then the most general function of $\bar{\eta}\eta$ is simply

$$f(\bar{\eta}\eta) = \sum_{n=0}^{\infty} a_n (\bar{\eta}\eta)^n = a_0 + a_1 \bar{\eta}\eta \quad . \tag{14.128}$$

Differentiation will be defined to correspond to the removal of one power of a Grassmann variable, but the derivative anticommutes with other Grassmann variables, so we have

$$\begin{aligned} \frac{\partial}{\partial \eta} f(\bar{\eta}\eta) &= a_1 \frac{\partial}{\partial \eta}(\bar{\eta}\eta) = -a_1 \bar{\eta} \frac{\partial}{\partial \eta} \eta = -a_1 \bar{\eta} \\ \frac{\partial}{\partial \bar{\eta}} f(\bar{\eta}\eta) &= a_1 \eta \quad . \end{aligned} \tag{14.129}$$

The rules for the integration of Grassmann numbers will be given shortly.

We now show how the anticommuting property of Grassmann numbers makes it possible to extend the path integral formalism to fermions. The first step is to find coherent states of the fermion annihilation operator, which we will denote by B_i (where i is the frequency of energy of the particle). This requires we find states with the following property:

$$B_i |b\rangle = b_i |b\rangle \quad . \tag{14.130}$$

At first it seems that it must be impossible to find a solution to this equation, because we know that fermion states can have at most *one particle in any quantum state*, and the coherent states (14.62) required a sum over states with an *arbitrarily large number of particles in each quantum state*. However, if the eigenvalue b_i is assumed to commute with the state $|b\rangle$,

$$b_i |b\rangle = |b\rangle b_i \ ,$$

then the anticommutation relations satisfied by the annihilation operators B_i imply that the eigenvalues b_i must be complex Grassmann numbers and that the eigenvalues and annihilation operators also anticommute,

$$\{B_i, b_{i'}\} = 0$$

(see Prob. 14.4). Hence, if we define

$$|b\rangle = \prod_i \left(1 - b_i B_i^\dagger\right) |0\rangle \ , \tag{14.131}$$

then

$$B_i|b\rangle = B_i \prod_{i'} \left(1 - b_{i'} B_{i'}^\dagger\right) |0\rangle$$
$$= b_i \prod_{i' \neq i} \left(1 - b_{i'} B_{i'}^\dagger\right) |0\rangle$$
$$= b_i \prod_{i'} \left(1 - b_{i'} B_{i'}^\dagger\right) |0\rangle = b_i|b\rangle \ , \tag{14.132}$$

where the property $b_i^2 = 0$ was used in the last step. These coherent states have a norm which is very similar to their Bose counterparts,

$$\langle b|b\rangle = \langle 0| \prod_{i'} (1 - B_{i'} b_{i'}^*) \prod_i \left(1 - b_i B_i^\dagger\right) |0\rangle$$
$$= \langle 0| \prod_i (1 + b_i^* b_i) |0\rangle$$
$$= e^{\sum_i b_i^* b_i} \ , \tag{14.133}$$

where it is convenient to introduce an exponential in the last step because of the property $e^x \, e^y = e^{x+y}$. Note the form of $\langle b|$, which is dictated by the requirement $\langle b|B_i^\dagger = \langle b|b_i^*$.

The form of (14.133) suggests *defining* definite integrals of Grassmann numbers over the interval $(-\infty, \infty)$ so that the resolution of unity assumes the familiar form

$$\mathbf{1} = \int \prod_i db_i^* \, db_i \, e^{-\sum_i b_i^* b_i} |b\rangle\langle b| \ , \tag{14.134}$$

where, by convention, the factors of $2\pi i$ are omitted for fermions. First, consider the implications of this equation for the case of *only one frequency*. We have

$$\mathbf{1} = \int db_i^* \, db_i \, e^{-b_i^* b_i} \left(1 - b_i B_i^\dagger\right) |0\rangle\langle 0| \left(1 - B_i b_i^*\right) \ . \tag{14.135}$$

Expanding both sides of this equation gives

$$\mathbf{1} = |0\rangle\langle 0| + |1_i\rangle\langle 1_i|$$
$$= \int db_i^* \, db_i \, (1 - b_i^* b_i) \left\{ |0\rangle\langle 0| - b_i|1_i\rangle\langle 0| - |0\rangle\langle 1_i|b_i^* + b_i|1_i\rangle\langle 1_i|b_i^* \right\}$$
$$= \int db_i^* \, db_i \left\{ (1 - b_i^* b_i) \, |0\rangle\langle 0| - b_i|1_i\rangle\langle 0| - |0\rangle\langle 1_i|b_i^* + b_i|1_i\rangle\langle 1_i|b_i^* \right\} \ . \tag{14.136}$$

The equality of the first and last lines implies that

$$\int db_i = 0 = \int db_i^* = 0$$
$$\int db_i \, b_i = 1 = \int db_i^* \, b_i^* \ . \tag{14.137}$$

We leave it as an exercise to show that these conditions are also sufficient to prove (14.134).

Note that these conditions also preserve the translational invariance of Grassmann integrals, which is essential for the reduction of path integrals. Translational invariance implies that

$$\int d\eta\, f(\bar{\eta}\eta) = \int d\eta\, f\left[\bar{\eta}(\eta + \eta')\right] \quad . \tag{14.138}$$

This identity can be proved by expanding out the function f and using $\int d\eta = 0$, which gives

$$\int d\eta\left[a_0 + a_1\bar{\eta}(\eta+\eta')\right] = \int d\eta\left[a_0 + a_1\bar{\eta}\eta\right] \quad . \tag{14.139}$$

Note also that the anticommutation relations require that

$$\int d\eta\, \bar{\eta}\, \eta = -\bar{\eta}\int d\eta\, \eta = -\bar{\eta} \tag{14.140}$$

whereas

$$\int d\bar{\eta}\; \bar{\eta}\eta = \eta \quad . \tag{14.141}$$

Now, consider a space spanned by two independent Grassmann generators. The Grassmann numbers in this space can be represented by a two-dimensional complex vector

$$\bar{\eta} = \begin{pmatrix} \bar{\eta}_1 & \bar{\eta}_2 \end{pmatrix} \qquad \text{and} \qquad \eta = \begin{pmatrix} \eta_1 \\ \eta_2 \end{pmatrix} \quad , \tag{14.142}$$

where η_1 and η_2 are complex numbers which multiply the independent Grassmann generators C_1 and C_2 (in applications, these numbers η_1 and η_2 will also have a four-dimensional Dirac vector structure which has nothing to do with the Grassmann space), and $\bar{\eta}_1$ and $\bar{\eta}_2$ are their complex conjugates (and Dirac conjugates). Our rules of integration imply that

$$\begin{aligned}
\int d\bar{\eta}\, d\eta\, e^{-\bar{\eta}\eta} &= \int d\bar{\eta}\, d\eta \left\{1 - (\bar{\eta}_1\eta_1 + \bar{\eta}_2\eta_2) + \tfrac{1}{2}\left[\bar{\eta}_1\eta_1\bar{\eta}_2\eta_2 + \bar{\eta}_2\eta_2\bar{\eta}_1\eta_1\right]\right\} \\
&= \int d\bar{\eta}\, d\eta \left\{1 - (\bar{\eta}_1\eta_1 + \bar{\eta}_2\eta_2) + \bar{\eta}_1\eta_1\bar{\eta}_2\eta_2\right\} \\
&= \int d\bar{\eta}_1 d\bar{\eta}_2\, d\eta_1 d\eta_2 \left\{1 - \bar{\eta}_1\eta_1 - \bar{\eta}_2\eta_2 + \bar{\eta}_1\eta_1\bar{\eta}_2\eta_2\right\} \\
&= -\int d\bar{\eta}_1\, \bar{\eta}_1 \int d\eta_1\, \eta_1 \int d\bar{\eta}_2\, \bar{\eta}_2 \int d\eta_2\, \eta_2 = -1 \, .
\end{aligned} \tag{14.143}$$

Now, suppose we generalize the quadratic form $\bar{\eta}\eta$ to $\bar{\eta}A\eta$, which can be written

$$\bar{\eta}A\eta = \bar{\eta}_i A_{ij}\eta_j \quad . \tag{14.144}$$

Then, the integral (14.143) generalizes to

$$\begin{aligned}
\int d\bar{\eta}_1 d\bar{\eta}_2 d\eta_1 d\eta_2 \, e^{-\bar{\eta}A\eta} &= \frac{1}{2}\int d\bar{\eta}_1 d\bar{\eta}_2 d\eta_1 d\eta_2 \, (\bar{\eta}_i A_{ij}\,\eta_j)\,(\bar{\eta}_k A_{k\ell}\,\eta_\ell) \\
&= \frac{1}{2}\int d\bar{\eta}_1 d\bar{\eta}_2 d\eta_1 d\eta_2 \, \bar{\eta}_i\,\eta_j\,\bar{\eta}_k\,\eta_\ell\,[A_{ij}A_{k\ell}] \\
&= -\frac{1}{2}\left[A_{22}A_{11} - A_{21}A_{12} + A_{22}A_{11} - A_{12}A_{21}\right] \\
&= -\det A \ ,
\end{aligned}$$

or, if an i is inserted in the exponent,

$$\boxed{\int [d\bar{\eta}\, d\eta]\; e^{-i\bar{\eta}A\eta} = -(i)^n \det A \ ,} \tag{14.145}$$

where n is the dimension of the matrix A. The result, which can be generalized to arbitrarily large matrices, will be very useful in the following sections. Note that this result for Grassmann integrals is very different from what would have been obtained from ordinary numbers:

$$\int [da^*\, da]\; e^{-ia^*Aa} = \frac{(2\pi)^n}{\det A} \quad . \tag{14.146}$$

Free Dirac Fields

Since we have successfully defined coherent states for Dirac annihilation operators and obtained a form for the resolution of unity which is identical to the one for scalar fields, the rest of the discussion for scalar fields may be carried over directly to Dirac fields. The generator for free Dirac fields is then

$$Z_0[\bar{\eta},\eta] = \int [d\bar{\psi}\, d\psi]\, e^{\,i\int d^4x \left\{\bar{\psi}(x)\left(\frac{i}{2}\gamma^\mu \overleftrightarrow{\partial}_\mu - m\right)\psi(x) + i\epsilon^2\bar{\psi}\psi + \bar{\eta}\psi + \bar{\psi}\eta\right\}} \ , \tag{14.147}$$

where $\bar{\psi}(x)$, $\psi(x)$, $\eta(x)$ and $\bar{\eta}(x)$ are all four-component Dirac vectors, which are also infinite-dimensional Grassmann variables, with

$$\psi(x) \longrightarrow \psi(x_i) = \psi_i C_i \quad . \tag{14.148}$$

We reduce Z_0 using the same steps carried out for the scalar field ϕ in Sec. 14.4. In particular, transform the integrals to momentum space using

$$\begin{aligned}
\psi(p) &= \int \frac{d^4x}{(2\pi)^2}\, e^{ip\cdot x}\, \psi(x) \\
\bar{\psi}(p) &= \int \frac{d^4x}{(2\pi)^2}\, e^{-ip\cdot x}\, \bar{\psi}(x) \ .
\end{aligned} \tag{14.149}$$

Integrating by parts,

$$\int d^4x\,\bar{\psi}(x)\overleftrightarrow{\partial}_\mu\psi(x) = 2\int d^4x\,\bar{\psi}(x)\overrightarrow{\partial}_\mu\,\psi(x)\ , \tag{14.150}$$

and substituting the Fourier transforms, we obtain

$$Z_0[\bar{\eta},\eta] = \int \left[d\bar{\psi}\,d\psi\right]\, e^{i\int d^4p\{\bar{\psi}(p)(\not{p}-m+i\epsilon)\psi(p)+\bar{\eta}(p)\psi(p)+\bar{\psi}(p)\eta(p)\}}. \tag{14.151}$$

Next, make the substitution

$$\begin{aligned}\psi(p) &= \psi^0(p) - (\not{p} - m + i\epsilon)^{-1}\,\eta(p)\\ \bar{\psi}(p) &= \bar{\psi}^0(p) - \bar{\eta}(p)\,(\not{p} - m + i\epsilon)^{-1}\end{aligned} \tag{14.152}$$

and use the translational invariance of the integrals to obtain

$$Z_0[\bar{\eta},\eta] = \int \left[d\bar{\psi}^0\,d\psi^0\right]\, e^{i\int d^4p\{\bar{\psi}^0(p)(\not{p}-m+i\epsilon)\psi^0(p)-\bar{\eta}(p)(\not{p}-m+i\epsilon)^{-1}\eta(p)\}}\ . \tag{14.153}$$

We have again separated Z_0 into a part independent of η and $\bar{\eta}$ and a part which depends on η and $\bar{\eta}$ only. Dividing by $Z_0(0)$, we have

$$\boxed{\overline{Z}_0[\bar{\eta},\eta] = e^{i\int d^4p\,\bar{\eta}(p)[m-\not{p}-i\epsilon]^{-1}\eta(p)}\ .} \tag{14.154}$$

The free propagator is the vacuum expectation value of the time-ordered product of field operators. In this case it is

$$\begin{aligned}iS_{\alpha\beta}(x_1,x_2) &= \langle 0|T\left(\psi_\alpha(x_1)\bar{\psi}_\beta(x_2)\right)|0\rangle\\ &= -(-i)^2\frac{\delta^2}{\delta\bar{\eta}_\alpha(x_1)\delta\eta_\beta(x_2)}\bar{Z}_0[\bar{\eta},\eta]\Bigg|_{\eta=\bar{\eta}=0}\end{aligned} \tag{14.155}$$

where the extra minus sign arises from the $\delta\eta$ differentiation, using

$$\frac{\delta}{\delta\eta}\bar{\psi}\,\eta = -\bar{\psi}\ . \tag{14.156}$$

Applied to the above form for $\overline{Z}_0$, this gives

$$iS_{\alpha\beta}(x_1,x_2) = \int \frac{d^4p}{(2\pi)^4}\,e^{ip\cdot(x_1-x_1)}\frac{-i}{(m-\not{p}-i\epsilon)_{\alpha\beta}}\ , \tag{14.157}$$

in precise agreement with Eq. (11.21b).

Fermion Loops

We close this discussion by considering fermion loops in a theory with a scalar meson coupled to a fermion. This is similar to the ϕ^3 case previously discussed, except that now there are two different kinds of particles. The interaction term is

$$\mathcal{L}_I = -g\,\bar{\psi}\psi\phi \ . \tag{14.158}$$

Replacing $\bar{\psi}\psi\phi$ by derivatives, using Eq. (14.156), gives

$$Z[J,\bar{\eta},\eta] = \exp\left\{(-i)^4\, g \int d^4x\, \frac{\delta}{\delta\bar{\eta}(x)}\,\frac{\delta}{\delta\eta(x)}\,\frac{\delta}{\delta J(x)}\right\} Z_0[J,\bar{\eta},\eta]\,. \tag{14.159}$$

The sign in the exponent depends on the order in which the η and $\bar{\eta}$ derivatives are written; for the order given in the above expression the sign is the same as for the generating function $Z[J]$ of ϕ^3 theory.

To illustrate the difference between closed fermion and boson loops and obtain the factor of -1 for closed fermion loops found previously in Sec. 11.5, we calculate the self-energy of a scalar particle which arises from a closed fermion loop, similar to that drawn in Fig. 14.6A (the drawing for the two cases is identical, but the meaning of the lines and vertices is different). The calculation is sufficiently different from that given in Sec. 14.5 that a new calculation is necessary. In order to track the difference between a charged scalar loop and a fermion loop, we will keep track of sign changes which arise from the interchange of the Grassmann numbers by multiplying by a factor of $\xi = -1$ for each interchange. Then, after the calculation is complete, we can recover the result for a charged scalar loop by changing $\xi \to 1$ (and making other changes which we will discuss below).

Drawing on our experience with the previous calculation, the second order contribution to the scalar particle self-energy comes from the term

$$\begin{aligned}\overline{Z}^{(2)} =& \frac{1}{2}g^2 \int \frac{d^4p_1'\, d^4p_1\, d^4k_1\, d^4p_2'\, d^4p_2\, d^4k_2}{(2\pi)^4}\,\delta^4(p_1'-p_1+k_1)\,\delta^4(p_2'-p_2+k_2)\\ &\times \frac{\delta}{\delta\bar{\eta}_\alpha(p_1)}\,\frac{\delta}{\delta\eta_\alpha(p_1')}\,\frac{\delta}{\delta J(k_1)}\,\frac{\delta}{\delta\bar{\eta}_\beta(p_2)}\,\frac{\delta}{\delta\eta_\beta(p_2')}\,\frac{\delta}{\delta J(k_2)}\overline{Z}_0[J,\bar{\eta},\eta]\,,\end{aligned} \tag{14.160}$$

where the momenta of the ψ fields are denoted by p_1 and p_2, the momenta of the $\bar{\psi}$ fields are denoted by p_1' and p_2', and momenta of both scalar fields are outgoing, so that their source terms are $J(k)$. Now, since the source terms occur only in the combinations $\bar{\eta}(p)\eta(p)$ and $J(k)J(-k)$, the only non-zero, *connected terms* come from the pairing of p_2 with p_1', p_1 with p_2', and k_1 and k_2 with the external momenta k. Hence the terms we need are

$$\overline{Z}^{(2)} \Rightarrow \frac{\xi^2 g^2}{2(2\pi)^4} \int d^4p_1'\, d^4p_1\, d^4k_1\, d^4p_2'\, d^4p_2\, d^4k_2$$

$$\times\, \delta^4(p_1' - p_1 + k_1)\delta^4(p_2' - p_2 + k_2)\frac{\delta}{\delta J(k_1)}\,\frac{\delta}{\delta J(k_2)}$$

$$\times \left\{\frac{\delta}{\delta\bar\eta_\alpha(p_1)}\,\frac{\delta}{\delta\eta_\alpha(p_1')}\,\bar\eta_\gamma(p_2')iS_{\gamma\beta}(p_2')iS_{\beta\delta}(p_2)\eta_\delta(p_2)\right\}\overline{Z}_0$$

$$= \frac{\xi^3 g^2}{2(2\pi)^4} \int d^4p_1'\, d^4p_1\, d^4k_1\, d^4p_2'\, d^4p_2\, d^4k_2$$

$$\times\, \delta^4(p_1' - p_1 + k_1)\delta^4(p_2' - p_2 + k_2)\frac{\delta}{\delta J(k_1)}\,\frac{\delta}{\delta J(k_2)}$$

$$\times\, \delta^4(p_2' - p_1)\delta^4(p_2 - p_1')\,\{iS_{\alpha\beta}(p_1)iS_{\beta\alpha}(p_2)\}\,\overline{Z}_0[J,\bar\eta,\eta]$$

$$= -\frac{\xi^2}{2}\int d^4k_1\, d^4k_2\, \delta^4(k_1+k_2)\frac{\delta}{\delta J(k_1)}\,\frac{\delta}{\delta J(k_2)}\left[-i\Sigma(-k_1)\right]\overline{Z}_0[J,\bar\eta,\eta]$$

$$= \frac{\xi^2}{2}\overline{Z}_0[J,\bar\eta,\eta]\int \frac{d^4k\, J(-k)J(k)}{[\mu^2 - k^2 - i\epsilon]^2}\left[-i\Sigma(-k)\right]\,, \tag{14.161}$$

where the extra factor of $\xi = -1$ arose in the second step when we needed to pass the $\delta/\delta\eta_\alpha$ derivative through $\bar\eta_\gamma$ to act on η_δ. [Two minus signs (ξ^2) also arose in the first step, but these do not change the overall sign.] The self-energy Σ introduced in the last step is

$$\Sigma(k) = -i\xi g^2 \int \frac{d^4p_1 d^4p_2}{(2\pi)^4}\,\delta^4(p_2 - p_1 - k)\,\{iS_{\alpha\beta}(p_1)iS_{\beta\alpha}(p_2)\}$$

$$= -ig^2 \int \frac{d^4p}{(2\pi)^4}\,\frac{\mathrm{tr}\,\{(m+\not p)(m+\not p+\not k)\}}{(m^2 - p^2 - i\epsilon)\,(m^2 - (p+k)^2 - i\epsilon)}\,. \tag{14.162}$$

Note that the final form of $\overline{Z}^{(2)}$ given in Eq. (14.161) has the same structure as Eq. (14.116), permitting us to identify $i\Sigma(k)$ as the loop contribution to the self-energy of the scalar particle, which we are seeking. Equation (14.162) exhibits features of fermion loops which were previously encountered in Sec. 11.5:

- a trace must be taken over the product of Dirac propagators.
- there is an additional factor of -1 which arises from the closed fermion loop.

The result for a loop involving charged *scalar* particles is easily obtained from (14.162). The calculation is the same except $\xi = 1$, and scalar propagators must be substituted for Dirac propagators. We obtain the result

$$\Sigma(k) = ig^2 \int \frac{d^4p}{(2\pi)^4}\,\frac{1}{(m^2 - p^2 - i\epsilon)\,(m^2 - (p+k)^2 - i\epsilon)}\,. \tag{14.163}$$

Note that there is no symmetry factor of $\frac{1}{2}$ if the scalar particles are charged.

This concludes our introductory discussion of the use of path integrals to quantize field theories. Note that the formalism allows us to use c-number fields (but Dirac fields must be described by anticommuting Grassmann numbers) and that the results are equivalent to the operator formalism we presented in Chapters 9–11. In the next chapter we use the techniques developed here to quantize QED and QCD, and we will discuss the Standard Model.

PROBLEMS

14.1 Obtain the path integral given in Eq. (14.81) from the path integral Eq. (14.77) by transforming from the coordinates $\{q_i, p_i\}$ to the coordinates $\{\phi_\alpha, \pi_\alpha\}$ using the transformations given in Eq. (14.78). This can be accomplished by working through the following steps:

(a) From the relations

$$\int_{L^3} \frac{d^3r}{L^3}\, e^{i(\boldsymbol{k}_i-\boldsymbol{k}_{i'})\cdot\boldsymbol{r}} = \delta_{ii'}$$

$$\sum_i \frac{e^{i\boldsymbol{k}_i\cdot(\boldsymbol{r}-\boldsymbol{r}')}}{L^3} = \delta^3(\boldsymbol{r}-\boldsymbol{r}') \ ,$$

prove the completeness and orthogonality relations given in Eq. (14.80).

(b) Prove that the volume integration is invariant by showing that, for each time "slice,"

$$\int \prod_i \frac{dq_i\, dp_i}{2\pi} = \eta \int \prod_\alpha \frac{d\phi_\alpha\, d\pi_\alpha}{2\pi} \ ,$$

where $|\eta|^2 = 1$ and can be ignored.

(c) The first term in the exponential in Eq. (14.77) may be symmetrized by integrating half of the expression by parts,

$$\int dt \sum_i \dot{q}_i p_i = \frac{1}{2}\int dt \sum_i (\dot{q}_i p_i - q_i \dot{p}_i) \ .$$

Show that

$$\frac{1}{2}\int dt \sum_i (\dot{q}_i p_i - q_i \dot{p}_i) = \frac{1}{2}\int dt \sum_\alpha \left(\dot{\phi}_\alpha \pi_\alpha - \phi_\alpha \dot{\pi}_\alpha\right) \to \int dt \sum_\alpha \dot{\phi}_\alpha \pi_\alpha \, .$$

14.2 Using the Feynman rules worked out in Chapter 11 or given in Appendix B, obtain the correct integrals corresponding to the two diagrams in Fig. 14.5. (Don't overlook the symmetry factors). Can the same results be obtained from Eq. (14.113)? (It is an interesting exercise to use the Feynman rules to construct these diagrams, but *do not forget that all such diagrams can be neglected* because they cancel when the generating function is renormalized.)

14.3 Using the generating function $W[J]$, defined in Eq. (14.125), calculate the propagator for a neutral scalar meson to second order in the coupling constant λ, and show that it contains no disconnected pieces. Compare your answer with the result given in Eq. (14.122).

14.4 From the eigenvalue condition (14.130) and the anticommutation relations $\{B_i, B_j\} = \delta_{ij}$, prove that the eigenvalues of the coherent states, b_i, must be Grassmann numbers and that the Grassmann numbers also anticommute with the annihilation operators.

CHAPTER 15

QUANTUM CHROMODYNAMICS AND THE STANDARD MODEL

In this chapter we apply the path integral formalism developed in the previous chapter to the quantization of gauge theories. The simple case of an Abelian theory (QED) is treated first, and then we discuss the quantization of non-Abelian gauge theories and obtain the new Feynman rules for *ghost* lines and vertices in QCD. Ghosts are particles which violate the connection between spin and statistics (in this case they are scalar particles which obey Fermi statistics), and hence they cannot exist in initial or final states but only appear as virtual particles inside of loop diagrams (hence the name "ghost"). We will show that one of their roles in QCD is to maintain unitarity. We conclude the chapter with a discussion of the current theory of the electroweak interactions, referred to as the *Standard Model.*

15.1 QUANTIZATION OF GAUGE THEORIES

The path integral formalism will now be used to quantize gauge field theories. The general results we obtain will first be applied to QED at the end of this section and to QCD in the next section.

We begin the discussion by considering the following generating function for the free electromagnetic field with source $\jmath_\mu$:

$$Z_0[\jmath] = \int \mathcal{D}[A_\mu] e^{\,i\int d^4x\{-\frac{1}{4}F_{\mu\nu}F^{\mu\nu}+\jmath_\mu A^\mu\}} \ . \tag{15.1}$$

Note that the source term is identical to a current interaction term, giving us the familiar identification of currents as sources of the electromagnetic field.

Recall from our discussion in Sec. 2.2 that quantization of the EM field presented a problem because $\partial A_0/\partial t$ was not contained in the Lagrangian, and hence A_0 was not a dynamical variable. This problem was "solved" by choosing a gauge (the Coulomb gauge) in which A_0 could be easily eliminated and the fields quantized. One constraint remained, the Coulomb gauge condition:

$$\nabla \cdot \boldsymbol{A} = 0 \ . \tag{15.2}$$

We have a similar problem with the generating function (15.1). The action is *invariant under the gauge transformation* $A'_\mu = A_\mu + \partial_\mu \Lambda$ (we assume the source current is conserved; $\partial_\mu j^\mu = 0$). This means that the field can be separated into two parts:

$$A_\mu = \{A_{D\mu}, A_{G\mu}\} \quad , \tag{15.3}$$

where $A_{G\mu}$ are the "gauge" components which leave the action invariant and $A_{D\mu}$ are the "dynamical" components upon which the action depends. Since the action is *independent of the gauge components,* they cannot be determined from the variational principle which gives the field equations and must be integrated out. One of the nice features of the path integral is that it is possible to express the integration over the gauge components as an overall factor which is *independent of the dynamical components* $A_{D\mu}$. This factor is infinite but has no effect on the dynamics because it is an overall multiplicative factor which can be absorbed into the normalization constant which is divided out when we evaluate propagators, S matrix elements, and other physical observables.

Before the gauge degrees of freedom can be separated from the dynamical degrees of freedom, we must define how the separation is to be made. This is done by imposing a constraint on the fields. The constraint, or gauge condition, defines the dynamical fields; all fields which satisfy the constraint are dynamical. The constraint is chosen so that fields which do not satisfy it differ from those which do by a gauge transformation which leaves the action invariant and therefore these additional fields are redundant. These fields which do not satisfy the constraint are gauge fields which must be integrated out.

An example, taken from Cheng and Li (1984), will illustrate this discussion. Suppose we have a complex field, $\psi = r\, e^{i\theta}$, and an action $\mathcal{A}(\psi^*\psi)$ which depends on $\psi^*\psi = r^2$ only. The action is therefore invariant under the gauge transformation

$$\psi' = e^{i\phi}\, \psi \ . \tag{15.4}$$

The path integral which determines the dynamics of this system is

$$Z = \int d\theta\, r dr\; e^{i\mathcal{A}(r^2)} \to \int r dr\; e^{i\mathcal{A}(r^2)} \quad , \tag{15.5}$$

where in the last expression we have integrated out the "gauge dependent" degrees of freedom by integrating over the redundant variable θ and dropped this factor because it is a constant. While elimination of the redundant degrees of freedom was trivial in this case, it is good to have a systematic method for carrying out the separation in the general case, and one method is to insert unity, written in the following form:

$$1 = \int d\phi\, \delta(\theta - \phi) \quad , \tag{15.6}$$

into the original integral. This gives

$$Z = \int d\phi \int d\theta\, r dr\, \delta(\theta - \phi)\, e^{iA(r^2)} = \int d\phi\, Z_\phi$$
$$= Z_\phi \int d\phi \to Z_\phi \ , \tag{15.7}$$

where the integral over ϕ can be separated out because Z_ϕ does not depend on ϕ (gauge invariance). This method separates the integral into a "dynamical" part, in which the fields are specified by the "gauge condition" $\theta = \phi$, and a "gauge" part which includes the redundant dependence on the gauge angle ϕ. The gauge group in this example is the group $U(1)$ of multiplications by a complex phase, and the integral over ϕ is an integral over all elements of the gauge group.

To prepare the way for application of these ideas to gauge theories, we will generalize the above example. A constraint which is more complicated that $\theta = \phi$ might be chosen to define the dynamical fields. Such a constraint can be written in the form

$$F(r, \theta) = 0 \ . \tag{15.8}$$

Then, in place of Eq. (15.6) we have a more general result,

$$\Delta^{-1}(r, \theta) = \int d\phi\, \delta\left[F(r, \theta - \phi)\right] \ , \tag{15.9}$$

where Δ can be evaluated directly,

$$\Delta(r, \theta) = \left.\frac{\partial F(r, \theta)}{\partial \theta}\right|_{F=0} \ . \tag{15.10}$$

It can also be shown that Δ is gauge invariant, which for this example means that it is independent of θ:

$$\Delta^{-1}(r, \theta + \phi') = \int d\phi\, \delta\left[F(r, \theta + \phi' - \phi)\right]$$
$$= \int d(\phi - \phi')\, \delta\left[F(r, \theta + \phi' - \phi)\right]$$
$$= \Delta^{-1}(r, \theta) \ . \tag{15.11}$$

The crucial step in this "proof" is the invariance of the measure for the group integration, referred to as a Hurwitz measure, which is expressed mathematically as $\int d\phi = \int d(\phi - \phi')$. This is trivially true in this example.

Using these results, the path integral for the general constraint (15.8) can be written

$$Z = \int d\phi \int d\theta\, r dr\, \Delta(r, \theta) \delta\left[F(r, \theta - \phi)\right]\, e^{iA(r^2)} = \int d\phi\, Z_\phi$$
$$= Z_\phi \int d\phi \to Z_\phi \ , \tag{15.12}$$

where

$$Z_\phi = \int d\theta\, r dr\, \Delta(r,\theta)\delta\left[F(r,\theta-\phi)\right]\, e^{iA(r^2)} \quad . \tag{15.13}$$

This new path integral includes the constraint and yet is still gauge invariant. To prove the latter, use the invariance of the group measure, of Δ, and of the action

$$\begin{aligned} Z_{\phi'} &= \int d\theta\, r dr\, \Delta(r,\theta)\delta\left[F(r,\theta-\phi')\right]\, e^{iA(r^2)} \\ &= \int d(\theta'-\phi+\phi')\, r dr\, \Delta(r,\theta'-\phi+\phi')\delta\left[F(r,\theta'-\phi)\right]\, e^{iA(r^2)} \\ &= \int d\theta'\, r dr\, \Delta(r,\theta')\delta\left[F(r,\theta'-\phi)\right]\, e^{iA(r^2)} = Z_\phi \quad . \end{aligned} \tag{15.14}$$

The path integral will now undergo one more transformation before it is in its final form. It is convenient to generalize the gauge fixing condition (15.8) by the more general constraint

$$F(r,\theta) = G \quad , \tag{15.15}$$

and then average over all values of G using the following integral:

$$1 = \sqrt{\frac{i}{2\pi\alpha}} \int_{-\infty}^{\infty} dG\, e^{-i\frac{1}{2\alpha}G^2} \quad , \tag{15.16}$$

where α is a constant referred to as the *gauge fixing parameter* (to be discussed below). Since G is independent of all the other variables, Δ will be unchanged by this substitution, and discarding the unimportant constant in Eq. (15.16) the path integral (15.13) can be replaced by

$$\begin{aligned} Z_\phi &= \int dG \int d\theta\, r dr\, \Delta(r,\theta)\delta\left[F(r,\theta-\phi)-G\right]\, e^{i\left(A(r^2)-\frac{1}{2\alpha}G^2\right)} \\ &= \int d\theta\, r dr\, \Delta(r,\theta)\, e^{i\left(A(r^2)-\frac{1}{2\alpha}F^2\right)} \quad . \end{aligned} \tag{15.17}$$

This is the form we will employ in our discussion of gauge theories. Note the following features

- The gauge degrees of freedom have been removed from the field integration.
- The new integral includes the gauge constraint in two places: the effective action includes a gauge fixing factor of F^2, and the integration measure includes the factor Δ. In spite of these factors, the overall expression is gauge invariant.

In the example we have been discussing, the initial gauge constraint was $\phi = \theta$, which is equivalent to the function $F = \theta$. In this simple case, $\Delta = 1$, and the gauge fixing term reduces to a constant,

$$\int d\theta\, e^{-i\frac{1}{2\alpha}\theta^2} = \text{constant} \quad ,$$

which can be dropped, showing that (15.17) reduces to (15.7).

We now return to our discussion of gauge theories. As discussed above, the gauge condition will be written in the following general form:

$$F(\mathbf{A}) = 0 \ , \tag{15.18}$$

where F is some function or operator which depends on the components of $\mathbf{A}_\mu = \frac{1}{2}\lambda_a A^a_\mu$ (where $\frac{1}{2}\lambda_a$ are the generators of the gauge transformation). Familiar choices are

$$\begin{aligned} \nabla \cdot \mathbf{A} &= 0 \qquad \text{Coulomb gauge} \\ \mathbf{A}_z &= 0 \qquad \text{axial gauge} \\ \partial_\mu \mathbf{A}^\mu &= 0 \qquad \text{Lorentz gauge} \ . \end{aligned} \tag{15.19}$$

In this chapter we will choose the Lorentz gauge because it is manifestly covariant.

The example presented above outlined how the path integral is to be constructed, but because of the greater complexity of the realistic problem, we will review the steps again here. This way of treating the gauge condition was invented by Faddeev and Popov [FP 67] and is referred to as the *Faddeev–Popov trick*. This trick is not required for the quantization of QED, but is useful in the quantization of QCD. The argument begins by considering the quantity

$$\Delta^{-1}(\mathbf{A}) = \int d\mathbf{U}\, \delta\left[F(\mathbf{A}^U)\right] \ , \tag{15.20}$$

where $d\mathbf{U}$ is an invariant integration over the elements $\mathbf{U}$ of the gauge group, defined in general by the transformation

$$\mathbf{A}'_\mu \equiv \mathbf{A}^U_\mu = \mathbf{U}\mathbf{A}_\mu\mathbf{U}^\dagger + \frac{i}{g}\left(\partial_\mu \mathbf{U}\right)\mathbf{U}^\dagger \tag{15.21}$$

[recall Eq. (13.27)]. For an Abelian gauge group, $\mathbf{A}_\mu = A_\mu$,

$$\mathbf{U} = e^{-ig\Lambda(x)} \ , \tag{15.22}$$

and we recover the familiar $A^U_\mu = A_\mu + \partial_\mu \Lambda(x)$. The invariant group integration (the Hurwitz measure) is defined by the requirement

$$\int d\mathbf{U} = \int d\left(\mathbf{U}'\mathbf{U}\right) \tag{15.23}$$

for any fixed element $\mathbf{U}'$ in the group. The idea behind this statement is that the sum (integral) over all elements $\mathbf{U}$ of a finite (continuous) group is the same as the sum (integral) over all elements $\mathbf{U}'\mathbf{U}$, because multiplication by $\mathbf{U}'$ maps the group into itself. For an Abelian gauge transformation we may take

$$\int d\mathbf{U} \Longrightarrow \int \mathcal{D}\left[\Lambda(x)\right] \ . \tag{15.24}$$

This definition satisfies the requirement (15.23)

$$\int d\left[\mathbf{U}'\mathbf{U}\right] \Longrightarrow \int \mathcal{D}\left[\Lambda'(x)+\Lambda(x)\right] = \int \mathcal{D}\left[\Lambda(x)\right] \tag{15.25}$$

because, at each point x_i (or more properly, in each volume α), $\Lambda(x_i) = \Lambda_i$ is integrated from $-\infty$ to ∞, and hence, if the integral exists,

$$\int_{-\infty}^{\infty} d\Lambda_i = \int_{-\infty}^{\infty} d\left(\Lambda_i + \Lambda_i'\right) \quad . \tag{15.26}$$

Returning to Eq. (15.20), note that $\Delta^{-1}(\mathbf{A})$ is gauge invariant because

$$\begin{aligned} \Delta^{-1}\left(\mathbf{A}^{U'}\right) &= \int d\mathbf{U}\,\delta\left[F\left(\mathbf{A}^{U'U}\right)\right] = \int d(\mathbf{U}'\mathbf{U})\,\delta\left[F\left(\mathbf{A}^{U'U}\right)\right] \\ &= \int d\mathbf{U}''\,\delta\left[F\left(\mathbf{A}^{U''}\right)\right] = \Delta^{-1}(\mathbf{A}) \quad . \end{aligned} \tag{15.27}$$

Now $\Delta(\mathbf{A})$ can be explicitly evaluated using fields defined in coordinate space, but it is more straightforward if the fields are treated in momentum space. Hence $\partial_\mu \mathbf{A}^\mu(x) \to -ik_\mu \mathbf{A}^\mu(k)$, and all the gauge conditions we have mentioned so far can be written

$$F(\mathbf{A}) = -in_\mu \mathbf{A}^\mu(k) = 0 \ , \tag{15.28}$$

where

$$\begin{aligned} n^\mu &= k^\mu && \text{Lorentz gauge} \\ n^\mu &= (0,0,0,1) && \text{axial gauge} \\ n^\mu &= (0,\hat{\boldsymbol{k}}) && \text{Coulomb gauge.} \end{aligned} \tag{15.29}$$

Now $\Delta(\mathbf{A})$ will be evaluated for the *general case of a non-Abelian gauge transformation* because it will be needed later. Working with the infinitesimal transformations given in Eq. (13.31), the gauge transformations are

$$\left(A^U\right)^a_\mu(k) = A^a_\mu(k) - ik_\mu\epsilon^a(k) - g\int \frac{d^4k'}{(2\pi)^2} f_{abc} A^b_\mu(k-k')\epsilon^c(k') \ , \tag{15.30}$$

where the transformations have been written in momentum space, the indices a, b, c are color indices, and f_{abc} are the structure constants of the group. Dividing space up into discrete cells, so that $A^a_\mu(k_i) \equiv A^a_{\mu i}$, the gauge transformation can be written

$$\left(A^U\right)^a_{\mu i} = A^a_{\mu i} + M^{ac}_{\mu ij}\,\epsilon^c_j \ , \tag{15.31}$$

where the gauge transformation matrix is

$$M^{ac}_{\mu ij} = -ik_{\mu i}\delta_{ij}\delta_{ac} - \frac{g}{(2\pi)^2} f_{abc} A^b_{\mu\ i-j} \quad . \tag{15.32}$$

Later, when we want to restore the continuum, we will use

$$\begin{aligned}(A^U)^a_\mu(k) &= A^a_\mu(k) + \int d^4k' M^{ac}_\mu(k,k')\epsilon^c(k') \\ M^{ac}_\mu(k,k') &= -ik_\mu\delta^4(k-k')\delta_{ac} - \frac{g}{(2\pi)^2} f_{abc}A^b_\mu(k-k') \ .\end{aligned} \tag{15.33}$$

The group integration is

$$\int d\mathbf{U} = \int \prod_{i,a} d\epsilon^a_i \ , \tag{15.34}$$

and hence the explicit form for $\Delta(\mathbf{A})$ is

$$\Delta^{-1}(\mathbf{A}) = \int \prod_{i,a} d\epsilon^a_i \ \delta\left[-in^\mu\left(A^a_{\mu i} + M^{ac}_{\mu ij}\,\epsilon^c_j\right)\right] \ . \tag{15.35}$$

Now, if we work "around" a field configuration $A^a_{\mu i}$ which satisfies the gauge condition, then $n^\mu A^a_{\mu i} = 0$, and we have

$$\Delta^{-1}(\mathbf{A}) = \int \prod_{i,a} d\epsilon^a_i \ \delta\left[-in^\mu M^{ac}_{\mu ij}\,\epsilon^c_j\right] \ . \tag{15.36}$$

This has the general form

$$\begin{aligned} I &= \int \prod_i dx_i \, \delta(\mathcal{O}_{ij}x_j) \\ &= \int \prod_i dy_i \, J\left(\frac{\partial x_i}{\partial y_j}\right) \delta(y_i) = J\left(\frac{\partial x_i}{\partial y_j}\right) = \frac{1}{J\left(\dfrac{\partial y_i}{\partial x_j}\right)} \\ &= \det{}^{-1}(\mathcal{O}) \ . \end{aligned} \tag{15.37}$$

Hence, finally we obtain the desired explicit form for Δ:

$$\boxed{\Delta(\mathbf{A}) = \det\left(-in^\mu M^{ac}_{\mu ij}\right) \ .} \tag{15.38}$$

For an Abelian gauge group, $\Delta(\mathbf{A}) = \Delta(A) = \Delta$, independent of A, and since all constant terms do not matter in the definition of the generating function, Δ can be discarded. This is true for QED, but is *not* true for QCD, where Δ depends on $\mathbf{A}$ and cannot be ignored. We will continue to include Δ in our discussion of QED, in order to be better prepared to include it later in our discussion of QCD.

Now we are almost ready to incorporate the constraint and implement the Faddeev–Popov trick. Before doing so, we generalize the gauge condition to

$$F(\mathbf{A}) - \mathbf{G}(x) = 0 \ , \tag{15.39}$$

where $\mathbf{G} = \frac{1}{2}\lambda_a G^a$ is an arbitrary Hermitian, traceless matrix in the gauge space. Then consider the more general relation

$$\Delta_G^{-1}(\mathbf{A}) = \int d\mathbf{U}\, \delta\left[F(\mathbf{A}^U) - \mathbf{G}(x)\right] \quad . \tag{15.40}$$

This quantity is still gauge invariant. Furthermore, working around the new field $\mathbf{A}_G$ which satisfies the new gauge condition

$$-in^\mu \mathbf{A}_{G\mu} = \mathbf{G} \,, \tag{15.41}$$

we can show that $\Delta(\mathbf{A})$ is independent of $\mathbf{G}$. To demonstrate this, return to Eq. (15.35) and generalize it to include $\mathbf{G}$:

$$\begin{aligned}\Delta_G^{-1}(\mathbf{A}) &= \int \prod_{ia} d\epsilon_i^a\, \delta\left[-in^\mu \left(A^a_{G\mu i} + M^{ac}_{\mu ij}\, \epsilon_j^c\right) - G_i^a\right] \\ &= \int \prod_{ia} d\epsilon_i^a\, \delta\left[-in^\mu M^{ac}_{\mu ij}\, \epsilon_j^c\right] = \Delta^{-1}(\mathbf{A})\end{aligned} \tag{15.42}$$

because $\mathbf{A}_{G\mu}$ satisfies the new gauge condition.

Finally, following the discussion leading to Eq. (15.17), we may insert the constraint imposed by the gauge condition in the form

$$\boxed{\mathbf{1} = \Delta(\mathbf{A})\mathcal{N}(\alpha) \int \mathcal{D}(G) \int d\mathbf{U}\, \delta\left[F(\mathbf{A}^U) - \mathbf{G}\right]\, e^{-i\frac{1}{\alpha}\,\mathrm{tr}\,\mathbf{G}^2} \,,} \tag{15.43}$$

where $\mathcal{N}(\alpha)$ is the (infinite) constant

$$\int \mathcal{D}(G)\, e^{-i\frac{1}{\alpha}\,\mathrm{tr}\,\mathbf{G}^2} = \int_{-\infty}^{\infty} \prod_{ia} dG_i^a\, e^{-\frac{i}{2\alpha}\sum_a (G_i^a)^2} = \prod_{ia} \sqrt{\frac{2\pi\alpha}{i}} = \mathcal{N}^{-1}(\alpha)\,. \tag{15.44}$$

This constant normalizes the integrals over each G_i^a and approaches infinity as the number of cells n, into which the spatial integrals are divided, approaches infinity. However, since it is a constant, it may be discarded leading to the following generating function for gauge theories:

$$\begin{aligned} Z_0[j] \cong \int \mathcal{D}[A_\mu G] d\mathbf{U}\, \Delta(\mathbf{A})\, \delta\left[F(\mathbf{A}^U) - \mathbf{G}\right] \\ \times\, e^{i\int d^4x\, \mathrm{tr}\left\{-\frac{1}{2}\mathbf{F}_{\mu\nu}\mathbf{F}^{\mu\nu} - \frac{1}{\alpha}\mathbf{G}^2 + 2\mathbf{j}_\mu \mathbf{A}^\mu\right\}} . \end{aligned} \tag{15.45}$$

Next, we simplify this by letting $\mathbf{A}_\mu \to \mathbf{A}_\mu^{(U^{-1})}$. Then, everything in the integral is explicitly gauge invariant except the $\delta[F(\mathbf{A}^U) - \mathbf{G}]$, and we get

$$\begin{aligned} Z_0[j] &= \int d\mathbf{U} \int \mathcal{D}\left[A_\mu^{(U^{-1})} G\right] \Delta\left(\mathbf{A}^{(U^{-1})}\right)\, \delta\left[F(\mathbf{A}) - \mathbf{G}\right] e^{i\mathcal{A}^{(U^{-1})}} \\ &= \underbrace{\int d\mathbf{U}}_{\substack{\text{integral over gauge} \\ \text{degrees of freedom}}} \quad \underbrace{\int \mathcal{D}[A_\mu G] \Delta(\mathbf{A})\, \delta\left[F(\mathbf{A}) - \mathbf{G}\right]\, e^{i\mathcal{A}}}_{\substack{\text{integral over degrees of} \\ \text{freedom fixed by the gauge}}} \ . \end{aligned} \tag{15.46}$$

This result accomplishes what we set out to do; it separates the integral into two parts: (i) an integration over the gauge degrees of freedom (now an infinite constant *explicitly* independent of $\mathbf{A}$) and (ii) an integration over all field variables constrained by the gauge condition $F(\mathbf{A}) = \mathbf{G}$. The infinite $\int d\mathbf{U}$ integration can now be dropped, and the integration over $\mathbf{G}$ carried out, removing the δ-functions and giving

$$Z_0[j] = \int \mathcal{D}[A_\mu]\Delta(\mathbf{A})\, e^{\,i\int d^4x\, \mathrm{tr}\left\{-\frac{1}{2}\mathbf{F}_{\mu\nu}\mathbf{F}^{\mu\nu} - \frac{1}{\alpha}[F(\mathbf{A})]^2 + 2\mathbf{j}_\mu \mathbf{A}^\mu\right\}} . \tag{15.47}$$

Discussion

(i) The new term $-\frac{1}{\alpha}\,\mathrm{tr}\,[F(\mathbf{A})]^2$ in the effective Lagrangian is the gauge fixing term we first discussed in Sec. 2.2. This one has a different structure and depends on a parameter α.

(ii) In QED, where the gauge transformation does not depend on A, $\Delta(A)$ is a constant and can be ignored. In this case we recover the same generating function we would have had if we had simply inserted the δ-function (and averaged over G).

(iii) Let us write out the effective Lagrangian (15.47) for QED in the generalized Lorentz gauge. In this case, the Lagrangian density in momentum space becomes

$$-\frac{1}{4}F_{\mu\nu}F^{\mu\nu} - \frac{1}{2\alpha}\left(\partial_\mu A^\mu\right)^2 \Rightarrow -\frac{1}{2}A^{*\mu}(k)\left[k^2 g_{\mu\nu} - k_\mu k_\nu\left(1 - \frac{1}{\alpha}\right)\right]A^\nu(k) \, . \tag{15.48}$$

The propagator is proportional to the inverse of the kinetic energy term. Hence, if $\Delta_{\mu\nu}$ is the propagator, then it must satisfy the equation

$$\Delta^{\mu\mu'}\left[-k^2 g_{\mu'\nu} + k_{\mu'}k_\nu\left(1 - \frac{1}{\alpha}\right)\right] = g^\mu{}_\nu \, . \tag{15.49}$$

From Lorentz invariance we know that the propagator must have the general form

$$\Delta^{\mu\nu} = \Delta_1\, g^{\mu\nu} + \Delta_2\, k^\mu k^\nu \quad , \tag{15.50}$$

where Δ_1 and Δ_2 are functions of k^2. Substituting (15.50) into (15.49) gives the following equations for Δ_1 and Δ_2:

$$\begin{aligned} &\Delta_1 k^2 = -1 \\ &\Delta_1\left(1 - \frac{1}{\alpha}\right) - \Delta_2 k^2\left(\frac{1}{\alpha}\right) = 0 \, . \end{aligned} \tag{15.51}$$

Hence the propagator is

$$\boxed{i\Delta_{\mu\nu} = \frac{i}{k^2}\left[-g_{\mu\nu} + \frac{k_\mu k_\nu}{k^2}(1-\alpha)\right] \quad .} \tag{15.52}$$

There are two common choices for the (arbitrary) gauge parameter α:

$$\begin{array}{ll} \alpha = 1 & \text{Feynman gauge} \\ \alpha = 0 & \text{Landau gauge} \quad . \end{array} \tag{15.53}$$

Note that $\alpha = 0$ corresponds to a limiting case where $G = 0$. The Feynman gauge is the one used in Chapters 10 and 11 and is used for many QED calculations.

(iv) The choice $\alpha = \infty$, which would eliminate the gauge fixing term, gives a singular propagator. A finite α is necessary and hence a gauge fixing term is necessary.

(v) In QED where current conservation takes on the simple form $k_\mu j^\mu = 0$ (for on-shell particles) the second term in the propagator vanishes and all results are independent of α explicitly. Gauge invariance (independence of α) and current conservation are again the same constraint.

We now turn to a discussion of the quantization of QCD.

15.2 GHOSTS AND THE FEYNMAN RULES FOR QCD

The final application of path integrals will be to the determination of the Feynman rules for QCD. We already have the basic ones, given in Fig. 13.1. Our focus here will be on gauge fixing and the appearance of ghosts in loop diagrams.

We have already done the bulk of the work necessary to obtain the Feynman rules for QCD. The candidate generating function is just the one given in Eq. (15.47). The principal differences are that:

- $\mathbf{F}_{\mu\nu}\mathbf{F}^{\mu\nu}$ now contains the covariant derivatives, which include $\mathbf{A}_\mu$ terms. When expanded out, they generate a kinetic energy term of the same form as in QED, plus three-gluon (A^3) and four-gluon (A^4) interaction terms. These generate the $3g$ and $4g$ couplings given in Eq. (13.57).

- $\Delta(\mathbf{A})$ *is now dependent on* $\mathbf{A}$. *This means that it can no longer be discarded,* as we did in QED. This Faddeev–Popov term generates new interactions which we want to discuss now.

A general result for the $\Delta(\mathbf{A})$ term was given in Eq. (15.38). In the generalized Lorentz gauges it is

$$\Delta(\mathbf{A}) = \det\left[-k_i^2\delta_{ij}\delta_{ab} + \frac{g}{(2\pi)^2} f_{acb}\, ik_i^\mu A_\mu^c(i-j)\right] \quad . \tag{15.54}$$

This determinant will make the volume integration over $\mathcal{D}[A]$ dependent on A and will influence the path densities in a non-uniform manner. *Any such influence is an effective interaction* which must be taken into account in all calculations. This can be done by writing the factor Δ as an exponential. Since Δ multiplies the rest of the path integral, its phase must be added to the other terms in the action, and these new terms will describe new interactions which can be calculated in the usual fashion. Fortunately, the technique for doing this has already been given in Eq. (14.145).

To convert the determinant into an exponential, we *introduce two new anti-commuting, colored, scalar fields*, denoted by $\bar{c}$ and c. Since such particles violate the connection between spin and statistics, they are referred to as ghosts, and we expect them to appear only in loops and not to exist outside the region of interactions. Using the identity relating a determinant to an integral over Grassmann variables, Eq. (14.145), we write

$$\begin{aligned} \Delta(\mathbf{A}) &= \int \mathcal{D}[\bar{c}c] \exp\left(-i\sum_{\substack{i,j\\a,b}}\left\{\bar{c}_i^a(-k_i^2)c_i^a + \frac{g}{(2\pi)^2} f_{acb}\bar{c}_i^a\, ik_i^\mu A_\mu^c(i-j)c_j^b\right\}\right) \\ &= \int \mathcal{D}[\bar{c}c]\, e^{\,i\int d^4p\,\bar{c}^a(p)p^2c^a(p) - i\frac{g}{(2\pi)^2}\int d^4p d^4p' f_{acb}\bar{c}^a(p')\,ip'^\mu A_\mu^c(p'-p)c^b(p)}\ . \end{aligned} \tag{15.55}$$

Fourier transforming the exponent to position space gives

$$\begin{aligned} &\frac{1}{(2\pi)^2}\int d^4p\, d^4p'\, \bar{c}(p')\, ip'^\mu A_\mu(p'-p)c(p) \\ &\quad = \int d^4p\, d^4p' \frac{d^4x\, d^4y\, d^4z}{(2\pi)^8} e^{-ip'\cdot x} e^{i(p'-p)\cdot y} e^{ip\cdot z} \left[\partial^\mu \bar{c}(x)\right] A_\mu(y)c(z) \\ &\quad = \int d^4x\, \left[\partial^\mu \bar{c}(x)\right] A_\mu(x)c(x) \end{aligned} \tag{15.56}$$

and hence

$$\boxed{\Delta(A) = \int \mathcal{D}[\bar{c}c]\, e^{\,i\int d^4x\left\{(\partial_\mu\bar{c}^a)(\partial^\mu c^a) - gf_{abc}(\partial^\mu\bar{c}^a)A_\mu^b c^c\right\}}.} \tag{15.57}$$

From the form of this expression, we deduce that ghost fields behave like massless scalar fields, with a propagator of the conventional form

$$i\Delta_c = \frac{i\delta_{ab}}{p^2 + i\epsilon}\ . \tag{15.58}$$

Furthermore, ghosts interact with the gluon fields with a ϕ^3 type interaction. Since the interaction with $\bar{c}$ is different than with c, the ghost lines should be oriented, and we have the following addition to Feynman Rule 1:

- the operator

$$g f_{abc}\,(p+k)^{\mu}$$

at each gluon ghost vertex, where the *outgoing* ghost (dotted in the diagram) has momentum $p+k$ and color a, the *incoming* gluon has polarization μ, color b, and momentum k, and the *incoming* ghost has momentum p and color c.

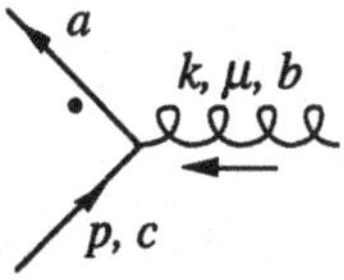

Fig. 15.1

The sign of this term follows from the observation that the term in the Feynman rules is $-i\mathcal{H}_{\rm int} = i\mathcal{L}_{\rm int}$. The momentum of the $\bar{c}$ field (outgoing) is $e^{ip'\cdot x}$ so that $\partial_\mu \to ip'_\mu$, giving

$$-i\,ip'^{\mu}\,gf_{abc} = p'^{\mu} g f_{abc}$$

and precisely the above result when $p' = p+k$ is substituted. The dot in the diagram of Fig. 15.1 tells which line has the momentum attached to it.

In constructing closed loops, $\bar{c}$ must pair with c, and since the momentum is associated with $\bar{c}$, we have the rule:

- a ghost line cannot be dotted at both ends.

Finally, since ghosts anticommute:

- a factor of (-1) multiplies each ghost loop.

The full set of Feynman rules for QCD is given in Appendix B [MP 78]. Any QCD diagram can be calculated from these rules, and they will be used in the next section and in Chapter 17.

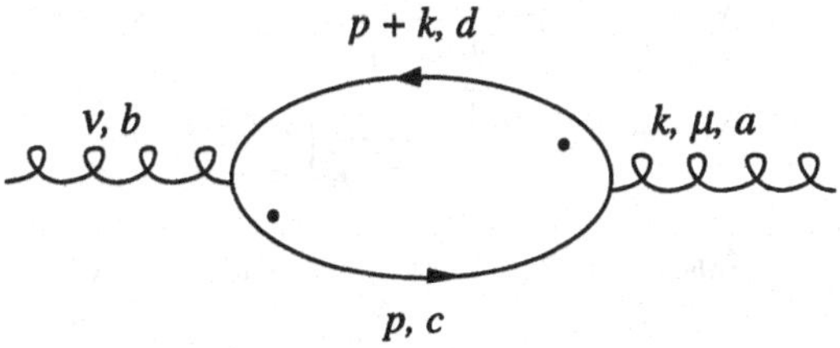

Fig. 15.2 The ghost loop contribution to the gluon self-energy showing that ghost lines are dotted at only one end.

As an illustration of the use of the Feynman rules for ghost loops, the second order ghost contribution to the gluon self-energy, given by the Feynman diagram shown in Fig. 15.2, is

$$\begin{aligned} i\Pi^{\mu\nu}_{ab} &= (-1)\, ig^2 f_{dac} f_{cbd} \int \frac{d^4p}{(2\pi)^4} \frac{(i)^2\,[(p+k)^\mu\, p^\nu]}{[p^2+i\epsilon]\,[(p+k)^2+i\epsilon]} \\ &= -ig^2 f_{acd} f_{bcd} \int \frac{d^4p}{(2\pi)^4} \frac{(p+k)^\mu\, p^\nu}{[p^2+i\epsilon]\,[(p+k)^2+i\epsilon]} \; . \end{aligned} \tag{15.59}$$

This diagram, and others which contribute to the gluon self-energy in second order, will be evaluated in Chapter 17, where it will be shown that the effective QCD coupling constant g approaches zero at high energies, a property referred to as *asymptotic freedom.*

The appearance of ghosts in QCD may appear a bit mysterious. In the next section it is shown that ghosts are necessary in order to preserve unitarity.

15.3 GHOSTS AND UNITARITY

Even though ghosts never appear in external states, ghost loops play an important role in QCD. Since they are a consequence of the quantization of a field with a gauge symmetry, it is expected that they will be needed in order to maintain gauge invariance, but it is perhaps less obvious that they are also needed to maintain unitarity. In this section we will look at a simple example which illustrates how ghosts help to maintain both gauge invariance and unitarity. [See also the discussions in Cheng and Li (1984) and Aitchison and Hey (1982).]

First, consider the annihilation of a quark–antiquark pair into two gluons. To second order in g^2 there are three diagrams which describe this process, as shown in Fig. 15.3. Omitting the polarization vectors of the final state gluons, these three diagrams are

$$\begin{aligned} \mathcal{M}^{\mu\nu}_{A\,ab} &= -g^2 \tfrac{1}{4} \lambda_b \lambda_a\, \bar{v}(\mathbf{p}_2, s_2)\gamma^\nu \left[\frac{1}{m - \not{a}_A}\right] \gamma^\mu u(\mathbf{p}_1, s_1) \\ \mathcal{M}^{\mu\nu}_{B\,ab} &= -g^2 \tfrac{1}{4} \lambda_a \lambda_b\, \bar{v}(\mathbf{p}_2, s_2)\gamma^\mu \left[\frac{1}{m - \not{a}_B}\right] \gamma^\nu u(\mathbf{p}_1, s_1) \\ \mathcal{M}^{\mu\nu}_{C\,ab} &= \left(\frac{ig^2}{K^2}\right) f_{abc} \tfrac{1}{2} \lambda_c\, \bar{v}(\mathbf{p}_2, s_2) \left[\gamma^\mu (K+k_1)^\nu - \gamma^\nu (K+k_2)^\mu \right. \\ &\qquad \left. + g^{\mu\nu} (\not{k}_2 - \not{k}_1)\right] u(\mathbf{p}_1, s_1) \; . \end{aligned} \tag{15.60}$$

Note that the sum of these three diagrams is symmetric under interchange of the two final state gluons.

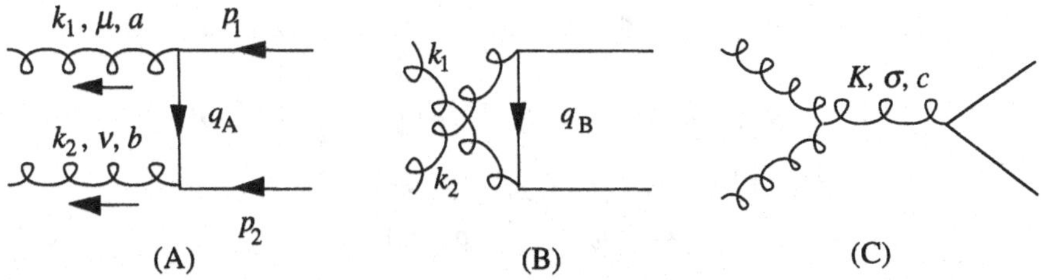

Fig. 15.3 The three Feynman diagrams which contribute to the production of two gluons from the annihilation of a quark–antiquark pair.

Now, the physical scattering amplitudes are obtained from the expressions (15.60) by contracting them with the polarization vectors of the outgoing gluons, which will be denoted by ϵ^μ (these vectors also carry a color index, which will be suppressed for simplicity). As we saw in Secs. 2.5 and 9.10, a massive spin one field has three independent polarization states defined by the requirement

$$p \cdot \epsilon = 0 \ , \tag{15.61}$$

where p is the four-momentum of the particle [recall Eq. (2.44)]. If the particle is massless, $p^2 = 0$, and this condition does not uniquely specify the polarization states. In particular, for a massless particle traveling in the $\hat{z}$ direction, the helicity states

$$\epsilon^{\pm\,\mu} = \mp\frac{1}{\sqrt{2}}(0, 1, \pm i, 0) \tag{15.62}$$

satisfy condition (15.61), but so does any vector of the form

$$\epsilon'^{\pm} = \epsilon^{\pm} + \alpha p \ , \tag{15.63}$$

where α is an arbitrary parameter. To uniquely define two transverse states of a massless particle, we must impose an additional condition which determines α. This condition is equivalent to fixing the gauge. In our discussion of the EM field in Part I, we chose the Coulomb gauge, which is equivalent to the requirement that $n \cdot \epsilon = 0$, where, for a particle moving in the $\hat{z}$-direction, $n^\mu = (0, 0, 0, 1)$. Returning to the problem under consideration, the requirement that the physics be *independent of the gauge* translates into the requirement that the amplitudes $\mathcal{M}$ give the same result for *any* polarization vector of the form (15.63). Hence the necessary and sufficient condition for a gauge invariant result is that the sum of the three diagrams in Fig. 15.3 satisfy the conditions $k_1^\mu \mathcal{M}_{\mu\nu} = 0$ and $k_2^\nu \mathcal{M}_{\mu\nu} = 0$. Because of the symmetry, it is only necessary to look at one of these relations; the other can be obtained by interchanging $k_1 \leftrightarrow k_2$ and $a \leftrightarrow b$.

The first two diagrams give

$$\begin{aligned}
k_{1\mu}\mathcal{M}^{\mu\nu}_{A+B\,ab} =& -g^2\tfrac{1}{4}\lambda_b\lambda_a\,\bar{v}(p_2,s_2)\gamma^\nu\left[\frac{1}{m-\not{q}_A}\right](\not{p}_1-\not{q}_A)u(p_1,s_1) \\
& -g^2\tfrac{1}{4}\lambda_a\lambda_b\,\bar{v}(p_2,s_2)(\not{p}_2+\not{q}_B)\left[\frac{1}{m-\not{q}_B}\right]\gamma^\nu u(p_1,s_1) \\
=& \; g^2\tfrac{1}{4}\left[\lambda_a,\lambda_b\right]\bar{v}(p_2,s_2)\gamma^\nu u(p_1,s_1) \\
=& \; ig^2 f_{abc}\tfrac{1}{2}\lambda_c\,\bar{v}(p_2,s_2)\gamma^\nu u(p_1,s_1) \; ,
\end{aligned} \tag{15.64}$$

where we used the Dirac equation to simplify the result. Note that this would be zero if the commutator vanished, which shows that, in QED (where diagram C does not exist), the two diagrams A and B are gauge invariant.

The contribution from diagram C is

$$\begin{aligned}
k_{1\mu}\mathcal{M}^{\mu\nu}_{C\,ab} =& \left(\frac{ig^2}{K^2}\right) f_{abc}\tfrac{1}{2}\lambda_c\,\bar{v}(p_2,s_2)\Big[\not{k}_1\,(K+k_1)^\nu - \gamma^\nu\,(K+k_2)\cdot k_1 \\
& \qquad + k_1^\nu\,(\not{k}_2-\not{k}_1)\Big]\,u(p_1,s_1) \\
=& \left(\frac{ig^2}{K^2}\right) f_{abc}\tfrac{1}{2}\lambda_c\,\bar{v}(p_2,s_2)\Big[\underbrace{(\not{k}_1+\not{k}_2)\,k_1^\nu}_{=0} - \gamma^\nu K^2 + k_2^\nu\,\not{k}_1\Big]u(p_1,s_1) \\
=& -\left(\frac{ig^2}{K^2}\right) f_{abc}\tfrac{1}{2}\lambda_c\,\bar{v}(p_2,s_2)\left[\gamma^\nu K^2 - k_2^\nu\,\not{k}_1\right]u(p_1,s_1) .
\end{aligned} \tag{15.65}$$

Note that the first term in the final expression cancels (15.64), so that the total result is simply

$$\begin{aligned}
k_{1\mu}\mathcal{M}^{\mu\nu}_{ab} &= ig^2 f_{abc}\tfrac{1}{2}\lambda_c\,\frac{k_2^\nu}{K^2}\,\bar{v}(p_2,s_2)\,\not{k}_1\,u(p_1,s_1) \\
&= k_2^\nu\,\mathcal{T} \; ,
\end{aligned} \tag{15.66}$$

where $\mathcal{T}$ is defined by this equation. Note that $\mathcal{T}$ is symmetric under interchange of the two gluons and that therefore $k_{2\nu}\mathcal{M}^{\mu\nu}_{ab} = k_1^\mu\,\mathcal{T}$. Now, because of the condition (15.61), the result (15.66) is zero when contracted with the polarization vector ϵ_2. Hence we conclude that the *physical* scattering amplitude is gauge invariant and that the gluon self-interaction diagram, C, is essential to this result. However, if some other vector, k_1 for example, is contracted into (15.66), the result is not zero, and this has implications for the unitarity relation, which will be examined now.

As we discussed in Sec. 12.8, the unitarity relation tells us that the imaginary part of a scattering amplitude must be equal to the integral over the product of the amplitudes which describe scattering from the initial state to an intermediate state and from the intermediate state to the final state. Symbolically,

$$Im\,\mathcal{M}_{fi} = -\int_k \mathcal{M}^*_{fk}\mathcal{M}_{ki} = -\int_k \mathcal{M}_{kf}\mathcal{M}_{ki} \; , \tag{15.67}$$

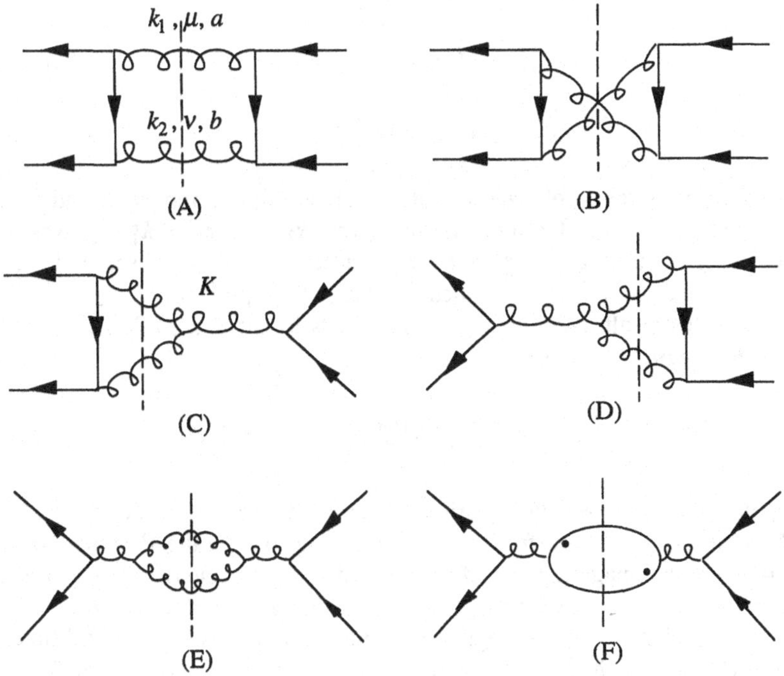

Fig. 15.4 Feynman diagrams (A–E) describe quark–antiquark scattering with a two-gluon intermediate state, and diagram (F) is the ghost loop contribution. All of these diagrams have a two-body cut indicated by the vertical dashed line.

where the integral is over all degrees of freedom of the intermediate state, which must be a physical state with both particles on the mass-shell. The $\int_k$ includes the two-body phase space factors discussed in Sec. 12.8 (the specific form of which are not needed in the subsequent discussion). This unitarity relation is a profound restriction on any theory and is directly related to the conservation of probability.

Let us see how this restriction applies to the elastic scattering of a $q\overline{q}$ pair through a two-gluon intermediate state, described by the six diagrams shown in Fig. 15.4. Unitarity tells us that the imaginary part of these diagrams must be equal (with suitable factors) to the "square" of the $q+\overline{q} \to 2g$ annihilation amplitudes which we have just studied, and in the remainder of this section we will show that this is true, but *only because of the presence of the ghost loop diagram*, F, which is needed to cancel unwanted contributions from the first five diagrams A–E.

In preparation for this demonstration, we note that the metric tensor can be decomposed as follows

$$\begin{aligned} g_{\mu\nu} &= g^T_{\mu\nu} + g^L_{\mu\nu} \\ g^T_{\mu\nu} &= -\sum_{i=1,2} \epsilon^i_\mu \epsilon^i_\nu \\ g^L_{\mu\nu} &= \tfrac{1}{2}\left(\hat{k}_{1\mu}\hat{k}_{2\nu} + \hat{k}_{1\nu}\hat{k}_{2\mu}\right) \;, \end{aligned} \tag{15.68}$$

where ϵ^i are transverse polarization states with zero time components and $\hat{k}_1$ and $\hat{k}_2$ are "unit" vectors in the direction of k_1 and k_2; i.e., $\hat{k}^\mu_1 = k^\mu_1/k_0$, where k_0 is the energy of either gluon in the overall center of mass frame (see Prob. 15.1). This decomposition is very convenient for the present problem.

Using this result, and drawing on the discussion in Sec. 12.8, the imaginary part of the diagrams A–E is

$$Im\,\mathcal{M}^{(A-E)}_{\text{scattering}} = -\frac{1}{2}\int_k \mathcal{M}^{\mu\nu}_{ab}\left[g_{\mu\mu'}g_{\nu\nu'}\right]\mathcal{M}^{\mu'\nu'}_{ab} \;, \tag{15.69}$$

where the symmetry factor of $\frac{1}{2}$ arises because the two gluons are identical and the factors of $g_{\mu\mu'}g_{\nu\nu'}$ are from the gluon propagators (in the Feynman gauge). In computing the imaginary part of the diagrams, the two intermediate gluons are fixed on their mass-shell, so that if k_1 and k_2 are their four-momenta, we have $k_1^2 = 0$ and $k_2^2 = 0$, just as for the original annihilation diagrams. The only difference between (15.69) and the correct unitarity relation is that the $g_{\mu\mu'}g_{\nu\nu'}$ term from the propagators *includes the contribution from the longitudinal polarization states*, the $g^L_{\mu\nu}$ term in (15.68). Unitarity requires that the sum be over *physical, transverse states only*, and hence we require

$$Im\,\mathcal{M}^{\text{total}}_{\text{scattering}} = -\frac{1}{2}\int_k \mathcal{M}^{\mu\nu}_{ab}\left[g^T_{\mu\mu'}g^T_{\nu\nu'}\right]\mathcal{M}^{\mu'\nu'}_{ab} \;. \tag{15.70}$$

The difference between (15.70) and (15.69) is the extent to which the imaginary parts of diagrams A–E violate unitarity and is

$$\begin{aligned} \delta\, Im\,\mathcal{M} &= Im\,\mathcal{M}^{(A-E)}_{\text{scattering}} - Im\,\mathcal{M}^{\text{total}}_{\text{scattering}} \\ &= -\frac{1}{2}\int_k \mathcal{M}^{\mu\nu}_{ab}\left[g^L_{\mu\mu'}g^T_{\nu\nu'} + g^T_{\mu\mu'}g^L_{\nu\nu'} + g^L_{\mu\mu'}g^L_{\nu\nu'}\right]\mathcal{M}^{\mu'\nu'}_{ab} \;. \end{aligned} \tag{15.71}$$

Note that each of the terms in this difference involves at least one factor of g^L, which in turn would be zero *if the annihilation amplitudes were gauge invariant.* Because $k_1^2 = k_2^2 = 0$, it follows from Eq. (15.66) that

$$k_{1\mu}k_{2\nu}\mathcal{M}^{\mu\nu}_{ab} = k_{1\mu}\epsilon^i_{2\nu}\mathcal{M}^{\mu\nu}_{ab} = \epsilon^i_{1\mu}k_{2\nu}\mathcal{M}^{\mu\nu}_{ab} = 0 \;, \tag{15.72}$$

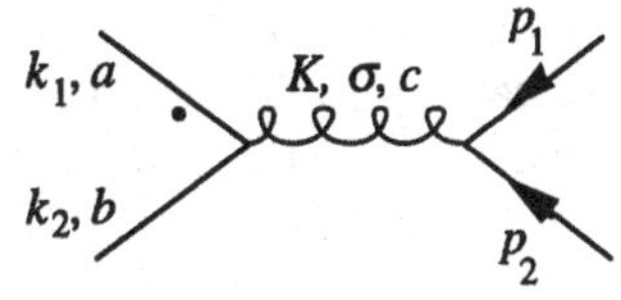

Fig. 15.5 The ghost production diagram used to calculate the imaginary part of diagram 15.4F.

and the only non-zero terms which contribute to (15.71) are

$$\hat{k}_{1\mu}\hat{k}_{1\nu}\mathcal{M}^{\mu\nu}_{ab} = \hat{k}_1 \cdot \hat{k}_2\,\mathcal{T} = 2\mathcal{T} = \hat{k}_{2\mu}\hat{k}_{2\nu}\mathcal{M}^{\mu\nu}_{ab} \ , \tag{15.73}$$

where $\mathcal{T}$ was defined in Eq. (15.66). Using these results, $\delta\, Im\, \mathcal{M}$ can be quickly reduced to

$$\begin{aligned}\delta\, Im\, \mathcal{M} &= -\frac{1}{2}\int_k \mathcal{M}^{\mu\nu}_{ab}\left[\frac{1}{4}\hat{k}_{1\mu}\hat{k}_{1\nu}\hat{k}_{2\mu'}\hat{k}_{2\nu'} + \frac{1}{4}\hat{k}_{2\mu}\hat{k}_{2\nu}\hat{k}_{1\mu'}\hat{k}_{1\nu'}\right]\mathcal{M}^{\mu'\nu'}_{ab} \\ &= -\int_k \mathcal{T}\mathcal{T} \ .\end{aligned} \tag{15.74}$$

Now the ghost loop diagram 15.4F, which we have ignored until now, will also contribute to the imaginary part of the $q\bar{q}$ scattering amplitude. The imaginary part of this diagram can be computed from the square of the ghost production diagram shown in Fig. 15.5, *even though this production diagram does not contribute to the annihilation amplitude* (because no ghosts may be in initial or final states). This diagram is

$$\begin{aligned}\mathcal{M}_{ab} &= ig^2 f_{abc}\tfrac{1}{2}\lambda_c\left(\frac{1}{K^2}\right)\bar{v}(p_2, s_2)\, \not{k}_1\, u(p_1, s_1) \\ &= ig^2 f_{bac}\tfrac{1}{2}\lambda_c\left(\frac{1}{K^2}\right)\bar{v}(p_2, s_2)\, \not{k}_2\, u(p_1, s_1) \\ &= \mathcal{T} \ .\end{aligned} \tag{15.75}$$

Recalling that ghost loops must always be multiplied by a minus sign and that there is no symmetry factor of $\frac{1}{2}$ in this case, the imaginary part of diagram 15.4F is

$$Im\, \mathcal{M}^{(F)}_{\text{scattering}} = \int_k \mathcal{T}\mathcal{T} \ . \tag{15.76}$$

Note that this contribution cancels the unitarity violating part given in Eq. (15.74). The scattering amplitude satisfies unitarity because the *imaginary part of the ghost loop cancels unphysical contributions from longitudinal polarization degrees of freedom in intermediate states.*

This cancellation seems miraculous, and if we had not shown that the Lorentz gauges which we are using require ghosts which satisfy the Feynman rules given in the previous section, it would seem to be an accident. As it is, the unitarity requirement is one way to see why such gauges require ghosts, and their presence was anticipated well before their quantization rules were known. Having the rules, we can regard this demonstration of unitarity as a confirmation that the rules are correct and that the theory is physical.

15.4 THE STANDARD ELECTROWEAK MODEL

Building on what has been learned, the standard Electroweak model, usually just referred to as the Standard Model, will be described next. This model was developed by Weinberg and Salam in the late 1960's [We 67, Sa 68] and is the first example of the modern unification of forces. In this model, the weak and electromagnetic forces are unified into a single force.

We will only describe one generation of the lepton sector of the Standard Model, but the other generations and the quark sectors are all very similar (for an elementary account, see [AL 73, La 81]). Briefly, the Standard Model can be described as a theory which has a local $SU(2)\times U(1)$ gauge symmetry spontaneously broken by the vacuum. Hence it combines features of the $SU(2)$ gauge group studied in Sec. 13.2 with spontaneous symmetry breaking studied in Secs. 13.6–13.8. There are two Lagrangians to be discussed. The first is the unbroken Lagrangian in which the gauge symmetry is manifest, and the second is the transformed Lagrangian expressed in terms of physical fields. The particles contained in each of these Lagrangians are summarized in Table 15.1, and they will be described in detail as we proceed.

We begin with the unbroken Lagrangian, which will be denoted by $\mathcal{L}$. It is constructed from a left-handed doublet of Fermi particles, a right-handed fermi singlet, a doublet of complex scalars, and four gauge bosons: three, A^i_μ, associated with the $SU(2)$ symmetry, and one, B_μ, associated with the $U(1)$ symmetry. All particles but the scalars are massless. The left-handed particles include the $\frac{1}{2}\left(1-\gamma^5\right)$ projection operator, and we use the notation

$$\psi_L = \begin{pmatrix} \psi_{\nu_L} \\ \psi_{e_L} \end{pmatrix} = \frac{1}{2}(1-\gamma^5)\begin{pmatrix} \psi_\nu \\ \psi_e \end{pmatrix} , \tag{15.77}$$

where ν and e are neutrino and electron fields. The right-handed singlet has the form

$$\psi_R = \psi_{e_R} = \tfrac{1}{2}\left(1+\gamma^5\right)\psi_e \ . \tag{15.78}$$

Under the gauge group, the lepton and scalar doublets transform as in Eq. (13.16):

$$\psi'_D = \mathbf{U}_2\mathbf{U}_1\psi_D \ , \tag{15.79}$$

where D is either doublet, and

$$\begin{aligned} \mathbf{U}_2 &= e^{-ig\frac{1}{2}\boldsymbol{\alpha}(x)} \\ \mathbf{U}_1 &= e^{-ig'\frac{1}{2}\beta(x)\mathbf{Y}} \ , \end{aligned} \tag{15.80}$$

Table 15.1 Fields and particles in one generation of the Standard Model.

	$\mathcal{L}$					$\mathcal{L}_{EW}$				
Field	Spin	Mass	I_W	Y	# of States	Field	Spin	Mass	Charge	# of States
$\begin{pmatrix}\psi_{\nu_L}\\ \psi_{e_L}\end{pmatrix}$	$\frac{1}{2}$	0	$\frac{1}{2}$	-1	2	ν_L	$\frac{1}{2}$	0	0	1
ψ_{eR}	$\frac{1}{2}$	0	0	-2	1	e	$\frac{1}{2}$	finite	-1	2
$\begin{pmatrix}\phi^+\\ \phi^0\end{pmatrix}$	0	finite	$\frac{1}{2}$	1	4	η_0	0	finite	0	1
						γ	1	0	0	2
A^i_μ	1	0	1	0	6	$W^\pm$	1	finite	±1	6
B_μ	1	0	0	1	2	Z^0	1	finite	0	3

where $\boldsymbol{\alpha}(x) = \tau_i\alpha_i(x)$, and τ_i are Pauli matrices which are the generators of *weak isospin*, I_W, and $\mathbf{Y}$ is the *hypercharge* operator, which has the values shown in Table 15.1 for each particle. There are two coupling constants:

$$\begin{array}{ll} g & SU(2) \text{ gauge group} \\ g' & U(1) \text{ gauge group} \ . \end{array} \tag{15.81}$$

Similarly, the right-handed singlet transforms as

$$\psi'_R = e^{-ig'\frac{1}{2}\beta(x)\mathbf{Y}}\,\psi_R = e^{ig'\beta(x)}\psi_R \tag{15.82}$$

because ψ_R has a hypercharge assignment of -2.

The unbroken Lagrangian consists of four parts:

$$\mathcal{L} = \mathcal{L}_{\text{lepton}} + \mathcal{L}_{\text{field}} + \mathcal{L}_{\text{scalar}} + \mathcal{L}_{\text{int}} \ , \tag{15.83}$$

where

$$\begin{aligned} \mathcal{L}_{\text{lepton}} &= \bar{\psi}_L\left[\frac{i}{2}\gamma^\mu \overleftrightarrow{D}_\mu(L)\right]\psi_L + \bar{\psi}_R\left[\frac{i}{2}\gamma^\mu \overleftrightarrow{D}_\mu(R)\right]\psi_R \\ \mathcal{L}_{\text{field}} &= -\frac{1}{2}\,\text{tr}\,(\mathbf{A}_{\mu\nu}\mathbf{A}^{\mu\nu}) - \frac{1}{4}B_{\mu\nu}B^{\mu\nu} \\ \mathcal{L}_{\text{scalar}} &= (D_\mu(\phi)\phi^\dagger)\,(D^\mu(\phi)\phi) - m^2|\phi|^2 - \lambda^2|\phi|^4 \\ \mathcal{L}_{\text{int}} &= -G_e\left[\bar{\psi}_R\,(\phi^\dagger\psi_L) + (\bar{\psi}_L\phi)\,\psi_R\right] \ . \end{aligned} \tag{15.84}$$

In these expressions,

$$\begin{aligned} \mathbf{A}_{\mu\nu} &= D_\mu \mathbf{A}_\nu - D_\nu \mathbf{A}_\mu \\ B_{\mu\nu} &= \partial_\mu B_\nu - \partial_\nu B_\mu \ , \end{aligned} \tag{15.85}$$

where $\mathbf{A}_\mu = \frac{1}{2}\tau_i A^i_\mu$, as in Eq. (13.25), and the D_μ are covariant derivatives, defined with the following properties:

$$\begin{aligned} \vec{D}'_\mu(L)\mathbf{U}_2\mathbf{U}_1\psi_L &= \mathbf{U}_2\mathbf{U}_1\vec{D}_\mu(L)\psi_L \\ \vec{D}'_\mu(\phi)\mathbf{U}_2\mathbf{U}_1\phi &= \mathbf{U}_2\mathbf{U}_1\vec{D}_\mu(\phi)\phi \\ \vec{D}'_\mu(R)\mathbf{U}_1\psi_R &= \mathbf{U}_1\vec{D}_\mu(R)\psi_R \ , \end{aligned} \tag{15.86}$$

where D'_μ is the gauge transformed version of D_μ. With these defining properties, the covariant derivatives have a form analogous to (13.32), including terms for both the $\mathbf{U}(2)$ and $\mathbf{U}(1)$ groups, when indicated. For example,

$$i\vec{D}_\mu(L)\psi_L = \left[i\vec{\partial}_\mu - g\mathbf{A}_\mu(x) + \tfrac{1}{2}g' B_\mu(x)\right]\psi_L \ , \tag{15.87}$$

where the positive sign in the B_μ term follows from the assignment $Y = -1$ to L. The gauge transformations for the gauge fields are

$$\begin{aligned} \mathbf{A}'_\mu &= \mathbf{U}_2\mathbf{A}_\mu\mathbf{U}_2^\dagger + \frac{i}{g}\left(\partial_\mu \mathbf{U}_2\right)\mathbf{U}_2^\dagger \\ B'_\mu &= B_\mu + \frac{i}{g'}\left(\partial_\mu \mathbf{U}_1\right)\mathbf{U}_1^\dagger = B_\mu + \mathbf{Y}\,\partial_\mu\beta(x) \ . \end{aligned} \tag{15.88}$$

With all these definitions, the gauge invariance of the Lagrangian (15.83) follows almost trivially. The only term requiring a demonstration is the interaction term, where the invariance under $SU(2)$ is again obvious, but invariance under $U(1)$ is a consequence of compatible hypercharge assignments,

$$\begin{aligned} \mathcal{L}'_{\text{int}} &= -G_e\bar{\psi}'_R\left(\phi'^\dagger\psi'_L\right) + \text{h.c.} \\ &= -G_e\underbrace{e^{-ig'\beta(x)}}_{\bar{R}}\ \underbrace{e^{+ig'\frac{1}{2}\beta(x)}}_{\phi^\dagger}\ \underbrace{e^{+ig'\frac{1}{2}\beta(x)}}_{L}\bar{\psi}_R\left(\phi^\dagger\psi_L\right) + \text{h.c.} \\ &= \mathcal{L}_{\text{int}} \ . \end{aligned} \tag{15.89}$$

We now arrange for this gauge symmetry to be spontaneously broken by making m^2 negative, as we did in Sec. 13.6. Then the energy density is a minimum when

$$|\phi| = \frac{v}{\sqrt{2}} = \frac{1}{\sqrt{2}}\sqrt{-\frac{m^2}{\lambda^2}} \ , \tag{15.90}$$

where v is the same constant first introduced in Sec. 13.6. We will choose the real part of the neutral component of ϕ to be the one which breaks the symmetry, and introduce fields which are small in the neighborhood of this point. Since the fields are now constrained, the gauge invariance is spontaneously broken, and the way in which the fields are defined is equivalent to "choosing a gauge". One choice of gauges, referred to as R_ξ gauges, is to introduce fields ξ_i' and η', where

$$\phi = \begin{pmatrix} \xi_1' + i\xi_2' \\ \frac{1}{\sqrt{2}}(v+\eta') + i\xi_3' \end{pmatrix} \ . \tag{15.91}$$

This choice is convenient for higher order calculations [see Cheng and Li (1984), for example], but here we will use the *unitary*, or U gauge, which gives a simple direct description of the particle content of the Standard Model. In this gauge ϕ is parameterized as follows:

$$\begin{aligned} \phi &= e^{-i\frac{1}{2v}\boldsymbol{\xi}}\ e^{+i\frac{1}{2v}\xi_3 \mathbf{Y}} \begin{pmatrix} 0 \\ \frac{1}{\sqrt{2}}(\eta + v) \end{pmatrix} \\ &= \mathcal{U}^{-1} \begin{pmatrix} 0 \\ \frac{1}{\sqrt{2}}(\eta + v) \end{pmatrix} = \mathcal{U}^{-1}\phi_0 \ , \end{aligned} \tag{15.92}$$

where $\boldsymbol{\xi} = \xi_i \tau_i$. The replacement of ϕ by the four real fields ξ_i and η is not dissimilar to the introduction of $\boldsymbol{\pi}$ in the non-linear sigma model discussed in Sec. 13.7, except that in this case we will be able to linearize the new Lagrangian.

We now transform the Lagrangian to its new form. Using the representation (15.92), we do a final gauge transformation to new fields defined as follows

$$\begin{aligned} \psi_L' &= \mathcal{U}\psi_L \\ \phi' &= \mathcal{U}\phi = \phi_0 \\ \psi_R' &= \mathcal{U}\psi_R \\ \mathbf{A}_\mu' &= \mathcal{U}\mathbf{A}_\mu\mathcal{U}^\dagger + \frac{i}{g}\left(\partial_\mu \mathcal{U}\right)\mathcal{U}^\dagger \\ B_\mu' &= B_\mu + \frac{i}{g'}\left(\partial_\mu \mathcal{U}\right)\mathcal{U}^\dagger \ . \end{aligned} \tag{15.93}$$

Because of the original gauge invariance of the Lagrangian, this transformation leaves the Lagrangian invariant in form, and the *practical result of the gauge transformation (15.93) is to replace ϕ by ϕ_0*. Accordingly, we will drop the primes on the new fields.

After the transformation (15.93), the new Lagrangian is no longer gauge invariant, because ϕ_0 breaks the gauge. This is a consequence of our decision to choose a particular gauge (the U gauge, as mentioned above) for the electroweak theory. However, there is still a gauge symmetry remaining, which was *not spontaneously broken* by the vacuum. These are the transformations which leave ϕ_0

invariant. Neither of the original $SU(2)$ nor $U(1)$ groups leave ϕ_0 invariant, but bearing in mind that $\mathbf{Y} = 1$ for the scalar fields, the generator

$$\left(\tfrac{1}{2}\tau_3 + \tfrac{1}{2}\mathbf{Y}\right)\phi_0 = \begin{pmatrix} 1 & 0 \\ 0 & 0 \end{pmatrix}\phi_0 = \phi_0 \tag{15.94}$$

does leave the vacuum value ϕ_0 unchanged, and hence it defines a gauge transformation which is unbroken by the vacuum field. We will call this the charge operator and write the new local gauge transformation in the following way:

$$U_Q = e^{-ie\gamma(x)\mathbf{Q}} \; , \tag{15.95}$$

where e is the electric charge and the charge assignment for individual particles is $e\mathbf{Q}$, where

$$\mathbf{Q} = \tfrac{1}{2}\tau_3 + \tfrac{1}{2}\mathbf{Y} \; . \tag{15.96}$$

Because this transformation is a *combination* of the "old" $SU(2)$ and $U(1)$ transformations, there is a new connection between these independent groups; there is a mixing, or *unification* of the two original groups. This unification shows up in the new gauge field, which must be a linear combination of the gauge fields A^3_μ and B_μ. We introduce *new fields* A_μ and Z_μ,

$$\begin{pmatrix} A_\mu \\ Z_\mu \end{pmatrix} = \begin{pmatrix} \sin\theta_W & \cos\theta_W \\ \cos\theta_W & -\sin\theta_W \end{pmatrix} \begin{pmatrix} A^3_\mu \\ B_\mu \end{pmatrix} \; , \tag{15.97}$$

and determine the mixing angle θ_W (referred to as the *Weinberg angle*) by choosing A_μ to be the new gauge field, and therefore require that Z_μ be unaffected by the gauge transformation.

Now, the transformation of the two original fields under the new gauge transformation can be found by substituting (15.94) into (15.88) (noting that $\mathbf{Y} = 0$ for A^3 and $\tau_3 = 0$ for B),

$$\begin{aligned} A'^3_\mu &= A^3_\mu + \frac{e}{g}\partial_\mu\gamma(x) \\ B'_\mu &= B_\mu + \frac{e}{g'}\partial_\mu\gamma(x) \; , \end{aligned} \tag{15.98}$$

because the subgroup generated by τ_3 is Abelian. Hence, the requirements that

$$\begin{aligned} A'_\mu &= A_\mu + \partial_\mu\gamma(x) \\ Z'_\mu &= Z_\mu \end{aligned} \tag{15.99}$$

give immediately

$$\begin{aligned} 1 &= \frac{e}{g}\sin\theta_W + \frac{e}{g'}\cos\theta_W \\ 0 &= \frac{e}{g}\cos\theta_W - \frac{e}{g'}\sin\theta_W \end{aligned} \tag{15.100}$$

which give the relations

$$\begin{aligned} \tan\theta_W &= \frac{g'}{g} \\ e &= \frac{gg'}{\sqrt{g^2+g'^2}} = g\sin\theta_W \quad . \end{aligned} \tag{15.101}$$

Hence the Weinberg angle and the electric charge are given by the two original coupling constants g and g'. Since these were both undetermined before, we have no prediction as of yet.

The final task is to express the original Lagrangian L in terms of ϕ_0 (v and η) and the new gauge fields A_μ (the photon) and Z_μ (the neutral vector boson). To make contact with the old phenomenology, we also define the charged vector boson field by

$$W_\mu = \frac{1}{\sqrt{2}}\left(A^1_\mu - iA^2_\mu\right) \quad . \tag{15.102}$$

Hence

$$\begin{aligned} \tfrac{1}{2}\left(\tau_1 A^1_\mu + \tau_2 A^2_\mu\right) &= \tfrac{1}{4}\left(\tau_1 + i\tau_2\right)\left(A^1_\mu - iA^2_\mu\right) + \tfrac{1}{4}\left(\tau_1 - i\tau_2\right)\left(A^1_\mu + iA^2_\mu\right) \\ &= \frac{1}{\sqrt{2}}\left(\tau_+ W_\mu + \tau_- W^\dagger_\mu\right) , \end{aligned} \tag{15.103}$$

where

$$\tau_\pm = \tfrac{1}{2}\left(\tau_1 \pm i\tau_2\right) \tag{15.104}$$

are the weak isospin raising and lowering operators. From Eqs. (15.77) and (15.84) we see that the operator $\tau_\pm$ connects neutrino and electron fields as follows:

$$\bar{\psi}_L \tau_+ \psi_L = \bar{\psi}_{\nu_L} \psi_{e_L} \tag{15.105}$$

and hence describes the creation of one unit of charge through processes like

$$e^- \rightarrow \nu_e \qquad \bar{\nu}_e \rightarrow e^+ \qquad 0 \rightarrow e^+ + \nu_e \qquad e^- + \bar{\nu}_e \rightarrow 0 \quad .$$

Hence the identification of W_μ with positively charged bosons is confirmed, in agreement with our discussion in Sec. 9.10. The fields A^3_μ and B_μ will be replaced by A_μ and Z_μ, where

$$\begin{aligned} A^3_\mu &= \sin\theta_W A_\mu + \cos\theta_W Z_\mu = \frac{g'A_\mu + gZ_\mu}{\sqrt{g^2+g'^2}} \\ B_\mu &= \cos\theta_W A_\mu - \sin\theta_W Z_\mu = \frac{gA_\mu - g'Z_\mu}{\sqrt{g^2+g'^2}} \quad . \end{aligned} \tag{15.106}$$

Hence, the lepton term in the original Lagrangian (15.84) becomes

$$\mathcal{L}_{\text{lepton}} = \bar{\psi}_e \left[\frac{i}{2} \overleftrightarrow{\not{\partial}} + e \not{A} + \frac{1}{\sqrt{2}} g_w \not{Z} \left([1 - 4\sin^2\theta_W] - \gamma^5 \right) \right] \psi_e$$
$$+ \bar{\psi}_{\nu_L} \left[\frac{i}{2} \overleftrightarrow{\not{\partial}} - \sqrt{2}\, g_w \not{Z} \right] \psi_{\nu_L} - g_w \cos\theta_W \left[\bar{\psi}_e \gamma^\mu (1-\gamma^5) \psi_\nu W^\dagger_\mu + \text{h.c.} \right] , \tag{15.107}$$

where, for convenience, a new combination of coupling constants has been introduced,

$$g_w = \frac{g}{2\sqrt{2}\cos\theta_W} = \frac{e}{\sqrt{2}\sin 2\theta_W} \ . \tag{15.108}$$

We will regard e and θ_W as the independent parameters and introduce g_w for convenience only. Note that the Lagrangian density (15.107) includes the new interactions shown in Fig. 15.6.

Next, look at the Lagrangian for the scalars. Reducing the covariant derivative gives

$$D_\mu \phi_0 = \left(\partial_\mu + ig\mathbf{A}_\mu + \frac{i}{2} g' B_\mu \right) \phi_0$$
$$= \frac{1}{\sqrt{2}} \begin{pmatrix} \frac{ig}{\sqrt{2}} W_\mu (\eta + v) \\ \partial_\mu \eta - \frac{ig}{2\cos\theta_W} Z_\mu (\eta + v) \end{pmatrix} \ . \tag{15.109}$$

Recalling that η and Z_μ are real fields, we see that

$$\left(D_\mu \phi_0^\dagger \right) (D^\mu \phi_0) = \frac{1}{2} \partial_\mu \eta \partial^\mu \eta + \frac{g^2}{4} W^\dagger_\mu W^\mu (\eta + v)^2 + \frac{1}{2} \frac{g^2}{4\cos^2\theta_W} Z_\mu Z^\mu (\eta + v)^2 \ . \tag{15.110}$$

We have the proper kinetic energy term for a scalar, called the Higgs, and interactions of the scalar with the gauge bosons. However, more significantly, we have *mass terms for the bosons.* The squares of these masses are the coefficients of the $W^\dagger_\mu W^\mu$ term and the $\frac{1}{2} Z_\mu Z^\mu$ terms. Hence

$$M_W^2 = \frac{g^2}{4} v^2$$
$$M_Z^2 = \frac{g^2}{4\cos^2\theta_W} v^2 \ . \tag{15.111}$$

Hence the *mass ratio is completely determined by the Weinberg angle,*

$$\frac{M_Z}{M_W} = \frac{1}{\cos\theta_W} \ . \tag{15.112}$$

These masses have been measured in colliding beam experiments. Recent values are $M_W = 80.22 \pm 0.26$ GeV and $M_Z = 91.173 \pm 0.020$ GeV. The average value of $\sin^2\theta_W$ extracted from many experiments is $0.2325 \pm\ 0.0008$ [RP 92].

- the operator

$$i\frac{g_W}{\sqrt{2}}\gamma^\mu\left([1-4\sin^2\theta_W]-\gamma^5\right)$$

at each vertex where a Z^0 weak boson with polarization μ is emitted from or absorbed by an electron.

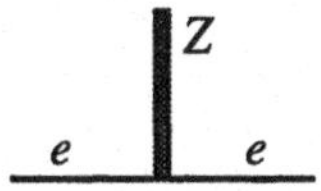

Fig. 15.6a

- the operator

$$-i\frac{g_W}{\sqrt{2}}\gamma^\mu\left(1-\gamma^5\right)$$

at each vertex where a Z^0 weak boson with polarization μ is emitted from or absorbed by a neutrino.

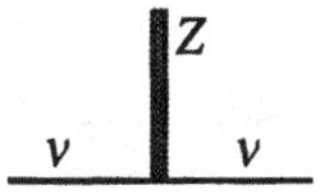

Fig. 15.6b

- the operator

$$-ig_W\cos\theta_W\,\gamma^\mu\left(1-\gamma^5\right)$$

at each vertex where a W^+ weak boson with polarization μ is emitted from or a W^- boson is absorbed by a neutrino, converting it to an electron. Electron to neutrino conversion is described by the same factor.

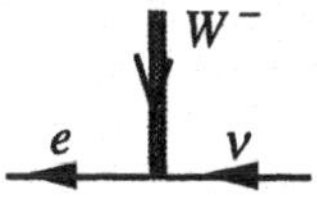

Fig. 15.6c

Fig. 15.6 Interactions and Feynman rules for the lepton sector of the Standard Model.

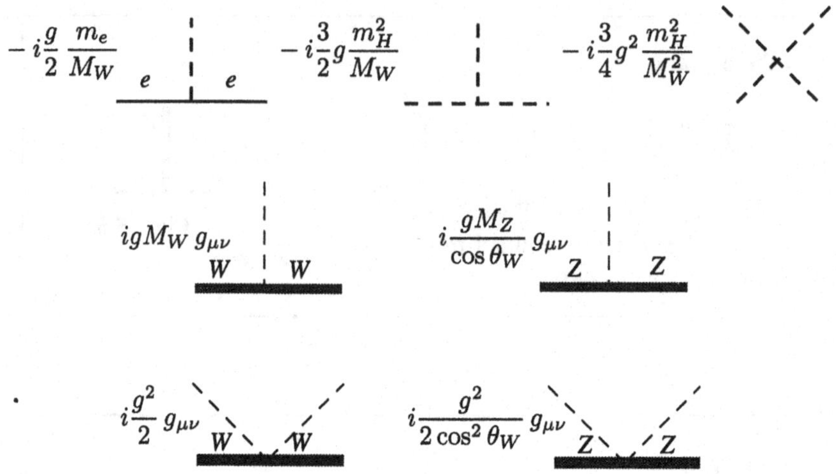

Fig. 15.7 Interactions and Feynman rules for the Higgs sector of the Standard Model.

The interactions of the Higgs particle come from the kinetic term (15.110), the scalar potential, and the interaction term. Expanding out the scalar potential gives

$$\begin{aligned} V' &= -\frac{m^2}{2}(\eta+v)^2 - \frac{\lambda^2}{4}(\eta+v)^4 \\ &= m^2\eta^2 - \lambda^2 v\eta^3 - \frac{\lambda^2}{4}\eta^4 + \text{constants} \\ &= -\frac{1}{2}m_H^2\eta^2 - \frac{1}{3!}(6\lambda^2 v)\eta^3 - \frac{1}{4!}6\lambda^2\eta^4 + \text{constants}\,. \end{aligned} \tag{15.113}$$

where

$$m_H^2 = -2m^2 \;. \tag{15.114}$$

Hence the Higgs has a mass of the correct sign (remember that $m^2 < 0$), and is independent of the other parameters in the lepton sector. The coupling of the Higgs to the electron and the electron mass come from the interaction term, which reduces to

$$\mathcal{L}'_{\text{int}} = -G_e\left(\frac{\eta+v}{\sqrt{2}}\right)\bar{\psi}_e\psi_e = -(\eta+v)\frac{m_e}{v}\bar{\psi}_e\psi_e \;. \tag{15.115}$$

The electron mass is an independent parameter which fixes G_e, and once G_e has been fixed, the coupling of the Higgs to the electron is also fixed and is proportional to the electron mass. The Feynman rules for all Higgs couplings are summarized in Fig. 15.7.

Finally, we expand the gauge field part of the Lagrangian. This generates many couplings of the photon, $W^\pm$ and Z and will ultimately confirm our charge assignments. First, consider the terms involving the square of $\partial_\mu A^i_\nu - \partial_\nu A^i_\mu$ and the $B_{\mu\nu}$ term. Since the transformation connecting A_μ and Z_μ to A^3_μ and B_μ is orthogonal, and because $\text{tr}\,(\tau_+\tau_-) = 1$, we have immediately

$$\mathcal{L}_{\text{field}}\Big|_{\text{KE terms}} = -\tfrac{1}{4}F_{\mu\nu}F^{\mu\nu} - \tfrac{1}{4}Z_{\mu\nu}Z^{\mu\nu} - \tfrac{1}{2}W^\dagger_{\mu\nu}W^{\mu\nu} \;, \tag{15.116}$$

where $F_{\mu\nu}$ is the usual EM field term, and

$$\begin{aligned} Z_{\mu\nu} &= \partial_\mu Z_\nu - \partial_\nu Z_\mu \\ W_{\mu\nu} &= \partial_\mu W_\nu - \partial_\nu W_\mu \;. \end{aligned} \tag{15.117}$$

Note the $W^\dagger_{\mu\nu}W^{\mu\nu}$ term is twice as large, as required of a complex vector field with two real components (recall Prob. 7.3).

Next, we calculate the four-gauge coupling terms. To reduce these, use

$$\begin{aligned} \epsilon_{3ij}A^i_\mu A^j_\nu &= \; i\left[W^\dagger_\mu W_\nu - W^\dagger_\nu W_\mu\right] \\ \frac{1}{\sqrt{2}}\left(\epsilon_{1ij}A^i_\mu A^j_\nu + i\epsilon_{2ij}A^i_\mu A^j_\nu\right) &= -iW^\dagger_\mu\left[\sin\theta_W A_\nu + \cos\theta_W Z_\nu\right] - (\mu \leftrightarrow \mu) \\ \frac{1}{\sqrt{2}}\left(\epsilon_{1ij}A^i_\mu A^j_\nu - i\epsilon_{2ij}A^i_\mu A^j_\nu\right) &= \; iW_\mu\left[\sin\theta_W A_\nu + \cos\theta_W Z_\nu\right] - (\mu \leftrightarrow \nu) \;. \end{aligned} \tag{15.118}$$

Hence the four-gauge couplings are

$$\begin{aligned} \mathcal{L}_{\text{field}}\Big|_{4g} =& -\tfrac{1}{2}g^2\left[W^\dagger_\mu W_\nu W^{\nu\dagger}W^\mu - W^\dagger_\mu W^{\mu\dagger}W_\nu W^\nu\right] \\ & - g^2 W^\dagger_\mu W^\mu\left[\sin^2\theta_W A_\nu A^\nu + \sin 2\theta_W A_\nu Z^\nu + \cos^2\theta_W Z_\nu Z^\nu\right] \\ & + g^2 W^\dagger_\mu W_\nu\left[\sin^2\theta_W A^\mu A^\nu + \tfrac{1}{2}\sin 2\theta_W\left(A^\nu Z^\mu + A^\mu Z^\nu\right)\right. \\ & \left. + \cos^2\theta_W Z^\mu Z^\nu\right] \;. \end{aligned} \tag{15.119}$$

Note that all of these couplings conserve charge and that all involve the charged bosons. Finally, the three-gauge couplings can also be found with the help of (15.118). We have

$$\begin{aligned} \mathcal{L}_{\text{field}}\Big|_{3g} =& \, ig\left[\sin\theta_W F_{\mu\nu} + cos\theta_W Z_{\mu\nu}\right]W^{\mu\dagger}W^\nu \\ & + ig\sin\theta_W A^\nu\left[W^\mu W^\dagger_{\mu\nu} - W^{\mu\dagger}W_{\mu\nu}\right] \\ & + ig\cos\theta_W Z^\nu\left[W^\mu W^\dagger_{\mu\nu} - W^{\mu\dagger}W_{\mu\nu}\right] \;. \end{aligned} \tag{15.120}$$

There are many terms, and it is helpful to separate out those which correspond to the expected electromagnetic couplings of the W_μ field. Recalling that $e = g\sin\theta$, the terms proportional to e and e^2 can be readily isolated from (15.120) and (15.119). These are also the terms proportional to A^2 in (15.119) and A in (15.120). These terms give:

$$\begin{aligned}\mathcal{L}_{\text{field}}\Big|_{EM} =& ieA^\nu\left[W^\mu W^\dagger_{\mu\nu} - W^{\mu\dagger}W_{\mu\nu}\right]\\ &- e^2\left[W^\dagger_\mu W^\mu A_\nu A^\nu - W^\dagger_\mu W_\nu A^\mu A^\nu\right] \ .\end{aligned} \tag{15.121}$$

If these terms are combined with the kinetic terms for W, we have

$$\begin{aligned}\mathcal{L}_{\text{field}}\Big|_{W} =& -\tfrac{1}{2}\left[W^\dagger_{\mu\nu} - ie\left(A_\mu W^\dagger_\nu - A_\nu W^\dagger_\mu\right)\right]\\ &\times\left[W^{\mu\nu} + ie\left(A^\mu W^\nu - A^\nu W^\mu\right)\right] .\end{aligned} \tag{15.122}$$

This is precisely the result expected from minimal substitution; if $\partial_\mu \to \partial_\mu + ieA_\mu$, we would obtain (15.122) from the free W Lagrangian in (15.116). Furthermore, the sign identifies W as a field with *plus* charge e. The electromagnetic couplings of the W^+ boson are given in Fig. 15.8, and the self-couplings of the W's and the Z are shown in Fig. 15.9.

At this time, we collect together all of the transformed terms into a Lagrangian density with four terms:

$$\mathcal{L}_{\text{EW}} = \mathcal{L}^{\text{EW}}_{\text{lepton}} + \mathcal{L}^{\text{EW}}_{\text{field}} + \mathcal{L}^{\text{EW}}_{\text{scalar}} + \mathcal{L}^{\text{EW}}_{\text{int}} \ . \tag{15.123}$$

These terms are combinations of terms from different parts of the original Lagrangian, collected together for convenience. The first term includes the original lepton part, (15.107), plus the electron mass term which came from the original interaction term. It is

$$\begin{aligned}\mathcal{L}^{\text{EW}}_{\text{lepton}} =& \bar\psi_{\nu_L}\left(\frac{i}{2}\overleftrightarrow{\partial\!\!\!/} - \sqrt{2}\,g_W\, Z\!\!\!/\right)\psi_{\nu_L} - g_W\cos\theta_W\left[\bar\psi_e\gamma^\mu(1-\gamma^5)\psi_\nu W^\dagger_\mu + \text{h.c.}\right]\\ &+\bar\psi_e\left(\frac{i}{2}\overleftrightarrow{\partial\!\!\!/} - m_e + e\,A\!\!\!/ + \frac{1}{\sqrt{2}}g_W\, Z\!\!\!/\left(\left[1-4\sin^2\theta_W\right]-\gamma^5\right)\right)\psi_e \ .\end{aligned} \tag{15.124}$$

The field part contains the free field terms together with the electromagnetic interactions and the boson mass term:

$$\begin{aligned}\mathcal{L}^{\text{EW}}_{\text{field}} =& -\tfrac{1}{4}F_{\mu\nu}F^{\mu\nu} - \tfrac{1}{4}Z_{\mu\nu}Z^{\mu\nu} + \tfrac{1}{2}M_Z^2\,Z_\mu Z^\mu + M_W^2\,W^\dagger_\mu W^\mu\\ &-\tfrac{1}{2}\left[W^\dagger_{\mu\nu} - ie\left(A_\mu W^\dagger_\nu - A_\nu W^\dagger_\mu\right)\right]\left[W^{\mu\nu} + ie\left(A^\mu W^\nu - A^\nu W^\mu\right)\right] .\end{aligned} \tag{15.125}$$

• the operator

$$ie\left[(p+p')_\mu\, g_{\nu\lambda} - (p'+q)_\nu\, g_{\mu\lambda} - (p-q)_\lambda\, g_{\mu\nu}\right]$$

at each vertex where a photon with momentum q and polarization μ is absorbed by a positively charged spin one boson with incoming momentum p and polarization ν and outgoing momentum p' and polarization λ.

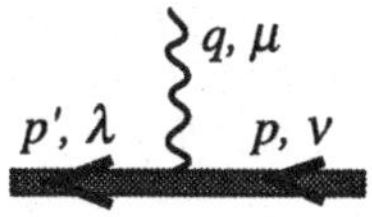

Fig. 15.8a

• the operator

$$-ie^2\left[2g_{\mu\nu}g_{\sigma\rho} - g_{\mu\sigma}g_{\nu\rho} - g_{\mu\rho}g_{\nu\sigma}\right]$$

at each vertex where two photons with polarization μ and ν are emitted from or absorbed by two weak charged bosons with polarizations σ and ρ.

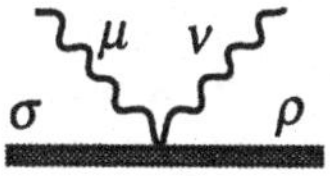

Fig. 15.8b

Fig. 15.8 Quantum Electrodynamics of a spin one boson.

The scalar part includes the Higgs kinetic energy piece and all Higgs interactions:

$$\begin{aligned}\mathcal{L}^{\rm EW}_{\rm scalar} =& \frac{1}{2}\partial_\mu\eta\partial^\mu\eta - \frac{1}{2}m_H^2\eta^2 - \frac{1}{3!}\left(\frac{3}{2}g\frac{m_H^2}{M_W}\right)\eta^3 - \frac{1}{4!}\left(\frac{3}{4}g^2\frac{m_H^2}{M_W^2}\right)\eta^4\\ &- \frac{g}{2}\frac{m_e}{M_W}\bar{\psi}_e\psi_e\eta + gM_W W_\mu^\dagger W^\mu\left(\eta + \frac{1}{2}\frac{g}{2M_W}\eta^2\right)\\ &+ \frac{1}{2}\frac{gM_Z}{\cos\theta_W}Z_\mu Z^\mu(\eta + \frac{1}{2}\frac{g}{2M_Z\cos\theta_W}\eta^2)\end{aligned} \qquad (15.126)$$

and finally the "interaction" part includes only the *weak interactions of gauge bosons*:

$$\begin{aligned}\mathcal{L}^{\rm EW}_{\rm int} =& \tfrac{1}{2}g^2\left[W_\mu^\dagger W^{\dagger\mu}W_\nu W^\nu - W_\mu^\dagger W_\nu W^{\nu\dagger}W^\mu\right]\\ &- g^2\cos^2\theta_W\left[W_\mu^\dagger W^\mu Z_\nu Z^\nu - W_\mu^\dagger W_\nu Z^\nu Z^\mu\right]\\ &- eg\cos\theta_W\left[2W_\mu^\dagger W^\mu A_\nu Z^\nu - W_\mu^\dagger W_\nu\left(A^\nu Z^\mu + Z^\nu A^\mu\right)\right]\\ &+ ig\cos\theta_W\left[W^{\mu\dagger}W^\nu Z_{\mu\nu} + W_{\mu\nu}^\dagger Z^\nu W^\mu - W_{\mu\nu}Z^\nu W^{\mu\dagger}\right].\end{aligned} \qquad (15.127)$$

- the operator

$$ig \cos\theta_W \left[g_{\nu\sigma} (p + p')_\mu - g_{\mu\nu} (p - q)_\sigma - g_{\mu\sigma} (q + p')_\nu \right]$$

at each vertex where a Z boson with momentum q and polarization μ is absorbed by a W boson with incoming momentum p and polarization ν and outgoing momentum p' and polarization σ.

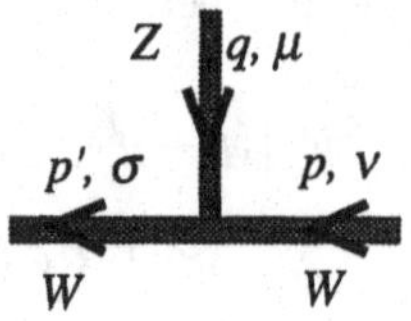

Fig. 15.9a

- the operator

$$ig^2 \left[2g_{\mu\rho} g_{\nu\sigma} - g_{\mu\nu} g_{\sigma\rho} - g_{\mu\sigma} g_{\nu\rho} \right]$$

at each vertex where two W bosons enter with polarization μ and ρ and leave with polarizations ν and σ.

Fig. 15.9b

- the operator

$$-ig^2 \cos^2\theta_W \left[2g_{\mu\nu} g_{\rho\sigma} - g_{\mu\sigma} g_{\nu\rho} - g_{\mu\rho} g_{\nu\sigma} \right]$$

at each vertex where W bosons enter and leave with polarizations μ and ν and Z bosons enter and leave with polarizations ρ and σ.

Fig. 15.9c

- the operator

$$-ieg \cos\theta_W \left[2g_{\mu\nu} g_{\rho\sigma} - g_{\mu\sigma} g_{\nu\rho} - g_{\mu\rho} g_{\nu\sigma} \right]$$

at each vertex where a W boson enters with polarization μ and leaves with polarization ν and a Z boson and a photon enter and leave with polarizations ρ and σ.

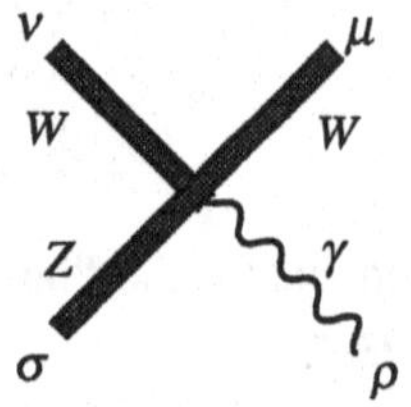

Fig. 15.9d

Fig. 15.9 Boson self-couplings in the Standard Model.

These Lagrangians account for the large number of interactions summarized in Figs. 15.6–15.9 and in Appendix B.

The particle content of the transformed Lagrangian was summarized in Table 15.1. There are five independent parameters. After the U gauge transformation, they are e, the absolute value of the electron charge, θ_W, the Weinberg angle, m_e, the electron mass, M_W, the mass of the charged boson, and m_H, the Higgs mass. These are related to the original parameters through

$$\begin{aligned} e &= \frac{gg'}{\sqrt{g^2+g'^2}} = g\sin\theta_W \\ \tan\theta_W &= \frac{g'}{g} \\ m_e &= \frac{G_e v}{\sqrt{2}} \\ M_W &= \frac{gv}{2} = \frac{g}{2}\sqrt{-\frac{m^2}{\lambda^2}} \\ m_H &= \sqrt{-2m^2} \ . \end{aligned} \tag{15.128}$$

The parameter g_w is only a shorthand for the combination (15.108).

We now look at the high energy behavior of the electroweak interactions.

15.5 UNITARITY IN THE STANDARD MODEL

We close this chapter with a short calculation which illustrates how the Standard Model solves a longstanding problem associated with the weak interactions. This calculation will also illustrate the new Feynman rules given in Figs. 15.6 and 15.9.

As discussed in Sec. 9.10, a massive vector particle (such as the $W^\pm$ or the Z) has three polarization states, owing to the fact that it can always be brought to rest and polarized in any of the three independent directions in space (recall that this is not true for a massless particle). The four-vectors $\epsilon^{i\mu}$ which describe these three states can be conveniently defined in the rest frame by the requirements

$$\begin{aligned} p_0 \cdot \epsilon^i &= 0 \\ \epsilon^i \cdot \epsilon^j &= -\delta_{ij} \ , \end{aligned} \tag{15.129}$$

where $p_0 = (m, 0, 0, 0)$ is the four-momentum of the particle at rest. These requirements are clearly satisfied by any three-space unit vector with a zero time component, and there are precisely three independent such vectors. The polarization states of a moving particle can then be obtained by boosting the rest frame polarization vectors. This will preserve the requirements (15.129), which can then be used to construct the polarization vectors in an arbitrary frame. For example, if a vector particle has momentum $p^\mu = (E, 0, 0, p)$, then three independent

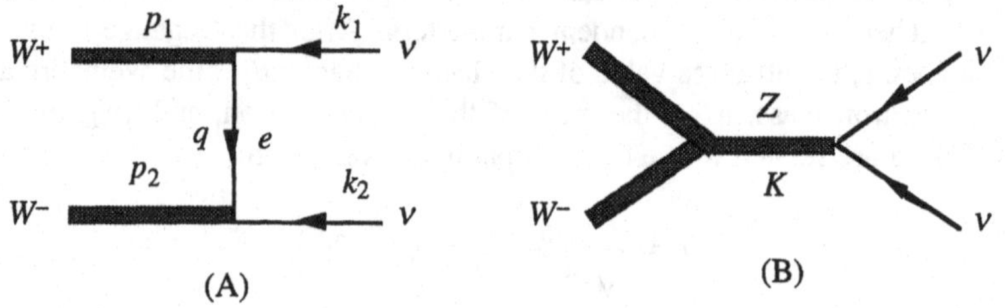

Fig. 15.10 Feynman diagrams which contribute to the lowest order production of W pairs by neutrinos.

polarization states which satisfy these requirements are

$$\begin{aligned} \epsilon^{\pm\,\mu} &= \mp\frac{1}{\sqrt{2}}(0,1,\pm i,0) \\ \epsilon^{3\mu} &= \frac{1}{m}(p,0,0,E) \ , \end{aligned} \tag{15.130}$$

where $\epsilon^{\pm}$ are the transverse states and ϵ^3 is the *longitudinal* state. Note that the individual components of the longitudinal polarization vector approach infinity as the momentum of the particle approaches infinity,

$$\epsilon^{3\mu} \underset{p\to\infty}{\longrightarrow} \frac{p}{m}(1,0,0,1) + \mathcal{O}\left(p^{-2}\right) \to \frac{p^\mu}{m} \ . \tag{15.131}$$

Because of this divergent behavior, theories with massive vector bosons generally give infinite results at very high energy, and such behavior is unphysical. A limit on the high energy behavior of scattering amplitudes can be obtained from the unitarity relation. In Sec. 12.8 we showed that one consequence of unitarity is that partial wave scattering amplitudes are bounded by a constant as $p \to \infty$. The most general bound we can obtain from the unitarity relation is somewhat less restrictive, but a growth which is linear in the energy is ruled out.

As an example of how the Standard Model controls the divergent behavior of longitudinal polarization states, consider the process $\nu + \bar{\nu} \to W^+ + W^-$ (this is a simple theoretical example, even though it is almost impossible to study in the laboratory). To second order in the weak coupling, only the two Feynman diagrams shown in Fig. 15.10 contribute. Suppose the two W bosons are produced in longitudinal states. Then the first of these diagrams gives

$$
\begin{aligned}
\mathcal{M}_A &= i(-i)\left(\frac{-ig}{2\sqrt{2}}\right)^2 \bar{v}(\boldsymbol{k}_2,s_2)\left[\frac{\not\epsilon_2\,(1-\gamma^5)(m_e+\not q)\not\epsilon_1\,(1-\gamma^5)}{m_e^2-q^2}\right] u(\boldsymbol{k}_1,s_1) \\
&\underset{E\to\infty}{\longrightarrow} \frac{4}{q^2}\left(\frac{g}{2\sqrt{2}}\right)^2\left(\frac{1}{M_W^2}\right)\bar{v}(\boldsymbol{k}_2,s_2)\;\not p_2\not q\not p_1\;u(\boldsymbol{k}_1,s_1) \\
&= \left(\frac{g^2}{2q^2M_W^2}\right)\bar{v}(\boldsymbol{k}_2,s_2)\,(\not p_2-\not k_2)\;\not q\;(\not p_1-\not k_1)\,u(\boldsymbol{k}_1,s_1) \\
&= -\left(\frac{g^2}{2M_W^2}\right)\bar{v}(\boldsymbol{k}_2,s_2)\;\not q\;u(\boldsymbol{k}_1,s_1)\;,
\end{aligned}
\tag{15.132}
$$

where, in the second step, $(1-\gamma^5)u = 2u$ and $\bar{v}(1+\gamma^5) = 2\bar{v}$ was used and we took the limit $E \to \infty$ (where E is the energy of the incoming neutrino and antineutrino) and, in the third step, terms involving $\not k_1\; u(\boldsymbol{k}_1,s_1) = 0$ were added to facilitate the reduction. Note that the final result goes to infinity as E^2. Before the development of the Standard Model, only the charge changing weak interactions were known, and it was assumed that they were mediated by a massive charged boson. In such a "theory," the production of W's would be described only by diagram A, which violates the unitarity bound. A theory with only W's has problems.

In the Standard Model this problem is eliminated because there is a neutral Z which gives a second diagram, Fig. 15.10B, which *exactly cancels* the divergent part of diagram A. This diagram is

$$
\begin{aligned}
\mathcal{M}_B &= i(i)\left(\frac{-ig}{4\cos\theta_W}\right)(ig\cos\theta_W)\,\bar{v}(\boldsymbol{k}_2,s_2)\,\gamma^\mu(1-\gamma^5)\,u(\boldsymbol{k}_1,s_1) \\
&\quad\times\left[\frac{g_{\mu\nu}-K_\mu K_\nu/M_Z^2}{M_Z^2-K^2}\right]\left[\epsilon_1\cdot\epsilon_2(p_1-p_2)^\nu+2\epsilon_2^\nu\epsilon_1\cdot p_2-2\epsilon_1^\nu\epsilon_2\cdot p_1\right] \\
&= -\frac{g^2}{2(M_Z^2-K^2)}\bar{v}(\boldsymbol{k}_2,s_2)\left[\epsilon_1\cdot\epsilon_2(\not p_1-\not p_2)+2\not\epsilon_2\,\epsilon_1\cdot p_2-2\not\epsilon_1\,\epsilon_2\cdot p_1\right. \\
&\qquad\qquad\left.-2\frac{\epsilon_2\cdot p_1\;\epsilon_1\cdot p_2}{M_Z^2}+2\frac{\epsilon_1\cdot p_2\;\epsilon_2\cdot p_1}{M_Z^2}\right]u(\boldsymbol{k}_1,s_1) \\
&\underset{E\to\infty}{\longrightarrow}\left(\frac{g^2\,p_1\cdot p_2}{2M_W^2K^2}\right)\bar{v}(\boldsymbol{k}_2,s_2)\,(\not p_2-\not p_1)\,u(\boldsymbol{k}_1,s_1) \\
&= \left(\frac{g^2}{4M_W^2}\right)\bar{v}(\boldsymbol{k}_2,s_2)\,(\not p_2-\not p_1)\,u(\boldsymbol{k}_1,s_1) \\
&= \left(\frac{g^2}{2M_W^2}\right)\bar{v}(\boldsymbol{k}_2,s_2)\;\not q\;u(\boldsymbol{k}_1,s_1)\;,
\end{aligned}
\tag{15.133}
$$

where, in the last step, we used $K^2 = 2M_W^2 + 2p_1\cdot p_2 \to 2p_1\cdot p_2$ as $E\to\infty$. This term cancels the contribution from diagram A, eliminating all terms which go like E^2 at large E. The exact result is therefore finite as $E \to \infty$.

This good behavior at high energy was obtained only through a delicate cancellation of diverging contributions from different terms. While this cancellation was obtained easily in this example, it presents a serious problem when these Feynman rules are used to calculate loop diagrams. Now the divergences due to the longitudinal polarization will show up in the bad high energy behavior of the spin one propagator; the $p_\mu p^\nu/M^2$ term in the propagator (9.138) gives strong ultraviolet divergences, which must be canceled by diverging contributions from other terms. This is an unfortunate feature of the Feynman rules in the U gauge (which are the ones given in this chapter), but there are gauges (the R_ξ gauges) in which the boson propagators do *not* diverge as $p \to \infty$ and in which all terms are individually well behaved at high momentum. The disadvantage of these R_ξ gauges is that they produce (several) ghosts and many ghost interactions which greatly increase the number of interactions and diagrams which must be calculated. However, it turns out that the advantage of the improved convergence for higher order loop corrections outweighs the disadvantage of having to use ghosts, and the R_ξ gauges are preferred when loop calculations are to be carried out. For a discussion of the Feynman rules for the R_ξ gauges, see Cheng and Li (1984).

PROBLEMS

15.1 Prove the relation (15.68). Show that it is true for $\hat{\boldsymbol{k}}_1 = \hat{z}$ and $\hat{\boldsymbol{k}}_2 = -\hat{z}$ and then generalize the result.

15.2 Compute the electron–neutrino elastic scattering amplitude to second order in the weak coupling constant g^2.

CHAPTER 16

RENORMALIZATION

In this chapter we return to the subject of renormalization, or the removal of infinities from field theories. An introductory discussion of this topic was presented in Chapter 11, but because this is a problem of central importance, we will now discuss it in somewhat greater depth.

The principal goal of this chapter is to introduce some of the main issues and develop the language to the point where the interested student is equipped to pursue the literature. After defining the problem and studying ϕ^3 theory as an example, we discuss the renormalization of QED, emphasizing the role played by gauge invariance in the form of the Ward-Takahashi identities. The chapter concludes with a brief discussion of the renormalization of QCD.

16.1 POWER COUNTING AND REGULARIZATION

Before defining what is meant by a renormalizable theory, look at the ultraviolet behavior of a typical Feynman diagram which arises from the perturbative expansion of the theory. Diverging diagrams will be evaluated using dimensional regularization, introduced in Chapter 11. For simplicity, we will first discuss theories with *no* derivatives in the interaction term; the discussion will be extended to derivative interactions later. Then a typical diagram will have

$$\begin{aligned} \ell &= \text{the number of internal loops} \\ n_B &= \text{the number of internal spin zero boson lines} \\ n_F &= \text{the number of internal fermion lines} \end{aligned} \tag{16.1}$$

and it is easy to see that if the momenta in all of its loops become large at *the same time*, then the overall divergence of all the integrals is determined by the quantity

$$D = \ell d - 2n_B - n_F \tag{16.2}$$

where d is the dimension of space–time. This quantity is sometimes referred to as the *superficial degree of divergence*, or as the *overall divergence* of the diagram,

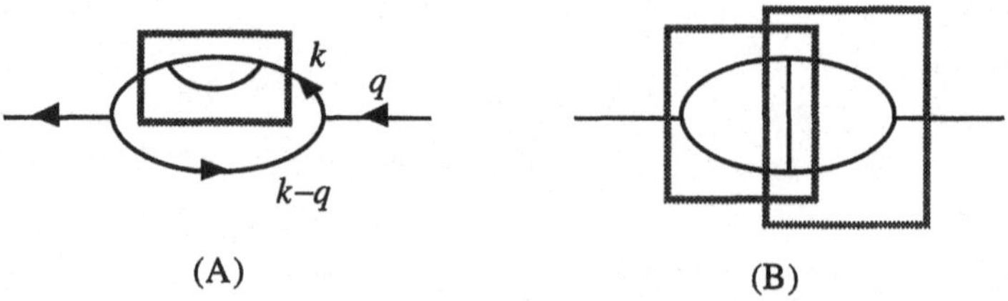

Fig. 16.1 Two graphs for the self-energy in ϕ^3 theory. The shaded boxes surround divergent subgraphs, as discussed in the text.

and is simply the number of powers of momenta in the numerator (counting each $d^d k_i$ as d) minus the number of powers of momenta in the denominator. Clearly, if $D \geq 0$, the diagram will diverge; if $D = 0$, it is superficially logarithmically divergent, and it would seem to converge if $D < 0$. However, since each loop momentum is an independent variable, a divergence can also occur if $D_i \geq 0$ for any loop (or combination of loops) in the diagram (which will be referred to as a subdiagram and denoted by i), even if the overall divergence $D < 0$. For the diagram to be finite, both D and *all* D_i associated with any subdiagram must be less than zero. This theorem is sometimes referred to as Weinberg's theorem [We 60].

To illustrate these ideas, consider the two loop diagrams for the self-energy in ϕ^3 theory, shown in Fig. 16.1. Both of these diagrams have a superficial divergence $D = 2d - 10$, which is less than zero for $d = 4$ dimensions, but the self-energy *insertion* in Fig. 16.1A (contained in the shaded rectangle) has a superficial divergence $D_\imath = d - 4$, showing that it diverges logarithmically in $d = 4$ dimensions. Diagram 16.1B has two *overlapping* vertex insertions, each with a superficial divergence $D_i = d - 6$, which converges in four dimensions. We will return to a more detailed discussion of these diagrams in Sec. 16.2 below.

To study the removal of infinities from a theory, we must express the superficial divergences D and $\{D_i\}$ of each Feynman diagram directly in terms of the properties of the interaction Lagrangian $\mathcal{L}_I$. In order to keep the discussion general, consider an interaction of the form

$$\mathcal{L}_I \simeq \left(\bar{\psi}\theta\psi\right)^{F/2} \phi^B \ ,$$

where F is the number of Fermi fields and B the number of boson fields which interact at each point in space–time. If N_B and N_F are the total number of bosons and fermions external to a Feynman diagram (the sum of those in *both* the initial *and* final states), and if there are n vertices (for an nth order Feynman diagram), then, since each internal line couples to two vertices,

$$\begin{aligned} N_B + 2n_B &= nB \\ N_F + 2n_F &= nF \ . \end{aligned} \tag{16.3}$$

Another constraint comes from momentum conservation. The number of loops ℓ is the same as the number of momenta unfixed by momentum conservation, which is the total number of internal momenta (lines) minus the number of vertices (each of which contributes one constraint) plus one (for the overall energy–momentum constraint which does not limit internal momenta). Hence

$$\ell = n_B + n_F - n + 1 \ . \tag{16.4}$$

These three constraints enable us to re-express the superficial divergence in terms of the number of external lines (which depends only on the physical process under consideration), B and F (which depend only on the theory), and n, the order of the diagram. The result can be written

$$D = nI + d - \tfrac{1}{2}(d-2)N_B - \tfrac{1}{2}(d-1)N_F \ , \tag{16.5}$$

where

$$I = \tfrac{1}{2}(d-2)B + \tfrac{1}{2}(d-1)F - d \tag{16.6}$$

is the *index of divergence* of the interaction Lagrangian $\mathcal{L}_I$.

Note that the index of divergence depends only on the theory, and not on any particular physical process, while the remaining quantities in (16.5) depend on the number and kind of the external particles (which are fixed for any physical process under consideration) and on the number of vertices n in the diagram. In four dimensions, (16.5) reduces to

$$D = nI + 4 - N_B - \tfrac{3}{2}N_F \ . \tag{16.7}$$

We will now distinguish three different classes of theories.

The first class has index $I > 0$. In this case, D will eventually become *greater than zero* as n increases, *regardless of the physical process* under consideration (i.e., for any N_B and N_F). Furthermore, as n increases, D becomes larger and larger. We conclude that if $I > 0$, there are always (higher order) Feynman diagrams which diverge regardless of the physical process under consideration. Such theories are called *non-renormalizable.*

Next, if $I = 0$, D is independent of n and in $d = 4$ dimensions is less than zero if $D_0 \equiv 4 - N_B - \frac{3}{2}N_F < 0$. For such theories, all but a finite number of elementary processes are superficially convergent. For example, in a theory with a $\bar{\psi}\theta\psi\phi$ interaction, $I = 0$, and processes involving five external bosons, such as boson production in boson–boson scattering ($B_1 + B_2 \rightarrow B_3 + B_4 + B_5$), or two external fermions and two bosons, such as boson–fermion elastic scattering ($B + F \rightarrow B + F$) or fermion annihilation ($F + \bar{F} \rightarrow B + B$), and all other more complex processes have no overall divergence; divergences which contribute to these cases come from simpler processes which occur as insertions inside of the Feynman diagrams. For QED, which has an index equal to zero, the only quantities which diverge are the electron self-energy, with $D_0 = -1$, the vacuum polarization, with $D_0 = -2$, and the γee vertex, with $D_0 = 0$. [The photon–photon scattering amplitude $(\gamma+\gamma \to \gamma+\gamma)$ also has $D_0 = 0$ but does not diverge because of gauge invariance.] We will refer to such theories as *superficially renormalizable* and save the word "renormalizable" (which is often used) for a different meaning given below.

Finally, if the index $I < 0$, then only a *finite* number of diagrams associated with a *finite* number of physical processes will diverge. An example is ϕ^3 theory in four dimensions, which has $I = -1$. In this theory only *one* diagram (or subdiagram) diverges: the *lowest order* meson self-energy, with $n = 2$ and $D = 0$. All other diagrams, except those containing this lowest order self-energy insertion, will converge (and after this self-energy is regularized, *all* other diagrams will converge). Such theories are said to be *super-renormalizable.*

Regularization Schemes

The term *regularization* is used quite generally to describe the process of removing infinities from any set of Feynman diagrams associated with any of the above classes of theory. These infinities are removed according to a definite procedure or prescription referred to as a *renormalization scheme.* Using this scheme, any Feynman diagram can be written as the sum of a finite and an infinite part, with the infinite part depending on a number of *infinite constants*, referred to as *renormalization constants*. These constants cannot be determined by the theory and are regarded as free parameters to be fixed by experiment. For non-renormalizable theories, the total number of required renormalization constants grows with the order n, and an *infinite* number are needed to remove all infinities to all orders. For theories which are superficially renormalizable, the number of renormalization constants is finite and can be absorbed into a finite number of parameters, such as the charges and masses of the particles in the theory. This process of redefining the theory by absorbing the renormalization constants into the parameters of the theory is referred to as *renormalization*, and the proof that this is possible is definitely non-trivial and is the subject of much of this chapter.

Using the ideas introduced here and in Chapter 11, we will now describe in more detail how Feynman diagrams are regularized and how infinities are systematically removed. To keep the discussion general, but not too abstract, consider ϕ^3 theory in d dimensions. Its index is

$$I_{\phi^3} = \frac{d}{2} - 3 \tag{16.8}$$

so that the theory is super-renormalizable in $d = 4$ dimensions, superficially renormalizable in $d = 6$ dimensions, and non-renormalizable in $d = 8$ dimensions.

Consider the lowest order ($n = 3$) vertex correction shown in Fig. 16.2A. This scalar vertex correction depends only on the square of the three external momenta, $\Lambda(p,q) = F\left(q^2, p^2, p \cdot q\right)$, and is divergent in $d \geq 6$ dimensions. The infinite part,

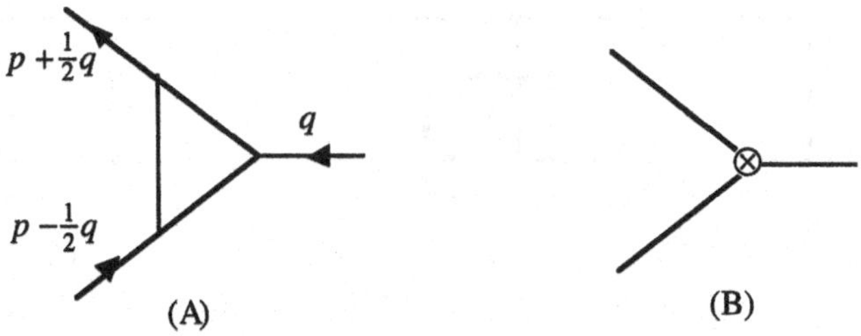

Fig. 16.2 The lowest order vertex correction in ϕ^3 theory is shown in (A), and the Feynman diagram generated by the corresponding counterterm Λ_0 is shown in (B).

which we will denote by Λ_0, can be removed by subtracting it from the diagram. Detailed examples of how these infinite parts are defined and subtracted will be presented in the next section. The *finite* part which remains after the subtraction will depend on precisely how the infinite part is defined and how the subtraction is carried out, and this dependence generally appears as a dependence on some *momentum scale*, which will be denoted by μ. Hence, the renormalized vertex correction Λ_R is written

$$\Lambda_R\,(p,q,\mu) = \Lambda\,(p,q) - \Lambda_0(\mu)\ , \tag{16.9}$$

where the dependence of Λ_R on the momentum scale μ, which inevitably arises in the subtraction, is shown explicitly. The infinite subtraction constant, Λ_0, can be treated in one of two equivalent ways. The first method, which we used in Chapter 11, is to absorb it into the lower order graphs (the coupling constant in this example), which are said to be *renormalized* by the subtraction. In this example, the three-point coupling to third order then becomes (recall that the coupling constant g_0 must multiply the vertex correction to give the diagram in Fig. 16.2)

$$\begin{aligned} g_R\Gamma &= g_0 + g_0\,(\Lambda_R + \Lambda_0) \\ &= g_R + g_0\Lambda_R \\ &\cong g_R + g_R\Lambda_R\ , \end{aligned} \tag{16.10}$$

where Γ is the full vertex function, the renormalized coupling is $g_R = g_0 + g_0\Lambda_0$, and $g_0\Lambda_R \cong g_R\Lambda_R$ to third order in perturbation theory. Defining the renormalization constant Z_1 by

$$\Lambda_0 = Z_1^{-1} - 1 \tag{16.11}$$

(as we did in Chapter 11) gives

$$g_R = \frac{g_0}{Z_1} = g_0\Gamma_0\ , \tag{16.12}$$

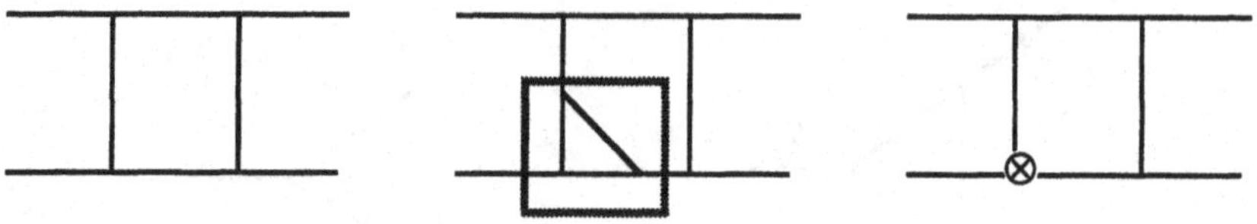

Fig. 16.3 Figure (A) shows the fourth order box diagram for scattering in ϕ^3 theory. In (B) the diagram has a divergent vertex insertion, which is renormalized by the box diagram with counterterm shown in (C).

showing that vertex contributions renormalize the coupling constant by a factor Z_1. In this method, the coupling constant is renormalized order-by-order in perturbation theory.

The second method, which we will employ in this chapter, is to keep the coupling unchanged and to remove the singular term Λ_0 by adding a counterterm to the Lagrangian. This is the method we used for handling mass renormalization in Chapter 11, and here we will extend it to the other renormalizations as well. For example, if the original ϕ^3 interaction term was

$$\mathcal{L}_I = -\frac{1}{3!} g_R \phi^3 \tag{16.13}$$

(for the symmetric case discussed in Chapter 9), then the infinity can be removed by adding a counterterm

$$\mathcal{L}'_I = -\frac{1}{3!} g_R \phi^3 - \frac{1}{3!} g_R (Z_1 - 1)\phi^3 \ . \tag{16.14}$$

Using (16.11), the counterterm can be written $g_R(Z_1 - 1) = -g_R Z_1 \Lambda_0$. This counterterm gives a new third order Feynman diagram, shown in Fig. 16.2B, so that when the calculation is carried out to third order we obtain

$$\begin{aligned} g_R \Gamma &= g_R + g_R \Lambda\,(p,q) - g_R Z_1 \Lambda_0 \\ &= g_R + g_R \Lambda_R\,(p,q,\mu) \ , \end{aligned} \tag{16.15}$$

because $Z_1 \Lambda_0 \cong \Lambda_0$ to second order in perturbation theory. In this method the counterterm is a new interaction specifically added to cancel the singularity in diagram 16.2A.

These two points of view are equivalent, but the introduction of counterterms is more general in that it allows us to discuss the removal of singularities from non-renormalizable theories. For example, consider the $\phi\phi$ scattering diagram shown in Fig. 16.3A. This diagram is finite in four and six dimensions but has a logarithmic divergence in $d = 8$ dimensions. To remove this singularity, it is necessary to add a counterterm to the ϕ^3 Lagrangian of the form

$$\mathcal{L}_4 = -\frac{1}{4!}\lambda_0\phi^4 \,, \tag{16.16}$$

where λ_0 is determined from the singular part of graph 16.3A (see Prob. 16.2). However, since a ϕ^4 term does not appear in the original Lagrangian, this counterterm cannot be absorbed into one of the original parameters of the theory, and its appearance changes the structure of the theory. This need not be a disaster; in this example λ_0 can be treated as a new parameter and determined from a measurement of $\phi\phi$ scattering at some fixed point, allowing us to predict the scattering at other points. But the appearance of a new counterterm certainly reduces the predictive power of the theory, and because the index I of this theory is positive (remember that $d = 8$), we can expect many new divergences to appear in higher order. This will introduce still more counterterms, further reducing the predictive power of the theory. In practice, non-renormalizable theories are useful only in cases where a good estimate can be obtained from the first few orders in perturbation theory. *Chiral perturbation theory*, based on the non-linear chiral models discussed in Chapter 13, is an example of a non-renormalizable theory which has enjoyed considerable success. For a discussion of effective Lagrangians, see Donoghue, Golowich, and Holstein (1992). We will not discuss non-renormalizable theories further.

Our discussion up to now has focused on how the infinities are removed in "lowest order." The central problem in the proof of renormalizability is to show that the addition of a finite number of counterterms is sufficient to remove *all* infinities from the theory. For example, return to the diagrams shown in Fig. 16.3. In $d = 6$ dimensions, 16.3A is finite, but 16.3B is infinite because of the diverging vertex subdiagram. A counterterm added to the Lagrangian renders this vertex correction finite, as we have discussed, and the same counterterm inserted in the $\phi\phi$ scattering box, shown in Fig. 16.3C, will also insure that the two diagrams 16.3B and 16.3C are finite. To prove renormalizability, we must show that such a procedure works for all diagrams to all orders.

The demonstration of the renormalizability of QED will be a major goal of this chapter. Before we discuss these problems further, it is helpful to consider a few more examples and to develop a technique for evaluating multi-loop diagrams.

16.2 ϕ^3 THEORY: AN EXAMPLE

To clarify some of the issues which will arise in the construction of a general proof of renormalizability, we look at ϕ^3 theory in six dimensions. As discussed above, this theory is superficially renormalizable (has an index equal to zero) and will provide a simple illustrative example.

First, consider the dimensions of the coupling constant in ϕ^3 theory. The action is dimensionless, so in d dimensions, the Lagrangian density must have the

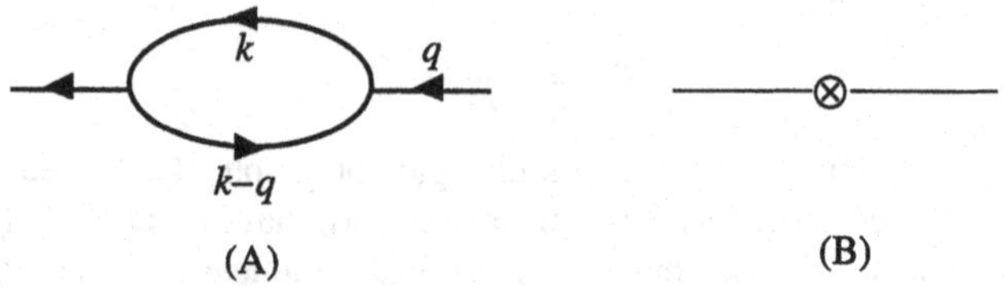

Fig. 16.4 The lowest order self-energy diagram in ϕ^3 theory with its corresponding counterterm.

dimension L^{-d} (where L is a length), and as the Lagrangian density contains the term $\partial_\mu\phi\partial^\mu\phi \sim L^{-2}\phi^2$, each field function $\phi(x)$ has dimension $L^{1-d/2} \to \mu^{d/2-1}$ (where μ is a mass), so that the coupling constant must have dimension $\mu^{3-d/2}$. Therefore, the coupling constant is dimensionless only in $d = 6$ dimensions, precisely the number of dimensions in which ϕ^3 theory is superficially renormalizable. In less than six dimensions, we will write

$$g_R = g\,\mu^{\epsilon/2}\,, \tag{16.17}$$

where $\epsilon = 6 - d$, g is a dimensionless coupling constant, and μ is an arbitrary mass scale. It is through equations like (16.17) that a mass scale associated with dimensional regularization enters the renormalization program. This subtle point was ignored in our less general treatment of Chapter 11.

For our first real example, we calculate the self-energy shown in Fig. 16.4A. This graph gives

$$\begin{aligned}\Sigma(q^2) &= i\frac{g_R^2}{2}\int\frac{d^dk}{(2\pi)^d}\,\frac{1}{(m^2-k^2)\,(m^2-(k-q)^2)}\\ &= i\frac{g_R^2}{2}\int_0^1 dx\int\frac{d^dk}{(2\pi)^d}\,\frac{1}{\left[m^2-k^2+2k\cdot qx-q^2x\right]^2}\end{aligned} \tag{16.18}$$

where we used Eq. (C.2) from Appendix C to combine the denominator into a single term and the extra factor of $\frac{1}{2}$ is the symmetry factor which accompanies bubble diagrams in symmetric ϕ^3 theory. Next, complete the square in the denominator by shifting $k \to k' + xq$, and carry out the d^dk' integration using Eq. (C.12):

$$\Sigma(q^2) = -\frac{g_R^2}{2(4\pi)^{d/2}}\Gamma\left(2-\frac{d}{2}\right)\int_0^1 dx\frac{1}{\left[m^2-q^2x(1-x)\right]^{2-d/2}}\,. \tag{16.19}$$

Next, we separate this into two parts, one proportional to q^2 and one finite as $q^2 \to 0$:

$$\Sigma(q^2) = m^2A(q^2) - q^2B(q^2)\,, \tag{16.20}$$

where

$$
\begin{aligned}
A(q^2) &= -\frac{g_R^2}{2(4\pi)^{d/2}}\Gamma\left(2-\frac{d}{2}\right)\int_0^1 dx \frac{1}{[m^2 - q^2 x(1-x)]^{3-d/2}} \\
B(q^2) &= -\frac{g_R^2}{2(4\pi)^{d/2}}\Gamma\left(2-\frac{d}{2}\right)\int_0^1 dx \frac{x(1-x)}{[m^2 - q^2 x(1-x)]^{3-d/2}} \quad .
\end{aligned}
\tag{16.21}
$$

The mass shift and wave function renormalization come from $A(q^2)$ and $B(q^2)$, as in Sec. 11.3.

We are interested in the behavior of each of the functions near $d = 6$ dimensions. Note that each is singular in the small parameter $\epsilon = 6 - d$ as $\epsilon \to 0$, because

$$
\Gamma\left(2-\frac{d}{2}\right) = \Gamma\left(\frac{\epsilon}{2}-1\right) = -\frac{\Gamma(1+\epsilon/2)}{\epsilon/2\,(1-\epsilon/2)} \quad . \tag{16.22}
$$

It is convenient to separate A and B into two parts: a singular part, which will be denoted by A_0 and B_0, and a finite part A_R and B_R. The separation between these two parts depends on the renormalization scheme, because the singular part can include any finite terms which it is desirable to include, and will also depend on the mass scale μ which enters through the substitution $g_R^2 = g^2\mu^\epsilon$ of Eq. (16.17). To make the results well-defined and unique, we must define, as part of the renormalization scheme, what finite terms are to be included in A_0 and B_0 and which are to remain in A_R and B_R. In this chapter we will adopt the somewhat unconventional scheme of including in A_0 and B_0 all finite terms which do not depend on momenta or on the scale parameter μ^2. With this choice, the finite terms emerge *only* from the expansion of the factor

$$
\left[\frac{\mu^2}{m^2 - q^2x(1-x)}\right]^{\epsilon/2} = 1 - \frac{\epsilon}{2}\log\left(\frac{m^2 - q^2x(1-x)}{\mu^2}\right) \tag{16.23}
$$

so that we have uniquely

$$
\begin{aligned}
A_0 &= \frac{g^2}{(4\pi)^{3-\epsilon/2}}\,\frac{\Gamma\left(1+\frac{\epsilon}{2}\right)}{\epsilon\left(1-\frac{\epsilon}{2}\right)} \\
B_0 &= \frac{g^2}{6(4\pi)^{3-\epsilon/2}}\,\frac{\Gamma\left(1+\frac{\epsilon}{2}\right)}{\epsilon\left(1-\frac{\epsilon}{2}\right)}
\end{aligned}
\tag{16.24}
$$

with corresponding finite terms

$$
\begin{aligned}
A_R(q^2) &= -\frac{g^2}{2(4\pi)^3}\int_0^1 dx \log\left[\frac{m^2 - q^2x(1-x)}{\mu^2}\right] \\
B_R(q^2) &= -\frac{g^2}{2(4\pi)^3}\int_0^1 dx\, x(1-x)\log\left[\frac{m^2 - q^2x(1-x)}{\mu^2}\right] \quad .
\end{aligned}
\tag{16.25}
$$

Note the explicit appearance of the scale parameter μ^2 in the finite terms. We will say no more about these finite pieces now.

The infinite parts (16.24) can be expanded in a Laurent series in the small parameter ϵ. Using the expansion of the Γ-function,

$$\Gamma\left(1+\frac{\epsilon}{2}\right) = 1 - \frac{\epsilon}{2}\gamma + \mathcal{O}(\epsilon^2)\,, \tag{16.26}$$

where $\gamma = 0.5772\cdots$ is Euler's constant, the terms which survive as $\epsilon \to 0$ are

$$\begin{aligned} A_0 &= \frac{g^2}{2(4\pi)^3}\left[\frac{2}{\epsilon} - \gamma + 1 + \log(4\pi)\right] \\ B_0 &= \frac{g^2}{12(4\pi)^3}\left[\frac{2}{\epsilon} - \gamma + 1 + \log(4\pi)\right] \quad . \end{aligned} \tag{16.27}$$

In the scheme which we will use in this chapter, these constants will become the counterterms discussed in the previous section. In some treatments, only the $1/\epsilon$ part of (16.27) are included in the counterterms, leaving the finite part to be combined with A_R and B_R. This is the *minimal subtraction* scheme [Ho 73], and in this scheme the counterterms are

$$\begin{aligned} A_{\rm MS} &= \frac{g^2}{(4\pi)^3}\,\frac{1}{\epsilon} \\ B_{\rm MS} &= \frac{g^2}{6(4\pi)^3}\,\frac{1}{\epsilon}\,, \end{aligned} \tag{16.28}$$

where the subscript MS refers to "minimal subtraction." Alternatively, the $\overline{\rm MS}$ scheme [BB 78] includes the $\log(4\pi)$ and Euler's constant γ in the counterterms, so that

$$\begin{aligned} A_{\overline{\rm MS}} &= \frac{g^2}{2(4\pi)^3}\left[\frac{2}{\epsilon} - \gamma + \log(4\pi)\right] \\ B_{\overline{\rm MS}} &= \frac{g^2}{12(4\pi)^3}\left[\frac{2}{\epsilon} - \gamma + \log(4\pi)\right] \quad . \end{aligned} \tag{16.29}$$

Since this combination of ϵ, γ, and $\log(4\pi)$ arising from the expansion of $(4\pi)^{\epsilon/2}\Gamma(\epsilon/2)$ occurs frequently, the $\overline{\rm MS}$ scheme is quite popular in QCD. In this chapter we include all of the terms in (16.27).

With this convention, our ϕ^3 Lagrangian becomes

$$\mathcal{L} = \tfrac{1}{2}\left[\partial_\mu\phi\partial^\mu\phi - m^2\phi^2\right] - \tfrac{1}{2}B_0\partial_\mu\phi\partial^\mu\phi + \tfrac{1}{2}A_0 m^2\phi^2 \tag{16.30}$$

where the A_0 and B_0 of Eq. (16.27) are precisely the correct factors required to cancel the divergence arising from diagram 16.4A. These counterterms will generate the Feynman diagram shown in Fig. 16.4B, as discussed in the previous section.

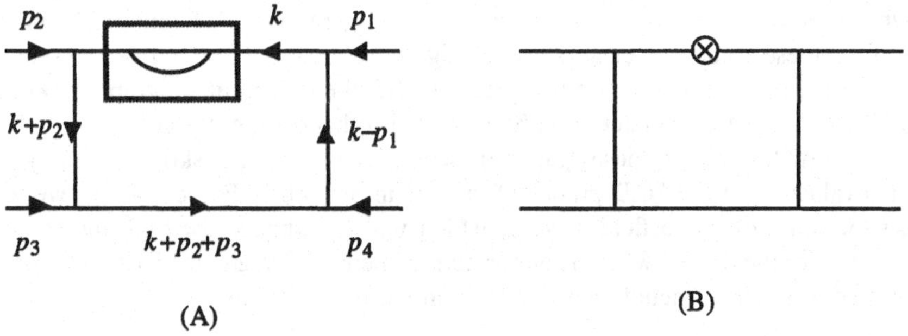

Fig. 16.5 A box diagram with a diverging self-energy insertion in the line with momentum k, and the corresponding diagram with counterterm.

To illustrate the usefulness of this procedure, consider the $\phi\phi$ scattering diagram shown in Fig. 16.5A. The *skeleton* of this graph (i.e., the graph which remains once all subgraphs have been removed) is convergent in $d = 6$ dimensions, but the self-energy insertion introduces a divergence. The graph 16.5A is

$$G_A = -i\, g_R^4 \int \frac{d^dk}{(2\pi)^d}\, \frac{m^2 A(k^2) - k^2 B(k^2)}{D\,(k, \{p_i\})}\,, \tag{16.31}$$

where

$$\begin{aligned} D\,(k, \{p_i\}) = {} & \left(m^2 - k^2 - i\epsilon\right)^2 \left(m^2 - (k+p_2)^2 - i\epsilon\right) \\ & \times \left(m^2 - (k+p_2+p_3)^2 - i\epsilon\right)\left(m^2 - (k-p_1)^2 - i\epsilon\right) \,. \end{aligned} \tag{16.32}$$

This is divergent because the quantities $A(k^2)$ and $B(k^2)$ contain divergent factors, as we have seen. However, the companion diagram, Fig. 16.5B, contains the counterterms which must accompany 16.5A, and adding this diagram to G_A gives

$$G_{A+B} = -i\, g_R^4 \int \frac{d^dk}{(2\pi)^d}\, \frac{\left[m^2\left(A(k^2) - A_0\right) - k^2\left(B(k^2) - B_0\right)\right]}{D\,(k, \{p_i\})} \,. \tag{16.33}$$

The differences $A_R(k^2) = A(k^2) - A_0$ and $B_R(k^2) = B(k^2) - B_0$ are finite as $\epsilon \to 0$ and go like $\log k^2$ at large k, and hence the integrand of (16.33) at large k^2 goes (for $d = 6$) like

$$\frac{k^6}{k^8} \log(k^2)\,,$$

which is convergent. The inclusion of the counterterms leads to a finite result for the two graphs in Fig. 16.5.

Fourth Order Self-Energy

Finally, we return to the graphs for the fourth order self-energy shown in Fig. 16.1, which illustrate the problems which can arise in the proof of renormalizability. Both of these graphs are examples of a singular subdiagram inserted into a skeleton which is itself divergent. And in graph 16.1B, the divergent subgraphs *overlap*. Will the method of counterterms be sufficient to handle these cases?

We will compute these graphs in detail, developing the skills necessary for the evaluation of the QED graphs of interest in Sec. 16.5. For simplicity, we will set the mass of the ϕ field to zero, which will not change the high momentum structure of the theory, which is our principal interest. The graph 16.1A (multiplied by two in order to include the insertion on the other line) is

$$\begin{aligned}\Sigma_A &= -ig_R^2 \int \frac{d^dk}{(2\pi)^d} \frac{B(k^2)}{(-k^2)\left[-(k-q)^2\right]} \\ &= \frac{i}{2}\frac{g^4}{(4\pi)^{d/2}}\Gamma\left(2-\tfrac{d}{2}\right)\int_0^1 dx \int \frac{d^dk}{(2\pi)^d}\frac{x^{d/2-2}(1-x)^{d/2-2}\mu^{2\epsilon}}{(-k^2)^{4-d/2}\left[-(k-q)^2\right]} \\ &= \frac{i}{2}\frac{g^4}{(4\pi)^{d/2}}\Gamma\left(2-\tfrac{d}{2}\right)B\left(\tfrac{d}{2}-1,\tfrac{d}{2}-1\right) \\ &\quad\times\int\frac{d^dk\,\mu^{2\epsilon}}{(2\pi)^d(-k^2)^{4-d/2}\left[-(k-q)^2\right]}\ ,\end{aligned} \tag{16.34}$$

where, in the last line, we have used the $B(\alpha,\beta)$-function, defined by

$$B(\alpha,\beta) \equiv \int_0^1 dx\, x^{\alpha-1}(1-x)^{\beta-1} = \frac{\Gamma(\alpha)\Gamma(\beta)}{\Gamma(\alpha+\beta)}\ . \tag{16.35}$$

Before we can evaluate this integral, we must extend our technique by introducing the following identity.

Identity

$$\begin{aligned}\frac{1}{A_1^{\alpha_1}A_2^{\alpha_2}\cdots A_n^{\alpha_n}} &= \frac{\Gamma(\alpha_1+\alpha_2+\cdots+\alpha_n)}{\Gamma(\alpha_1)\Gamma(\alpha_2)\cdots\Gamma(\alpha_n)}\int_0^1 dx_1\,dx_2\cdots dx_n\,\delta\left(1-\sum_i x_i\right) \\ &\quad\times\frac{x_1^{\alpha_1-1}x_2^{\alpha_2-1}\cdots x_n^{\alpha_n-1}}{[A_1x_1+A_2x_2+\cdots+A_nx_n]^{\alpha_1+\alpha_2+\cdots+\alpha_n}}\ .\end{aligned} \tag{16.36}$$

This is a generalization of the identities (11.79) introduced in Sec. 11.6.

Proof: The proof begins with the integral representation for the Γ-function,

$$\Gamma(\alpha) = \int_0^\infty t^{\alpha-1}e^{-t}dt\ .$$

Changing variables gives

$$\frac{\Gamma(\alpha_1)}{A_1^{\alpha_1}} = \int_0^\infty dt_1\, t_1^{\alpha_1-1} e^{-A_1 t_1} \quad .$$

Then

$$\frac{\Gamma(\alpha_1)}{A_1^{\alpha_1}} \frac{\Gamma(\alpha_2)}{A_2^{\alpha_2}} = \int_0^\infty dt_1\, dt_2\, t_1^{\alpha_1-1} t_2^{\alpha_2-1} e^{-(A_1 t_1 + A_2 t_2)} \quad .$$

Next, use the identity

$$1 = \int_0^\infty dt\, \delta\,(t - t_1 - t_2)$$

to get

$$\frac{\Gamma(\alpha_1)\Gamma(\alpha_2)}{A_1^{\alpha_1} A_2^{\alpha_2}} = \int_0^\infty dt\, \delta\,(t - t_1 - t_2)\, dt_1\, dt_2\, t_1^{\alpha_1-1} t_2^{\alpha_2-1} e^{-(A_1 t_1 + A_2 t_2)}$$

and then scale the integral by introducing $t_1 = tx_1$, $t_2 = tx_2$. This gives

$$\begin{aligned} \frac{\Gamma(\alpha_1)\Gamma(\alpha_2)}{A_1^{\alpha_1} A_2^{\alpha_2}} &= \int_0^\infty dx_1\, dx_2\, \delta(1 - x_1 - x_2) x_1^{\alpha_1-1} x_2^{\alpha_2-1} \\ &\quad \times \int_0^\infty dt\, t^{\alpha_1+\alpha_2-1} e^{-t(A_1 x_1 + A_2 x_2)} \\ &= \Gamma\,(\alpha_1 + \alpha_2) \int_0^\infty \frac{dx_1\, dx_2\, \delta(1 - x_1 - x_2) x_1^{\alpha_1-1} x_2^{\alpha_2-1}}{[A_1 x_1 + A_2 x_2]^{\alpha_1+\alpha_2}} \quad . \end{aligned}$$

This gives the desired result for two factors, and we see from the proof that the result is readily extended to any number of factors. ∎

With the identity (16.36), the denominator in (16.34) can be combined and the integral carried out:

$$\begin{aligned} \Sigma_A &= \frac{ig^4}{2(4\pi)^{d/2}} \Gamma\left(2 - \tfrac{d}{2}\right) B\left(\tfrac{d}{2} - 1, \tfrac{d}{2} - 1\right) \frac{\Gamma\left(5 - \frac{d}{2}\right)}{\Gamma\left(4 - \frac{d}{2}\right)} \\ &\quad \times \int_0^1 dx \int \frac{d^d k\, \mu^{2\epsilon}\, (1 - x)^{3-d/2}}{(2\pi)^d \left(-k^2 + 2k \cdot qx - q^2 x\right)^{5-d/2}} \\ &= q^2 \frac{g^4}{2(4\pi)^d} \frac{\Gamma\left(2 - \frac{d}{2}\right)\Gamma(5 - d)}{\Gamma\left(4 - \frac{d}{2}\right)} B\left(\tfrac{d}{2} - 1, \tfrac{d}{2} - 1\right) \\ &\quad \times \int_0^1 dx \frac{(1 - x)^{3-d/2}}{[x(1 - x)]^{5-d}} \left(\frac{\mu^2}{-q^2}\right)^\epsilon , \end{aligned} \qquad (16.37)$$

where the integral was evaluated by completing the square of the denominator by shifting $k \to k + xq$ and using (C.12). Note that $\Sigma_A \simeq q^2$, showing that the massless condition will survive the final renormalization. Equation (16.37) can be written

$$\Sigma_A = q^2 \frac{g^4}{2(4\pi)^d} \frac{\Gamma(1+\epsilon) B\left(\frac{d}{2}-1, \frac{d}{2}-1\right) B\left(d-4, \frac{d}{2}-1\right)}{(\epsilon-1)\epsilon\left(\epsilon/2-1\right)\epsilon/2} \left(\frac{\mu^2}{-q^2}\right)^{\epsilon}, \tag{16.38}$$

which displays a double pole at $\epsilon = 0$ and, what is more critical, a singular logarithmic term proportional to $\epsilon^{-1} \log\left(-q^2/\mu^2\right)$. Keeping only these two terms, (16.38) becomes

$$\Sigma_A \cong q^2 \frac{g^4}{36\,(4\pi)^6} \left(\frac{1}{\epsilon^2} - \frac{1}{\epsilon} \log\left(-\frac{q^2}{\mu^2}\right) + \cdots\right). \tag{16.39}$$

The $1/\epsilon^2$ term (and other constant terms proportional to $1/\epsilon$) can be removed by redefining the counterterms, but the $\epsilon^{-1} \log\left(-q^2/\mu^2\right)$ term cannot be removed in this way, and the renormalization program would *fail at this point* if this term were not canceled by the counterterms (16.24). However, as we will now show, it is canceled.

The contribution from the counterterm [which gives a Feynman integral identical to Eq. (16.34) but with $B(k^2)$ replaced by $-B_0$] is

$$\begin{aligned}
\Sigma_A^{\rm ct} &= i \frac{g^4}{6(4\pi)^{d/2}} \frac{\Gamma\left(1+\frac{\epsilon}{2}\right)}{\epsilon\left(1-\frac{\epsilon}{2}\right)} \int \frac{d^d k}{(2\pi)^d} \frac{\mu^\epsilon}{[-k^2]\left[-(k-q)^2\right]} \\
&= -\frac{g^4}{6(4\pi)^d} \frac{\Gamma\left(1+\frac{\epsilon}{2}\right)}{\epsilon\left(1-\frac{\epsilon}{2}\right)} \Gamma\left(2-\tfrac{d}{2}\right) \int_0^1 dx \frac{\mu^\epsilon}{\left[-q^2 x(1-x)\right]^{2-d/2}},
\end{aligned} \tag{16.40}$$

which can be written

$$\Sigma_A^{\rm ct} = -q^2 \frac{g^4}{6(4\pi)^d} \frac{2\Gamma^2\left(1+\frac{\epsilon}{2}\right) B\left(\frac{d}{2}-1, \frac{d}{2}-1\right)}{\left(\frac{\epsilon}{2}-1\right)^2 \epsilon^2} \left(\frac{\mu^2}{-q^2}\right)^{\epsilon/2}. \tag{16.41}$$

We expand this, keeping the $1/\epsilon^2$ and $\epsilon^{-1} \log\left(-q^2/\mu^2\right)$ terms only:

$$\Sigma_A^{\rm ct} \cong -q^2 \frac{g^4}{36\,(4\pi)^6} \left(\frac{2}{\epsilon^2} - \frac{1}{\epsilon} \log\left(-\frac{q^2}{\mu^2}\right) \cdots\right). \tag{16.42}$$

Adding this to Σ_A, we find that the $\epsilon^{-1} \log\left(-q^2/\mu^2\right)$ term cancels

$$\Sigma_A + \Sigma_A^{\rm ct} = -q^2 \frac{g^4}{36\,(4\pi)^6} \left(\frac{1}{\epsilon^2} + \frac{1}{\epsilon}\,[\text{constant}] + \text{finite}\right). \tag{16.43}$$

The singular part of this is now of an acceptable form (i.e., a constant times q^2) to serve as a new addition to the counterterm of order g^4.

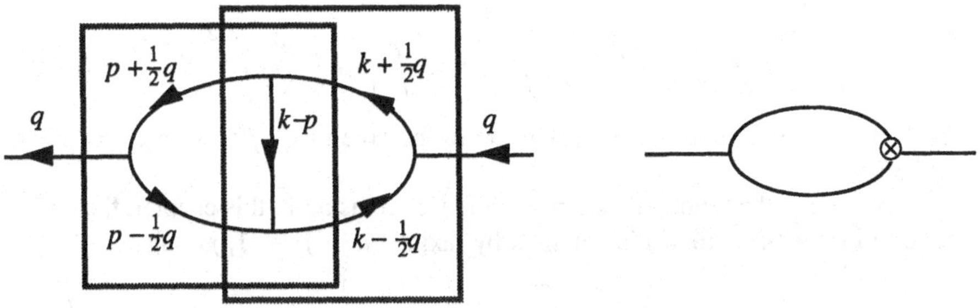

(A) (B)

Fig. 16.6 Figure (A) is the same diagram given in Fig. 16.1B, with the momenta labeled, and (B) is the corresponding skeleton diagram with counterterm.

Overlapping Divergences

Next, look at the graph 16.1B, which contains overlapping divergences. This graph is redrawn in Fig. 16.6A, which shows the labeling of momenta which we will use. To evaluate this, first evaluate the vertex contribution shown in the shaded box on the right-hand side of the diagram. We have

$$\begin{aligned}\Lambda(p,q) &= -ig_R^2 \int \frac{d^dk}{(2\pi)^d}\,\frac{1}{\underbrace{A_+(k)}_{x}\,\underbrace{A_-(k)}_{y}\,A_0}\\ &= -2ig_R^2\int_0^1 dx\int_0^{1-x} dy\int\frac{d^dk}{(2\pi)^d}\,\frac{1}{[X^2-k^2]^3}\ ,\end{aligned} \tag{16.44}$$

where we will frequently use the notation

$$A_\pm(k) = -(k \pm \tfrac{1}{2}q)^2 \qquad\qquad A_0 = -(k-p)^2\ . \tag{16.45}$$

In going from the first to the second line, Eq. (16.36) was used (with $\alpha_1 = \alpha_2 = \alpha_3 = 1$ and the Feynman parameters x and y associated with the denominators A_+ and A_- as indicated) and X^2 is found by completing the square in the denominator, which requires the shift $k \to k + p(1-x-y) - \frac{1}{2}q(x-y)$:

$$X^2 = \left[p(1-x-y) - \tfrac{1}{2}q(x-y)\right]^2 - p^2(1-x-y) - \tfrac{1}{4}q^2(x+y)\ . \tag{16.46}$$

It is convenient to map the x, y integration into $\xi = x + y$ and $\xi\eta = \frac{1}{2}(x-y)$ as we did in Eq. (11.124). This gives a simpler expression for X^2

$$\begin{aligned}X^2 &= -(1-\xi)\xi\,(p+\eta q)^2 - q^2\xi\left(\tfrac{1}{4}-\eta^2\right)\\ &= \ (1-\xi)\xi\,Y^2\ ,\end{aligned} \tag{16.47}$$

where Y, defined by this equation, will be used below. Carrying out the integral over k then gives

$$\Lambda(p,q) = \frac{g^2}{(4\pi)^{d/2}}\,\Gamma\left(3-\frac{d}{2}\right)\int_0^1 \xi\,d\xi \int_{-\frac{1}{2}}^{\frac{1}{2}} d\eta \left(\frac{\mu^2}{X^2}\right)^{\epsilon/2} . \tag{16.48}$$

In $d=6$ dimensions, this is singular due to the pole in $\Gamma(\epsilon/2)$, where $\epsilon = 6-d$ as before.

Following the renormalization scheme we are using in this chapter, the counterterm implied by (16.48) is obtained by taking $(\mu/X)^\epsilon \to 1$, giving

$$\Lambda_0 = \frac{g^2}{(4\pi)^{d/2}}\,\Gamma\left(1+\frac{\epsilon}{2}\right)\frac{1}{\epsilon} , \tag{16.49}$$

and the finite term is

$$\Lambda_R(p,q) = -\frac{g^2}{2(4\pi)^3}\int_0^1 \xi\,d\xi \int_{-\frac{1}{2}}^{\frac{1}{2}} d\eta \log\left(\frac{X^2}{\mu^2}\right) . \tag{16.50}$$

We now calculate the full diagram 16.6A. The Feynman integral is

$$\begin{aligned}\Sigma_B &= i\frac{g_R^2}{2}\int \frac{d^dp}{(2\pi)^d}\,\frac{\Lambda(p,q)}{A_+(p)\,A_-(p)} \\ &= \frac{ig^4}{2(4\pi)^{d/2}}\Gamma\left(\frac{\epsilon}{2}\right)\int_0^1 d\xi \int_{-\frac{1}{2}}^{\frac{1}{2}} d\eta \int \frac{d^dp}{(2\pi)^d}\,\frac{\xi^{1-\epsilon/2}(1-\xi)^{-\epsilon/2}\,(\mu^2)^\epsilon}{A_+(p)\,A_-(p)\,(Y^2)^{\epsilon/2}} .\end{aligned} \tag{16.51}$$

It is convenient to improve the convergence of the dp integral by integrating over η by parts, giving

$$\begin{aligned}\int_{-\frac{1}{2}}^{\frac{1}{2}} d\eta \frac{1}{A_+(p)\,A_-(p)\,(Y^2)^{\epsilon/2}} &= \frac{1}{2}\left\{\frac{1}{[A_+(p)]^{1+\epsilon/2}A_-(p)} + \frac{1}{A_+(p)[A_-(p)]^{1+\epsilon/2}}\right\} \\ &\quad - \frac{\epsilon}{2}\int_{-\frac{1}{2}}^{\frac{1}{2}} \eta\,d\eta \frac{2p\cdot q + 2\eta q^2 - \frac{2\eta q^2}{1-\xi}}{A_+(p)\,A_-(p)\,(Y^2)^{1+\epsilon/2}} .\end{aligned} \tag{16.52}$$

The first two terms in the { } give equal contributions to (16.51) and integrate straightforwardly to

$$\begin{aligned}\Sigma_B^{(1)} &= \frac{ig^4}{2(4\pi)^{d/2}}\Gamma\left(\tfrac{\epsilon}{2}\right) B\left(2-\tfrac{\epsilon}{2},1-\tfrac{\epsilon}{2}\right)\int \frac{d^dp}{(2\pi)^d}\,\frac{(\mu^2)^\epsilon}{[A_+(p)]^{1+\epsilon/2}A_-(p)} \\ &= \frac{ig^4}{2(4\pi)^{d/2}}\Gamma\left(\tfrac{\epsilon}{2}\right) B\left(2-\tfrac{\epsilon}{2},1-\tfrac{\epsilon}{2}\right)\frac{\Gamma\left(2+\frac{\epsilon}{2}\right)}{\Gamma\left(1+\frac{\epsilon}{2}\right)} \\ &\qquad \times \int_0^1 dx \int \frac{d^dp}{(2\pi)^d}\,\frac{(\mu^2)^\epsilon}{[-p^2-q^2x(1-x)]^{2+\epsilon/2}} \\ &= q^2\frac{g^4}{2(4\pi)^d}\left(\frac{2}{\epsilon}\right) B\left(2-\tfrac{\epsilon}{2},1-\tfrac{\epsilon}{2}\right) B\left(2-\tfrac{\epsilon}{2},2-\epsilon\right)\Gamma(\epsilon-1)\left(\frac{\mu^2}{-q^2}\right)^\epsilon ,\end{aligned} \tag{16.53}$$

where in the second step we combined the denominators and shifted $p \to p - q(1-2x)$.

This result displays the $1/\epsilon^2$ singularity. Neglecting all terms of the form [constant]$/\epsilon$ and using

$$\Gamma(\epsilon - 1) = \frac{\Gamma(1+\epsilon)}{\epsilon(\epsilon-1)} \to -\frac{1}{\epsilon}\,,$$

Eq. (16.53) reduces quickly to

$$\begin{aligned}\Sigma_B^{(1)} &\simeq -q^2 \frac{g^4}{12\,(4\pi)^6}\left[\frac{1}{\epsilon^2}\left(\frac{\mu^2}{-q^2}\right)^{\epsilon} + \cdots\right] \\ &\to -q^2 \frac{g^4}{12\,(4\pi)^6}\left[\frac{1}{\epsilon^2} - \frac{1}{\epsilon}\log\left(-\frac{q^2}{\mu^2}\right) + \cdots\right].\end{aligned} \tag{16.54}$$

This result clearly has a structure similar to Σ_A.

We complete the calculation of Σ_B by evaluating the contributions from the second term in (16.52). In preparation for this evaluation, first obtain a more symmetrical form by shifting $p \to p - \eta q$, so that

$$\begin{aligned}\Sigma_B^{(2)} &= -\frac{ig^4}{2(4\pi)^{d/2}}\Gamma\left(1+\tfrac{\epsilon}{2}\right)\int_0^1 d\xi \int_{-\frac{1}{2}}^{\frac{1}{2}} \eta\, d\eta\ \xi^{1-\epsilon/2}(1-\xi)^{-\epsilon/2} \\ &\times \int \frac{d^dp}{(2\pi)^d}\frac{(\mu^2)^{\epsilon}\left[2p\cdot q - \frac{2\eta q^2}{1-\xi}\right]}{\left[-\left(p+\left(\frac{1}{2}-\eta\right)q\right)^2\right]\left[-\left(p-\left(\frac{1}{2}+\eta\right)q\right)^2\right]\left(-p^2 - \frac{q^2\left(\frac{1}{4}-\eta^2\right)}{1-\xi}\right)^{1+\epsilon/2}}.\end{aligned} \tag{16.55}$$

The only contribution of order $1/\epsilon^2$ from this term comes from the singularity at $\xi = 1$, but it is hard to see this without doing the p integration first. The first step is to combine denominators using the identity (16.36) with $\alpha_1 = \alpha_2 = 1$ and $\alpha_3 = 1 + \epsilon/2$. We have

$$\begin{aligned}I &= \int \frac{d^dp}{(2\pi)^d}\frac{(\mu^2)^{\epsilon}\left[2p\cdot q - \frac{2\eta q^2}{1-\xi}\right]}{\underbrace{\left[-\left(p+\left(\frac{1}{2}-\eta\right)q\right)^2\right]}_{x}\underbrace{\left[-\left(p-\left(\frac{1}{2}+\eta\right)q\right)^2\right]}_{y}\left(-p^2 - \frac{q^2\left(\frac{1}{4}-\eta^2\right)}{1-\xi}\right)^{1+\epsilon/2}} \\ &= -\frac{\Gamma\left(3+\frac{\epsilon}{2}\right)}{\Gamma\left(1+\frac{\epsilon}{2}\right)}\int_0^1 dx \int_0^{1-x} dy\,(1-x-y)^{\epsilon/2}\, 2q^2\,(\mu^2)^{\epsilon} \\ &\qquad \times \int \frac{d^dp}{(2\pi)^d}\frac{\left[\frac{1}{2}(x-y) - \eta(x+y) + \frac{\eta}{1-\xi}\right]}{[-p^2 - Zq^2]^{3+\epsilon/2}}\,,\end{aligned} \tag{16.56}$$

where the Feynman parameters x and y are associated with the denominators as shown, which requires the shift $p \to p - q\left[\frac{1}{2}(x-y) - \eta(x+y)\right]$, and

$$Z = \frac{1}{4}\left(x+y-(x-y)^2\right) + \left[\eta^2(x+y) - \eta(x-y) + \frac{\left(\frac{1}{4}-\eta^2\right)}{1-\xi}\right](1-x-y)$$
$$= a_Z + \frac{b_Z}{1-\xi}\,, \tag{16.57}$$

where a_Z and b_Z are defined by this expression. Completing the dp integration gives

$$I = -i\frac{\Gamma(\epsilon)}{(4\pi)^{d/2}\Gamma\left(1+\frac{\epsilon}{2}\right)}\int_0^1 dx\int_0^{1-x} dy\,(1-x-y)^{\epsilon/2}$$
$$\times\, 2q^2\left[\frac{1}{2}(x-y) - \eta(x+y) + \frac{\eta}{1-\xi}\right]\left(\frac{\mu^2}{-Zq^2}\right)^{\epsilon}. \tag{16.58}$$

Note that this has a $1/\epsilon$ singularity from the $\Gamma(\epsilon)$ and the integrand also goes like $(1-\xi)^{\epsilon-1}$, which generates another singularity as $\xi \to 1$. Keeping only this singular term (the last term in the square bracket) and including additional factors of ξ from Eq. (16.55) we integrate by parts:

$$\int dI = \int_0^1 d\xi\,\xi^{(1-\epsilon/2)}(1-\xi)^{-\epsilon/2}\,I(\xi)$$
$$= -i\frac{2q^2}{(4\pi)^{d/2}}\left(\frac{\mu^2}{-q^2}\right)^{\epsilon}\frac{\Gamma(\epsilon)}{\Gamma\left(1+\frac{\epsilon}{2}\right)}\,\eta\int_0^1 d\xi\,\xi^{1-\epsilon/2}(1-\xi)^{\epsilon/2-1}$$
$$\times\int_0^1 dx\int_0^{1-x} dy\,(1-x-y)^{\epsilon/2}\left[(1-\xi)\,a_Z + b_Z\right]^{-\epsilon}$$
$$= -i\frac{2q^2}{(4\pi)^{d/2}}\left(\frac{\mu^2}{-q^2}\right)^{\epsilon}\frac{\Gamma(\epsilon)}{\Gamma\left(1+\frac{\epsilon}{2}\right)}\frac{2\eta}{\epsilon}\int_0^1 d\xi(1-\xi)^{\epsilon/2}$$
$$\times\int_0^1 dx\int_0^{1-x} dy(1-x-y)^{\epsilon/2}\left\{\left(1-\frac{\epsilon}{2}\right)\xi^{-\epsilon/2}\left[(1-\xi)\,a_Z+b_Z\right]^{-\epsilon}\right.$$
$$\left. + \epsilon\xi^{(1-\epsilon/2)}a_Z\left[(1-\xi)\,a_Z+b_Z\right]^{-(1+\epsilon)}\right\},$$

where a_Z and b_Z were defined in (16.57). Now the second term in the $\{\ \}$ goes only like $1/\epsilon$ and can be ignored. In the first term we will keep the $1/\epsilon^2$ terms only, which means we can take $\epsilon \to 0$ everywhere except in the $\left(-\mu^2/q^2\right)^{\epsilon}$ term. We get

$$\int dI \cong -i\frac{4q^2}{(4\pi)^3}\left(\frac{1}{\epsilon}\right)^2\left(\frac{\mu^2}{-q^2}\right)^{\epsilon}\eta\int_0^1 dx\int_0^{1-x} dy$$
$$= -i\frac{2q^2}{(4\pi)^3}\left(\frac{1}{\epsilon}\right)^2\left(\frac{\mu^2}{-q^2}\right)^{\epsilon}\eta\,. \tag{16.59}$$

Fig. 16.7 The second counterterm required to renormalize the overlapping self energy diagram Fig. 16.1B (or 16.6A). This diagram is equal to diagram 16.6B.

Inserting this into (16.55) gives

$$\Sigma_B^{(2)} \simeq -q^2 \frac{g^4}{(4\pi)^6} \int_{-\frac{1}{2}}^{\frac{1}{2}} \eta^2 d\eta \left[\left(\frac{1}{\epsilon}\right)^2 \left(\frac{\mu^2}{-q^2}\right)^{\epsilon} + \cdots \right]$$

$$\simeq -q^2 \frac{g^4}{12(4\pi)^6} \left[\frac{1}{\epsilon^2} - \frac{1}{\epsilon} \log\left(-\frac{q^2}{\mu^2}\right) + \cdots \right] . \tag{16.60}$$

The leading result for diagram 16.1B is therefore

$$\Sigma_B = \Sigma_B^{(1)} + \Sigma_B^{(2)}$$

$$\cong -q^2 \frac{g^4}{6(4\pi)^6} \left[\frac{1}{\epsilon^2} - \frac{1}{\epsilon} \log\left(-\frac{q^2}{\mu^2}\right) + \cdots \right] . \tag{16.61}$$

Now we compute the effect of the counterterm, Fig. 16.6B. This integral is

$$\Sigma_B^{\text{ct}} = -\frac{ig^2}{2} \int \frac{d^d p}{(2\pi)^d} \frac{\Lambda_0 \left(\mu^2\right)^{\epsilon/2}}{A_+(p)\, A_-(p)}$$

$$= -q^2 \frac{g^2}{2(4\pi)^{d/2}} \Lambda_0\, \Gamma\left(\frac{\epsilon}{2} - 1\right) B\left(2 - \frac{\epsilon}{2}, 2 - \frac{\epsilon}{2}\right) \left(\frac{\mu^2}{-q^2}\right)^{\epsilon/2}$$

$$= \quad q^2 \frac{g^4}{(4\pi)^d} \frac{\Gamma^2\left(1 + \frac{\epsilon}{2}\right)}{\epsilon^2 \left(1 - \frac{\epsilon}{2}\right)} B\left(2 - \frac{\epsilon}{2}, 2 - \frac{\epsilon}{2}\right) \left(\frac{\mu^2}{-q^2}\right)^{\epsilon/2} \tag{16.62}$$

where the result for the second step follows from (16.21), with $m^2 = 0$. Extracting the $1/\epsilon^2$ term, we have

$$\Sigma_B^{\text{ct}} = q^2 \frac{g^4}{6(4\pi)^6} \left[\frac{1}{\epsilon^2} - \frac{1}{2\epsilon} \log\left(-\frac{q^2}{\mu^2}\right) + \cdots \right] . \tag{16.63}$$

Note that this term *does not cancel* the $\epsilon^{-1} \log(-q^2/\mu^2)$ term in (16.61); it is precisely $\frac{1}{2}$ the size needed for a cancellation.

The problem is that we have only included *one* vertex counterterm, Fig. 16.6B. There should be a *second counterterm,* corresponding to subtraction at the left-hand vertex (shown in Fig. 16.7). This term is equal to (16.63), increasing

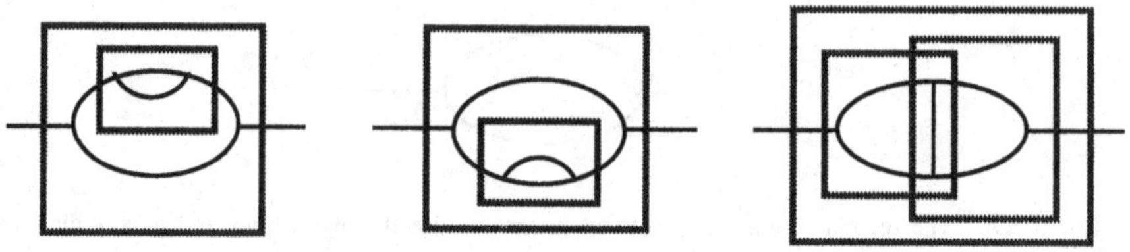

(A) (B) (C)

Fig. 16.8 The three diagrams which contribute to the fourth order self-energy, with their diverging subdiagrams enclosed in a shaded box. Diagrams (A) and (B) have nested divergences, while (C) has overlapping divergences.

the subtraction by a factor of two as needed, giving

$$\Sigma_B + 2\Sigma_B^{\rm ct} = q^2 \frac{g^4}{6(4\pi)^6}\left[\frac{1}{\epsilon^2} + \frac{1}{\epsilon}(\text{constant}) + \cdots\right] . \tag{16.64}$$

The troublesome $\epsilon^{-1}\log(-q^2/\mu^2)$ term has been canceled, the infinite part is constant, and the renormalization program can be carried through.

Review of the Fourth Order Calculation

We conclude this section with a brief review of the fourth order self-energy calculation which we have just completed. Evaluation of the self-energy requires the calculation of the diagrams shown in Fig. 16.8. The subdiagrams with divergences are enclosed by a shaded box, and in diagrams A and B these subdiagrams (with divergences arising from an internal self-energy) are *nested* within the overall divergence of the skeleton diagram. Each of those subdivergences is removed by counterterms associated with them, drawn in Fig. 16.9A and B. The final sum of these four diagrams does not completely reduce the divergence to the strength of the original skeleton, which is $\sim 1/\epsilon$, but it does cancel the serious

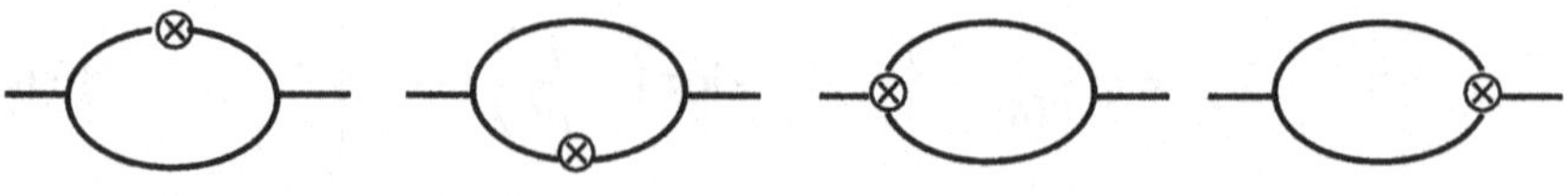

(A) (B) (C) (D)

Fig. 16.9 The graphs with counterterms which must be added to the graphs shown in Fig. 16.8 in order to give a finite result.

$\epsilon^{-1}\log(-q^2/\mu^2)$ term, giving a new singular self-energy of the form

$$\Sigma^{(4)}(q^2) = -q^2 B^{(4)}(q^2) \ ,$$

with a new *constant* counterterm of the form

$$B_0^{(4)} = \frac{c_1}{\epsilon^2} + \frac{c_2}{\epsilon} + c_3 \ ,$$

where the c_i are constants.

The third diagram, Fig. 16.8C, has an overlapping structure; the subdivergences are enclosed by shaded boxes which *overlap*. There are two subdivergences associated with *either* $p \to \infty$, k finite, or $k \to \infty$, p finite (in the notation of Fig. 16.6). Because there are two subdivergences (even though there is only a *single* diagram), we must remove *each* of these subdivergences, which requires the *two* counterterms shown in Fig. 16.9C and D. It is the generation of *two* (or more) counterterms by a *single* diagram associated with the presence of overlapping divergences which complicates the general proof of renormalizability. In the next section we discuss some features of the problem in the general case. After this, we turn to a discussion of QED.

16.3 PROVING RENORMALIZABILITY

We now return to the general discussion which was interrupted at the end of Sec. 16.1. As we observed there, the proof of renormalizability requires that we demonstrate, order-by-order in perturbation theory, that a finite number of counterterms can be introduced (one for each term in the original Lagrangian) which will render all Feynman diagrams of that order finite. If this procedure can be shown to work to all orders, we will have *proved* that a theory is renormalizable.

The first step in a general proof is to identify those diagrams or *subdiagrams* which are divergent. For superficially renormalizable theories with a ϕ^3 structure (i.e., with an interaction Lagrangian which is a product of only three fields), the divergent diagrams are usually limited to the self-energy and vertex corrections, and the counterterms will be defined by these graphs. An explicit example of how these terms are defined was given for ϕ^3 theory in the last section. In fact, if the counterterms are to have the same structure as the original terms which occurred in the Lagrangian, they *must* be limited to self-energies [which generate counterterms associated with the kinetic energy and mass terms in the free Lagrangian; recall Eq. (16.30)] and vertex corrections [which generate counterterms associated with the ϕ^3 interaction term; recall Eq. (16.14)]. If there are ϕ^4-type terms, as there are in QCD, additional four-point counterterms can be expected to be present.

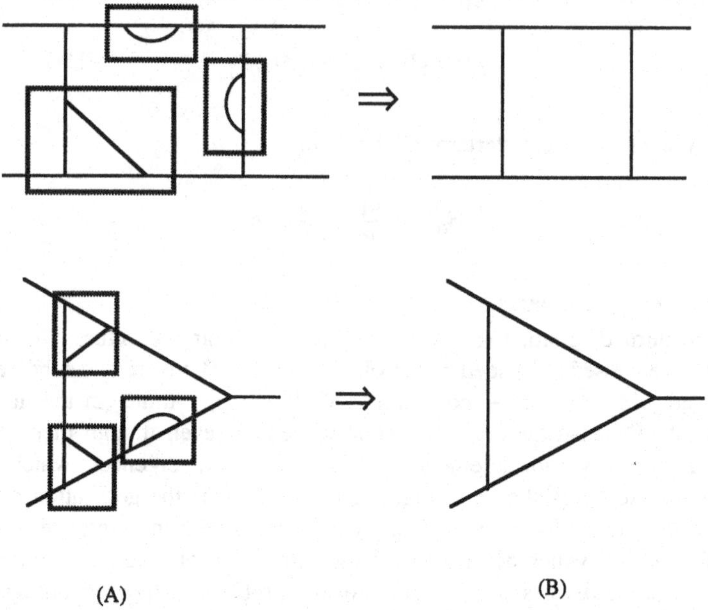

Fig. 16.10 Examples of graphs with insertions and their corresponding skeletons.

The proof that a superficially renormalizable theory with a ϕ^3 structure is, in fact, renormalizable proceeds in two stages. First, we discuss all diagrams with three or more external particles (i.e., diagrams which are not self-energies), and then we consider self-energies. Examples of nth order diagrams with three or more external particles are given in Fig. 16.10 and an example of a sixth order self-energy diagram is given in Fig. 16.11. From the previous discussion, we know that the divergences can only come from *subdiagrams which contain vertex or self-energy insertions* or from the overall diagram if it is a vertex correction or self-energy. In both of these figures, these diverging parts are enclosed by a shaded box. For graphs with three or more external particles, the singularities from vertex insertions and self-energies *do not overlap*, and the graphs have a well-defined *skeleton*, which is the diagram with all self-energy or vertex insertions removed (or collapsed to a point). Examples of diagrams (A) and their skeletons (B) are shown in Fig. 16.10. The proof that such diagrams cannot have overlapping divergences (except for those contained inside of self-energy insertions) will not be given; its truth may seem evident from the examination of many examples, and a discussion can be found in Vol. 2 of Bjorken and Drell (1964). However, as previously emphasized, self-energy diagrams can have *overlapping divergences*. In the example shown in Fig. 16.11, there are four diverging subgraphs and several

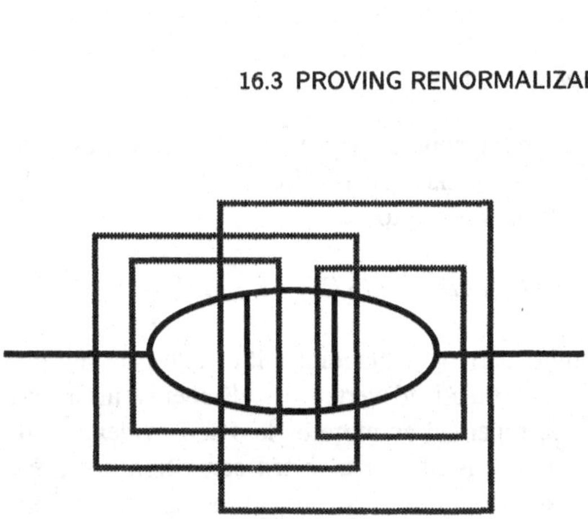

Fig. 16.11 A graph with several overlapping divergences. Each shaded box encloses a subgraph which is divergent, and many boxes overlap.

overlaps. Overlapping divergences require special discussion, and it is for this reason that self-energies must be treated separately.

Diagrams with No Overlapping Divergences

All diagrams with no overlapping divergences (three or more external particles) can be written in the form

$$G(\Gamma,\Delta) = G_S\left(\Gamma',\Delta'\right) \ , \tag{16.65}$$

where G is the original graph with bare vertices Γ and bare propagators Δ and G_S is the skeleton with dressed vertices Γ' and dressed propagators Δ'. [The arguments of G and G_S can refer to each propagator and vertex function individually or to all propagators and vertex functions collectively, in which case $G_S(\Gamma',\Delta')$ actually refers to a whole class of graphs, only one of which is shown in Fig. 16.10A.] Note that the use of Eq. (16.65) depends on the vertex and self-energy insertions being *disjoint* or *nested* as just discussed. If the insertions are overlapping, then the use of (16.65) would be ambiguous.

With Eq. (16.65), applied to graphs G with disjoint or nested insertions only, we describe how the graph is made finite by the subtraction of the counterterms. First, for propagators (for simplicity, we will only give explicit formulae for spin zero particles, but all of this discussion can be extended to include Dirac particles and massless vector bosons) it is convenient to introduce the notation

$$\Delta' = \Delta(\Sigma) = \frac{1}{m^2 - p^2 + \Sigma(p^2)} \tag{16.66}$$

so that in our previous notation the bare propagator $\Delta = \Delta(0)$. We will also use the notation

$$\Sigma_0(p^2) = m^2 A_0 - p^2 B_0 \ , \tag{16.67}$$

where A_0 and B_0 are the counterterms [calculated to lowest order in ϕ^3 theory in Eq. (16.24)]. Then, if the skeleton of G is not a vertex correction, the finite part of the graph G will be *defined* to be

$$G_R(\Gamma, \Delta) = G_S\left(\Gamma' - \Gamma_0, \Delta\left(\Sigma - \Sigma_0\right)\right) \quad . \tag{16.68}$$

In words: the finite part of a diagram with vertex or self-energy insertions is obtained by using renormalized vertex or self-energy insertions. The proof that G_R is finite will be regarded as more or less self-evident: if the skeleton G_S is finite (which it is if G_S is neither a vertex correction nor a self-energy), then it is permissible to take the limit $\epsilon \to 0$ before doing the final integration, and the finiteness of G_R follows from the finite behavior of $\Gamma' - \Gamma_0$ and $\Sigma - \Sigma_0$.

However, if G_S is a vertex correction, G_R as defined in (16.68) may require a *new additional* overall subtraction, and in this case the finite part will be *defined* to be

$$G_R(\Gamma, \Delta) = G_S\left(\Gamma' - \Gamma_0, \Delta\left(\Sigma - \Sigma_0\right)\right) - G_0 \quad , \tag{16.69}$$

where G_0 is a new, higher order counterterm. The proof that this G_R is finite for vertex functions is less obvious. It requires that we show that the infinite part of $G_S\left(\Gamma' - \Gamma_0, \Delta\left(\Sigma - \Sigma_0\right)\right)$ is a *constant* (independent of momenta) and hence is a legitimate counterterm. In view of the calculations carried out in Sec. 16.2 it may be useful to sketch a more general demonstration of this point.

Consider the insertion of a self-energy bubble in an nth order single-loop diagram, as shown in Fig. 16.12A. The corresponding skeleton (with counterterm) is shown in Fig. 16.12B. In massless ϕ^3 theory, these two diagrams have the form

$$G = \int \frac{d^d p}{(2\pi)^d} \int dy \frac{N}{[-p^2]\,[A(y)]^{n-1}} \left[\frac{f(\epsilon)}{\epsilon}\left(\frac{\mu^2}{-p^2}\right)^{\epsilon/2} - \frac{f(\epsilon)}{\epsilon}\right] , \tag{16.70}$$

where $\epsilon = 6 - d$ and $f(\epsilon)$ can be read from Eq. (16.24)

$$f(\epsilon) = \frac{g^2}{6(4\pi)^{3-\epsilon/2}} \frac{\Gamma\left(1 + \frac{\epsilon}{2}\right)}{1 - \frac{\epsilon}{2}} \quad ,$$

p is the four-momentum of the line with the insertion, and $\int dy$ is the Feynman parameterized integral over the other $n-1$ internal lines. Using (16.36), we can write

$$G = \int \frac{d^d p}{(2\pi)^d} \int dy \int_0^1 dx \frac{N x^{n-2}}{\Gamma(n-1)} \frac{f(\epsilon)}{\epsilon}$$
$$\times \left[\frac{(1-x)^{\epsilon/2}\Gamma\left(n + \frac{\epsilon}{2}\right)\left(\mu^2\right)^{\epsilon/2}}{\Gamma\left(1 + \frac{\epsilon}{2}\right)\left[-p^2 + Y_n^2\right]^{n+\epsilon/2}} - \frac{\Gamma(n)}{\left[-p^2 + Y_n^2\right]^n}\right] ,$$

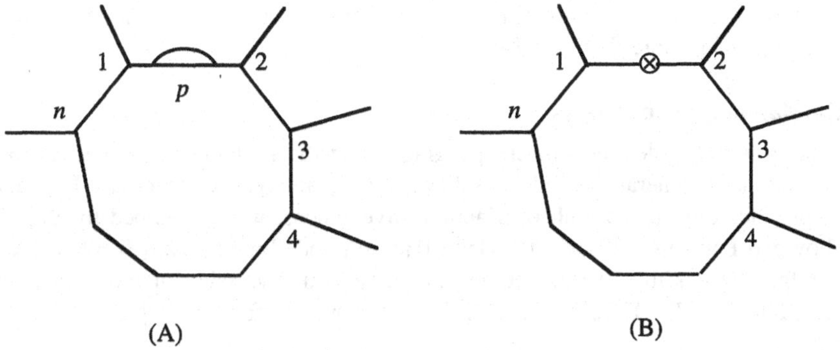

Fig. 16.12 A graph for the n-point function with a self-energy insertion and the graph with the corresponding counterterm. [The lines for the external particles 5 to $n-1$ are not drawn on the graphs.]

where Y_n^2 is obtained by shifting p to complete the square of $xA^2(y) - p^2(1-x)$ (which is the same for both terms). Now carrying out the p integration gives

$$G = \frac{1}{(4\pi)^{d/2}} \int dy \int_0^1 dx \frac{Nx^{n-2}}{\Gamma(n-1)} \frac{f(\epsilon)}{\epsilon} \left(\frac{1}{Y_n^2}\right)^{n-3+\epsilon/2}$$
$$\times \left[\frac{(1-x)^{\epsilon/2}}{\Gamma\left(1+\frac{\epsilon}{2}\right)} \Gamma\left(n + \frac{\epsilon}{2} - \frac{d}{2}\right)\left(\frac{\mu^2}{Y_n^2}\right)^{\epsilon/2} - \Gamma\left(n - \frac{d}{2}\right)\right].$$

Now note that if $n > 3$, the Γ-functions have no singularities, and the limit $\epsilon \to 0$ gives a finite result. However, if $n = 3$, the Γ-functions have a pole at $\epsilon = 0$, and the terms in the square bracket become

$$\begin{aligned}
\frac{1}{\epsilon}\Big[\quad\Big]\left(\frac{\mu^2}{Y_n^2}\right)^{\epsilon/2} &= \frac{\Gamma(\epsilon/2)}{\epsilon}\left[\frac{(1-x)^{\epsilon/2}\Gamma(\epsilon)}{\Gamma\left(1+\frac{\epsilon}{2}\right)\Gamma(\epsilon/2)}\left(\frac{\mu^2}{Y_n^2}\right)^{\epsilon} - \left(\frac{\mu^2}{Y_n^2}\right)^{\epsilon/2}\right] \\
&= \frac{\Gamma(\epsilon/2)}{\epsilon}\left[\frac{(1-x)^{\epsilon/2}\Gamma(1+\epsilon)}{2\Gamma^2\left(1+\frac{\epsilon}{2}\right)}\left(\frac{\mu^2}{Y_n^2}\right)^{\epsilon} - \left(\frac{\mu^2}{Y_n^2}\right)^{\epsilon/2}\right] \\
&= \frac{\Gamma(\epsilon/2)}{\epsilon}\left[\frac{1}{2} - \frac{\epsilon}{2}\log\left(\frac{Y_n^2}{\mu^2}\right) + \epsilon\alpha_1 + \epsilon^2 g_1\left(\frac{Y_n^2}{\mu^2}\right) + \cdots\right. \\
&\qquad \left. - \left(1 - \frac{\epsilon}{2}\log\left(\frac{Y_n^2}{\mu^2}\right) + \epsilon^2 g_2\left(\frac{Y_n^2}{\mu^2}\right) + \cdots\right)\right] \\
&= \frac{\Gamma(\epsilon/2)}{\epsilon}\left[-\frac{1}{2} + \epsilon\alpha_1 + \epsilon^2(g_1 - g_2) + \cdots\right] + \ .
\end{aligned}$$

Note that the $\epsilon \log \left(Y_n^2/\mu^2\right)$ term cancels, insuring that the singular term, while of order $1/\epsilon^2$, is nevertheless a constant, and hence a legitimate counterterm. To complete the proof (which will not be done here) the argument must be extended to graphs with more than one loop.

Overlapping Divergences

Finally, we consider the self-energy graphs and the problem of overlapping divergences. A general method for subtracting the divergences from any Feynman graph, including those with overlapping divergences, was developed by Bogoliubov and Parasiuk [BP 57, 80], Hepp [He 66], and Zimmerman [Zi 69, 71] and is referred to as the BPHZ method. For a general discussion of these methods, see Muta (1987). Here we will illustrate the results for the diagram shown in Fig. 16.11.

A good way to see how the BPHZ results are obtained is to begin with the Dyson equations for the self-energy. These equations are illustrated diagrammatically in Fig. 16.13. They are coupled, non-linear equations for the self-energy Σ and dressed vertex function Γ' expressed in terms of the dressed propagator Δ' (which depends on Σ) and the Bethe–Salpeter scattering amplitude, $\mathcal{M}$, discussed in Chapter 12. However, this form of the equations is not convenient for the renormalization program because, for example, substitution of the renormalized vertex function into 16.13A would suggest only *one* subtraction of the counterterm Λ_0, which we know from our discussion in the previous section is incorrect. *Two* factors of Λ_0 are needed. However, it is not necessary to write the equation for the self-energy in the form 16.13A; a completely equivalent form of this equation is shown diagrammatically in Fig. 16.14, where $\mathcal{V}$ is the kernel of the BS equation which the scattering amplitude $\mathcal{M}$ satisfies. (Recall the discussion in Chapter 12.) It is not difficult to prove that 16.14 is equivalent to 16.13A if Eq. (12.50) is used, and this proof is left as an exercise (Prob. 16.3).

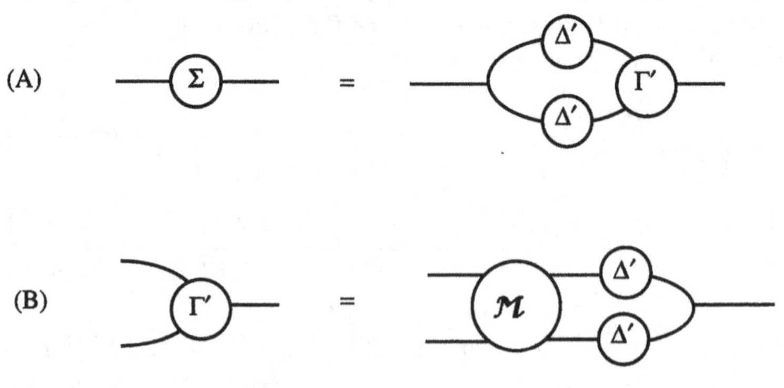

Fig. 16.13 Diagrammatic representation of the Dyson equations for the vertex function and the self-energy (included in the propagator). These equations permit the determination of Σ and Γ' if $\mathcal{M}$ is known.

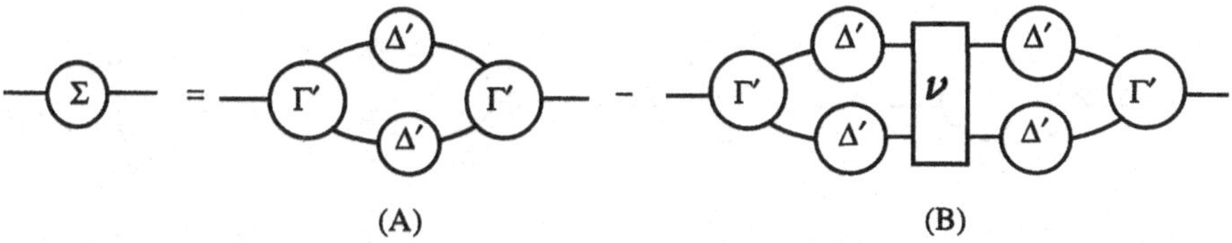

Fig. 16.14 An alternative (and equivalent) equation for the self-energy which is more suitable for applications to renormalization.

Using Fig. 16.14, it is easy to see why the two subtraction terms shown in Fig. 16.9C and D are required in order to renormalize the fourth order self-energy. The kernel $\mathcal{V}$ must be at least of second order, and hence, to fourth order, 16.14B can only generate the overlapping diagram 16.8C (with a negative sign). However, Fig. 16.14A gives *two* of the overlapping diagrams 16.8C (one obtained from the product of the third order Γ in the left vertex and the first order Γ in the right vertex and another with these contributions interchanged) so that the sum of the two contributions is correct. But the counterterms are already of third order, so they can come only from Fig. 16.14A, and there are thus two such terms.

Next, note that the sixth order diagram in Fig. 16.11 arises, in the language of the Dyson equation, from *three* terms generated by Fig. 16.14A (two contributions of a fifth order Γ denoted by $\Gamma^{(5)}$ and shown in Fig. 16.15, with a first order Γ, and one product of two third order Γ's) minus *two* terms generated by Fig. 16.14B (the two possible products of a third order $\Gamma^{(3)}$ with a first order Γ). The counterterms required for graph 16.11 follow from a consideration of the renormalization of $\Gamma^{(5)}$, as illustrated in Fig. 16.15. The full diagram 16.15A contains a subdivergence, which is regulated by the diagram with a counterterm shown in Fig 16.15B. These two diagrams then have an overall divergence, which requires the two subtractions illustrated in Figs. 16.15C and D (these two subtractions could be represented as a

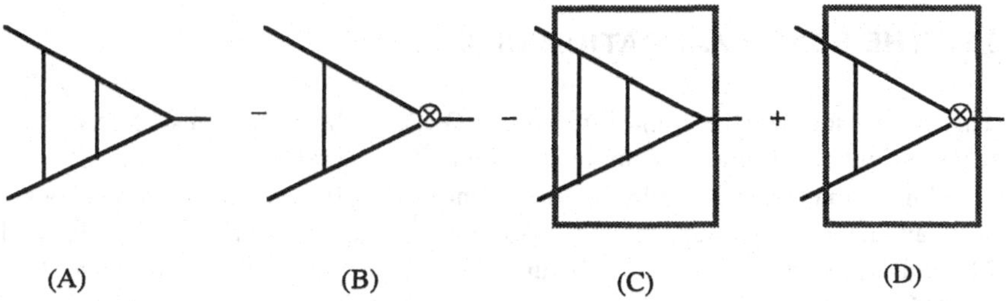

Fig. 16.15 The 5th order vertex (A) with the graph containing the third order counter term (B). The counterterms generated by graphs (A) and (B) are shown in figures (C) and (D); the shaded box encloses the overall divergence.

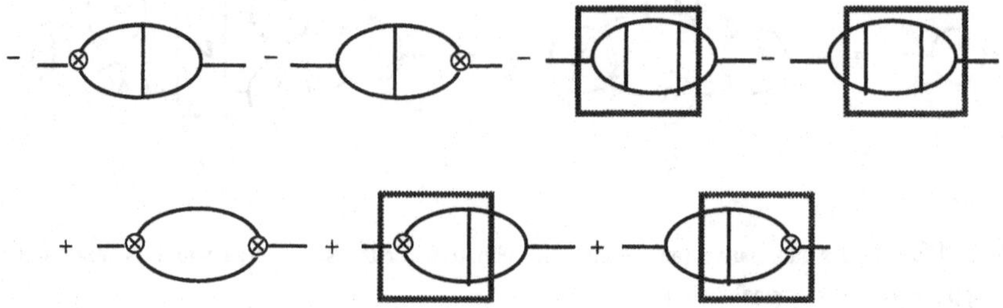

Fig. 16.16 Diagrammatic representation of the seven counterterms needed to renormalize the fifth order vertex correction shown in Fig. 16.11.

single term). Inserting all of these terms, and the one which arises in third order, into Fig. 16.14 and keeping careful track of the cancellation between 16.14A and B give seven counterterm subtractions required to remove all subdivergences from graph 16.11 (see Prob. 16.4). These are illustrated in Fig. 16.16. After these have been taken into account, an *overall* subtraction, corresponding to a new counterterm $\Sigma_0^{(6)}$, is required in order to obtain a finite $\Sigma^{(6)}$.

The discussion of overlapping divergences, and hence the proof of renormalizability, is in general a complicated business, and we will conclude our discussion of the general problem here. The aim of the last three sections has been to introduce some of the issues and to give the student a good starting point for further study. For additional discussion, see Muta (1987).

We now turn to the specific case of QED. Fortunately, because of the Ward-Takahashi identities, the general problem of renormalization can be stated in terms of diagrams with *no overlapping divergence*, and the proof can be completed more easily.

16.4 THE RENORMALIZATION OF QED

This discussion will be limited to spinor QED, with one species of fermion, a massless quark, but most of the results are quite general.

First, note that the index of divergence of QED in $d = 4$ dimensions is zero, and hence it is a superficially renormalizable theory, as defined in Sec. 16.1. The only interaction has a ϕ^3 structure, with $\mathcal{L}_I = -\bar{\psi}\gamma^\mu\psi A_\mu$, and hence, from Eq. (16.6)

$$I = B + \tfrac{3}{2}F - 4 = 0 \ . \tag{16.71}$$

The only graphs or subgraphs which diverge are the self-energies (the self-energy of the photon, or vacuum polarization, and the self-energy of the quark) and the vertex corrections. In lowest order these are all regularized by counterterms derivable from the results given in Chapter 11. There is no mass counterterm, because massless QED is chirally symmetric, and no mass can be generated without spontaneous symmetry breaking (which we assume does not happen here). Hence there are only three counterterms, related to Z_1, Z_2, and Z_3, and the Lagrangian density can be written

$$\begin{aligned}\mathcal{L} =& \frac{i}{2}\bar{\psi}\overleftrightarrow{\partial\!\!\!/}\psi - \frac{1}{4}F_{\mu\nu}F^{\mu\nu} - e_R\bar{\psi}\gamma^\mu\psi A_\mu \\ &+ \frac{i}{2}(Z_2-1)\bar{\psi}\overleftrightarrow{\partial\!\!\!/}\psi - \frac{1}{4}(Z_3-1)F_{\mu\nu}F^{\mu\nu} - e_R(Z_1-1)\bar{\psi}\gamma^\mu\psi A_\mu \\ =& \frac{i}{2}Z_2\bar{\psi}\overleftrightarrow{\partial\!\!\!/}\psi - \frac{Z_3}{4}F_{\mu\nu}F^{\mu\nu} - e_R Z_1\bar{\psi}\gamma^\mu\psi A_\mu \ . \end{aligned} \tag{16.72}$$

In this expression, keep in mind that the counterterms are added to *remove* the infinities, and hence ψ, A^μ, and the renormalized charge e_R are *finite* quantities. [This is why the vertex counterterm, with the form $e_R(Z_1-1)$, is consistent with (11.128)]. Absorbing the infinities back into the fields, by defining the *bare* quantities

$$\begin{aligned}\psi^{(0)} &= \sqrt{Z_2}\,\psi \\ A^{(0)}_\mu &= \sqrt{Z_3}\,A_\mu \ ,\end{aligned} \tag{16.73}$$

gives

$$\mathcal{L}^{(0)} = \frac{i}{2}\bar{\psi}^{(0)}\overleftrightarrow{\partial\!\!\!/}\psi^{(0)} - \frac{1}{4}F^{(0)}_{\mu\nu}F^{(0)\mu\nu} - \left(\frac{e_R Z_1}{Z_2\sqrt{Z_3}}\right)\bar{\psi}^{(0)}\gamma^\mu\psi^{(0)}A^{(0)}_\mu, \tag{16.74}$$

leading to the identification of the bare charge, e_0, as

$$e_0 = \frac{e_R Z_1}{Z_2\sqrt{Z_3}} = \frac{e_R}{\sqrt{Z_3}} \ , \tag{16.75}$$

in agreement with Eq. (11.130).

Now, as we discussed in the last section, the proof of renormalizability is straightforward for diagrams which do not have overlapping divergences, which are all those except some self-energy diagrams, examples of which are given in Fig. 16.17. Both of these diagrams reduce to Fig. 16.1B in scalar ϕ^3 theory.

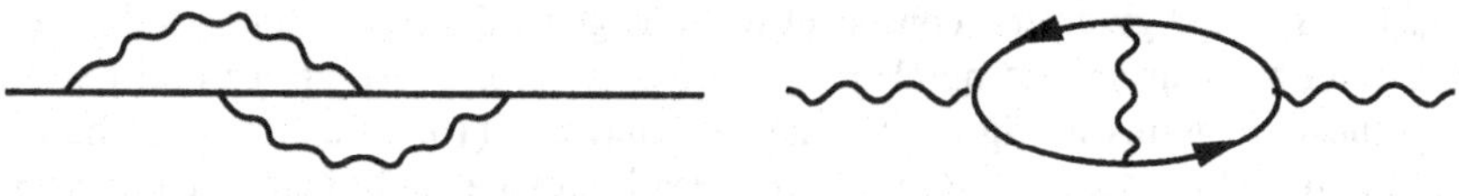

(A) (B)

Fig. 16.17 Diagrams in QED with overlapping divergences.

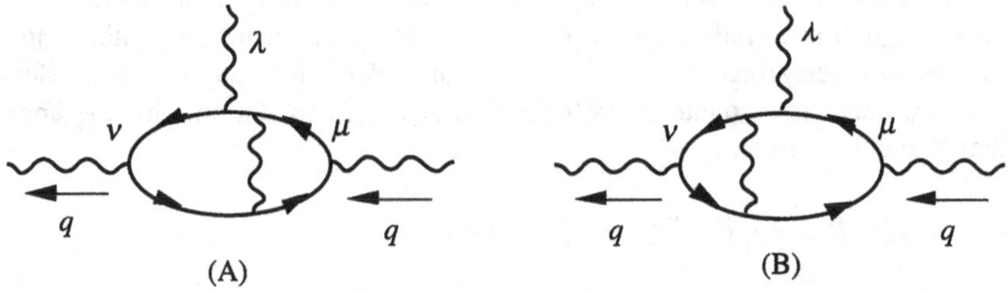

Fig. 16.18 Diagrams which can be used to calculate the vacuum polarization 16.17B without overlapping divergences. Note that the photon with index λ does not couple to the antiquark.

However, because of the Ward-Takahashi identities, we do not need to consider self-energy diagrams explicitly. The Ward-Takahashi identity for the quark self-energy relates it directly to the quark vertex,

$$\Lambda^\mu(p,p) = -\frac{\partial}{\partial p_\mu}\Sigma(p) ,$$

and hence, if we integrate the vertex $\Lambda^\mu(p,p)$ we can obtain the self-energy. Since the vertex *contains no overlapping divergences*, the proof of renormalizability for the vertex is straightforward, and once the vertex has been renormalized, the self-energy can be obtained from it.

A similar argument can be used for the vacuum polarization if we introduce a new amplitude $\Delta^{\mu\nu\lambda}$ related to the vacuum polarization by [Wa 50]

$$\Delta^{\mu\nu\lambda}(q,q) = -\frac{\partial}{\partial q_\lambda}\Pi^{\mu\nu}(q) \quad . \tag{16.76}$$

The two diagrams for $\Delta^{\mu\nu\lambda}$ which enable us to calculate graph 16.17B are shown in Fig. 16.18. Note that the "third" photon (with polarization index λ) couples only to the circulating fermion (*not* the anti-fermion). The diagrams in which the third photon couples to the anti-fermion are of the opposite sign (because of the opposite charge of the anti-fermion), and when added to the two graphs shown in Fig. 16.18 would give zero, consistent with the general theorem (Furry's theorem) that a vertex with an odd number of photons is zero. If we are careful to keep only those diagrams in Fig. 16.18, then we may use (16.76) and exploit the fact that the *pseudovertex* $\Delta^{\mu\nu\lambda}$ (*pseudo* because it includes only half of the possible couplings of the photon with polarization λ) has *no overlapping divergences* (in common with other vertex functions) to complete the proof of renormalizability.

The proof then proceeds as in the previous section. Consider any graph G in QED and its corresponding skeleton G_S. Since G has no overlapping divergences [self-energies need not be considered because of the Ward-Takahashi identities (11.137) and (16.76)], we may use a generalization of the relation (16.65),

$$G\left(\gamma^{\mu}, D, S, e_0\right) = G_S\left(\Gamma^{\mu}, D', S', e_0\right) \ , \tag{16.77}$$

where $D = D^{\mu\nu}$ and S are the lowest order photon and quark propagators and Γ^{μ}, $D' = D'^{\mu\nu}$, and S' are the dressed vertex function and photon and electron propagators. With the counterterms included, the finite parts of Γ^{μ}, D', and S', denoted by $\tilde{\Gamma}^{\mu}$, $\tilde{D}$, and $\tilde{S}$, are related to Γ^{μ}, D', and S' by the renormalization constants

$$\begin{aligned} \tilde{S} &= \frac{1}{Z_2} S' \\ \tilde{D}^{\mu\nu} &= \frac{1}{Z_3} D'^{\mu\nu} \\ \tilde{\Gamma}^{\mu} &= Z_1 \Gamma^{\mu} \ . \end{aligned} \tag{16.78}$$

These constants can then be absorbed into the charge (as we did in Chapter 11), and if the graph G has N_F external quark lines and N_B external photon lines, then

$$G\left(\gamma^{\mu}, D, S, e_0\right) = \left(Z_3\right)^{-N_B/2} \left(Z_2\right)^{-N_F/2} G_S\left(\tilde{\Gamma}^{\mu}, \tilde{D}, \tilde{S}, e_R\right) \ . \tag{16.79}$$

This equation expresses G in terms of finite quantities and an overall multiplicative infinite renormalization constant. The finite part of G, G_R, is then defined to be

$$\begin{aligned} G_R\left(\gamma^{\mu}, D, S, e_0\right) &= \left(Z_3\right)^{N_B/2} \left(Z_2\right)^{N_F/2} G\left(\gamma^{\mu}, D, S, e_0\right) \\ &= G_S\left(\tilde{\Gamma}^{\mu}, \tilde{D}, \tilde{S}, e_R\right) \ . \end{aligned} \tag{16.80}$$

This completes our proof that QED is renormalizable. Once it has been demonstrated that self-energy diagrams need not be explicitly considered and that therefore any diagram can be made finite by subtracting the infinities from its singular subdiagrams (self-energies or vertex parts which do not overlap) using counterterms defined in a lower order calculation, the proof is merely a matter of showing that the factors Z_1, Z_2, and Z_3 can always be absorbed into the charge, except for some remaining overall factor associated with the external lines. This redefinition of the charge was already discussed in Chapter 11.

As an illustration of the usefulness of the Ward-Takahashi identity (16.76), in the next section we compute the vacuum polarization to fourth order. In addition to being an interesting example, the result is of practical importance. The final section of this chapter will discuss the renormalization of QCD.

16.5 FOURTH ORDER VACUUM POLARIZATION

As an illustration of the techniques we have developed, we will calculate the fourth order vacuum polarization in QED. There are three fourth order contributions,

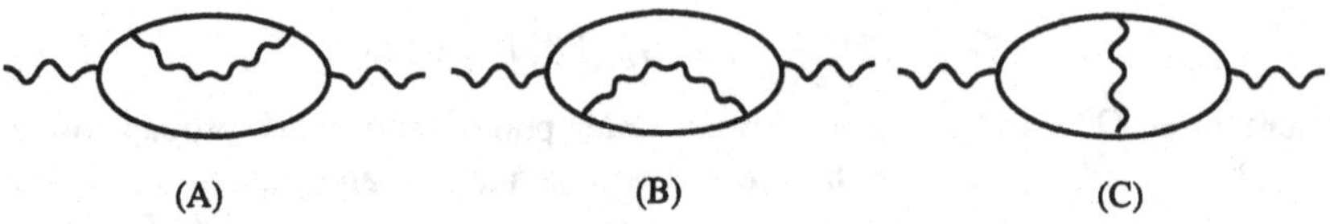

Fig. 16.19 Graphs which contribute to the fourth order vacuum polarization in QED.

shown in Fig. 16.19. The two diagrams A and B are equal, and diagram C has the familiar overlapping divergence. To avoid this overlapping divergence, we can use the Ward-Takahashi identity (16.76) to obtain the fourth order vacuum polarization from the six diagrams shown in Fig 16.20. In this case A = B, and D = F. All of these diagrams have the same skeleton, which will be labeled as shown in Fig 16.21. We will only calculate the parts of the diagrams which are *singular*, or which *diverge as* $q^2 \to \infty$. These contributions will be referred to as the "leading" contributions.

The vacuum polarization diagrams can be expressed in terms of a scalar function $\Pi(q^2)$ as was done in Eq. (11.69),

$$\Pi^{\mu\nu}(q) = \left(g^{\mu\nu}q^2 - q^\mu q^\nu\right)\Pi(q^2) \ . \tag{11.69}$$

It turns out that the leading terms can be calculated from the Ward-Takahashi identity (16.76) by ignoring the derivatives of $\Pi(q^2)$ (as we will see once we

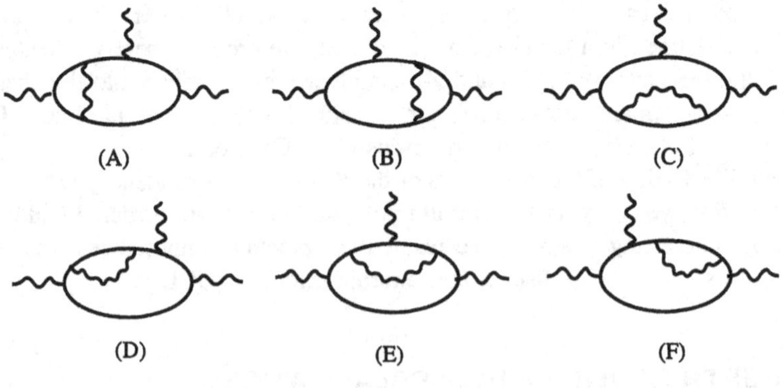

Fig. 16.20 Diagrams which can be used to calculate the fourth order vacuum polarization without overlapping divergences. Note again that the third photon does not couple to the antiquark.

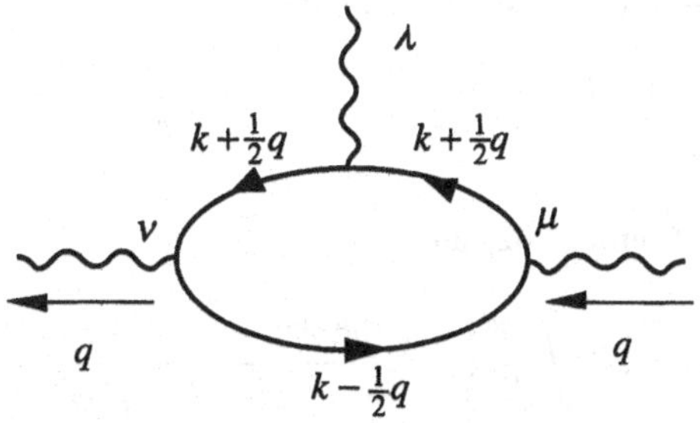

Fig. 16.21 The skeleton diagram for the fourth order calculation, with momenta labeled.

have the answer), and hence

$$\Delta^{\mu\nu\lambda}(q) = \left(q^\mu g^{\nu\lambda} + q^\nu g^{\mu\lambda} - 2q^\lambda g^{\mu\nu}\right)\Pi(q^2) \ . \tag{16.81}$$

We will find $\Pi(q^2)$ by calculating $\Delta^{\mu\nu\lambda}$ from the six diagrams in Fig 16.20.

As a warm-up, we first redo the calculation of the lowest order vacuum polarization. This is obtained from the skeleton graph 16.21, which is (don't forget the minus sign for a closed fermion loop)

$$\Delta^{\mu\nu\lambda}(q) = i\,e_R^2 \int \frac{d^dk}{(2\pi)^d}\,\frac{\mathrm{tr}\left\{\gamma^\nu\left(\not k + \frac{1}{2}\not q\right)\gamma^\lambda\left(\not k + \frac{1}{2}\not q\right)\gamma^\mu\left(\not k - \frac{1}{2}\not q\right)\right\}}{\left[-\left(k+\frac{1}{2}q\right)^2\right]^2\left[-\left(k-\frac{1}{2}q\right)^2\right]} \ . \tag{16.82}$$

The trace is readily reduced. Let $k_\pm = k \pm \frac{1}{2}q$. Then

$$\begin{aligned} N = \mathrm{tr}\{\ \} &= 2k_+^\lambda\,\mathrm{tr}\left\{\gamma^\nu \not k_+\,\gamma^\mu \not k_-\right\} - k_+^2\,\mathrm{tr}\left\{\gamma^\nu\gamma^\lambda\gamma^\mu \not k_-\right\} \\ &= 8k_+^\lambda\left[k_+^\nu k_-^\mu - g^{\mu\nu}k_+\cdot k_- + k_-^\nu k_+^\mu\right] - 4k_+^2\left[g^{\nu\lambda}k_-^\mu - g^{\mu\nu}k_-^\lambda + g^{\mu\lambda}k_-^\nu\right] \ . \end{aligned}$$

The denominator is combined into a single term, with the Feynman parameter x associated with k_+^2 and $1-x$ with k_-^2. Then k is shifted,

$$k \to k + \tfrac{1}{2}q(1-2x) \ , \tag{16.83}$$

to complete the square in the denominator. Shifting k in the numerator, dropping the odd powers of k (which integrate to zero), and neglecting terms which go like q^3 in the numerator (they are finite and cannot give rise to the leading terms we are keeping), the numerator reduces to

$$\begin{aligned} N \to & 8k^\lambda\left[\left(q^\nu k^\mu + q^\mu k^\nu\right)(1-2x) - g^{\mu\nu}(k\cdot q)(1-2x)\right] \\ & + 8q^\lambda(1-x)\left[2k^\nu k^\mu - k^2 g^{\mu\nu}\right] + 4k^2 x\left[g^{\nu\lambda}q^\mu - g^{\nu\mu}q^\lambda + g^{\mu\lambda}q^\nu\right] \\ & - 8k\cdot q(1-x)\left[g^{\nu\lambda}k^\mu + g^{\mu\lambda}k^\nu - g^{\mu\nu}k^\lambda\right] \ . \end{aligned}$$

Next, the numerator can be simplified using Eq. (C.9) from Appendix C, which allows us to make the replacement $k^\mu k^\nu \to k^2 g^{\mu\nu}/d$:

$$N \to 4k^2 \left(\frac{d-2}{d}\right) \left[\left(q^\mu g^{\nu\lambda} + q^\nu g^{\mu\lambda}\right) x - q^\lambda g^{\mu\nu}(2-x) \right] .$$

Hence the leading terms from (16.82) are

$$\Delta^{\mu\nu\lambda}(q) = 8i\, e_{_R}^2 \frac{(d-2)}{d} \int_0^1 x\, dx \int \frac{d^d k}{(2\pi)^d}\, k^2 \frac{\left(q^\mu g^{\nu\lambda} + q^\nu g^{\mu\lambda}\right) x - q^\lambda g^{\mu\nu}(2-x)}{\left[-k^2 - q^2 x(1-x)\right]^3} .$$

Keeping the leading terms from the k integration gives

$$\begin{aligned} \Delta^{\mu\nu\lambda}(q) =& \frac{8e^2(d-2)}{(4\pi)^{d/2} d} \Gamma\left(2 - \tfrac{d}{2}\right) \\ &\times \int_0^1 x\, dx \frac{\left(q^\mu g^{\nu\lambda} + q^\nu g^{\mu\lambda}\right) x - q^\lambda g^{\mu\nu}(2-x)}{\left[x(1-x)\right]^{2-d/2}} \left(\frac{\mu^2}{-q^2}\right)^{\epsilon/2}, \end{aligned} \tag{16.84}$$

where $\epsilon = 4 - d$ is a small quantity, and we have replaced

$$e_{_R}^2 = e^2 \left(\mu^2\right)^{\epsilon/2} . \tag{16.85}$$

Since we are keeping the leading terms only, we let

$$\Gamma\left(2 - \frac{d}{2}\right) = \Gamma\left(\frac{\epsilon}{2}\right) \cong \frac{2}{\epsilon} \tag{16.86}$$

and let $\epsilon \to 0$ everywhere except in the q^2 term, which will give the leading large q^2 behavior. In this case, the x integrations give

$$\int_0^1 x^2 dx = \frac{1}{3} \qquad \int_0^1 (2-x)x\, dx = \frac{2}{3}$$

and $\Pi(q^2)$ can be extracted from (16.84), giving ($\alpha = e^2/4\pi$)

$$\begin{aligned} \Pi(q^2) &\cong \frac{2\alpha}{3\pi} \frac{1}{\epsilon} \left(\frac{\mu^2}{-q^2}\right)^{\epsilon/2} + \cdots \\ &= \frac{2\alpha}{3\pi} \left[\frac{1}{\epsilon} - \frac{1}{2} \log\left(-\frac{q^2}{\mu^2}\right) + \cdots\right] . \end{aligned} \tag{16.87}$$

Note that this result agrees with the results obtained in Sec. 11.6.

Having demonstrated how the method works, we now calculate the fourth order self-energy. The calculation requires the second order quark self-energy and the third order vertex correction. In doing this calculation, we will keep only the terms which diverge as $\epsilon \to 0$, or which go like $\log(-q^2/\mu^2)$ at large q^2. Our previous experience with dimensional regularization tells us there can be no terms which go linearly with q^2; the leading terms must be of the form

$$\left(\frac{1}{\epsilon}\right)^{n_1}\left(\frac{\mu^2}{-q^2}\right)^{\epsilon/n_2} \cong \left(\frac{1}{\epsilon}\right)^{n_1} - \frac{1}{n_2}\left(\frac{1}{\epsilon}\right)^{n_1-1}\log\left(-\frac{q^2}{\mu^2}\right) + \cdots , \qquad (16.88)$$

where n_1 and n_2 are integers. Hence the $\log(-q^2/\mu^2)$ terms will be obtained automatically at the same time the singular terms are obtained.

The second order quark self-energy is

$$\Sigma(p) = -ie_R^2 \int \frac{d^dk}{(2\pi)^d} \frac{\gamma^\alpha \not{k}\, \gamma_\alpha}{[-k^2]\left[-(k-p)^2\right]} . \qquad (16.89)$$

To reduce this, note that in d dimensions $\gamma^\alpha\gamma_\alpha = d$, and hence

$$\gamma^\alpha \not{k}\, \gamma_\alpha = (2-d)\not{k} , \qquad (16.90)$$

and completing the calculation as we did in Sec. 16.2 gives

$$\Sigma(\not{p}) = -\not{p}\, B(p^2)$$

with

$$B(p^2) = \frac{e^2}{(4\pi)^{d/2}}\Gamma\left(\frac{\epsilon}{2}\right)(d-2)B\left(2-\frac{\epsilon}{2}, 1-\frac{\epsilon}{2}\right)\left(\frac{\mu^2}{-p^2}\right)^{\epsilon/2} . \qquad (16.91)$$

The second order vertex correction for finite q is $\Lambda^\mu(p,q)$, where p and q are defined in Fig. 16.22. [A similar labeling was used in Chapter 11; compare with Fig. 11.11.] If $q = 0$ (which is the case at the λ vertex of $\Delta^{\mu\nu\lambda}$), then Λ^μ can be obtained from the Ward-Takahashi identity,

$$\Lambda^\lambda(p,0) = \gamma^\lambda B(p^2) + 2\not{p}\, p^\lambda \frac{d}{dp^2}B(p^2) . \qquad (16.92)$$

For finite q, the vertex correction is obtained following the method used in Sec. 16.2 which led up to Eq. (16.48). Recalling that Λ^μ is defined with one factor of e_R omitted, the vertex correction is

$$\Lambda^\mu(p,q) = i\, e_R^2 \int \frac{d^dk}{(2\pi)^d} \frac{\gamma^\alpha \not{k}_+ \gamma^\mu \not{k}_- \gamma_\alpha}{\underbrace{A_+(k)}_{x}\ \underbrace{A_-(k)}_{y}\ A_0} , \qquad (16.93)$$

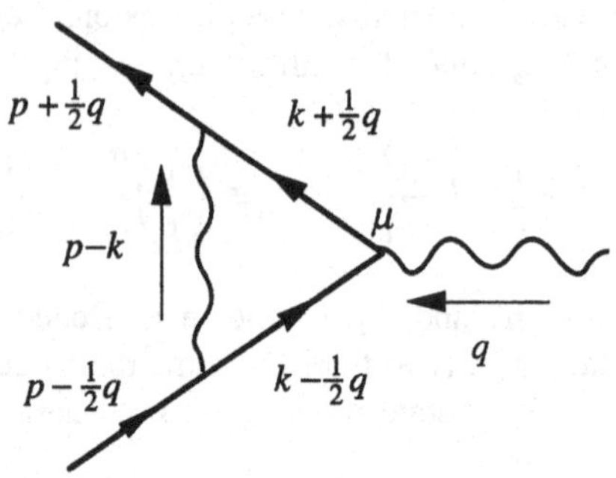

Fig. 16.22 The lowest order vertex correction, with momenta labeled.

where the A's in the denominator were introduced in Eq. (16.45). We combine the denominator into a single term, using the Feynman parameters indicated. These steps were already carried out in the work leading to Eq. (16.48); here the only additional work is the reduction of the numerator. Recalling that the shift in k was $k \to k + p(1-\xi) - q\xi\eta$, the numerator in (16.93) becomes (dropping terms proportional to two powers of q, which will not contribute to the leading terms, and terms odd in k, which integrate to zero)

$$\begin{aligned} N =& \gamma^\alpha \not{k}_+ \gamma^\mu \not{k}_- \gamma_\alpha \\ =& \gamma^\alpha \not{k} \gamma^\mu \not{k} \gamma_\alpha + \gamma^\alpha \not{p} \gamma^\mu \not{p} \gamma_\alpha (1-\xi)^2 \\ & + \gamma^\alpha \not{q} \gamma^\mu \not{p} \gamma_\alpha \left(\tfrac{1}{2} - \xi\eta\right)(1-\xi) - \gamma^\alpha \not{p} \gamma^\mu \not{q} \gamma_\alpha \left(\tfrac{1}{2} + \xi\eta\right)(1-\xi) . \end{aligned} \tag{16.94}$$

Only the first term diverges; terms proportional to $\not{p}p$ and $p\,q$ will give a finite contribution but must be retained for reasons which will be apparent soon. In the first term we use (C.9) and (16.90) twice to obtain

$$\gamma^\alpha \not{k} \gamma^\mu \not{k} \gamma_\alpha \Rightarrow \frac{k^2}{d} \gamma^\alpha \gamma^\beta \gamma^\mu \gamma_\beta \gamma_\alpha = \frac{k^2}{d} (2-d)^2 \gamma^\mu \ . \tag{16.95}$$

Similarly, the second term in (16.94) becomes

$$\gamma^\alpha \not{p} \gamma^\mu \not{p} \gamma_\alpha = (2-d)\left[2 \not{p}\, p^\mu - p^2 \gamma^\mu\right] \ . \tag{16.96}$$

Hence the contributions of the first two terms in (16.94) to Λ^μ are then of the form

$$\Lambda^\mu_{\text{first two}}(p,q) = \gamma^\mu \Lambda_1(p,q) + 2p^\mu \not{p}\, \Lambda_2(p,q) \ , \tag{16.97}$$

where Λ_1 and Λ_2 are

$$\begin{aligned} \Lambda_1(p,q) &= -2i\, e_R^2 (d-2) \int \frac{d^d k}{(2\pi)^d} \int_0^1 \xi\, d\xi \int_{-\frac{1}{2}}^{\frac{1}{2}} d\eta \frac{\left[-k^2\left(1-\frac{2}{d}\right) - (1-\xi)^2 p^2\right]}{\left[-k^2 + (1-\xi)\xi\, Y^2\right]^3} \\ \Lambda_2(p,q) &= -2i\, e_R^2 (d-2) \int \frac{d^d k}{(2\pi)^d} \int_0^1 \xi\, d\xi \int_{-\frac{1}{2}}^{\frac{1}{2}} d\eta \frac{(1-\xi)^2}{\left[-k^2 + (1-\xi)\xi\, Y^2\right]^3} \ , \end{aligned}$$

where Y was defined in Eq. (16.47). Carrying out the dk integration gives

$$\begin{aligned}
\Lambda_1(p,q) =& \frac{2e^2}{(4\pi)^{d/2}}(d-2)\,\Gamma\left(\frac{\epsilon}{2}\right)\int_0^1 d\xi\, \xi^{1-\epsilon/2}(1-\xi)^{-\epsilon/2} \\
&\times \int_{-\frac{1}{2}}^{\frac{1}{2}} d\eta \left(\frac{\mu^2}{Y^2}\right)^{\epsilon/2}\left[1-\frac{2}{d}-\frac{\Gamma\left(1+\frac{\epsilon}{2}\right)}{2\Gamma\left(\epsilon/2\right)}\left\{1-\frac{2}{d}+\frac{1-\xi}{\xi}\frac{p^2}{Y^2}\right\}\right] \\
\Lambda_2(p,q) =& \frac{e^2}{(4\pi)^{d/2}}(d-2)\,\Gamma\left(1+\frac{\epsilon}{2}\right)\int_0^1 d\xi\, \xi^{-\epsilon/2}(1-\xi)^{1-\epsilon/2} \\
&\times \int_{-\frac{1}{2}}^{\frac{1}{2}} d\eta\, \frac{1}{Y^2}\left(\frac{\mu^2}{Y^2}\right)^{\epsilon/2} .
\end{aligned} \tag{16.98}$$

Note that Λ_2 is finite as $\epsilon \to 0$ and may therefore be simplified by setting $\epsilon = 0$ in most places (i.e., everywhere except in the $(\mu^2/Y^2)^{\epsilon/2}$ term, where the infinitesimal power gives extra convergence at large p^2 needed later for the dp integral) giving

$$\Lambda_2(p,q) = \frac{\alpha}{2\pi}\int_0^1 d\xi(1-\xi)\int_{-\frac{1}{2}}^{\frac{1}{2}} d\eta\left(\frac{1}{Y^2}\right)\left(\frac{\mu^2}{Y^2}\right)^{\epsilon/2} . \tag{16.99}$$

The expression for Λ_1 is singular as $\epsilon \to 0$ and must be evaluated very carefully. Keeping only p^2 and $p\cdot q$ terms in the numerator (because the q^2 terms will not contribute to the leading term), the third term in the [ ] of (16.98) can be further reduced by replacing $p^2 \to -Y^2 - 2\eta\, p\cdot q$, giving

$$\begin{aligned}
\Lambda_1(p,q) =& \frac{2e^2}{(4\pi)^{d/2}}(d-2)\,\Gamma\left(\frac{\epsilon}{2}\right)\int_0^1 d\xi\, \xi^{-\epsilon/2}(1-\xi)^{-\epsilon/2} \\
&\times \int_{-\frac{1}{2}}^{\frac{1}{2}} d\eta\left(\frac{\mu^2}{Y^2}\right)^{\epsilon/2}\left[\frac{1}{2}\xi+\frac{\epsilon}{4}(1-2\xi)+\frac{\epsilon}{2}\eta(1-\xi)\frac{p\cdot q}{Y^2}\right] .
\end{aligned}$$

The first two terms in the square bracket may be simplified by integrating over η by parts [as we did in Eq. (16.52)]. In carrying out this integration, the second term proportional to $1-2\xi$ integrates to zero, and Λ_1 can be written as

$$\Lambda_1(p,q) = \frac{1}{2}\left[B\left(p_+^2\right)+B\left(p_-^2\right)\right]+\overline{\Lambda}_1(p,q) ,$$

where B is the singular self-energy term given in Eq. (16.91), and $\overline{\Lambda}_1$ is a term which is finite as $\epsilon \to 0$ and may therefore be simplified by setting $\epsilon = 0$ everywhere except in the $(\mu^2/Y^2)^{\epsilon/2}$ term (as we did for Λ_2), giving

$$\overline{\Lambda}_1(p,q) = \frac{\alpha}{\pi}(p\cdot q)\int_0^1 d\xi\,(1-2\xi)\int_{-\frac{1}{2}}^{\frac{1}{2}} d\eta\,\eta\left(\frac{1}{Y^2}\right)\left(\frac{\mu^2}{Y^2}\right)^{\epsilon/2} . \tag{16.100}$$

The last two terms in (16.94) complete our description of Λ^μ. These are also finite, and letting $d=4$ in the numerator permits us to use Eq. (11.118) to reduce it to

$$N_{\text{last two}} = \Big\{ -(\not p\,\gamma^\mu\,\not q)\,[1-2\xi\eta] + \not q\,\gamma^\mu\,\not p\,(1+2\xi\eta)\Big\}(1-\xi)\ .$$

Hence, the vertex correction to third order is

$$\Lambda^\mu(p,q) = \gamma^\mu\Big(\tfrac{1}{2}\left[B\left(p_+^2\right)+B\left(p_-^2\right)\right] + \overline{\Lambda}_1(p,q)\Big)$$
$$+\,2p^\mu\,\not p\,\Lambda_2(p,q) + \Lambda_3^\mu(p,q) \qquad (16.101)$$

where $\overline{\Lambda}_1$ was given in Eq. (16.100), Λ_2 in Eq. (16.99), and using (A.13) Λ_3 is

$$\Lambda_3(p,q) = \frac{e^2}{(4\pi)^2}\int_0^1 d\xi \int_{-\frac{1}{2}}^{\frac{1}{2}} d\eta \left(\frac{\mu^2}{Y^2}\right)^{\epsilon/2} \frac{[\not p\,\gamma^\mu\,\not q\,(1-2\xi\eta) - \not q\,\gamma^\mu\,\not p\,(1+2\xi\eta)]}{Y^2}\,. \qquad (16.102)$$

As a test of these results, note that $\Lambda_3(p,0) = \overline{\Lambda}_1(p,0) = 0$, and

$$\Lambda_2(p,0) = -\frac{\alpha}{2\pi}\left(\frac{1}{2}\right)\left(\frac{\mu^2}{-p^2}\right)^{\epsilon/2}\left(\frac{1}{p^2}\right) = \frac{d}{dp^2}B(p^2) = B'(p^2) \qquad (16.103)$$

and hence (16.101) is consistent with the Ward-Takahashi identity, Eq. (16.92).

We are now prepared to tackle a calculation of the six diagrams in Fig. 16.20. Their combined contribution is

$$\begin{aligned}
\Delta^{\mu\nu\lambda} &= i\,e_R^2 \int \frac{d^dk}{(2\pi)^d}\frac{1}{A_+^2 A_-} \\
&\times \Big[N^{\nu\lambda\mu}\Big\{ \underbrace{2B(k_+^2) + B(k_-^2) + \overline{\Lambda}_1(k,q) + \overline{\Lambda}_1(k,-q)}_{\text{(a)}} - \underbrace{2B(k_+^2) - B(k_-^2)}_{\text{(b)}} \Big\} \\
&+2k^\mu\Lambda_2(k,q)\,\mathrm{tr}\left\{\gamma^\nu\,\not k_+\gamma^\lambda\,\not k_+\not k\,\not k_-\right\} + 2k^\nu\Lambda_2(k,-q)\,\mathrm{tr}\left\{\not k\,\not k_+\gamma^\lambda\,\not k_+\gamma^\mu\,\not k_-\right\} \\
&+\mathrm{tr}\left\{\gamma^\nu\,\not k_+\,\gamma^\lambda\,\not k_+\,\Lambda_3^\mu(k,q)\,\not k_-\right\} + \mathrm{tr}\left\{\Lambda_3^\nu(k,-q)\,\not k_+\,\gamma^\lambda\,\not k_+\,\gamma^\mu\,\not k_-\right\}\Big] \\
&- i\,e_R^2 \int \frac{d^dk}{(2\pi)^d}\frac{2k_+^\lambda \Lambda_2(k_+,0)}{A_+A_-}\,\mathrm{tr}\left\{\gamma^\nu\,\not k_+\,\gamma^\mu\,\not k_-\right\} \qquad (16.104)
\end{aligned}$$

where $N^{\nu\lambda\mu} = \mathrm{tr}\left\{\gamma^\nu\,\not k_+\,\gamma^\lambda\,\not k_+\,\gamma^\mu\,\not k_-\right\}$, $A_\pm = -k_\pm^2$, and the (a) terms come from the γ^μ contributions to the three vertex functions, the (b) terms are from the three electron self-energies, and the remaining terms are additional vertex corrections. Note that the vertex for the *outgoing* photon with polarization ν is $\Lambda^\nu(k,-q)$.

Now, note that the singular $B(k^2)$ parts of the (a) and (b) terms cancel *exactly*, and hence *only finite* contributions to the integrand remain. This tells us immediately that the fourth order result will only go like $1/\epsilon$ (there are *no* $1/\epsilon^2$ terms as there were in ϕ^3 theory), so that the leading terms have the structure of Eq. (16.88) with $n_1 = 1$. Hence the leading $\log(-q^2/\mu^2)$ terms are finite. Accordingly, we can evaluate these terms by letting $\epsilon \to 0$ in all factors which multiply the $1/\epsilon$ singularity, *except* for the $\left(-\mu^2/q^2\right)^\epsilon$ term, which gives the log. The final result therefore comes from only four terms: the $\overline{\Lambda}_1$ terms, two Λ_2 terms, and the Λ_3 terms. These will each be calculated in turn.

Consider the contribution from one of the $\overline{\Lambda}_1$ terms first. From Eq. (16.100), we see that

$$\overline{\Lambda}_1(k,q) \underset{k\to\infty}{\longrightarrow} \frac{k\cdot q}{\left(k^2\right)^{1+\epsilon/2}}$$

and hence the $\overline{\Lambda}_1$ integral will diverge in four dimensions only if multiplied by a term which goes like k^3/k^6. Precisely such a term arises from the k^3 term in $N^{\nu\lambda\mu}$ (which did not contribute to the second order calculation because it was odd in k), which now gives

$$\begin{aligned}\Delta_1^{\mu\nu\lambda} =& \frac{i\,e_R^2\alpha}{\pi}\int\frac{d^dk}{(2\pi)^d}\frac{\mathrm{tr}\left\{\gamma^\nu\, \not{k}\,\gamma^\lambda\, \not{k}\,\gamma^\mu\, \not{k}\right\}(k\cdot q)}{A_+^2A_-}\\ &\times\int_0^1 d\xi\,(1-2\xi)\int_{-\frac12}^{\frac12}\eta\,d\eta\,\frac{1}{Y^2}\left(\frac{\mu^2}{Y^2}\right)^{\epsilon/2}. \end{aligned} \tag{16.105}$$

Combining the denominators into a single term, as we did in obtaining Eq. (16.56), will give a contribution of the form (remember that $\epsilon = 0$ except for the $(\mu^2/Y^2)^{\epsilon/2}$ term)

$$\begin{aligned}\Delta_1^{\mu\nu\lambda} =& \frac{i\,e_R^2\alpha}{\pi}\int_0^1 d\xi\,(1-2\xi)\int_{-\frac12}^{\frac12}\eta\,d\eta\int_0^1 x\,dx\int_0^{1-x}dy\\ &\times\int\frac{d^dk}{(2\pi)^d}\frac{\mathrm{tr}\left\{\gamma^\nu\, \not{k}\,\gamma^\lambda\, \not{k}\,\gamma^\mu\, \not{k}\right\}k\cdot q}{\left(-k^2-Zq^2\right)^{4+\epsilon/2}}, \end{aligned} \tag{16.106}$$

where Z, defined in Eq. (16.57), is the factor which emerged when the square of the denominator was completed (we will not need to know its precise form here). To complete the square in the denominator, we had to shift $k \to k + \gamma q$ (where γ depends on the parameters x, y, and η), but since four powers of k must be retained in the numerator in order to get the divergent result we are seeking, the numerator is unaffected by this shift. After carrying out the trace and averaging over the direction of k, the numerator will be proportional to k^4, and after the k

integral is done, we will obtain a factor of

$$\left(\frac{1}{Zq^2}\right)^{\epsilon} \simeq \left(\frac{1}{q^2}\right)^{\epsilon} ,$$

where the factor of $Z^{-\epsilon}$ can be set to unity, since departures of Z from unity make only a finite contribution. Hence there are no additional contributions from the Feynman parameters x, y, ξ, and η, and the integrations over the Feynman parameter ξ (or η) gives zero. Hence the divergent part of Δ_1 is *zero*. This conclusion holds for both of the terms proportional to Λ_1.

Next, calculate the terms proportional to Λ_2. It will turn out that these are zero also. First, evaluate the last term proportional to $\Lambda_2(k_+, 0)$. Doing the trace gives

$$\Delta_2^{\mu\nu\lambda} = -\frac{i\,\alpha e^2}{2\pi} \int \frac{d^d k}{(2\pi)^d} \frac{(\mu^2)^{\epsilon}}{\underbrace{A_-\ {A_+}^{2+\epsilon/2}}_{x}} 4k_+^{\lambda}\left[k_+^{\nu}k_-^{\mu} + k_-^{\nu}k_+^{\mu} - g^{\nu\mu}k_+\cdot k_-\right] . \tag{16.107}$$

Combining the denominator and completing the square using the now familiar shift $k \to k + \frac{1}{2}q(1-2x)$, Eq. (16.83), give

$$\begin{aligned}\Delta_2^{\mu\nu\lambda} = -i\,4\alpha^2 \int \frac{d^d k}{(2\pi)^d} \int_0^1 dx\, x \frac{k^2\left(\mu^2\right)^{\epsilon}}{\left[-k^2 - q^2 x(1-x)\right]^{3+\epsilon/2}} \\ \times \left[\left(q^{\mu}g^{\nu\lambda} + q^{\nu}g^{\mu\lambda}\right)(1-2x) - g^{\mu\nu}q^{\lambda}(3-4x)\right]\end{aligned} \tag{16.108}$$

where the $k^{\mu}k^{\nu}$ terms were reduced to $g^{\mu\nu}k^2/4$ using (C.9). Doing the k integration [setting $\epsilon = 0$ everywhere except in the $\left(-\mu^2/q^2\right)$ factor] gives the following leading term:

$$\Delta_2^{\mu\nu\lambda} = \frac{\alpha^2}{24\pi^2}\,\frac{1}{\epsilon}\left(\frac{\mu^2}{-q^2}\right)^{\epsilon}\left[q^{\mu}g^{\nu\lambda} + q^{\nu}g^{\mu\lambda} + g^{\mu\nu}q^{\lambda}\right] . \tag{16.109}$$

Note that this does not have the gauge invariant form of Eq. (16.81), and hence it must be canceled.

The cancellation comes from the other terms proportional to Λ_2. Using $k = \frac{1}{2}\left(k_+ + k_-\right)$, the numerator for the first of these terms reduces to

$$\begin{aligned} N_1 &= 2k^{\mu}\,\mathrm{tr}\left\{\gamma^{\nu}\,\not{k}_+\,\gamma^{\lambda}\,\not{k}_+\not{k}\,\not{k}_-\right\} \\ &= k^{\mu}k_+^2\,\mathrm{tr}\left\{\gamma^{\nu}\,\not{k}_+\,\gamma^{\lambda}\,\not{k}_-\right\} + k^{\mu}k_-^2\,\mathrm{tr}\left\{\gamma^{\nu}\,\not{k}_+\,\gamma^{\lambda}\,\not{k}_+\right\} \\ &= 2k^{\mu}k^2\,\mathrm{tr}\left\{\gamma^{\nu}\,\not{k}_+\,\gamma^{\lambda}\,\not{k}\right\} + \mathcal{O}(q^2) \end{aligned} \tag{16.110}$$

where terms of $\mathcal{O}(q^2)$ can be dropped because they are not of leading order. Since $\Lambda_2(k,q) = \Lambda_2(k,-q)$, the second of these terms, which differs only by $\mu \leftrightarrow \nu$, may be added, and after the trace has been taken, the first two Λ_2 terms become

$$\Delta_2'^{\mu\nu\lambda} = \frac{i\,e^2\alpha}{\pi}\int_0^1 d\xi\,(1-\xi)\int_{-\frac{1}{2}}^{\frac{1}{2}} d\eta \int \frac{d^dk}{(2\pi)^d}\,\frac{4k^2\,(\mu^2)^\epsilon}{A_+^2 A_-\,(Y^2)^{1+\epsilon/2}}$$
$$\times\Big[4k^\mu k^\nu k^\lambda + \tfrac{1}{2}\left(k^\mu k^\lambda q^\nu + k^\nu k^\lambda q^\mu + 2k^\mu k^\nu q^\lambda\right)$$
$$- \left(g^{\nu\lambda}k^\mu + g^{\mu\lambda}k^\nu\right)\left(k^2 + \tfrac{1}{2}q\cdot k\right)\Big]\,.$$

Now combine the denominators as we did leading to Eq. (16.56), shift $k \to k+\gamma q$, where $\gamma = -\eta - \frac{1}{2}(x-y) + \eta(x+y)$, drop all terms odd in k, and keep only terms linear in q. We obtain

$$\Delta_2'^{\mu\nu\lambda} = 24\frac{i\,e^2\alpha}{\pi}\int_0^1 d\xi\,(1-\xi)\int_{-\frac{1}{2}}^{\frac{1}{2}} d\eta\int_0^1 x\,dx\int_0^{1-x} dy$$
$$\times\int \frac{d^dk}{(2\pi)^d}\,\frac{N_2^{\mu\nu\lambda}\,(\mu^2)^\epsilon}{(-k^2 - Zq^2)^{4+\epsilon/2}}\,,$$

where

$$N_2^{\mu\nu\lambda} = 8\gamma\,k\cdot q\,k^\mu k^\nu k^\lambda + k^2\left[\left(4\gamma+\tfrac{1}{2}\right)\left(k^\mu k^\lambda q^\nu + k^\nu k^\lambda q^\mu\right) + (4\gamma+1)k^\mu k^\nu q^\lambda\right]$$
$$- k^2(k\cdot q)\left[g^{\nu\lambda}k^\mu + g^{\mu\lambda}k^\nu\right]\left(4\gamma+\tfrac{1}{2}\right) - k^4\gamma\left[g^{\mu\lambda}q^\nu + g^{\nu\lambda}q^\mu\right]\,.$$

Reduce this numerator by averaging over k, using (C.9) from Appendix C, and a new identity (C.10):

$$\int \frac{d^dk}{(2\pi)^d}\,\frac{k^\mu k^\nu k^\lambda k^\delta}{D(k^2)} = \frac{\left[g^{\mu\nu}g^{\lambda\delta} + g^{\mu\lambda}g^{\nu\delta} + g^{\mu\delta}g^{\nu\lambda}\right]}{d(d+2)}\int \frac{d^dk}{(2\pi)^d}\,\frac{k^4}{D(k^2)}\,.$$

The proof of (C.10) is left as an exercise. Using these identities in four dimensions gives

$$N_2^{\mu\nu\lambda} \to \frac{k^4}{4}\left\{-\frac{8}{3}\gamma\left[g^{\mu\lambda}q^\nu + g^{\nu\lambda}q^\mu\right] + \left(1+\frac{16}{3}\gamma\right)g^{\mu\nu}q^\lambda\right\} \equiv \frac{k^4}{4}\,n_2^{\mu\nu\lambda}\,. \tag{16.111}$$

Doing the k integral and replacing the $Z^{-\epsilon}$ factor by unity give

$$\Delta_2'^{\mu\nu\lambda} = -6\frac{\alpha^2}{4\pi^2}\,\frac{1}{\epsilon}\left(\frac{\mu^2}{-q^2}\right)^\epsilon\int_0^1 d\xi\,(1-\xi)\int_{-\frac{1}{2}}^{\frac{1}{2}} d\eta\int_0^1 x\,dx\int_0^{1-x} dy\;n_2^{\mu\nu\lambda}$$
$$= -\frac{\alpha^2}{24\pi^2}\,\frac{1}{\epsilon}\left(\frac{\mu^2}{-q^2}\right)^\epsilon\left[q^\mu g^{\nu\lambda} + q^\nu g^{\mu\lambda} + q^\lambda g^{\mu\nu}\right]\,. \tag{16.112}$$

As advertised, this term cancels (16.109), giving zero.

The final term, and the only non-zero contribution to $\Delta^{\mu\nu\lambda}$, comes from the Λ_3 terms, which are already linear in q and therefore quite easy to reduce. To reduce them, recall that Y is unchanged if $q \to -q$ and $\eta \to -\eta$, and hence $\Lambda_3^\nu(k,-q)$ can be written in a form similar to $\Lambda_3^\mu(k,q)$ provided $\eta \to -\eta$ in the integral which defines it [recall Eq. (16.102)]. Hence, retaining terms linear in q only, the numerator of the $\Lambda_3^{\mu\nu\lambda}$ contributions can be combined as follows:

$$
\begin{aligned}
N_3 = &\operatorname{tr}\left\{\gamma^\nu \not{k}_+ \gamma^\lambda \not{k}_+ \left[\not{k}\,\gamma^\mu \not{q}\,(1-2\xi\eta) - \not{q}\,\gamma^\mu \not{k}\,(1+2\xi\eta)\right] \not{k}_-\right\} \\
&- \operatorname{tr}\left\{\left[\not{k}\,\gamma^\nu \not{q}\,(1+2\xi\eta) - \not{q}\,\gamma^\nu \not{k}\,(1-2\xi\eta)\right] \not{k}_+ \gamma^\lambda \not{k}_+ \gamma^\mu \not{k}_-\right\} \\
\to\; &(1-2\xi\eta)k^2\left[\operatorname{tr}\left(\gamma^\nu \not{k}\,\gamma^\lambda\gamma^\mu \not{q}\,\not{k}\right) + \operatorname{tr}\left(\not{q}\,\gamma^\nu\gamma^\lambda \not{k}\,\gamma^\mu \not{k}\right)\right] \\
&- (1+2\xi\eta)k^2\left[\operatorname{tr}\left(\gamma^\nu \not{k}\,\gamma^\lambda \not{k}\,\not{q}\,\gamma^\mu\right) + \operatorname{tr}\left(\gamma^\nu \not{q}\,\not{k}\,\gamma^\lambda \not{k}\,\gamma^\mu\right)\right] \ .
\end{aligned}
$$

Again, as above, we may use (C.9) and $\gamma^\alpha\gamma^\mu\gamma_\alpha = -2\gamma^\mu$ immediately to get

$$
N_3 \to -4k^4(1-2\xi\eta)\left[q^\nu g^{\lambda\mu} + q^\mu g^{\lambda\nu} - g^\lambda g^{\mu\nu}\right] + 4k^4(1+2\xi\eta)q^\lambda g^{\mu\nu}.
$$

Hence the Λ_3 contributions are

$$
\begin{aligned}
\Delta_3^{\mu\nu\lambda} = &-\frac{i\,e^2\alpha}{\pi}\int_0^1 d\xi \int_{-\frac{1}{2}}^{\frac{1}{2}} d\eta \left[\left(g^{\mu\lambda}q^\nu + g^{\nu\lambda}q^\mu\right)(1-2\xi\eta) - 2q^\lambda g^{\mu\nu}\right] \\
&\times \int \frac{d^dk}{(2\pi)^d} \frac{\left(\mu^2\right)^\epsilon k^4}{A_+^2 A_- \left(Y^2\right)^{1+\epsilon/2}} \ .
\end{aligned}
$$

Combining denominators and doing the k integral gives

$$
\begin{aligned}
\Delta_3^{\mu\nu\lambda} &= 6\,\frac{\alpha^2}{4\pi^2}\frac{1}{\epsilon}\left(\frac{\mu^2}{-q^2}\right)^\epsilon \int_0^1 d\xi \int_{-\frac{1}{2}}^{\frac{1}{2}} d\eta \int_0^1 x\,dx \int_0^{1-x} dy \\
&\qquad \times \left[\left(g^{\mu\lambda}q^\nu + g^{\nu\lambda}q^\mu\right)(1-2\xi\eta) - 2q^\lambda g^{\mu\nu}\right] \\
&= \frac{\alpha^2}{4\pi^2}\,\frac{1}{\epsilon}\left(\frac{\mu^2}{-q^2}\right)^\epsilon \left[g^{\mu\lambda}q^\nu + g^{\nu\lambda}q^\mu - 2q^\lambda g^{\mu\nu}\right] . \qquad (16.113)
\end{aligned}
$$

This is the final result for the fourth order self-energy, and combined with our previous results gives

$$
\boxed{\Pi(q^2) = \frac{2\alpha}{3\pi}\,\frac{1}{\epsilon} + \frac{\alpha^2}{4\pi^2}\,\frac{1}{\epsilon} - \left(\frac{\alpha}{3\pi} + \frac{\alpha^2}{4\pi^2}\right)\log\left(-\frac{q^2}{\mu^2}\right) + \cdots .} \qquad (16.114)
$$

The result (16.114) shows how the renormalization constant Z_3 must be redefined in fourth order. The fact that the singular term is constant is another example of the requirements of renormalizability. In addition, we have obtained the finite, high q^2 part of the vacuum polarization correction to fourth order. Note that it adds to the second order corrections, further enhancing the effects already discussed in Chapter 11. In the next chapter we will show how this result may be used to calculate the QCD corrections to R, which were discussed in Sec. 10.4.

The calculation of (16.114) is long, but note that none of the singular contributions arises from singularities in the integrals over the Feynman parameters, as they did in the ϕ^3 example discussed in Sec. 16.2 above. This is because we calculated (16.114) from diagrams with *no overlapping divergences.*

16.6 THE RENORMALIZATION OF QCD

We conclude this chapter with a brief discussion of the renormalization of QCD. First, consider the four types of interactions which can occur in QCD (with quarks). There are three-gluon ($3g$) and four-gluon ($4g$) vertices, a quark-gluon vertex ($gq\bar{q}$), and the ghost-gluon vertex (gcc). Two of these vertices have derivative operators, which introduce an extra power of momentum into the vertex and add a single unit to the formula for the index of divergence (see Prob. 16.1). The new formula for the index is

$$I = B + \tfrac{3}{2}F + n_D - 4 \ ,$$

where n_D is the number of derivatives at the vertex. Evaluating this formula in four dimensions for each vertex gives the following computations:

$$\begin{aligned} 3g&: \quad 3 + \tfrac{3}{2}(0) + 1 - 4 = 0 \\ 4g&: \quad 4 + \tfrac{3}{2}(0) + 0 - 4 = 0 \\ gq\bar{q}&: \quad 1 + \tfrac{3}{2}(2) + 0 - 4 = 0 \\ gcc&: \quad 3 + \tfrac{3}{2}(0) + 1 - 4 = 0 \ . \end{aligned} \tag{16.115}$$

In every case the index of divergence is zero. Note that ghosts are treated like bosons for this estimate, since their propagator goes like p^{-2}. Hence, by the definition given in Sec. 16.1, QCD is superficially renormalizable.

In QCD, the divergent subgraphs are seven in number: the gluon, ghost and quark self-energies, renormalized by the constants Z_3, $\tilde{Z}_3$, and Z_3^F, and the three-gluon coupling, with renormalization constant Z_1, the gcc coupling with constant $\tilde{Z}_1$, the $gq\bar{q}$ coupling with constant Z_1^F, and the four-gluon vertex with constant Z_4. In addition, there is a mass shift for the quark, which we shall ignore.

A central issue in QCD is to demonstrate that the $SU(3)$ local gauge invariance is preserved by renormalization. Since the freedom to choose α, the gauge parameter in the gluon propagator, is a consequence of this freedom, it should be true that this parameter is also unaffected by renormalization. We will discuss the implication of these remarks now.

Recall that the QCD Lagrangian, including ghost fields and gauge fixing terms, can be written (assuming, for simplicity, that the quarks are massless)

$$\mathcal{L}_F = \mathcal{L}_g + \mathcal{L}_f + \mathcal{L}_q + \mathcal{L}_c + \mathcal{L}_{3g} + \mathcal{L}_{4g} + \mathcal{L}_{gq\bar{q}} + \mathcal{L}_{gcc} \,, \tag{16.116}$$

where the "free" Lagrangians and interaction terms are

$$\begin{aligned}
\mathcal{L}_g &= -\tfrac{1}{4}\left(\partial_\mu A^a_\nu - \partial_\nu A^a_\mu\right)\left(\partial^\mu A^{a\nu} - \partial^\nu A^{a\mu}\right)\\
\mathcal{L}_f &= -\frac{1}{2\alpha}\left(\partial_\mu A^{a\mu}\right)\left(\partial_\nu A^{a\nu}\right)\\
\mathcal{L}_q &= \frac{i}{2}\bar{\psi}\overleftrightarrow{\not\partial}\psi\\
\mathcal{L}_c &= \partial_\mu \bar{c}^a \partial^\mu c^a\\
\mathcal{L}_{3g} &= \tfrac{1}{2}g_R f_{abc}\, A^{b\mu}\, A^{c\nu}\left(\partial_\mu A^a_\nu - \partial_\nu A^a_\mu\right)\\
\mathcal{L}_{4g} &= -\tfrac{1}{4}g_R^2 f_{abe} f_{cde} A^a_\mu A^b_\nu A^{c\mu} A^{d\nu}\\
\mathcal{L}_{gq\bar{q}} &= -g_R\bar{\psi}\gamma^\mu \tfrac{1}{2}\lambda_a\psi\, A^a_\mu\\
\mathcal{L}_{gcc} &= -g_R f_{abc}\partial_\mu \bar{c}^a c^c A^{b\mu} \;.
\end{aligned} \tag{16.117}$$

These were discussed in Chapters 13 and 15. As in our discussion of QED, the fields and coupling constant g_R are *assumed to be finite, renormalized quantities.*

Counterterms must be added to the Lagrangian to cancel infinities which arise from the seven types of diverging graphs or subgraphs. Since the gauge fixing term, $\mathcal{L}_f$, is associated with the gluon propagator, and the gauge fixing parameter α is unchanged, counterterms are introduced as follows:

$$\begin{aligned}
\mathcal{L}_C =&\,(Z_3-1)\left(\mathcal{L}_g+\mathcal{L}_f\right) + \left(Z_3^F-1\right)\mathcal{L}_q + \left(\tilde{Z}_3-1\right)\mathcal{L}_c\\
&+(Z_1-1)\,\mathcal{L}_{3g} + (Z_4-1)\,\mathcal{L}_{4g} + \left(Z_1^F-1\right)\mathcal{L}_{gq\bar{q}} + \left(\tilde{Z}_1-1\right)\mathcal{L}_{gcc}\,.
\end{aligned} \tag{16.118}$$

Hence the bare Lagrangian, which includes the counterterms, is

$$\mathcal{L}^{(0)}_{\rm QCD} = \mathcal{L}_R + \mathcal{L}_C \;. \tag{16.119}$$

Absorbing the infinite renormalization constants into the fields gives the following *bare* fields, denoted by a superscript (0):

$$\begin{aligned}
A^{a(0)}_\mu &= \sqrt{Z_3}\,A^a_\mu\\
c^{a(0)} &= \sqrt{\tilde{Z}_3}\,c^a\\
\psi^{(0)} &= \sqrt{Z_3^F}\,\psi\,,
\end{aligned} \tag{16.120}$$

and we see that the *bare coupling constant*, g_0, depends on four different combinations of renormalization constants:

$$
\begin{aligned}
3g \Rightarrow g_0 &= \frac{Z_1}{(Z_3)^{3/2}} g_R = Z_g g_R \\
4g \Rightarrow g_0^2 &= \frac{Z_4}{(Z_3)^2} g_R^2 \\
gq\bar{q} \Rightarrow g_0 &= \frac{Z_1^F}{Z_3^F \sqrt{Z_3}} g_R \\
gcc \Rightarrow g_0 &= \frac{\tilde{Z}_1}{\tilde{Z}_3 \sqrt{Z_3}} g_R \quad .
\end{aligned} \tag{16.121}
$$

Clearly, *each of these quantities must be identical* if the $SU(3)$ gauge symmetry is to survive the renormalization. Hence, we must have generalizations of the Ward-Takahashi identity, referred to as the Slavnov–Taylor identities [Ta 71, Sl 72]:

$$
\frac{Z_1}{Z_3} = \frac{Z_1^F}{Z_3^F} = \frac{\tilde{Z}_1}{\tilde{Z}_3} = \frac{Z_4}{Z_1} \quad . \tag{16.122}
$$

For a further discussion of these identities, the reader is referred to the literature [La 81, MP 78]. We will not pursue the discussion of the renormalizability of QCD further.

In the next chapter we will assume that QCD is renormalizable and the identities (16.122) can be proved and show that QCD is an *asymptotically free* theory.

PROBLEMS

16.1 Consider a theory with an interaction Lagrangian with n_D space–time derivatives. For example, pseudovector πN coupling has the form

$$
\mathcal{L}_I = -g\bar{\psi}\gamma^5\gamma^\mu\psi\partial_\mu\phi
$$

with $n_D = 1$ space–time derivative. Using the methods worked out in Sec. 16.1, show that Eq. (16.5) is unchanged but that the index of divergence of the theory becomes

$$
I = \tfrac{1}{2}(d-2)B + \tfrac{1}{2}(d-1)F + n_D - d \quad .
$$

Discuss the significance of this result.

16.2 Compute the counterterm λ_0 which arises from the four-point function shown in Fig. 16.3A when calculated in $d = 8$ dimensions.

16.3 Using the BS equation [in particular, Eq. (12.50) may be helpful], prove that the Dyson equation illustrated in Fig. 16.13A is equivalent to the equation illustrated in Fig. 16.14. (You will also need to use the equation in Fig. 16.13B.)

16.4 Show that the counterterms needed to remove the subdivergences from the sixth order self-energy, graph 16.11, are as shown in Fig, 16.16. You may use Figs. 16.14 and 16.15 to construct your argument. (Hint: Don't forget the contribution from $\Gamma^{(3)}$ and its counterterms.)

CHAPTER 17

THE RENORMALIZATION GROUP AND ASYMPTOTIC FREEDOM

We now exploit one of the most powerful consequences of renormalization: the fact that the final results for renormalized scattering amplitudes cannot depend on the choice of the renormalization scale μ^2. If it is really true that the choice of μ^2 does not change the final results, then there is a symmetry associated with this invariance, i.e., a group of transformations involving changes in μ^2 which leave all physical results unchanged. These transformations are referred to as the *renormalization group*. Following our discussion of the renormalization group, we will discuss asymptotic freedom and show that QCD is asymptotically free.

17.1 THE RENORMALIZATION GROUP EQUATIONS

We will use the notation of Chapter 16, where g_0 denotes the *unrenormalized* coupling constant and g_R the *renormalized* coupling constant. These two constants are related through renormalization,

$$g_0 = Z_g g_R \ , \tag{17.1}$$

where Z_g is a renormalization constant. In d dimensions, both of these coupling constants have the dimensions of $\epsilon = 4-d$, and the renormalized coupling constant g_R is related to the *dimensionless* coupling constant g as in Eq. (16.17),

$$g_R = g\mu^{\epsilon/2} \ ,$$

where μ is the (arbitrary) mass scale used in the dimensional regularization scheme discussed in Chapter 16. Now, the unrenormalized coupling constant g_0 is the only one of these constants which is *independent of the renormalization scale* μ (for the simple reason that the scale does not enter into its definition), and hence the dimensionless coupling constant g depends on the renormalization scale μ through a combination of Eqs. (17.1) and (16.17),

$$g = \mu^{-\epsilon/2} Z_g^{-1} g_0 \ . \tag{17.2}$$

The implication of this dependence of g on renormalization scale will be studied in this chapter.

Now, as a specific example of how these ideas can be extended to scattering amplitudes, consider the *unrenormalized* scattering matrix for the interaction of n gluons (the n-point function). This amplitude depends on g_0 and ϵ, but not on the mass scale μ. Hence

$$\mu \frac{d}{d\mu} \mathcal{M}^{(n)} \left(\{p_i\}, g_0, \epsilon \right) = 0 \ , \tag{17.3}$$

where $\{p_i\}$ are the external momenta upon which $\mathcal{M}$ depends and the dependence of $\mathcal{M}$ on ϵ is shown explicitly. The validity of (17.3) is regarded as self-evident; $\mathcal{M}$ cannot depend on μ since none of its arguments do. It is a mathematical statement of the physical fact that the physics must be independent of the renormalization point μ.

Now, the *renormalized* n-point function is obtained from the unrenormalized n-point function by multiplying by a factor of $Z_3^{1/2}$ for each external gluon, so that

$$\mathcal{M}_R^{(n)} \left(\{p_i\}, g, \mu \right) = Z_3^{n/2} \left(g, \epsilon, \mu \right) \mathcal{M}^{(n)} \left(\{p_i\}, g_0, \epsilon \right) \ , \tag{17.4}$$

where we have displayed the fact that $\mathcal{M}_R$ depends on μ and will ignore any dependence of $\mathcal{M}$ on quark masses. The renormalized amplitude *does* depend on μ through $Z_3^{n/2}$, an essentially "trivial" dependence which is associated with the dimension of the amplitude (see below).

An equation describing the dependence of $\mathcal{M}_R$ on μ can be readily obtained from (17.3). We have

$$\mu \frac{d}{d\mu} \mathcal{M}_R^{(n)} \left(\{p_i\}, g, \mu \right) = \mu \frac{n}{2} \frac{d}{d\mu} \left(\log Z_3 \right) \mathcal{M}_R^{(n)} \left(\{p_i\}, g, \mu \right) \ . \tag{17.5}$$

If we now express the total derivative on the left-hand side in terms of partial derivatives, we obtain the Callan–Symanzik equation [Ca 70, Sy 70]

$$\left[\mu \frac{\partial}{\partial \mu} + \beta(g) \frac{\partial}{\partial g} - n\gamma(g) \right] \mathcal{M}_R^{(n)} \left(\{p_i\}, g, \mu \right) = 0 \ , \tag{17.6}$$

where β and γ are taken to be functions of g and

$$\begin{aligned} \beta(g) &= \mu \frac{dg}{d\mu} \\ \gamma(g) &= \frac{\mu}{2} \frac{d}{d\mu} \left(\log Z_3 \right) \ . \end{aligned} \tag{17.7}$$

Equation (17.6) is an example of a renormalization group equation. It describes an arbitrariness in the behavior of $\mathcal{M}_R$ which arises from the choice of the renormalization mass μ. If we choose a different μ, $\mathcal{M}_R$ will still be finite, but it will

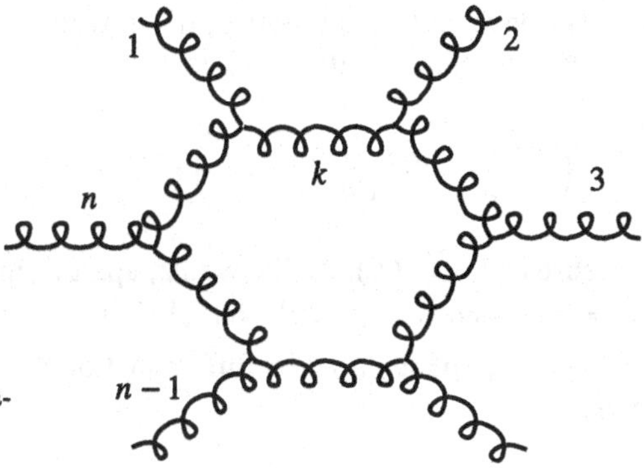

Fig. 17.1 One-loop graph for the n-point function.

have a different value, and the behavior of $\mathcal{M}_R$ as μ is varied is constrained by Eq. (17.6).

At first glance, it may seem that the dependence of $\mathcal{M}_R$ on the renormalization mass scale μ is an amusing fact of no physical importance. However, as we will now see, this dependence gives a powerful new way to study the behavior of scattering at large momenta.

17.2 SCATTERING AT LARGE MOMENTA

We can use Eq. (17.6), together with dimensional analysis, to deduce the behavior of $\mathcal{M}_R^{(n)}$ as the momenta $\{p_i\} \to \infty$. We will continue to consider the simplest case in which there is *no quark mass dependence*. Then, from an analysis of the simplest Feynman graph which contributes to $\mathcal{M}^{(n)}$, the one-loop graph with n $3g$ couplings shown in Fig. 17.1, we can determine that the dimension of $\mathcal{M}^{(n)}$ is

$$\mathcal{M}^{(n)} \simeq \int d^4k \frac{(k^n)}{(k^2)^n} \to (\mu)^{4-n} \tag{17.8}$$

and that therefore $\mathcal{M}^{(n)}$ can be rescaled as follows:

$$\mathcal{M}_R^{(n)}\left(\{\lambda p_i\}, g, \mu\right) = \mu^{4-n} \mathcal{M}_R^{(n)}\left(\left\{\frac{\lambda p_i}{\mu}\right\}, g, 1\right) , \tag{17.9}$$

where we have also replaced $\{p_i\}$ by $\{\lambda p_i\}$, where λ is a dimensionless parameter which is convenient to introduce, and μ is the mass scale which entered through the renormalization of $\mathcal{M}$. Note that Eq. (17.9) tells us that the variation of $\mathcal{M}_R^{(n)}$ with respect to the momentum scale (as measured by the parameter λ) is related to the variation of $\mathcal{M}_R^{(n)}$ with respect to the renormalization scale μ, and hence the renormalization group equations can be used to study the dependence of the scattering amplitudes on momentum scale.

To develop this connection, begin with the Callan–Symanzik equation (17.6) for the amplitude (17.9), which is

$$\left(\mu\frac{\partial}{\partial\mu} + \beta(g)\frac{\partial}{\partial g} - n\gamma(g)\right)\mu^{4-n}\mathcal{M}_R^{(n)}\left(\left\{\frac{\lambda p_i}{\mu}\right\}, g, 1\right) = 0 \ . \tag{17.10}$$

Because of Eq. (17.9), we know that, apart from a "trivial" factor derived from the overall dependence of $\mathcal{M}_R^{(n)}$ on μ^{4-n}, the action of the partial derivative $\mu\partial/\partial\mu$ on $\mathcal{M}_R^{(n)}$ can be replaced by differentiation by $-\lambda\partial/\partial\lambda$. Making this substitution gives

$$\left(4 - n - \lambda\frac{\partial}{\partial\lambda} + \beta(g)\frac{\partial}{\partial g} - n\gamma(g)\right)\mathcal{M}_R^{(n)}\left(\{\lambda p_i\}, g, \mu\right) = 0 \ . \tag{17.11}$$

This equation now relates the variation of $\mathcal{M}_R^{(n)}$ on λ (the momentum scale) to its variation on g and will permit us, under certain conditions, to estimate the behavior of $\mathcal{M}_R$ for large λ (large momenta).

We will now show that the solution to this equation can be written in the following form:

$$\mathcal{M}_R^{(n)}\left(\{\lambda p_i\}, g, \mu\right) = S\,\mathcal{M}_R^{(n)}\left(\{p_i\}, g_r(\lambda, g), \mu\right) \ , \tag{17.12}$$

where S is an overall scaling factor

$$S = \lambda^{4-n}\exp\left\{-n\int_1^\lambda \frac{dx}{x}\,\gamma\left[g_r(x,g)\right]\right\} \ , \tag{17.13}$$

and $g_r(\lambda, g)$ is the *running coupling constant* which depends on λ and g and which is interpreted as the effective coupling constant for $\lambda \geq 1$. The running coupling constant is *defined* by the equations

$$g_r(1, g) = g \tag{17.14a}$$

$$\lambda\frac{\partial}{\partial\lambda}g_r(\lambda, g) = \beta(g_r) \ . \tag{17.14b}$$

Before we show that (17.12) really does satisfy Eq. (17.11), note that it tells us that $\mathcal{M}$ at large momenta $\{\lambda p_i\}$ with $\lambda \to \infty$ *can be obtained from* $\mathcal{M}$ *at some fixed momenta* $\{p_i\}$, *provided we replace the coupling constant g by the running coupling constant* g_r and multiply by a scaling factor dependent on γ. We see that knowledge of the running coupling constant is sufficient to determine much of the behavior of the scattering at high momenta.

To prove that (17.12) is a solution to Eq. (17.11), begin by noting that Eq. (17.14b) can be integrated if we regard g as a parameter independent of λ. Then

$$\frac{dg_r}{\beta(g_r)} = \frac{d\lambda}{\lambda} ,$$

and using the condition (17.14a), we obtain the following integral equation for the running coupling constant:

$$\int_g^{g_r(\lambda,g)} \frac{dx}{\beta(x)} = \log\lambda . \tag{17.15}$$

This equation gives an implicit solution for the running coupling constant $g_r(\lambda, g)$ which will be discussed shortly. But first note that this equation can be used to show that the running coupling constant satisfies the following equation:

$$\left(\lambda\frac{\partial}{\partial\lambda} - \beta(g)\frac{\partial}{\partial g}\right) g_r(\lambda, g) = 0 . \tag{17.16}$$

This can be readily demonstrated by differentiating (17.15) with respect to g, which gives

$$\frac{1}{\beta(g_r)}\frac{\partial g_r}{\partial g} - \frac{1}{\beta(g)} = 0 .$$

From this we conclude that

$$\beta(g)\frac{\partial g_r}{\partial g} = \beta(g_r) ,$$

and comparing this with Eq. (17.14b) gives (17.16).

The proof that (17.12) solves Eq. (17.11) can now be completed. Differentiating (17.12) gives

$$\lambda\frac{\partial}{\partial\lambda}\mathcal{M}_R^{(n)}\left(\{\lambda p_i\}, g, \mu\right) = S\left\{\lambda\frac{\partial g_r}{\partial\lambda}\frac{\partial}{\partial g_r} + 4 - n - n\gamma(g_r)\right\}\mathcal{M}_R^{(n)}\left(\{p_i\}, g_r, \mu\right)$$

$$\frac{\partial}{\partial g}\mathcal{M}_R^{(n)}\left(\{\lambda p_i\}, g, \mu\right) = S\left\{\frac{\partial g_r}{\partial g}\frac{\partial}{\partial g_r} - n\int_1^\lambda \frac{dx}{x}\frac{\partial\gamma}{\partial g_r}\frac{\partial g_r(x,g)}{\partial g}\right\} \times \mathcal{M}_R^{(n)}\left(\{p_i\}, g_r, \mu\right) , \tag{17.17}$$

where S is the scaling factor of Eq. (17.13). Combining these equations and using (17.16) give

$$\left(4 - n - \lambda\frac{\partial}{\partial\lambda} + \beta(g)\frac{\partial}{\partial g}\right)\mathcal{M}_R^{(n)}\left(\{\lambda p_i\}, g, \mu\right)$$
$$= n\left\{\gamma(g_r) - \beta(g)\int_1^\lambda \frac{dx}{x}\frac{\partial\gamma}{\partial g_r}\frac{\partial g_r(x,g)}{\partial g}\right\}\mathcal{M}_R^{(n)}\left(\{\lambda p_i\}, g, \mu\right) . \tag{17.18}$$

Now we may use (17.16) again to carry out the integral over x giving

$$
\begin{aligned}
\gamma(g_r) - \beta(g) \int_1^\lambda \frac{dx}{x} \frac{\partial \gamma}{\partial g_r} \frac{\partial g_r(x,g)}{\partial g} \\
= \gamma(g_r) - \int_1^\lambda \frac{dx}{x} \frac{\partial \gamma}{\partial g_r} x \frac{\partial g_r(x,g)}{\partial x} \\
= \gamma(g_r) - [\gamma(g_r) - \gamma(g)] = \gamma(g)
\end{aligned}
\tag{17.19}
$$

and hence Eq. (17.11) is obtained.

An understanding of scattering at large momenta is greatly facilitated by an understanding of how the running coupling constant behaves as $\lambda \to \infty$, and we will discuss this in the next section.

Before turning to this discussion, rewrite the solution (17.12) for the case when $\gamma(\infty, g) = \gamma_\infty$ is finite. In this case it is convenient to subtract the integral over γ as follows:

$$
\int_1^\lambda \frac{dx}{x} \gamma \left[g_r(x,g)\right] = \int_1^\lambda \frac{dx}{x} \left\{ \gamma \left[g_r(x,g)\right] - \gamma_\infty \right\} + \gamma_\infty \log \lambda \ , \tag{17.20}
$$

and write the solution in the following form:

$$
\mathcal{M}_R^{(n)} \left(\{\lambda p_i\}, g, \mu\right) = S_+ \mathcal{M}_R^{(n)} \left(\{p_i\}, g_r(\lambda, g), \mu\right) \ , \tag{17.21}
$$

where

$$
S_+ = \lambda^{4-n-n\gamma_\infty} \exp \left\{ -n \int_1^\lambda \frac{dx}{x} \left[\gamma \left[g_r(x,g)\right] - \gamma_\infty\right] \right\} \ . \tag{17.22}
$$

Note that the integral over x now converges to a finite value as $\lambda \to \infty$, so that the *dimension* of $\mathcal{M}_R$ has changed from $4 - n$ to $4 - n - n\gamma_\infty$. The quantity γ_∞ is referred to as the *anomalous dimension* of the field, since it indicates how the scaling behavior of $\mathcal{M}_R$ departs from that predicted by simple dimensional analysis.

17.3 BEHAVIOR OF THE RUNNING COUPLING CONSTANT

From the defining Eqs. (17.14), it is clear that the behavior of the running coupling constant g_r depends on $\beta(g_r)$, and from the implicit solution, Eq. (17.15), it is clear that the zeros of β play a special role. As the upper limit of the integral in (17.15) approaches a value of g at which $\beta(g) = 0$, the integral will diverge, and hence λ must either approach ∞ or 0, the only two points at which $\log \lambda$ also diverges. Clearly, the behavior of $\log \lambda$ depends on whether or not the integral is positive or negative as the zero is approached. We are therefore led to distinguish the two possibilities shown in Fig. 17.2. In the first case, the zero will be denoted by g_+, and such a zero is referred to as an *ultraviolet fixed point.* In this case

$$
\beta(g) \begin{cases} < 0 & \text{if } g > g_+ \\ > 0 & \text{if } g < g_+ \end{cases} \ . \tag{17.23}
$$

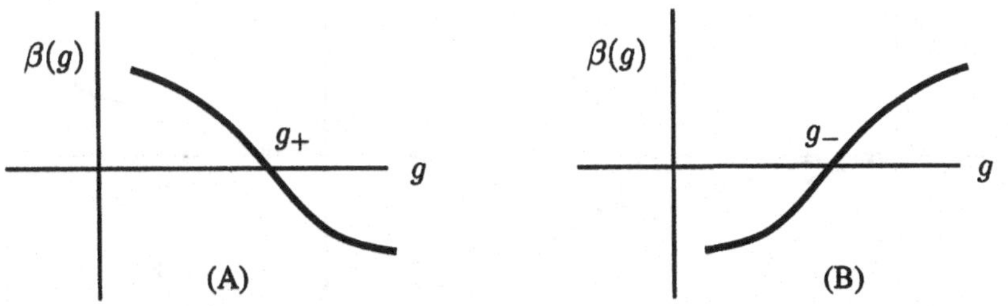

Fig. 17.2 Behavior of the β-function near (A) an ultraviolet fixed point and (B) an infrared fixed point.

In the second case, the zero is denoted by g_- and is referred to as an *infrared fixed point.* In this case

$$\beta(g)\begin{cases} >0 & \text{if } g > g_- \\ <0 & \text{if } g < g_- \end{cases} . \tag{17.24}$$

To understand the behavior of g_r near these points, suppose that g_r is less than $g_\pm$ but close enough to $g_\pm$ so that the zero at $g_\pm$ is the only zero under consideration. Then, as $g_r \to g_\pm$ from below, the integral will diverge to + infinity at g_+ and − infinity at g_-, and hence $\lambda \to \infty$ as $g \to g_+$ at $\lambda \to 0$ as $g_r \to g_-$. The same conclusion is reached if $g_r > g_\pm$; in this case the sign of the integral is changed, but the sign of β is also changed, so the same argument holds. We conclude that the two solutions $g_r(\lambda, g)$ are

$$\begin{aligned} g_+ &= g_r(\infty, g) && \text{ultraviolet fixed point} \\ g_- &= g_r(0, g) && \text{infrared fixed point} \ . \end{aligned}$$

In such a case, g_+ is the *effective coupling constant at infinite momenta,* and g_- is the *effective constant at low momenta.*

The function β will usually approach zero as some power of g when $g \to 0$. The two ways it can approach zero are illustrated in Fig. 17.3. In the first case, $\beta(g)$ is *negative* for small g; in the second case it is positive. From the previous discussion, we see that, in the first case,

$$g_r(\infty, g) = 0 \qquad \text{asymptotic freedom} \ .$$

This case is referred to as *asymptotic freedom.* The name comes from the fact that the *effective coupling constant of the theory approaches zero as the momenta approach infinity.* This means that the scattering amplitudes at large momenta can be *calculated using perturbation theory,* and all of the wonderful perturbative methods which we have described can be applied to high energy calculations,

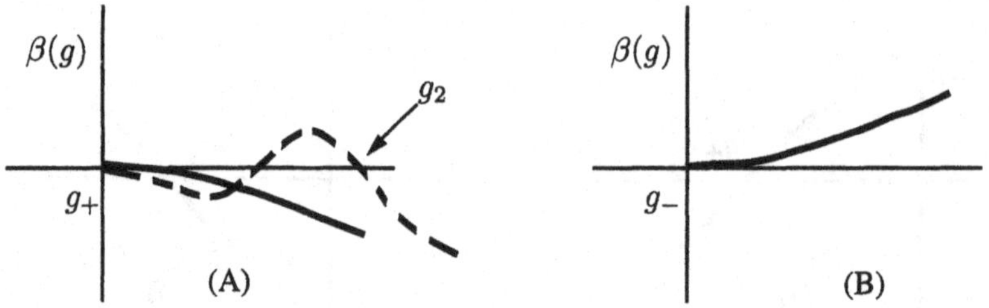

Fig. 17.3 Behavior of the β-function for small coupling. In (A) the theory is asymptotically free, but for a theory with a β-function given by the dashed curve, the asymptotically free region cannot be physically realized if $g > g_2$.

even if the effective coupling constant at moderate energies is large. As we will show in the next section, this remarkable property is a feature of non-Abelian gauge theories and of QCD in particular.

In order to prove that a theory is asympotically free, *it is sufficient to demonstrate that $\beta(g) < 0$ as $g \to 0$ in lowest order perturbation theory* [GW 73, Po 73]. This is because such a demonstration insures that g_r is small at large momenta, and hence also justifies using perturbation theory to estimate $\beta(g)$. However, such a demonstration does not guarantee that the theory as physically realized by nature is asymptotically free. For example, if the theory has a β-function with two other zeros at finite g (as shown by the dashed line in Fig. 17.3A), and if the physical value of $g > g_2$, then the coupling will evolve to g_2 as $\lambda \to \infty$, and we will never reach the asymptotically free region. Alternatively, even if there are no other zeros, the coupling constant may run so slowly that it does not become small until the momenta are so large that they are inaccessible experimentally. For QCD, neither of these situations seems to hold, and it appears that the asymptotically free region of QCD is physical accessible.

Anticipating the results we will obtain in the next section, we assume that in lowest order perturbation theory

$$\beta(g) = -\beta_0\, g^3 \,, \tag{17.25}$$

where β_0 is a positive number. In this case we have an asymptotically free theory where $g_+ = 0$, and the effective coupling constant at large λ is obtained by solving Eq. (17.15). Substituting (17.25) into (17.15) gives

$$-\int_g^{g_r} \frac{dx}{\beta_0 x^3} = \log\lambda = \frac{1}{2\beta_0}\left[\frac{1}{g_r^2} - \frac{1}{g^2}\right] \tag{17.26}$$

from which we obtain

$$g_r^2(\lambda, g) = \frac{g^2}{1 + 2\beta_0 g^2 \log\lambda} = \frac{g^2}{1 + 2\beta_0 g^2 \log\left(\frac{Q}{\Lambda}\right)} \,, \tag{17.27}$$

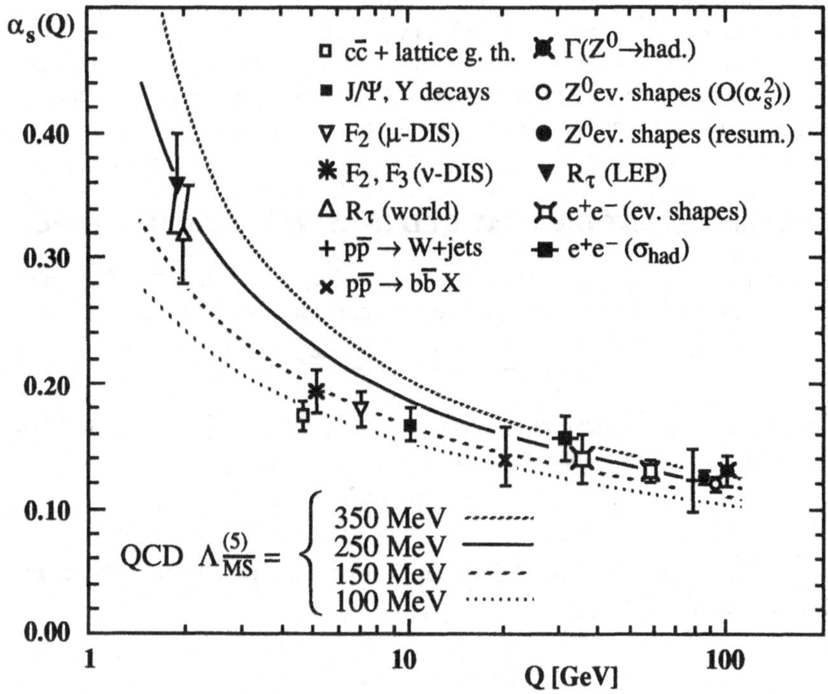

Fig. 17.4 Value of the strong fine structure constant as extracted from a variety of experiments at different momentum scales Q. Note that the fine structure constant grows with decreasing Q as predicted by Eq. (17.28). The empirically determined value of Λ is between 150 and 250 MeV. (*Courtesy of S. Bethke* [Be 92].)

where λ has been replaced by Q/Λ, with Λ being the momentum scale at which the coupling has the initial value of g and Q being the momentum scale at which it is g_r. Letting $Q \gg \Lambda$, we may write the running coupling constant in the following form:

$$\boxed{\alpha_S(Q) = \frac{g_r^2(Q)}{4\pi} \simeq \frac{1}{8\pi\beta_0 \log{(Q/\Lambda)}}} \quad . \tag{17.28}$$

Since β_0 can be calculated from perturbation theory, the above formula shows that the coupling constant at large momenta really depends on only one parameter, Λ. This prediction can be tested by extracting the effective coupling constant from a variety of experiments carried out over a range of momentum scales. Unfortunately, the expected log variation is slow, and it has not been until recently that enough evidence has been collected to convince a skeptic that the coupling constant does indeed "run" according to Eq. (17.28). A recent determination of how the QCD coupling constant "runs" is shown in Fig. 17.4 [BP 92]. From these results a value of $\Lambda \simeq 150 - 250$ MeV is obtained.

Note that the running coupling constant (17.28) approaches zero very slowly as $Q \to \infty$, and hence the force between two quarks at very short distance does not really go to zero, as is sometimes assumed.

17.4 DEMONSTRATION THAT QCD IS ASYMPTOTICALLY FREE

We will now show that QCD is an asymptotically free theory. Combining Eqs. (17.2) and (17.7), the demonstration requires that we compute

$$\beta(g) = \mu \frac{dg}{d\mu} = -\frac{\epsilon}{2} g - \frac{g}{Z_g} \frac{dZ_g}{dg} \beta(g) \ , \tag{17.29}$$

where Z_g is a renormalization constant defined in Eq. (16.121), which in lowest order perturbation theory has the form

$$Z_g = 1 - g^2 \kappa \left(\frac{1}{\epsilon} \right) \ , \tag{17.30}$$

where κ is a constant to be determined below. Hence Eq. (17.29) has the following solution to third order in g:

$$\begin{aligned} \beta(g) &= \frac{-\epsilon g}{2\left(1 - \dfrac{2g^2\kappa}{Z_g \epsilon}\right)} \\ &\cong -\frac{\epsilon}{2} g - g^3 \kappa \ , \end{aligned} \tag{17.31}$$

which, in the limit $\epsilon \to 0$, gives $\kappa = \beta_0$, where β_0, defined in Eq. (17.25), determines how the coupling constant "runs". If κ is positive, the theory is asymptotically free.

The key to the demonstration is therefore the calculation of Z_g. From Eq. (16.121), there are three equivalent forms for Z_g:

$$Z_g = \frac{Z_1}{(Z_3)^{3/2}} = \frac{Z_1^F}{Z_3^F \sqrt{Z_3}} = \frac{\tilde{Z}_1}{\tilde{Z}_3 \sqrt{Z_3}} \ , \tag{17.32}$$

where we ignore the possibility of using Z_4 (why?). The diagrams which contribute to the six renormalization constants which enter (17.32) are shown in Fig. 17.5. We will choose the last combination in (17.32) which is comparatively simple to calculate.

We begin with the lowest order calculation of $\tilde{Z}_3$, the renormalization constant for the ghost propagator, labeled in Fig. 17.6. Using the Feynman gauge ($\alpha = 1$) and the Feynman rules for QCD, summarized in Appendix B, we obtain

$$\tilde{\Pi}_{ab} = i g_R^2 f_{acd} f_{bdc} \int \frac{d^d k}{(2\pi)^d} \frac{\left(k + \frac{1}{2}q\right) \cdot q}{A_+ A_-} \ , \tag{17.33}$$

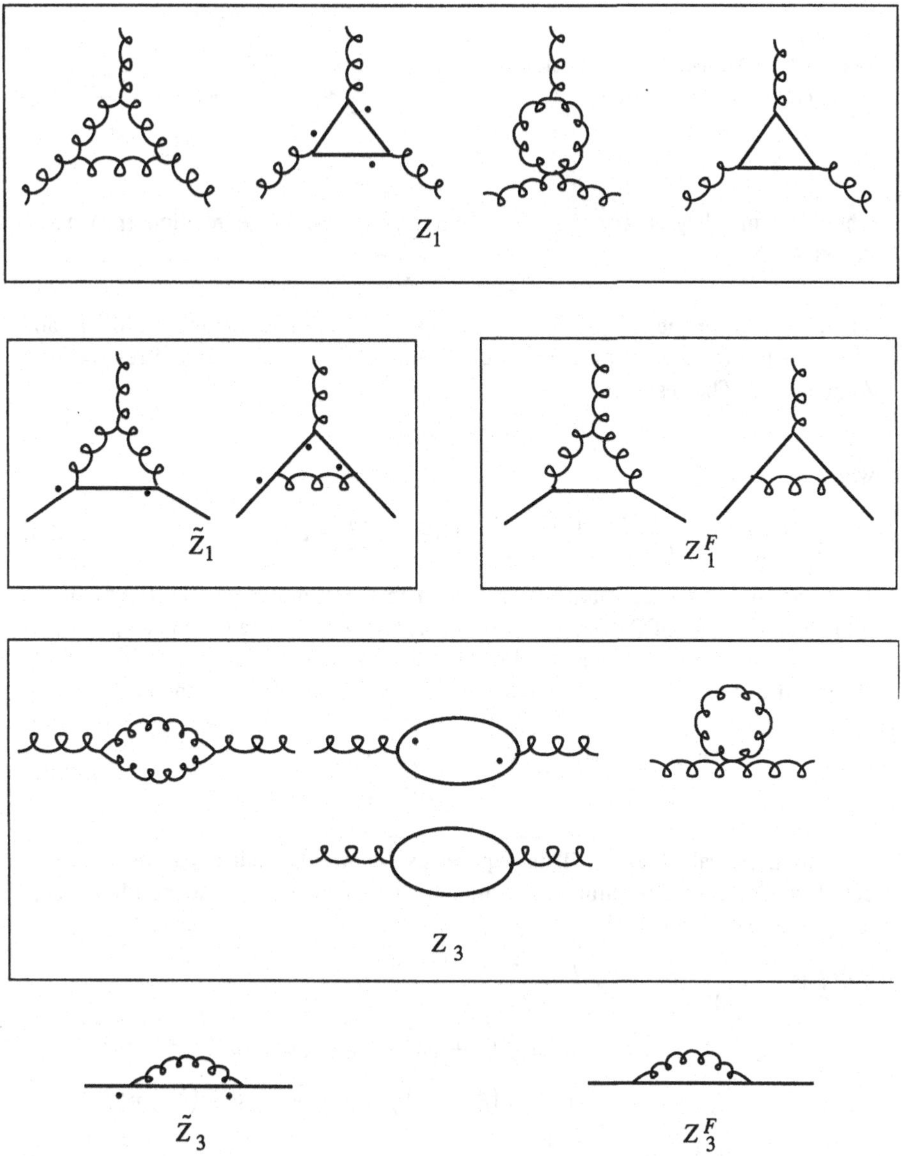

Fig. 17.5 Feynman graphs which contribute to the calculation of seven of the renormalization constants in QCD. Both ghost and quark lines are solid, but ghost lines have dots at one end, and gluons are corkscrews.

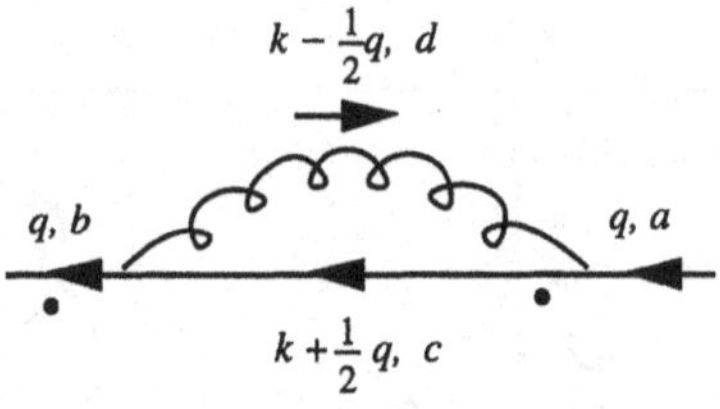

Fig. 17.6 Feynman graph for calculation of the renormalization constant $\tilde{Z}_3$.

where, as in Chapter 16, $A_\pm = -\left(k \pm \frac{1}{2}q\right)^2$. Using the relation (proved in Appendix D)

$$f_{acd} f_{bcd} = N\,\delta_{ab} \tag{17.34}$$

where $N = 3$ for the $SU(3)$ color group, we combine denominators, shift k, and carry out the integral using techniques developed in Chapter 16 and reviewed in Appendix C. The result is

$$\tilde{\Pi}_{ab} = q^2\,\delta_{ab}\,\tilde{\Pi}(q^2) \tag{17.35}$$

where

$$\tilde{\Pi}(q^2) = \frac{g^2 N}{2(4\pi)^{d/2}} \left(\frac{\mu^2}{-q^2}\right)^{\epsilon/2} \Gamma\left(\tfrac{\epsilon}{2}\right) B\left(\tfrac{d}{2}-1, \tfrac{d}{2}-1\right) \ . \tag{17.36}$$

The renormalization constant $\tilde{Z}_3$ is related to the singular part of $\tilde{\Pi}(q^2)$ in the same way that the QED renormalization constant Z_2 is related to the singular part of electron self energy Σ [recall Eq. (11.49)] and using MS renormalization (as discussed in Sec. 16.2), the counterterm obtained from $\tilde{\Pi}(q^2)$ is therefore

$$\boxed{\tilde{Z}_3 - 1 = \frac{g^2 N}{(4\pi)^2}\,\frac{1}{\epsilon}} \ . \tag{17.37}$$

Next, we calculate Z_3. This requires evaluating the four diagrams, drawn in detail in Fig. 17.7. Recalling the symmetry factor of $\frac{1}{2}$ for a closed gluon loop, the gluon loop diagram A is

$$\begin{aligned}
\Pi^{\mu\nu}_{ab}(A) = &-\frac{i}{2} g_R^2 f_{acd} f_{bcd} \int \frac{d^dk}{(2\pi)^d} \frac{1}{A_+A_-} \\
&\times \left[-g^{\mu\lambda}\left(k+\tfrac{3}{2}q\right)^\sigma + g^{\lambda\sigma}\,2k^\mu - g^{\sigma\mu}\left(k-\tfrac{3}{2}q\right)^\lambda\right] \\
&\times \left[\ g^\nu{}_\lambda\left(k+\tfrac{3}{2}q\right)_\sigma - g_{\lambda\sigma}\,2k^\nu + g^\nu{}_\sigma\left(k-\tfrac{3}{2}q\right)_\lambda\right] \\
= &\ i g_R^2 N\,\delta_{ab} \int \frac{d^dk}{(2\pi)^d} \frac{1}{A_+A_-} \\
&\times \left[g^{\mu\nu}\left(k^2 + \tfrac{9}{4}q^2\right) + k^\mu k^\nu(2d-3) - \tfrac{9}{4}q^\mu q^\nu\right] ,
\end{aligned} \tag{17.38}$$

where (17.34) and $g_{\sigma\lambda}g^{\sigma\lambda} = d$ have been used and terms in the numerator which are odd in k have been dropped, because A_+A_- is symmetric as $k \to -k$. We will

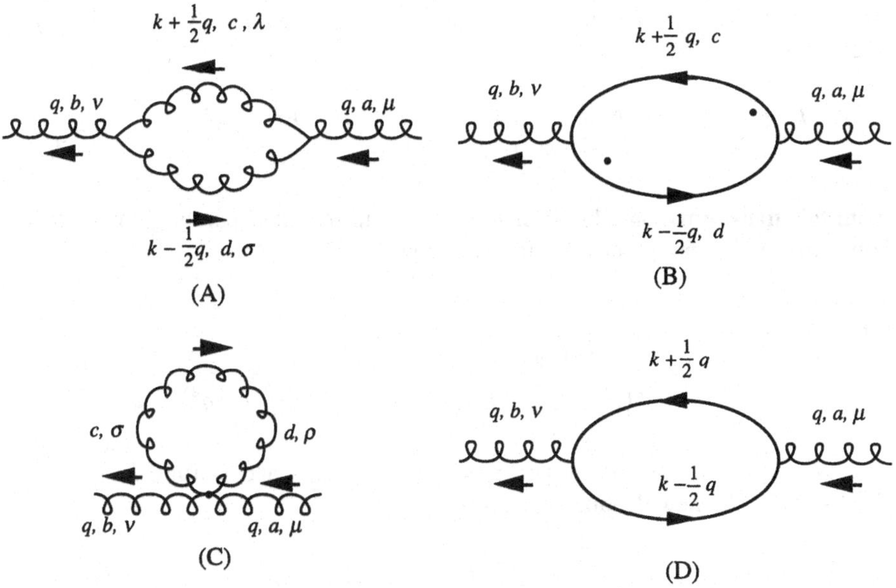

Fig. 17.7 The four Feynman graphs which are needed for a calculation of the renormalization constant Z_3. (A) The gluon loop graph, (B) the ghost loop, (C) the gluon tadpole, and (D) the quark loop contributions.

postpone further reduction of this expression until the other gluonic contributions are included. The ghost loop diagram, with the factor of -1 for a closed "fermion" loop, is

$$\begin{aligned}\Pi^{\mu\nu}_{ab}(B) &= -ig_R^2 f_{acd}f_{bcd}\int \frac{d^dk}{(2\pi)^d}\,\frac{1}{A_+A_-}\left(k+\tfrac{1}{2}q\right)^\mu\left(k-\tfrac{1}{2}q\right)^\nu \\ &= -ig_R^2 N\delta_{ab}\int \frac{d^dk}{(2\pi)^d}\,\frac{1}{A_+A_-}\left(k^\mu k^\nu - \tfrac{1}{4}q^\mu q^\nu\right) \quad . \end{aligned} \qquad (17.39)$$

Finally, the contribution C from the gluon four-point vertex, with its factor of $\frac{1}{2}$, is

$$\begin{aligned}\Pi^{\mu\nu}_{ab}(C) =& i\frac{1}{2}g_R^2\int \frac{d^dk}{(2\pi)^d}\left(\frac{1}{-k^2}\right)\delta^{cd}g_{\sigma\rho} \\ &\times\Big[f_{abe}f_{cde}\left(g^{\mu\sigma}g^{\nu\rho}-g^{\mu\rho}g^{\nu\sigma}\right)+f_{ace}f_{bde}\left(g^{\mu\nu}g^{\sigma\rho}-g^{\mu\rho}g^{\nu\sigma}\right) \\ &\quad+f_{ade}f_{cbe}\left(g^{\mu\sigma}g^{\nu\rho}-g^{\mu\nu}g^{\sigma\rho}\right)\Big] \\ =& ig_R^2 N\delta_{ab}\,(d-1)\int \frac{d^dk}{(2\pi)^d}\left(\frac{1}{-k^2}\right)g^{\mu\nu} \quad . \end{aligned} \qquad (17.40)$$

This integral can be shown to be zero in dimensional regularization (see Prob. 17.1), but we will find it convenient to transform it and add it to the other gluonic contributions. To this end, shift $k \to k + \frac{1}{2}q$, and multiply and divide by A_-. Then

$$\int \frac{d^dk}{(2\pi)^d} \frac{1}{A_+} = \int \frac{d^dk}{(2\pi)^d} \frac{A_-}{A_+A_-} = \int \frac{d^dk}{(2\pi)^d} \frac{-k^2 - \frac{1}{4}q^2}{A_+A_-} \ . \tag{17.41}$$

With this transformation, (17.40) may be added to the other purely gluonic contributions to Π (from graphs A and B), giving

$$\Pi^{\mu\nu}_{ab}(\text{gluon}) = -ig_R^2 N\delta_{ab} \int \frac{d^dk}{(2\pi)^d} \frac{1}{A_+A_-} \times \left\{2\left(q^\mu q^\nu - g^{\mu\nu}q^2\right) + (d-2)\left[g^{\mu\nu}\left(k^2 + \tfrac{1}{4}q^2\right) - 2\,k^\mu k^\nu\right]\right\}. \tag{17.42}$$

We now evaluate this by combining denominators and shifting $k \to k + \frac{1}{2}q(1-2x)$. The new denominator becomes

$$D = -k^2 - q^2x(1-x) \ , \tag{17.43}$$

and using the replacement $k^\mu k^\nu \to k^2 g^{\mu\nu}/d$, we obtain

$$\Pi^{\mu\nu}_{ab}(\text{gluon}) = -ig_R^2 N\delta_{ab} \int_0^1 dx \int \frac{d^dk}{(2\pi)^d} \frac{1}{D^2} \times \left\{\left(q^\mu q^\nu - g^{\mu\nu}q^2\right)\left[2 - \left(\tfrac{d}{2} - 1\right)(1-2x)^2\right] + g^{\mu\nu}(d-2)\left[k^2\left(1 - \tfrac{d}{2}\right) + q^2x(1-x)\right]\right\}. \tag{17.44}$$

First look at the gauge violating $g^{\mu\nu}$ term. We can show that it is identically zero by direct integration:

$$\begin{aligned}
-i\int_0^1 dx \int \frac{d^dk}{(2\pi)^d} \frac{1}{D^2}\left[k^2\left(1 - \tfrac{2}{d}\right) + q^2x(1-x)\right]
&= -i\int_0^1 dx \int \frac{d^dk}{(2\pi)^d}\left[-\left(1 - \tfrac{2}{d}\right)\frac{1}{D} + \tfrac{2}{d}\,\frac{q^2x(1-x)}{D^2}\right] \\
&= \frac{1}{(4\pi)^{d/2}}\left(\frac{1}{-q^2}\right)^{1-d/2} B\left(\tfrac{d}{2}, \tfrac{d}{2}\right)\left[-\left(1 - \tfrac{2}{d}\right)\Gamma\left(1 - \tfrac{d}{2}\right) - \tfrac{2}{d}\Gamma\left(2 - \tfrac{d}{2}\right)\right] \\
&= \frac{1}{(4\pi)^{d/2}}\left(\frac{1}{-q^2}\right)^{1-d/2} B\left(\tfrac{d}{2}, \tfrac{d}{2}\right)\Gamma\left(1 - \tfrac{d}{2}\right)\left[\tfrac{2}{d} - 1 - \tfrac{2}{d}\left(1 - \tfrac{d}{2}\right)\right] = 0 \ .
\end{aligned}$$

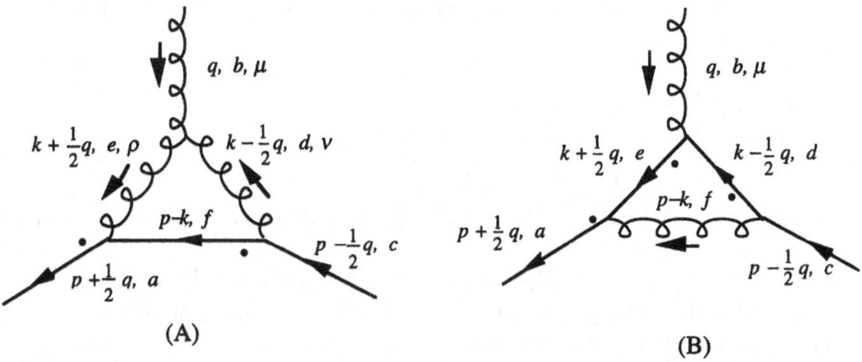

Fig. 17.8 The two Feynman graphs which are needed for a calculation of the renormalization constant $\tilde{Z}_1$.

In a similar way the gauge invariant term can be integrated, giving

$$\Pi^{\mu\nu}_{ab}(\text{gluon}) = \delta_{ab}\left(g^{\mu\nu}q^2 - q^\mu q^\nu\right)\Pi_{\text{gluon}}(q^2) \tag{17.45}$$

with singular part

$$\Pi_{\text{gluon}}(q^2) = -\frac{g^2 N}{(4\pi)^2}\left(\frac{\mu^2}{-q^2}\right)^{\epsilon/2}\frac{1}{\epsilon}\,\frac{10}{3}\,. \tag{17.46}$$

Using the MS renormalization scheme and remembering that the general connection between the renormalization constant constant Z_3 and the singular part of $\Pi(q^2)$ is similar to that for the photon, Eq. (11.73), we obtain

$$Z_3(\text{gluon}) - 1 = \frac{g^2}{(4\pi)^2}\,\frac{1}{\epsilon}\left[N\frac{10}{3}\right]\,. \tag{17.47}$$

This must be supplemented by the last diagram (D), the $q\bar{q}$ contribution to the gluon vacuum polarization. Using

$$\text{tr}\,(\lambda_a\lambda_b) = 2\delta_{ab} \tag{17.48}$$

and the results for the photon vacuum polarization, Eq. (16.87), we have immediately

$$Z_3(\text{quark}) - 1 = \frac{g^2}{(4\pi)^2}\,\frac{1}{\epsilon}\left[-\frac{4}{3}N_f\right]\,, \tag{17.49}$$

where N_f is the number of quark *flavors* $(u, d, s, \ldots)$, as reviewed in Appendix D. Combining (17.49) and (17.47) gives

$$\boxed{Z_3 - 1 = \frac{g^2}{(4\pi)^2}\,\frac{1}{\epsilon}\left[N\frac{10}{3} - \frac{4}{3}N_f\right]}\,. \tag{17.50}$$

Finally, to complete the calculation, we must evaluate the *gcc* vertex corrections. These come from the two diagrams shown in Fig. 17.8. The first of these diagrams is

$$\begin{aligned} g_R \tilde{\Lambda}^\mu_{abc}(A) = & - g_R^3 f_{aef} f_{cfd} f_{bde} \\ & \times \int \frac{d^d k}{(2\pi)^d} \frac{1}{A_+ A_- A_0} \left(p + \tfrac{1}{2} q\right)_\rho (p-k)_\nu \\ & \times \left[g^{\mu\nu} \left(k - \tfrac{3}{2} q\right)^\rho - g^{\nu\rho} 2k^\mu + g^{\mu\rho} \left(k + \tfrac{3}{2} q\right)^\nu \right] , \end{aligned} \tag{17.51}$$

where $A_0 = -(p-k)^2$. We need to calculate only the divergent part, which comes from the terms proportional to two powers of k in the numerator. The shift in k which we will eventually make will not change these terms, and using

$$f_{aef} f_{cfd} f_{bde} = \tfrac{1}{2} N f_{acb} = -\tfrac{1}{2} N f_{abc} \tag{17.52}$$

(which is proved in Appendix D), we obtain immediately

$$g_R \tilde{\Lambda}^\mu_{abc}(A) \cong -g_R^3 f_{abc} \frac{N}{2} \int \frac{d^d k}{(2\pi)^d} \frac{1}{A_+ A_- A_0} \left[\left(p + \tfrac{1}{2} q\right)^\mu k^2 - k^\mu \left(p + \tfrac{1}{2} q\right) \cdot k \right] . \tag{17.53}$$

Using (17.52), the second diagram (B) gives

$$\begin{aligned} g_R \tilde{\Lambda}^\mu_{abc}(B) &= -g_R^3 f_{abc} \frac{N}{2} \int \frac{d^d k}{(2\pi)^d} \frac{1}{A_+ A_- A_0} \left(k + \tfrac{1}{2} q\right)^\mu \left(p + \tfrac{1}{2} q\right) \cdot \left(k - \tfrac{1}{2} q\right) \\ &\cong -g_R^3 f_{abc} \frac{N}{2} \int \frac{d^d k}{(2\pi)^d} \frac{1}{A_+ A_- A_0} k^\mu \left(p + \tfrac{1}{2} q\right) \cdot k , \end{aligned} \tag{17.54}$$

where we have again kept only the divergent part (terms proportional to two powers of k in the numerator). Adding diagrams A and B together gives

$$g_R \tilde{\Lambda}^\mu_{abc} = -g_R^3 f_{abc} \frac{N}{2} \int \frac{d^d k}{(2\pi)^d} \frac{1}{A_+ A_- A_0} \left(p + \tfrac{1}{2} q\right)^\mu k^2 . \tag{17.55}$$

Now, combining the denominators and shifting k, as we did in the evaluation of Eqs. (16.44)–(16.48), give

$$g_R \tilde{\Lambda}^\mu_{abc} = i g_R f_{abc} \left(p + \tfrac{1}{2} q\right)^\mu \tilde{\Lambda} , \tag{17.56}$$

where

$$\begin{aligned} \tilde{\Lambda} &= i g_R^2 N \int_0^1 \xi \, d\xi \int_{-\frac{1}{2}}^{\frac{1}{2}} d\eta \int \frac{d^d k}{(2\pi)^d} \frac{k^2}{\left(-k^2 + (1-\xi)\xi Y^2\right)^3} \\ &= \frac{g^2 N}{(4\pi)^{d/2}} \Gamma\left(\frac{\epsilon}{2}\right) \int_0^1 \xi \, d\xi \int_{-\frac{1}{2}}^{\frac{1}{2}} d\eta \left(\frac{\mu^2}{(1-\xi)\xi Y^2} \right)^{\epsilon/2} . \end{aligned} \tag{17.57}$$

In evaluating this, $k^\mu k^\nu$ was replaced by $k^2 g^{\mu\nu}/d$, and Y^2 was defined in Eq. (16.48). Recalling the analogous connection Eq. (1.128), the MS counterterm implied by (17.57) is

$$\boxed{\tilde{Z}_1 - 1 = -\frac{g^2}{(4\pi)^2}\,\frac{1}{\epsilon}\,N\,,} \tag{17.58}$$

Now, combining the final results (17.37), (17.50), and (17.58) gives

$$Z_g = \frac{\tilde{Z}_1}{\tilde{Z}_3\sqrt{Z_3}} \cong 1 - \frac{g^2}{(4\pi)^2}\,\frac{1}{\epsilon}\left[11 - \frac{2}{3}N_f\right] \quad . \tag{17.59}$$

Noting that $g_R \sim g$ in lowest order and comparing with (17.30), we see that

$$\beta_0 = \lambda = \frac{1}{(4\pi)^2}\left[11 - \frac{2}{3}N_f\right] \tag{17.60}$$

is positive provided

$$\frac{33}{2} > N_f\,, \tag{17.61}$$

which is easily satisfied by the Standard Model, which has only $N_f = 6$ flavors of quarks. Note that it is the *factor* of 11, *coming from the non-Abelian color gauge group*, which results in a positive β_0 and hence an asymptotically free theory. It has been shown that the *only theories which exhibit asymptotic freedom are non-Abelian gauge theories.*

Substituting (17.60) into (17.28) gives our final result for the running coupling constant of QCD:

$$\boxed{\frac{g_r^2(Q)}{4\pi} = \frac{2\pi}{\left[11 - \frac{2}{3}N_f\right]\log\left(Q/\Lambda\right)} \quad .} \tag{17.62}$$

In deriving this, we assumed that the quark mass is small compared to the momentum scale Q, and hence the value of N_f which should be used in (17.62) is the number of quarks with masses less that Q. The coupling also depends on the (unknown) scale parameter Λ, which can be determined experimentally, as shown in Fig. 17.4.

17.5 QCD CORRECTIONS TO THE RATIO R

We conclude this chapter with a calculation of the correction to the ratio of hadronic production in e^+e^- annihilation to $\mu^+\mu^-$ production, the ratio R introduced and discussed in Sec. 10.4. This will serve as a simple application of some of the ideas we have developed in the last two chapters.

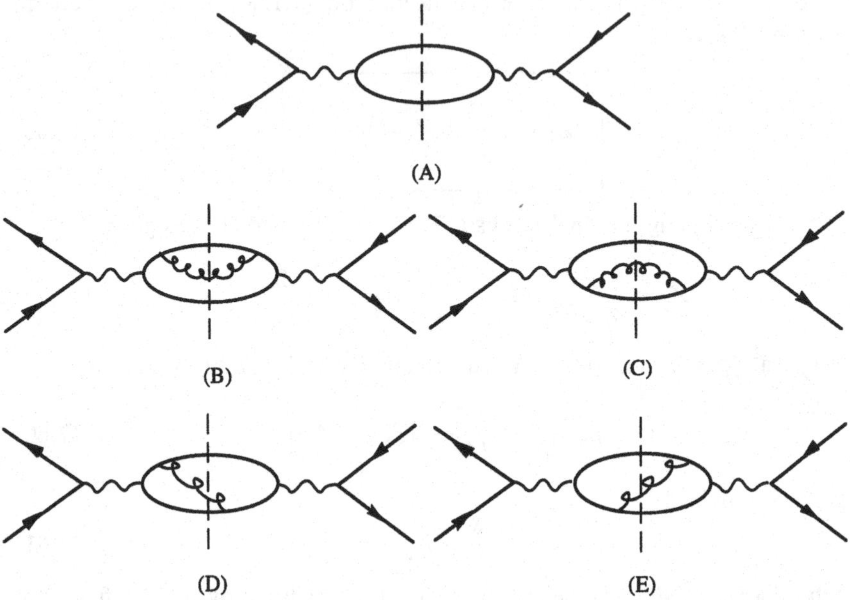

Fig. 17.9 The square of the e^+e^- annihilation amplitude to second order in the strong coupling constant g consists of 5 contributions. The leading term, of order g^0, is the square of the $e^+ + e^- \to q + \bar{q}$ amplitude and is shown in (A). The corrections of order g^2 come from the square of each $e^+ + e^- \to q + \bar{q} + g$ amplitude, diagrams (B) and (C), and the interference of the two amplitudes, diagrams (D) and (E). The vertical dashed line indicates that the intermediate particles are on their mass-shell in these diagrams.

As we discussed in Chapter 10, the total hadronic e^+e^- cross section depends on the square of the matrix elements given in Figs. 10.7 and 10.8. The square of these matrix elements are shown in Fig. 17.9. In these figures the vertical dotted line which cuts the diagram means that all particles cut by this line are on their mass-shell, and from the general unitarity principle discussed in Sec. 11.7, we know that these contributions are just the *imaginary parts* of the corresponding diagrams with vacuum polarization insertions. In fact, because the external e^+e^-

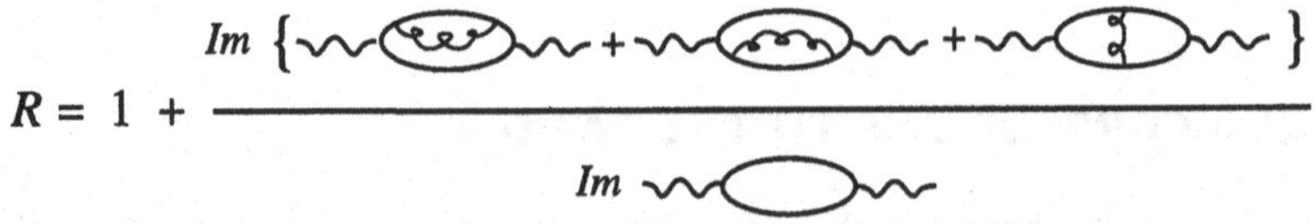

Fig. 17.10 The correction to the ratio R is equal to the ratio of the imaginary part of the fourth order self-energy to the imaginary part of the second order self-energy.

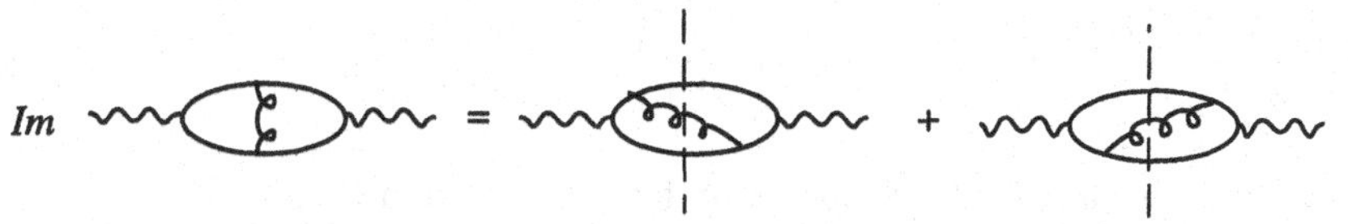

Fig. 17.11 The diagram with overlapping divergences has two cuts, giving the contributions of both Fig. 17.9D and E. Hence Fig. 17.10 is consistent with Fig. 17.9.

couplings are the same in all graphs, the ratio R is just the ratio of the imaginary parts of the vacuum polarization diagrams themselves, as illustrated in Fig. 17.10. Here there are only three fourth order graphs, while there are four contributions shown in Fig. 17.9, but the imaginary part of the overlapping diagram has *two cuts*, as illustrated in Fig. 17.11, so the counting is correct.

In Chapter 16 we calculated the leading fourth order contributions to the vacuum polarization. From Eq. (16.114), the finite part which dominates at high q^2 is

$$\Pi(q^2) = -\left(\frac{\alpha}{3\pi} + \frac{\alpha^2}{4\pi^2}\right) \log\left(-\frac{q^2}{\mu^2}\right) \; . \tag{17.63}$$

This was a purely electromagnetic result but can be modified to apply to the problem under consideration if we multiply each fourth order graph by the correct color factors, which are simply

$$\text{tr}\,\left(\tfrac{1}{2}\lambda_a\,\tfrac{1}{2}\lambda_a\right) = 4 \; . \tag{17.64}$$

The lowest order graph has a color factor of $\text{tr}\,\mathbf{1} = 3$. Hence the result (17.63) is modified for QCD to

$$\Pi^R(q^2) = -\frac{\alpha}{\pi}\left[1 + \frac{\alpha_s}{\pi}\right] \log\left(-\frac{q^2}{\mu^2}\right) , \tag{17.65}$$

where $\alpha_s = g_r^2/4\pi$ is the strong fine structure constant, which can be expected to run as shown in Eq. (17.62). The imaginary part of the log for large positive q^2 is simply

$$Im\,\log\left(-\frac{q^2}{\mu^2}\right) = \pi \; , \tag{17.66}$$

and hence for each flavor of quark the hadronic production cross section is proportional to

$$Im\,\Pi^R(q^2) = -\alpha\left[1 + \frac{\alpha_s}{\pi}\right] \; . \tag{17.67}$$

Multiplying this by the square of the charge of each quark flavor, Q_i^2, summing over all flavors produced, and dividing by the same factor for $\mu^+\mu^-$, which is simply $-\alpha$, give

$$R = \sum_i Q_i^2 \left[1 + \frac{\alpha_s}{\pi}\right] \quad . \tag{17.68}$$

We have obtained the QCD correction factor reported in Chapter 10.

This calculation was very simple because the difficult work of obtaining Eq. (16.114) had already been carried out. The calculation illustrates how QED calculations may, in some cases, be extended to QCD.

PROBLEM

17.1 In Eq. (17.41) we transformed the integral

$$\int \frac{d^dk}{(2\pi)^d} \left(\frac{1}{-k^2}\right) = -\int \frac{d^dk}{(2\pi)^d} \frac{k^2 + \frac{1}{4}q^2}{A_+A_-} \quad .$$

Evaluate the right-hand side of this equation by combining denominators and shifting k, and thereby show that it is identically zero. Use the same method to prove that

$$\int \frac{d^dk}{(2\pi)^d} \left(\frac{1}{-k^2}\right)^2 = 0 \quad .$$

[In dimensional regularization, it is often assumed that

$$\int \frac{d^dk}{(2\pi)^d} \left(\frac{1}{-k^2}\right)^\alpha = 0$$

for any $\alpha > 0$. Can you find an argument to justify this? (See Muta (1987), Sec. 2.5.5.)]

APPENDIX A

RELATIVISTIC NOTATION

In this Appendix we summarize the notation for relativistic four-vectors and Dirac matrices and spinors.

A.1 VECTORS AND TENSORS

In the natural system of units, the speed of light, c, is equal to unity, so that the space–time four-vector is denoted

$$\begin{aligned} x^\mu &= (t, \boldsymbol{r}) = (t, x, y, z) = (t, r^i) \\ x_\mu &= (t, -\boldsymbol{r}) = (t, -x, -y, -z) = (t, -r^i) \ , \end{aligned} \tag{A.1}$$

where the Greek index μ varies from 0 to 3 and the Roman indices on three-vectors vary from 1 to 3. Other frequently encountered four-vectors are

energy–momentum:	$p^\mu = (E, \boldsymbol{p})$
four-gradient:	$\partial^\mu = (\frac{\partial}{\partial t}, -\nabla)$
four-current:	$j^\mu = (\rho, \boldsymbol{j})$
four-vector potential:	$A^\mu = (A_0, \boldsymbol{A}) = (\phi, \boldsymbol{A})$

The invariant length of the four-vector is written

$$x^2 = x \cdot x = g_{\mu\nu} x^\mu x^\nu = x_\mu x^\mu = t^2 - \boldsymbol{r}^2 = t^2 - x^2 - y^2 - z^2 \ , \tag{A.2}$$

where a sum over repeated indices is always assumed and $g_{\mu\nu} = g^{\mu\nu}$ is the *metric tensor*

$$G = \{g_{\mu\nu}\} = \left(\begin{array}{c|ccc} 1 & & & \\ \hline & -1 & & \\ & & -1 & \\ & & & -1 \end{array} \right) = \left(\begin{array}{c|c} 1 & \\ \hline & -\mathbf{1} \end{array} \right) .$$

The symbol p is used to denote either the energy–momentum four-vector or the magnitude of the momentum three-vector (they can be distinguished from each other by the context in which they are used). The four-divergence of a vector field is

$$\partial_\mu V^\mu = \frac{\partial}{\partial t} V^0 + \nabla \cdot \boldsymbol{V} \ . \tag{A.3}$$

In manipulating three-vectors and tensors, we use the Kronecker δ_{ij} function and the antisymmetric symbol ϵ_{ijk}, which is antisymmetric in any pair of indices and normalized to $\epsilon_{123} = 1$. Useful identities are

$$\begin{aligned} \epsilon_{ijk}\,\epsilon_{jki'} &= 2\delta_{ii'} \\ \epsilon_{ijk}\,\epsilon_{i'j'k} &= \delta_{ii'}\delta_{jj'} - \delta_{ij'}\delta_{ji'} \ . \end{aligned} \tag{A.4}$$

When manipulating four-vectors and tensors we will sometimes need the four-dimensional antisymmetric symbol $\epsilon_{\mu\nu\lambda\delta}$, which, when *all indices are down*, is antisymmetric under the interchange of any pair of indices and normalized to $\epsilon_{0123} = 1$. Useful identities are

$$\begin{aligned} \epsilon_{\mu\nu\lambda\delta}\epsilon^{\mu\nu\lambda\delta} &= 24 \\ \epsilon_{\mu\nu\lambda\delta}\,\epsilon_{\mu'}{}^{\nu\lambda\delta} &= 6\,g_{\mu\mu'} \ . \end{aligned} \tag{A.5}$$

A.2 DIRAC MATRICES

The Dirac matrices γ^μ satisfy the following anticommutation relations:

$$\{\gamma^\mu, \gamma^\nu\} = \gamma^\mu\gamma^\nu + \gamma^\nu\gamma^\mu = 2g^{\mu\nu} \ . \tag{A.6}$$

The 4×4 representation of these matrices used in this book is

$$\gamma^0 = \begin{pmatrix} 1 & 0 \\ 0 & -1 \end{pmatrix} \qquad \gamma^i = \begin{pmatrix} 0 & \sigma_i \\ -\sigma_i & 0 \end{pmatrix} \ , \tag{A.7}$$

where each element in the above expressions is a 2×2 matrix and the $\boldsymbol{\sigma}$ are the Pauli matrices

$$\begin{gathered} \boldsymbol{\sigma} = (\sigma_1, \sigma_2, \sigma_3) \\ \sigma_1 = \begin{pmatrix} 0 & 1 \\ 1 & 0 \end{pmatrix} \qquad \sigma_2 = \begin{pmatrix} 0 & -i \\ i & 0 \end{pmatrix} \qquad \sigma_3 = \begin{pmatrix} 1 & 0 \\ 0 & -1 \end{pmatrix} \ . \end{gathered} \tag{A.8}$$

Note that

$$\begin{gathered} \gamma^{0\dagger} = \gamma^0 \qquad \gamma^{i\dagger} = -\gamma^i \\ \gamma^0\gamma^{\mu\dagger}\gamma^0 = \gamma^\mu \ . \end{gathered} \tag{A.9}$$

Other matrices which are related to or constructed from the γ-matrices, are $\beta = \gamma^0$, $\boldsymbol{\alpha} = \gamma^0\boldsymbol{\gamma}$, $\gamma^5 = i\gamma^0\gamma^1\gamma^2\gamma^3$, $\sigma^{\mu\nu} = \frac{i}{2}[\gamma^\mu, \gamma^\nu]$, the charge congugation matrix $C = -i\alpha_2$, and time inversion matrix $T = C\gamma^5$. Their explicit 2×2 block form is

$$
\begin{aligned}
\boldsymbol{\alpha} = \gamma^0\boldsymbol{\gamma} &= \begin{pmatrix} 0 & \boldsymbol{\sigma} \\ \boldsymbol{\sigma} & 0 \end{pmatrix} & \gamma^5 &= \begin{pmatrix} 0 & 1 \\ 1 & 0 \end{pmatrix} \\
\sigma^{0i} = i\alpha_i = i &\begin{pmatrix} 0 & \sigma_i \\ \sigma_i & 0 \end{pmatrix} & C &= \begin{pmatrix} 0 & -i\sigma_2 \\ -i\sigma_2 & 0 \end{pmatrix} \\
\sigma^{ij} = \gamma^5\alpha_k &= \begin{pmatrix} \sigma_k & 0 \\ 0 & \sigma_k \end{pmatrix} & T &= \begin{pmatrix} -i\sigma_2 & 0 \\ 0 & -i\sigma_2 \end{pmatrix} ,
\end{aligned}
\tag{A.10}
$$

where ijk are in cyclic order. Note that $C = -C^\dagger = -C^{-1}$, $\left(\gamma^5\right)^2 = 1$, and that

$$
\begin{aligned}
C\gamma^\mu C^{-1} &= -\gamma^{\mu\mathsf{T}} \\
\gamma^5\gamma^\mu &= -\,\gamma^\mu\gamma^5 \quad .
\end{aligned}
\tag{A.11}
$$

Using the notation

$$
\not{p} = p^\mu\gamma_\mu \quad ,
$$

the following identities hold for the γ-matrices:

$$
\begin{aligned}
\gamma^\mu\gamma_\mu &= 4 \\
\gamma^\mu \not{a}\ \gamma_\mu &= -2\not{a} \\
\gamma^\mu \not{a}\not{b}\ \gamma_\mu &= \ 4\,a\cdot b \\
\gamma^\mu \not{a}\not{b}\not{c}\ \gamma_\mu &= -2\not{c}\not{b}\not{a} \quad .
\end{aligned}
\tag{A.12}
$$

In d dimensions, these identities generalize to

$$
\begin{aligned}
\gamma^\alpha\gamma_\alpha &= d \\
\gamma^\alpha \not{a}\ \gamma_\alpha &= (2-d)\not{a} \\
\gamma^\alpha \not{a}\not{b}\ \gamma_\alpha &= 4a\cdot b-(4-d)\not{a}\not{b} \\
\gamma^\alpha \not{a}\not{b}\not{c}\ \gamma_\alpha &= -2\not{c}\not{b}\not{a} + (4-d)\not{a}\not{b}\not{c} \\
&= -(6-d)\not{c}\not{b}\not{a} + 2(4-d)\left[\not{a}\ b\cdot c - \not{b}\ c\cdot a + \not{c}\ a\cdot b\right] \quad .
\end{aligned}
\tag{A.13}
$$

Extending the definition of γ^5 to $d \neq 4$ dimensions poses special problems related to the existence of the chiral anomaly. This is not discussed in this book; for a recent modern treatment, see [KN 92].

Trace Theorems

The trace of an odd number of γ-matrices is zero. Other traces are

$$\begin{aligned} \mathrm{tr}\,\{\not{a}\not{b}\} &= 4\,a\cdot b \\ \mathrm{tr}\,\{\not{a}\not{b}\not{c}\not{d}\} &= 4\,(a\cdot b c\cdot d - a\cdot c b\cdot d + a\cdot d b\cdot c) \\ \mathrm{tr}\,\left\{\gamma^5\,\not{a}\not{b}\right\} &= 0 \\ \mathrm{tr}\,\left\{\gamma^5\,\not{a}\not{b}\not{c}\not{d}\right\} &= 4i\epsilon_{\mu\nu\lambda\rho}a^\mu b^\nu c^\lambda d^\rho \quad . \end{aligned} \tag{A.14}$$

A.3 DIRAC SPINORS

The four-component Dirac particle u and antiparticle v spinors are defined by the relations

$$\begin{aligned} u\,(\boldsymbol{p},s) &= \sqrt{E_p+m}\begin{pmatrix} 1 \\ \dfrac{\boldsymbol{\sigma}\cdot\boldsymbol{p}}{E_p+m} \end{pmatrix}\chi^{(s)} \\ v(\boldsymbol{p},s) &= \sqrt{E_p+m}\begin{pmatrix} \dfrac{\boldsymbol{\sigma}\cdot\boldsymbol{p}}{E_p+m} \\ 1 \end{pmatrix}\left[-i\sigma_2\chi^{(s)}\right] \quad , \end{aligned} \tag{A.15}$$

where $E_p = \sqrt{m^2+\boldsymbol{p}^2} = \sqrt{m^2+p^2}$ is the relativistic energy of the particle and the two-component spinors χ are

$$\chi^{(\frac{1}{2})} = \begin{pmatrix} 1 \\ 0 \end{pmatrix} \qquad \chi^{(-\frac{1}{2})} = \begin{pmatrix} 0 \\ 1 \end{pmatrix} \quad . \tag{A.16}$$

The antiparticle two-component spinor is sometimes denoted by $\eta^{(s)}$, where $\eta^{(-s)} = -i\sigma_2\chi^{(s)}$. Hence

$$\eta^{(-\frac{1}{2})} = -i\sigma_2\chi^{(+\frac{1}{2})} = \begin{pmatrix} 0 \\ +1 \end{pmatrix} \qquad \eta^{(+\frac{1}{2})} = -i\sigma_2\chi^{(-\frac{1}{2})} = \begin{pmatrix} -1 \\ 0 \end{pmatrix} \quad . \tag{A.17}$$

Note the sign (phase) of $\eta^{(+\frac{1}{2})}$. This phase convention is introduced so that the spinors are charge conjugates of one another (see below).

The adjoint Dirac spinors are

$$\begin{aligned} \bar{u}\,(\boldsymbol{p},s) &= u^\dagger\,(\boldsymbol{p},s)\;\gamma^0 \\ \bar{v}\,(\boldsymbol{p},s) &= v^\dagger\,(\boldsymbol{p},s)\;\gamma^0 \quad . \end{aligned} \tag{A.18}$$

With this definition, the spinors satisfy the following normalization and orthogonality relations:

$$\begin{aligned} \bar{u}(\boldsymbol{p},s)\,u(\boldsymbol{p},s') &= 2m\,\delta_{ss'} & \bar{v}(\boldsymbol{p},s)\,u(\boldsymbol{p},s') &= 0 \\ \bar{v}(\boldsymbol{p},s)\,v(\boldsymbol{p},s') &= -2m\,\delta_{ss'} & \bar{u}(\boldsymbol{p},s)\,v(\boldsymbol{p},s') &= 0 \ . \end{aligned} \tag{A.19}$$

The completeness relations are expressed in terms of the positive and negative energy projection operators

$$\begin{aligned} \sum_s u(\boldsymbol{p},s)\,\bar{u}(\boldsymbol{p},s) &= \not{p} + m = 2m\Lambda_+(p) \\ \sum_s v(\boldsymbol{p},s)\,\bar{v}(\boldsymbol{p},s) &= \not{p} - m = -2m\Lambda_-(p) \ . \end{aligned} \tag{A.20}$$

The u and v spinors are related by charge conjugation:

$$C\,\bar{v}^{\mathsf{T}}(\boldsymbol{p},s) = u(\boldsymbol{p},s) \qquad C\,\bar{u}^{\mathsf{T}}(\boldsymbol{p},s) = v(\boldsymbol{p},s) \ . \tag{A.21}$$

APPENDIX B

FEYNMAN RULES

In this Appendix we collect together all of the rules for the calculation of relativistic cross sections and decay rates. The rules fall into two parts. There are rules for the calculation of the cross section and decay rates from the relativistic scattering matrix, $\mathcal{M}$, and then there are rules for calculating $\mathcal{M}$ in a given theory. The former are quite general, but the latter, referred to as the Feynman rules, depend on the specific theory.

B.1 DECAY RATES AND CROSS SECTIONS

The rules for calculation of relativistic decay rates and cross sections were derived in Secs. 9.2 and 9.3.

The differential n-body decay rate, dW_n, for a particle with energy E is obtained from the following factors:

- a factor of $(2\pi)^4\delta^4(p_f - p_i)$, where p_f is the total four-momentum of the n decay products and p_i is the four-momentum of the decaying particle,
- a factor of
$$\frac{d^3k_i}{(2\pi)^3 2E_{k_i}}$$
for each *particle in the final state*, where $\boldsymbol{k}_i$ and E_{k_i} are the momentum and energy of the ith particle,
- a factor of $1/2E$ for the initial particle which is decaying, and
- the absolute square of the $\mathcal{M}$-matrix.

The differential decay rate is then

$$dW_n = (2\pi)^4\delta^4(p_f - p_i)\,\frac{1}{2E}\prod_{i=1}^{i=n}\frac{d^3k_i}{(2\pi)^3 2E_{k_i}}\;|\mathcal{M}|^2 \quad . \tag{B.1}$$

The total decay rate is obtained by integrating Eq. (B.1) over all outgoing momenta and summing over all outgoing spins:

$$W = \int \sum_{\text{spins}} dW_n \ . \tag{B.2}$$

The differential cross section for the production of n particles (elastic scattering occurs when $n = 2$ *and* the final particles are identical to the initial ones) is obtained from the following factors:

- a factor of $(2\pi)^4\delta^4(p_f - p_i)$, where p_f is the total four-momentum of the n particles in the final state and p_i is the total four-momentum of the two initial particles,
- a factor of

$$\frac{d^3k_i}{(2\pi)^3 2E_{k_i}}$$

  for each *particle in the final state*, where $\mathbf{k}_i$ and E_{k_i} are the momentum and energy of the ith particle,
- a factor of

$$\frac{1}{4EE'} \ ,$$

  where E and E' are the energies of the two particles in the initial state,
- a factor of $1/v$, where v is the *flux*, or relative velocity of the two (colinear) colliding particles, equal to

$$v = \frac{p}{E} + \frac{p'}{E'} \ ,$$

  where p and p' are the magnitudes of their momenta, and
- the absolute square of the $\mathcal{M}$-matrix.

The differential cross section is therefore

$$d\sigma = (2\pi)^4\delta^4(p_f - p_i)\,\frac{1}{4EE'\,v}\prod_{i=1}^{i=n}\frac{d^3k_i}{(2\pi)^3 2E_{k_i}}\,|\mathcal{M}|^2 \ . \tag{B.3}$$

The unpolarized cross section for scattering into some final state in the phase volume $\Delta\Omega$ is therefore obtained by integrating Eq. (B.3) over all outgoing momenta in $\Delta\Omega$, summing over all final spins, and *averaging* over initial spins:

$$\Delta\sigma = \int_{\Delta\Omega} \frac{1}{(2s+1)(2s'+1)} \sum_{\text{spins}} d\sigma \ , \tag{B.4}$$

where s and s' are the spins of the initial particles.

Finally, in calculating both decay rates and differential cross sections, for *each* set of m identical particles in the final state, the integrals over momenta must *either* be divided by $m!$ or limited to the restricted cone $\theta_1 < \theta_2 < \ldots < \theta_m$.

B.2 GENERAL RULES

The Feynman rules for calculation of the $\mathcal{M}$-matrix depend of the theory used to do the calculation. The basic rules are given first, and then the forms required for specific theories. Any diagram will either be a *tree* diagram (with no loops) or will have one or more closed *loops*.

- The diagrams consist of *lines* and *vertices*.
- Each *internal* line represents the propagation of a particular particle from one space–time point to another, and the vertices are the points in space–time where particles are created or destroyed, as described by the interaction Lagrangian of the theory.
- Label the momenta of each external particle, and use energy–momentum conservation to determine the four-momentum of each internal line. *Tree* diagrams have no closed loops, and each internal momentum can be fixed in terms of the external momenta. *Loop* diagrams have momenta which cannot be uniquely specified, and these must be integrated over. There will be one undetermined four-momentum for *each* loop.

The Feynman rules tell how to associate a number with each Feynman diagram. There are several basic rules from which the number is constructed:

Rule 0: a factor of i.

Rule 1: an operator for each vertex, the precise form of which depends on the theory and the particular particles involved.

Rule 2: a propagator for each internal line with four-momentum k, the precise form of which depends on the particle propagating. For spin zero bosons with isospin indices i, j, for fermions with Dirac indices α, β, and for photons or massive vector bosons with polarization indices μ, ν, the forms are:

$$i\Delta_{ij}(k) = \frac{-i\delta_{ij}}{m^2 - k^2 - i\epsilon} \qquad \text{spin zero}$$

$$iS_{\alpha\beta}(k) = \frac{-i(m + \not{k})_{\alpha\beta}}{m^2 - k^2 - i\epsilon} \qquad \text{spin } \tfrac{1}{2}$$

$$i\Delta_{\mu\nu}(k) = \frac{i}{-k^2 - i\epsilon}\left[g_{\mu\nu} - \frac{k_\mu k_\nu}{k^2}(1 - \alpha)\right] \qquad \text{photon or gluon}$$

$$i\Delta_{\mu\nu}(k) = \frac{i\left[g_{\mu\nu} - k_\mu k_\nu/m^2\right]}{m^2 - k^2 - i\epsilon} \qquad \text{vector boson.}$$

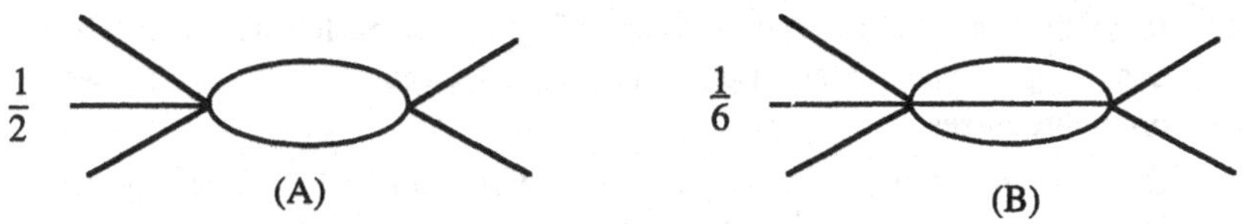

Fig. B.1 Symmetry factors for bubbles with identical neutral bosons.

In all cases, k is constrained by momentum conservation, and for the photon or gluon, α is the gauge parameter (which could be unity).

Rule 3: for fermions, assemble the incoming fermion spinors, vertex operators, propagators, and outgoing fermion spinors in *order* along each fermion line to make a well-formed matrix element. In particular:

- multiply from the *left* by $\bar{u}(\boldsymbol{p}_-, s_-)$ for each *outgoing fermion* with momentum $\boldsymbol{p}_-$ and spin s_-.
- multiply from the *right* by $u(\boldsymbol{k}_-, s_-)$ for each *incoming fermion* with momentum $\boldsymbol{k}_-$ and spin s_-.
- multiply from the *right* by $v(\boldsymbol{p}_+, s_+)$ for each *outgoing antifermion* with momentum $\boldsymbol{p}_+$ and spin s_+.
- multiply from the *left* by $\bar{v}(\boldsymbol{k}_+, s_+)$ for each *incoming antifermion* with momentum $\boldsymbol{k}_+$ and spin s_+.

for photons and vector bosons, construct well-formed vector products by saturating any free vector polarization indices μ on current operators γ^μ by:

- multiplying by ϵ^*_μ for each *outgoing particle* with polarization index μ.
- multiplying by ϵ_μ for each *incoming particle* with polarization index μ.

Rule 4: • symmetrize between identical bosons in the initial or final state.

• antisymmetrize between identical fermions in the initial or final state.

Rule 5: integrate over each internal four-momentum k not fixed by energy–momentum conservation with a weight

$$\int \frac{d^4k}{(2\pi)^4} \ .$$

Rule 6: for each closed fermion loop, a minus sign.

Rule 7: multilpy by the proper symmetry factor, which is $\frac{1}{2}$ for bubbles with *two identical neutral bosons* of the type shown in Fig. B.1A and 1/3! for bubbles with *three identical neutral bosons* of the type shown in Fig. B.1B.

Rule 8: renormalize with a factor of $\sqrt{Z_2}$ for each external fermion and $\sqrt{Z_3}$ for each external photon. (These factors are unity in all lowest order processes.)

Rule 9: for each particle with a mass which could be shifted by self-interactions, a mass counterterm $i\delta m$ is added to remove the mass shift. (This is zero to lowest order.)

B.3 SPECIAL RULES

The operators specified in **Rule 1** which are associated with each vertex depend on the theory, and their form is derived from the interaction Hamiltonian. For the following theories, the operator at the right is to be substituted for every vertex of the form shown at the left:

Symmetric ϕ^3 theory: $\mathcal{H}_{\text{int}}(x) = \frac{\lambda}{3!} : \phi^3(x): \Longrightarrow$

$-i\lambda$ 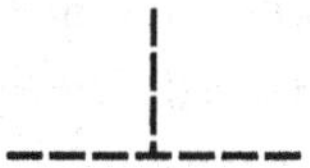 (B.5)

Charged ϕ^3 theory: $\mathcal{H}_{\text{int}}(x) = \lambda : \Phi_1^\dagger(x)\Phi_1(x) : \phi(x) \Longrightarrow$

$-i\lambda$ 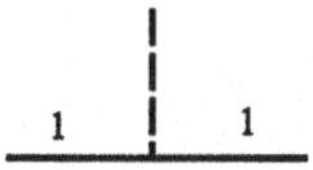

(B.6)

Pion–nucleon interaction with chiral contact term:

$$\mathcal{H}_{\text{int}}(x) = ig : \bar{\psi}(x)\, \gamma^5\, \tau_i\, \psi(x) : \phi_i(x) + i\frac{g^2}{2m} : \bar{\psi}(x)\, \delta_{ij}\, \psi(x)\, \phi_i(x)\phi_j(x) : \Longrightarrow$$

$g\gamma^5\tau_i$ 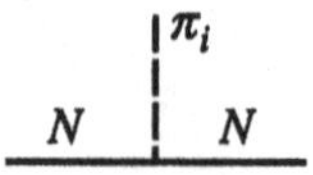

$i\frac{g^2}{m}\delta_{ij}$ 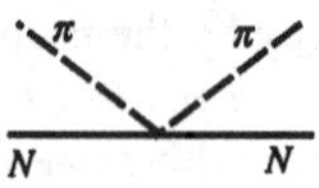

(B.7)

Spinor QED (positive charge):

$$H_I = \frac{1}{8\pi} \int \frac{d^3r\, d^3r'}{|r - r'|} J^0(r,t) J^0(r',t) - \int d^3r\, \vec{J}(r,t) \cdot \vec{A}(r,t)$$

with $J^\mu = e\!: \left[\bar{\psi}\gamma^\mu\psi\right] : \implies$

$$-ie\,\gamma_\mu \qquad \text{(B.8)}$$

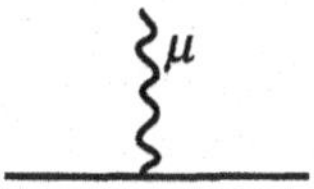

Scalar QED (positive charge):

$$-ie\,(p + p')_\mu$$

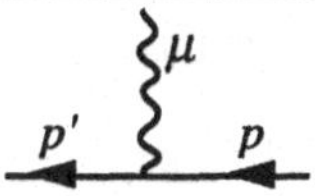

$$2ie^2\, g_{\mu\nu} \qquad \text{(B.9)}$$

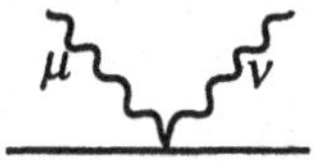

QED for a vector boson (positive charge):

$$ie\left[(p + p')_\mu\, g_{\nu\lambda} - (p' + q)_\nu\, g_{\mu\lambda} - (p - q)_\lambda\, g_{\mu\nu}\right]$$

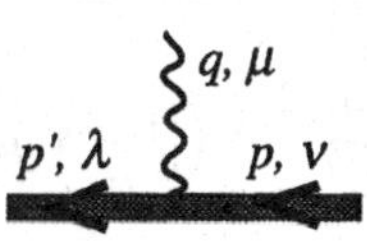

$$-ie^2\left[2g_{\mu\nu}g_{\sigma\rho} - g_{\mu\sigma}g_{\nu\rho} - g_{\mu\rho}g_{\nu\sigma}\right] \qquad \text{(B.10)}$$

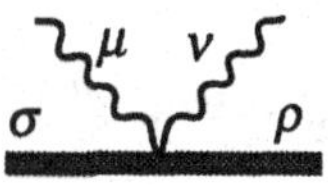

QCD:

$$-ig\,\gamma^\mu\,\tfrac{1}{2}\lambda_a$$

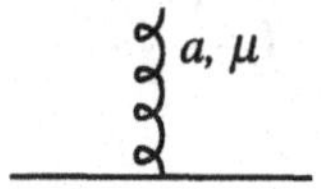

$$g\,f_{abc}\left[g_{\mu\nu}\,(q-k)_\sigma + g_{\nu\sigma}\,(r-q)_\mu + g_{\sigma\mu}\,(k-r)_\nu\right]$$

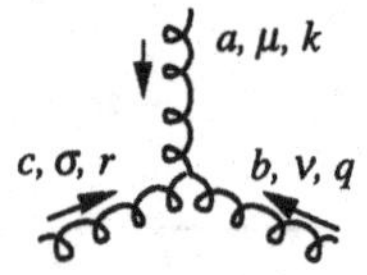

$$-\,ig^2\Big[\,f_{abe}f_{cde}\,(g_{\mu\sigma}g_{\nu\rho}-g_{\mu\rho}g_{\nu\sigma}) + f_{ace}f_{bde}\,(g_{\mu\nu}g_{\sigma\rho}-g_{\mu\rho}g_{\nu\sigma}) + f_{ade}f_{cbe}\,(g_{\mu\sigma}g_{\nu\rho}-g_{\mu\nu}g_{\rho\sigma})\Big] \qquad \text{(B.11)}$$

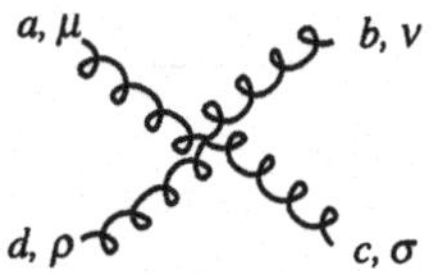

$$gf^{abc}\,(p+k)^\mu$$

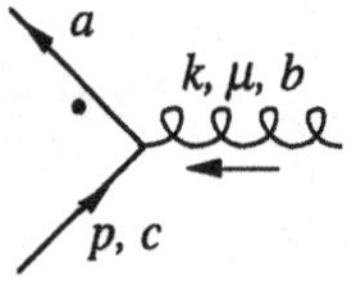

Standard Model (lepton sector): $g_W = \dfrac{g}{2\sqrt{2}\cos\theta_W}$

$$i\frac{g_W}{\sqrt{2}}\,\gamma^\mu\,\left(\left[1-4\sin^2\theta_W\right]-\gamma^5\right)$$

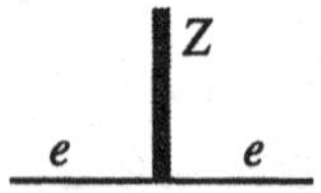

$$-i\frac{g_W}{\sqrt{2}}\,\gamma^\mu\,\left(1-\gamma^5\right) \qquad \text{(B.12)}$$

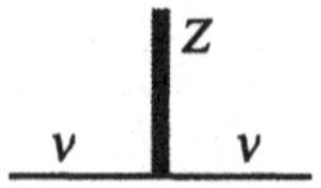

$$-ig_W\,\cos\theta_W\,\gamma^\mu\,\left(1-\gamma^5\right)$$

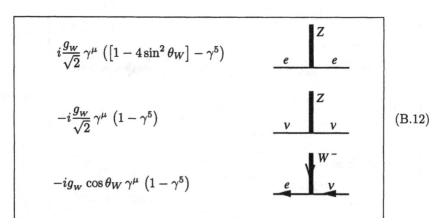

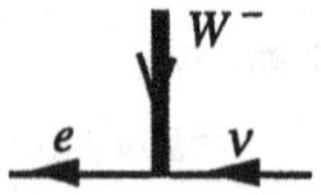

Standard Model (boson self-couplings and Higgs couplings in unitary gauge):

$$ig\cos\theta_W\left[g_{\nu\sigma}(p+p')_\mu - g_{\mu\nu}(p-q)_\sigma - g_{\mu\sigma}(q+p')_\nu\right]$$

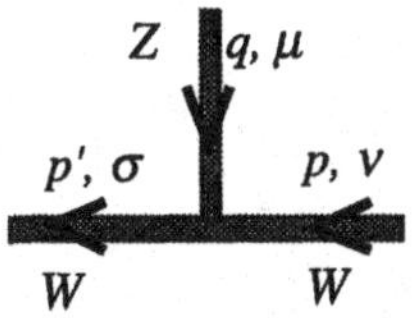

$$ig^2\left[2g_{\mu\rho}g_{\nu\sigma} - g_{\mu\nu}g_{\sigma\rho} - g_{\mu\sigma}g_{\nu\rho}\right]$$

$$-ig^2\cos^2\theta_W\left[2g_{\mu\nu}g_{\rho\sigma} - g_{\mu\sigma}g_{\nu\rho} - g_{\mu\rho}g_{\nu\sigma}\right]$$

$$-ieg\cos\theta_W\left[2g_{\mu\nu}g_{\rho\sigma} - g_{\mu\sigma}g_{\nu\rho} - g_{\mu\rho}g_{\nu\sigma}\right]$$

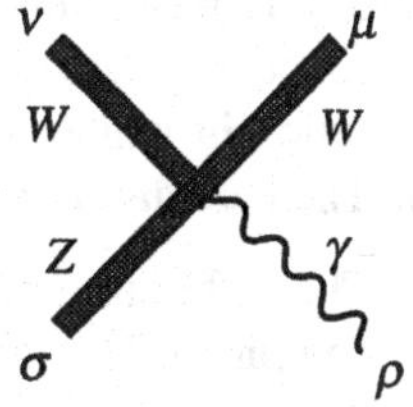

$$-i\frac{g}{2}\frac{m_e}{M_W}$$

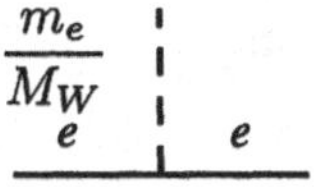

$$-i\frac{3}{2}g\frac{m_H^2}{M_W}$$

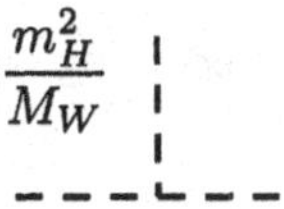

$$-i\frac{3}{4}g^2\frac{m_H^2}{M_W^2}$$

$$igM_W\, g_{\mu\nu}$$

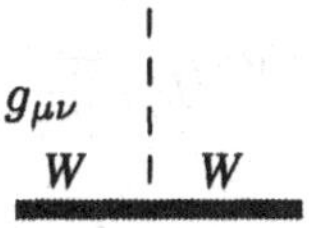

$$i\frac{gM_Z}{\cos\theta_W}g_{\mu\nu}$$

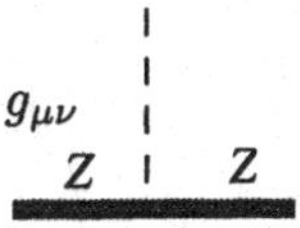

$$i\frac{g^2}{2}g_{\mu\nu}$$

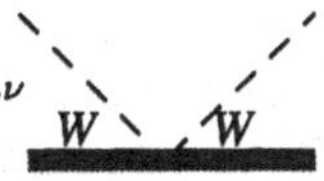

$$i\frac{g^2}{2\cos^2\theta_W}g_{\mu\nu}$$

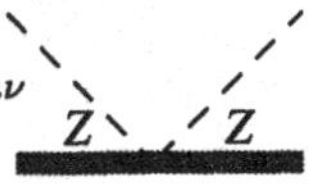

APPENDIX C

EVALUATION OF LOOP DIAGRAMS

In this Appendix we collect together all of the formulae for the evaluation of Feynman loop diagrams.

Using dimensional regularization (explained in Sec. 11.6), the general one-loop Feynman integral is of the following form:

$$I = \int \frac{d^d k}{(2\pi)^d} \frac{N}{A_1^{\alpha_1} A_2^{\alpha_2} \cdots A_n^{\alpha_n}} , \qquad \text{(C.1)}$$

where d is the number of space–time dimensions (not necessarily integer) and the numbers α_i are integers unless the above integral results from a multi-loop calculation in which some of the loops have already been evaluated (see Chapter 16). The calculation of this Feynman integral is carried out in two steps.

- The different denominators are combined into a single denominator and the combined denominator reduced to standard form by translating, or shifting, the internal momenta.
- The integral is then evaluated using an integral identity.

The first step makes use of identities of the form

$$\begin{aligned} \frac{1}{A_1 A_2} &= \int_0^1 dz \frac{1}{[A_1 z + A_2(1-z)]^2} = \int_0^1 dz \frac{1}{[D(z)]^2} \\ \frac{1}{A_1 A_2 A_3} &= 2\int_0^1 dz_1 \int_0^{1-z_1} dz_2 \frac{1}{[A_1 z_1 + A_2 z_2 + A_3(1-z_1-z_2)]^3} \\ &= 2\int_0^1 dz_1 \int_0^{1-z_1} dz_2 \frac{1}{[D(z_1, z_2)]^3} . \end{aligned} \qquad \text{(C.2)}$$

The integration variables z_i are referred to as Feynman parameters. A generalization of these identities which will work for any case was proved in Chapter 16 [following Eq. (16.36)]:

$$\frac{1}{A_1^{\alpha_1} A_2^{\alpha_2} \cdots A_n^{\alpha_n}} = \frac{\Gamma(\alpha_1 + \alpha_2 + \cdots + \alpha_n)}{\Gamma(\alpha_1)\Gamma(\alpha_2)\cdots\Gamma(\alpha_n)} \int_0^1 dx_1\, dx_2 \cdots dx_n\, \delta\Big(1 - \sum x_i\Big) \times \frac{x_1^{\alpha_1-1} x_2^{\alpha_2-1} \cdots x_n^{\alpha_n-1}}{[A_1x_1 + A_2x_2 + \cdots + A_nx_n]^{\alpha_1+\alpha_2+\cdots+\alpha_n}} , \tag{C.3}$$

where $\Gamma(\alpha)$ is the generalization of $(\alpha - 1)!$ to non-integer numbers. It has the properties

$$\begin{aligned} \Gamma(\alpha+1) &= \alpha\Gamma(\alpha) \\ \Gamma(2) &= \Gamma(1) = 1 \\ \Gamma\left(1+\frac{\epsilon}{2}\right) &= 1 - \frac{\epsilon}{2}\gamma + \mathcal{O}(\epsilon^2) , \end{aligned} \tag{C.4}$$

where $\gamma = 0.5772\cdots$ is Euler's constant and the last relation holds when ϵ is infinitesimal. We will also have use for the B-function, defined as follows:

$$B(\alpha, \beta) = \int_0^1 dx\, x^{\alpha-1}(1-x)^{\beta-1} = \frac{\Gamma(\alpha)\Gamma(\beta)}{\Gamma(\alpha+\beta)} . \tag{C.5}$$

To complete the reduction to standard form (the first step), note that the combined denominator D always has the form

$$D = k^2 + 2k\cdot Q + B^2 , \tag{C.6}$$

where k is the internal loop momentum and Q is a vector function of the external momenta and the Feynman parameters. Thus the square of the denominator can always be completed by shifting $k = k' - Q$, which gives

$$D \to D' = k'^2 + B^2 - Q^2 . \tag{C.7}$$

This shift must also be carried out in the numerator N, which assumes the general form

$$N = N_0 + k'_\mu N_1^\mu + k'_\mu k'_\nu N_2^{\mu\nu} + k'_\mu k'_\nu k'_\sigma N_3^{\mu\nu\sigma} + k'_\mu k'_\nu k'_\sigma k'_\lambda N_4^{\mu\nu\sigma\lambda} + \cdots , \tag{C.8}$$

where the N_i are tensors which do not depend on k'. Since the denominator is even in k' (in fact, it depends on k'^2 only), all of the odd terms reduce to zero and the even ones can be simplified using the following identities:

$$\int \frac{d^dk}{(2\pi)^d} \frac{k^\mu k^\nu}{D(k^2)} = \frac{g^{\mu\nu}}{d} \int \frac{d^dk}{(2\pi)^d} \frac{k^2}{D(k^2)} \tag{C.9}$$

and

$$\int \frac{d^dk}{(2\pi)^d} \frac{k^\mu k^\nu k^\lambda k^\delta}{D(k^2)} = \frac{[g^{\mu\nu}g^{\lambda\delta} + g^{\mu\lambda}g^{\nu\delta} + g^{\mu\delta}g^{\nu\lambda}]}{d(d+2)} \int \frac{d^dk}{(2\pi)^d} \frac{k^4}{D(k^2)} , \tag{C.10}$$

where d is the number of space–time dimensions. If the integrals are finite, we may set $d = 4$ immediately.

After the first step has been carried out, the integral will have the standard form

$$I = \int \frac{d^d k}{(2\pi)^d} \frac{(k^2)^n}{[C^2 - k^2 - i\epsilon]^\alpha} , \tag{C.11}$$

where n is an integer. These integrals can be reduced to a sum of integrals of the form

$$\int \frac{d^d k}{(2\pi)^d} \frac{1}{[C^2 - k^2 - i\epsilon]^\alpha} = \frac{i}{(4\pi)^{d/2}} \frac{\Gamma\left(\alpha - \frac{d}{2}\right)}{\Gamma(\alpha)} \left(\frac{1}{C^2}\right)^{\alpha - d/2} . \tag{C.12}$$

This identity was proved in Sec. 11.6. A convenient combination of Eqs. (C.9) and (C.12) is

$$\int \frac{d^d k}{(2\pi)^d} \frac{k^\mu k^\nu}{[C^2 - k^2 - i\epsilon]^\alpha} = -\frac{i g^{\mu\nu}}{2(4\pi)^{d/2}} \frac{\Gamma\left(\alpha - 1 - \frac{d}{2}\right)}{\Gamma(\alpha)} \left(\frac{1}{C^2}\right)^{\alpha - 1 - d/2} . \tag{C.13}$$

APPENDIX D

QUARKS, LEPTONS, AND ALL THAT

In this Appendix we summarize the current model of the fundamental forces and particles of nature and compute some color factors needed in Chapter 17.

D.1 FUNDAMENTAL PARTICLES AND FORCES

There are three kinds of fundamental particles. These are the *fermions*, which have spin $\frac{1}{2}$, the *gauge bosons*, which have spin 1 and are the carriers of the fundamental forces, and a scalar particle called the *Higgs*. As of the spring of 1993 (the time this book was finished), the fundamental forces were believed to be three in number: the strong forces mediated by gluons, the electroweak forces mediated by the photon and the three intermediate vector bosons ($W^{\pm}$, and Z), and gravity, which is not discussed at all in this book. (There are expectations that the strong and electroweak forces, and possibly gravity, are really a single unified force.)

The fundamental particles are summarized in Table D.1. The only particles in this table which have not been observed to date are the Higgs and the top quark t. Discovery of these two particles is one of the missions of the Superconducting Super Collider (SSC). The Higgs is an essential part of the verification of the standard electroweak model, which predicts its existence (see Sec. 15.4), and the top quark is expected on the basis of recent measurements which suggest that there are *only three* generations (denoted by G in the table) of particles.

The fermions can be classified according to which forces they experience. The *leptons* interact only through the electroweak (EW) forces described by the Standard Model, while the *quarks* also experience the strong forces described by QCD. The leptons in the Standard Model consist of the three neutral (and maybe massless) *neutrinos* and three charged leptons which include the electron, muon, and τ. All of these particles have distinct antiparticles, for a total of $2 \times 6 = 12$ particles. Because these particles do not see the strong forces, which are sensitive to an intrinsic property referred to as *color*, they can be said to be colorless. The quarks are colored; each comes in three colors corresponding to the three degrees

Table D.1 The fundamental particles of nature.

- **Fermions: (spin 1/2, charge q)**

G	quarks (QCD and EW)					leptons (EW only)		
				q	mass		q	mass
1	u_R	u_G	u_B	2/3	$\sim 5\,\text{MeV}$	e	-1	$\sim 0.5\,\text{MeV}$
	d_R	d_G	d_B	$-1/3$	$\sim 7\,\text{MeV}$	ν_e	0	$< 18\,\text{eV}$
2	c_R	c_G	c_B	2/3	~ 1500	μ	-1	$\sim 105\,\text{MeV}$
	s_R	s_G	s_B	$-1/3$	~ 200	ν_μ	0	$< 0.25\,\text{MeV}$
3	t_R	t_G	t_B	2/3	?	τ	-1	$\sim 1780\,\text{MeV}$
	b_R	b_G	b_B	$-1/3$	~ 4500	ν_τ	0	$< 35\,\text{MeV}$

- **Gauge Bosons (spin 1)**

  g — gluons (eight colors) — QCD

  $\gamma, W^\pm, Z^0$ — photon and intermediate vector bosons — EW

- **Scalar (spin 0)**

  H — Higgs — EW

of freedom of the $SU(3)$ gauge group. The quarks also experience the standard electroweak interactions, which (except for electromagnetic) are not described in this book [see Cheng and Li (1984)].

It has been found that the strong color "charge" goes to zero as the momentum flowing through the interaction vertex becomes very large (a property known as *asymptotic freedom*; see Chapter 17). This means that perturbation theory can be used to study high energy interactions. Conversely, at low momenta (large distances) the strong forces become very strong, and colored particles (quarks and gluons) are therefore *confined* to colorless clusters. All observed strongly interacting particles, or *hadrons*, are believed to be composites of quarks and gluons.

There are two major types of hadrons: *mesons*, which are bosons with integral spin and believed to be colorless composites of gluons and quark–antiquark pairs, and *baryons*, which are fermions with half-integral spin and are colorless composites of three quarks (one of each color, denoted by q_R, q_B, or q_G for red, green, or blue) in a sea of gluons and $q\bar{q}$ pairs. To produce a colorless composite, the color wave function must be fully antisymmetric (the only scalar which can be formed from three vectors), so that the remaining part of their wave function must be symmetric (because quarks are fermions with a fully antisymmetric wave function). This explains the existence of states like the charged Δ^{++}, which is a spin $\frac{3}{2}$ composite of three "up" (u) quarks and has a fully symmetric spatial and spin wave function.

The structure and spectrum of composite hadrons, and the forces between them ("nuclear" forces), are yet to be understood in terms of the fundamental interactions between quarks and gluons. One of the missions of the Continuous Electron Beam Accelerator Facility (CEBAF) is to carry out the experimental studies which will be needed for such an understanding.

D.2 COMPUTATION OF COLOR FACTORS

In this section we will summarize the properties of the $SU(3)$ color group and obtain the color factors needed in Chapter 17. $SU(3)$ is the group of unitary transformations with unit determinant and is discussed in many references [see, for example, Carruthers (1966)]. The transformations of this continuous group, given in Sec. 13.3, are specified by a set of continuous real parameters ϵ_a where

$$\mathbf{U} = e^{-ig\frac{1}{2}\lambda_a \epsilon_a(x)} \quad , \tag{D.1}$$

and the matrices $\frac{1}{2}\lambda_a$ are the *generators* of the group. The transformations $\mathbf{U}$ will be unitary only if the generators are Hermitian matrices, and the requirement that the $\det|\mathbf{U}| = 1$ leads, from an examination of the infinitesimal transformations, to the requirement that the λ_a be traceless. There are precisely eight independent Hermitian 3×3 traceless matrices $(9 - 1)$, and hence the sum in (D.1) will run from 1 to 8. The standard choice for the eight independent λ_a are

$$\lambda_1 = \begin{pmatrix} 0 & 1 & 0 \\ 1 & 0 & 0 \\ 0 & 0 & 0 \end{pmatrix} \qquad \lambda_2 = \begin{pmatrix} 0 & -i & 0 \\ i & 0 & 0 \\ 0 & 0 & 0 \end{pmatrix}$$

$$\lambda_4 = \begin{pmatrix} 0 & 0 & 1 \\ 0 & 0 & 0 \\ 1 & 0 & 0 \end{pmatrix} \qquad \lambda_5 = \begin{pmatrix} 0 & 0 & -i \\ 0 & 0 & 0 \\ i & 0 & 0 \end{pmatrix}$$

$$\lambda_6 = \begin{pmatrix} 0 & 0 & 0 \\ 0 & 0 & 1 \\ 0 & 1 & 0 \end{pmatrix} \qquad \lambda_7 = \begin{pmatrix} 0 & 0 & 0 \\ 0 & 0 & -i \\ 0 & i & 0 \end{pmatrix}$$

$$\lambda_3 = \begin{pmatrix} 1 & 0 & 0 \\ 0 & -1 & 0 \\ 0 & 0 & 0 \end{pmatrix} \qquad \lambda_8 = \frac{1}{\sqrt{3}} \begin{pmatrix} 1 & 0 & 0 \\ 0 & 1 & 0 \\ 0 & 0 & -2 \end{pmatrix} \quad .$$

Note that these matrices satisfy the relations

$$\mathrm{tr}\,(\lambda_a\lambda_b) = 2\delta_{ab} \tag{D.2a}$$

$$\sum_a \lambda_a\lambda_a = \frac{16}{3} \ . \tag{D.2b}$$

The space of all traceless 3×3 Hermitian matrices, **A**, is an eight-dimensional linear vector space spanned by the eight matrices λ_a. A scalar product can be defined by $\mathbf{A}\cdot\mathbf{B} = \mathrm{tr}\,\{\mathbf{A}\,\mathbf{B}\}$, in which case Eq. (D.2a) shows that the basis "vectors" λ are orthogonal and normalized to 2. Since the commutator $-i\,[\frac{1}{2}\lambda_a, \frac{1}{2}\lambda_b]$ is also Hermitian and traceless, it can be expanded in terms of the complete set λ_a. The expansion is written

$$[\tfrac{1}{2}\lambda_a, \tfrac{1}{2}\lambda_b] = i f_{abc}\,\tfrac{1}{2}\lambda_c \ . \tag{D.3}$$

The f_{abc} are the *structure constants* of the group and, from the above definition, are antisymmetric in the first two indices $f_{abc} = -f_{bac}$. They are also antisymmetric in the last two indices (and hence are fully antisymmetric in all indices). This can be shown from the orthogonality property (D.2) and the commutation relations (D.3):

$$\begin{aligned} 4i f_{abe} &= 2i\,\mathrm{tr}\,\{f_{abc}\lambda_c\lambda_e\} \\ &= \mathrm{tr}\,\{\lambda_e\lambda_a\lambda_b - \lambda_e\lambda_b\lambda_a\} \\ &= \mathrm{tr}\,\{\lambda_b\lambda_e\lambda_a - \lambda_b\lambda_a\lambda_e\} = 4i f_{eab} \\ &= -4i f_{aeb} \ . \end{aligned}$$

The explicit values of f_{abc} can be found by direct computation:

$$\begin{aligned} f_{123} &= 1 & f_{246} &= \tfrac{1}{2} & f_{367} &= -\tfrac{1}{2} \\ f_{147} &= \tfrac{1}{2} & f_{257} &= \tfrac{1}{2} & f_{458} &= \tfrac{\sqrt{3}}{2} \\ f_{156} &= -\tfrac{1}{2} & f_{345} &= \tfrac{1}{2} & f_{678} &= \tfrac{\sqrt{3}}{2} \ . \end{aligned}$$

The structure constants f_{abc} can be used to construct another representation of $SU(3)$. Define the 8×8 matrices

$$\mathbf{F}_a = (F_a)_{bc} = -i f_{abc} \ . \tag{D.4}$$

These matrices are clearly Hermitian and traceless, and we will show that they satisfy the commutation relations

$$[\mathbf{F}_a, \mathbf{F}_b] = i f_{abc}\mathbf{F}_c \ . \tag{D.5}$$

Therefore, they have the correct algebraic properties to represent the group. This representation is referred to as the *regular* representation, and because it is eight dimensional, it is the correct representation for color transformations of the gluons.

To prove (D.5), we use the Jacobi identy

$$[\lambda_a, [\lambda_b, \lambda_c]] + [\lambda_b, [\lambda_c, \lambda_a]] + [\lambda_c, [\lambda_a, \lambda_b]] = 0 \quad , \tag{D.6}$$

which holds for *any* three matrices (write out the terms and see that they cancel identically). Applying the identity (D.6) to the λ-matrices and using the commutation relations give

$$\begin{aligned} &[\lambda_a, f_{bce}\lambda_e] + [\lambda_b, f_{cae}\lambda_e] + [\lambda_c, f_{abe}\lambda_e] = 0 \\ &\Rightarrow f_{bce}f_{aed} + f_{cae}f_{bed} + f_{abe}f_{ced} = 0 \quad . \end{aligned} \tag{D.7}$$

This identity is easy to recall and use if we note that the indices e and d are fixed and that a, b, c are cyclically permuted in the three terms. Multiplying by $(-i)^2$ and substituting (D.4) when possible give

$$(F_b)_{ce}(F_a)_{ed} - (F_a)_{ce}(F_b)_{ed} + i f_{abe}(F_e)_{cd} = 0 \quad .$$

But this is just the (c, d) element of the matrix equation (D.5), proving that (D.5) is indeed satisfied.

In Chapter 17, we will need the relations

$$f_{acd}f_{bcd} = N\,\delta_{ab} \tag{D.8a}$$

$$f_{ade}f_{bef}f_{cfd} = \tfrac{1}{2}N f_{abc} \quad , \tag{D.8b}$$

where $N = 3$. These relations can be proved with the help of the Jacobi identity.

To prove (D.8a), begin by showing that the 8×8 matrix $F^2_{cd} = \text{tr}\,\{\mathbf{F}_c\mathbf{F}_d\}$ commutes with all matrices $\mathbf{F}_a$. This is shown by direct evaluation of the commutator

$$\begin{aligned} \left[\mathbf{F}^2, \mathbf{F}_a\right]_{cd} &= F_{cgf}F_{efg}F_{aed} - F_{ace}F_{efg}F_{dgf} \\ &= -F_{cgf}\left(F_{gae}F_{fed} + F_{afe}F_{ged}\right) + \left(F_{cge}F_{aef} + F_{gae}F_{cef}\right)F_{dgf} \\ &= \text{tr}\,\{-\mathbf{F}_c\mathbf{F}_d\mathbf{F}_a + \mathbf{F}_c\mathbf{F}_a\mathbf{F}_d - \mathbf{F}_c\mathbf{F}_a\mathbf{F}_d + \mathbf{F}_a\mathbf{F}_c\mathbf{F}_d\} = 0 \quad , \end{aligned}$$

where the Jacobi identity was used in going to the second line and in the third line the sums were written as traces, making use of the antisymmetry of the f's. Since $\mathbf{F}^2$ commutes with all $\mathbf{F}_a$, it also commutes with all the $SU(3)$ matrices, and by Schur's Lemma it must be a multiple of the identity. Hence

$$\begin{aligned} \text{tr}\,\{\mathbf{F}_a\mathbf{F}_b\} = N\delta_{ab} &= -f_{acd}f_{bdc} \\ &= f_{acd}f_{bcd} \quad . \end{aligned}$$

This establishes the general structure of Eq. (D.8a); it remains only to find N. This is easily done using the explicit numbers for the f_{abc}. Taking the trace of both sides of (D.8a) gives

$$8N = f_{acd}f_{acd} = 6\left(1 + 6\left(\frac{1}{2}\right)^2 + 2\left(\frac{\sqrt{3}}{2}\right)^2\right) = 24$$

which gives

$$N = 3 \ .$$

To evaluate the second identity (D.8b), we first prove that

$$\begin{aligned}
&\mathrm{tr}\,\{\mathbf{F}_a\,[\mathbf{F}_b\mathbf{F}_c + \mathbf{F}_c\mathbf{F}_b]\} \\
&\quad = F_{ade}F_{bef}F_{cfd} + \underbrace{F_{ade}F_{cef}F_{bfd}}_{F_{aed}F_{bdf}F_{cef}} \\
&\quad = -\left(F_{dbe}F_{aef} + F_{bae}F_{def}\right)F_{cfd} - \left(F_{ebd}F_{adf} + F_{bad}F_{edf}\right)F_{cef} \\
&\quad = \mathrm{tr}\,\{\mathbf{F}_b\mathbf{F}_a\mathbf{F}_c\} + F_{bae}\,\mathrm{tr}\,\{\mathbf{F}_e\mathbf{F}_c\} - \mathrm{tr}\,\{\mathbf{F}_b\mathbf{F}_a\mathbf{F}_c\} - F_{bad}\,\mathrm{tr}\,\{\mathbf{F}_d\mathbf{F}_c\} \\
&\quad = 0 \ ,
\end{aligned}$$

where the Jacobi identity was again used in the second step and in the last step we used (D.8a). Armed with this result it is an easy matter to prove (D.8b) by multiplying both sides of the commutation relation (D.5) by $\mathbf{F}_d$ and taking the trace:

$$\begin{aligned}
\mathrm{tr}\,\{[\mathbf{F}_a, \mathbf{F}_b]\ \mathbf{F}_d\} &= iNf_{abd} \\
&= \mathrm{tr}\,\{2\mathbf{F}_a\mathbf{F}_b\mathbf{F}_d - (\mathbf{F}_a\mathbf{F}_b + \mathbf{F}_b\mathbf{F}_a)\,\mathbf{F}_d\} \\
&= \mathrm{tr}\,\{2\mathbf{F}_a\mathbf{F}_b\mathbf{F}_d\} \\
&= 2if_{aef}f_{bfg}f_{dge} \ .
\end{aligned}$$

Dividing by 2 gives the result (D.8b).

Because the Jacobi identity holds for all matrices, these results can be generalized to higher dimensional representations.

REFERENCES

BOOKS

Classics

Bjorken, J. D. and S. D. Drell (1964), *Relativistic Quantum Mechanics* and *Relativistic Quantum Fields*, McGraw-Hill, New York.

Bogoliubov, N. N. and D. V. Shirkov (1959), *Introduction to the Theory of Quantized Fields*, Interscience, New York.

Mandl, A. (1966), *Introduction to Quantum Field Theory*, Wiley-Interscience, New York.

Sakurai, J. J. (1967), *Advanced Quantum Mechanics*, Addison-Wesley, Reading, MA.

New Books

Aitchison, I. J. R. and A. J. G. Hey (1982), *Gauge Theories in Particle Physics*, Adam Hilger, Philadelphia.

Cheng, T-P. and L-F. Li (1984), *Gauge Theory of Elementary Particle Physics*, Oxford University Press, Oxford.

Itzykson, C. and J-B. Zuber (1980) *Quantum Field Theory*, McGraw-Hill, New York.

Muta, T. (1987), *Foundations of Quantum Chromodynamics*, World Scientific, Singapore.

Ramond, P. (1981), *Field Theory, A Modern Primer*, Benjamin, Reading, MA.

Ryder, L. H. (1985), *Quantum Field Theory*, Cambridge University Press, Cambridge.

A Few Elementary Texts

Feynman, R. P. (1961), *Quantum Electrodynamics*, Benjamin, Reading, MA.

Gottfried, K. (1966), *Quantum Mechanics, Vol. I: Fundamentals*, Benjamin, Reading, MA.

Merzbacher, E. (1970), *Quantum Mechanics*, Wiley, New York.

Sakurai, J. J. (1985), *Modern Quantum Mechanics*, Addison-Wesley, Reading, MA.

Schiff, L. I. (1968), *Quantum Mechanics*, McGraw-Hill, New York.

Winter, R. G. (1986), *Quantum Physics*, Faculty Publishing, Davis, CA.

Related or Specialized Topics

Abramowitz, M. and I. A. Stegun, Eds. (1964), *Handbook of Mathematical Functions*, U.S. National Bureau of Standards, U.S. Government Printing Office, Washington, D.C.

Barton, G. (1965), *Introduction to Dispersion Techniques in Field Theory*, Benjamin, Reading, MA.

Bethe, H. A. and E. E. Salpeter (1957), *Quantum Mechanics of One and Two Electron Atoms*, Academic Press, New York.

Bhaduri, R. K. (1988), *Models of the Nucleon*, Addison-Wesley, Reading, MA.

Carruthers, P. A. (1966), *Introduction to Unitary Symmetry*, Wiley-Interscience, New York.

Das, T. P. (1973), *Relativistic Quantum Mechanics of Electrons*, Harper and Row, New York.

Donoghue, J. F., E. Golowich, and B. R. Holstein (1992), *Dynamics of the Standard Model*, Cambridge University Press, Cambridge.

Fetter, A. L. and J. D. Walecka (1971), *Quantum Theory of Many-Particle Systems*, McGraw-Hill, New York.

Feynman, R. P. and A. R. Hibbs (1965), *Quantum Mechanics and Path Integrals*, McGraw-Hill, New York.

Goldstein, H. (1977), *Classical Mechanics*, Addison-Wesley, Reading, MA.

Landau, R. H. (1990), *Quantum Mechanics II*, Wiley-Interscience, New York.

Negele, J. W. and H. Orland (1988), *Quantum Many Particle Systems*, Addison-Wesley, Reading, MA.

Rose, M. E. (1957), *Elementary Theory of Angular Momentum*, Wiley, New York.

Streater, R. F. and A. Wightman (1964), *PCT, Spin and Statistics, and All That*, Benjamin, New York.

Todorov, I. T. (1971), *Analytic Properties of Feynman Diagrams in Quantum Field Theory*, Oxford University Press, Oxford.

ARTICLES

[AL 73] Abers, E. S. and B. W. Lee, *Physics Reports* **9**, 1 (1973).

[BB 78] Bardeen, W. A., A. J. Buras, D. W. Duke, and T. Muta, *Phys. Rev. D* **18**, 3998 (1978).

[Be 47] Bethe, H. A., *Phys. Rev.* **72**, 339 (1947).

[Be 92] Bethke, S. "Tests of QCD", in Proc. of the XXVI International Conference on High Energy Physics, Dallas, Texas, 6-12 August 1992; and Heidelberg preprint HD-PY 92/13.

[BG 87] Bhatt, G. C. and H. Grotch, *Ann. of Phys.* **178**, 1 (1987).

[Bl 50] Bleuler, K., *Helv. Phys. Acta* **23**, 567 (1950).

[BL 50] Bethe, H. A. and C. Longmire, *Phys. Rev.* **77**, 647 (1950).

[BP 57] Bogoluibov, N. N. and O. S. Parasiuk, *Acta Math.* **97**, 277 (1957).

[BS 66] Blankenbecler, R. and R. Sugar, *Phys. Rev.* **142**, 1051 (1966).

[Ca 70] Callan, C. G., *Phys. Rev. D* **2**, 154 (1970).

[CG 35] Chadwick, J. and M. Goldhaber, *Nature* **134**, 237 (1935).

[CJ 74] Chodos, A., R. L. Jaffe, K. Johnson, C. B. Thorn, and V. F. Weisskopf, *Phys. Rev. D* **9**, 3471 (1974); Chodos, A., R. L. Jaffe, K. Johnson, and C. B. Thorn, *Phys. Rev. D* **10**, 2599 (1974).

[Co 89] Coester, F. (1989), in *Quarks, Mesons and Nuclei II: Electroweak Interactions*, W. -Y. P. Hwang and E. M. Henley, Eds., World Scientific, Singapore, p. 124.

[Cu 54] Cutkosky, R. E., *Phys. Rev.* **96**, 1135 (1954).

[Cu 61] Cutkosky, R. E., *Rev. Mod. Phys.* **33**, 448 (1961).

[De 85] De Pascale, M. P., *et. al.*, *Phys. Rev. C* **32**, 1830 (1985).

[Di 28] Dirac, P. A. M., *Proc. Roy. Soc. (London)* **A117**, 610 (1928); **A118**, 351 (1928).

[DW 69] Dashen, R., *Phys. Rev.* **183**, 1245 (1969); R. Dashen and M. Weinstein, ibid, 1291.

[EK 91] Eides, M. I., S. G. Karshenboim, and V. A. Shelyuto, *Ann. Phys.* **205**, 231 (1991); 291 (1991).

[Fe 48] Feynman, R. P., *Rev. Mod. Phys.* **20**, 367 (1948).

[Fe 49] Feynman, R. P., *Phys. Rev.* **76**, 749 (1949); 769 (1949).

[FP 67] Faddeev, L. D. and V. N. Popov, *Phys. Lett.* **25B**, 29 (1967).

[Fr 91] Frank, M., Ph.D. Thesis, Kent State University (1991).

[FT 75] Fleischer, J. and J. A. Tjon, *Nucl. Phys.* **B84**, 375 (1975); *Phys. Rev. D* **15**, 2537 (1977); **21**, 87 (1980).

[FV 58] Feshbach, H. and F. Villars, *Rev. Mod. Phys.* **30**, 24 (1958).

[FW 50] Foldy, L. L. and S. A. Wouthuysen, *Phys. Rev.* **78**, 29 (1950).

[GL 60] Gell-Mann, M. and M. Lévy, *Nuovo Cim.* **16**, 705 (1960).

[Go 61] Goldstone, J., *Nuovo Cim.* **19**, 154 (1961).

[Gr 69] Gross, F., *Phys. Rev.* **186**, 1448 (1969).

[Gr 82] Gross, F., *Phys. Rev. C* **26**, 2203 (1982).

[Gu 50] Gupta, S. N., *Proc. Phys. Soc. (London)* **A63**, 681 (1950).

[GV 92] Gross, F., J. W. Van Orden, and K. Holinde, *Phys. Rev. C* **45**, 2095 (1992).

[GW 73] Gross, D. J. and F. Wilczek, *Phys. Rev. Lett.* **30**, 1343 (1973); *Phys. Rev. D* **8**, 3633 (1973).

[He 66] Hepp, K., *Comm. Math. Physics* **2**, 301 (1966).

[Ho 73] 't Hooft, G., *Nucl. Phys.* **B61**, 445 (1973).

[JW 59] Jacob, M. and G. C. Wick, *Ann. of Physics* **7**, 404 (1957).

[KL 90] Kinoshita, T. and W. B. Lindquist, *Phys. Rev. D* **42**, 636 (1990).

[KN 92] Krewald, S. and K. Nakayama, *Ann. of Physics* **216**, 201 (1992).

[KP 90] Keister, B. D. and W. N. Polyzou, *Adv. Nucl. Phys.* **20**, 1 (1990).

[KS 84] Kinoshita, T. and J. Sapirstein, in *Proc. of the 9th Int. Conf. on Atomic Phys.*, Univ. of Washington, Seattle, WA, July 1984.

[La 51] Lamb, W. E., *Rep. Prog. Phys.* **14**, 19 (1951).

[La 81] Langacker, P., *Physics Reports* **72**, 185 (1981).

[LP 86] Lundeen, S. P. and F. M. Pipkin, *Phys. Rev. Lett.* **46**, 232 (1981); *Metrologia* **22**, 9 (1986).

[LR 47] Lamb, W. E. and R. C. Retherford, *Phys. Rev.* **72**, 241 (1947).

[Lu 57] Lüders, G., *Ann. of Phys.* **2**, 1 (1957).

[MP 78] Marciano, W. and H. Pagels, *Physics Reports* **36**, 137 (1978).

[NJ 61] Nambu, Y. and G. Jona-Lasinio, *Phys. Rev.* **122**, 345 (1961).

[Po 73] Politzer, H. D., *Phys. Rev. Lett.* **30**, 1346 (1973).

[PS 83] Pal'chikov, V. G., Yu. L. Sokolov, and V. P. Yakovlev, *Lett. J. Tech. Phys.* **38**, 347 (1983).

[RP 92] *Review of Particle Properties*, *Phys. Rev. D* **45** (1992).

[Sa 68] Salam, A., in *Elementary Particle Physics (Nobel Symp. No. 8)*, ed. N. Svartholm, Almqvist and Wilsell, Stockholm (1968).

[SB 51] Salpeter, E. E. and H. A. Bethe, *Phys. Rev.* **84**, 1232 (1951).

[Sc 48] Schwinger, J., *Phys. Rev.* **73**, 416 (1948).

[Sl 72] Slavnov, A. A., *Teor. Mat. Fiz.* **10**, 153 (1972) [*Theor. Math. Phys.* **10**, 99 (1973)].

[Sy 70] Symanzik, K., *Comm. Math. Phys.* **18**, 227 (1970).

[Ta 57] Takahashi, Y., *Nuovo Cim.* **6**, 371 (1957).

[Ta 71] Taylor, J. C., *Nucl. Phys.* **B33**, 436 (1971).

[VS 87] Van Dyck, R. S. Jr., P. B. Schwinberg, and H. G. Dehmelt, *Phys. Rev. Lett.* **59**, 26 (1987).

[Wa 50] Ward, J. C., *Phys. Rev.* **78**, 182 (1950).

[We 60] Weinberg, S., *Phys. Rev.* **118**. 838 (1960).

[We 67] Weinberg, S., *Phys. Rev. Lett.* **19**, 1264 (1967).

[We 68] Weinberg, S., *Phys. Rev. Lett.* **18**, 188 (1967); *Phys. Rev.* **166**, 1568 (1968).

[Wi 39] Wigner, E. P., *Ann. of Math.* **40**, 149 (1939).

[Wi 54] Wick, G. C., *Phys. Rev.* **96**, 1124 (1954).

[Wi 59] Winter, R. G., *Am. J. Phys.* **27**, 355 (1959).

[Zi 69] Zimmerman, W., *Comm. Math. Physics* **15**, 208 (1969).

[Zi 71] Zimmerman, W., in *Lectures on Elementary Particles and Quantum Field Theory*, Proceedings of the 1970 Brandeis Summer Institute in Theoretical Physics, S. Deser, M. Grisaru and H. Pendleton, eds. (MIT Press, Cambridge, MA, 1971), p. 396.

INDEX

Letters in *italic* are references to the problems (*p*) or to the appendices (*a*).

9 780471 353867